ENCYCLOPEDIA OF COMPUTER SCIENCE AND TECHNOLOGY

HARRY HENDERSON

Facts On File, Inc.

ENCYCLOPEDIA OF COMPUTER SCIENCE AND TECHNOLOGY

For information contact:

Facts On File, Inc.
132 West 31st Street
New York NY 10001

Library of Congress Cataloging-in-Publication Data
Henderson, Harry, 1951–
Encyclopedia of computer science and technology / Harry Henderson.
p. cm.
Includes bibliographical references and index.
ISBN 0-8160-4373-6
1. Computer science—Encyclopedias. 2. Computers—Encyclopedias. I. Title.
QA76.15.H43 2003
004'.03—dc21 2002006796

Facts On File books are available at special discounts when purchased in bulk quantities for businesses, associations, institutions, or sales promotions. Please call our Special Sales Department in New York at (212) 967-8800 or (800) 322-8755.

You can find Facts On File on the World Wide Web at http://www.factsonfile.com

Text design by Erika K. Arroyo
Cover design by Cathy Rincon
Illustrations by Sholto Ainslie

Printed in the United States of America

VB JT 10 9 8 7 6 5 4 3 2 1

This book is printed on acid-free paper.

CONTENTS

INTRODUCTION

Facts On File's *Encyclopedia of Computer Science and Technology* is intended to provide an overview and reference for general readers, technically inclined persons, students who are taking computer science and programming courses, and professionals who want to learn more about less familiar topics.

This book can be used in several ways. As with any encyclopedia, you can look up specific entries by consulting the Table of Contents, referring from topics in the Index, or simply by browsing. The entries in this work are intended to read like "mini-essays," giving not just the bare definition of a topic but also developing its significance for the use of computers and its relationship to other topics. Related topics are shown in this SPECIAL FONT. At the end of each entry is a list of books, articles, and/or websites for further exploration of the topic.

The appendices provide further information for reference and exploration. They include a guide to finding books and publications in the computing field; a chronology of significant events in computing; a listing of achievements in computing as recognized in major awards; and finally, brief descriptions and contact information for some important organizations in the computer field.

To obtain an overview of particular areas in computing, you should read groups of related entries. The following listing groups the entries by category.

FUNDAMENTAL IDEAS OF COMPUTING

Church, Alonzo
computer science
computability and complexity
hexadecimal system
information theory
mathematics of computing
measurement units used in computing
Turing, Alan M.
von Neumann, John
Wiener, Norbert

DEVELOPMENT OF COMPUTERS

Aiken, Howard
analog and digital
analog computer
Atanasoff, John Vincent
Babbage, Charles
calculator
Eckert, J. Presper
history of computing
Hollerith, Hermann
Mauchly, John William
mainframe
minicomputer
Zuse, Konrad

COMPUTER ARCHITECTURE

addressing
arithmetic logic unit (ALU)
bits and bytes
buffering
bus
cache
computer engineering
concurrent programming
Cray, Seymour
device driver

distributed computing
embedded system
parallel port
reduced instruction set computer (RISC)
serial port
supercomputer
USB

General Hardware Components

CD-ROM and DVD-ROM
disk array
hard disk
flat-panel display
floppy disk
keyboard
monitor
motherboard
optical computing
printers
punched cards and paper tape
scanner
tape drives

Personal Computer Components

BIOS
boot sequence
chip
chipset
clock speed
CPU (central processing unit)
green PC
IBM PC
microprocessor
personal computer
plug and play
portable computers

User Interface and Support

Engelbart, Douglas
ergonomics of computing
help systems
installation of software
Jobs, Steven Paul
Kay, Alan
Macintosh
mouse
technical support
technical writing
user groups
user interface
WYSIWYG

Data Types and Algorithms

algorithm
array
binding
bitwise operations
Boolean operators
branching statements
characters and strings
class
constants and literals
data
data abstraction
data structures
data types
encapsulation
enumerations and sets
heap (data structure)
Knuth, Donald
loop
list processing
numeric data
pointers and indirection
queue
recursion
sorting and searching
stack
tree
variable

Other Program Language Concepts

arithmetic operators and expressions
assembler
authoring systems
automatic programming
Backus-Naur Form
compiler
finite state machine
flag
functional languages
interpreter
nonprocedural languages
parsing
procedures and functions
programming languages
random number generation
real time processing
scheduling and prioritization
scripting languages
Stroustrup, Bjarne
template
Wirth, Niklaus

COMPUTER LANGUAGES

Ada
Algol
APL
awk
BASIC
C
C++
COBOL
Forth
FORTRAN
Java
LISP
Logo
Pascal
Perl
PL/1
Prolog
Python
RPG
Simula
Smalltalk

OPERATING SYSTEMS

demon
emulation
file
Input/Output (I/O)
job control language
kernel
memory
memory management
message passing
Microsoft Windows
MS-DOS
multiprocessing
multitasking
operating system
regular expression
Ritchie, Dennis
shell
software agent
Stallman, Richard
system administrator
Torvalds, Linus
UNIX

SOFTWARE DEVELOPMENT AND ENGINEERING

application program interface (API)
bugs and debugging
CASE
Dijkstra, Edsger
documentation of program code
documentation, user
document model
error handling
flowchart
Hopper, Grace Murray
library, program
macro
object-oriented programming
open source movement
plug-in
programming as a profession
programming environment
pseudocode
quality assurance, software
shareware
simulation
software engineering
structured programming
systems programming
template

DATABASES

database administration
database management system (DBMS)
data conversion
data dictionary
data mining
data security
data warehouse
hashing
information retrieval
SQL
XML

BUSINESS APPLICATIONS

application software
application suite
auctions, on-line
auditing in data processing
banking and computers
business applications of computers
desktop publishing
enterprise computing
font
groupware
home office
management information system
middleware

office automation
Optical Character Recognition
personal information manager (PIM)
presentation software
project management software
smart card
spreadsheet
systems analyst
telecommuting
text editor
transaction processing
word processing

Graphics and Multimedia

animation, computer
bitmapped image
color in computing
computer graphics
graphics card
graphics formats
graphics tablet
image processing
multimedia
music, computer
photography, digital
sound file formats
streaming
video editing, digital

Other Applications

art and the computer
cartography, computer-assisted
computer-aided design and manufacturing (CAD/CAM)
computer-aided instruction (CAI)
computer games
education and computers
film industry and computing
financial software
fractals in computing
journalism and computers
law enforcement and computers
libraries and computing
linguistics and computing
mathematics software
medical applications of computers
military applications of computers
scientific computing applications
space exploration and computers
statistics and computing
workstation

Data Communications and Networking (General)

bandwidth
broadband
client-server computing
data acquisition
data communications
data compression
error correction
file server
file transfer protocols
local area network (LAN)
modem
network
Shannon, Claude E.
telecommunications
terminal
wireless computing

On-line Services, Internet, and World Wide Web

application service provider (ASP)
Berners-Lee, Tim
bulletin board systems (BBS)
Bush, Vannevar
Cerf, Vinton G.
certificate, digital
CGI
chat, on-line
conferencing systems
cyberspace and cyber culture
digital cash
digital convergence
Domain Name System (DNS)
e-commerce
e-mail
HTML
hypertext and hypermedia
information superhighway
Internet
Internet appliance
Internet applications programming
Internet Service Provider (ISP)
Internet telephony
intranet
Licklider, J. C. R.
netiquette
netnews and newsgroups
on-line research
on-line services
portal
search engine

TCP/IP
videoconferencing
virtual community
virtual reality
Web browser
Web cam
Web filter
webmaster
webpage design
Web server
World Wide Web
XML

AI and Robotics

artificial intelligence
artificial life
cellular automata
cognitive science
computer vision
Dreyfus, Hubert L.
expert systems
Feigenbaum, Edward
fuzzy logic
genetic algorithms
handwriting recognition
knowledge representation
McCarthy, John
Minsky, Marvin
neural network
robotics
speech recognition and synthesis
telepresence

Future Computing

biology and computing
Dertouzos, Michael
Joy, Bill
molecular computing
nanotechnology
quantum computing

Computer Industry

certification of computer professionals
compatibility and portability
computer industry
education in the computer field
employment in the computer field
entrepreneurs in computing
Gates, William III (Bill)
Grove, Andrew
journalism and the computer industry
marketing of software
research laboratories in computing
standards in computing
Wozniak, Steven

Computer Security and Risks

authentication
backup and archive systems
biometrics
computer crime and security
computer virus
copy protection
encryption
firewall
hackers and hacking
risks of computing
Y2K Problem

Legal and Social Issues

computer literacy
digital divide
disabled persons and computing
intellectual property and computing
popular culture and computing
privacy in the digital age
social impact of computing
women in computing

ENTRIES
A-Z

A

Ada

Starting in the 1960s, the U.S. Department of Defense (DOD) began to confront the growing unmanageability of its software development efforts. Whenever a new application such as a communications controller (see EMBEDDED SYSTEM) was developed, it typically had its own specialized programming language. With more than 2,000 such languages in use, it had become increasingly costly and difficult to maintain and upgrade such a wide variety of incompatible systems. In 1977, a DOD working group began to formally solicit proposals for a new general-purpose programming language that could be used for all applications ranging from weapons control and guidance systems to bar-code scanners for inventory management. The winning language proposal eventually became known as Ada, named for 19th-century computer pioneer Ada Lovelace (see also BABBAGE, CHARLES). After a series of reviews and revisions of specifications, the American National Standards Institute officially standardized Ada in 1983, and this first version of the language is sometimes called Ada-83.

LANGUAGE FEATURES

In designing Ada, the developers adopted basic language elements based on emerging principles (see STRUCTURED PROGRAMMING) that had been implemented in languages developed during the 1960s and 1970s (see ALGOL and PASCAL). These elements include well-defined control structures (see BRANCHING STATEMENT and LOOP) and the avoidance of the haphazard jump or "goto" directive.

Ada combines standard structured language features (including control structures and the use of subprograms) with user-definable data type "packages" similar to the classes used later in C++ and other languages (see CLASS and OBJECT-ORIENTED PROGRAMMING). As shown in this simple example, an Ada program has a general form similar to that used in Pascal. (Note that words in boldface type are language keywords.)

```
with Ada.Text_IO; use Ada.Text_IO;
procedure Get_Name is
Name : String (1..80);
Length : Integer;

begin
Put ("What is your first name? ");
Get_Line (Name, Length);
New_Line;
Put ("Nice to meet you, ");
Put (Name (1..Length));
end Get_Name;
```

The first line of the program specifies what "packages" will be used. Packages are structures that combine data types and associated functions, such as those needed for getting and displaying text. The Ada.Text.IO package, for example has a specification that includes the following:

```
package Text_IO is
type File_Type is limited private;
type File_Mode is (In_File, Out_File,
Append_File);
procedure Create (File : in out File_Type;
Mode : in File_Mode := Out_File;
Name : in String := "");
procedure Close (File : in out File_Type);
procedure Put_Line (File : in File_Type; Item :
in String);
procedure Put_Line (Item : in String);
end Text_IO;
```

The package specification begins by setting up a data type for files, and then defines functions for creating and closing a file and for putting text in files. As with C++ classes, more specialized packages can be derived from more general ones.

In the main program **Begin** starts the actual data processing, which in this case involves displaying a message using the Put function from the Ada.Text.IO function and getting the user response with Get_Line, then using Put again to display the text just entered.

Ada is particularly well suited to large, complex software projects because the use of packages hides and protects the details of implementing and working with a data type. A programmer whose program uses a package is restricted to using the visible interface, which specifies what parameters are to be used with each function. Ada compilers are carefully validated to ensure that they meet the exact specifications for the processing of various types of data (see DATA TYPES), and the language is "strongly typed," meaning that types must be explicitly declared, unlike the case with C, where subtle bugs can be introduced when types are automatically converted to make them compatible.

Because of its application to embedded systems and real-time operations, Ada includes a number of features designed to create efficient object (machine) code, and the language also makes provision for easy incorporation of routines written in assembly or other high-level languages. The latest version, Ada 95, also emphasizes support for parallel programming (see MULTIPROCESSING). The future of Ada is unclear, however, because the Department of Defense no longer requires use of the language in government contracts.

Further Reading

"Ada 95 Lovelace Tutorial" http://www.adahome.com/Tutorials/Lovelace/lovelace.htm

Ada 95 On-line Reference Manual (hypertext) http://www.adahome.com/rm95/rm9x-toc.html

English, John. *Ada 95: The Craft of Object-Oriented Programming*. Upper Saddle River, N.J.: Prentice Hall, 1996.

Feldman, Michael B., and Elliott B. Koffman. *Ada: Problem-Solving and Program Design*. 3rd ed. Reading, Mass.: Addison-Wesley, 1999.

addressing

In order for computers to manipulate data, they must be able to store and retrieve it on demand. This requires a way to specify the location and extent of a data item in memory. These locations are represented by sequential numbers, or addresses.

Physically, a modern RAM (random access memory) can be visualized as a grid of address lines that crisscross with data lines. Each line carries one bit of the address, and together, they specify a particular location in memory (see MEMORY). Thus a machine with 32 address lines can handle up to 32 bits, or 4 gigabytes (billions of bytes) worth of addresses. However the amount of memory that can be addressed can be extended through indirect addressing, where the data stored at an address is itself the address of another location where the actual data can be found. This allows a limited amount of fast memory to be used to point to data stored in auxiliary memory or mass storage thus extending addressing to the space on a hard disk drive.

Some of the data stored in memory contains the actual program instructions to be executed. As the processor executes program instructions, an instruction pointer accesses the location of the next instruction. An instruction can also specify that if a certain condition is met the processor will jump over intervening locations to fetch the next instruction. This implements such control structures as branching statements and loops.

ADDRESSING IN PROGRAMS

A variable name in a program language actually references an address (or often, a range of successive addresses, since most data items require more than one byte of storage). For example, if a program includes the declaration

```
Int Old_Total, New_Total;
```

when the program is compiled, storage for the variables Old_Total and New_Total is set aside at the next available addresses. A statement such as

```
New_Total = 0;
```

is compiled as an instruction to store the value 0 in the address represented by New_Total. When the program later performs a calculation such as:

```
New_Total = Old_Total + 1;
```

the data is retrieved from the memory location designated by Old_Total and stored in a register in the CPU, where 1 is added to it, and the result is stored in the memory location designated by New_Total.

Although programmers don't have to work directly with address locations, programs can also use a special type of variables to hold and manipulate memory addresses for more efficient access to data (see POINTERS AND INDIRECTION).

Addressing

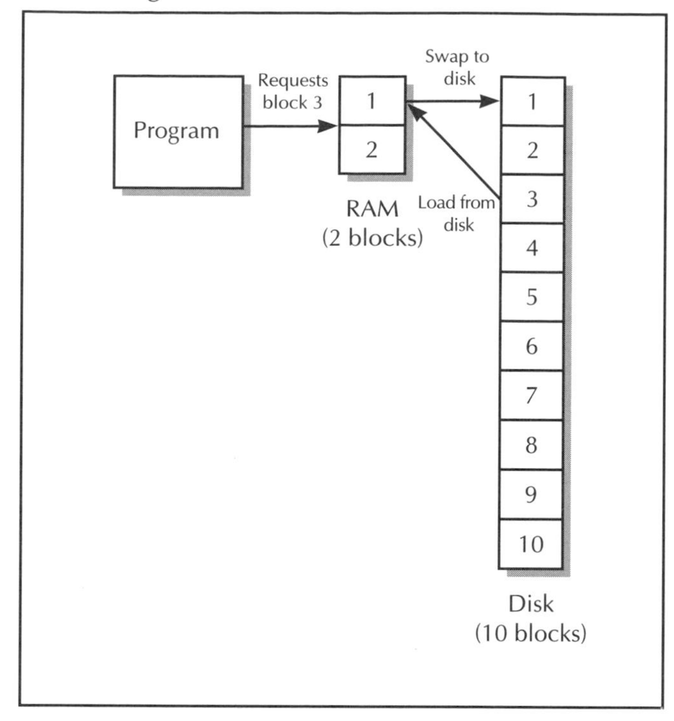

Virtual memory uses indirect addressing. When a program requests data from memory, the address is looked up in a table that keeps track of each block's actual location. If the block is not in RAM, one or more blocks in RAM are copied to the swap file on disk, and the needed blocks are copied from disk into the vacated area in RAM.

Further Reading

"Computer Architecture Tutorial." http://www.cs.iastate.edu/~prabhu/Tutorial/title.html

Murdocca, Miles J. and Vincent P. Heuring. "Principles of Computer Architecture." http://www.cs.rutgers.edu/~murdocca/POCA/poca.html

Aiken, Howard

(1900–1973)
American
Electrical Engineer

Howard Hathaway Aiken was a pioneer in the development of automatic calculating machines. Born on March 8, 1900, in Hoboken, New Jersey, he grew up in Indianapolis, Indiana, where he pursued his interest in electrical engineering by working at a utility company while in high school. He earned a B.A. in electrical engineering in 1923 at the University of Wisconsin.

By 1935, Aiken was involved in theoretical work on electrical conduction that required laborious calculation. Inspired by work a hundred years earlier (see BABBAGE, CHARLES), Aiken began to investigate the possibility of building a large-scale, programmable, automatic computing device (see CALCULATOR). As a doctoral student at Harvard, Aiken aroused interest in his project, particularly from Thomas Watson, Sr., head of International Business Machines (IBM). In 1939, IBM agreed to underwrite the building of Aiken's first calculator, the Automatic Sequence Controlled Calculator, which became known as the Harvard Mark I.

MARK I AND ITS PROGENY

Like Babbage, Aiken aimed for a general-purpose programmable machine rather than an assembly of special-purpose arithmetic units. Unlike Babbage, Aiken had access to a variety of tested, reliable components, including card punches, readers, and electric typewriters from IBM and the mechanical electromagnetic relays used for automatic switching in the telephone industry. His machine used decimal numbers (23 digits and a sign) rather than the binary numbers of the majority of later computers. Sixty registers held whatever constant data numbers were needed to solve a particular problem. The operator turned a rotary dial to enter each digit of each number. Variable data and program instructions were entered via punched paper tape. Calculations had to be broken down into specific instructions similar to those in later low-level programming languages such as "store this number in this register" or "add this number to the number in that register" (see ASSEMBLER). The results (usually tables of mathematical function values) could be printed by an electric typewriter or output on punched cards. Huge (about 8 feet high by 51 feet long), slow, but reliable, the Mark I worked on a variety of problems during World War II, ranging from equations used in lens design and radar to the designing of the implosive core of an atomic bomb.

Aiken completed an improved model, the Mark II, in 1947. The Mark III of 1950 and Mark IV of 1952, however, were electronic rather than electromechanical, replacing relays with vacuum tubes.

Compared to later computers such as the ENIAC and UNIVAC, the sequential calculator, as its name suggests, could only perform operations in the order specified. Any looping had to be done by physically creating a repetitive tape of instructions. (After all, the program as a whole was not stored in any sort of memory, and so previous instructions could not be re-accessed.) Although Aiken's machines soon slipped out of the mainstream of computer development, they did include the modern feature of parallel processing, because different calculation units could work on different instructions at the same time. Further, Aiken recognized the value of maintaining a library of frequently needed routines that could be reused in new programs—another fundamental of modern software engineering.

Aiken's work demonstrated the value of large-scale automatic computation and the use of reliable, available technology. Computer pioneers from around the world came to Aiken's Harvard computation lab to debate many issues that would become staples of the new discipline of computer science. The recipient of many awards including the Edison Medal of the IEEE and the Franklin Institute's John Price Award, Howard Aiken died on March 14, 1973, in St. Louis, Missouri.

Further Reading

Cohen, I. B. *Howard Aiken: Portrait of a Computer Pioneer.* Cambridge, Mass.: MIT Press, 1999.

Cohen, I. B., R. V. D. Campbell, and G. Welch, eds. *Makin' Numbers: Howard Aiken and the Computer.* Cambridge, Mass.: MIT Press, 1999.

Algol

The 1950s and early 1960s saw the emergence of two high-level computer languages into widespread use. The first was designed to be an efficient language for performing scientific calculations (see FORTRAN). The second was designed for business applications, with an emphasis on data processing (see COBOL). However many programs continued to be coded in low-level languages (see ASSEMBLER) designed to take advantages of the hardware features of particular machines.

In order to be able to easily express and share methods of calculation (see ALGORITHM), leading programmers began to seek a "universal" programming language that was not designed for a particular application or hardware platform. By 1957, the German GAMM (Gesellschaft für angewandte Mathematik und Mechanik) and the American ACM (Association for Computing Machinery) had joined forces to develop the specifications for such a language. The result became known as the Zurich Report or Algol-58, and it was refined into the first widespread implementation of the language, Algol-60.

LANGUAGE FEATURES

Algol is a block-structured, procedural language. Each variable is declared to belong to one of a small number of kinds of data including integer, real number (see DATA TYPES), or a series of values of either type (see ARRAY). While the number of types is limited and there is no facility for defining new types, the compiler's type checking (making sure a data item matches the variable's declared type) introduced a level of security not found in most earlier languages.

An Algol program can contain a number of separate procedures or incorporate externally defined procedures (see LIBRARY, PROGRAM), and the variables with the same name in different procedure blocks do not interfere with one another. A procedure can call itself (see RECURSION). Standard control structures (see BRANCHING STATEMENTS and LOOP) were provided.

The following simple Algol program stores the numbers from 1 to 10 in an array while adding them up, then prints the total:

```
begin
  integer array ints[1:10];
  integer counter, total;
  total := 0;
  for counter :=1 step 1 until counter > 10
  do
    begin
    ints [counter] := counter;
    total := total + ints[counter];
    end;
  printstring "The total is: ";
  printint (total);
  end
```

ALGOL'S LEGACY

The revision that became known as Algol-68 expanded the variety of data types (including the addition of boolean, or true/false values) and added user-defined types and "structs" (records containing fields of different types of data). Pointers (references to values) were also implemented, and flexibility was added to the parameters that could be passed to and from procedures.

Although Algol was used as a production language in some computer centers (particularly in Europe), its relative complexity and unfamiliarity impeded its acceptance, as did the widespread corporate backing for the rival languages FORTRAN and especially COBOL. Algol achieved its greatest success in two respects: for a time it became the language of choice for describing new algorithms for computer scientists, and its structural features would be adopted in the new procedural languages that emerged in the 1970s (see PASCAL and C).

Further Reading

"Algol 68 Home Page." http://www.nunan.fsnet.co.uk/algol68/

Backus, J. W., and others. "Revised Report on the Algorithmic Language Algol 60." Originally published in *Numerische Mathematik,* the *Communications of the ACM,* and the *Journal of the British Computer Society.* Available on-line at http://www.masswerk.at/algol60/report.htm

algorithm

When people think of computers, they usually think of silicon chips and circuit boards. Moving from relays to vacuum tubes to transistors to integrated circuits has vastly increased the power and speed of computers, but the essential idea behind the work computers do remains the algorithm. An algorithm is a reliable, definable procedure for solving a problem. The idea of the algorithm goes back to the beginnings of mathematics and elementary school students are usually taught a variety of algorithms. For example, the procedure for long division by successive division, subtraction, and attaching the next digit is an algorithm. Since a bona fide algorithm is guaranteed to work given the specified type of data and the rote following of a series of steps, the algorithmic approach is naturally suited to mechanical computation.

ALGORITHMS IN COMPUTER SCIENCE

Just as a cook learns both general techniques such as how to sauté or how to reduce a sauce and a repertoire of specific recipes, a student of computer science learns both general problem-solving principles and the details of common algorithms. These include a variety of algorithms for organizing data (see SORTING AND SEARCHING), for numeric problems (such as generating random numbers or finding primes), and for the manipulation of data structures (see LIST PROCESSING and QUEUE).

A working programmer faced with a new task first tries to think of familiar algorithms that might be applicable to the current problem, perhaps with some adaptation. For example, since a variety of well-tested and well-understood sorting algorithms have been developed, a programmer is likely to apply an existing algorithm to a sorting problem rather than attempt to come up with something entirely new. Indeed, for most widely used programming languages there are packages of modules or procedures that implement commonly needed data structures and algorithms (see LIBRARY, PROGRAM).

If a problem requires the development of a new algorithm, the designer will first attempt to determine whether the problem can, at least in theory, be solved (see COMPUTABILITY AND COMPLEXITY). Some kinds of problems have been shown to have no guaranteed answer. If a new algorithm seems feasible, principles found to be effective in the past will be employed, such as breaking complex problems down into component parts and building up from the simplest case to generate a solution (see RECURSION). For example, the merge-sort algorithm divides the data to be sorted into successively smaller portions until they are sorted, and then merges the sorted portions back together.

Another important aspect of algorithm design is choosing an appropriate way to organize the data. (See DATA STRUCTURES.) For example, a sorting algorithm that uses a branching (tree) structure would probably use a data structure that implements the nodes of a tree and the operations for adding, deleting, or moving them (see CLASS).

Once the new algorithm has been outlined (see PSEUDOCODE), it is often desirable to demonstrate that it will work for any suitable data. Mathematical techniques such as the finding and proving of loop invariants (where a true assertion remains true after the loop terminates) can be used to demonstrate the correctness of the implementation of the algorithm.

PRACTICAL CONSIDERATIONS

It is not enough that an algorithm be reliable and correct, it must also be accurate and efficient enough for its intended use. A numerical algorithm that accumulates too much error through rounding or truncation of intermediate results may not be accurate enough for a scientific application. An algorithm that works by successive approximation or convergence on an answer may require too many iterations even for today's fast computers, or may consume too much of other computing resources such as memory. On the other hand, as computers become more and more powerful and processors are combined to create more powerful supercomputers (see SUPERCOMPUTER and CONCURRENT PROGRAMMING), algorithms that were previously considered impracticable might be reconsidered. Code profiling (analysis of which program statements are being executed the most frequently) and techniques for creating more efficient code can help in some cases. It is also necessary to keep in mind special cases where an otherwise efficient algorithm becomes much less efficient (for example, a tree sort may work well for random data but will become badly unbalanced and slow when dealing with data that is already sorted or mostly sorted).

Sometimes an exact solution cannot be mathematically guaranteed or would take too much time and resources to calculate, but an approximate solution is acceptable. A so-called "greedy algorithm" can proceed in stages, testing at each stage whether the solution is "good enough." Another approach is to use an algorithm that can produce a reasonable if not optimal solution. For example, if a group of tasks must be apportioned among several people (or computers) so that all tasks are completed in the shortest possible time, the time needed to find an exact solution rises exponentially with the number of workers and tasks. But an algorithm that first sorts the tasks by decreasing length and then distributes them among the workers by "dealing" them one at a time like cards at a bridge table will, as demonstrated by Ron Graham, give an allocation guaranteed to be within 4/3 of the optimal result—quite suitable for most applications. (A procedure that can produce a practical, though not perfect solution is actually not an algorithm but a heuristic.)

An interesting approach to optimizing the solution to a problem is allowing a number of separate programs to "compete," with those showing the best performance surviving and exchanging pieces of code ("genetic material") with other successful programs (see GENETIC ALGORITHMS). This of course mimics evolution by natural selection in the biological world.

Further Reading

Berlinksi, David. *The Advent of the Algorithm: The Idea that Rules the World.* New York: Harcourt, 2000.

Cormen, T. H., C. E. Leiserson, and R. L. Rivest. *Introduction to Algorithms.* Cambridge, Mass.: MIT Press, 1990.

Knuth, Donald E. *The Art of Computer Programming.* Vol. 1: *Fundamental Algorithms.* 3rd ed. Addison-Wesley, 1997. Vol. 2: *Seminumerical Algorithms.* 3rd ed. Addison-Wesley, 1997. Vol. 3: *Searching and Sorting.* 2nd ed. Addison-Wesley, 1998.

analog and digital

The word *analog* (derived from Greek words meaning "by ratio") denotes a phenomenon that is continuously variable, such as a sound wave. The word *digital,* on the other hand, implies a discrete, exactly countable value that can be represented as a series of digits (numbers). Sound recording provides familiar examples of both approaches. Recording a phonograph record involves electromechanically transferring a physical signal (the sound wave) into an "analogous" physical representation (the continuously varying peaks and dips in the record's surface). Recording a CD, on the other hand, involves sampling (measuring) the sound level at thousands of discrete instances and storing the results in a physical representation of a numeric format that can in turn be used to drive the playback device.

Virtually all modern computers depend on the manipulation of discrete signals in one of two states denoted by the numbers 1 and 0. Whether the 1 indicates the presence of an electrical charge, a voltage level, a magnetic state, a pulse of light, or some other phenomenon, at a given point there is either "something" (1) or "nothing" (0). This is the most natural way to represent a series of such states.

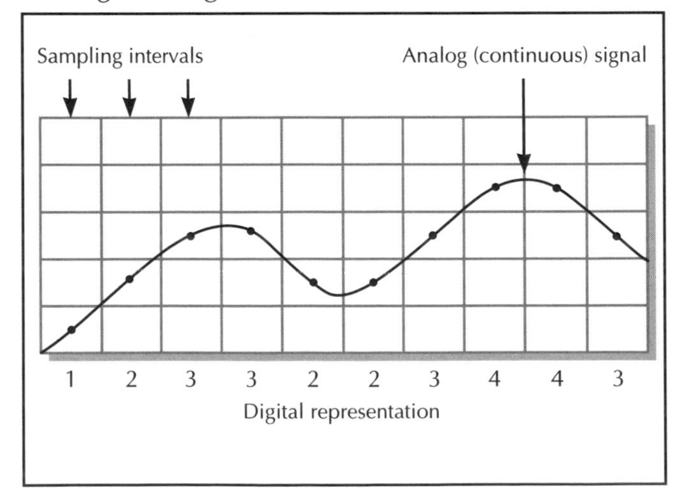

Most natural phenomena such as light or sound intensity are analog values that vary continuously. To convert such measurements to a digital representation, "snapshots" or sample readings must be taken at regular intervals. Sampling more frequently gives a more accurate representation of the original analog data, but at a cost in memory and processor resources.

Digital representation has several advantages over analog. Since computer circuits based on binary logic can be driven to perform calculations electronically at ever-increasing speeds, even problems where an analog computer better modeled nature can now be done more efficiently with digital machines (see ANALOG COMPUTER). Data stored in digitized form is not subject to the gradual wear or distortion of the medium that plagues analog representations such as the phonograph record. Perhaps most important, because digital representations are at base simply numbers, an infinite variety of digital representations can be stored in files and manipulated, regardless of whether they started as pictures, music, or text.

CONVERTING BETWEEN ANALOG AND DIGITAL REPRESENTATIONS

Because digital devices (particularly computers) are the mechanism of choice for working with representations of text, graphics, and sound, a variety of devices are used to digitize analog inputs so the data can be stored and manipulated. Conceptually, each digitizing device can be thought of as having three parts: a component that scans the input and generates an analog signal, a circuit that converts the analog signal from the input to a digital format, and a component that stores the resulting digital data for later use. For example, in the ubiquitous flatbed scanner a moving head reads varying light levels on the paper and converts them to a varying level of current (see SCANNER). This analog signal is in turn converted into a digital reading by an analog-to-digital converter, which creates numeric information that represents discrete spots (pixels) representing either levels of gray or of particular colors. This information is then written to disk using the formats supported by the operating system and the software that will manipulate them.

Further Reading

Chalmers, David J. "Analog vs. Digital Computation." http://www.u.arizona.edu/~chalmers/notes/analog.html

Hoeschele, David F. *Analog-to-Digital and Digital-to-Analog Conversion Techniques.* 2nd ed. New York: Wiley-Interscience, 1994.

analog computer

Most natural phenomena are analog rather than digital in nature (see ANALOG AND DIGITAL). But just as mathematical laws can describe relationships in nature, these relationships in turn can be used to construct a model in which natural forces generate mathematical solutions. This is the key insight that leads to the analog computer.

The simplest analog computers use physical components that model geometric ratios. The earliest known analog computing device is the Antikythera Mechanism. Constructed by an unknown scientist on the island of Rhodes around 87 B.C., this device used a precisely

Analog and Digital (2)

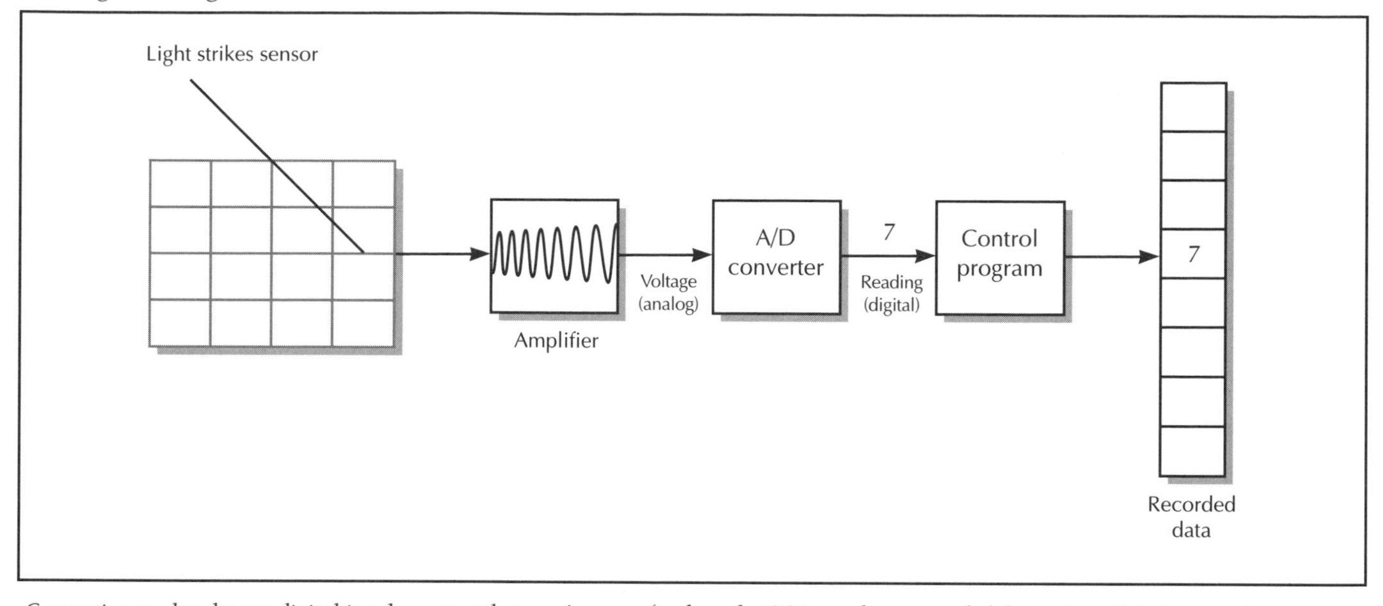

Converting analog data to digital involves several steps. A sensor (such as the CCD, or charge-coupled device in a digital camera) creates a varying electrical current. An amplifier can strengthen this signal to make it easier to process, and filters can eliminate spurious spikes or "noise." The "conditioned" signal is then fed to the analog to digital (A/D) converter, which produces numeric data that is usually stored in a memory buffer from which it can be processed and stored by the controlling program.

crafted differential gear mechanism to mechanically calculate the interval between new moons (the synodic month). (Interestingly, the differential gear would not be rediscovered until 1877.)

Another analog computer, the slide rule, became the constant companion of scientists, engineers, and students until it was replaced by electronic calculators in the 1970s. Invented in simple form in the 17th century, the slide rule's movable parts are marked in logarithmic proportions, allowing for quick multiplication, division, the extraction of square roots, and sometimes the calculation of trigonometric functions.

The next insight involved building analog devices that set up dynamic relationships between mechanical movements. In the late 19th century two British scientists, James Thomson and his brother Sir William Thomson (later Lord Kelvin) developed the mechanical integrator, a device that could solve differential equations. An important new principle used in this device is the closed feedback loop, where the output of the integrator is fed back as a new set of inputs. This allowed for the gradual summation or integration of an equation's variables. In 1931, VANNEVAR BUSH completed a more complex machine that he called a "differential analyzer." Consisting of six mechanical integrators using specially shaped wheels, disks, and servomechanisms, the differential analyzer could solve equations in up to six independent variables. As the usefulness and applicability of the device became known, it was quickly replicated in various forms in scientific, engineering, and military institutions.

These early forms of analog computer are based on fixed geometrical ratios. However, most phenomena that scientists and engineers are concerned with, such as aerodynamics, fluid dynamics, or the flow of electrons in a circuit, involve a mathematical relationship between forces where the output changes smoothly as the inputs are changed. The "dynamic" analog computer of the mid-20th century took advantage of such force relationships to construct devices where input forces represent variables in the equation, and nature itself "solves" the equation by producing a resulting output force.

In the 1930s, the growing use of electronic circuits encouraged the use of the flow of electrons rather than mechanical force as a source for analog computation. The key circuit is called an operational amplifier. It generates a highly amplified output signal of opposite polarity to the input, over a wide range of frequencies. By using components such as potentiometers and feedback capacitors, an analog computer can be programmed to set up a circuit in which the laws of electronics manipulate the input voltages in the same way the equation to be solved manipulates its variables. The results of the calculation are then read as a series of voltage values in the final output.

Starting in the 1950s, a number of companies marketed large electronic analog computers that contained

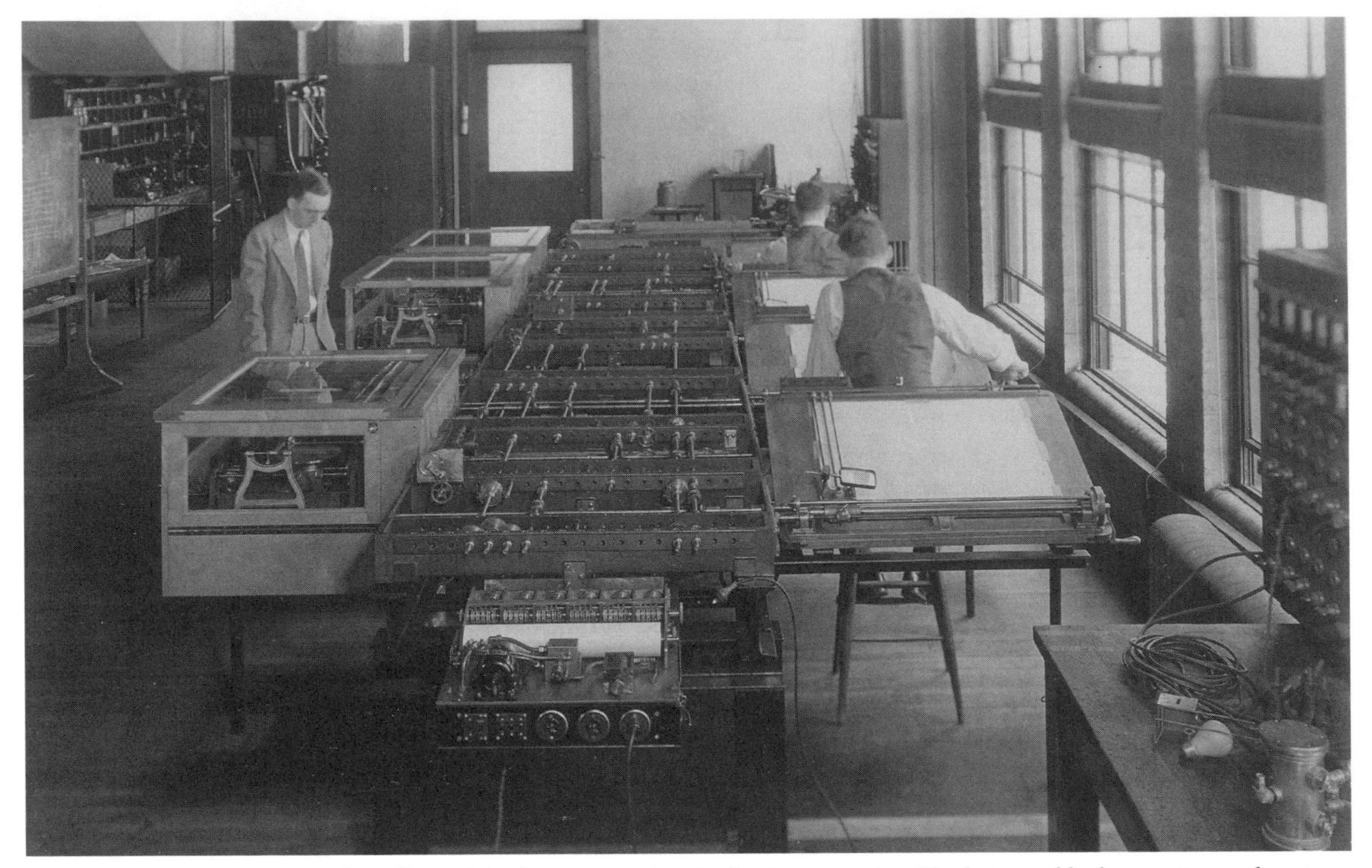

Completed in 1931, Vannevar Bush's Differential Analyzer was a triumph of analog computing. The device could solve equations with up to six independent values. (PHOTO COURTESY OF MIT MUSEUM)

many separate computing units that could be harnessed together to provide "real time" calculations in which the results could be generated at the same rate as the actual phenomena being simulated. In the early 1960s, NASA set up training simulations for astronauts using analog real-time simulations that were still beyond the capability of digital computers.

Gradually, however, the use of faster processors and larger amounts of memory enabled the digital computer to surpass its analog counterpart even in the scientific programming and simulations arena. In the 1970s, some hybrid machines combined the easy programmability of a digital "front end" with analog computation, but by the end of that decade the digital computer had rendered analog computers obsolete.

Further Reading

"Doug Coward's Analog Computer Museum and History Center." http://www.best.com/~dcoward/analog/

Hoeschele, David F., Jr. *Analog-to-Digital and Digital-to-Analog Conversion Techniques*. 2nd ed. New York: John Wiley, 1994.

Vassos, Basil H., and Galen Ewing, eds. *Analog and Computer Electronics for Scientists*. 4th ed. New York: John Wiley, 1993.

animation, computer

Ever since the first hand-drawn cartoon features entertained moviegoers in the 1930s, animation has been an important part of the popular culture. Traditional animation uses a series of hand-drawn frames that, when shown in rapid succession, create the illusion of lifelike movement.

COMPUTER ANIMATION TECHNIQUES

The simplest form of computer animation (illustrated in games such as *Pong*) involves drawing an object, then erasing it and redrawing it in a different location. A somewhat more sophisticated approach can create motion in a scene by displaying a series of pre-drawn images called *sprites*—for example, there could be a series of sprites showing a sword-wielding troll in different positions.

Since there are only a few intermediate images, the use of sprites doesn't convey truly lifelike motion. Modern animation uses a modern version of the traditional drawn animation technique. The drawings are "keyframes" that capture significant movements by the characters. The keyframes are later filled in with transitional frames in a process called *tweening*. Since it is pos-

Animation

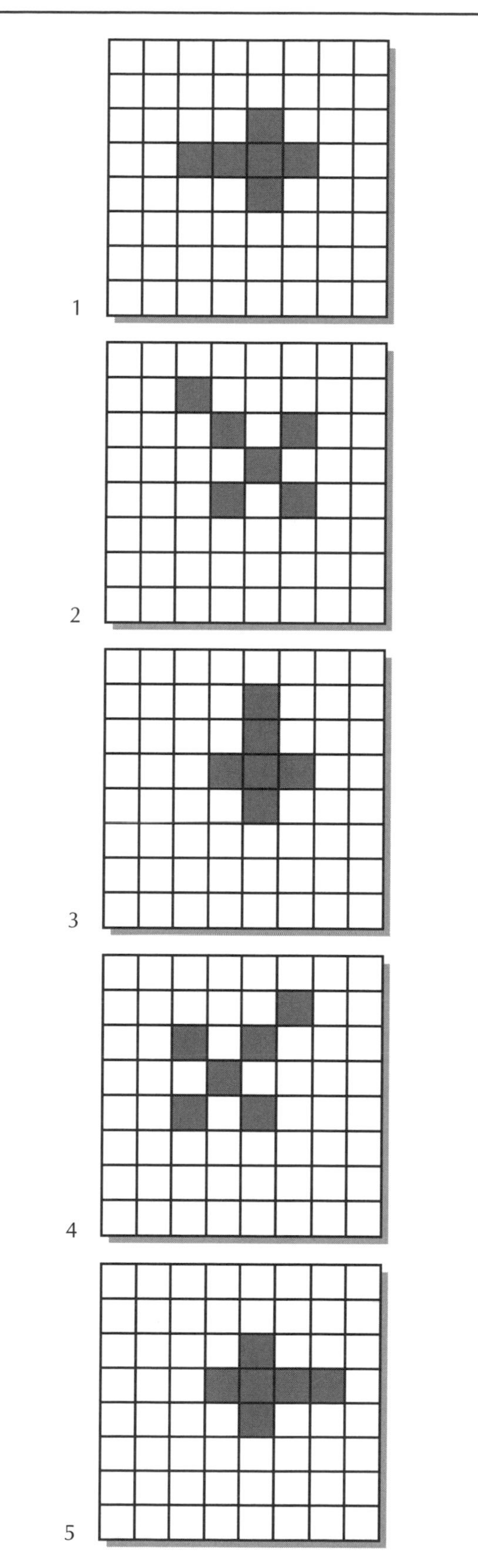

Sprite graphics use the same principle as old-style cartoons—rapidly running through a set of slightly different images to create the illusion of motion. In this very simplified example a cross-shaped pattern (perhaps representing an airplane) is turned 180 degrees in the course of five images. Sprites were used extensively in early videogames where memory and processing power were too limited for more sophisticated graphics techniques.

sible to create algorithms that describe the optimal in-between frames, the advent of sufficiently powerful computers has made computer animation both possible and desirable. Today computer animation is used not only for cartoons but also for video games and movies. The most striking use of this technique is morphing, where the creation of plausible intermediate images between two strikingly different faces creates the illusion of one face being transformed into the other.

Algorithms that can realistically animate people, animals, and other complex objects require the ability to create a model that includes the parts of the object that can move separately (such as a person's arms and legs). Because the movement of one part of the model often affects the positions of other parts, a treelike structure is often used to describe these relationships. (For example, an elbow moves an arm, the arm in turn moves the hand, which in turn moves the fingers). Alternatively, live actors performing a repertoire of actions or poses can be digitized using wearable sensors and then combined to portray situations, such as in a video game.

Less complex objects (such as clouds or rainfall) can be treated in a simpler way, as a collection of "particles" that move together following basic laws of motion and gravity. Of course when different models come into contact (for example, a person walking in the rain), the interaction between the two must also be taken into consideration.

While realism is always desirable, there is inevitably a tradeoff between the resources available. Computationally intensive physics models might portray a very realistic spray of water using a high-end graphics workstation, but simplified models have to be used for a program that runs on a game console or desktop PC. The key variables are the frame rate (higher is smoother) and the display resolution. The amount of available video memory is also a consideration: many desktop PCs sold today have 32MB or 64MB of video memory.

APPLICATIONS

Computer animation is used extensively in many feature films, such as for creating realistic dinosaurs (*Jurassic Park*) or buglike aliens (*Starship Troopers*). Computer games combine animation techniques with other techniques (see COMPUTER GRAPHICS) to provide smooth action within a vivid 3D landscape. Simpler forms of animation are now a staple of website design, often written in Java or with the aid of animation scripting programs such as Macromedia's Flash.

The intensive effort that goes into contemporary computer animation suggests that the ability to fascinate the human eye that allowed Walt Disney to build an empire is just as compelling today.

Further Reading

"3-D Animation Workshop." http://www.webreference.com/3d/indexa.html

Comet, Michael B. "Character Animation: Principles and Practice." http://www.comet-cartoons.com/toons/3ddocs/charanim/

Hamlin, J. Scott. *Effective Web Animation: Advanced Techniques for the Web.* Reading, Mass.: Addison-Wesley, 1999.

O'Rourke, Michael. *Principles of Three-Dimensional Computer Animation: Modeling, Rendering, and Animating with 3D Computer Graphics.* New York: Norton, 1998.

Parent, Rick. *Computer Animation: Algorithms and Techniques.* San Francisco: Morgan Kaufmann, 2002.

APL (a programming language)

This programming language was developed by Harvard (later IBM) researcher Kenneth E. Iverson in the early 1960s as a way to express mathematical functions clearly and consistently for computer use. The power of the language to compactly express mathematical functions attracted a growing number of users, and APL soon became a full general-purpose computing language.

Like many versions of BASIC, APL is an interpreted language, meaning that the programmer's input is evaluated "on the fly," allowing for interactive response (see INTERPRETER). Unlike BASIC or FORTRAN, however, APL has direct and powerful support for all the important mathematical functions involving arrays or matrices (see ARRAY).

APL has over 100 built-in operators, called "primitives." With just one or two operators the programmer can perform complex tasks such as extracting numeric or trigonometric functions, sorting numbers, or rearranging arrays and matrices. (Indeed, APL's greatest power is in its ability to manipulate matrices directly without resorting to explicit loops or the calling of external library functions.)

To give a very simple example, the following line of APL code:

```
X [ ΔX]
```

sorts the array X. In most programming languages this would have to be done by coding a sorting algorithm in a dozen or so lines of code using nested loops and temporary variables.

However, APL has also been found by many programmers to have significant drawbacks. Because the language uses Greek letters to stand for many operators, it requires the use of a special type font that was generally not available on non-IBM systems. A dialect called J has been devised to use only standard ASCII characters, as well as both simplifying and expanding the language. Many programmers find mathematical expressions in APL to be cryptic, making programs hard to maintain or revise. Nevertheless, APL Special Interest Groups in the major computing societies testify to continuing interest in the language.

Further Reading

ACM Special Interest Group for APL and J Languages. http://www.acm.org/sigapl/

"Why APL?" http://www.acm.org/sigapl/whyapl.htm

"APL Frequently Asked Questions." Available from various sites including http://www.izap.com/~sirlin/apl/apl.faq.html

Gilman, Leonard, and Allen J. Rose. *APL: An Interactive Approach.* 3rd ed. (reprint). Malabar, Fla.: Krieger, 1992.

application program interface (API)

In order for an application program to function, it must interact with the computer system in a variety of ways, such as reading information from disk files, sending data to the printer, and displaying text and graphics on the monitor screen (see USER INTERFACE). The program may need to find out whether a device is available or whether it can have access to an additional portion of memory. In order to provide these and many other services, an operating system such as Microsoft Windows includes an extensive application program interface (API). The API basically consists of a variety of functions or procedures that an application program can call upon, as well as data structures, constants, and various definitions needed to describe system resources.

Applications programs use the API by including calls to routines in a program library (see LIBRARY, PROGRAM and PROCEDURES AND FUNCTIONS). In Windows, "dynamic link libraries" (DLLs) are used. For example, this simple function puts a message box on the screen:

```
MessageBox (0, "Program Initialization Failed!",
"Error!", MB_ICONEXCLAMATION | MB_OK | MB_SYS-
TEMMODAL);
```

In practice, the API for a major operating system such as Windows contains hundreds of functions, data structures, and definitions. In order to simplify learning to access the necessary functions and to promote the writing of readable code, compiler developers such as Microsoft and Borland have devised frameworks of C++ classes that package related functions together. For example, in the Microsoft Foundation Classes (MFC), a program generally begins by deriving a class representing the application's basic characteristics from the MFC class CWinApp. When the program wants to display a window, it derives it from the CWnd class, which has the functions common to all windows, dialog boxes, and controls. From CWnd is derived the specialized class for each type of window:

Application Programming Interface

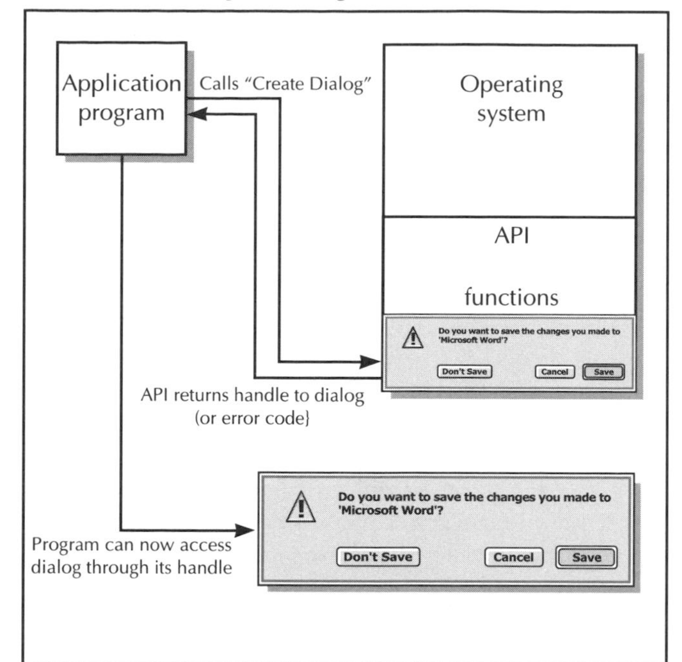

Modern software uses API calls to obtain interface objects such as dialog boxes from the operating system. Here the application calls the CreateDialog API function. The operating system returns a pointer (called a handle) that the application can now use to access and manipulate the dialog.

for example, CFrameWnd implements a typical main application window, while CDialog would be used for a dialog box. Thus in a framework such as MFC or Borland's OWL, the object-oriented concept of encapsulation is used to bundle together objects and their functions, while the concept of inheritance is used to relate the generic object (such as a window) to specialized versions that have added functionality (see OBJECT-ORIENTED PROGRAMMING and ENCAPSULATION INHERITANCE).

Programmers using languages such as Visual Basic can take advantage of a further level of abstraction. Here the various kinds of windows, dialogs, and other controls are provided as building blocks that the developer can insert into a form designed on the screen, and then settings can be made and code written as appropriate to control the behavior of the objects when the program runs. While the programmer will not have as much direct control or flexibility, avoiding the need to master the API means that useful programs can be written more quickly.

Further Reading

"DevCentral Tutorials: MFC and Win32." http://devcentral.iftech.com/learning/tutorials/submfc.asp

Petzold, Charles. *Programming Windows: the Definitive Guide to the Win32 API.* 5th ed. Redmond, Wash.: Microsoft Press, 1999.

"Windows API Guide." http://www.vbapi.com/

Application Service Provider (ASP)

The Application Service Provider offers a new model for delivery of software applications to end-users. Traditionally, users buy a program such as Microsoft Office on CD (see APPLICATION SUITE) and install it on a desktop PC. The up-front cost is typically in the hundreds of dollars. Often, such suites include many features (or entire programs) that are not actually needed in the user's working environment, but whose cost is included in that of the total package.

An Applications Service Provider (ASP) essentially rents access to software. The program is run on the provider's server, not on the user's desktop. Advocates of the ASP business model point to several advantages. Since the software resides on the ASP's server, there is no need to update numerous desktop installations every time a new version of the software is released. Software piracy (unauthorized copying) is also eliminated. There is no need to ship CDs or stock store shelves, and both the ASP and the software developer receive a steady stream of revenue rather than one that peaks after a new release and then declines. Meanwhile, the end-user doesn't have to come up with a large up-front cost to furnish an office with software, but rather can pay a more manageable monthly subscription fee.

CONCERNS AND PROSPECTS

Critics of the ASP model point to several drawbacks. Most critically, providing large full-featured applications suites such as MS Office over an Internet connection is problematic unless both the ASP and user have a broadband connection such as DSL or cable (see BROADBAND and INTERNET SERVICE PROVIDER). Providing enough capacity on the server to support simultaneous use of the program is also an issue. A server breakdown or communications problems could bring work in an office to a halt and imperil the user's business. Users also raise privacy and security issues involved with the direct access between the ASP's servers and desktop PCs containing sensitive data.

By 1999, one forecaster, Forester Research, was predicting that the ASP market would, in a few years, almost equal the current $7 billion market for software suites. Another forecaster, International Data, predicted a more modest (but still quite significant) $2 billion market for ASPs by 2003.

Microsoft has responded to the ASP model by beginning to incorporate it into the marketing of some of its software, especially that related to Web servers. ASPs may also find niche markets by offering more economical access to expensive specialized software, supporting new applications such as e-commerce support and wireless networking, and in providing an integrated package that includes security (anti-virus) service and technical support. As Internet Service Providers (ISPs) come under increasing pressure to differentiate their offerings, the addition of ASP has become an attractive option.

Further Reading

Lee, Jonathan and Krappe, Kirk. *ASP Revolution: Winning Business Strategies Using Application Service Providers.* Berkeley, Calif.: Osborne-McGraw-Hill, 2000.

ZDNet ASP Resource Center. http://www.zdnet.com/enterprise/filters/resources/0,,6011014,00.html

application software

Application software consists of programs that enable computers to perform useful tasks, as opposed to programs that are concerned with the operation of the computer itself (see OPERATING SYSTEM and SYSTEMS PROGRAMMING). To most users, applications programs *are* the computer: They determine how the user will accomplish tasks.

The following table gives a selection of representative applications:

GENERAL AREA	APPLICATIONS	EXAMPLES
Business Operations	payroll, accounts receivable, inventory	specialized business software, general spreadsheets and databases
Education	school management, curriculum reinforcement, reference aids, curriculum expansion or supplementation, training	attendance and grade book management, drill-and-practice software for reading or arithmetic, CD or online encyclopedias, educational games or simulations, collaborative and Web-based learning, corporate training programs
Engineering	design and manufacturing	computer-aided design (CAD), computer-aided manufacturing (CAM)
Entertainment	games, music, and video	desktop and console games, online games, digitized music distribution (MP3 files), streaming video (including movies)
Government	administration, law enforcement, military	tax collection, criminal records and field support for police, legal citation databases, combat information and weapons control systems
Health Care	hospital administration, health care delivery	hospital information and billing systems, medical records management, medical imaging, computer-assisted treatment or surgery
Internet and World Wide Web	web browser, search tools, e-commerce	browser and plug-in software for video and audio, search engines, e-commerce support and secure transactions
Libraries	circulation, cataloging, reference	automated book check-in systems, cataloging databases, CD or online bibliographic and full-text databases
Office Operations	e-mail, document creation	e-mail clients, word processing, desktop publishing
Science	statistics, modeling	mathematical and statistical software, modeling of molecules, gene typing, weather forecasting

DEVELOPING AND DISTRIBUTING APPLICATIONS

Applications can be divided into three categories based on how they are developed and distributed. Commercial applications such as word processors, spreadsheets, and general-purpose Database Management Systems (DBMS) are developed by companies specializing in such software and distributed to a variety of businesses and individual users (see WORD PROCESSING, SPREADSHEET, and DATABASE MANAGEMENT SYSTEM). Niche or specialized applications (such as hospital billing systems) are designed for and marketed to a particular industry (see MEDICAL APPLICATIONS OF COMPUTERS). These programs tend to be much more expensive and usually include extensive technical support. Finally, in-house applications are developed by programmers within a business or other institution for their own use. Examples might include employee training aids or a Web-based product catalog (although such applications could also be developed using commercial software such as multimedia or database development tools).

While each application area has its own needs and priorities, the discipline of software development (see SOFTWARE ENGINEERING and PROGRAMMING ENVIRONMENT) is generally applicable to all major products. Software developers try to improve speed of development as well as program reliability by using software development tools that simplify the writing and testing of computer code, as well as the manipulation of graphics, sound, and other resources used by the program. An applications developer must also have a good understanding of the features and limitations of the relevant operating system. The developer of commercial software must work closely with the marketing department to work out issues of feature selection, timing of releases, and anticipation of trends in software use (see MARKETING OF SOFTWARE).

Further Reading

"Business Software Buyer's Guide." http://businessweek.buyerzone.com/software/business_software/buyers_guide1.html

ZDnet Buyer's Guide to Computer Applications. http://www.zdnet.com/computershopper/edit/howtobuy/

application suite

An application suite is a set of programs designed to be used together and marketed as a single package. For example, a typical office suite might include word processing, spreadsheet, database, personal information manager, and e-mail programs.

While an operating system such as Microsoft Windows provides basic capabilities to move text and graphics from one application to another (such as by cutting and pasting), an application suite such as Microsoft Office makes it easier to, for example, launch a Web browser from a link within a word processing document or embed a spreadsheet in the document. In addition to this "interoperability," an application suite generally offers a consistent set of commands and features across the different applications, speeding up the learning process. The use of the applications in one package from one vendor simplifies technical support and upgrading. (The development of comparable applications suites for Linux is likely to increase that operating system's acceptance on the desktop.)

Applications suites have some potential disadvantages as compared to buying a separate program for each application. The user is not necessarily getting the best program in each application area, and he or she is also forced to pay for functionality that may not be needed or desired. Due to their size and complexity, software suites may not run well on older computers. Despite these problems, software suites sell very well and are ubiquitous in today's office. (For an alternative model for delivering software, see APPLICATION SERVICE PROVIDER.)

Further Reading

Villarosa, Joseph. "How Suite It Is: One-stop shopping for software can save you both time and money." *Forbes* magazine on-line. http://www.forbes.com/buyers/070.htm

Arithmetic Logic Unit (ALU)

The Arithmetic Logic Unit is the part of a computer system that actually performs calculations and logical comparisons on data. It is part of the central processing unit (CPU), and in practice there may be separate and multiple arithmetic and logic units (see CPU).

The ALU works by first retrieving a code that represents the operation to be performed (such as ADD). The code also specifies the location from which the data is to be retrieved and to which the results of the operation are to be stored. (For example, addition of the data from memory to a number already stored in a special accumulator register within the CPU, with the result to be stored back into the accumulator.) The operation code can also include a specification of the format of the data to be used (such as fixed or floating-point numbers)—the operation and format are often combined into the same code.

In addition to arithmetic operations, the ALU can also carry out logical comparisons, such as bitwise operations that compare corresponding bits in two data words, corresponding to Boolean operators such as AND, OR, and XOR (see BITWISE OPERATIONS and BOOLEAN OPERATORS).

The data or operand specified in the operation code is retrieved as words of memory that represent numeric data, or indirectly, character data (see MEMORY, NUMERIC DATA, and CHARACTERS AND STRINGS). Once the operation is performed, the result is stored (typically in a register in the CPU). Special codes are also stored in registers to indicate characteristics of the result (such as whether it is positive, negative, or zero). Other special conditions called exceptions indicate a problem with the processing. Common exceptions include overflow, where the result fills more bits than are available in the register, loss of precision (because there isn't room to store the necessary number of decimal places), or an attempt to divide by zero. Exceptions are typically indicated by setting a FLAG in the machine status register.

THE BIG PICTURE

Detailed knowledge of the structure and operation of the ALU is not needed by most programmers. Programmers who need to directly control the manipulation of data in the ALU and CPU write programs in assembly language (see ASSEMBLER) that specify the sequence of operations to be performed. Generally only the lowest-level operations involving the physical interface to hardware devices require this level of detail (see DEVICE DRIVER). Modern compilers can produce optimized machine code that is almost as efficient as directly-coded assembler. However, understanding the architecture of the ALU and CPU for a particular chip can help predict its advantages or disadvantages for various kinds of operations.

Further Reading

Kleitz, William. *Digital and Microprocessor Fundamentals: Theory and Applications.* Upper Saddle River, N.J.: Prentice Hall, 1999.

arithmetic operators and expressions

All programming languages provide operators to specify arithmetic functions. Some of them, such as addition +, subtraction –, multiplication *, and division /, are familiar from elementary school arithmetic (although the asterisk rather than the traditional x is used for multiplication in program code, to avoid confusion with the letter *x*). Additional operators found in languages such as C, C++ and Java include % (modulus, or remainder after division), ++ (adds one and stores the result back into the operand), and – – (decrement; subtracts one and stores the result back into the operand).

Operands are data items such as variables, constants, or literals (actual numbers) that are operated on by the operator. An operator is called unary if it takes just one operand (the increment operator ++ is an example). An operator that takes two operands is considered to be binary, and this is true of most arithmetic operations such as addition, multiplication, subtraction, and division.

A combination of operands and operators constitutes an arithmetic expression that evaluates to a particular value when the program runs. Thus in the C statement:

```
Total = SubTotal + Tax * Tax_Rate;
```

the value of the variable Tax is multiplied by the value of the variable Tax_Rate, the result is added to the value of SubTotal, and the result of the entire expression is stored in the variable Total. Compilers generally parse arithmetic expressions by converting them from an "infix" form (as in A + B) to a "postfix" form (as in + A B), resolving them into a simple form that is ready for conversion to machine code.

OPERATOR PRECEDENCE

The preceding example raises an important question. How does one know that the tax is to be multiplied by the tax rate and then the result added to the subtotal, as opposed to adding the subtotal and tax and multiplying the result by the tax rate? The former procedure is intuitively correct to human observers, but since computers lack intuition, specific rules of precedence are defined for operators. These rules, which are similar for all computer languages, tell the compiler that when code is generated for arithmetic operations, multiplications and divisions are carried out first (moving from left to right), and then additions and subtractions are resolved in the same way. The rules of precedence do become more complex when the relational, logical, and assignment operators are included. Finally, expressions can be enclosed in parenthesis to overrule precedence and force them to be evaluated. Thus in the expression (A + B) * C the addition will be carried out before the multiplication.

Generally speaking, the levels of precedence for most languages are as follows:

1. scope resolution operators (specify local vs. global versions of a variable)
2. invoking a method from a class; array subscript; function call; increment or decrement
3. size of (gets number of bytes in an object), address and pointer dereference, other unary operators (such as "not" and complement); creation and deallocation functions; type casts
4. class member selection through a pointer
5. multiplication, division, and modulus
6. addition and subtraction
7. left and right shift operators
8. less than and greater than
9. equal and not equal operators
10. bitwise operators (AND, then exclusive OR, inclusive OR)
11. logical operators (AND, then OR)
12. assignment statements

The basic arithmetic operators are built into each programming language, but many of the newer object-oriented languages such as C++ allow for programmer-defined operators and a process called overloading in which the same operator can be defined to work with several different kinds of data. Thus the + operator can be extended so that if it is given character strings instead of numbers, it will "add" the strings by combining (concatenating) them.

Further Reading

Stroustrup, Bjarne C. *The C++ Programming Language.* 3rd ed. Reading, Mass.: Addison-Wesley, 1997. See chap. 6 "Expressions and Statements" and chap. 11 "Operator Overloading."

"Summary of Operators in Java." http://sunsite.ccu.edu.tw/java/tutorial/java/nutsandbolts/opsummary.html

Arithmetic Operations and Expressions

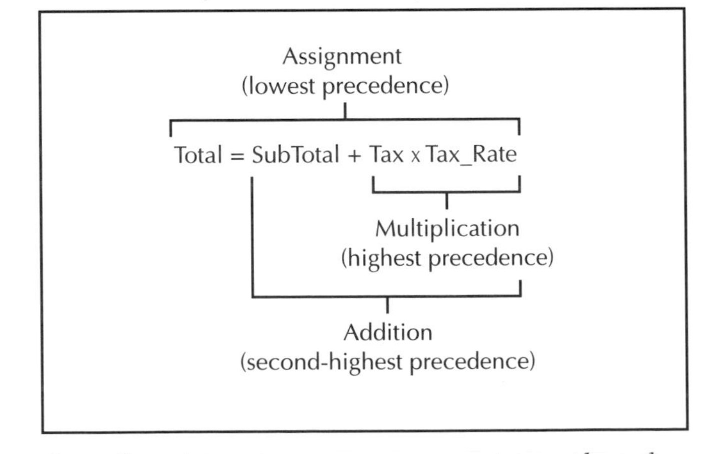

A compiler or interpreter processes a program statement by applying its operators in order of precedence. Here the multiplication is done first, and then its result is used in the addition. The assignment (=) operator has the lowest precedence and is applied last, assigning the value of the entire expression to the variable Total.

array

An array stores a group of similar data items in consecutive order. Each item is an *element* of the array, and it can be retrieved using a *subscript* that specifies the item's location relative to the first item. Thus in the C language, the statement

```
int Scores (10);
```

sets up an array called *Scores*, consisting of 10 integer values. The statement

```
Scores [5] = 93;
```

stores the value 93 in array element number 5. One subtlety, however, is that in languages such as C, the first element of the array is [0], so [5] represents not the fifth but the sixth element in *Scores*. (Many version of BASIC allow for setting either 0 or 1 as the first element of arrays.)

In languages such as C that have pointers, an equivalent way to access an array is to declare a pointer and store the address of the first element in it (see POINTERS AND INDIRECTION):

```
int * ptr;
ptr = &Scores [0];
```

(See POINTER.)

Arrays are useful because they allow a program to work easily with a group of data items without having to use separately named variables. Typically, a program uses a loop to traverse an array, performing the same operation on each element in order (see LOOP). For example, to print the current contents of the *Scores* array, a C program could do the following:

```
int index;
for (index = 0; i < 10; i++)

  printf ("Scores [%d] = %d \n", index, Scores
  [index]);
```

This program might print a table like this:

```
Scores [0] = 22
Scores [1] = 28
Scores [2] = 36
```

and so on. Using a pointer, a similar loop would increment the pointer to step to each element in turn.

An array with a single subscript is said to have one dimension. Such arrays are often used for simple data lists, strings of characters, or vectors. Most languages also support multidimensional arrays. For example, a two-dimensional array can represent X and Y coordinates, as on a screen display. Thus the number 16 stored at Colors[10][40] might represent the color of the point at X=10, Y=40 on a 640 by 480 display. A matrix is also a two-dimensional array, and languages such as APL provide built-in support for mathematical operations on such arrays. A four-dimensional array might hold four test scores for each person.

Array

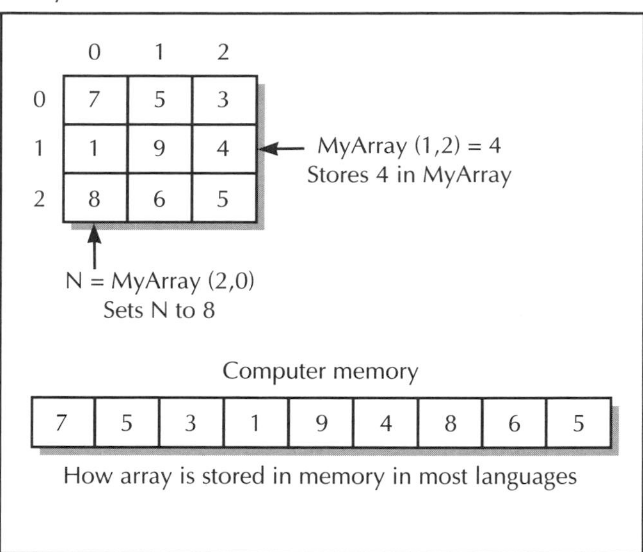

A two-dimensional array can be visualized as a grid, with the array subscripts indicating the row and column in which a particular value is stored. Here the value 4 is stored at the location (1,2), while the value at (2,0), which is 8, is assigned to N. As shown, the actual computer memory is a one-dimensional line of successive locations. In most computer languages the array is stored row by row.

Some languages such as FORTRAN 90 allow for defining "slices" of an array. For example, in a 3 x 3 matrix, the expression MAT(2:3, 1:3) references two 1 x 3 "slices" of the matrix array. Pascal allows defining a subrange, or portion of the subscripts of an array.

ASSOCIATIVE ARRAYS

It can be useful to explicitly associate pairs of data items within an array. In an *associative array* each data element has an associated element called a *key*. Rather than using subscripts, data elements are retrieved by passing the key to a hashing routine (see HASHING). In the Perl language, for example, an array of student names and scores might be set up like this:

```
%Scores = ("Henderson" => 86, "Johnson" => 87,
"Jackson" => 92);
```

The score for Johnson could later be retrieved using the reference:

```
$Scores ("Johnson")
```

Associative arrays are handy in that they facilitate look-up tables or can serve as small databases. However, expanding the array beyond its initial allocation requires rehashing all the existing elements.

PROGRAMMING ISSUES

To avoid error, any reference to an array must be within its declared *bounds*. For example, in the earlier example, Scores[9] is the last element, and a reference to Scores[10] would be out of bounds. Attempting to reference an out-of-bounds value gives an error message in some languages such as Pascal, but in others such as standard C and C++, it simply retrieves whatever happens to be in that location in memory.

Another issue involves the allocation of memory for the array. In a *static* array, such as that used in FORTRAN 77, the necessary storage is allocated before the program runs, and the amount of memory cannot be changed. Static arrays use memory efficiently and reduce overhead, but are inflexible, since the programmer has to declare an array based on the largest number of data items the program might be called upon to handle. A *dynamic* array, however, can use a flexible structure called a *heap* to allocate memory. The program can change the size of the array at any time while it is running. C and C++ programs can create dynamic arrays and allocate memory using special functions (malloc and free in C) or operators (new and delete in C++).

In the early days of microcomputer programming, arrays tended to be used as an all-purpose data structure for storing information read from files. Today, since there are more structured and flexible ways to store and retrieve such data, arrays are now mainly used for small sets of data (such as look-up tables).

Further Reading

Jensen, Ted. "A Tutorial on Pointers and Arrays in C." http://pw2.netcom.com/~tjensen/ptr/pointers.htm

Sebesta, Robert W. *Concepts of Programming Languages*. 4th ed. Reading, Mass.: Addison-Wesley, 1999.

art and the computer

While the artistic and technical temperaments are often viewed as opposites, the techniques of artists have always shown an intimate awareness of technology, including the physical characteristics of the artist's tools and media. The development of computer technology capable of generating, manipulating, displaying, or printing images has offered a variety of new tools for existing artistic traditions, as well as entirely new media and approaches.

Computer art began as an offshoot of research into image processing or the simulation of visual phenomena, such as by researchers at Bell Labs in Murray Hill, N.J., during the 1960s. One of these researchers, A. Michael Noll, applied computers to the study of art history by simulating techniques used by painters Piet Mondrian and Bridget Riley in order to gain a better understanding of them. In addition to exploring existing realms of art, experimenters began to create a new genre of art, based on the ideas of Max Bense, who coined the terms "artificial art" and "generative esthetics." Artists such as Manfred Mohr studied computer science because they felt the computer could provide the tools for an esthetic strongly influenced by mathematics and natural science. For example, Mohr's *P-159/A* (1973) used mathematical algorithms and a plotting device to create a minimalistic yet rich composition of lines. Other artists working in the minimalist, neoconstructivist, and conceptual art traditions found the computer to be a compelling tool for exploring the boundaries of form.

By the 1980s, the development of personal computers made digital image manipulation available to a much wider group of people interested in artistic expression, including the more conventional realms of representational art and photography. Programs such as Adobe Photoshop blend art

Air, *created by Lisa Yount with the popular image editing program Adobe Photoshop, is part of a group of photocollages honoring the ancient elements of earth, air, water, and fire. The "wings" in the center are actually the two halves of a mussel shell.*

and photography, making it possible to combine images from many sources and apply a variety of transformations to them. The use of computer graphics algorithms make realistic lighting, shadow, and fog effects possible to a much greater degree than their approximation in traditional media. Fractals can create landscapes of infinite texture and complexity. The computer has thus become a standard tool for both "serious" and commercial artists.

IMPLICATIONS AND PROSPECTS

While traditional artistic styles and genres can be reproduced with the aid of a computer, the computer has the potential to change the basic paradigms of the visual arts. The representation of all elements in a composition in digital form makes art fluid in a way that cannot be matched by traditional media, where the artist is limited in the ability to rework a painting or sculpture. Further, there is no hard-and-fast boundary between still image and animation, and the creation of art works that change interactively in response to their viewer becomes feasible. Sound, too, can be integrated with visual representation, in a way far more sophisticated than that pioneered in the 1960s with "color organs" or laser shows. Indeed, the use of virtual reality technology makes it possible to create art that can be experienced "from the inside," fully immersively (see VIRTUAL REALITY). The use of the Internet opens the possibility of huge collaborative works being shaped by participants around the world.

The growth of computer art has not been without misgivings. Many artists continue to feel that the intimate physical relationship between artist, paint, and canvas cannot be matched by what is after all only an arrangement of light on a flat screen. However, the profound influence of the computer on contemporary art is undeniable.

Further Reading

Computer-Generated Visual Arts (Yahoo). http://dir.yahoo.com/Arts/Visual_Arts/Computer_Generated/

Ashford, Janet. *Arts and Crafts Computer: Using Your Computer as an Artist's Tool.* Berkeley, Calif.: Peachpit Press, 2001.

Popper, Frank. *Art of the Electronic Age.* New York: Thames & Hudson, 1997.

Rush, Michael. *New Media in Late 20th-Century Art.* New York: Thames & Hudson, 1999.

artificial intelligence

The development of the modern digital computer following World War II led naturally to the consideration of the ultimate capabilities of what were soon dubbed "thinking machines" or "giant brains." The ability to perform calculations flawlessly and at superhuman speeds led some observers to believe that it was only a matter of time before the intelligence of computers would surpass human levels. This belief would be reinforced over the years by the development of computer programs that could play chess with increasing skill, culminating in the match victory of IBM's Deep Blue over world champion Garry Kasparov in 1997.

However, the quest for artificial intelligence would face a number of enduring challenges, the first of which is a lack of agreement on the meaning of the term *intelligence,* particularly in relation to such seemingly different entities as humans and machines. While chess skill is considered a sign of intelligence in humans, the game is deterministic in that optimum moves can be calculated systematically, limited only by the processing capacity of the computer. Human chess masters use a combination of pattern recognition, general principles, and selective calculation to come up with their moves. In what sense could a chess-playing computer that mechanically evaluates millions of positions be said to "think" in the way humans do? Similarly, computers can be provided with sets of rules that can be used to manipulate virtual building blocks, carry on conversations, and even write poetry. While all these activities can be perceived by a human observer as being intelligent and even creative, nothing can truly be said about what the computer might be said to be experiencing.

In 1950, computer pioneer Alan M. Turing suggested a more productive approach to evaluating claims of artificial intelligence in what became known as the Turing test (see TURING, ALAN). Basically, the test involves having a human interact with an "entity" under conditions where he or she does not know whether the entity is a computer or another human being. If the human observer, after engaging in teletyped "conversation" cannot reliably determine the identity of the other party, the computer can be said to have passed the Turing test. The idea behind this approach is that rather than attempting to precisely and exhaustively define intelligence, we will engage human experience and intuition about what intelligent behavior is like. If a computer can successfully imitate such behavior, then it at least may become problematic to say that it is *not* intelligent.

Computer programs have been able to pass the Turing test to a limited extent. For example, a program called ELIZA written by Joseph Weizenbaum can carry out what appears to be a responsive conversation on themes chosen by the interlocutor. It does so by rephrasing statements or providing generalizations in the way that a nondirective psychotherapist might. But while ELIZA and similar programs have sometimes been able to fool human interlocutors, an in-depth probing by the humans has always managed to uncover the mechanical nature of the response.

Although passing the Turing test could be considered evidence for intelligence, the question of whether a computer might have consciousness (or awareness of self) in the sense that humans experience it might be impossible to answer. In practice, researchers have had to confine themselves to producing (or simulating) intelligent *behavior,* and they have had considerable success in a variety of areas.

TOP-DOWN APPROACHES

The broad question of a strategy for developing artificial intelligence crystallized at a conference held in 1956 at Dartmouth College. Four researchers can be said to be founders of the field: Marvin Minsky (founder of the AI Laboratory at MIT), John McCarthy (at MIT and later, Stanford), and Herbert Simon and Allen Newell (developers of a mathematical problem-solving program called Logic Theorist at the Rand Corporation, who later founded the AI Laboratory at Carnegie Mellon University). The 1950s and 1960s were a time of rapid gains and high optimism about the future of AI (see MINSKY, MARVIN and MCCARTHY, JOHN).

Most early attempts at AI involved trying to specify rules that, together with properly organized data, can enable the machine to draw logical conclusions. In a *production system* the machine has information about "states" (situations) plus rules for moving from one state to another—and ultimately, to the "goal state." A properly implemented production system cannot only solve problems, it can give an explanation of its reasoning in the form of a chain of rules that were applied.

The program SHRDLU, developed by Marvin Minsky's team at MIT, demonstrated that within a simplified "microworld" of geometric shapes a program can solve problems and learn new facts about the world. Minsky later developed a more generalized approach called "frames" to provide the computer with an organized database of knowledge about the world comparable to that which a human child assimilates through daily life. Thus, a program with the appropriate frames can act as though it understands a story about two people in a restaurant because it "knows" basic facts such as that people go to a restaurant to eat, the meal is cooked for them, someone pays for the meal, and so on.

While promising, the frames approach seemed to founder because of the sheer number of facts and relationships needed for a comprehensive understanding of the world. During the 1970s and 1980s, however, expert systems were developed that could carry out complex tasks such as determining the appropriate treatment for infections (MYCIN) and analysis of molecules (DENDRAL). Expert systems combined rules of inference with specialized databases of facts and relationships. Expert systems have thus been able to encapsulate the knowledge of human experts and make it available in the field (see EXPERT SYSTEMS and KNOWLEDGE REPRESENTATION).

BOTTOM-UP APPROACHES

Several "bottom-up" approaches to AI were developed in an attempt to create machines that could learn in a more humanlike way. The one that has gained the most practical success is the neural network, which attempts to emulate the operation of the neurons in the human brain. Researchers believe that in the human brain perceptions or the acquisition of knowledge leads to the reinforcement of particular neurons and neural paths, improving the brain's ability to perform tasks. In the artificial neural network a large number of independent processors attempt to perform a task. Those that succeed are reinforced or "weighted," while those that fail may be negatively weighted. This leads to a gradual improvement in the overall ability of the system to perform a task such as sorting numbers or recognizing patterns (see NEURAL NETWORK).

The approach characterized as "artificial life" adds a genetic component in which the successful components pass on program code "genes" to their offspring. Thus, the power of evolution through natural selection is simulated, leading to the emergence of more effective systems (see ARTIFICIAL LIFE and GENETIC ALGORITHMS).

In general the top-down approaches have been more successful in performing specialized tasks, but the bottom-up approaches may have greater general application, as well as leading to cross-fertilization between the fields of artificial intelligence, cognitive psychology, and research into human brain function.

APPLICATION AREAS

While powerful artificial intelligence is not yet ubiquitous in everyday computing, AI principles are being successfully used in a number of application areas. These areas, which are all covered separately in this book, include

- Knowledge representation: development of expert systems for diagnosis and analysis in chemistry, medicine, and other fields.
- Natural language processing: creation of conversational approaches to querying databases, customer service calls, and other areas.
- Computer vision: developing the ability of a computer to "understand" the objects in a scene and their relationships.
- Robotics: Perhaps the greatest challenge is to develop an artificial intelligence that can interact with the complex everyday world that human beings encounter. A successful robot would have to master many disciplines of AI.
- Software agents: Sometimes called "bots," software agents perform autonomous tasks in cyberspace, such as looking for and evaluating competing offerings of merchandise.

PROSPECTS

The field of AI has been characterized by successive waves of interest in various approaches, and ambitious projects have often failed. However, expert systems and, too a lesser extent, neural networks have become the basis for viable products. Robotics and computer vision offer a significant potential payoff in industrial and military applications. The creation of software agents to help users navigate the complexity of the Internet is now of great commercial interest. The growth of AI has turned out to be a steeper and more complex path than originally anticipated. One view suggests steady progress. Another, shared by science fiction writers such as Vernor Vinge, suggests a breakthrough, perhaps arising from artificial life research, might someday create a true—but truly alien—intelligence.

Further Reading

American Association for Artificial Intelligence. "Welcome to AI Topics." http://www.aaai.org/Pathfinder/html/welcome.html

Feigenbaum, E. A. and J. Feldman, eds. *Computers and Thought.* New York: McGraw-Hill, 1963.

"An Introduction to the Science of Artificial Intelligence." http://library.thinkquest.org/2705/

Jain, Sanjay [and others]. *Systems that Learn: an Introduction to Learning Theory.* 2nd ed. Cambridge, Mass: MIT Press, 1999.

Kurzweil, Ray. *The Age of Spiritual Machines: When Computers Exceed Human Intelligence.* New York: Viking, 1999.

McCorduck, Pamela. *Machines Who Think.* San Francisco: W. H. Freeman, 1979.

Shapiro, Stuart C. *Encyclopedia of Artificial Intelligence.* 2nd ed. New York: Wiley, 1992.

artificial life (AL)

This is an emerging field that attempts to simulate the behavior of living things in the realm of computers and robotics. The field overlaps artificial intelligence (AI) since intelligent behavior is an aspect of living things. The design of a self-reproducing mechanism by John von Neumann in the mid-1960s was the first model of artificial life (see VON NEUMANN, JOHN). The field was expanded by the development of cellular automata as typified in John Conway's Game of Life in the 1970s, which demonstrated how simple components interacting according to a few specific rules could generate complex emergent patterns. A program by Craig Reynolds uses this principle to model the flocking behavior of simulated birds, called "boids." (See CELLULAR AUTOMATA.)

The development of genetic algorithms by John Holland added selection and evolution to the act of reproduction. This approach typically involves the setting up of numerous small programs with slightly varying code, and having them attempt a task such as sorting data or recognizing patterns. Those programs that prove most "fit" at accomplishing the task are allowed to survive and reproduce. In the act of reproduction, biological mechanisms such as genetic mutation and crossover are allowed to intervene (see GENETIC ALGORITHMS). A rather similar approach is found in the neural network, where those nodes that succeed better at the task are given greater "weight" in creating a composite solution to the problem (see NEURAL NETWORK).

A more challenging but interesting approach to AL is to create actual robotic "organisms" that navigate in the physical rather than the virtual world. Roboticist Hans Moravec of the Stanford AI Laboratory and other researchers have built robots that can deal with unexpected obstacles by improvisation, much as people do, thanks to layers of software that process perceptions, fit them to a model of the world, and make plans based on goals. But such robots, built as full-blown designs, share few of the characteristics of artificial life. As with AI, the bottom-up approach offers a different strategy that has been called "fast, cheap, and out of control"—the production of numerous small, simple, insectlike robots that have only simple behaviors, but are potentially capable of interacting in surprising ways. If a meaningful genetic and reproductive mechanism can be included in such robots, the result would be much closer to true artificial life (see ROBOTICS).

The philosophical implications arising from the possible development of true artificial life are similar to those involved with "strong AI." Human beings are used to viewing themselves as the pinnacle of a hierarchy of intelligence and creativity. However, artificial life with the capability of rapid evolution might quickly outstrip human capabilities, perhaps leading to a world like that portrayed by science fiction writer Gregory Benford, where flesh-and-blood humans become a marginalized remnant population.

Further Reading

"ALife On-line 2.0." http://alife.org/

"Karl Sims Retrospective." http://www.biota.org/ksims/

Langton, Christopher G., ed. *Artificial Life: an Overview.* Cambridge, Mass.: MIT Press, 1995.

Levy, Stephen. *Artificial Life: the Quest for a New Creation.* New York: Pantheon Books, 1992.

assembler

All computers at bottom consist of circuits that can perform a repertoire of mathematical or logical operations. The earliest computers were programmed by setting switches for operations and manually entering numbers in working storage, or memory. A major advance in the flexibility of computers came with the idea of stored programs, where a set of instructions could be read in and held in the machine in the same way as other data. These instructions were in machine language, consisting of numbers representing instructions (operations to be performed) and other numbers representing the address of data to be manipulated (or an address containing the address of the data, called indirect addressing—see ADDRESSING). Operations include basic arithmetic (such as addition), the movement of data between storage (memory) and special processor locations called registers, and the movement of data from an input device (such as a card reader) and an output device (such as a printer).

Writing programs in machine code is obviously a tedious and error-prone process, since each operation must be specified using a particular numeric instruction code together with the actual addresses of the data to be used. It soon became clear, however, that the computer could itself be used to keep track of binary codes and actual addresses, allowing the programmer to use more human-friendly names for instructions and data variables. The program that translates between symbolic language and machine language is the assembler.

With a symbolic assembler, the programmer can give names to data locations. Thus, instead of saying (and having to remember) that the quantity Total will be in location &H100, the program can simply define a two-byte chunk of memory and call it Total:

```
Total DB
```

The assembler will take care of assigning a physical memory location and, when instructed, retrieving or storing the data in it.

Most assemblers also have macro capability. This means that the programmer can write a set of instructions (a procedure) and give it a name. Whenever that name is used in the program, the assembler will replace it with the actual code for the procedure and plug in whatever variables are specified as operands (see MACRO).

APPLICATIONS

In the mainframe world of the 1950s, the development of assembly languages represented an important first step toward symbolic programming; higher-level languages such as FORTRAN and COBOL were developed so that programmers could express instructions in language that was more like mathematics and English respectively. High-level languages offered greater ease of programming and source code that was easier to understand (and thus to maintain). Gradually, assembly language was reserved for systems programming and other situations where efficiency or the need to access some particular hardware capability required the exact specification of processing (see SYSTEMS PROGRAMMING and DEVICE DRIVER).

During the 1970s and early 1980s, the same evolution took place in microcomputing. The first microcomputers typically had only a small amount of memory (perhaps 8–64K), not enough to compile significant programs in a high-level language (with the partial exception of some versions of BASIC.) Applications such as graphics and games in particular were written in assembly language for speed. As available memory soared into the hundreds of kilobytes and then megabytes, however, high level languages such as C and C++ became practicable, and assembly language began to be relegated to systems programming, including device drivers and other programs that had to interact directly with the hardware.

Assembler

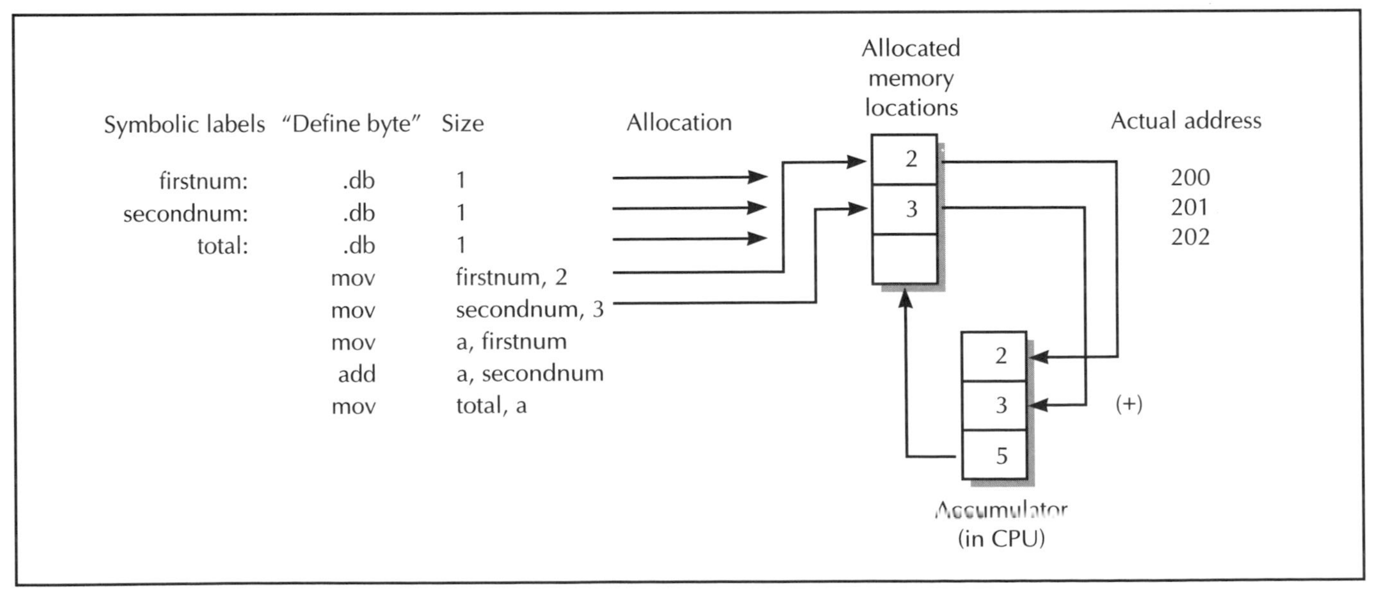

In this assembly language example, the "define byte" (.db) directive is used to assign one memory byte to the symbolic names (variables) firstnum, secondnum, and total. The two mov commands then load 2 and 3 into firstnum and secondnum respectively. Firstnum is then loaded into the processor's accumulator (a), and secondnum is then added to it. Finally, the sum is moved into the memory location labeled total.

While many people learning programming today receive little or no exposure to assembly language, some understanding of this detailed level of programming is still useful because it illustrates fundamentals of computer architecture and operation.

Further Reading

Abel, Peter. *IBM PC Assembly Language and Programming*. 5th ed. Upper Saddle River, N.J.: Prentice Hall, 2001.

Duntemann, Jeff. *Assembly Language Step by Step: Programming with DOS and Linux*. 2nd ed. New York: Wiley, 2000.

Miller, Karen. *An Assembly Language Introduction to Computer Architecture Using the Intel Pentium*. New York: Oxford University Press, 1999.

Atanasoff, John Vincent

(1903–1995)
American
Computer Engineer

John V. Atanasoff is considered by many historians to be the inventor of the modern electronic computer. He was born October 4, 1903, in Hamilton, New York. As a young man, Atanasoff showed considerable interest in and a talent for electronics. His academic background (B.S. in electrical engineering, Florida State University, 1925; M.S. in mathematics, Iowa State College, 1926; and Ph.D. in experimental physics, University of Wisconsin, 1930) well equipped him for the design of computing devices. He taught mathematics and physics at Iowa State until 1942, and during that time, he conceived the idea of a fully electronic calculating machine that would use vacuum tubes for its arithmetic circuits and would store binary numbers on a rotating drum memory that used high and low charges on capacitors. Atanasoff and his assistant Clifford E. Berry built a successful computer called ABC (Atanasoff-Berry Computer) using this design in 1942. (By that time he had taken a wartime research position at the Naval Ordnance Laboratory in Washington, D.C.)

The ABC was a special-purpose machine designed for solving up to 29 simultaneous linear equations using an algorithm based on Gaussian elimination to eliminate a specified variable from a pair of equations. Because of inherent unreliability in the system that punched cards to hold the many intermediate results needed in such calculations, the system was limited in practice to solving five or fewer sets of equations.

Despite its limitations, the ABC's design proved the feasibility of fully electronic computing, and similar vacuum tube switching and regenerative memory circuits were soon adopted in designing the ENIAC and EDVAC, which unlike the ABC, were general-purpose electronic computers. Equally important was Atanasoff's use of capacitors to store data in memory electronically: The descendent of his capacitors can be found in the DRAM chips in today's computers.

According to a federal court it was John Atanasoff, not John Mauchly and Presper Eckert, who built the first digital computer. At any rate, the "ABC" or Atanasoff-Berry Computer represented a pioneering achievement in the use of binary logic circuits for computation. (PHOTO COURTESY OF IOWA STATE UNIVERSITY)

When Atanasoff returned to Iowa State in 1948, he discovered that the ABC computer had been dismantled to make room for another project. Only a single memory drum and a logic unit survived. Iowa State granted him a full professorship and the chairmanship of the physics department, but he never returned to that institution. Instead, he founded the Ordnance Engineering Corporation in 1952, which grew to a 100-person workforce before he sold the firm to Aerojet General in 1956. He then served as a vice president at Aerojet until 1961.

Atanasoff then semi-retired, devoting his time to a variety of technical interests (he had more than 30 patents to his name by the time of his death). However, when Sperry Univac (owner of Eckert and Mauchly's computer patents) began demanding license fees from competitors in the mid-1960s, the head lawyer for one of these competitors, Honeywell, found out about Atanasoff's work on the ABC and enlisted his aid as a witness in an attempt to overturn the patents. After prolonged litigation, Judge Earl Richard

Larson ruled in 1973 that the two commercial computing pioneers had learned key ideas from Atanasoff's apparatus and writings and that their patent was invalid because of this "prior art."

Atanasoff received numerous awards for his work for the Navy on acoustics and for his pioneering computer work. These awards included the IEEE Computer Pioneer Award (1984) and the National Medal of Technology (1990). In addition, he had both a hall at Iowa State University and an asteroid (3546-Atanasoff) named in his honor. John Atanasoff died on June 15, 1995, in Monrovia, Maryland.

Further Reading

Burks, A. R., and A. W. Burks. *The First Electronic Computer: the Atanasoff Story.* Ann Arbor, Mich: University of Michigan Press, 1988.

Lee, J. A. N. *Computer Pioneers.* Los Alamitos, Calif.: IEEE Computer Society Press, 1995.

"Reconstruction of the Atanasoff-Berry Computer (ABC)." http://www.scl.ameslab.gov/ABC/ABC.html

auctions, on-line

By the late 1990s, millions of computer users had discovered a new way to buy and sell an immense variety of items ranging from traditional collectibles to the exotic (such as a working German Enigma encoding machine.) By 1999, the on-line auction market was estimated at $4.5 billion, almost tripling the previous year's revenue. The largest site, eBay, founded in 1995, has more than 5 million items on offer at any given time. The giant online bookstore Amazon.com and the Web portal Yahoo! have also established on-line auction services.

PROCEDURES

On-line auctions differ from traditional auctions in several ways. Traditional auction firms generally charge the seller and buyer a commission of around 10 percent of the sale or "hammer" price. On-line auctions charge the buyer nothing, and the seller typically pays a fee of about 3–5 percent of the amount realized. On-line auctions can charge much lower fees because unlike traditional auctions, there is no live auctioneer, no catalogs to produce, and little administration, since all payments pass from buyer to seller directly.

An on-line auction is like a mail bid auction in that bids can be posted at any time during the several days a typical auction runs. A buyer specifies a maximum bid and if he or she becomes the current high bidder, the high bid is adjusted to a small increment over the next highest bid. As with a "live" auction, however, bidders can revise their bids as many times as they wish until the close of the auction. An important difference between online and traditional live auctions is that a traditional auction ends as soon as no one is willing to top the current high bid. With an on-line auction, the bidding ends at the posted ending time. This has led to a tactic known as "sniping," where some bidders submit a bid just over the current high bid just before the auction ends, such that the previous high bidder has no time to respond.

FUTURE AND IMPLICATIONS

On-line auctions have become very popular, and an increasing number of people run small businesses by selling items through auctions. The markets for traditional collectibles such as coins and stamps have been considerably affected by on-line auctions. Knowledgeable buyers can often obtain items for considerably less than a dealer would charge, or sell items for more than a dealer would pay. However, many items are overpriced compared to the normal market, and faked or ill-described items can be a significant problem. Attempts to hold the auction service legally responsible for such items are met with the response that the auction service is simply a facilitator for the seller and buyer and doesn't play the role of traditional auctioneers who catalog items and provide some assurance of authenticity. If courts or regulators should decide that on-line auctions must bear this responsibility, the cost of using the service may rise or the variety of items that can be offered may be restricted.

Further Reading

"AW Looks at the Year of the On-line Auction." http://www.auctionwatch.com/awdaily/features/yearin/index.html

Cohen, Adam. *The Perfect Store: Inside eBay.* Boston: Little, Brown, 2002.

Kovel, Ralph M., and Terry H. Kovel. *Kovels' Bid, Buy, and Sell On-line: Basic Auction Information and Tricks of the Trade.* New York: Three Rivers Press, 2001.

Prince, Dennis L. *Starting Your On-line Auction Business.* Roseville, Calif.: Prima Tech, 2000.

auditing in data processing

The tremendous increase in the importance and extent of information systems for all aspects of commerce and industry has made it imperative that businesses be able to ensure the accuracy and integrity of their accounting systems and corporate databases. Errors can result in loss of revenue and even exposure to legal liability.

Auditing involves the analysis of the security and accuracy of software and the procedures for using it. For example, sample data can be extracted using automated scripts or other software tools and examined to determine whether correct and complete information is being entered into the system, and whether the reports on which management relies for decision making are accurate. Auditing is also needed to confirm that data reported to regulatory agencies meets legal requirements.

In addition to confirming the reliability of software and procedures, auditors must necessarily also be concerned with issues of security, since attacks or fraud involving computer systems can threaten their integrity or reliability (see COMPUTER CRIME AND SECURITY). The safeguarding of customer privacy has also become a sensitive concern (see PRIVACY IN THE DIGITAL AGE). To address such issues, the auditor must have a working knowledge of basic psychology and human relations, particularly as they affect large organizations.

Auditors recommend changes to procedures and practices to minimize the vulnerability of the system to both human and natural threats. The issues of backup and archiving and disaster recovery must also be addressed (see BACKUP AND ARCHIVE SYSTEMS). As part accountant and part systems analyst, the information systems auditor represents a bridging of traditional practices and rapidly changing technology.

Further Reading

Hall, James. *Information Systems Auditing and Assurance.* Cincinnati, Ohio: South Western College, 2000.

"Information Systems Auditing and Control Association and Foundation." http://www.isaca.org/isacafx.htm

Krist, Martin A. *A Standard for Auditing Computer Applications.* Boca Raton, Fla.: Auerbach, 1999.

Taylor, Donald H., and G. William Glezen. *Auditing: an Assertions Approach.* 7th ed. New York: Wiley, 1997.

authentication

This process by which two parties in a communication or transaction can assure each other of their identity is a fundamental requirement for any transaction not involving cash, such as the use of checks or credit or debit cards. (In practice, for many transactions, authentication is "one way"—the seller needs to know the identity of the buyer or at least have some way of verifying the payment, but the buyer need not confirm the identity of the seller—except, perhaps in order to assure proper recourse if something turns out to be wrong with the item purchased.)

Traditionally, authentication involves paper-based identification (such as driver's licenses) and the making and matching of signatures. Since such identification is relatively easy to fake, there has been growing interest in the use of characteristics such as voice, facial measurements, or the patterns of veins in the retina that can be matched uniquely to individuals (see BIOMETRICS). Biometrics, however, requires the physical presence of the person before a suitable device, so it is primarily used for guarding entry into high-security areas.

AUTHENTICATION IN ON-LINE SYSTEMS

Since many transactions today involve automated systems rather than face-to-face dealings, authentication systems generally involve the sharing of information unique to the parties. The PIN used with ATM cards is a common example: It protects against the physical diversion of the card by requiring information likely known only to the legitimate owner. In e-commerce, there is the additional problem of safeguarding sensitive information such as credit card numbers from electronic eavesdroppers or intruders. Here a system is used by which information is encrypted before it is transmitted over the Internet. Encryption can also be used to verify identity through a digital signature, where a message is transformed using a "one-way function" such that it is highly unlikely that a message from any other sender would have the same encrypted form (see ENCRYPTION). The most widespread system is public key cryptography, where each person has a public key (known to all interested parties) and a private key that is kept secret. Because of the mathematical relationship between these two keys, the reader of a message can verify the identity of the sender or creator.

The choice of technology or protocol for authentication depends on the importance of the transaction, the vulnerability of information that needs to be protected, and the consequences of failing to protect it. A website that is providing access to a free service in exchange for information about users will probably not require authentication beyond perhaps a simple user/password pair. An on-line store, on the other hand, needs to provide a secure transaction environment both to prevent losses and to reassure potential customers that shopping on-line does not pose an unacceptable risk.

Authentication ultimately depends on a combination of technological and social systems. For example, cryptographic keys or "digital certificates" can be deposited with a trusted third party such that a user has reason to believe that a business is who it says it is.

Further Reading

Oppliger, Rolf. *Authentication Systems for Secure Networks.* Boston: Artech House, 1996.

Tung, Brian. *Kerberos: a Network Authentication System.* Reading, Mass.: Addison Wesley, 1999.

authoring systems

Multimedia presentations such as computer-based-training (CBT) modules are widely used in the corporate and educational arenas. Programming such a presentation in a high-level language such as C++ (or even Visual Basic) involves writing code for the detailed arrangement and control of graphics, animation, sound, and user interaction. Authoring systems offer an alternative way to develop presentations or courses. The developer specifies the sequence of graphics, sound, and other events, and the authoring system generates a finished program based on those specifications.

Authoring systems can use a variety of models for organizing presentations. Some use a scripting language that specifies the objects to be used and the actions to be performed (see SCRIPTING LANGUAGES). A scripting language uses many of the same features as a high-level programming language, including the definition of variables and the use of control structures (decision statements and loops). Programs such as the ubiquitous Hypercard (for the Macintosh) and Asymetrix Toolbook for Windows organize presentations into segments called "cards," with instructions fleshed out in a scripting language.

As an alternative, many modern authoring systems such as Discovery Systems' CourseBuilder use a graphical approach to organizing a presentation. The various objects (such as graphics) to be used are represented by icons, and the icons are connected with "flow lines" that describe the sequence of actions, serving the same purpose as control structures in programming languages. This "iconic" type of authoring system is easiest for less experienced programmers to use and makes the creation of small presentations fast and easy. Such systems may become more difficult to use for lengthy presentations (due to the number of symbols and connectors involved), and speed of the finished program can be a problem. Other popular models for organizing presentations include the "timeline" of Macromedia Flash, which breaks the presentation into "movies" and specifies actions for each frame, as well as providing multiple layers to facilitate animation. With the migration of many presentations to the Internet, the ability of authoring systems to generate HTML (or DHTML) code is becoming important.

Further Reading

Makedon, Fillia, and Samuel A. Rebelsky, ed. *Electronic Multimedia Publishing: Enabling Technologies and Authoring Issues.* Boston: Kluwer Academic, 1998.

"Multimedia Authoring Systems FAQ." http://www.tiac.net/users/jasiglar/MMASFAQ.HTML

Murray, T. "Authoring Intelligent Tutoring Systems: An analysis of the state of the art." *International J. of Artificial Intelligence in Education* 10 (1999), 98–129.

Wilhelm, Jeffrey D., Paul Friedman, and Julie Erickson. *Hyperlearning: where Projects, Inquiry, and Technology Meet.* York, Me.: Stenhouse, 1998.

automatic programming

From the beginning of the computer age, computer scientists have grappled with the fact that writing programs in any computer language, even relatively high-level ones such as FORTRAN or C, requires painstaking attention to detail. While language developers have responded to this challenge by trying to create more "programmer friendly" languages such as COBOL with its English-like syntax, another approach is to use the capabilities of the computer to automate the task of programming itself. It is true that any high-level language compiler does this to some extent (by translating program statements into the underlying machine instructions), but the more ambitious task is to create a system where the programmer would specify the problem and the system would generate the high-level language code. In other words, the task of programming, which had already been abstracted from the machine code level to the assembler level and from that level to the high-level language, would be abstracted a step further.

During the 1950s, researchers began to apply artificial intelligence principles to automate the solving of mathematical problems (see ARTIFICIAL INTELLIGENCE). For example, in the 1950s Anthony Hoare introduced the definition of pre-conditions and post-conditions to specify the states of the machine as it proceeds toward an end state (the solution of the problem). The program Logic Theorist demonstrated that a computer could use a formal logical calculus to solve problems from a set of conditions or axioms. Techniques such as deductive synthesis (reasoning from a set of programmed principles to a solution) and transformation (step-by-step rules for converting statements in a specification language into the target programming language) allowed for the creation of automated programming systems, primarily in mathematical and scientific fields (see also PROLOG).

The development of the expert system (combining a knowledge base and inference rules) offered yet another route toward automated programming (see EXPERT SYSTEMS). Herbert Simon's 1963 Heuristic Compiler was an early demonstration of this approach.

APPLICATIONS

Since many business applications are relatively simple in logical structure, practical automatic principles have been used in developing application generators that can create, for example, a database management system given a description of the data structures and the required reports. While some systems output code in a language such as C, others generate scripts to be run by the database management software itself (for example, Microsoft Access).

To simplify the understanding and specification of problems, a visual interface is often used for setting up the application requirements. Onscreen objects can represent items such as data files and records, and arrows or other connecting links can be dragged to indicate data relationships.

The line between automated program generators and modern software development environments is blurry. A programming environment such as Visual Basic encapsulates a great deal of functionality in objects called *controls,* which can represent menus, lists, buttons, text input boxes, and other features of the Windows interface, as well as other functionalities (such as a Web browser). The Visual Basic programmer can design an application by assembling the appropriate interface objects and processing tools, set properties (characteristics), and write whatever additional code is necessary. While not completely automating programming, much of the same effect can be achieved.

Further Reading

Andrews, James H. *Logic Programming: Operational Semantics and Proof Theory.* New York: Cambridge University Press, 1992.

"Automatic Programming Server." http://www.cs.utexas.edu/users/novak/cgi/apserver.cgi

"Programming and Problem Solving by Connecting Diagrams." http://www.cs.utexas.edu/users/novak/cgi/vipdemo.cgi

Tahid, Walid, ed. *Semantics, Applications and Implementation of Program Generation.* New York: Springer-Verlag, 2000.

awk

This is a scripting language developed under the UNIX operating system (see SCRIPTING LANGUAGES) by Alfred V. Aho, Brian W. Kernighan, and Peter J. Weinberger in 1977. (The name is an acronym from the their last initials.) The language builds upon many of the pattern matching utilities of the operating system and is designed primarily for the extraction and reporting of data from files. A number of variants of awk have been developed for other operating systems such as DOS.

As with other scripting languages, an awk program consists of a series of commands read from a file by the awk interpreter. For example the following UNIX command line:

```
awk -f MyProgram > Report
```

reads awk statements from the file MyProgram into the awk interpreter and sends the program's output to the file Report.

LANGUAGE FEATURES

An awk statement consists of a *pattern* to match and an *action* to be taken with the result (although the pattern can be omitted if not needed). Here are some examples:

```
{print $1} # prints the first field of every
   line of # input (since no pattern is speci-
   fied)
/debit/ {print $2} # print the second field of
   every # line that contains the word "debit"
if ( Code == 2 ) # if Code equals 2, print third
   field
print $3 # of each line
```

Pattern matching uses a variety of regular expressions familiar to UNIX users. Actions can be specified using a limited but adequate assortment of control structures similar to those found in C. There are also built-in variables (including counters for the number of lines and fields), arithmetic functions, useful string functions for extracting text from fields, and arithmetic and relational operators. Formatting of output can be accomplished through the versatile (but somewhat cryptic) print function familiar to C programmers.

Awk became popular for extracting reports from data files and simple databases on UNIX systems. For more sophisticated applications it has been supplanted by Perl, which offers a larger repertoire of database-oriented features (see PERL).

Further Reading

Aho, Alfred V., Brian Kernighan, and Peter J. Weinberger. *The awk Programming Language.* Reading, Mass.: Addison-Wesley, 1998.

"Awk Quick Reference." http://www.geocities.com/ResearchTriangle/Lab/1059/awkqref.zip

B

Babbage, Charles

(1791–1871)
British
Mathematician, Inventor

Charles Babbage made wide-ranging applications of mathematics to a variety of fields including economics, social statistics, and the operation of railroads and lighthouses. Babbage is best known, however, for having conceptualized the key elements of the general-purpose computer nearly a century before the dawn of electronic digital computing.

As a student at Trinity College, Cambridge, Babbage was already making contributions to the reform of calculus, championing new European methods over the Newtonian approach still clung to by British mathematicians. But Babbage's interests were shifting from the theoretical to the practical. Britain's growing industrialization as well as its worldwide interests increasingly demanded accurate numeric tables for navigation, actuarial statistics, interest rates, and engineering parameters. All tables had to be hand-calculated, a long process that inevitably introduced numerous errors. Babbage began to consider the possibility that the same mechanization that was revolutionizing industries such as weaving could be turned to the automatic calculation of numeric tables.

Starting in 1820, Babbage began to build a mechanical calculator called the difference engine. This machine used series of gears to accumulate additions and subtractions (using the "method of differences") to generate tables. His small demonstration model worked well, so Babbage undertook the full-scale "Difference Engine Number One," a machine that would have about 25,000 moving parts and would be able to calculate up to 20 decimal places. Unfortunately, Babbage was unable, despite financial support from the British government, to overcome the difficulties inherent in creating a mechanical device of such complexity with the available machining technology.

Undaunted, Babbage turned in the 1830s to a new design that he called the Analytical Engine. Unlike the Difference Engine, the new machine was to be programmable using instructions read in from a series of punch cards (as in the Jacquard loom). A second set of cards would contain the variables, which would be loaded into the "store"—a series of wheels corresponding to memory in a modern computer. Under control of the instruction cards, numbers could be moved between the store and the "mill" (corresponding to a modern CPU) and the results of calculations could be sent to a printing device.

Collaborating with Ada Lovelace (who translated his lecture transcripts by L. F. Menebrea) Babbage wrote a series of papers and notes that explained the workings of the proposed machine, including a series of "diagrams" (programs) for performing various sorts of calculations.

Building the Analytical Engine would have been a far more ambitious task than the special-purpose Difference Engine, and Babbage made little progress in the actual construction of the device. Although Babbage's ideas would remain obscure for nearly a century, he would then be recognized as having designed most of the key elements of the modern computer: the central processor, memory, instructions, and data organization. Only in the lack of a capability to manipulate memory addresses did the design fall short of a modern computer.

Further Reading

"The Analytical Engine: the First Computer." http://www.fourmilab.ch/babbage/

Babbage, Henry Prevost, ed. *Babbage's Calculating Engines: A Collection of Papers.* With a new introduction by Allan G. Bromley. Los Angeles: Tomash, 1982.

Campbell-Kelly, M., ed. *The Works of Charles Babbage.* 11 vols. London: Picerking and Chatto, 1989.

[Charles Babbage Resources.] http://atschool.eduweb.co.uk/jralston/frbabb.html

Swade, Doron D. "Redeeming Charles Babbage's Mechanical Computer." *Scientific American,* February 1993.

If it had been built, Charles Babbage's Analytical Engine, although mechanical rather than electrical, would have had most of the essential features of modern computers. These included punched card input, a processor, a memory (store), and a printer. (PHOTO COURTESY OF NMPFT/SCIENCE & SOCIETY PICTURE LIBRARY)

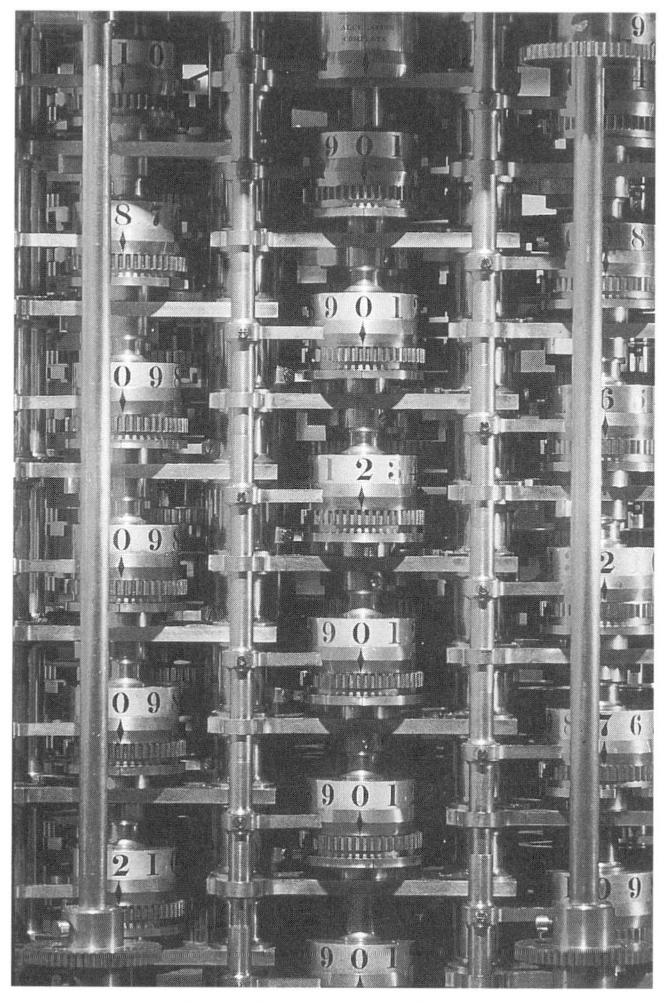

A reconstruction of part of the Babbage Difference Engine, displayed at the London Science Museum. (PHOTO COURTESY OF NMPFT/SCIENCE & SOCIETY PICTURE LIBRARY)

backup and archive systems

The need to create backup copies of data has become increasingly important as dependence on computers has grown and the economic value of data has increased. Potential threats to data include bugs in the operating system or software applications, malicious acts such as the introduction of computer viruses, theft, hardware failure (such as in hard disk drives), power outages, fire, and natural disasters such as earthquakes and floods.

A variety of general principles must be considered in devising an overall strategy for creating and maintaining backups:

Reliability: Is there assurance that the data is stored accurately on the backup medium, and will automatic backups run reliably as scheduled? Can the data be accurately retrieved and restored if necessary?

Physical storage: Is the backed-up data stored securely and organized in a way to make it easy to retrieve particular disks or tapes? Is the data stored at the site where it is to be used, or off-site (guarding against fire or other disaster striking the workplace).

Ease of Use: To the extent backups must be set up or initiated by human operators, is the system easy to understand and use with minimal training? Ease of use both promotes reliability (because users will be more likely to perform the backups), and saves money in training costs.

Economy: How does a given system compare to others in terms of the cost of the devices, software, media (such as tapes or cartridges), training, and administration?

CHOICE OF METHODS

The actual choice of hardware, software, and media depends considerably on how much data must be backed up (and how often) as well as whether the data is being generated on individual PCs or being stored at a central location. (See FILE SERVER, DATA WAREHOUSE.)

Backups for individual PCs can be accomplished using the backup software that comes with various versions of Microsoft Windows or through third-party software. The most commonly used media are disks such as Zip or Jazz (each storing up to a few hundred megabytes), removable cartridge drives (up to about 2GB per cartridge), various types of tape, or CD-ROMs (since CD read/write drives are now inexpensive). See CD-ROM, HARD DISK, and TAPE DRIVES.)

In addition to backing up documents or other data generated by users, the operating system and applications software is often backed up to preserve configuration information that would otherwise be lost if the program were reinstalled. (The backing up of software for archival purposes is different from making additional copies of a program for multiple users, which usually requires a license from the software manufacturer.) There are utilities for Microsoft Windows and other operating systems that simplify the backing up of configuration information by identifying and backing up only those files (such as the Windows Registry) that contain information particular to the installation.

The widespread use of local area networks makes it easier to back up data automatically from individual PCs and to store data at a central location (see LOCAL AREA NETWORK and FILE SERVER). However, having all data eggs in one basket increases the importance of building reliability and redundancy into the storage system, including the use of RAID (multiple disk arrays), "mirrored" disk drives, and uninterruptible power supplies (UPS). Despite such measures, the potential risk in centralized storage has led to advocacy of a "replication" system, preferably at the operating system level, that would automatically create backup copies of any given object at multiple locations on the network.

Another alternative of growing interest is the use of the Internet to provide remote (off-site) backup services. According to market researcher International Data Corp., the market for on-line backup services will grow from $140 million in 2000 to $4.8 billion in 2003. On-line backup offers ease of use (the backups can be run automatically and the service is particularly handy for laptop computer users on the road) and the security of off-site storage, but raise questions of privacy and security of sensitive information, particularly if encryption is not built into the process. On-line data storage is also provided to individual users by a variety of Internet service providers seeking to expand the stagnant ISP business. Application Service Providers (ASPs) also have a natural entry into the on-line storage market since they already host the applications their users use to create data. (See APPLICATION SERVICE PROVIDER.)

A practice that still persists in some mainframe installations is the tape library, which maintains an archive of data on tape that can be retrieved and mounted as needed.

ARCHIVING

Although using much of the same technology as making backups, archiving of data is different in its objectives and needs. An archive is a store of data that is no longer needed for routine current use, but must be retrievable upon demand, such as the production of bank records or e-mail as part of a legal process. (Data may also be archived for historical or other research purposes.) Since archives may have to be maintained for many years (even indefinitely), the ability of the medium (such as tape) to maintain data in readable condition becomes an important consideration. Besides physical deterioration, the obsolescence of file formats can also render archived data unusable.

MANAGEMENT CONSIDERATIONS

If backups must be initiated by individual users, the users must be trained in the use of the backup system and motivated to make backups, a task that is easy to put off to another time. Even if the backup is fully automated, sample backup disks or tapes should be checked periodically to make sure that data could be restored from them. Backup practices should be coordinated with disaster recovery and security policies.

Further Reading

Hodge, Gail M. "Best Practices for Digital Archiving: an Information Life-Cycle Approach," *D-Lib Magazine*, vol. 6, no. 1, January 2000.http://www.dlib.org/dlib/january00/01hodge.html

McMains, John R., and Bob Chronister. *Windows NT Backup & Recovery.* Berkeley, Calif.: Osborne McGraw Hill, 1998.

Preston, W. Curtis, and Gigi Esterbrook. *UNIX Backup and Recovery.* Sebastopol, Calif.: O'Reilly, 1999.

Backus-Naur Form

As the emerging discipline of computer science struggled with the need to precisely define the rules for new programming languages, the Backus-Naur Form (BNF) was devised as a notation for describing the precise grammar of a computer language. BNF represents the unification of separate work by John W. Backus and Peter Naur in 1958, when they were trying to write a specification for the Algol language.

A series of BNF statements defines the syntax of a language by specifying the combinations of symbols that constitute valid statements in the language.

Thus in a hypothetical language a program can be defined as follows:

```
<program> ::= program
    <declaration_sequence>
  begin
    <statements_sequence>
  end ;
```

Here the symbol ::= means "is defined as" and items in brackets <> are *metavariables* that represent placeholders for valid symbols. For example, <declaration_sequence> can consist of a number of different statements defined elsewhere.

Statements in square brackets [] indicate optional elements. Thus the If statement found in most programming languages is often defined as:

```
<if_statement> ::= if <boolean_expression> then
    <statement_sequence>
  [ else
<statement_sequence> ]
  end if ;
```

This can be read as "an If statement consists of a boolean_expression (something that evaluates to "true" or "false") followed by one or more statements, followed by an optional *else* that in turn is followed by one or more statements, followed by the keywords end if." Of course each item in angle brackets must be further defined—for example, a boolean_expression.

Curly brackets {} specify an item that can be repeated one or more times. For example, in the definition

```
<identifier> ::= <letter> { <letter> | <digit> }
```

An identifier is defined as a letter followed by one or more instances of either a letter or a digit.

An extended version of BNF (EBNF) offers operators that make definitions more concise yet easier to read. The preceding definition in EBNF would be:

```
Identifier = Letter

  {Letter | Digit}
```

Backus-Naur Form

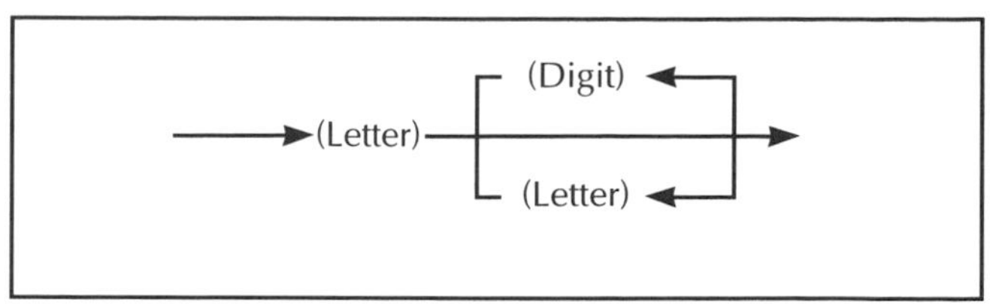

This "railroad diagram" indicates that an identifier must begin with a letter, which can be followed by a digit or another letter. The tracks curving back indicate that an element can appear more than once.

EBNF statements are sometimes depicted visually in railroad diagrams, so called because the lines and arrows indicating the relationship of symbols resemble railroad tracks. The definition of <identifier> expressed in a railroad diagram is depicted in the above figure.

BNF and EBNF are useful because they can provide unambiguous definitions of the syntax of any computer language that is not context-dependent (which is to say, nearly all of them). It can thus serve as a reference for introduction of new languages (such as scripting languages) and for developers of parsers for compilers.

Further Reading

Jensen, K., N. Wirth [et al.] *Pascal User Manual and Report: ISO Pascal Standard.* New York: Springer-Verlag, 1985.

Sebesta, Robert W. *Concepts of Programming Languages.* Reading, Mass.: Addison-Wesley, 1999.

bandwidth

In its original sense, bandwidth refers to the range of frequencies that a communications medium can effectively transmit. (At either end of the bandwidth, the transmission becomes too attenuated to be received reliably.) For a standard voice telephone, the bandwidth is about 3KHz.

In digital networks, bandwidth is used in a rather different sense to mean the amount of data that can be transmitted in a given time—what is more accurately described as the information transfer rate. A common measurement is Mb/sec (megabits per second). For example, a fast Ethernet network may have a bandwidth of 100 Mb/sec while a home phone-line network might have a bandwidth of from 1 to 10 Mb/sec and a DSL or cable modem runs at about 1 Mb/sec. (By comparison, a typical dial-up modem connection has a bandwidth of about 28-56 Kb/sec, roughly 20 times slower than even a slow home network.)

The importance of bandwidth for the Internet is that it determines the feasibility of delivering new media such as sound (MP3), streaming video, and digital movies over the network, and thus the viability of business models based on such products. The growth of high-capacity access to the Internet (see BROADBAND) is changing the way people use the network.

Further Reading

Gilder, George. *Telecosm: How Infinite Bandwidth Will Revolutionize Our World.* New York: Free Press, 2000.

Lu, Cary. *The Race for Bandwidth: Understanding Data Transmission.* Redmond, Wash.: Microsoft Press, 1998.

banking and computers

Beginning in the 1950s, banks undertook extensive automation of operations, starting with electronic funds transfer (EFT) systems. Check clearing (the sending of checks for payment to the bank on which they are drawn) was facilitated by the development of magnetic ink character recognition (MICR) that allowed checks to be automatically sorted and tabulated. Today an automated clearing house (ACH) network processes checks and other payments through regional clearinghouses.

Starting in the 1960s, the use of credit cards became an increasingly popular alternative to checks, and they were soon joined by automatic teller machine (ATM) networks and the use of debit cards (cards for transferring funds from a checking account at the point of sale).

Direct deposit of payroll and benefit checks has also been promoted for its safety and convenience. Credit card, ATM, and debit card systems rely upon large data processing facilities operated by the issuing financial institution. Because of the serious consequences of system failure both in immediate financial loss and customer goodwill, these fund transfer systems must achieve a high level of reliability and security. Reliability is promoted through the use of fault-tolerant hardware (such as redundant systems that can take over for one another in the event of a problem). The funds transfer messages must be provided a high level of security against eavesdropping or tampering through the use of algorithms such as the long-established DES (Data Encryption Standard)—see ENCRYPTION. Designers of EFT systems also face the challenge of providing a legally acceptable paper trail. Electronic signatures are increasingly accepted as an alternative to written signatures for authorizing fund transfers.

ON-LINE BANKING

The new frontier of electronic banking is the on-line bank, where customers can access many banking functions via the Internet, including balance queries, transfers, automatic payments, and loan applications. For the consumer, on-line banking offers greater convenience and access to information than even the ATM, albeit without the ability to obtain cash.

From the bank's point of view, on-line banking offers a new way to reach and serve customers while relieving the strain on the ATM hardware and network. However, use of the Internet increases vulnerability to hackers and raises issues of privacy and the handling of personal information similar to those found in other e-commerce venues (see COMPUTER CRIME AND SECURITY and PRIVACY IN THE DIGITAL AGE). Nevertheless, more than 20 new Internet banking venues were launched in 2000 (most affiliated with traditional banks, but a few found solely on the Internet). A particularly attractive feature of on-line banking is the ability to integrate bank services with popular personal finance software such as Quicken. (On-line banking is increasingly popular with businesses, with an estimated 33 percent of business accounts being used on-line in 2001.)

Credit card issuers are also starting to turn to the Internet to provide additional services. Citigroup, for example, reported in October 2000 that 3 million of its 40 million cardholders were using the bank's website to view their transaction information, and about 25 percent of the on-line users were paying their monthly credit card bills on-line.

Further Reading

"Bankrate.com" http://www.bankrate.com/brm/default.asp

"On-line Banking Report." http://www.on-linebankingreport.com/home/Resource

Reid, Joy-Ann Lomena. *On-line Banking.* Norwalk, Conn.: Business Communication, 2000.

BASIC

The BASIC (Beginner's All-purpose Symbolic Instruction Code) language was developed by J. Kemeny and T. Kurtz at Dartmouth College in 1964. At the time, the college was equipped with a time-shared computer system linked to terminals throughout the campus, an innovation at a time when most computers were programmed from a single location using batches of punch cards. John G. Kemeny and Thomas Kurtz wanted to take advantage of the interactivity of their system by providing an easy-to-learn computer language that could compile and respond immediately to commands typed at the keyboard. This was in sharp contrast to the major languages of the time, such as COBOL, Algol, and FORTRAN in which programs had to be completely written before they could be tested.

Unlike the older languages used with punch cards, BASIC programs did not have to have their keywords typed in specified columns. Rather, statements could be typed like English sentences, but without punctuation and with a casual attitude toward spacing. In general, the syntax for decision and control structures is simpler than other languages. For example, a for loop counting from 1 to 10 in C looks like this:

```
for (i = 1; i <= 10; i++)
  printf("%d", i);
```

The same loop in BASIC reads as follows:

```
for i = 1 to 10
  print i
next i
```

BASIC AND MICROCOMPUTERS

During the 1960s and 1970s BASIC was used on a growing number of time-sharing computers. The language's simplicity and ease of use made it useful for writing short utility programs and for teaching basic principles of computing, particularly to noncomputer science majors. When the first personal computers became widely available in the early 1980s, they typically had memory capacities of 8KB–64KB, not enough to run the editor, compiler, and other utilities needed for a language such as C. However, a simple interpreter version of BASIC could be put on a read-only memory (ROM) chip, as was done with the Apple II, the early IBM PC, and dozens of other microcomputers. More advanced versions of BASIC (including compilers) could be loaded from tape (the first sales by a young entrepreneur named Bill Gates consisted of such products).

As a consequence of the adopting of BASIC for a variety of microcomputers, numerous dialects of the language came into existence. Commands for generating simple graphics and for manipulating memory and hardware directly (PEEK and POKE) made many BASIC programs platform-specific.

Gradually, as microcomputers gained in memory capacity and processing power, languages such as Pascal (especially with the integrated development environment created at the University of California at San Diego) and C (from the UNIX community) began to supplant BASIC for the development of more complex microcomputer software.

CRITIQUE AND PROSPECTS

Most versions of BASIC used line numbers (a legacy of the early text editors that worked on a line-by-line basis) and a Goto statement could be used to make program control jump to a given line. While the language had simple subroutines (reached by a Gosub statement), it lacked the ability to explicitly pass variables to a procedure as in Pascal and C. Indeed, all variables were global, meaning that they could be accessed from anywhere in the program, leading to the danger of their values being unintentionally changed.

As interest in the principles of structured programming grew (see STRUCTURED PROGRAMMING), BASIC's structural shortcomings made it poorly regarded among computer scientists, who preferred Pascal as a teaching language and C for systems programming. In 1984, BASIC's original developers responded to what they saw as the problems of "street Basic" by introducing True BASIC, a modern, well-structured version of the language, and the 1988 ANSI BASIC standard incorporated similar features. These efforts had only limited impact. However, Microsoft introduced new BASIC development systems (Quick BASIC in the 1980s and Visual Basic in the 1990s) that also featured improved control structures and data types and that dispensed with the need for cumbersome line numbers. Visual Basic in particular has achieved considerable success, offering a combination of the interactivity of traditional BASIC and access to powerful pre-packaged "controls" that provide menus, dialog boxes, and other features of the Windows user interface. Recent versions of Visual Basic have become increasingly object-oriented, using classes similar to those in C++.

While BASIC in its newer forms continues to have a significant following, it can be argued that what was most distinctive about the original BASIC (the quick, interactive approach to programming) is no longer much in evidence. The writing of short utility programs is now more likely to be undertaken in any of a variety of scripting languages.

Further Reading

Kemeny, J. G., and Thomas E. Kurtz. *Back to Basic: the History, Corruption, and Future of the Language.* Reading, Mass.: Addison-Wesley, 1985.

"Top 219 Q(uick) Basic Sites." http://www.top219.org/qbasic/

Lomax, Paul, and Ron Petrusha. *VB and VBA in a Nutshell: the Languages.* Sebastopol, Calif.: O'Reilly, 1998.

Berners-Lee, Tim

(1955–)
British
Computer Scientist

A graduate of Oxford University, Tim Berners-Lee created what would become the World Wide Web in 1989 while working at CERN, the giant European physics research institute. At CERN, he struggled with organizing the dozens of incompatible computer systems and software that had been brought to the labs by thousands of scientists from around the world. With existing systems each requiring a specialized access procedure, researchers had little hope of finding out what their colleagues were doing or of learning about existing software tools that might solve their problems.

Berners-Lee's solution was to bypass traditional database systems and to consider text on all systems as "pages" that would each have a unique address, a Universal Document Identifier (later known as a Uniform Resource Locator, or URL). He and his assistants used existing ideas of hypertext to link words and phrases on one page to another page (see HYPERTEXT AND HYPERMEDIA), and adapted existing hypertext editing software for the NeXT computer to create the first World Wide Web pages, a server to provide access to the pages and a simple browser, a program that could be used to read pages and follow the links as the reader desired (see WEB SERVER and WEB BROWSER). But while existing hypertext systems were confined to browsing a single file or at most, the contents of a single computer system, Berners-Lee's World Wide Web used the emerging Internet to provide nearly universal access.

Between 1990 and 1993, word of the Web spread throughout the academic community as Web software was written for more computer platforms (see WORLD WIDE WEB). As demand grew for a body to standardize and shape the evolution of the Web, Berners-Lee founded the World Wide Web Consortium (W3C) in 1994. Together with his colleagues, he has struggled to maintain a coherent vision of the Web in the face of tremendous growth and commercialization, the involvement of huge corporations with conflicting agendas, and contentious issues of censorship and privacy. His general approach has been to develop tools that would empower the user to make the ultimate decision about the information he or she would see or divulge.

Berners-Lee now works at the Massachusetts Institute of Technology Computer Laboratory for Computer Science. In his original vision for the Web, users would create webpages as easily as they could read them, using software no more complicated than a word processor. While there are programs today that hide the details of HTML coding and allow easier Web page creation, Berners-Lee feels the Web must become even easier to use if it is to be a truly interactive, open-ended knowledge system. He is also interested in developing software that can take better advantage of the rich variety of information on the Web, creating a "semantic" Web of meaningful connections that would allow for logical analysis and permit human beings and machines not merely to connect, but to actively collaborate (see also XML).

Further Reading

Berners-Lee, Tim. Home page with biography and links: http://www.w3.org/People/Berners-Lee/

———, Tim. Papers on Web design issues. http://www.w3.org/DesignIssues/

———, Tim. Proposal for the World Wide Web, 1989. http://www.w3.org/History/1989/proposal.html

Berners-Lee, Tim and Mark Fischetti. *Weaving the Web*. San Francisco: HarperSanFrancisco, 1999.

Henderson, Harry. *Pioneers of the Internet*. San Diego, Calif.: Lucent Books, 2002.

binding

Designers of program compilers are faced with the question of when to translate a statement written in the source language into final instructions in machine language (see also ASSEMBLER). This can happen at different times depending on the nature of the statement and the decision of the compiler designer.

Many programming languages use formal data types (such as integer, floating point, double, string, and so on) that result in allocation of an exact amount of storage space to hold the data (see DATA TYPES). A statement that declares a variable with such a type can be effectively bound immediately (that is, a final machine code statement can be generated). This is also called compile-time binding.

However, there are a variety of statements for which binding must be deferred until more information becomes available. For example, it is common for programmers to use libraries of precompiled routines. A statement that calls such a routine cannot be turned immediately into machine language because the compiler doesn't know the actual address where the routine will be embedded in the final compiled program. (That address will be determined by a program called a linker that links the object code from the source program to the library routines called upon by that code.)

Another aspect of binding arises when there is more than one object in a program with the same name. In languages such as C or Pascal that use a nested block structure, lexical binding can determine that a name refers to the closest declaration of that name—that is, the smallest scope that contains that name (see VARIABLE). In a few languages such as Lisp, however, the reference for a name depends on how (or for what) the function is being called, so binding can be done only at run time.

BINDING AND OBJECT-ORIENTED LANGUAGES

The use of polymorphism in object-oriented languages such as C++ raises a similar issue. Here there can be a base class and a hierarchy of derived classes. A function in the base class can be declared to be virtual, and versions of the same function can be declared in the derived classes. In this case a statement containing a pointer to the function in the base class cannot be bound until run time, because only then will it be known which version of the virtual function is being called. However, compilers for object-oriented languages can be written so they do early binding on statements for which it is safe (such as those involving static data types), but do dynamic binding when necessary.

From the point of view of efficiency, early binding is better because memory can be allocated efficiently. Dynamic binding provides greater flexibility, however, and facilitates debugging—for example, because the name of a variable is normally lost once it is bound and the machine code is generated.

Further Reading

Pratt, T. W., and M. V. Zeikowitz. *Programming Languages: Design and Implementation*. Upper Saddle River, N.J.: Prentice Hall, 1996.

biology and computing

The attempt to apply mathematical and information science techniques to biology (sometimes called *bioinformatics*) is inherently difficult because a living organism represents a complex interaction of chemical processes. Understanding any one process in isolation gives little understanding of the role it plays in physiology.

Today, however, the ability of increasingly powerful computer systems to create and visualize models of living processes is allowing life scientists to explore them at an unprecedented level of detail. Instead of drastically simplifying a model to make it manageable, additional factors can be added and a static model can be turned into a dynamic simulation. Further, the use of computer graphics to depict processes (such as enzyme reactions or protein synthesis) makes it much easier for the researcher to comprehend and manipulate the simulation.

APPLICATIONS

The computer-assisted mathematical approach to biology has achieved remarkable success in a variety of areas. In particular, the use of computers and robots in sequencing the human genome resulted in the project being completed years ahead of schedule. In 2000, IBM announced that a successor to its Deep Blue processor (which had defeated world chess champion Garry Kasparov in 1997) would undertake a simulation of the detailed process by which genetic information results in the synthesis of proteins. At the same time, a growing database of information and simulation in general molecular biology is gradually enabling researchers to replace the relatively hit-and-miss search for medically effective molecules (drugs) with an approach directed by the analysis of similar molecules. Further, analysis of the detailed structure of pathogens such as HIV could well lead to the tailoring of agents that could block the organism's ability to penetrate target cells or to reproduce.

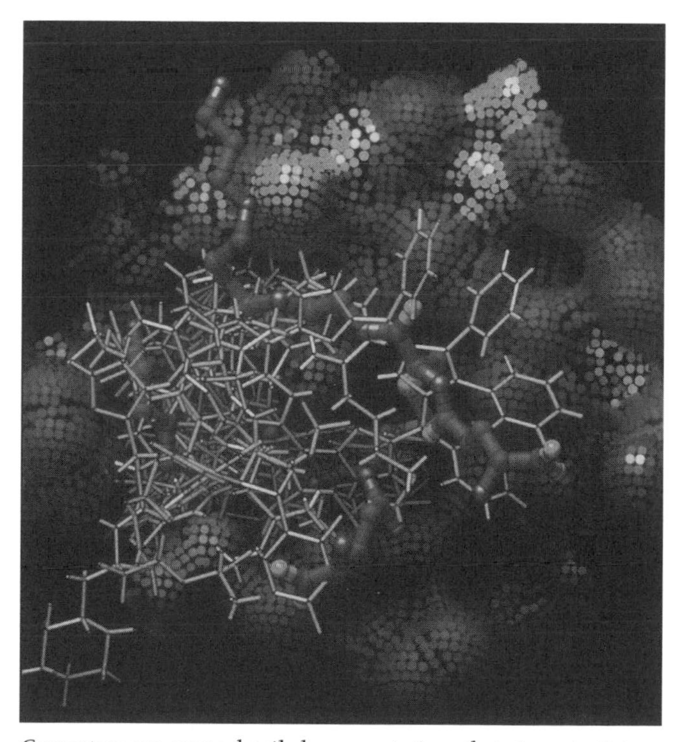

Computers can create detailed representations that give scientists unprecedented ability to visualize nature's most intricate structures. This is a computer model of Trypanathione Reductase, a protein crystal. (NASA PHOTO; MARSHALL SPACE FLIGHT CENTER IMAGE EXCHANGE)

The treatment of disease also promises to be revolutionized by the development of powerful models of the fundamental processes within the cell, including the reaction of the cell to pathogens such as HIV, the development of cancer, and the processes that cumulatively result in aging. Models of organs such as the heart and kidneys are also being created. Ultimately, a highly detailed model of the human body as a whole may enable researchers and doctors to accurately predict or test the effects of candidate drugs or treatments.

The study of biology extends of course from the consideration of individual organisms to the study of populations and interaction with the environment (ecology). Here, too, computer models are helping researchers understand the spread of disease (epidemiology), the growth and decline of populations, and the effects of changes in the balance of predators and prey.

Ultimately, the whole in biology is always greater than the sum of the parts. The ability to digitize and manipulate data from observations and make it available in databases that can be used by researchers means that the growing web of information can be accessed and augmented at an ever-increasing rate as it approaches the complexity and richness of life itself.

Further Reading

Misener, Stephen, and Stephen A. Krawetz, eds. *Bioinformatics: Methods and Protocols.* Totowa, N.J.: Humana Press, 2000.

Rashidi, Hooman H., and Lukas K. Buehler. *Bioinformatics Basics: Applications in Biological Science and Medicine.* Grand Rapids, Mich.: CRC Press, 1999.

biometrics

The earliest use of biometrics was probably the development by Alphonse Bertillon in 1882 of anthropometry, a system of classification by physical measurements and description. While this was soon supplanted by the discovery that fingerprints could serve as an easier to use means of unique identification of persons, the need for a less invasive means of physical identification has led to the development of a variety of biometric scanners that take Bertillon's ideas to a much more detailed level.

TECHNOLOGIES

In general, biometric scanning involves four steps: the capture of an image using a camera or other device, the extraction of key features from the image, the creation of a template that uniquely characterizes the person being scanned, and the matching of the template to stored templates in order to identify the person.

There are several possible targets for biometric scanning, including

FACIAL SCANNING

Facial scanning uses cameras and image analysis software that looks at areas of the human face that change little during the course of life and are not easily alterable, such as the upper outline of the eye sockets and the shape of the cheekbones. Researchers at MIT developed a series of about 125 grayscale images called eigenfaces from which features can be combined to characterize any given face. The template resulting from a scan can be compared with the one on file for the claimed identity, and coefficients expressing the degree of similarity are calculated. Variance above a specified level results in the person being rejected. Facial scanning is often viewed as less intrusive than the use of fingerprints, and it can also be applied to surveillance images.

FINGER SCANNING

Finger scanning involves the imaging and automatic analysis of the pattern of ridges on one or more fingertips. Unlike traditional fingerprinting, the actual fingerprint is not saved, but only enough key features are retained to provide a unique identification. This information can be stored in a database and also compared with full fingerprints stored in existing databases (such as that maintained by the Federal Bureau of Investigation). Finger scanning can meet with resistance because of its similarity to fingerprinting and the association of the latter with criminality.

HAND GEOMETRY

This technique measures several characteristics of the hand, including the height of and distance between the fingers and the shape of the knuckles. The person being scanned places the hand on the scanner's surface, aligning the fingers to five pegs. Hand-scanning is reasonably accurate in verifying an individual compared to the template on file, but not accurate enough to identify a scan from an unknown person.

IRIS AND RETINA SCANNING

These techniques take advantage of many unique individual characteristics of these parts of the eye. The scanned characteristics are turned into a numeric code similar to a bar code. Retina scanning can be uncomfortable because it involves shining a bright light into the back of the eye, and has generally been used only in high-security installations. However, iris scanning involves the front of the eye and is much less intrusive, and the person being scanned needs only to look into a camera.

VOICE SCANNING

Voice scanning and verification systems create a "voiceprint" from a speech sample and compare it to the voice of the person being verified. It is a quick and nonintrusive technique that is particularly useful for remote transactions such as telephone access to banking information.

APPLICATIONS OF BIOMETRICS

Due to the expense of the equipment and the time involved in scanning, biometrics were originally used primarily in verifying identity for people entering high-security installations. However, the development of faster and less intrusive techniques, combined with the growing need to verify users of banking (ATM) and other networks has led to a growing interest in biometrics. For example, a pilot program in the United Kingdom has used iris scanning to replace the PIN (personal identification number) as a means of verifying ATM users.

The general advantage of biometrics is that it does not rely on cards or other artifacts that can be stolen or otherwise transferred from one person to another, and in turn, a person needing to identify him or herself doesn't have to worry about forgetting or losing a card. However, while workers at high-security installations can simply be required to submit to biometric scans, citizens and consumers have more choice about whether to accept techniques they may view as uncomfortable, intrusive, or threatening to privacy.

Further Reading

Ashborn, Julian D. M. *Biometrics: Advanced Identity Verification, the Complete Guide.* New York: Springer-Verlag, 2000.

"Biometrics Overview." http://www.biometricgroup.com/a_bio1/technology/research_a_technology.htm

Harreld, Heather. "Biometrics Points to Greater Security." *Federal Computer Week,* July 22, 1999. Available on-line at http://www.cnn.com/TECH/computing/9907/22/biometrics.idg/index.html

BIOS

With any computer system a fundamental design problem is how to provide for the basic communication between the processor (see CPU) and the devices used to obtain or display data, such as the video screen, keyboard, and parallel and serial ports.

In personal computers, the BIOS (Basic Input-Output System) solves this problem by providing a set of routines for direct control of key system hardware such as disk drives, the keyboard, video interface, and serial and parallel ports. In PCs based on the IBM PC architecture, the

BIOS is divided into two components. The fixed code is stored on a PROM (programmable read-only memory) chip commonly called the "ROM BIOS" or "BIOS chip." This code handles interrupts (requests for attention) from the peripheral devices (which can include their own specialized BIOS chips). During the boot sequence the BIOS code runs the POST (Power-On Self Test) and queries various devices to make sure they are functional. (At this time the PC's screen will display a message giving the BIOS manufacturer, model, and other information.) Once DOS is running, routines in the operating system kernel can access the hardware by making calls to the BIOS routines. In turn, application programs can call the operating system, which passes requests on to the BIOS routines.

The BIOS scheme has some flexibility in that part of the BIOS is stored in system files (in IBM PCs, IO.SYS and IBMIO.COM). Since this code is stored in files, it can be upgraded with each new version of DOS. In addition, separate device drivers can be loaded from files during system startup as directed by DEVICE commands in CONFIG.SYS, a text file containing various system settings.

For further flexibility in dealing with evolving device capabilities, PCs also began to include CMOS (Complementary Metal Oxide Semiconductor) chips that allow for the storage of additional parameters, such as for the configuration of memory and disk drive layouts. The data on these chips is maintained by a small on-board battery so settings are not lost when the main system power is turned off.

Additionally, modern PC BIOS chips use "flash memory" (EEPROM or "Electrically Erasable Programmable Read-only Memory") to store the code. These chips can be "flashed" or reprogrammed with newer versions of the BIOS, enabling the support of newer devices without having to replace any chips.

BEYOND THE BIOS

While the BIOS scheme was adequate for the earliest PCs, it suffered from a lack of flexibility and extensibility. The routines were generic and thus could not support all the functions of newer devices. Because BIOS routines for such tasks as graphics tended to be slow, applications programmers often bypassed the BIOS and dealt with devices directly or created device drivers specific to a particular model of device. This made the life of the PC user more complicated because programs (particularly games) may not work with some video cards, for example, or at least required an updated device driver.

While both the main BIOS and the auxiliary BIOS chips on devices such as video cards are still essential to the operation of the PC, modern operating systems such as Microsoft Windows XP and applications written for them generally do not use BIOS routines and employ high performance device drivers instead. (By the mid-1990s BIOSes included built-in support for "Plug and Play," a system for automatically loading device drivers as needed. Thus, the BIOS is now usually of concern only if there is a hardware failure or incompatibility.)

Further Reading

"System BIOS Function and Operation." http://www.pcguide.com/ref/mbsys/bios/func.htm

bitmapped image

A bitmap is a series of bits (within a series of bytes in memory) in which the bits represent the pixels in an image. In a monochrome bitmap, each pixel can be represented by one bit, with a 1 indicating that the pixel is "on." For grayscale or color images several bits must be used to store the information for each pixel. The pixel value bits are usually preceded by a data structure that describes various characteristics of the image.

For example, in the Microsoft Windows BMP format, the file for an image begins with a BITMAPFILEHEADER that includes a file type, size, and layout. This is followed by a BITMAPINFOHEADER that gives information about the image itself (dimensions, type of compression, and color format). Next comes a color table that describes each color found in the image in terms of its RGB (red, green, blue) components. Finally comes the consecutive bytes representing the bits in each line of the image, starting from lower left and proceeding to the upper right.

The actual number of bits representing each pixel depends on the dimensions of the bitmap and the number of colors being used. For example, if the bitmap has a maximum of 256 colors, each pixel value must use one

Bitmapped Image

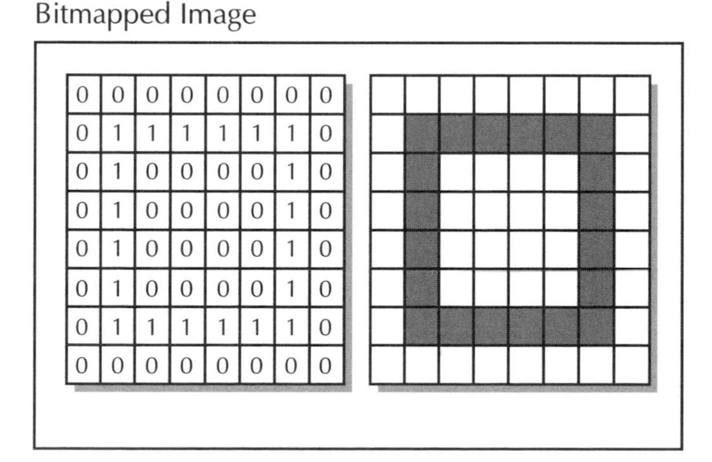

In a monochrome bitmapped image, a one is used to represent a pixel that is turned on, while the empty pixels are represented by zeroes. Color bitmaps must use many more bits per pixel to store color numbers.

byte to store the index that "points" to that color in the color table. However, an alternative format stores the actual RGB values of each pixel in three consecutive bytes (24 bits), thus allowing for a maximum of 224 (16,777,216) colors (see also COLOR IN COMPUTING).

SHORTCOMINGS AND ALTERNATIVES

The relationship between number of possible colors and amount of storage needed for the bitmap means that the more realistic the colors, the more space is needed to store an image of a given size, and generally, the more slowly the bitmap can be displayed. Various techniques have been used to shrink the required space by taking advantage of redundant information resulting from the fact that most images have areas of the same color (see DATA COMPRESSION).

Vector graphics offer an alternative to bitmaps, particularly for images that can be constructed from a series of lines. Instead of storing the pixels of a complete image, vector graphics provides a series of vectors (directions and lengths) plus the necessary color information. This can make for a much smaller image, as well as making it easy to scale the image to any size by multiplying the vectors by some constant.

Further Reading

Indiana University. Center for Innovative Computer Applications. "List of Image File Formats." http://www.cica.indiana.edu/graphics/image.formats.html

Rimmer, Steve. *Windows and OS/2 Bitmapped Graphics.* 2nd Ed. New York: McGraw Hill, 1996.

bits and bytes

Computer users soon become familiar with the use of bits (or more commonly bytes) as a measurement of the capacity of computer memory (RAM) and storage devices such as disk drives. They also speak of such things as "16-bit color," referring to the number of different colors that can be specified and generated by a video display.

In the digital world a bit is the smallest discernable piece of information, representing one of two possible states (indicated by the presence or absence of something such as an electrical charge or magnetism, or by one of two voltage levels). Bit is actually short for "binary digit," and a bit corresponds to one digit or place in a binary (base two) number. Thus an 8 bit value of

11010101

corresponds, from right to left, to $(1 * 2^0) + (0 * 2^1) + (1 * 2^2) + (0 * 2^3) + (1 * 2^4) + (0 * 2^5) + (1 * 2^6) + (1 * 2^7)$, or 213 in terms of the familiar decimal system.

With regard to computer architectures the number of bits is particularly relevant to three areas: (1) The size of the basic "chunk" of data or instructions that can be fetched, processed, or stored by the central processing unit (CPU); (2) The "width" of the data bus over which data is sent between the CPU and other devices—given the same processor speed, a 32-bit bus can transfer twice as much data in a given time as a 16-bit bus; and (3) The width of the address bus (now generally 32 bits), which determines how many memory locations can be addressed, and thus the maximum amount of directly usable RAM.

Bits and Bytes

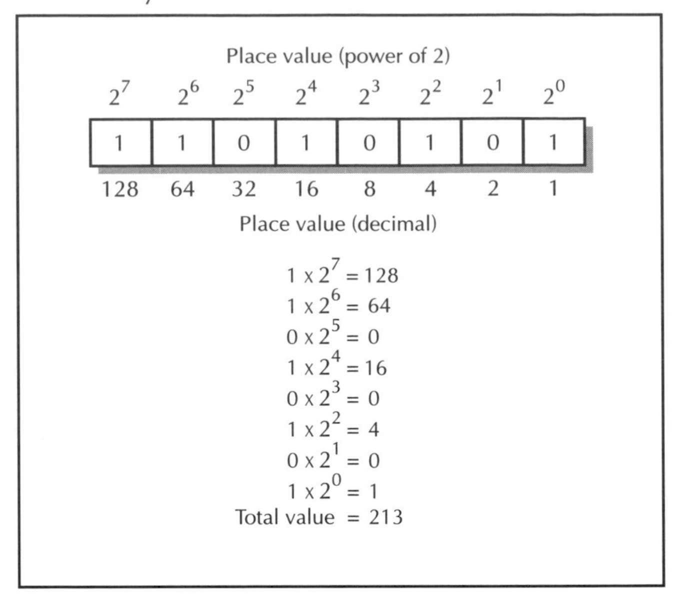

One byte in memory can store an 8-bit binary number. Just as each place to the left in a decimal number represents the next higher power of 10, the places in the byte increase as powers of two. Here the places with ones in them add up to a total decimal value of 213.

Measurement	Number of Bytes		Examples of Use
byte	1		small integer, character
kilobyte	2^{10}	1,024	RAM (PCs in the 1980s)
megabyte	2^{20}	1,048,576	hard drive (PCs to mid-1990s) RAM (modern PCs)
gigabyte	2^{30}	1,073,741,824	hard drive (modern PCs)
terabyte	2^{40}	1,099,511,627,776	large drive arrays

The first PCs used 8-bit or 16-bit processors, while today's PC processors and operating systems often use 32-bits at a time, with 64-bit processors on the near horizon. Besides the "width" of data transfer, the number of bits can also be used to specify the range of available values. For example, the range of colors that can be displayed by a video card is often expressed as 16 bit (65,536 colors) or 32 bit (16,777,777,216 colors, because only 24 of the bits are used for color information).

Since multiple bits are often needed to specify meaningful information, memory or storage capacity is often expressed in terms of bytes. A byte is 8 bits or binary digits, which amounts to a range of from 0 to 255 in terms of decimal (base ten) numbers. A byte is thus enough to store a small integer or a character code in the standard ASCII CHARACTER set. Common multiples of a byte are a kilobyte (thousand bytes), megabyte (million bytes), gigabyte (billion bytes), and occasionally terabyte (trillion bytes). The actual numbers represented by these designations are actually somewhat larger, as indicated in the accompanying table.

Further Reading

"How Bits and Bytes work." http://www.howstuffworks.com/bytes.htm

bitwise operations

Since each bit of a number (see BITS AND BYTES) can hold a truth value (1 = true, 0 = false), it is possible to use individual bits to specify particular conditions in a system, and to compare individual pairs of bits using special operators that are available in many programming languages.

Bitwise operators consist of logical operators and shift operators. The logical operators, like Boolean operators in general (see BOOLEAN OPERATORS), perform logical comparisons. However, as the name suggests, bitwise logical operators do a bit-for-bit comparison rather than comparing the overall value of the bytes. They compare the corresponding bits in two bytes (called source bits) and write result bits based on the type of comparison.

The **AND** operator compares corresponding bits and sets the bit in the result to one if both are one. Otherwise, it sets it to zero.

Example: 10110010 AND 10101011 = 10100010

The **OR** operator compares corresponding bits and sets the bit in the result to one if either or both of the bits are ones.

Example: 10110110 OR 10010011 = 10110111

The **XOR** ("exclusive OR") operator works like OR except that it sets the result bit to one only if either (not both) of the source bits are ones.

Example: 10110110 XOR 10010011 = 00100101

The **COMPLEMENT** operator switches all the bits to their opposites (ones for zeroes and zeroes for ones).

Example: COMPLEMENT 11100101 = 00011010

The shift operators simply shift all the bits left (**LEFT SHIFT**) or right (**RIGHT SHIFT**) by the number of places specified after the operator. Thus

00001011 **LEFT SHIFT 2** = 00101100

and

00001011 **RIGHT SHIFT 2** = 00000010 (bits that shift off the end of the byte simply "drop off" and are replaced with zeroes).

While we have used words in our general description of these operators, actual programming languages often use special symbols that vary somewhat with the language. The operators used in the C language are typical:

& AND

| OR

^ Exclusive OR

~ Complement

>> Right Shift

<< Left Shift

MASKING

There are a number of programming tasks where the contents of individual bits must be read or manipulated. Operating systems and network protocols often have data structures where several separate pieces of information are stored in a single byte in order to save space. (For example, in IBM architecture PC's interrupts are often enabled or disabled by setting particular bits in a mask register.) Operations using BITMAPPED IMAGES can also involve bit manipulation.

Suppose the right three bits of a byte contain a desired piece of information. The byte is ANDed with a prepared byte called a mask in which the desired bits are set to one and the rest of the bits are zero: in this case it would be 00000111. Thus if the byte contains 11010110:

11010110 AND 00000111 = 000000110

The result contains only the value of the right three bits. Similarly, if one wants to set a particular bit to zero, one simply ANDs the byte with a byte that has a zero in that position and ones in the rest of the byte. Thus to "zero out" the second bit from the left in 11010110:

11010110 AND 10111111 = 10010110

Further Reading

"Bitwise Operations Overview." http://www.harper.cc.il.us/bus-ss/cis/166/mmckenzi/lect19/l19.htm

Boolean operators

In 1847, British mathematician George Boole proposed a system of algebra that could be used to manipulate propositions, that is, assertions that could be either true or false. In his system, called propositional calculus or Boolean Algebra, propositions can be combined using the "and" and "or" operators (called Boolean operators), yielding a new proposition that is also either true or false. For example:

"A cat is an animal" AND "The sun is a star" is true because both of the component propositions are true.

"A square has four sides" AND "The Earth is flat" is false because only one of the component propositions is true.

However "A square has four sides" OR "The Earth is flat" is true, because *at least one* of the component propositions is true.

A chart called a truth table can be used to summarize the AND and OR operations. Here 1 means true and 0 means false, and you read across from the side and down from the top to see the result of each combination.

AND TABLE

	0	1
0	0	0
1	0	1

OR TABLE

	0	1
0	0	1
1	1	1

A variant of the OR operator is the "exclusive OR," sometimes called "XOR" operator. The XOR operator yields a result of true (1) if *only one* of the component propositions is true:

XOR TABLE

	0	1
0	0	1
1	1	0

Additionally, there is a NOT operator that simply reverses the truth value of a proposition. That is, NOT 1 is 0 and NOT 0 is 1.

APPLICATIONS

Note the correspondence between the two values of Boolean logic and the binary number system in which each digit can have only the values of 1 or 0. Electronic digital computers are possible because circuits can be designed to follow the rules of Boolean logic, and logical operations can be harnessed to perform arithmetic calculations.

Besides being essential to computer design, Boolean operations are also used to manipulate individual bits in memory (see BITWISE OPERATIONS), storing and extracting information needed for device control and other purposes. Computer programs also use Boolean logic to make decisions using branching statements such as If and loop statements such as While. For example, the Basic loop

```
While (Not Eof()) OR (Line = 50)
   Read (Line$)
   Print (Line$)
   Line = Line + 1
Endwhile
```

will read and print lines from the previously opened file until *either* the Eof (end of file) function returns a value of True *or* the value of Line reaches 50. (In some programming languages different symbols are used for the operators. In C, for example, AND is &&, OR is ||, and NOT is !.)

Users of databases and Web search engines are also familiar with the use of Boolean statements for defining search criteria. In many search engines, the search phrase "computer science" AND "graduate" will match sites that have both the phrase "computer science" and the word "graduate," while sites that have only one or the other will either not be listed or will be listed after those that have both (see SEARCH ENGINE).

Further Reading

University at Albany Libraries. "Boolean Searching on the Internet." http://www.albany.edu/library/internet/boolean.html

Whitesitt, J. E. *Boolean Algebra and Its Applications.* New York: Dover, 1995.

boot sequence

All computers are faced with the problem that they need instructions in order to be able to read in the instructions they need to operate. The usual solution to this conundrum is to store a small program called a "loader" in a ROM (read-only memory) chip. When the computer is switched on, this chip is activated and runs the loader. The loader program has the instructions needed to be able to access the disk containing the full operating system. This process is called booting (short for "bootstrapping").

BOOTING A PC

While the details of the boot sequence vary with the hardware and operating system used, a look at the booting of a "Wintel" machine (IBM architecture PC running DOS and Microsoft Windows) can serve as a practical example.

When the power is turned on, a chip called the BIOS (Basic Input-Output System) begins to execute a small program. The first thing it does is to run a routine called the POST (Power-On Self Test) that sends a query over the system BUS (see BUS) to each of the key devices (memory, keyboard, video display, and so on) for a response that indicates it is functioning properly. If an error is detected, the system generates a series of beeps, the number of which indicates the area where the problem was found, and then halts.

Assuming the test runs successfully (sometimes indicated by a single beep), the BIOS program then queries the devices to see if they have their own BIOS chips, and if so, executes their programs to initialize the devices, such as the video card and disk controllers. At this point, since the video display is available, informational and error messages can be displayed as appropriate. The BIOS also sets various parameters such as the organization of the disk drive, using information stored in a CMOS chip. (There is generally a way the user can access and change these information screens, such as when installing additional memory chips.)

The BIOS now looks for a disk drive that is bootable—that is, that contains files with the code needed to load the operating system. This is generally a hard drive, but could be a floppy disk or even a CD-ROM. (The order in which devices are checked can be configured.) On a hard drive, the code needed to start the operating system is found in a "master boot record."

The booting of the operating system (DOS) involves the determination of the disk structure and file system and the loading of the operating system kernel (found in files called IO.SYS and MSDOS.SYS), and a command interpreter (COMMAND.COM). The latter can then read the contents of the files AUTOEXEC.BAT and CONFIG.SYS, which specify system parameters, device drivers, and other programs to be loaded into memory at startup. If the system is to run Microsoft Windows, that more elaborate operating system will then take over, building upon or replacing the foundation of DOS.

Further Reading

PC Guide. "System Boot Sequence." http://www.pcguide.com/ref/mbsys/bios/bootSequence-c.html

branching statements

The simplest calculating machines (see CALCULATOR) could only execute a series of calculations in an unalterable sequence. Part of the transition from calculator to full computer is the ability to choose different paths of execution according to particular values—in some sense, to make decisions.

Branching statements (also called decision statements or selection statements) give programs the ability to choose one or more different paths of execution depending on the results of a logical test. The general form for a branching statement in most programming languages is

```
if (Boolean expression)

statement

else statement
```

For example, a blackjack game written in C might have a statement that reads:

```
if ((Card_Count + Value(This_Card)) > 21)
  printf ("You're busted!");
```

Here the Boolean expression in parenthesis following the if keyword is evaluated. If it is true, then the following statement (beginning with printf) is executed. (The Boolean expression can be any combination of expressions, function calls, or even assignment statements, as long as they evaluate to true or false—see also BOOLEAN OPERATORS.)

The else clause allows the specification of an alternative statement to be executed if the Boolean expression is *not* true. The preceding example could be expanded to:

```
if (Card_Count + Value (This_Card) > 21)
  printf ("You're busted!");
else
  printf("Do you want another card?");
```

In most languages if statements can be nested so that a second if statement is executed only if the first one is true. For example:

```
if (Turn > Max_Turns)
  {
  if (Winner() )
    PrintScore();
  }
```

Here the first if test determines whether the maximum number of turns in the game has been exceeded. If it has, the second if statement is executed, and the Winner() function is called to determine whether there is a winner. If there is a winner, the PrintScore() function is called. This example also illustrates the general rule in most languages that wherever a single statement can be used a block of statements can also be used. (The block is delimited by braces in the C family of languages, while Pascal uses Begin . . . End.)

The switch or case statement found in many languages is a variant of the if statement that allows for easy testing of several possible values of a condition. One could write:

```
if (Category = = "A")
  AStuff();
else if (Category = = "B")
  BStuff();
else if (Category = = "C")
  CStuff();
else
  printf "(None of the above\n");
```

However, C, Pascal, and many other languages provide a more convenient multiway branching statement (called switch in C and case in Pascal). Using a switch statement, the preceding test can be rewritten in C as:

```
switch (Category) {
case "A":
  AStuff();
  break;
case "B":
  BStuff();
  break;
case "C"
  CStuff();
  break;
default:
  printf ("None of the above\n");
}
```

(Here the break statements are needed to prevent execution from continuing on through the other alternatives when only one branch should be followed.)

Further Reading

Sebesta, Robert W. *Concepts of Programming Languages.* Reading, Mass.: Addison-Wesley, 1999.

broadband

Technically, broadband refers to the carrying of multiple communications channels in a single wire or cable. In the broader sense used here, broadband refers to high-speed data transmission over the Internet using a variety of technologies (see also DATA COMMUNICATIONS and TELECOMMUNICATIONS). This can be distinguished from the relatively slow (56 Kbps or slower) dial-up phone connections used by most home, school, and small business users until the late 1990s. A quantitative change in speed results in a qualitative change in the experience of the Web, making continuous multimedia (video and sound) transmissions possible.

BROADBAND TECHNOLOGIES

The earliest broadband technology to be developed consists of dedicated point-to-point telephone lines designated T1, T2, and T3, with speeds of 1.5, 6.3, and 44.7 Mbps respectively. These lines provide multiple data and voice channels, but cost thousands of dollars a month, making them practicable only for large companies or institutions.

Two other types of phone line access offer relatively high speed at relatively low cost. The earliest, ISDN (Integrated Services Digital Network) in typical consumer form offers two 64 Kbps channels that can be combined for 128 Kbps. (Special services can combine more channels, such as a 6 channel 384 Kbps configuration for videoconferencing.) The user's PC is connected via a digital adapter rather than the usual analog-to-digital modem.

The more common (and generally faster) phone line broadband technology today is DSL (Digital Subscriber Line). Unlike T1, DSL sends digital packets over ordinary phone lines, making it much less expensive. However, the transmission rate falls off with distance from the phone company's central office, with a maximum distance of about 18,000 feet. A typical speed for a consumer DSL service is about 1 Mbps; however, because consumer DSL is generally asymmetric (ADSL), the "upstream" speed (from the user's PC to the Internet Service Provider) is only about 128 Kbps. (Since most users don't upload or e-mail many large files, this is usually not a problem from the consumer's point of view.) As of 2002, there were over 3 million DSL subscribers in the United States.

Fiber optic cable can provide connection speeds of 4.5 to 9 Mbps (about three to four times that of DSL or cable). Until recently, fiber optic was too expensive for consumer use, and was generally limited to high-bandwidth connections within local area networks. However, a pilot program was recently rolled out in Palo Alto, California, home of many technological "early adopters."

The primary consumer alternative to phone-based broadband is the cable modem, which uses the television cable available in many (but not all) communities. This service is convenient in that it doesn't require a second phone line (and doesn't interfere with simultaneous use of the cable for television viewing). Speeds are generally comparable to DSL, but whereas DSL slows down at longer distances, cable slows down as more people are connected to a given circuit. As of 2002, there were about twice as many (7 million) cable users as DSL users.

Recently, various forms of wireless Internet access have attracted attention (see WIRELESS COMPUTING). While most radio and cellular services are not designed to carry large amounts of data, a service called 2.5G (and eventually, 3G), promises data transmission speeds of 2.4 Mbps or higher, comparable to cable or DSL. By 2002, there were only about 300,000 people connected to the Internet through wireless services.

Satellite-based Internet access is also available, though it is expensive for consumer use and is limited to downloading (outgoing data must be sent over a conventional modem). There were slightly more than 50,000 people with satellite Internet service by early 2002.

Wireless methods are likely to become more popular with the growing use of palm computers or PDAs (Personal Digital Assistants) with wireless connectivity, and the incorporation of wireless e-mail and Web browsing capability into high-end cell phones.

APPLICATIONS

The promise of broadband is in the delivery of high-quality multimedia, video, movies, and sound. The use of buffering techniques (see STREAMING) makes it possible to deliver a smooth flow of data shortly after connection, while smoothing away some of the vagaries of Internet traffic. (In practice, however, the available infrastructure at the ISP and major regional nodes has not kept up with the demand for rich media, often resulting in sporadic performance.)

Many news and other sites now routinely offer video to supplement their coverage of current news stories. Software such as RealPlayer provides an easy-to-use interface for playing and controlling video and sound.

Starting in the late 1990s, much industry "hype" centered on the provision of "channels" (also called "Push" technology). Channels are continuously updated multimedia content analogous to TV broadcast channels. However, while a variety of channels are now available, they have not become a ubiquitous part of the Internet experience.

An interesting application of streaming sound is the several thousand radio stations as of 2000 that were making their live broadcasts available to Internet users. If bandwidth problems can be solved, television stations may soon join their radio counterparts on the Net.

Perhaps the most popular (and controversial) application of broadband is the downloading and sharing of MP3 sound files containing high-quality recordings of popular music. The Napster program, which allows users to find and download MP3s from other users' hard drives, found millions of users but was eventually shut down by litigation for allegedly facilitating the violation of copyright and the depriving of record producers and artists of revenue.

TRENDS

The investment of major companies in integrating and delivering broadband connectivity has been impressive. AT&T, for example, has spent over $100 billion in acquiring cable systems and expanding DSL services. However, relatively high cost (about $40–$60 per month in 2000) and lack of availability in many rural and even poorer urban areas mean that broadband access is not yet ubiquitous. There is some danger of those Internet users who are limited to dial-up modem access being relegated to second-class citizenship on an Internet that increasingly offers sites rich in graphics and multimedia. On the other hand, with the recent failure of many businesses based upon the delivery of rich Internet content, some analysts have suggested that broadband access may be overrated and a "tough sell" to the mass consumer market.

An additional obstacle to the development of broadband Internet access is the difficulty in developing viable business plans that would recoup the cost of providing the content that drives consumers to seek and use broadband. While advertising (in the form of banners and pop-up windows, for example) is an obvious revenue source, it is limited—in part by the difficulty in persuading advertisers they are getting their money's worth. Another obvious strategy, subscription-based services, has met with consumer resistance.

Despite these obstacles and misgivings, the growing availability of broadband communications combined with the ever-increasing processing power of the desktop PC as well as the growth of mobile computing is likely to make multimedia experience pervasive and increasingly immersive.

The "always on" nature of most broadband technologies also impacts the way people use the Internet. Without the need to explicitly dial and wait for a connection, it becomes easier and more natural to resort to the Net for breaking news, the satisfying of routine and esoteric information needs, and on-demand entertainment. On the other hand, always-on systems are more vulnerable to hacking and intrusion, requiring that users obtain firewall and antivirus software and become more educated about potential security threats (see FIREWALL).

Further Reading

Abe, George, and Alicia Buckley. *Residential Broadband.* 2nd ed. Indianapolis, Ind.: Cisco Press, 2000.

Bates, Regis J. *Broadband Telecommunications Handbook.* New York: McGraw Hill, 1999.

buffering

Computer designers must deal with the way different parts of a computer system process data at different speeds. For example, text or graphical data can be stored in main memory (RAM) much more quickly than it can be sent to a printer, and in turn data can be sent to the printer faster than the printer is able to print the data. The solution to this problem is the use of a buffer (sometimes called a spool), or memory area set aside for the temporary storage of data. Buffers are also typically used to store data to be displayed (video buffer), to collect data to be transmitted to (or received from) a modem, for transmitting audio or video content (see STREAMING) and for many other devices (see INPUT/OUTPUT). Buffers can also be used for data that must be reorganized in some way before it can be further processed. For example, character data is stored in a communications buffer so it can be serialized for transmission.

BUFFERING TECHNIQUES

The two common arrangements for buffering data are the pooled buffer and the circular buffer. In the pool buffer, multiple buffers are allocated, with the buffer size being equal to the size of one data record. As each data record is received, it is copied to a free buffer from the pool. When it is time to remove data from the buffer for processing, data is read from the buffers in the order in which it had been stored (first in, first out, or FIFO). As a buffer is read, it is marked as free so it can be used for more incoming data.

In the circular buffer there is only a single buffer, large enough to hold a number of data records. The buffer is set up as a queue (see QUEUE) to which incoming data records are written and from which they are read as needed for processing. Because the queue is circular, there is no "first" or "last" record. Rather, two pointers (called In and Out) are maintained. As data is stored in the buffer, the In pointer is incremented. As data is read back from the buffer, the Out pointer is incremented. If either pointer reaches around back to the beginning, it begins to wrap around. The software managing the buffer must make sure that if the In pointer goes past the Out pointer, then the Out pointer must not go past In. Similarly, if Out goes past In, then In must not go past Out.

The fact that programmers sometimes fail to check for buffer overflows has resulted in a seemingly endless series of security vulnerabilities, such as in earlier versions of the UNIX sendmail program. In one technique, attackers can use a too-long value to write data, or worse, commands into the areas that control the program's execution, possibly taking over the program (see also COMPUTER CRIME AND SECURITY).

Buffering is conceptually related to a variety of other techniques for managing data. A disk cache is essentially a special buffer that stores additional data read from a disk in anticipation that the consuming program may soon request it. A processor cache stores instructions and data in anticipation of the needs of the CPU. Streaming of multimedia (video or sound) buffers a portion of the content so it can be played smoothly while additional content is being received from the source.

Depending on the application, the buffer can be a part of the system's main memory (RAM) or it can be a separate memory chip or chips onboard the printer or other device. Decreasing prices for RAM have led to increases in the typical size of buffers. Moving data from main memory to a peripheral buffer also facilitates the multitasking feature found in most modern operating systems, by allowing applications to buffer their output and continue processing.

Further Reading

Festa, Paul. "Study Says 'Buffer Overflow' is Most Common Security Bug." CNET News, November 23, 1999. http://news.com.com/2100-1001-233483.html?legacy=cnet

"Secure Programming for Linux and UNIX HOWTO: Chapter 5, Avoid Buffer Overflow." http://www.linuxvoodoo.com/howto/HOWTO/Secure-Programs-HOWTO/buffer-overflow.html

bugs and debugging

In general terms a bug is an error in a computer program that leads to unexpected and unwanted behavior. (Lore has it that the first "bug" was a burnt moth found in the relays of the early Mark I computer in the 1940s; however, as early as 1878 Thomas Edison had referred to "bugs" in the design of his new inventions.)

Computer bugs can be divided into two categories: syntax errors and logic errors. A syntax error results from failing to follow a language's rules for constructing statements, or from using the wrong symbol. For example, each statement in the C language must end with a semicolon. This sort of syntax error is easily detected and reported by modern compilers, so fixing it is trivial.

A logic error, on the other hand, is a syntactically valid statement that does not do what was intended. For example, if a C programmer writes:

```
if Total = 100
```

instead of

```
if Total = = 100
```

the programmer may have intended to test the value of Total to see if it is 100, but the first statement actually *assigns* the value of 100 to Total. That's because a single equals sign in C is the assignment operator; testing for equality requires the double equals sign. Further, the error will result in the if statement always being true, because the truth value of an assignment is the value assigned (100 in this case) and any nonzero value is considered to be "true." (See BRANCHING STATEMENTS.)

Loops and pointers are frequent sources of logical errors (see LOOP and POINTERS AND INDIRECTION). The boundary condition of a loop can be incorrectly specified (for example, < 10 when < = 10 is wanted). If a loop and a pointer or index variable are being used to retrieve data from an array, pointing beyond the end of the array will retrieve whatever data happens to be stored out there.

Errors can also be caused in the conversion of data of different types (see DATA TYPES). For example, in many language implementations the compiler will automatically convert an integer value to floating point if it is to be assigned to a floating point variable. However, while an integer can retain at least nine decimal digits of precision, a float may only be able to guarantee seven. The result could be a loss of precision sufficient to render the program's results unreliable, particularly for scientific purposes.

DEBUGGING TECHNIQUES

The process of debugging (identifying and fixing bugs) is aided by the debugging features integrated into most modern programming environments. Some typical features include the ability to set a *breakpoint* or place in the code where the running program should halt so the values of key variables can be examined. A *watch* can be set on specified certain variables so their changing values will be displayed as the program executes. A *trace* highlights the source code to show what statements are being executed as the program runs. (It can also be set to follow execution into and through any procedures or subroutines called by the main code.)

During the process of software development, debugging will usually proceed hand in hand with software testing. Indeed, the line between the two can be blurry. Essentially, debugging deals with fixing problems so that the program is doing what it intends to do, while testing determines whether the program's performance adequately meets the needs and objectives of the end user.

Further Reading

Robbins, John. *Debugging Applications.* Redmond, Wash.: Microsoft Press, 2000.

Rosenberg, Jonathan B. *How Debuggers Work: Algorithms, Data Structures, and Architecture.* New York: Wiley, 1996.

bulletin board systems (BBS)

An electronic bulletin board is a computer application that lets users access a computer (usually with a modem and phone line) and read or post messages on a variety of topics. The messages are often organized by topic, resulting in *threads* of postings, responses, and responses to the responses. In addition to the message service, many bulletin boards provide files that users can download, such as games and other programs, text documents, pictures, or sound files. Some bulletin boards expect users to upload files to contribute to the board in return for the privilege of downloading material.

The earliest form of bulletin board appeared in the late 1960s in government installations and a few universities participating in the Defense Department's ARPANET (the ancestor to the Internet). As more universities came on-line in the early 1970s, the Netnews (or Usenet) system offered a way to use UNIX file-transfer programs to store messages in topical newsgroups (see NETNEWS AND NEWSGROUPS). The news system automatically propagated messages (in the form of a "news feed") from the site where they were originally posted to regional nodes, and from there throughout the network.

By the early 1980s, a significant number of personal computer users were connecting modems to their PCs. Bulletin board software was developed to allow an operator (called a "sysop") to maintain a bulletin board on his or her PC. Users (one or a few at a time) could dial a phone number to connect to the bulletin board. In 1984, programmer Tom Jennings developed the Fido BBS software, which allowed participating bulletin boards to propagate postings in a way roughly similar to the distribution of UNIX Netnews messages.

DECLINE OF THE BBS

In the 1990s, two major developments led to a drastic decline in the number of bulletin boards. The growth of major services such as America On-line and CompuServe (see ON-LINE SERVICES) offered users a friendlier user interface, a comprehensive selection of forums and file downloads, and richer content than bulletin boards with their character-based interface and primitive graphics. An even greater impact resulted from the development of the World Wide Web and Web browsing software, which offered access to a worldwide smorgasbord of services in which each Web home page had the potential of serving as a virtual bulletin board and resource center (see WORLD WIDE WEB and WEB BROWSER). As the 1990s progressed, increasingly rich multimedia content became available over the Internet in the form of streaming video, themed "channels," and the sharing of music and other media files.

Traditional bulletin boards are now found mostly in remote and underdeveloped areas (where they can provide users who have only basic phone service and perhaps obsolescent PCs with an e-mail gateway to the Internet). However the BBS contributed much to the grassroots on-line culture, providing a combination of expansive reach and a virtual small-town atmosphere (see also VIRTUAL COMMUNITY). Venues such as The Well (see CONFERENCING SYSTEMS) retain much of the "feel" of the traditional bulletin board system.

Further Reading

"The BBS Corner." http://www.dmine.com/bbscorner/

Byrant, Alan D. *Growing and Maintaining a Successful BBS: The Sysop's Handbook.* Reading, Mass.: Addison-Wesley, 1995.

bus

A computer bus is a pathway for data to flow between the central processing unit (CPU), main memory (RAM), and various devices such as the keyboard, video, disk drives, and communications ports. Connecting a device to the bus allows it to communicate with the CPU and other components without there having to be a separate set of wires for each device. The bus thus provides for flexibility and simplicity in computer architecture.

Mainframe computers and large minicomputers typically have proprietary buses that provide a wide multipath connection that allows for data transfer rates from about 3 MB/sec. to 10 MB/sec or more. This is in keeping with the use of mainframes to process large amounts of data at high speeds (see MAINFRAME).

MICROCOMPUTER BUSES

The bus played a key role in the development of the modern desktop computer in the later 1970s and 1980s. In the microcomputer, the bus is fitted with connectors called expansion slots, into which any expansion card that meets connection specifications can be inserted. Thus the S-100 bus made it possible for microcomputer pioneers to build a variety of systems with cards to expand the memory and add serial and parallel ports, disk controllers, and other devices. (The Apple II had a similar expansion capability.) In 1981, when IBM announced its first PC, it also defined an 8-bit expansion bus that became known as the ISA (Industry Standard Architecture) as other companies rushed to "clone" IBM's hardware.

In the mid-1980s, IBM advanced the industry with the AT (Advanced Technology) machine, which had the 16-bit Intel 80286 chip and an expanded bus that could transmit data at up to 2 MB/sec. The clone manufacturers soon matched and exceeded these specifications, however. IBM responded by trying both to improve the microcomputer bus and to define a proprietary standard that it could control via licensing. The result was called the Micro-Channel Architecture (MCA), which increased data throughput to 20 MB/sec with full 32-bit capability. This bus had other advanced features such as a direct connection to the video system (Video Graphics Array) and the ability to configure cards in software rather than having to set physical switches. In addition, cards could now incorporate their own processors and memory in a way similar to that of their powerful mainframe counterparts (this is called *bus mastering*). Despite these advantages, however, the proprietary nature of the MCA and the fact that computers using this bus could not use any of the hundreds of ISA cards led to a limited market share for the new systems.

Instead of paying IBM and adopting the new standard, nine major clone manufacturers joined to develop the EISA (Extended ISA) bus. EISA was also a 32-bit bus, but its maximum transfer rate of 33 MB/sec made it considerably faster than the MCA. It was tailored to the new Intel 80386 and 80486 processors, which supported the synchronous transfer of data in rapid bursts. The EISA matched and exceeded the MCA's abilities (including bus mastering and no-switch configuration), but it also retained the ability to use older ISA expansion cards. The EISA soon became the industry standard as the Pentium family of processors were introduced.

A standard ISA bus PC expansion card. This "open architecture" allowed dozens of companies to create hundreds of add-on devices for IBM-compatible personal computers.

However, the endless hunger for more data-transfer capability caused by the new graphics-oriented operating systems such as Microsoft Windows led to the development of *local buses*. A local bus is connected to the processor's memory bus (which typically runs at half the processor's external speed rather than the much slower system bus speed), a considerable advantage in moving data (such as graphics) from main memory to the video card.

Two of these buses, the VESA (or VL) bus and the PCI bus came into widespread use in higher-end machines, with the PCI becoming dominant. The PCI bus runs at 33 MHz and supports features such as Plug and Play (the ability to automatically configure a device, supported in Windows 98 and later) and Hot Plug (the ability to connect or reconnect devices while the PC is running). The PCI retains compatibility with older 8-bit and 16-bit ISA expansion cards. At the end of the 1990s, PC makers were starting to introduce even faster buses such as the AGP (accelerated graphics port), which runs at 66 MHz.

Two important auxiliary buses are designed for the connection of peripheral devices to the main PC bus. The older SCSI (Small Computer Systems Interface) was announced in 1986 (with the expanded SCSI-2 in 1994). SCSI is primarily used to connect disk drives and other mass storage devices (such as CD-ROMs), though it can be used for scanners and other devices as well. SCSI-2 can transfer data at 20 MB/sec over a 16-bit path, and SCSI-3 (still in development) will offer a variety of high-speed capabilities. SCSI was adopted as the standard peripheral interface for many models of Apple Macintosh computers as well as UNIX workstations. On IBM architecture PCs SCSI is generally used for servers that require large amounts of mass storage. Multiple devices can be connected in series (or "chained").

The newer USB (Universal Serial Bus) is relatively slow (12 MB/sec) but convenient because devices do not need separate cards to connect to the bus. A simple plug can be inserted directly into a USB socket on the system board or the socket can be connected to a USB hub to which several devices can be connected. In 2002, USB 2.0 is gradually entering the marketplace. It offers 480 MB/sec data transfer speed.

It is uncertain whether the next advance will be the adoption of a 64-bit PCI bus or the development of an entirely different bus architecture. The latter is attractive as a way to get past certain inherent bottlenecks in the PCI design, but the desire for downward compatibility with the huge number of existing ISA, EISA, and PCI devices is also very strong.

Further Reading

PC Guide. "System Buses." http://www.pcguide.com/ref/mbsys/buses/index.htm

Bush, Vannevar

(1890–1974)
American
Engineer and Inventor

Vannevar Bush, grandson of two sea captains and son of a clergyman, was born in Everett, Massachusetts, just outside of Boston. Bush earned his B.S. and M.S. degrees in engineering at Tufts University, and received a joint doctorate from Harvard and MIT in 1916. He went on to full professorship at MIT and became Dean of its Engineering School in 1932.

Bush combined an interest in mathematics with the design of mechanical devices to automate calculations. During his undergraduate years he invented an automatic surveying machine using two bicycle wheels and a recording instrument. His most important invention was the differential analyzer, a special type of computer that used combinations of rotating shafts and cams to incrementally add or subtract the differences needed to arrive at a solution to the equation (see also ANALOG COMPUTER). His improved model (Rockefeller Differential Analyzer, or RDA2) replaced the shafts and gears with an electrically-driven system, but the actual integrators were still mechanical. Several of these machines were built in time for World War II, when they served for such purposes as calculating tables of ballistic trajectories for artillery.

Later, Bush turned his attention to problems of information processing. Together with John H. Howard (also of MIT), he invented the Rapid Selector, a device that could retrieve specific information from a roll of microfilm by scanning for special binary codes on the edges of the film. His most far-reaching idea, however, was what he called the "Memex"—a device that would link or associate pieces of information with one another in a way similar to the associations made in the human brain. Bush visualized this as a desktop workstation that would enable its user to explore the world's information resources by following links, the basic principle of what would later become known as hypertext (see HYPERTEXT AND HYPERMEDIA).

In his later years, Bush wrote books that became influential as scientists struggled to create large-scale research teams and to define their roles and responsibilities in the cold war era. He played the key role in establishing the National Science Foundation in 1950, and served on its advisory board from 1953 to 1956. He then became CEO of the drug company Merck (1955–1962) as well as serving as chairman (and then honorary chairman) of the MIT Corporation (1957–1974).

Bush would receive numerous honorary degrees and awards that testified to the broad range of his interests and achievements not only in electrical and mechanical engineering, but also in social science. In 1964, he received the National Medal of Science. Bush died on June 28, 1974, in Belmont, Massachusetts.

Further Reading

Bush, Vannevar. *Pieces of the Action*. New York: William Morrow, 1970.

———. *Science: the Endless Frontier.* U.S. Government Printing Office, 1945.

Nyce, J. M., and P. Kahn. *From Memex to Hypertext: Vannevar Bush and the Mind's Machine*. Boston: Academic Press, 1991. [Includes two essays by Bush: "As We May Think" and "Memex II."]

business applications of computers

Efficient and timely data processing is essential for businesses of all sizes from corner shop to multinational corporation. Business applications can be divided into the broad categories of Administration, Accounting, Office, Production, and Marketing and Sales.

Administrative applications deal with the organization and management of business operations. This includes personnel-related matters (recruiting, maintenance of personnel records, payroll, pension plans, and the provision of other benefits such as health care). It also includes management information or decision support systems, communications (from simple e-mail to teleconferencing), and the administration of the data processing systems themselves.

The Accounting category includes databases of accounts receivable (money owed to the firm) and payable (such as bills from vendors). While this software is decidedly unglamorous, in a large corporation small inefficiencies can add up to significant costs or lost revenue. (For example, paying a bill before it is due deprives the firm of the "float" or interest that can be earned on the money, while paying a bill too late can lead to a loss of discounts or the addition of penalties.) A variety of reports must be regularly generated so management can spot such problems and so taxes and regulatory requirements can be met.

The Office category involves the production and tracking of documents (letters and reports) as required for the day-to-day operation of the business. Word processing, desktop publishing, presentation and other software can be used for this purpose (see APPLICATION SUITE, WORD PROCESSING, SPREADSHEET, and PRESENTATION SOFTWARE).

Production is a catchall term for the actual product or service that the business provides. For a manufacturing business this may require specialized design and manufacturing programs (see COMPUTER-AIDED DESIGN AND MANUFACTURING CAD/CAM) as well as software for tracking and scheduling the completion of tasks. For a business that markets already produced goods the primary applications will be in the areas of transportation (tracking the shipping of goods), inventory and warehousing, and distribution. Service businesses will need to establish accounts for customers and keep track of the services performed (on an hourly basis or otherwise).

Marketing and Sales includes market research, advertising, and other programs designed to make the public aware of and favorably disposed to the product or service. Once people come to the store to buy something, the actual retail transaction must be provided for, including the point-of-sale terminal (formerly "cash register") with its interface to the store inventory system and the verification of credit cards or other forms of payment.

CHANGING ROLE OF COMPUTERS

Computer support for business functions can be provided in several forms. During the 1950s and 1960s (the era of mainframe dominance), only the largest firms had their own computer facilities. Many medium- to small-sized businesses contracted with agencies called service bureaus to provide computer processing for such functions as payroll processing. Service bureaus and in-house data processing facilities often developed their own software (typically using the COBOL language).

The development of the minicomputer (and in the 1980s, the desktop microcomputer) allowed more businesses to undertake their own data processing, in the expectation (not always fulfilled) that they would be able both to save money and to create systems better tailored to their needs. Areas such as payroll and accounts payable/receivable generally still relied upon specialized software packages. However, the growing availability of powerful database software (such as dBase and its descendants) as well as spreadsheet programs enabled businesses to maintain and report on a variety of information.

During the 1980s, the daily life of the office began to change in marked ways. The specialized word processing machines gave way to programs such as WordStar, WordPerfect, and Microsoft Word running on desktop computers. Advanced word processing and desktop publishing software moved more of the control of the appearance of documents into the hands of office personnel. The local area network (LAN) made it possible to share resources (such as the new laser printers and databases on a powerful file server PC) as well as providing for communication in the form of e-mail.

As the Internet and the World Wide Web came into prominence in the later 1990s, another revolution was soon under way. Every significant organization is now expected to have its own website or sites. These webpages serve a Janus-like function. On the one hand, they present the organization's face to the world, providing announcements, advertising, catalogs, and the capability for on-line purchasing (e-commerce). On the other hand, many organizations now put their databases and other records on websites (in secured private networks) so that employees can readily access and update them. The growth in mobile computing and readily available Internet connections (including wireless services) increasingly enables traveling businesspersons to effectively take the office and its resources with them on the road.

Further Reading

Bodnar, George H,. and William S. Hopwood. *Accounting Information Systems.* Upper Saddle River, N.J.: Prentice Hall, 2000.

Cortada, James W. *21st Century Business: Managing and Working in the New Digital Economy.* Upper Saddle River, N.J.: Prentice Hall, 2000.

O'Brien, James A. *Introduction to Information Systems.* New York: McGraw Hill, 2000.

C

C

The C programming language was developed in the early 1970s by Dennis Ritchie, who based it on the earlier languages BCPL and B. C was first used on DEC PDP-11 computers running the newly developed UNIX operating system, where the language provided a high-level alternative to the use of PDP Assembly language for development of the many utilities that give UNIX its flexibility. Since the 1980s, C and its descendent, C++, have become the most widely used programming languages.

LANGUAGE FEATURES

Like the earlier Algol and the somewhat later Pascal, C is a procedural language that reflects the philosophy of structured programming that was gradually taking shape during the 1970s. In general, C's approach can be described as providing the necessary features for real world computing in a compact and efficient form. The language provides the basic control structures such as if and switch (see BRANCHING STATEMENTS) and while, do, and for (see LOOP). The built-in data types provide for integers (int, short, and long), floating-point numbers (float and double), and characters (char). An array of any type can be declared, and a string is implemented as an array of char (see DATA TYPES and CHARACTERS AND STRINGS).

Pointers (references to memory locations) are used for a variety of purposes, such as for storing and retrieving data in an array (see POINTERS AND INDIRECTION). While the use of pointers can be a bit difficult for beginners to understand, it reflects C's emphasis as a systems programming language that can "get close to the hardware" in manipulating memory.

Data of different types can be combined into a record type called a struct. Thus, for example:

```
struct Employee_Record {
  char [10] First_Name;
  char [1] Middle_Initial
  char [20] Last_Name
  int Employee_Number
} ;
```

(There is also a union, which is a struct where the same structure can contain one of two different data items.)

The standard mathematical and logical comparison operators are available. There are a couple of quirks: the equals comparison operator is = =, while a single equal sign = is an assignment operator. This can create a pitfall for the wary, since the condition

```
if (Total = 10)
  printf ("Finished!");
```

always prints Finished, since the assignment Total = 10 returns a value of 10 (which not being zero, is "true" and satisfies the if condition).

C also features an increment ++ and decrement – – operator, which is convenient for the common operation of raising or lowering a variable by one in a counting loop. In C the following statements are equivalent:

```
Total = Total + 1;
Total += 1;
Total ++;
```

Unlike Pascal's two separate kinds of procedures (func, or function, which returns a value, and proc, or procedure, which does not), C has only functions. Arguments are passed to functions by value, but can be passed by reference by using a pointer. (See PROCEDURES AND FUNCTIONS.)

SAMPLE PROGRAM

The following is a brief example program:

```
#include <stdio.h>
float Average (void);
main () {
printf ("The average is: %f", Average() );
}
float Average (void) {
int NumbersRead = 0;
int Number;
int Total = 0;
while (scanf("%d\n", &Number) == 1)
  {
    Total = Total + Number;
    NumbersRead = NumbersRead + 1;
  }
return (Total / NumbersRead);
}
}
```

Statements at the beginning of the program that begin with # are preprocessor directives. These make changes to the source code before it is compiled. The #include directive adds the specified source file to the program. Unlike many other languages, the C language itself does not include many basic functions, such as input/output (I/O) statements. Instead, these are provided in standard libraries. (The purpose of this arrangement is to keep the language itself simple and portable while keeping the implementation of functions likely to vary on different platforms separate.) The stdio.h file here is a "header file" that defines the I/O functions, such as printf() (which prints formatted data) and scanf() (which reads data into the program and formats it).

The next part of the program declares any functions that will be defined and used in the program (in this case, there is only one function, Average). The function declaration begins with the type of data that will be returned by the function to the calling statement (a floating point value in this case). After the function name comes declarations for any parameters that are to be passed to the function by the caller. Since the Average function will get its data from user input rather than the calling statement, the value (void) is used as the parameter.

Following the declaration of Average comes the main() function. Every C program must have a main function. Main is the function that runs when the program begins to execute. Typically, main will call a number of other functions to perform the necessary tasks. Here main calls Average within the printf statement, which will print the average as returned by that function. (Calling functions within other statements is an example of C's concise syntax.)

Finally, the Average function is defined. It uses a loop to read in the data numbers, which are totaled and then divided to get the average, which is sent back to the calling statement by the return statement.

A programmer could create this program on a UNIX system by typing the code into a source file (test.c in this case) using a text editor such as vi. A C compiler (gcc in this case) is then given the source code. The source code is compiled, and linked, creating the executable program file a.out. Typing that name at the command prompt runs the program, which asks for and averages the numbers.

```
% gcc test.c
% a.out
5
7
9
.
```

The average is: 7.000000

SUCCESS AND CHANGE

In the three decades after its first appearance, C became one of the most successful programming languages in history. In addition to becoming the language of choice for most UNIX programming, as microcomputers became capable of running high-level languages, C became the language of choice for developing MS-DOS, Windows, and Macintosh programs. The application programming interface (API) for Windows, for example, consists of hundreds of C functions, structures, and definitions (see APPLICATION PROGRAMMING INTERFACE and MICROSOFT WINDOWS).

However, C has not been without its critiques among computer scientists. Besides containing idioms that can encourage cryptic coding, the original version of C (as defined in Kernighan and Ritchie's *The C Programming Language*) did not check function parameters to make sure they matched the data types expected in the function definitions. This problem led to a large number of hard-to-catch bugs. However, the development of ANSI standard C with its stricter requirements, as well as type checking built into compilers has considerably ameliorated this problem. At about the same time, C++ became available as an object-oriented extension and partial rectification of C. While C++ and Java have considerably supplanted C for developing new programs, C programmers have a relatively easy learning path to the newer languages and the extensive legacy of C code will remain useful for years to come.

Further Reading

Kernighan, B. W., and D. M. Ritchie. *The C Programming Language.* 2nd ed. Upper Saddle River, N.J.: Prentice-Hall, 1988.

Ritchie, D. M. "The Development of the C Language," in *History of Programming Languages II,* ed. T. J. Bergin and R. G. Gibson, 678–698. Reading, Mass.: Addison-Wesley, 1995.

C++

The C++ language was designed by Bjarne Stroustrup at AT&T's Bell Labs in Murray Hill, New Jersey, starting in 1979. By that time the C language had become well established as a powerful tool for systems programming (see C). However Stroustrup (and others) believed that C's limited data structures and function mechanism were proving inadequate to express the relationships found in increasingly large software packages involving many objects with complex relationships.

Consider the example of a simple object: a stack onto which numbers can be "pushed" or from which they can be "popped" (see STACK). In C, a stack would have to be implemented as a struct to hold the stack data and stack pointer, and a group of separately declared functions that could access the stack data structure in order to, for example "push" a number onto the stack or "pop" the top number from it. In such a scheme there is no direct, enforceable relationship between the object's data and functions. This means, among other things, that parts of a program could be dependent on the internal structure of the object, or could directly access and change such internal data. In a large software project with many programmers working on the code, this invites chaos.

An alternative paradigm already existed (see OBJECT-ORIENTED PROGRAMMING) embodied in a few new languages (see SIMULA and SMALLTALK). These languages allow for the structuring of data and functions together in the form of objects (or classes). Unlike a C struct, a class can contain both the data necessary for describing an object and the functions needed for manipulating it (see CLASS). A class "encapsulates" and protects its private data, and communicates with the rest of the program only through calls to its defined functions.

Further in object-oriented languages, the principle of inheritance could be used to proceed from the most general, abstract object to particular versions suited for specific tasks, with each object retaining the general capabilities and revising or adding to them. Thus, a "generic" list foundation class could be used as the basis for deriving a variety of more specialized lists (such as a doubly-linked list).

While attracted to the advantages of the object-oriented approach, Stroustrup also wanted to preserve the C language's ability to precisely control machine behavior needed for systems programming. He thus decided to build a new language on C's familiar syntax and features with object-oriented extensions. Stroustrup wrote the first version, called "C with Classes" as his Ph.D. thesis at Cambridge University in England. This gradually evolved into C++ through the early 1980s.

C++ FEATURES

The fundamental building block of C++ is the class. A class is used to create objects of its type. Each object contains a set of data and can carry out specified functions when called upon by the program. For example, the following class defines an array of integers and declares some functions for working with the array. Typically, it would be put in a header file (such as stack.h):

```
const int Max_size=20; // maximum elements in
Stack

class Stack { // Declare the Stack class
public: // These functions are available
outside
Stack(); // Constructor to create Stack
objects
void push (int); // push int on Stack
int pop(); // remove top element
private: // This data can only be used in
class
int index;
int Data[Max_size];
};
```

Next, the member functions of the Stack class are defined. The definitions can be put in a source file Stack.cpp:

```
#include "Stack.h" // bring in the declarations
Stack::Stack() { index=0;} // set zero for new
  stack
void Stack::push (int item){ // put a number on
  stack
Data[index++] = item;
}
int Stack::pop(){ // remove top number
return Data [index–];
}
```

Now a second source file (Stacktest.cpp) can be written. It includes a main() function that creates a Stack object and tests some of the class functions:

```
#include "Stack.cpp" // include the Stack class
#include <iostream.h> // include standard I/O
  library
main() {
Stack S; // Create a Stack object called S
int index;
for (index = 1; index <= 5; index++)
  S.push(index); // put numbers 1-5 on stack
for (index = 1; index <=5; index++)
  cout < S.pop(); // print the stack
}
```

The stack implementation is completely separate from any program code that uses stack objects. Thus, a programmer could revise the stack class (perhaps using an improved algorithm or generalizing it to work with different data types). As long as the required parameters for the member functions aren't changed, programs that use stack objects won't need to be changed.

In addition to classes and inheritance, C++ has some other important features. The data types for function parameters can be fully defined, and types checked automatically (although programmers can bypass this type checking if they really want or need to). New operators can be added to a class by defining special operator functions, and the same operator can be given different meanings when working with different data types. (This is called overloading.) Thus, the + operator can be defined with a String class to combine (concatenate) two strings. The operator will still mean "addition" when used with numeric data.

An abstract object (one with no actual implementation) can be used as the basis for virtual functions. These functions can be redefined in each derived object so that whenever an object of that type is encountered the compiler will automatically search "downward" from the base class and find the correct derived class function.

Later versions of C++ include a related concept called templates. A template is an abstract specification that can be used to generate class definitions for data types passed to it (see TEMPLATE). Thus, a list template could be passed a vector and a 2D array and it will create a list class definition for each of these types. Templates are generally used when there is no hierarchical inheritance relationship between the types (in that case the virtual base class is a better approach).

C++ provides object-oriented alternatives to the standard libraries. For example, input/output uses a stream model, and I/O operators can be overloaded so they'll work with new classes. There is also an improved error-handling mechanism using appropriate objects.

GROWTH OF C++

During the late 1980s and 1990s, C++ became a very popular language for a variety of applications ranging from systems programming to business applications and games. The growth of the language coincided with the development of more powerful desktop computers and the release of inexpensive, easy-to-use but powerful development environments from Microsoft, Borland, and others. Since these compilers could also handle traditional C code, programmers could "port" existing code and use the object-oriented techniques of C++ as they mastered them. By the late 1990s, however, C++, although still dominant in many areas, was being challenged by Java, a language that simplified some of the more complex features of C++ and that was designed particularly for writing software to run on Web servers and browsers (see JAVA).

Further Reading

"C++ Archive." http://www.austinlinks.com/CPlusPlus/

"Complete C++ Language Tutorial." http://www.cplusplus.com/doc/tutorial/

Stroustrup, Bjarne. "A History of C++: 1979–1991." In *History of Programming Languages II*, edited by Thomas J. Bergin, Jr., and Richard G. Gibson, Jr. New York: ACM Press; Reading, Mass.: Addison-Wesley, 1996, 699–755.

———. *The C++ Programming Language*. 3rd ed. Reading, Mass.: Addison-Wesley, 1997.

cache

A basic problem in computer design is how to optimize the fetching of instructions or data so that it will be ready when the processor (CPU) needs it. One common solution is to use a cache. A cache is an area of relatively fast-access memory into which data can be stored in anticipation of its being needed for processing. Caches are used mainly in two contexts: the processor cache and the disk cache.

CPU CACHE

The use of a processor cache is advantageous because instructions and data can be fetched more quickly from the cache (static memory chips next to or within the CPU) than they can be retrieved from the main memory (usually dynamic RAM). An algorithm analyzes the instructions currently being executed by the processor and tries to anticipate what instructions and data are likely to be needed in the near future. (For example, if the instructions call for a possible branch to one of two sets of instructions, the cache will load the set that has been used most often or most recently. Since many programs loop over and over again through the same instructions until some condition is met, the cache's prediction will be right most of the time.)

These predicted instructions and data are transferred from main memory to the cache while the processor is still executing the earlier instructions. If the cache's prediction was correct, when it is time to fetch these instructions and data they are already waiting in the high-speed cache memory. The result is an effective increase in the CPU's speed despite there being no increase in clock rate (the rate at which the processor can cycle through instructions).

The effectiveness of a processor cache depends on two things: the mix of instructions and data being processed and the location of the cache memory. If a program uses long sequences of repetitive instructions and/or data, caching will noticeably speed it up. A cache located within the CPU itself (called an L1 cache) is faster (albeit more expensive) than an L2 cache, which is a separate set of chips on the motherboard.

Cache (1)

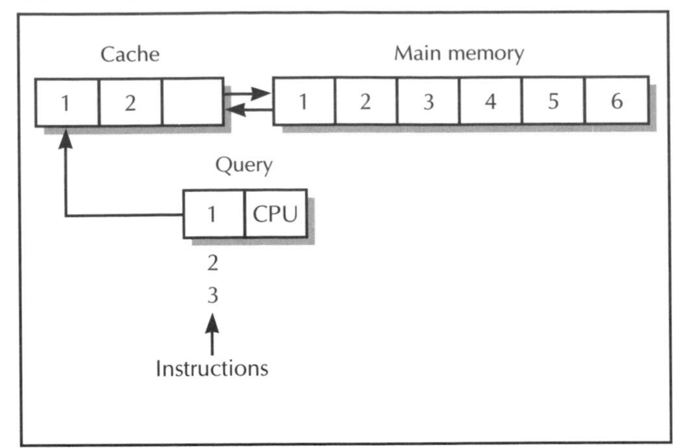

A cache system attempts to ensure that data is available in fast-access memory before it is needed. Here when the processor (CPU) is looking for cell 1, it finds it is already available in the cache. If cell 3 was wanted, it would have to be transferred from main memory rather than the cache, but at that time cells 4 and 5 might then be stored in the cache in anticipation of being needed for the next instruction.

Changes made to data by the CPU are normally written back to the cache, not to main memory, until the cache is full. In multiprocessor systems, however, designers of processor caches must deal with the issue of cache coherency. If, for example, several processors are executing parts of the same code and are using a shared main memory to communicate, one processor may change the value of a variable in memory but not write it back immediately (since its cache is not yet full). Meanwhile, another processor may load the old value from the cache, unaware that it has been changed. This can be prevented by using special hardware that can detect such changes and automatically "write through" the new value to the memory. The processors, having received a hardware or software "signal" that data has been changed, can be directed to reread it.

DISK CACHE

A disk cache uses the same general principle as a processor cache. Here, however, it is RAM (either a part of main memory or separate memory on the disk drive) that is the faster medium and the disk drive itself that is slower. When an application starts to request data from the disk, the cache reads one or more complete blocks or sectors of data from the disk rather than just the data record being requested. Then, if the application continues to request sequential data records, these can be read from the high-speed memory on the cache rather than from the disk drive. It follows that disk caching is most effective when an application, for example, loads a database file that is stored sequentially on the disk.

Similarly, when a program writes data to the disk, the data can be accumulated in the cache and written back to the drive in whole blocks. While this increases efficiency, if a power outage or other problem erases or corrupts the cache contents, the cache will no longer be in synch with the drive. This can cause corruption in a database.

NETWORK CACHE

Caching techniques can be used in other ways. For example, most Web browsers are set to store recently read pages on disk so that if the user directs the browser to go back to such a page it can be read from disk rather than having to be retransmitted over the Internet (generally a slower process). Web servers and ISPs (such as cable services) can also cache popular pages so they can be served up quickly.

Further Reading

Nottingham, Mark. "Caching Tutorial for Web Authors and Webmasters." http://www.wdvl.com/Internet/Cache/index.html

"System Cache." http://www.pcguide.com/ref/mbsys/cache/

Peir, J.-K., W. Hsu, and A. J. Smith. "Implementation Issues in Modern Cache Memories." *IEEE Transactions on Computers,* 48, no. 2, 100–110.

calculator

The use of physical objects to assist in performing calculations begins in prehistory with such practices as counting with pebbles or making what appears to be counting marks on pieces of bone. Nor should such simple manipulations be despised: In somewhat more sophisticated form it yielded the abacus, whose operators regularly outperformed mechanical calculators until the advent of electronics.

Generally, however, the term *calculator* is used to refer to a device that is able to store a number, add it to another number, and mechanically produce the result, taking care of any carried digits. In 1623, astronomer Johannes Kepler commissioned such a machine from Wilhelm Schickard. The machine combined a set of "Napier's bones" (slides marked with logarithmic intervals, the ancestor of the slide rule) and a register consisting of a set of toothed wheels that could be rotated to displays the digits 0 to 9, automatically carrying one place to the left. This ingenious machine was destroyed in a fire before it could be delivered to Kepler.

In 1642, French philosopher and mathematician Blaise Pascal invented an improved mechanical calculator. Its mechanism used a carry mechanism with a weight that would drop when a carry was reached, pulling the next wheel into position. This avoided having to use excessive force to carry a digit through several places. Pascal produced a number of his machines and tried to market them to accountants, but they never really caught on.

Schikard's and Pascal's calculators could only add, but in 1674 German mathematician Gottfried Wilhelm Leibniz invented a calculator that could work with all the digits of a number at once, rather than carrying from digit to digit. It worked by allowing a variable number of gear teeth to be engaged in each digit wheel. The operator could, for example, set the wheels to a number such as 215, and then turn a crank three times to multiply it by three, giving a result of 645. This mechanism, gradually improved, would remain fundamental to mechanical calculators for the next three centuries.

The first calculator efficient enough for general business use was invented by an American, Dorr E. Felt, in 1886. His machine, called a Comptometer, used the energy transmitted through the number-setting mechanism to perform the addition, considerably speeding up the calculating process. Improved machines by William Burroughs and others would replace the arm of the operator with an electric motor and provide a printing tape for automatically recording input numbers and results.

ELECTRONIC CALCULATORS

The final stage in the development of the calculator would be characterized by the use of electronics to replace mechanical (or electromechanical) action. The use of logic circuits to perform calculations electronically was first seen in the giant computers of the late 1940s, but this was obviously impractical for desktop office use. By the late 1960s, however, transistorized calculators comparable in size to mechanical desktop calculators came into use. By the 1970s, the use of integrated circuits made it possible to shrink the calculator down to palm-size and smaller. These calculators use a microprocessor with a set of "microinstructions" that enable them to perform a repertoire of operations ranging from basic arithmetic to trigonometric, statistical, or business-related functions.

The most advanced calculators are programmable by their user, who can enter a series of steps (including perhaps decisions and branching) as a stored program, and then apply it to data as needed. At this point the calculator can be best thought of as a small, somewhat limited computer. However, even these limits are constantly stretched: During the 1990s it became common for students to use graphing calculators to plot equations. Calculator use is now generally accepted in schools and even in the taking of the Scholastic Aptitude Test (SAT). However, some educators are concerned that overdependence on calculators may be depriving students of basic numeracy, including the ability to estimate the magnitude of results.

Further Reading

Aspray, W., ed. *Computing Before Computers*. Ames: Iowa State University Press, 1989.

"Calculator Related Links." http://www.geocities.com/SiliconValley/Park/7227/links.html

cartography, computer-assisted

Cartography, or the art of mapmaking, has been transformed in many ways by the use of computers. Traditionally, mapmaking was a tedious process of recording, compiling, and projecting or plotting information about the location, contours, elevation, or other characteristics of natural geographic features or the demographic or political structure of human communities.

Instead of being transcribed from the readings of surveying instruments, geographic information can be acquired and digitized by sensors such as cameras aboard orbiting satellites. The availability of such extensive, detailed information would overwhelm any manual system of transcribing or plotting. Instead, the Geographic Information System (GIS, first developed in Canada in the 1960s) integrates sensor input with scanning and plotting devices, together with a database management system to compile the geographic information.

The format in which the information is stored is dependent on the scope and purpose of the information system. A detailed topographical view, for example, would have physical coordinates of latitude, longitude, and elevation. On the other hand, a demographic map of an urban area might have regions delineated by ZIP code or voting precinct, or by individual address.

Geographic data can be stored as either a raster or a vector representation. A raster system divides the area into a grid and assigns values to each cell in the grid. For example, each cell might be coded according to its highest point of elevation, the amount of vegetation (ground cover) it has, its population density, or any other factor of interest. The simple grid system makes raster data easy to manipulate, but the data tends to be "coarse" since there is no information about variations within a cell.

Unlike the arbitrary cells of the raster grid, a vector representation is based upon the physical coordinates of actual points or boundaries around regions. Vector representation is used when the actual shape of an entity is important, as with property lines. Vector data is harder to manipulate than raster data because geometric calculations must be made in order to yield information such as the distance between two points.

The power of geographic information systems comes from the ability to integrate data from a variety of sources, whether aerial photography, census records, or even scanned paper maps. Once in digital form, the data can be represented in a variety of ways for various purposes. A sophisticated GIS can be queried to determine, for example, how much of a proposed development would have a downhill gradient and be below sea level (such that flooding might be a problem). These results can in turn be used by simulation programs to determine, for example, whether a release of chemicals into the groundwater from a proposed plant site might affect a particular town two miles away. Geographic information systems are thus vital

Cartography, Computer-Assisted

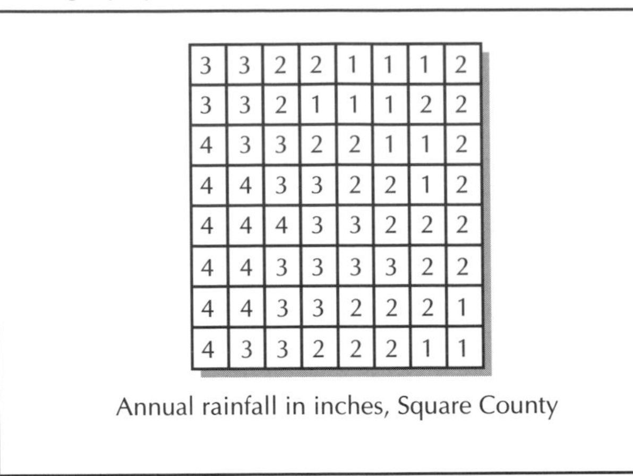

A raster grid showing annual rainfall totals in inches for mythical Square County. Raster data is easy to work with, but the "coarseness" of the grid means that it doesn't capture much local variation or detail.

for the management of a variety of complex systems that are distributed over a geographical area, such as water and sewage systems, power transmission grids, and traffic control systems. Other applications include emergency planning (and evacuation routes) and the long-term study of the effects of global warming trends.

FROM INFORMATION TO NAVIGATION

The earliest use of maps was for facilitating navigation. The development of the Global Positioning System (GPS) made it possible for a device to triangulate readings from three of 24 satellites to pinpoint the user's position on the Earth's surface within a few meters (or even closer in military applications). The mobile navigation systems that have now become a consumer product essentially use the current physical coordinates to look up information in the onboard geographical information system. Depending on the information stored and the user's needs, the resulting display can range from a simple depiction of the user's location on a highway or city street map to the generating of detailed driving directions from the present location to a desired location. As these systems are fitted with increasingly versatile natural language systems (and perhaps voice recognition capabilities), the user will be able to ask questions such as "Where's the nearest gas station?" or even "Where's the nearest French restaurant rated at least three stars?"

Further Reading

Burrough, P. A., and R. McDonnell. *Principles of Geographical Information Systems.* 2nd ed. New York: Oxford University Press, 1998.

Demers, M. N. *Fundamentals of Geographic Information Systems.* New York: Wiley, 1997.

Longley, P. A., (and others), eds. *Geographical Information Systems: Principles, Techniques, Management and Applications.* New York: Wiley, 1998.

Ramadan, K. "The Use of GPS for GIS Applications." http://www.geogr.muni.cz/lgc/gis98/proceed/RAMADAN.html

U.S. Geological Survey. "Geographic Information Systems." http://www.usgs.gov/research/gis/title.html

CASE (computer-aided software engineering)

During the late 1950s and 1960s, software rapidly grew more complex—especially operating system software and large business applications. With the typical program consisting of many components being developed by different programmers, it became difficult both to see the "big picture" and to maintain consistent procedures for transferring data from one program module to another. As computer scientists worked to develop sounder principles (see STRUCTURED PROGRAMMING) it also occurred to them that the power of the computer to automate procedures could be used to create tools for facilitating program design and managing the resulting complexity. CASE, or computer-aided software engineering, is a catchall phrase that covers a variety of such tools involved with all phases of development.

DESIGN TOOLS

The earliest design tool was the flowchart, often drawn with the aid of a template that could be used to trace the symbols on paper (see FLOWCHART). With its symbols for the flow of execution through branching and looping, the flowchart provides a good tool for visualizing how a program is intended to work. However large and complex programs often result in a sea of flowcharts that are hard to relate to one another and to the program as a whole. Starting in the 1960s, the creation of programs for manipulating flow symbols made it easier both to design flowcharts and to visualize them in varying levels of detail.

Another early tool for program design is pseudocode, a language that is at a higher level of abstraction than the target programming language, but that can be refined by adding details until the actual program source code has been specified (see PSEUDOCODE). This is analogous to a writer outlining the main topics of an essay and then refining them into subtopics and supporting details. Attempts were made to create a well-defined pseudocode that could be automatically parsed and transformed into compilable language statements, but they met with only limited success.

During the 1980s and 1990s, the graphics capabilities of desktop computers made it attractive to use a visual rather than linguistic approach to program design. Symbols (sometimes called "widgets") represent program functions such as reading data from a file or creating various kinds of charts. A program can be designed by connecting the widgets with "pipes" representing data flow and by setting various characteristics or properties.

CASE principles can also be seen in mainstream programming environments such as Microsoft's Visual Basic and Visual C++, Borland's Delphi and Turbo C++, and others (see also PROGRAMMING ENVIRONMENT). The design approach begins with setting up forms and placing objects (controls) that represent both user interface items (such as menus, lists, and text boxes) and internal processing (such as databases and Web browsers). However these environments do not in themselves provide the ability of full CASE tools to manage complex projects with many components.

ANALYSIS TOOLS

Once a program has been designed and implementation is under way, CASE tools can help the programmers maintain consistency across their various modules. One such tool (now rather venerable) is the data dictionary, which is a database whose records contain information about the definition of data items and a list of program components that use each item (see DATA DICTIONARY). When the definition of a data item is changed, the data dictionary can provide a list of affected components. Database technology is also applied to software design in the creation of a database of objects within a particular program, which can be used to provide more extensive information during debugging.

INTEGRATION AND TRENDS

A typical CASE environment integrates a variety of tools to facilitate the flow of software development. This process may begin with design using visual flowcharting, "rapid prototyping," or other design tools. Once the overall design is settled, the developer proceeds to the detailed specification of objects used by the program and perhaps creates a data dictionary or other databases with information about program objects. During the coding process, source control or versioning facilities help log and keep track of the changes to code and the succession of new versions ("builds"). While testing the program, an integrated debugger (see BUGS AND DEBUGGING) can use information from the program components database to help pinpoint errors. As the code is finished, other tools can automatically generate documentation and other supporting materials (see also TECHNICAL WRITING and DOCUMENTATION OF PROGRAM CODE).

Just as some early proponents of the English-like COBOL language proclaimed that professional programmers would no longer be needed for generating business applications, CASE tools have often been hyped as a panacea for all the ills of the software development cycle. Rather than causing the demise of the programmer, however, CASE tools have played an important role in keeping software development viable.

CASE

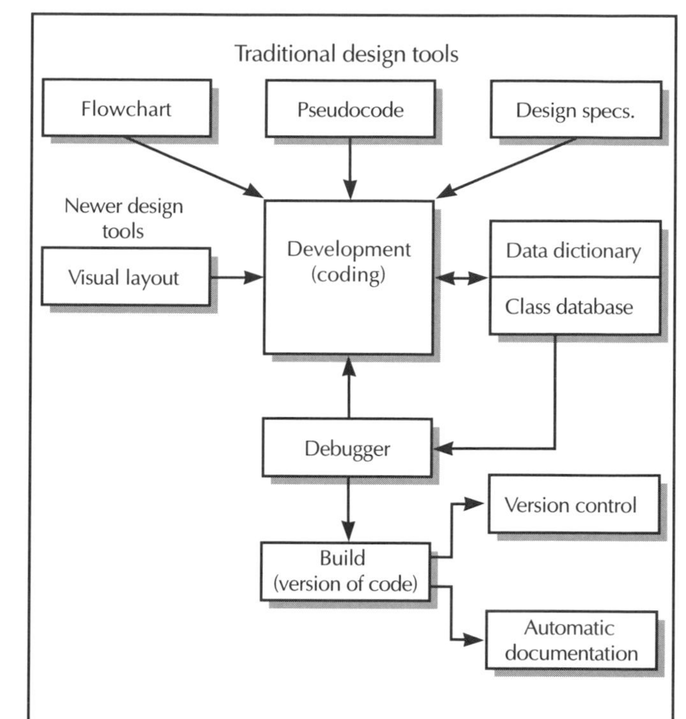

Many tools are used today to aid the complex endeavor of software engineering. Design tools include the traditional flowchart, pseudocode, and design specifications document. Additionally, many systems today use interactive, visual layout tools. During the coding and debugging phase, a data dictionary and/or class database can be used to describe and verify relationships and characteristics of objects in the program. Once the code is "built," a version control system keeps track of what was changed, and various automatic documentation features can be used to obtain listings of classes, functions, and other program elements.

Further Reading

Bergin, Thomas, ed. *Computer-Aided Software Engineering: Issues and Trends for the 1990s and Beyond.* Harrisburg, Penn.: Idea Group, 1993.

CASE (Computer-Aided Software Engineering) Home Page. http://osiris.sunderland.ac.uk/sst/casehome.html

Dixon, Robert L. *Winning with Case: Managing Modern Software Development.* New York: McGraw Hill, 1991.

Fisher, Alan S. *CASE: Using Software Development Tools.* 2nd ed. New York: Wiley, 1991.

Muller, Hausi, Ronald J. Norman, and Jacob Slonim, eds. *Computer-Aided Software Engineering.* Boston: Kluwer Academic Publishers, 1996.

"Software Engineering Archives." http://www.qucis.queensu.ca/Software-Engineering/

CD-ROM and DVD-ROM

CD-ROM (Compact Disk Read-Only Memory) is an optical data storage system that uses a disk coated with a thin layer of metal. In writing data, a laser etches billions of tiny pits in the metal. The data is encoded in the pattern of pits and spaces between them (called "lands"). Unlike the case with a magnetic hard or floppy disk, the data is written in a single spiral track that begins at the center of the disk. The CD-ROM drive uses another laser to read the encoded data (which is read from the other side as "bumps" rather than pits). The drive slows down as the detector (reading head) moves toward the outer edge of the disk. This maintains a constant linear velocity and allows for all sectors to be the same size. This system was adapted from the one used for the audio CDs that largely supplanted phonograph records during the 1980s.

A CD can hold about 650MB of data. By the early 1990s, the CD had become inexpensive and ubiquitous, and it has now largely replaced the floppy disk as the medium of software distribution. The relatively large capacity meant that one CD could replace multiple floppies for a distribution of products such as Microsoft Windows or Word, and it also made it practical to give users access to the entire text of encyclopedias and other reference works. Further, the CD was essential for the delivery of multimedia (graphics, video, and sound) to the desktop, since such applications require far more storage than is available on 1.44MB floppy disks. CD drives declined in price from several hundred dollars to about $50, while their speeds have increased by a factor of 30 or more, allowing them to keep up with games and other software that needs to read data quickly from the disk.

CD-ROM

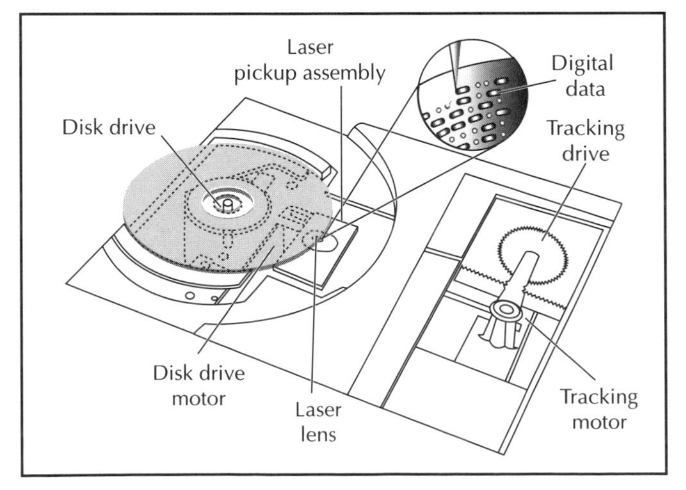

Schematic of the components of a CD drive. The tracking drive and tracking motor move the laser pickup assembly across the spinning disk drive to position it to the correct track. The laser beam hits the disk surface, reflecting differently from the pits and flat areas (lands). This pattern of differences encodes the data as ones and zeroes.

RECORDABLE CDS

In the late 1990s, a new consumer technology enabled users to create their own CDs with data or audio tracks. The cheapest kind, CD-R (Compact Disk Recordable) uses a layer of a dyed material and a thin gold layer to reflect the laser beam. Data is recorded by a laser beam hitting the dye layer in precise locations and marking it (in one of several ways, depending on technology). The lengths of marked ("striped") track and unmarked track together encode the data. This kind of drive is also known generically as a WORM (Write Once, Read Many times) device, and it is often used for making archival or backup copies of data or software.

A more versatile alternative is the CD-RW (Compact Disk, Readable/Writeable), which can be recorded on, erased, and re-recorded many times. These disks have a layer made from a mixture of such materials as silver, antimony, and rare earths such as indium and tellurium. The mixture forms many tiny crystals. To record data, an infrared laser beam is directed at pinpoint spots on the layer. The heat from the beam melts the crystals in the target spot into an amorphous mass. Because the amorphous state has lower reflectivity than the original crystals, the reading laser can distinguish the marked "pits" from the surrounding lands. Because of a special property of the material, a beam with a heat level lower than the recording beam can reheat the amorphous material to a point at which it will, upon cooling, revert to its original crystal form. This permits repeated erasing and re-recording.

DVD-ROM

The DVD (alternatively, Digital Video Disc or Digital Versatile Disc) is similar to a CD, but uses laser light with a shorter wavelength. This means that the size of the pits and lands will be considerably smaller, which in turns means that much more data can be stored on the same size disk. A DVD disk typically stores up to 4.7GB of data, equivalent to about 6 CDs. This capacity can be doubled by using both sides of the disk, though current readers can read only one side at a time.

The high capacity of DVD-ROMs (and their recordable equivalent, DVD-RAMs) makes them useful for storing feature-length movies or videos, very large games and multimedia programs, or large illustrated encyclopedias. Although DVD is now included or is an available option with many newer PCs, only a limited number of products currently make use of it. (In the movie area, its chief competitor is the analog videodisc or laserdisc.)

Further Reading

About.com "Home Recording: Burning CDs." http://homerecording.about.com/cs/burningcds/

White, Ron. *How Computers Work*. Millennium ed. Indianapolis, Ind.: Que, 1999. 286–295.

cellular automata

In the 1970s, British mathematician John H. Conway invented a pastime called the Game of Life, which was popularized in Martin Gardner's column in *Scientific American.* In this game (better termed a simulation), each cell in a grid "lived" or "died" according to the following rules:

1. A living cell remains alive if it has either two or three living neighbors.
2. A dead cell becomes alive if it has three living neighbors.
3. A living cell dies if it has other than two or three living neighbors.

Investigators created hundreds of starting patterns of living cells and simulated how they changed as the rules were repeatedly applied. (Each application of the rules to the cells in the grid is called a *generation.*) They found, for example, that a simple pattern of three living cells in a row "blinked" or switched back and forth between a horizontal and vertical orientation. Other patterns, called "glider guns" ejected smaller patterns (gliders or spaceships) that traveled across the grid.

The Game of Life is an instance of the general class called cellular automata. Each cell operates like a tiny computer that takes as input the states of its neighbors and produces its own state as the output. (See also FINITE STATE MACHINE.) The cells can be arranged in one (linear), two (grid), or three dimensions, and a great variety of sets of rules can be applied to them, ranging from simple variants of Life to exotic rules that can take into account how long a cell has been alive, or subject it to various "environmental" influences.

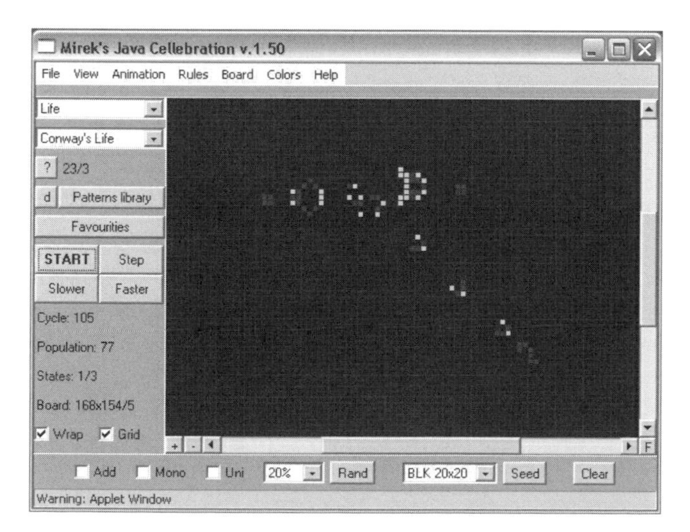

A screen from a Game of Life simulator called Mirek's Cellebration. (This version runs as a Web browser–accessible Java applet.) This and other programs make it easy to experiment with a variety of Life patterns and track them across hundreds of "generations."

APPLICATIONS

Cellular automata theory has been applied to a variety of fields that deal with the complex interrelationships of components, including biology (microbe growth and population dynamics in general), ecology (including forestry), and animal behavior, such as the flight of birds. (The cues that a bird identifies in its neighbors are like the input conditions for a cell in a cellular automaton. The "output" would be the bird's flight behavior.)

The ability of cellular automatons to generate a rich complexity from simple components and rules mimics the development of life from simple components, and thus cellular automation is an important tool in the creation and study of ARTIFICIAL LIFE. This can be furthered by combining a set of cellular automation rules with a GENETIC ALGORITHM, including a mechanism for inheritance of characteristics. Cellular automation principles can also be applied to engineering in areas such as pattern or image recognition.

In 2002, computer scientist and mathematician Stephen Wolfram (developer of the *Mathematica* program) published a book titled *A New Kind of Science* that undertakes the modest project of explaining the fundamental structure and behavior of the universe using the principles of cellular automation. Time will tell whether this turns out to be simply an idiosyncratic (albeit interesting) approach or a generally useful paradigm.

Further Reading

ALife On-line. "Frequently Asked Questions About Cellular Automata." http://alife.santafe.edu/alife/topics/cas/ca-faq/ca-faq.html

Gutowitz, Howard, ed. *Cellular Automata.* Cambridge, Mass.: MIT Press, 1991. [MCell: A Game of Life Simulation for Windows.] http://www.mirwoj.opus.chelm.pl/

"Useful Resources for Cellular Automata on the Web." http://www.scs.carleton.ca/~roytenbe/cell.html

Wolfram, S. *A New Kind of Science.* Champaign, Ill.: Wolfram Media, 2002.

———. *Theory and Applications of Cellular Automata.* Singapore: World Scientific, 1986.

Cerf, Vinton D.

(1943–)

American

Computer Scientist

Vinton (Vint) Cerf is a key pioneer in the development of the packet-switched networking technology that is the basis for the Internet. In high school, Cerf distinguished himself from his classmates by wearing a jacket and a tie and carrying a large brown briefcase, which he later described as "maybe a nerd's way of being different." He has a lifelong love for fantasy and science fiction, both of which explore difference. Finally, Cerf was set apart by being hearing-impaired as a result of a birth defect. He would overcome this handicap through a combination of hearing aids and communications strategies. And while he was fascinated by chemistry and rocketry, it would be communications, math, and computer science that would form his lifelong interest.

After graduating from Stanford in 1965 with a B.S. in mathematics, Cerf worked at IBM as an engineer on its time-sharing systems, while broadening his background in computer science. At UCLA he earned on M.S. and then a Ph.D. in computer science while working on technology that could link one computer to another. Soon he was working with Len Kleinrock's Network Measurement Center to plan the ARPA network, a government-sponsored computer link. In designing software to simulate a network that as yet existed only on paper, Cerf and his colleagues had to explore the issues of network load, response time, queuing, and routing, which would prove fundamental for the real-world networks to come.

By the summer of 1968, four universities and research sites (UCLA, UC Santa Barbara, the University of Utah, and SRI) as well as the firm BBN (Bolt Beranek and Newman) were trying to develop a network. At the time, a custom combination of hardware and software had to be devised to connect each center's computer to the other. The hardware, a refrigerator-sized interface called an IMP, was still in development.

By 1970, the tiny four-node network was in operation, cobbled together with software that allowed a user on one machine to log in to another. This was a far cry from a system that would allow any computer to seamlessly communicate with another, however. What was needed on the software end was a universal, consistent language—a protocol—that any computer could use to communicate with any other computer on the network.

In a remarkable display of cooperation, Cerf and his colleagues in the Network Working Group set out to design such a system. The fundamental idea of the protocol is that data to be transmitted would be turned into a stream of "packets." Each packet would have addressing information that would enable it to be routed across the network and then reassembled back into proper sequence at the destination. Just as the Post Office doesn't need to know what's in a letter to deliver it, the network doesn't need to know whether the data it is handling is e-mail, a news article, or something else entirely. The message could be assembled and handed over to a program that would know what to do with it.

With the development of what eventually became TCP/IP (Transmission Control Protocol/Internet Protocol) Vint Cerf and Bob Kahn essentially became the fathers of the Internet we know today (see TCP/IP). As the on-line world began to grow in the 1980s, Cerf worked with MCI in the development of its electronic mail system, and then set up systems to coordinate Internet researchers.

In later years, Cerf undertook new initiatives in the development of the Internet. He was a key founder and the first president of the Internet Society in 1992, serving in that post until 1995 and then as chairman of the board, 1998–1999. This group seeks to plan for expansion and change as the Internet becomes a worldwide phenomenon. Cerf's interest in science fiction came full circle in 1998 when he joined an effort at the Jet Propulsion Laboratory (JPL) in Pasadena, California. There they are designing an "interplanetary Internet" that would allow a full network connection between robot space probes, astronauts, and eventual colonists on Mars and elsewhere in the solar system. Cerf has received numerous honors, including the IEEE Kobayashi Award (1992), International Telecommunications Union Silver Medal (1995), and the National Medal of Technology (1997).

Further Reading

"Cerf's Up." Personal Perspectives. http://www.worldcom.com/about_the_company/cerfs_up/personal_perspective/index.phtml

Hafner, Katie and Matthew Lyon. *Where Wizards Stay Up Late: the Origins of the Internet.* New York: Simon & Schuster, 1996.

Henderson, Harry. *Pioneers of the Internet.* San Diego, Calif.: Lucent Books, 2002.

"Vint Cerf" [Insider Profile] http://www.netinsider.com/profile/vcerf/profile

certificate, digital

The ability to use public key encryption over the Internet makes it possible to send sensitive information (such as credit card numbers) to a website without electronic eavesdroppers being able to decode it and use it for criminal purposes (see ENCRYPTION and COMPUTER CRIME AND SECURITY). Any user can send information by using a person or organization's public key, and only the owner of the public key will be able to decode that information.

However, the user still needs assurance that a site actually belongs to the company that it says it does, rather than being an imposter. This assurance can be provided by a trusted third party certification authority (CA), such as VeriSign, Inc. The CA verifies the identity of the applicant and then provides the company with a digital certificate, which is actually the company's public key encrypted together with a key used by the CA and a text message. (This is sometimes called a digital signature.) When a user queries the website, the user's browser uses the CA's public key to decrypt the certificate holder's public key. That public key is used in turn to decrypt the accompanying message. If the message text matches, this proves that the certificate is valid (unless the CA's private key has somehow been compromised).

Certificate, Digital

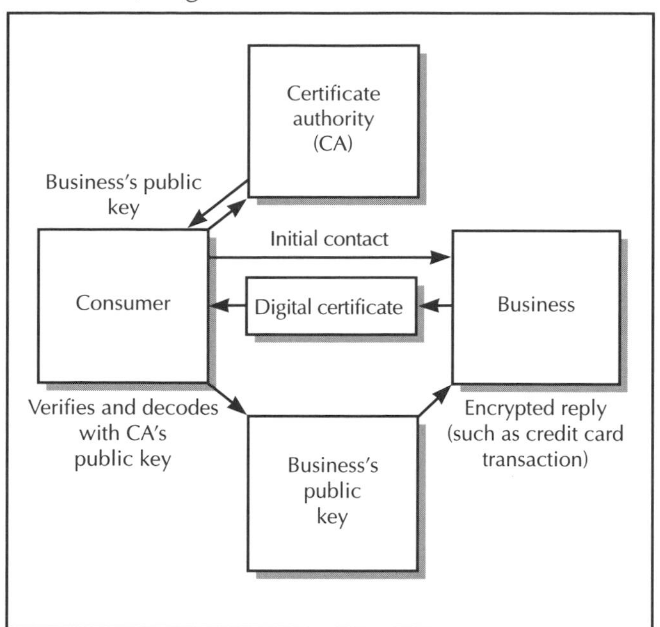

Digital Certification relies upon public key cryptography and the existence of a trusted third party, the Certificate Authority (CA). First a business properly identifies itself to the CA and receives a digital certificate. A consumer can obtain a copy of the business's digital certificate, and use it to obtain the business's public key from the CA. The consumer can now send encrypted information (such as a credit card number) to the business.

The supporting technology for digital certification is included in a standard called Secure Sockets Layer (SSL), which is a protocol for sending encrypted data across the Internet. SSL is supported by leading browsers such as Microsoft Internet Explorer and Netscape. As a result, digital certification is usually transparent to the user, unless the user is notified that a certificate cannot be verified.

Digital certificates are often attached to software such as browser plug-ins so the user can verify before installation that the software actually originates with its manufacturer and has not been tampered with (such as by introduction of a virus).

The use of digital certification is expanding. For example, VeriSign and the federal General Services Administration (GSA) have begun an initiative called ACES (Access Certificates for Electronic Services) that will allow citizens a secure means to send information (such as loan applications) and to view benefits records. The IRS has a pilot program for accepting tax returns that are digitally certified and signed.

Further Reading

Brands, Stefan A. *Rethinking Public Key Infrastructures and Digital Certificates.* Cambridge, Mass.: MIT Press, 2000.

"Digital Certificates Secure On-line Data: Creating Trust through Identity Verification." *Computing,* vol. 8, no. 1, Jan. 2000. Available on-line at www.smartcomputing.com

Feghhi, Jalal, and Peter Williams. *Digital Certificates: Applied Internet Security.* Reading, Mass.: Addison-Wesley, 1998.

certification of computer professionals

Unlike medicine, the law, or even civil engineering, the computer-related fields do not have legally required certification. Given society's critical dependence on computer software and hardware for areas such as infrastructure management and medical applications, there have been persistent attempts to require certification or licensing of software engineers. However, the fluid nature of the information science field would make it difficult to decide which application areas should have entry restrictions.

At present, a variety of academic degrees, professional affiliations, and industry certificates may be considered in evaluating a candidate for a position in the computing field.

ACADEMIC AND PROFESSIONAL CREDENTIALS

The field of computer science has the usual levels of academic credentials (baccalaureate, master's, and doctoral degrees), and these are often considered prerequisites for an academic position or for industry positions that involve research or development in areas such as ROBOTICS or ARTIFICIAL INTELLIGENCE. For business-oriented IT positions, a bachelor's degree in computer science or information systems may be required or preferred, and candidates who also have a business-oriented degree (such as an MBA) may be in a stronger position. However, degrees are generally viewed only as a minimum qualification (or "filter") before evaluating experience in the specific application or platform in question. While not a certification, membership in the major professional organizations such as the Association for Computing Machinery (ACM) and Institute for Electrical and Electronic Engineers (IEEE) can be viewed as part of professional status. Through special interest groups and forums, these organizations provide computer professionals with a good way to track emerging technical developments or to broaden their knowledge.

In the early years of computing and again, in the microcomputer industry of the 1980s, programming experience and ability were valued more highly than academic credentials. (Bill Gates, for example, had no formal college training in computer science.) In general, degree or certification requirements tend to be imposed as a sector of the information industry becomes well defined and established in the corporate world. For example, as local area networks came into widespread use in the 1980s, certifications were developed by Microsoft, Novell, and others. In turn, colleges and trade schools can train technicians, using the certificate examinations to establish a curriculum, and numerous books and packaged training courses have been marketed.

In a newly emerging sector there is less emphasis on credentials (which are often not yet established) and more emphasis on being able to demonstrate knowledge through having actually developed successful applications. Thus, in the late 1990s, a high demand for webpage design and programming emerged, and a good portfolio was more important than the holding of some sort of certificate. However as e-commerce and the Web became firmly established in the corporate world, the cycle is beginning to repeat itself as certification for webmastering and e-commerce applications is developed.

INDUSTRY CERTIFICATIONS

As of 2002, several major industry certifications have achieved widespread acceptance.

Since 1973, the Institute for Certification of Computing Professionals (ICCP) has offered certification based on general programming and related skills rather than mastery of particular platforms or products. The Associate Computing Professional (ACP) certificate is offered to persons who have a basic general knowledge of information processing and who have mastered one major programming language. The more advanced Certified Computing Professional (CCP) certificate requires several years of documented experience in areas such as programming or information systems management. Both certificates also require passing an examination.

A major trade group, the Computing Technology Industry Association (CompTIA) offers the A+ Certificate for computer technicians. It is based on passing a Core Service Technician exam focusing on general hardware-related skills and a DOS/Windows Service Technician exam that emphasizes knowledge of the operating system. The exams are updated regularly based on required job skills as assessed through industry practices.

Networking vendor Novell offers the Certified NetWare Engineer (CNE) certificate indicating mastery of the installation, configuration, and maintenance of its networking products or its GroupWise messaging system. The Certified NetWare Administrator (CNA) certificate emphasizes system administration.

Microsoft offers a variety of certificates in its networking and applications development products. The best known is the Microsoft Certified System Engineer (MCSE) certificate. It is based on a series of required and elective exams that cover the installation, management, configuration, and maintenance of Windows 2000 and other Microsoft networks.

A number of other vendors including Cisco Systems and Oracle offer certification in their products. Given the ever-changing marketplace, it is likely that most computer professionals will acquire multiple certificates as their career advances.

Further Reading

[CompTIA Certification Page] http://www.comptia.org/certification/index.htm

Institute for Certification of Computing Professionals. http://www.iccp.org

"MCSE Guide." http://www.mcseguide.com/

Novell Education Page http://www.novell.com/education/index.html

CGI

By itself, a webpage coded in HTML is simply a "static" display that does not interact with the user (other than for the selection of links). (See HTML.) Many Web services, including on-line databases and e-commerce transactions, require that the user be able to interact with the server. For example, an on-line shopper may need to browse or search a catalog of CD titles, select one or more for purchase, and then complete the transaction by providing credit card and other information. These functions are provided by "gateway programs" on the server that can access databases or other facilities.

One way to provide interaction with (and through) a webpage is to use the CGI (Common Gateway Interface). CGI is a facility that allows Web browsers and other client programs to link to and run programs stored on a website. The stored programs, called scripts, can be writ-

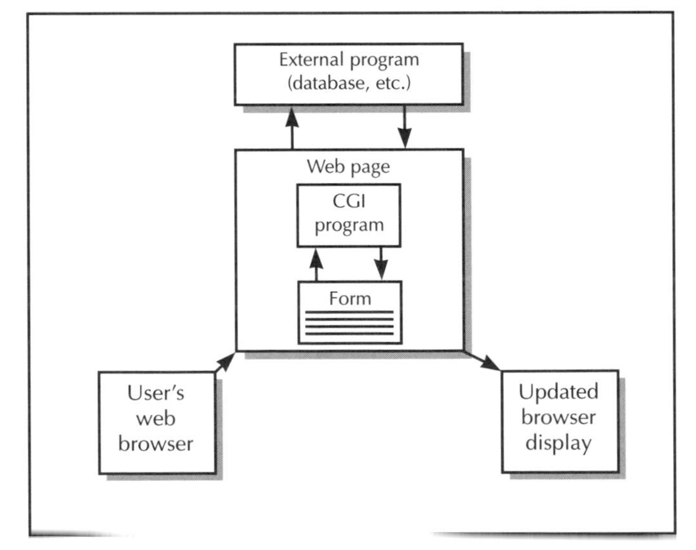

CGI or Common Gateway Interface allows a program linked to a webpage to obtain data from databases and use it to generate forms to be shown on users' Web browsers. For example, a CGI program can link a Web user to a "shopping cart" and inventory system for online purchases.

ten in C or various languages such as Perl or Tcl (see also SCRIPTING LANGUAGES) and placed in a cgi-bin folder on the Web server.

The CGI script is referenced by an HTML hyperlink on the webpage, such as

```
<A HREF="http://www.MyServer.com/cgi-bin/
  MyScript">MyScript</A>
```

Or more commonly, it is included in an HTML form that the user fills in, then clicks the Submit button. In either case, the script executes. The script can then process the information the user provided on the form, and return information to the user's Web browser in the form of an HTML document. The script can perform additional functions such as logging the user's query for marketing purposes.

In the early to mid-1990s, CGI was widely used in creating the "sinews" of functionality for websites, and CGI scripts continue to be pervasive, particularly in connecting websites to traditional databases. However, several alternative routes to dynamic webpage authoring have emerged, including the use of applets (see JAVA), JavaScript or Perl programs (see PERL) and DHTML (dynamic HTML, which combines advanced HTML features with JavaScript).

Note: the acronym CGI can also stand for "Computer-Generated Imagery" (see COMPUTER GRAPHICS).

Further Reading

"A Guide to HTML and CGI Scripts." http://snowwhite.it.brighton.ac.uk/~mas/mas/courses/html/html.html

Hamilton, Jacqueline D. *CGI Programming 101*. Houston, Tex.: CGI101.com, 2000. (First six chapters are available free on-line at http://www.cgi101.com/class/)

"The Most Simple Intro to CGI." http://bignosebird.com/prcgi.shtml

characters and strings

While the attention of the first computer designers focused mainly on numeric calculations, it was clear that much of the data that business people and others would want to manipulate with the new machines would be textual in nature. Billing records, for example, would have to include customer names and addresses, not just balance totals.

The "natural" representation of data in a computer is as a series of two-state (binary) values, interpreted as binary numbers. The solution for representing text (letters of the alphabet, punctuation marks, and other special symbols) is to assign a numeric value to each text symbol. The result is a character code, such as ASCII (American Standard Code for Information Interchange), which is the scheme used most widely today. (Another system, EBCDIC (Extended Binary-Coded Decimal Interchange Code) was used during the heyday of IBM mainframes, but is seldom used today.)

The seven-bit ASCII system is compact (using one byte of memory to store each character), and was quite suitable for early microcomputers that required only the basic English alphabet, punctuation, and a few control characters (such as carriage return). In an attempt to use characters to provide simple graphics capabilities, an

TABLE OF 7-BIT ASCII CHARACTER CODES

The following are control (nonprinting) characters:

0 Null (nothing)
7 Bell (rings on an old teletype; beeps on most PCs)
8 Backspace
9 Tab
10 Line feed (goes to next line without changing column position)
13 Carriage return (positions to beginning of next line)
26 End of file
27 [Esc] (Escape key)

The characters with codes from 32 to 127 produce printable characters.

32	[space]	64	@	96	`
33	!	65	A	97	a
34	"	66	B	98	b
35	#	67	C	99	c
36	$	68	D	100	d
37	%	69	E	101	e
38	&	70	F	102	f
39	'	71	G	103	g
40	(	72	H	104	h
41	)	73	I	105	i
42	*	74	J	106	j
43	+	75	K	107	k
44	,	76	L	108	l
45	-	77	M	109	m
46	.	78	N	110	n
47	/	79	O	111	o
48	0	80	P	112	p
49	1	81	Q	113	q
50	2	82	R	114	r
51	3	83	S	115	s
52	4	84	T	116	t
53	5	85	U	117	u
54	6	86	V	118	v
55	7	87	W	119	w
56	8	88	X	120	x
57	9	89	Y	121	y
58	:	90	Z	122	z
59	;	91	[	123	{
60	<	92	\	124	|
61	=	93	]	125	}
62	>	94	^	126	~
63	?	95	_	127	[delete]

"extended ASCII" was developed for use on IBM-compatible PCs. This used eight bits, increasing the number of characters available from 128 to 256. However, the use of bitmapped graphics in Windows and other operating systems made this version of ASCII unnecessary. Instead, the ANSI (American National Standards Institute) eight-bit character set used the additional character positions to store a variety of special symbols (such as fractions and the copyright symbol) and various accent marks used in European languages.

As computer use became more widespread internationally, even 256 characters proved to be inadequate. A new standard called Unicode can accommodate all of the world's alphabetic languages including Arabic, Hebrew, and Japanese (Kana, as opposed to ideograms). Unicode also includes many sets of internationally used symbols such as those used in mathematics and science. In order to accommodate this wealth of characters, Unicode uses 16 bits to store each character, allowing for 65,535 different characters at the expense of requiring twice the memory storage.

PROGRAMMING WITH STRINGS

Before considering how characters are actually manipulated in the computer, it is important to realize that what the binary value such as 1000001 (decimal 65) stored in a byte of memory actually represents depends on the context given to it by the program accessing that location. If the program declares an integer variable, then the data is numeric. If the program declares a character (char) value, then the data will be interpreted as an uppercase "A" (in the ASCII system).

Most character data used by programs actually represents words, sentences, or longer pieces of text. Multiple characters are represented as a *string*. For example, in traditional BASIC the statement:

```
NAME$ = "Homer Simpson"
```

declares a string variable called NAME$ (the $ is a suffix indicating a string) and sets its value to the character string "Homer Simpson." (The quotation marks are not actually stored with the characters.)

Some languages (such as BASIC) store a string in memory by first storing the number of characters in the string, followed by the characters, with one in each byte of memory. In the family of languages that includes C, however, there is no string type as such. Instead, a string is stored as an array of char. Thus, in C the preceding example might look like this:

```
char Name [20] = "Homer Simpson";
```

This declares Name as an array of up to 20 characters, and initializes it to the string literal "Homer Simpson."

An alternative (and equivalent) form is:

```
char * Name = "Homer Simpson";
```

Here Name is a pointer that returns the memory location where the data begins. The string of characters "Homer Simpson" is stored starting at that location.

Unlike the case with BASIC, in the C languages, the number of characters is not stored at the beginning of the data. Rather, a special "null" character is stored to mark the end of the string.

Programs can test strings for equality or even for greater than or less than. However, programmers must be careful to understand the collating sequence, or the order given to characters in a character set such as ASCII. For example the test

```
If State = "CA"
```

will fail if the current value of State is "ca." The lowercase characters have different numeric values than their uppercase counterparts (and indeed must, if the two are to be distinguished). Similarly, the expression:

```
"Zebra" < "aardvark"
```

is true because uppercase Z comes before lowercase "a" in the collating sequence.

Programming languages differ considerably in their facilities for manipulating strings. BASIC includes built-in functions for determining the length of a string (LEN) and for extracting portions of a string (substrings). For example given the string Test consisting of the text "Test Data," the expression Right$ (Test, 4) would return "data."

Following their generally minimalist philosophy, the C and C++ languages contains no string facilities. Rather, they are provided as part of the standard library, which can be included in programs as needed. In the following little program:

```
#include <iostream.h>
#include <string.h>
void main ()
{
char String1[20];
char String2[20];
strcpy (String1, "Homer");
strcpy (String2, " Simpson");
//Concatenate string2 to the end of string1
strcat (String1, String2);
cout String1 <<endl;
}
```

Here the strcpy function is used to initialize the two strings, and then the strcat (string concatenate) function is used to combine the two strings and store the result back in string1, which is then sent to the output.

As an alternative, one can take advantage of the object orientation of C++ and define a string class. The addition operator (+) can then be extended, or "overloaded" so that it will concatenate strings. Then, the preceding program, instead of using the strcat function, can use the more natural syntax:

```
cout << String1 + String2
```

to display the combined strings.

STRING-ORIENTED LANGUAGES

Sophisticated string processing (such as parsing and pattern matching) tends to be awkward to express in traditional number-oriented programming languages. Several languages have been designed especially for manipulating textual data. Snobol, designed in the early 1960s, is best known for its sophisticated pattern-matching and pattern processing capabilities. A similar language, Icon, is widely used for specialized string-processing tasks today. Many programmers working with textual data in the UNIX environment have found that the awk and Perl languages are easier to use than C for extracting and manipulating data fields. (See AWK and PERL.)

Further Reading

"ANSI Character Set." http://webopedia.lycos.com/Standards/Data_Formats/ANSI_Character_Set.html

Sebern, Mark J. "ANSI String Class." [for C++] http://www.msoe.edu/eecs/ce/courseinfo/stl/string.htm

"String Processing Algorithms." http://www.cee.hw.ac.uk/~alison/ds98/node70.html

Unicode Consortium. *Unicode Standard Version 3.0 with CD-ROM*. Reading, Mass.: Addison-Wesley, 2000.

chat, on-line

In general terms, to "chat" is to communicate in real time by typing messages to other on line users who can immediately type messages in reply. It is this conversational immediacy that distinguishes chat services from conferencing systems or bulletin boards.

COMMERCIAL SERVICES

Many PC users have become acquainted with chatting through participating in "chat rooms" operated by on-line services such as America On-line (AOL). A chat room is a "virtual space" in which people meet either to socialize generally or to discuss particular topics. At their best, chat rooms can develop into true communities whose participants develop long-term friendships and provide one another with information and emotional support (see VIRTUAL COMMUNITY).

However, the essentially anonymous character of chat (where participants often use "handles" rather than real names) that facilitates freedom of expression can also provide a cover for mischief or even crime. Chat rooms have acquired a rather lurid reputation in the eyes of the general public. There has been considerable public concern about children becoming involved in inappropriate sexual conversation. This has been fueled by media stories (sometimes exaggerated) about children being recruited into face-to-face meetings with pedophiles. AOL and other on-line services have tried to reduce such activity by restricting on-line sex chat to adults, but there is no reliable mechanism for a service to verify its user's age. A chat room can also be supervised by a host or moderator who tries to prevent "flaming" (insults) or other behavior that the on-line service considers to be inappropriate.

DISTRIBUTED SERVICES

For people who find commercial on-line services to be too expensive or confining, there are alternatives available for just the cost of an Internet connection. The popular Internet Relay Chat (IRC) was developed in Finland by Jarkko Oikarinen in the late 1980s. Using one of the freely available client programs, users connect to an IRC server, which in turn is connected to one of dozens of IRC networks. Users can create their own chat rooms (called channels). There are thousands of IRC channels with participants all over the world. To participate, a user simply joins a channel and sees all messages currently being posted by other users of the channel. In turn, the user's messages are posted for all to see. While IRC uses only text, there are now enhanced chat systems (often written in Java to work with a Web browser) that add graphics and other features.

A service called "instant messaging" is growing in popularity. It works by having users set up a list of friends. Anytime a person on the list is on-line, he or she will be automatically notified when someone else on the list is on-line. They can then have a two or many-way chat. AOL has the most popular instant messaging service, but the service has been criticized for its unwillingness to establish a common instant messaging protocol that could be used to communicate with nonmembers.

As with chat, there are noncommercial alternatives for instant messaging. The most popular is ICQ ("I Seek You"), which is a freely downloadable program that can also be used to create lists of users who can automatically notify one another when they are on-line. ICQ has become a popular way for people to keep in touch while playing games or working on projects that require real-time coordination.

There are many other technologies that can be used for conversing via the Internet. Some chat services (such as Cu-SeeMe) enable participants to transmit their images (see VIDEOCONFERENCING and WEB CAM). Voice can also be transmitted over an Internet connection (see INTERNET TELEPHONY). "Text messaging" via cellphone has also become an international fad that by 2002 was catching on in the United States.

Further Reading

[AOL Chat Resources.] http://aol.about.com/internet/aol/cs/chat/

ICQ, Inc. "An Introduction to ICQ." http://www.icq.com/products/whatisicq.html

Pioch, Nicolas. "A Short IRC Primer." http://www.irchelp.org/irchelp/ircprimer.html

Weverka, Peter. *Mastering ICQ: the Official Guide.* Dulles, Va.: ICQ Press, 2001.

chip

As early as the 1930s, researchers had begun to investigate the electrical properties of materials such as silicon and germanium. Such materials, dubbed "semiconductors," were neither a good conductor of electricity (such as copper) nor a good insulator (such as rubber). In 1939, one researcher, William Shockley, wrote in his notebook "It has today occurred to me that an amplifier using semiconductors rather than vacuum [tubes] is in principle possible." In other words, if the conductivity of a semiconductor could be made to vary in a controlled way, it could serve as an electronic "valve" in the same way that a vacuum tube can be used to amplify a current or to serve as an electronic switch.

The needs of the ensuing wartime years made it evident that a solid-state electronic device would bring many advantages over the vacuum tube: compactness, lower power usage, higher reliability. Increasingly complex electronic equipment, ranging from military fire control systems to the first digital computers, further underscored the inadequacy of the vacuum tube.

In 1947, William Shockley, along with John Bardeen and Walter Brattain, invented the transistor, a solid-state electronic device that could replace the vacuum tube for most low-power applications, including the binary switching that is at the heart of the electronic digital computer. But as the computer industry strove to pack more processing power into a manageable volume, the transistor itself began to appear bulky.

Starting in 1958, two researchers, Jack Kilby of Texas Instruments and Robert Noyce of Fairchild Semiconductor, independently arrived at the next stage of electronic miniaturization: the integrated circuit (IC). The basic idea of the IC is to make semiconductor resistors, capacitors, and diodes, combine them with transistors, and assemble them into complete, compact solid-state circuits. Kilby did this by embedding the components on a single piece of germanium called a substrate. However, this method required the painstaking and expensive hand-soldering of the tiny gold wires connecting the components. Noyce soon came up with a superior method: Using a lithographic process, he was able to print the pattern of wires for the circuit onto a board containing a silicon substrate. The components could then be easily connected to the circuit. Thus was born the ubiquitous PCB (printed circuit board). This technology would make the minicomputer (a machine that was roughly refrigerator-sized rather than room-sized) possible during the 1960s and 1970s. Besides the PCBs being quite reliable compared to hand-soldered connections, a failed board could be easily "swapped out" for a replacement, simplifying maintenance.

FROM IC TO CHIP

The next step to the truly integrated circuit was to form the individual devices onto a single ceramic substrate (much smaller than the printed circuit board) and encapsulate them in a protective polymer coating. The device then functioned as a single unit, with input and output leads to connect it to a larger circuit. However, the speed of this "hybrid IC" is limited by the relatively large distance between components. The modern IC that we now call the "computer chip" is a monolithic IC. Here the devices, rather than being attached to the silicon substrate, are formed by altering the substrate itself with tiny amounts of impurities (a process called "doping"). This creates regions with an excess of electrons (n-type, for negative) or a deficit (p-type for positive). The junction between a p and an n region functions as a diode. More complex arrangements of p and n regions form transistors. Layers of transistors and other devices can be formed on top of one another, resulting in a highly compact integrated circuit. Today this is generally done using optical lithography techniques, although as the separation between components approaches 100 NM (nanometers, or billionths of a meter) it becomes limited by the wavelength of the light used. On the horizon is the possibility of using nanotechnology to create circuits whose components consist of only a few atoms apiece.

In computers, the IC chip is used for two primary functions: logic (the processor) and memory. The MICROPROCESSORS of the 1970s were measured in thousands of transistor equivalents, while chips such as the Pentium and Athlon being marketed by the late 1990s are meas-

Chip

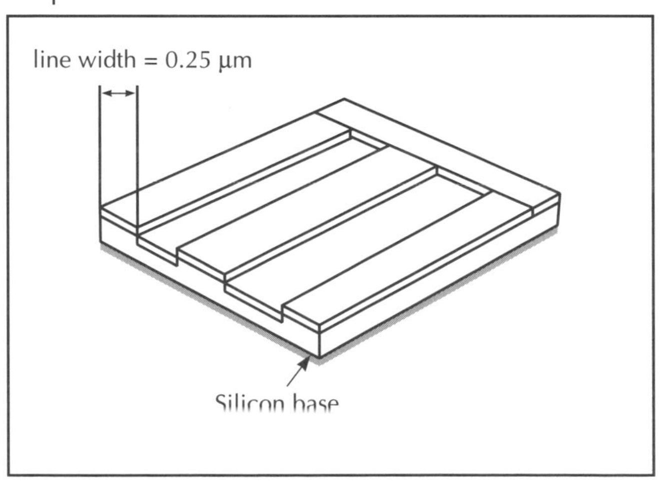

Cross section of an integrated circuit (IC). The components and their connections are so tiny that they are measured in microns, or millionths of a meter.

ured in tens of millions of transistors. Meanwhile, memory chips have increased in capacity from the 4K and 16K common around 1980 to 256MB and more. In what became known as "Moore's Law," Gordon Moore has observed that the number of transistors per chip has doubled roughly every 18 months.

Further Reading

Baker, R. Jacob, Harry W. Li, and David E. Boyce. *CMOS Circuit Design, Layout and Simulation*. New York: IEEE Press, 1998.

"Chip Directory." http://www.embeddedlinks.com/chipdir/

Saint, Christopher and Judy Saint. *IC Layout Basics*. New York: McGraw Hill, 2001.

Semiconductor Industry Association. http://www.semichips.org/

chipset

In personal computers a chipset is a group of integrated circuits that together perform a particular function. System purchasers generally think in terms of the processor itself (such as a Pentium III, Pentium IV, or competitive chips from AMD or Cyrix). However they are really buying a *system chipset* that includes the microprocessor itself (see MICROPROCESSOR) and often a memory cache (which may be part of the microprocessor or a separate chip—see CACHE) as well as the chips that control the memory bus (which connects the processor to the main memory on the motherboard). (See BUS.) The overall performance of the system depends not just on the processor's architecture (including data width, instruction set, and use of instruction pipelines) but also on the type and size of the cache memory, the memory bus (RDRAM or "Rambus" being the most recent) and the speed with which the processor can move data to and from memory.

In addition to the system chipset, other chipsets on the motherboard are used to support functions such as graphics (the AGP, or Advanced Graphics Port, for example), drive connection (EIDE controller), communication with external devices (see PARALLEL PORT, SERIAL PORT, and USB), and connections to expansion cards (the PCI bus).

At the end of the 1990s, the PC marketplace had chipsets based on two competing architectures. Intel, which originally developed an architecture called Socket 7, has switched to the more complex Slot-1 architecture, which is most effective for multiprocessor operation but offers the advantage of including a separate bus for accessing the cache memory. Meanwhile, Intel's main competitor, AMD, has enhanced the Socket 7 into "Super Socket 7" and is offering faster bus speeds. On the horizon around 2004 may be completely new architecture. In choosing a system, consumers are locked into their choice because the microprocessor pin sockets used for each chipset architecture are different.

Further Reading

"Intel Chipsets." http://developer.intel.com/design/chipsets/

"Motherboard Chipset Database." http://www.motherboards.org/chipsets.html

Risley, David. "Chipsets: Introduction." http://www.hardwarecentral.com/hardwarecentral/tutorials/46/1/

Church, Alonzo

(1903–1995)
American
Mathematician

Born in Washington, D.C., mathematician and logician Alonzo Church made seminal contributions to the fundamental theory of computation. Church was mentored by noted geometer Oswald Veblen and graduated from Princeton with an A.B. in mathematics in 1924. Veblen encouraged Church to devote his graduate thesis to the investigation of the fundamental problem of computability. At the time, mathematician David Hilbert and his followers were attempting to create a formal way to express mathematical propositions.

In 1927, Church received his Ph.D. from Princeton for a dissertation on the axiom of choice in set theory. During the 1930s, Church developed the lambda calculus, which provided rules for substituting bound variables in generating mathematical functions. The Church Thesis (also called the Church-Turing thesis, because Alan Turing [see TURING, ALAN] approached the same conclusion from a different angle) stated that every calculable function in number theory could be defined in lambda calculus and was also computable in Turing's sense (see COMPUTABILITY AND COMPLEXITY). This provided the theoretical confidence that given appropriate technology, computers could tackle a variety of problems reliably. At the same time, another of Church's achievements, the Church Theorem, proved that there were theorems that could not be proven by any computer.

Church's lambda calculus became important for the design and verification of computer languages, and the LISP language in particular was based on lambda expressions. Computer scientists working with problems in list processing and the use of recursion also have owed much to Church's pioneering work.

Church taught at Princeton for many years. In 1961, he received the title of Professor of Mathematics and Philosophy. In 1967, he took the same position at UCLA, where he was active until 1990. He received numerous honorary degrees, and in 1990 an international symposium was held in his honor at the State University of New York at Buffalo.

Further Reading

Barendregt, H. "The Impact of the Lambda Calculus in Logic and Computer Science." *The Bulletin of Symbolic Logic*, 3, 181–215.

Church, Alonzo. *Introduction to Mathematical Logic*. Princeton, N.J.: Princeton University Press, 1956.

"The Church-Turing Thesis." http://www.alanturing.net/pages/Reference%20Articles/The%20TuringChurch%20Thesis.html

Davis, M. *The Undecidable: Basic Papers on Undecidable Propositions, Unsolvable Problems, and Computable Functions.* Hackett, N.Y.: Raven Press, 1965.

class

A class is a data type that combines both a data structure and methods for manipulating the data. For example, a string class might consist of an array to hold the characters in the string and methods to compare strings, combine strings, or extract portions of a string (see CHARACTERS AND STRINGS).

As with other data types, once a class is declared, objects (sometimes called instances) of the class can be created and used. This way of structuring programs is called object-oriented programming because the class object is the basic building block (see OBJECT-ORIENTED PROGRAMMING).

Object-oriented programming and classes provide several advantages over traditional block-structured languages. In a traditional BASIC or even Pascal program, there is no particular connection between the data structure and the procedures or functions that manipulate it. In a large program one programmer might change the data structure without alerting other programmers whose code assumes the original structure. On the other hand, someone might write a procedure that directly manipulates the internal data rather than using the methods already provided. Either transgression can lead to hard-to-find bugs.

With a class, however, data and procedures are bound together, or encapsulated. This means that the data in a class object can be manipulated only by using one of the methods provided by the class. If the person in charge of maintaining the class decides to provide an improved implementation of the data structure, as long as the data parameters expected by the class methods do not change, code that uses the class objects will continue to function properly.

Most languages that use classes also allow for inheritance, or the ability to create a new class that derives data and methods from a "parent" class and then modifies or extends them. For example, a class that provides support for 3D graphics could be derived from an existing class for 2D graphics by adding data items such as a third (Z) coordinate and replacing a method such as "line" with a version that works with three coordinates instead of two.

In designing classes, it is important to identify the essential features of the physical situation you are trying to model. The most general characteristics can be put in the "base class" and the more specialized characteristics would be added in the inherited (derived) classes.

Class

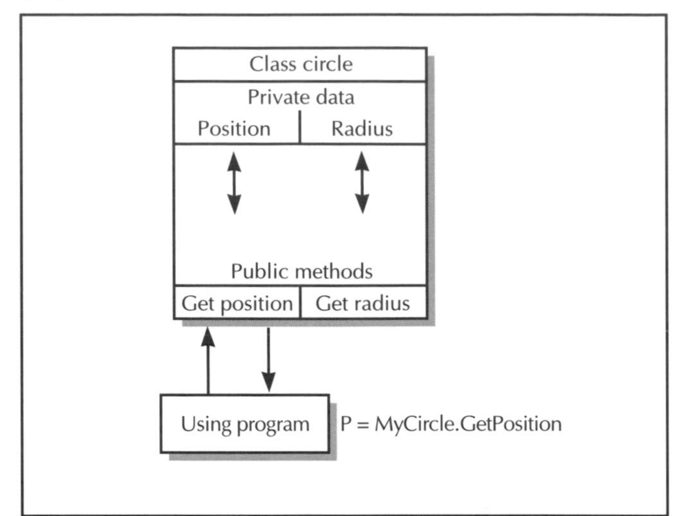

A class encapsulates (or hides) its internal information from the rest of the program. When the program calls MyCircle.GetPosition, the GetPosition member function of the MyCircle Circle class object retrieves the private Position data and sends it back to the calling statement, where it is assigned to the variable P. Private data cannot be directly accessed or changed by an outside caller.

CLASSES AND C++

Classes first appeared in the Simula 67 language, which introduced the terms class and object (see SIMULA). As the name suggests, the language was used mainly for simulation and modeling, but its object-oriented ideas would prove influential. The Smalltalk language developed at Xerox PARC in the 1970s ran on the Alto computer, which pioneered the graphic user interface that would become popular with the Macintosh in the 1980s. Smalltalk used classes to build a seamless and extensible operating system and environment (see SMALLTALK).

However it was Bjarne Stroustrup's C++ language that brought classes into the programming mainstream (see C++). C++ essentially builds its classes by extending the C struct so that it contains both methods (class functions) and data. An access mechanism allows class variables to be designated as completely accessible (public), which is rare, accessible only by derived classes (protected), or accessible only within the class itself (private). The creation of a new object of the class is specified by a constructor function, which typically allocates memory for the object and sets initial default values. The corresponding destructor function frees up the memory when the object no longer exists.

C++ allows for multiple inheritance, meaning that a class can be derived from more than one parent or base class. The language also provides two powerful mechanisms for extending functionality. The first, called virtual functions, allows a base class and its derived classes to have functions based on the same *interface*. For example, a base graphics class might have virtual line, circle, set-color, and other functions that would be implemented in derived classes for 3D objects, 3D solid objects, and so on. When the program calls a method in a virtual class, the COMPILER automatically searches the class's "family tree" until it finds the class that corresponds to the actual data type of the object.

A template specifies how to create a class definition based on the type of data to be used by the class. In other words, where a regular procedure takes and manipulates data parameters and returns data, a template takes data parameters and returns a definition of a class for working with that data (see TEMPLATE).

Other languages of the 1980s and later have embraced classes. Examples include descendants of the Algol family of languages (see PASCAL, ADA, and C++'s close cousin—JAVA), and Microsoft's Visual Basic, which has gradually added classes and other object-oriented features with each release. (There is even a version of COBOL with classes.)

The use of class frameworks such as the Microsoft Foundation Classes (MFC), the C++ STL (Standard Template Library), and various Java implementations has provided a superior way to organize the complexities of data access and operating system functions.

Further Reading

Sebesta, Robert W. *Concepts of Programming Languages*. 4th ed. Reading, Mass.: Addison-Wesley, 1999.

Stroustrup, Bjarne. *The C++ Programming Language*. 3rd ed. Reading, Mass.: Addison-Wesley, 1997.

client-server computing

It is often more efficient to have a large, relatively expensive computer provide an application or service to users on many smaller, inexpensive computers that are linked to it by a network connection. The term server can apply to both the application providing the service and the machine running it. The program or machine that receives the service is called the client.

A familiar example is browsing the Web. The user runs a Web browser, which is a client program. The browser connects to the Web server that hosts the desired website. Another example is a corporate server that runs a database. User's client programs connect to the database over a local area network (LAN). Many retail transactions are also handled using a client-server arrangement. Thus, when a travel or theater booking agent sells a ticket, the agent's client program running on a PC or terminal connects to the server containing the database that keeps track of what seats are available (see also TERMINAL).

There are several advantages to using the client-server model. Having most of the processing done by one or more servers means that these powerful and more costly machines can be used to the greatest efficiency. If more processing capacity is needed, more servers can be brought on-line without having to revamp the whole system. Users, on the other hand, only need PCs (or terminals) that are powerful enough to run the smaller client program to connect to the server.

Keeping the data in a central location helps ensure its integrity: If a database is on a server, transactions can be committed in an orderly way to ensure that, for example, the same ticket isn't sold to two people. A client-server model also offers flexibility to users. Any client program that meets the standards supported by the server can be used to make a connection. (The marketplace generally decides which clients will be supported: for example most websites today support both Microsoft Internet Explorer and Netscape, although they may cater to some features unique to one or the other and other browsers may also work to some extent.)

Client-server computing does have potential disadvantages. If there is only one server, a failure of the server (whether from a hardware failure, a bug, or a hacker attack) brings the whole system to a halt, since the client has no ability to complete transactions on its own. The clients' access to the server is also dependent on the network that connects them. A network failure or traffic bottleneck will also prevent the client from getting any work done.

EXTENDING THE MODEL

One way used in larger organizations to improve the efficiency of the client-server model is to introduce an intermediary between the client and the server. The intermediary program can cache frequently requested data so it can be supplied immediately rather than having to be retrieved from the server (see CACHE). The intermediary can also act as a "traffic cop" to route client requests to the server that currently has the least load or the fastest network access.

Another design consideration is the distribution of processing between the client and the server. At one extreme is the "thin client," where the client machine may only display forms and transmit information to and display information from the server. A POS (Point of Sale) terminal typifies this approach. On the other hand, a "fat client" running on a full-featured desktop PC may perform functions such as verifying the completeness and validity of data before sending it to the server, or use information from the server to generate graphics (this is typical with on-line games, where limiting the amount of information that must be sent over the network can be crucial to speed).

The ultimate extension of the client-server model is "distributed object computing." This is an application of object-oriented programming principles to the organization of the resources needed for data processing. In this

Client-Server

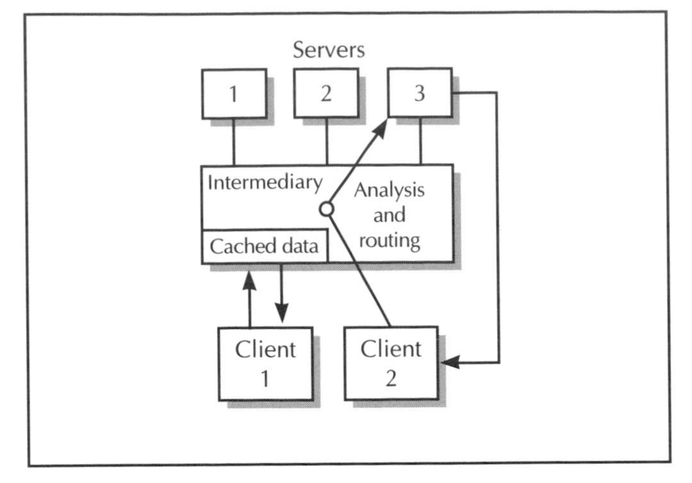

In this schematic an intermediary program receives requests from clients and routes them to available servers. Since the data needed by Client 1 has already been cached, no server connection is necessary. Client 2, on the other hand, is connected to Server 3 to receive data.

model each object (such as a database table) is accessible throughout the network by all other objects, regardless of their physical location. This scheme provides the ultimate in flexibility, because objects can be moved freely among physical machines in order to even out the load. One popular implementation of distributed object computing is CORBA (Common Object Request Broker Architecture). For Windows-based programs, Microsoft has developed the DCOM (Distributed Component Object Model), which allows controls (that is, objects with functional interfaces) written using ActiveX to communicate with each other in a networked environment. (For example, an Excel spreadsheet in an ActiveX control can be embedded in a Word document, and instructed to update itself regularly by obtaining data from a Microsoft Access database table on another machine.) The new Microsoft .NET initiative is also geared toward creating applications that can fluidly interoperate over the Internet.

Further Reading

Fox, Dan. *Building Distributed Applications with Visual Basic .NET*. Indianapolis, Ind.: Sams, 2002.

Goodyear, Mark, ed. *Enterprise System Architectures*. Grand Rapids, Mich.: CRC Press, 1999.

Graham, Steve [and others]. *Building Web Services with Java: Making Sense of XML, SOAP, WSDL and UDDI*. Indianapolis, Ind.: Sams, 2001.

Mathers, Tod W. *Windows NT/2000 Thin Client Solutions: Implementing Terminal Services and Citrix MetaFrame*. Indianapolis, Ind.: New Riders, 2000.

"Network Design Manual: Client-Server Fundamentals." http://www.networkcomputing.com/netdesign/1005part1a.html

Sinclair, Joseph T., and Mark S. Merkow. *Thin Clients Clearly Explained*. San Francisco: Morgan Kaufmann, 2000.

clock speed

The transfer of data within the microprocessor and between the microprocessor and memory must be synchronized to ensure that the data needed to execute each instruction is available when the flow of execution has reached an appropriate point. This synchronization is accomplished by moving data in intervals that correspond to the pulses of the system clock (a quartz crystal). This is done by sending control signals that tell the components of the processor and memory when to send or wait for data. Thus, if the microprocessor is the heart of the computer, the clock is the heart's pacemaker. Because most devices cannot run at the same pace as the processor, circuits in various parts of the motherboard create secondary control signals that run at various ratios of the actual system clock speed.

The following table shows the speed of various system components in relation to the system clock rate. Although the example uses a 600 MHz clock, the ratios will generally hold for faster processors.

Device	Speed	Relationship
Processor	600	System bus * 4.5
System (Memory) Bus	133	(depends on multiplier)
Level 2 Cache	300	Processor / 2
AGP	66	System bus / 2
PCI bus	33	System bus / 4

Microprocessors are rated according to the frequency (that is, number of pulses per second) of their associated clock. For example, a 1.2 GHz Pentium IV processor has 1.2 billion (giga-) pulses per second. It follows that all other things being equal, the higher a processor's clock frequency, the more instructions it can process per second. An alternative way to rate processors is according to the number of a standard type of instruction that it can process per second, hence MIPS (millions of instructions per second).

The relationship between clock speed and processor performance is not as simple as the preceding might imply, however. Each processor is designed with circuits that can move data at a certain rate. In some cases a processor can be run at a higher clock rate than specified (this is called overclocking), but then reliability comes into question. Also, the actual processing power of a processor depends on many other factors. If a processor implements instructions in its microcode that are more efficient for handling certain operations (such as floating point math or graphics rendering), applications that depend on these operations may run faster on one processor than on another, even if the two processors run at the same clock speed. The speed of the system bus (which connects the processor to the RAM memory) also affects the speed at which data can be fetched, processed, and stored. A processor with a clock speed of 733 MHz should perform better on a motherboard with a bus speed of 133 MHz than on one with a bus speed of only 100 MHz.

Speed is "sexy" in marketing terms, so the major chip manufacturers always tout their fastest chips. However, the difference in speed between, for example, as 2.2 GHz version of a processor and a 2.0 GHz version may be unnoticeable to the user of all but the most processor-intensive applications (such as image processing). Indeed, if the system with the slower chip has a faster bus, faster memory (such as RDRAM), or a larger processor cache (see CACHE) it may well outperform the one with a faster chip.

In PCs the term "clock" can also refer to the battery-powered "real-time" clock that provides a timing interval that can be accessed by the operating system and applications.

Further Reading

"Signaling, Clocks, and Synchronous Data Transfer."http://www.pcguide.com/intro/fun/clock-i.htm

COBOL

Common Business-Oriented Language was developed under the impetus of a 1959 Department of Defense initiative to create a common language for developing business applications that centered on the processing of data from files. (The military, after all, was a "business" whose inventory control and accounting needs dwarfed those of all but the largest corporations.) At the time, the principal business-oriented language for mainframe computers was FLOW-MATIC, a language developed by Grace Hopper's team at Remington-Rand UNIVAC and limited to that company's computers (see HOPPER, GRACE MURRAY). The first COBOL compilers became available in 1960, and the American National Standards Institute (ANSI) issued a standard specification for the language in 1968. Expanded standards were issued in 1974 and 1985 (COBOL-74 and COBOL-85) with a new standard under way as of 2000.

The committee that outlined the language that would become COBOL focused on making program statements resemble declarative English sentences rather than the mathematical expressions used by FORTRAN for scientific programming. COBOL's designers hoped that accountants, managers, and other business professionals could quickly master the language, reducing if not removing the need for professional programmers. (This theme of "programming without programmers" would recur with regard to other languages such as RPG, BASIC, and various database systems, always with limited success.)

PROGRAM STRUCTURE

A COBOL program as a whole resembles a business form in that it is divided into specific sections called divisions, each with required and optional items.

The Identification division simply identifies the programmer and gives some information about the program:

```
IDENTIFICATION DIVISION.
PROGRAM-ID WEEKLY REPORT.
AUTHOR JAMES BRADLEY.
DATE-WRITTEN DECEMBER 10, 2000.
DATE-COMPILED DECEMBER 12, 2000.
REMARKS THIS IS AN EXAMPLE PROGRAM.
```

The Environment division contains specifications about the environment (hardware) for which the program will be compiled. In some cases (for example, microcomputer versions of COBOL) it may not be needed. In other cases, it might simply have a Configuration section that specifies the machine to be used:

```
ENVIRONMENT DIVISION.
CONFIGURATION SECTION.
SOURCE-COMPUTER IBM-370.
OBJECT-COMPUTER IBM-370.
```

(The reason for the separate source and object computers is that programs were sometimes compiled on one computer for use on another, often smaller, one.)

In some cases, the Environment Division must also include an Input-Output section that specifies devices and files that will be used by the program. For example:

```
INPUT-OUTPUT SECTION.
FILE-CONTROL.
   SELECT STUDENT-FILE ASSIGN TO READER
   SELECT STUDENT-LISTING ASSIGN TO LOCAL-
   PRINTER
```

The Data division gives a description of the data records and other items that will be processed by the program. It is roughly comparable to the declarations of variables in languages such as Pascal, C, or BASIC. Since COBOL focuses on the processing of file records and the formatting of reports, it tends to have fewer data types than many other languages, but it makes it easier to describe the kinds of data structures commonly used in business applications. For example, it is easy to describe records that have fields and subfields by using level numbers to indicate the relationship:

```
DATA DIVISION.
FILE SECTION.
FD INFILE
     LABEL RECORDS ARE OMITTED.
```

```
01 STUDENT-DATA.
   02 STUDENT-ID PIC 999999.
   02 STUDENT-NAME.
      03 LAST-NAME PIC X(15).
      03 INITIAL PIC X.
      03 FIRST-NAME PIC X(10).
   02 GPA PIC 9.99
```

The "PIC" or picture clause specifies the type of data (using 9's and a decimal point for numbers and X for text) and the length. In addition to specifying the input records, the Data division often includes items that specify the format of the lines of output that are to be printed.

The Procedure division provides the statements that perform the actual data manipulation. Procedures can be organized as subroutines (roughly equivalent to procedures or functions on other languages). Some sample procedure statements are:

```
READ STUDENT-DATA INTO STUDENT-WORK-RECORD
   AT END MOVE 'E' TO PROC-FLAG-ST
      GO TO EXIT-PRINT
ADD 1 TO TOTAL-STUDENT-RECORDS
```

Mathematical expressions can be computed using a Compute statement:

```
COMPUTE GPA = TOTAL-GRADES / CLASSES
```

Branching (if) statements are available, and looping is provided by the Perform statement, for example:

```
PERFORM 100-PRINT-LINE
   UNTIL LINES-FL IS EQUAL TO 'E'
```

(As with older versions of BASIC, subroutines are numbered.)

IMPACT AND PROSPECTS

From the 1960s through the 1980s, COBOL became the workhorse language for business applications for mainframe and mid-size computers, and it is still widely used today. (The concerns about possible problems at the end of the century often involved older programs written in COBOL, see Y2K PROBLEM.) The main line of programming language evolution bypassed COBOL and went through Algol (a contemporary of COBOL) and on into Pascal, C, and other block-structured languages (see also STRUCTURED PROGRAMMING).

Some modern versions of COBOL have incorporated later developments in structured programming (such as modularization) and even object-oriented design. Nevertheless, usage of COBOL continues to decline slowly as developers increasingly turn to languages such as C or Visual Basic or database development systems.

Further Reading

Baroudi, Carol. *Mastering COBOL*. Alameda, Calif.: Sybex, 1999.

"COBOL 2000 Draft Standard." http://www.microfocus.com/Standards/Adobe/cd12all.pdf

Sammet, J. E. "The Early History of COBOL," in *History of Programming Languages*. Wexelblat, R. L., ed., 199–276. New York: Academic Press, 1985.

cognitive science

Cognitive science is the study of mental processes such as reasoning, memory, and the processing of perception. It is necessarily an interdisciplinary approach that includes fields such as psychology, linguistics, and neurology. The importance of the computer to cognitive science is that it offers a potential nonhuman model for a thinking entity. The attempts at artificial intelligence over the past 50 years have used the insights of cognitive science to help devise artificial means of reasoning and perception. At the same time, the models created by computer scientists (such as the neural network and Marvin Minsky's idea of "multiple intelligent agents") have in turn been applied to the study of human cognition (see MINSKY, MARVIN LEE and NEURAL NETWORK).

Since the late 19th century, technological metaphors have been used to describe the human mind. The neurons and synapses of the brain were compared to the multitude of switches in a telephone company central office. The invention of digital computers seemed to offer an even more compelling correspondence between neurons and their electrochemical states and the binary state of a vacuum tube or transistor. It is only a small further step to assert that human mental processes can be reduced in principle to computation, albeit a very complex tapestry of computation. Various schools of popular psychology and personal improvement have offered simplistic images of the human mind suffering from "bad programming" that can be debugged or manipulated through various processes. The simulation of some forms of reasoning and language construction by AI programs certainly suggests that there are fruitful analogies between human and machine cognition, but construction of a detailed model that would be applicable to both human and artificial intelligences seemed almost as distant in the science fictional year of 2001 as it was when Alan Turing and other AI pioneers first considered such questions in the early 1950s (see TURING, ALAN MATHISON).

SYMBOLISTS AND CONNECTIONISTS

Unlike standard computer memory cells, neurons can have hundreds of potential connections (and thus states). If a human being is a computer, it must be to a considerable extent an analog computer, with input in the form of levels of various chemicals and electrical impulses. Yet in the 1980s, Allen Newell and Herbert Simon suggested that the "output" of human mental experience can be effectively mapped as relationships between symbols (words, images, and so forth) that correspond to physical states (this is called the Physical Symbol System Hypothesis). If so, then such a symbol system would be "computable" in the Turing-Church sense (see COMPUTABILITY AND COMPLEXITY). Working from the computer end, AI researchers have created a variety of programs that seem to "understand" restricted universes of discourse such as a table with variously shaped blocks upon it or "story frames" based upon common human activities such as eating in a restaurant. Thus, symbol manipulators can at least appear to be intelligent.

The "connectionists," however, argue that it is not symbolic representations that are significant, but the structure within the mind that generates them. By designing neural networks (or distributed processor networks) the connectionists have been able to create systems that produce apparently intelligent behavior (such as pattern recognition) without any reference to symbolic representation.

Critiques have also come from philosophers. Herbert Dreyfus has pointed out that computers lack the body, senses, and social milieu that shape human thought. That machines can generate symbolic representations according to some sort of programmed rules doesn't make the machine truly intelligent, at least not in the way experienced by human beings. John Searle responded to the famous Turing test (which states that if a human being can't distinguish a computer's conversation from a human's, the computer is arguably intelligent). Searle's "Chinese Room" imagines a room in which an English-speaking person who knows no Chinese is equipped with a program that lets him manipulate Chinese words in such a way that a Chinese observer would think he knows Chinese. Similarly, Searle argues, the computer might act "intelligently," but it doesn't really understand what it is doing.

Advances in cognitive science will both influence and depend on developments in brain research (especially the connection between physical states and cognition) and in artificial intelligence.

Further Reading

Bechtel, William and Adele Abrahamson. *Connectionism and the Mind: Parallel Processing, Dynamics, and Evolution in Networks.* 2nd ed. Cambridge, Mass.: Blackwell, 2000.

"Cognitive and Psychological Sciences on the Internet." http://www-psych.stanford.edu/cogsci/

Horgan, Terence and John Tienson. *Connectionism and the Philosophy of Psychology.* Cambridge, Mass.: MIT Press, 1996.

Indiana State University Cognitive Science Program. "Cognitive Science Resources." http://www.psych.indiana.edu/cogsci.html

color in computing

With the exception of a few experimental systems, color graphics first became widely available only with the beginnings of desktop computers in the late 1970s. The first microcomputers were able to display only a few colors (some, indeed, displayed only monochrome or grayscale). Today's PC video hardware has the potential to display millions of colors, though of course the human eye cannot directly distinguish colors that are too close together. There are several important schemes that are used to define a "color space"—that is, a range of values that can be associated with physical colors.

RGB

One of the simplest color systems displays colors as varying intensities of red, green, and blue. This corresponds to the electronics of a standard color computer monitor, which uses three electron guns that bombard red, green, and blue phosphors on the screen. A typical RGB color scheme uses 8 bits to store each of the red, green, and blue components for each pixel, for a total of 24 bits (16,777,216 colors). 32-bit color provides the same number of colors but includes 8 bits for *alpha,* or the level of transparency. The number of bits per pixel is also called the bit depth or color depth.

CMYK

CMYK stands for cyan, magenta, yellow, and black. This four component color system is standard for most types of color printing, since black is an ink color in printing but is simply the absence of color in video. One of the more difficult tasks to be performed by DESKTOP PUBLISHING software is to properly match a given RGB screen color to the corresponding CMYK print color. Recent ver-

Colors in Computing (1)

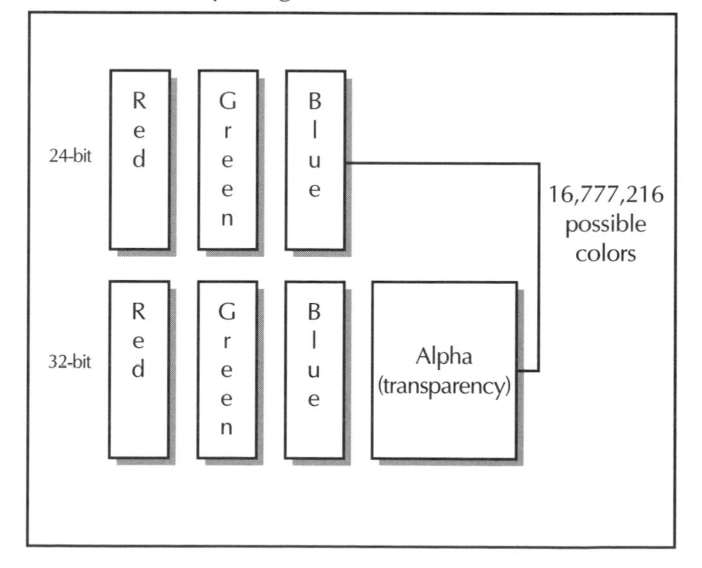

Both 24-bit and 32-bit RGB formats use one byte (8 bits) each to store the red, green, and blue colors, for a total of 16,777,216 possible colors. The 32-bit format, however, adds 8 more bits to store transparency (alpha) data.

sions of Microsoft Windows and the Macintosh operating system include a CMS (color matching system) to support color matching.

PALETTES

Although most color schemes now support thousands or millions of colors, it would be wasteful and inefficient to use three or four bytes to store the color of each pixel in memory. After all, any given application is likely to need only a few dozen colors. The solution is to set up a *palette*, which is a table of (usually 256) color values currently in use by the program. (A palette is also sometimes called a CLUT, or color look-up table.) The color of each pixel can then be stored as an index to the corresponding value in the palette.

The user of a paint program can select a palette from the full range of colors available from the operating system. Many color graphics image formats such as GIF (graphic interchange format) store a palette of the colors used by the image. When converting an image that has more colors that the palette can hold, various algorithms can be used to choose a palette that preserves as much of the color range as possible.

Color in Computing (2)

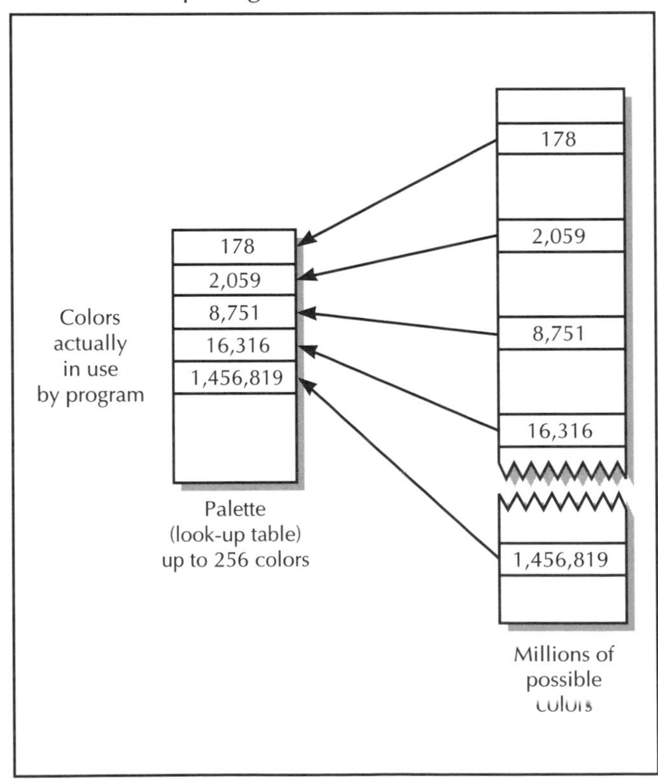

A color look-up table (CLUT) or palette can be used to store the colors actually being used by an image. Here up to 256 colors can be selected out of millions of possibilities.

Further Reading

Apple Computer. "Introduction to Color and Color Management Systems." http://devworld.apple.com/techpubs/macos8/MultimediaGraphics/ColorSyncManager/ManagingColorWithColorSync/ColorSync.9.html
"Color." http://www.webopedia.com/Graphics/Color/
International Color Consortium. http://www.color.org/
"Colour-Related Links." http://mcs.open.ac.uk/ik7/colour.html

compatibility and portability

The computers of the 1940s were each hand built and unique. When the first commercial models were developed, such as the UNIVAC and the first IBM mainframes, the question of compatibility was born. Broadly speaking, compatibility is the degree to which a program or hardware device designed for one system can work with or run on another.

The designers of high-level languages usually intend that a source program written using the proper language syntax will compile and run on any system for which a compiler is available. However, there are many factors that can destroy compatibility. For example, if one machine stores the bytes of a numeric value from least significant to most significant while another does it in the opposite order, program code that depends on directly referencing memory locations will give the wrong results on one machine or another. Similarly, standard data sizes such as "integer" might be 16 bits on one system and 32 bits on another.

Language designers can minimize such problems by separating hardware-related issues from the language itself, as is the case with C and C++. A program is then linked with standard libraries implemented for each hardware or operating system environment.

Manufacturers often design newer models of their computers so they are "upwardly compatible" with existing models. This means that a program written for the smaller machine should run correctly on the new, larger one. This is of obvious benefit to users who do not want to have to rewrite their software every time they upgrade their machine. Often, however, such systems are not "downwardly compatible"—a program written for the new, larger machine may rely on features or architectural characteristics that are not available on the older, smaller machines. Sometimes a "compatibility mode" can be specified for a compiler or operating system. This restricts the use of features to those available on the older system.

Compatibility is also important with regard to software. Generally speaking, a newer version of a program such as a word processor will be able to read files that were originally created by a previous version, although this may not be true for more than a few versions back. However, files saved from the newest version may well be incompatible with older versions, because they contain formatting or other information that is not understandable by the earlier version. Sometimes an intermediate format (such as RTF, or Rich Text Format) can be used to transfer files between otherwise incompatible systems.

Compatibility between vendors can be an important competitive issue. If a developer wants to enter a market where one or two products are viewed as industry standards, the new product will have to be compatible with at least most files created by the dominant products. A technically superior product can thus be a market disaster if it is not compatible with the industry standard. In areas (such as graphics file formats) where there are many alternatives in widespread use, most programs will support multiple formats.

PORTABILITY

Portability is the ability to adapt software or hardware to a wide variety of platforms (that is, computer systems or operating systems). Developers want their products to be portable so they can adapt to an often rapidly changing marketplace. A typical strategy for portability is to choose a language that is in widespread use and a COMPILER that is certified as meeting the ANSI or other standard for the language. The program should be written in such a way that it makes as few assumptions as possible about hardware-dependent matters such as how data is stored in memory. It is also sometimes possible to use standard frameworks that provide the same functions in several different operating systems such as Windows, Macintosh, and UNIX.

However, there is a tradeoff: The more "generic" a program is made in order to be portable, the less optimized it will be for any given hardware or operating environment. The program will also not be able to take advantage of the special features of a given operating system, which may put it at a competitive disadvantage compared to the "native version" of a program. (This is particularly true with Windows, given that operating system's dominance in personal computing.)

The Internet has in general been a force for portability. The Java language, in particular, is designed to be platform-independent. A Java program is compiled into an intermediate language called byte code, which is interpreted or compiled by a "virtual machine" program running on each platform. Thus, the same Java program should run in a browser under Windows, Macintosh, or UNIX (see JAVA).

Further Reading

Henderson, John. *Software Portability.* Brookfield, Vt.: Gower Publishing, 1988.

"Software Portability Home Page." http://www.cs.wvu.edu/~jdm/research/portability/home.html

compiler

A compiler is a program that takes as input a program written in a source language and produces as output an equivalent program written in another (target) language. Usually the input program is in a high-level language such as C++ and the output is in assembly language for the target machine (see ASSEMBLER).

Compilers are useful because programming directly in low-level machine instructions (as had to be done with the first computers) is tedious and prone to errors. Use of assembly language helps somewhat by allowing substitution of symbols (variable names) for memory locations and the use of mnemonic names for operations (such as "add" for addition, rather than some binary instruction code). An ASSEMBLER is essentially a compiler that needs to make only relatively simple translations, because assembly language is still at a relatively low level.

Moving to higher-level languages with relatively English-like statements makes programming easier and makes programs easier to read and maintain. However, the task of translating high-level statements to machine-level code becomes a more complex multistep process.

THE COMPILATION PROCESS

Compilers are traditionally thought of as having a "front end" that analyzes the source code (high-level language statements) and a "back end" that generates the appropriate low-level code. The front end processing begins with *lexical analysis*. The compiler scans the source program looking for matches to valid tokens as defined by the language. A token is any word or symbol that has meaning in the language, such as a keyword (reserved word) such as if or while. Next, the tokens are *parsed* or grouped according to the rules of the language. The result of parsing is a "parse tree" that resolves statements into their component parts. For example, an assignment statement may be parsed into an identifier, an assignment operator (such as =), and a value to be assigned. The value in turn may be an arithmetic expression that consists of operators and operands.

Parsing can be done either "bottom up" (finding the individual components of the statement and then linking them together) or "top down" (identifying the type of statement and then breaking it down into its component parts). A set of grammatical rules specifies how each construct (such as an arithmetic expression) can be broken into (or built up from) its component parts.

The next step is *semantic analysis*. During this phase the parsed statements are analyzed further to make sure they don't violate language rules. For example, most languages require that variables must be declared before they are referenced by the program. Many languages also have rules for which data types may be converted to other types when the two types are used in the same operation.

The result of front-end processing is an *intermediate representation* somewhere between the source statements

Compiler (1)

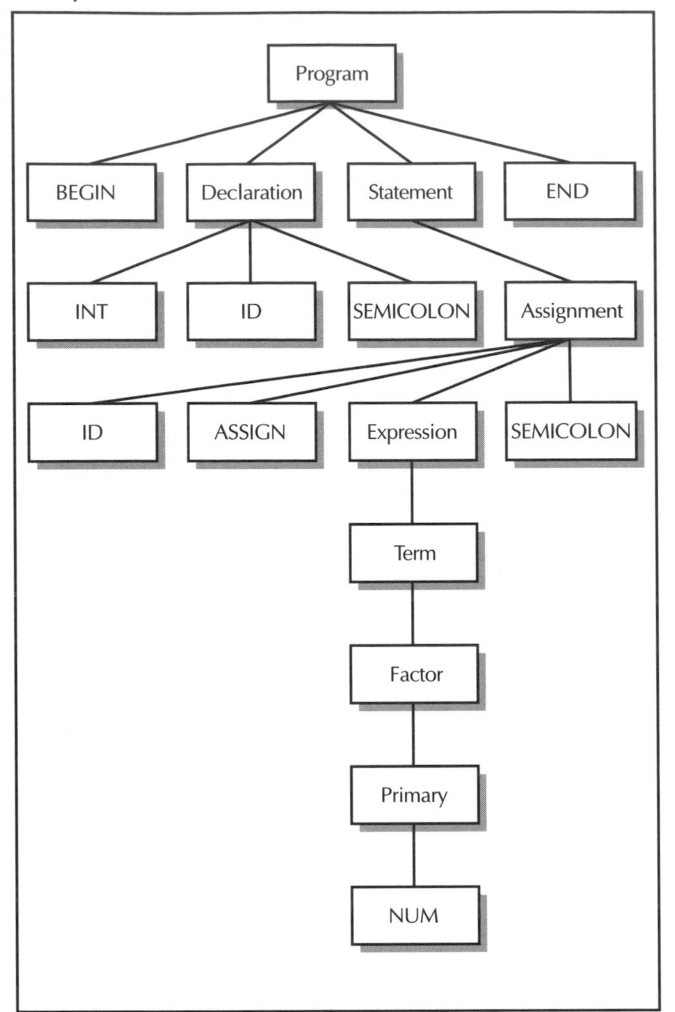

A parse tree showing how an assignment statement in Pascal can be broken down into its component parts. Here ID stands for a variable name, or identifier. An expression can be broken all the way down to a single number.

and machine-level statements. The intermediate representation is then passed to the back end.

CODE GENERATION AND OPTIMIZATION

The process of code generation usually involves multiple passes that gradually substitute machine-specific code and data for the information in the parse tree. An important consideration in modern compilers is *optimization,* which is the process of substituting equivalent (but more efficient) constructs for the original output of the front end. For example, an optimizer can replace an arithmetic expression with its value so that it need not be repeatedly calculated while the program is running. It can also "hoist out" an invariant expression from a loop so that it is calculated only once before the loop begins. On a larger scale, optimization can also improve the communication between different parts (procedures) of the program.

The compiler must attempt to "prove" that the change it is making in the program will never cause the program to operate incorrectly. It can do this, for example, by tracing the possible paths of execution through the program (such as through branching and loops) and verifying that each possible path yields the correct result. A compiler that is too "aggressive" in making assumptions can produce subtle program errors. (Many compilers allow the user to control the level of optimization, and whether to optimize for speed or for compactness of program size.) During development, a compiler is often set to include special debugging code in the output. This code preserves potentially important information that can help the debugging facility better identify program bugs. After the program is working correctly, it will be recompiled without the debugging code.

The final code generation is usually accomplished by using templates that match each intermediate construction with a construction in the target (usually assembly) language, plugging items in as specified by the template. Often a final step, called *peephole optimization,* examines the assembly code and identifies redundancies or, if possible, replaces a memory reference so that a faster machine register is used instead.

In most applications the assembly code produced by the compiler is linked to code from other source files. For example, in a C++ applications class definitions and code that use objects from the classes may be compiled separately. Also most languages (such as C and C++) have operating system-specific libraries that contain commonly used support functions.

As an alternative to bringing the external code into the final application file, code can be "dynamically linked" to libraries that will be accessed only while the program is being run. This eliminates the waste that would occur if several running applications are all using the same standard library code (see LIBRARY, PROGRAM).

In mainframes compilers were usually invoked as part of a batch file using some form of JCL (job control language). With operating systems such as UNIX and MS-DOS a program called *make* is typically used with a file that specifies the compiler, linker, and other options to be used to compile the program. Modern visually-oriented development environments (such as those provided by products such as Visual C++) allow options to be set via menus or simply by selecting from a variety of typical configurations.

Compiler design has become a highly complex field. A modern compiler is developed using a variety of tools (including packaged parsers and lexical analyzers), and involves a large team of programmers. Nevertheless, the principles of compiler design are emphasized in the general computer science curriculum because when a student understands even a simplified compiler in detail, he or she has become acquainted both with important ideas (such as

Compiler (2)

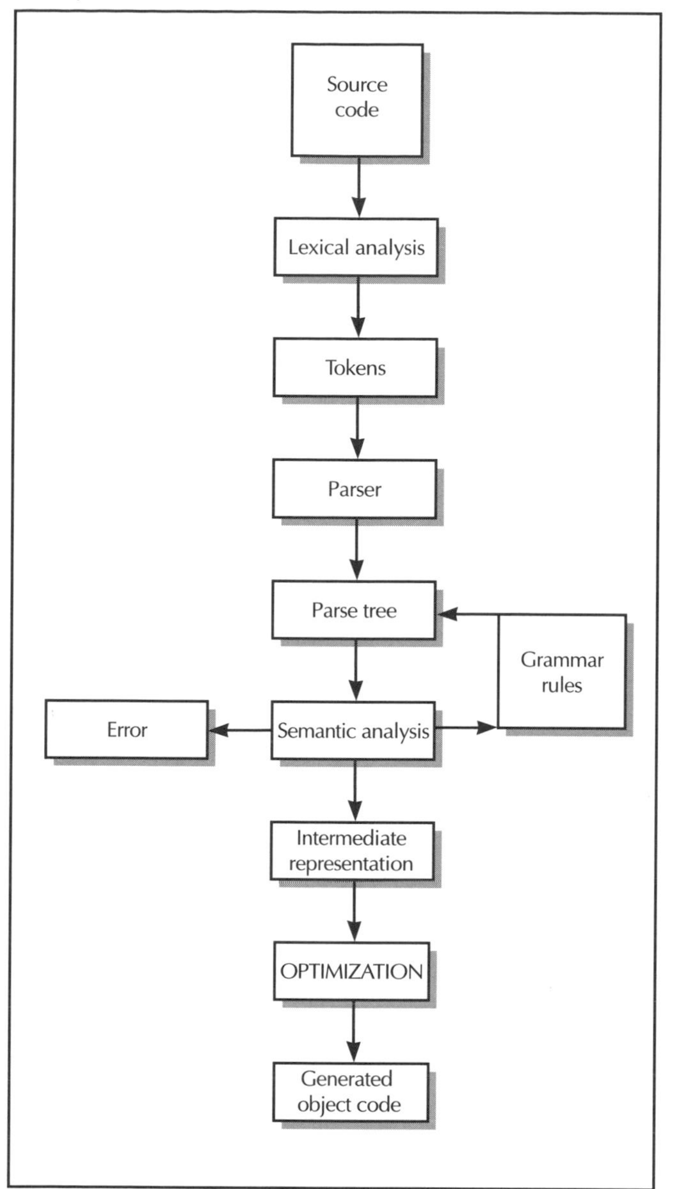

Compilation is a multistep process. Lexical analysis breaks statements down into tokens, which are then parsed and subjected to semantic analysis. The resulting intermediate representation can be optimized before the final object code is generated.

language grammar, parsing, and optimization) and with many levels of understanding computer architecture.

Further Reading

Aho, Alfred V., Ravi Sethi, and Jeffrey D. Ullman. *Compiler Design: Principles, Techniques, and Tools*. Reading, Mass.: Addison-Wesley, 1985.

"Compiler Connection: a Resource for Compiler Developers and Those Who Use Their Products and Services." http://www.compilerconnection.com/

Parsons, Thomas W. *An Introduction to Compiler Construction*. New York: W. H. Freeman, 1992.

computability and complexity

Interestingly, one of the important discoveries of 20th century-mathematics is that certain kinds of problems were not computable. The Turing Machine and Alonzo Church's lambda calculus provided equivalent models that could be used to determine what was computable (see TURING, ALAN MATHISON and CHURCH, ALONZO). Thus far, the equivalence between the Turing Machine and actual computers has held. That is, any decision problem (a problem with a "yes" or "no" answer) that can in theory be solved with a Turing Machine can in theory be solved by any actual computer. Conversely, if a problem can't be solved by a Turing Machine, it cannot be solved by a computer, no matter how powerful.

THE HALTING PROBLEM

The Halting Problem is a classic example of an undecidable problem (or proposition). The problem is this: Given any computer program, can you determine whether the program will halt (end) given any input? There are specific programs that can be shown to halt on particular inputs. For example, this program:

```
If Input = 99 then end.
```

will obviously halt on an input of 99. But to decide whether a determination can be made for any program for any input, it is only necessary to construct a logical paradox. Assume that there is a program P that halts if and only if it receives input D. (Further assume that the program can print something to let you know that it has halted.)

Since the input can be anything, you can let it be a copy of the program itself. The question then becomes: Will the program halt if it is given a copy of itself? Create a procedure (or subroutine) called HaltTest, and define it as:

```
If P halts then print "Halted"
  else print "Didn't Halt."
```

Now create another program called Main. It calls HaltTest and is programmed to do the opposite of what HaltTest indicates.

```
If HaltTest (Main) prints "Yes" then loop for-
ever else halt;
```

But what happens when Main is run? It calls HaltTest, giving itself (Main) as input. If HaltTest halts, then Main loops forever. But if HaltTest doesn't halt, then Main halts. But this means that Main halts if it doesn't halt, and doesn't halt if it halts. This paradox shows that whether Main halts is undecidable.

The undecidability of the Halting Problem has some interesting implications. For example, it means that there is no way a computer can reliably determine that a program doesn't contain an infinite loop. Also, because a mathematical function f(x) is equivalent to a computer program with input x, similar proofs by contradiction can be written to show that it can't be decided whether a program will halt on all inputs (which is equivalent to f(x) being defined for all x.) Nor can it be decided whether two different programs (or mathematical functions) are equivalent for all x.

It is important to realize that a program (or function) being undecidable in all cases doesn't necessarily mean that it can't be decided for some cases (or inputs). Indeed, the answer of the Halting Problem for any given input can be determined by feeding that particular input to the program, which will either halt or run forever.

COMPLEXITY

If a problem turns out to be computable, we then enter the realm of complexity—the analysis of how much computation will be required (see ALGORITHM). Sometimes a designer can devise a significantly faster algorithm for a given problem (such as finding prime factors or sorting). However, other problems appear to have complexity based on an exponential expression, meaning that they become more complex much more rapidly as the input increases. An example is the Traveling Salesman Problem, which is to find the most efficient route for a person traveling to a number of cities to visit each of the cities.

Mathematicians therefore categorize the complexity of problems as P (solvable in a polynomial period of time), EXP (requiring an exponential time), or an intermediate class NP, which means "nondeterministic polynomial." An NP problem is one that can be solved in polynomial time if one is able to guess (and then verify) the answer. The Traveling Salesman Problem is believed to be in the NP class.

While abstruse, the study of computability and complexity has important implications for practical applications. For example, determining the complexity of a cryptographic algorithm can help determine whether the resulting encryption is strong enough to withstand the efforts of a feasible attacker.

Further Reading

Bovet, Daniel P. *Introduction to the Theory of Complexity.* New York: Prentice Hall, 1994.

Hemaspaandra, Lane and Mitsunori Ogihara. *The Complexity Theory Companion.* New York: Springer-Verlag, 2001.

Papadimitriou, C. H. *Computational Complexity.* Reading, Mass.: Addison-Wesley, 1994.

Sipser, Michael. *Introduction to the Theory of Computation.* Boston: PWS Pub. Co., 1997.

computer-aided design and manufacturing (CAD/CAM)

The use of computers in the design and manufacturing of products revolutionized industry in the last quarter of the 20th century. Although computer-aided design (CAD) and computer-aided manufacturing (CAM) are different

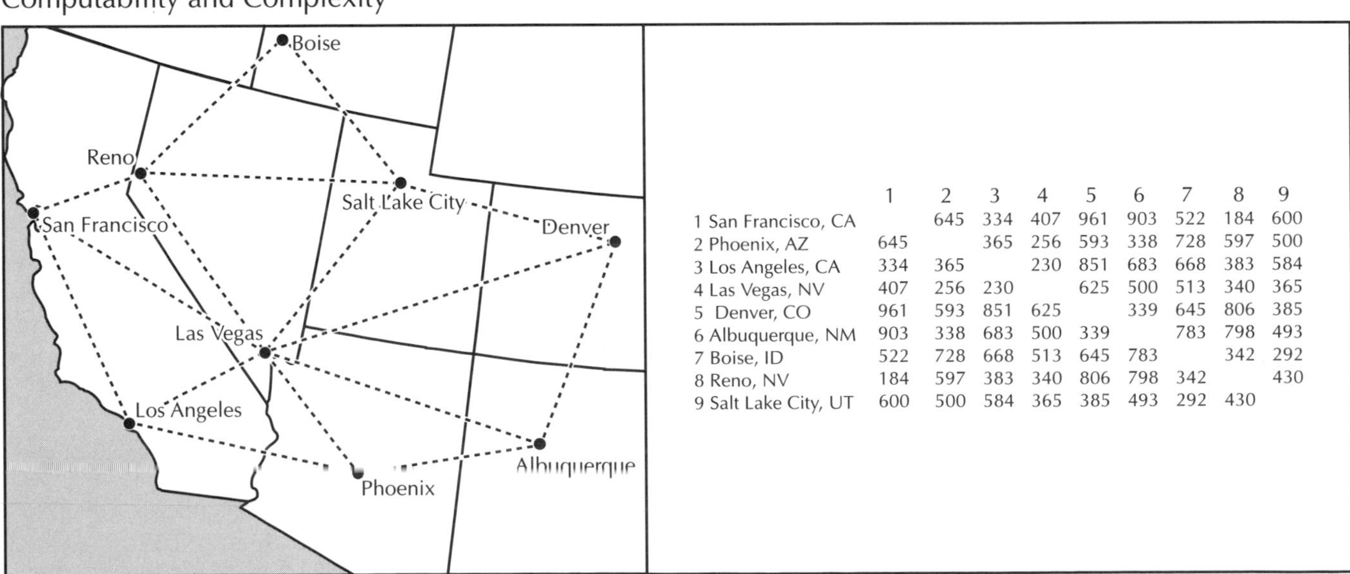

	1	2	3	4	5	6	7	8	9
1 San Francisco, CA		645	334	407	961	903	522	184	600
2 Phoenix, AZ	645		365	256	593	338	728	597	500
3 Los Angeles, CA	334	365		230	851	683	668	383	584
4 Las Vegas, NV	407	256	230		625	500	513	340	365
5 Denver, CO	961	593	851	625		339	645	806	385
6 Albuquerque, NM	903	338	683	500	339		783	798	493
7 Boise, ID	522	728	668	513	645	783		342	292
8 Reno, NV	184	597	383	340	806	798	342		430
9 Salt Lake City, UT	600	500	584	365	385	493	292	430	

A representation of the Traveling Salesman Problem using some cities in the western United States. The table shows the actual distance between pairs of cities.

areas of activity, they are now so closely integrated that they are often discussed together as CAD/CAM.

COMPUTER-AIDED DESIGN

In 1950, science fiction writer Robert Heinlein had his future inventor create "Drafting Dan," an automated drafting system that would enable designers to turn their ideas into manufacturing plans in a fraction of the time required for the hand preparation of schematics and parts lists. By the 1960s, engineers had developed the first computer-assisted design programs, running on terminals attached to mainframe computers.

The activity of a CAD workstation centers on the creation of geometrical models (first 2D, then 3D). With the aid of models, a virtual representation of the product being designed can be built up. With its knowledge of geometrical and physical relationships, routines in the CAD system can perform not only measurement of dimensions and mass but also structural analysis. (In some cases CAD can be interfaced with systems that provide full-blown simulation of the effects of stresses, heat, and other factors.)

The growth of desktop computing power in the 1980s and 1990s moved CAD from the mainframe to the high-end workstation (such as those built by Sun Microsystems) and even to high-end personal computers. The growing processing power also meant that the geometric models could become more sophisticated, including solid models with realistically rendered surfaces rather than just wireframes. The model of surfaces can include such factors as reflectivity, friction, or even aerodynamic characteristics. In designing a product (or a subsystem of a product), engineers can now use simulation software to determine how well a group of parts in a complex assembly (such as a car's steering mechanism) will perform. The ability to get detailed data in real time means that the CAD operator can work in a feedback loop in which the design is incrementally refined until the required parameters are met.

This growing modeling capability has been combined with the use of detailed databases containing the standard parts used in a particular industry or application. Libraries of templates allow the designer to "plug in" standard assemblies of parts and then modify them. The databases can also be used with algorithms that can assist the designer in optimizing the design for some desired characteristic, such as strength, light weight, or lower cost. Recent systems even have the capability to set "strategic" design goals for a whole family of products and to identify particular optimizations that would help each part or subsystem achieve those goals.

COMPUTER-AIDED MANUFACTURING

The automated fabricating of products on the factory floor originally developed independently of computer-aided design. Numerically controlled machine tools and lathes can be programmed using specialized languages such as APT (Automatically Programmed Tool) or more recently, through a system that uses a graphical interface. Advances in pattern recognition and other artificial intelligence techniques have been used to improve the ability of the automatic tool to identify particular features (such as holes into which bolts are to be inserted) and to properly orient surfaces. At some point the programmability and flexibility of the system with regard to its ability to manipulate the environment gives it the characteristics of a robot (see ROBOTICS).

INTEGRATION OF CAD AND CAM

As CAD systems became more capable, it soon became evident that there could be substantial benefits to be gained from integrating the design and manufacturing process.

The CAD software can also output detailed parts and assembly specifications that can be fed into the CAM process. In turn, manufacturing considerations can be applied to the selection of parts during the design process.

The integration of design, simulation, and manufacturing continues. The goal is to give the engineer a seamless way to "tweak" a design and have a number of simulation modules automatically depict the effects of the design change. In essence, the designer or engineer would be working in a virtual world that accurately reflects the physical constraints that the product will face in the real world.

The automation of the design and manufacturing process has been mainly responsible for the increasing productivity of modern factories. Factories using traditional methods in producing complex products such as automobiles or consumer electronics have generally had to refit for CAD/CAM in order to remain competitive. Low-skill but relatively high-paying factory jobs characteristic of the earlier industrial era have given way to smaller numbers of more technical jobs. This has meant a greater emphasis on education and specialized training for the industrial workforce.

Further Reading

"CAD On the Web." http://www.cadontheweb.com/

"Go Cadcam.com." http://www.gocadcam.com/

Rehg, J. A. *Introduction to Robotics in CIM Systems.* 3rd ed. Upper Saddle River, N.J.: Prentice Hall, 1997.

Semerson, J., and K. Curran. *Computer Numerical Control: Operation and Programming.* Upper Saddle River, N.J.: Prentice Hall, 1996.

computer-aided instruction (CAI)

Also called computer based training (CBT), computer-aided instruction (CAI) is the use of computer programs to provide instruction or training. (See EDUCATION AND COMPUTERS for a more comprehensive discussion of the use of computers for teaching and learning.)

The American reaction to Soviet space achievements led to many attempts to modernize the educational system. While the high cost and limited capabilities of 1950s computing technology allowed only for theoretical research by IBM and some universities, by the 1960s more powerful solid-state computers were starting to make what were then called "teaching machines" practicable. The first large-scale initiative was the PLATO teaching system designed by the Computer-based Educational Research Laboratory at the University of Illinois, Urbana. PLATO used a large timesharing system to provide educational software to about a thousand users at terminals throughout the university. PLATO pioneered the use of graphics and what would later be called multimedia, and was eventually marketed by Control Data Corporation, a leading manufacturer of high-end mainframe computers. Stanford University also began a large-scale initiative to deliver computerized instruction.

The early CAI systems required expensive hardware, however, and generally could be sustained only by research funding or where they met the growing training needs of the military, the aerospace industry, or other specialized users. However, the advent of the personal computer in the late 1970s provided both a new technology for delivering educational software and a potential market. With its color graphics and astute marketing the Apple II had became a staple of classrooms by the mid-1980s, when its successor, the Macintosh, brought more advanced graphics (see MACINTOSH) and a program called Hypercard that made it easy for educators to create simple interactive presentations (see HYPERTEXT AND HYPERMEDIA). The Intel-based IBM PC and its "clones" also gained a foothold in the classroom, and Microsoft Windows brought a graphical interface similar to that on the Macintosh.

APPLICATIONS

The simplest (and probably least interesting) form of CAI is often called "drill and practice" programs. Such programs (usually found in the elementary grades) repetitively present math problems, reading vocabulary, or other exercises and test the user's understanding. (Teaching keyboard skills to young students is another common application.) In an attempt to hold the student's interest, many such programs provide a gamelike atmosphere and offer periodic rewards or reinforcement for success.

More sophisticated programs allow the student more creative scope, such as by letting the student program and test virtual "robots" as a means of mastering a programming language. Many computer games, while not designed explicitly for instruction, provide simulations that exercise thinking and planning skills (see COMPUTER GAMES). (For example, the strategy game *Civilization* incorporates concepts such as resource management, labor allocation, and a balanced economy.)

Industry remains a large market for computer-based training. A variety of CBT packages are available for introducing and teaching programming languages such as C++ and Java as well as for preparing students to earn industry certificates such as the A+ certificate for computer technicians.

TRENDS

Two continuing trends in CAI are the growing use of graphics and multimedia, including video or movies, and the increasing delivery of training via the Internet. Some training software can be accessed directly over the Internet through a Web browser, without requiring special software on the user's PC. Increasingly, even products delivered on CD and run from the user's PC include links to supplemental material on the Web.

Further Reading

Ko, Suasan Schor and Steve Rossen. *Teaching On-line: A Practical Guide*. Boston: Houghton Mifflin, 2001.

Rosenberg, Marc J. *E-Learning: Strategies for Delivering Knowledge in the Digital Age*. New York: McGraw Hill, 2001.

Sloman, Martyn. *The E-learning Revolution: How Technology is Driving a New Training Paradigm*. New York: AMACOM, 2002.

Smith, Reesa. "Computer-Based Training." http://www.gsu.edu/~mstswh/courses/it7000/papers/computer1.htm

Web-Based Training Information Center http://www.filename.com/wbt/index.html

computer crime and security

The growing economic value of information, products, and services accessible through computer systems has attracted increased attention from opportunistic criminals. In particular, the many potential vulnerabilities of on-line systems and the Internet have made computer crime attractive and pose significant challenges to professionals whose task it is to secure such systems.

The motivations of persons who use computer systems in unauthorized ways vary. Some hackers primarily seek detailed knowledge of systems, while others (often teenagers) seek "bragging rights." Other intruders have the more traditional criminal motive of gaining access to information such as credit card numbers and personal identities that can be used to make unauthorized purchases. (The FBI estimates that by the end of the 1990s there were about 350,000 instances of such "identity theft," though only a small portion were actually exploited by criminals.) Computer access can also be used for purposes of harassment, extortion, espionage, sabotage, or terrorism. (Large-scale attacks coordinated by governments or terrorist groups often come under the heading of "information warfare.") The accidentally propagated "Internet Worm" of 1988 provided a foretaste of the vulnerability of an increasingly networked world.

The new emphasis on the terrorist threat following September 11, 2001, has included some additional attention to cyberterrorism, or the attack on computers controlling key infrastructure (including banks, water and power systems, air traffic control, and so on). So far ideologically inspired attacks on computer systems have mainly amounted to simple electronic vandalism of websites. Internal systems belonging to federal agencies and the military tend to be relatively protected and isolated from direct contact with the Internet. However, the possibility of a crippling attack or electronic hijacking cannot be ruled out. Commercial systems may be more vulnerable to denial-of-service attacks (see below) that cause economic losses by preventing consumers from accessing services.

FORMS OF ATTACK

Surveillance-based attacks involve scanning Internet traffic for purposes of espionage or obtaining valuable information. Not only businesses but also the growing number of Internet users with "always-on" Internet connections (see BROADBAND) are vulnerable to "packet-sniffing" software that exploits vulnerabilities in the networking software or operating system. The main line of defense against such attacks is the software or hardware firewall, which both "hides" the addresses of the main computer or network and identifies and blocks packets associated with the common forms of attack (see FIREWALL).

In the realm of harassment or sabotage, a "denial of service" (DOS) attack can flood the target system with packets that request acknowledgment (an essential feature of network operation). This can tie up the system so that a Web server, for example, can no longer respond to user requests, making the page inaccessible. More sophisticated DOS attacks can be launched by first using viruses to insert programs in a number of computers, and then instructing the programs to simultaneously launch attacks from a variety of locations.

Computer viruses can also be used to randomly vandalize computers, impeding operation or destroying data (see COMPUTER VIRUS). But a virus can also be surreptitiously inserted as a "Trojan Horse" into a computer's operating system where it can intercept passwords and other information, sending them to the person who planted the virus. Viruses were originally spread through infected floppy disks (often "bootleg" copies of software). Today, however, the Internet is the main route of access, with viruses embedded in e-mail attachments. This is possible because many e-mail and other programs have the ability to execute programs (scripts) that they receive. The main defense against viruses is regular use of antivirus software, turning off scripting capabilities unless absolutely necessary, and making a policy of not opening unknown or suspicious-looking e-mail attachments.

COMPUTER SECURITY

Because there are a variety of vulnerabilities of computer systems and of corresponding types of attacks, computer security is a multifaceted discipline. The vulnerability of computer systems is not solely technical in nature. Sometimes the weakest link in a system is the human link. Hackers are often adept at a technique they call "social engineering." This involves tricking computer operators into giving out sensitive information (such as passwords) by masquerading as a colleague or someone else who might have a legitimate need for the information.

Since computer crimes and attacks can take so many forms, the best defense is layered or in depth. It includes appropriate software (firewalls and antivirus programs, and network monitoring programs for larger installations). It emphasizes proper training of personnel, ranging from security investigators to clerical users. Finally, if information is compromised, the use of strong encryption can make it much less likely to be usable (see ENCRYPTION).

While the flexibility and speed of the Internet can aid attackers, it can also facilitate defense. Emergency response networks and major vendors of antivirus software can quickly disseminate protective code or "patches" that close vulnerabilities in operating systems or applications.

Individual consumers can reduce their vulnerability by ensuring that they do not give out personal information without verifying both the requester and the need for the data. Use of secure websites for credit card transactions has become standard. Generally speaking, vulnerability to computer crime is inversely proportional to the degree of privacy individuals have with regard to their personal information (see PRIVACY IN THE DIGITAL AGE). Public concern about privacy and security has led to recent laws and initiatives aimed at disclosure of organizations' privacy policy and limiting the redistribution of information once collected.

Further Reading

Center for Strategic and International Studies. *Cybercrime, Cyberterrorism, Cyberwarfare: Averting an Electronic Waterloo.* Washington, D.C.: CSIS Press, 1998. Summary on-line at http://www.csis.org/pubs/cyberfor.html

Koyacich, Gerald L. and William C. Boni. *High Technology Crime Investigator's Handbook.* Boston: Butterworth-Heinemann, 1999.

Stoll, Clifford. *The Cuckoo's Egg.* Garden City, N.Y.: Doubleday, 1989.

Sterling, Bruce. *The Hacker Crackdown: Law and Disorder on the Computer Frontier.* New York: Bantam, 1992.

computer engineering

Computer engineering involves the design and implementation of all aspects of computer systems. It is the practical complement to computer science, which focuses on the study of the theory of the organization and processing of information (see COMPUTER SCIENCE). Because hardware requires software (particularly operating systems) in order to be useful, computer engineering overlaps into software design, although the latter is usually considered to be a separate field (see SOFTWARE ENGINEERING).

To get an idea of the scope of computer engineering, consider the range of components commonly found in today's desktop computers:

PROCESSOR

The design of the microprocessor includes the number and width of registers, method of instruction processing (pipelining), the chipset (functions to be integral to the package with the microprocessor), the amount of cache, the connection to memory bus, the possible use of multiple processors, the order in which data will be moved and stored in memory (low or high-order byte first?), and the clock speed. (See MICROPROCESSOR, CHIPSET, CACHE, BUS, MULTIPROCESSING, MEMORY, and CLOCK SPEED.)

MEMORY

The design of memory includes the type (static or dynamic) and configuration of RAM, the maximum addressable memory, and the use of parity for error detection (see MEMORY, ADDRESSING and ERROR CORRECTION). Besides random-access memory, other types of memory include ROM (read-only memory) and CMOS (rewritable persistent memory).

MOTHERBOARD

The motherboard is the platform and data transfer infrastructure for the computer system. It includes the main data bus and secondary buses (such as for high-speed connection between the processor and video subsystem—see BUS). The designer must also decide which components will be integral to the motherboard, and which provided as add-ons through ports of various kinds.

PERIPHERAL DEVICES

Peripheral devices include fixed and removable disk drives; CD and DVD-ROM drives, tape drives, scanners, printers, and modems.

DEVICE CONTROL

Each peripheral device must have an interface circuit that receives commands from the CPU and returns data (see GRAPHICS CARD).

INPUT/OUTPUT AND PORTS

A variety of standards exist for connecting external devices to the motherboard (see PARALLEL PORT, SERIAL PORT, and USB). Designers of devices in turn must decide which connections to support.

There are also a variety of input devices to be handled, including the keyboard, mouse, joystick, track pad, graphics tablet, and so on.

Of course this discussion isn't limited to the desktop PC; similar or analogous components are also used in larger computers (see MAINFRAME, MINICOMPUTER, and WORKSTATION).

NETWORKING

Networking adds another layer of complexity in controlling the transfer of data between different computer systems, using various typologies and transport mechanisms (such as Ethernet); interfaces to connect computers to the network; routers, hubs, and switches (see NETWORK).

OTHER CONSIDERATIONS

In designing all the subsystems of the modern computer and network, computer engineers must consider a variety of factors and tradeoffs. Hardware devices must be designed with a form factor (size and shape) that will fit efficiently into a crowded computer case. For devices that require their own source of power, the capacity of the available power supply and the likely presence of other power-consuming devices must be taken into account. Processors and other circuits generate heat, which must be dissipated. Heat and other forms of stress affect reliability. And in terms of how a device processes input data or commands, the applicable standards must be met. Finally, cost is always an issue.

Moving beyond hardware to operating system (OS) design, computer engineers must deal with many additional questions, including the file system, how the OS will communicate with devices (or device drivers), and how applications will obtain data from the OS (such as the contents of input buffers). Today's operating systems include hundreds of system functions. Since the 1980s, the provision of all the objects needed for a standard user interface (such as windows, menus, and dialog boxes) has been considered to be part of the OS design.

TRENDS

In the early days of mainframe computing (and again at the beginning of microcomputing) many distinctive system architectures entered the market in rapid succession. For example, the Apple II (1977), IBM PC (1981), and Apple Macintosh (1984) (see IBM PC and MACINTOSH). Because architectures are now so complex (and so much has been invested in legacy hardware and software), wholly new architectures seldom emerge today. Because of the complexity and cost involved in creating system architectures, development tends to be incremental, such as adding PCI card slots to the IBM PC architecture while retaining older ISA slots, or replacing IDE controllers with EIDE.

The growing emphasis on networks in general and the Internet in particular has probably diverted some effort and resources from the design of stand-alone PCs to network and telecommunications engineering. At the same time, new categories of personal computing devices have emerged over the years, including the suitcase-size "transportable" PC, the laptop, the book-sized notebook PC, the handheld PDA (personal digital assistant), as well as network-oriented PCs and "appliances." (See PORTABLE COMPUTERS.)

As computing capabilities are built into more traditional devices (ranging from cars to home entertainment centers), computer engineering has increasingly overlapped other fields of engineering and design. This often means thinking of devices in nontraditional ways: a car that is able to plan travel, for example, or a microwave that can keep track of nutritional information as it prepares food (see EMBEDDED SYSTEM). The computer engineeer must consider not only the required functionality but the way the user will access the functions (see USER INTERFACE).

Further Reading

IEEE Computer Society. http://www.computer.org

Patterson, D. A. and J. L. Hennessy. *Computer Organization and Design*. 2nd ed. San Francisco: Morgan Kaufmann, 1998.

"PC Guide." www.pcguide.com

Stallings, W. *Computer Organization and Architecture*. 4th ed. Upper Saddle River, N.J.: Prentice Hall, 1996.

computer games

In the early days of computing, interest in computer games was theoretical rather than aimed at the development of commercial products. Computer scientists saw game playing as a way to develop techniques that could demonstrate intelligent machine behavior (see ARTIFICIAL INTELLIGENCE). Today, computer games are an important sector of the software industry, and rival television's attention share among youth.

DETERMINISTIC GAMES: COMPUTER CHESS

Simple, deterministic games such as tick-tack-toe provided a well-bounded arena that could be managed by the limited capacity computers of the early 1950s. (Deterministic means that there is no element of chance, and that it is always possible to find a best move in any situation.) In 1953, Arthur Samuel unveiled a program that could play a more sophisticated game, checkers, on the new IBM 701 mainframe.

Chess, a game with simple rules but virtually infinite complexity, posed a much greater challenge. In 1956, a group of programmers at the Los Alamos Atomic Energy Laboratory developed a chess program for the MANIAC –I computer that used a reduced (6 x 6 rather than 8 x 8) board. It could look ahead two moves, and played a bit better than a human beginner. Starting in the 1970s, regular computer-only chess tournaments provided a benchmark for measuring the growing strength of chess programs such as Chess 3.0, written by David Slate and Larry Atkins. A later program called Belle, written by Ken Thompson and Joe Condon at Bell Laboratories, received a Master rating from the U.S. Chess Federation. In 1983, Belle and other contenders were defeated by Cray Blitz, which ran on a Cray XMP supercomputer. This program, its rival, Carnegie Mellon's Hitech, and 1989 championship winner Deep Thought, used specialized circuitry to examine around 200,000 moves per second. In the mid-1990s, IBM developed Deep Blue, which used a special parallel processing circuit and could examine 200,000,000 moves per second. World Champion Garry Kasparov defeated the program in a 1996 match by a score of 4–2 games, but was defeated 3.5–2.5 in a 1997 rematch. A computer had arguably become the chess champion of the world.

Many observers equate chess-playing skill with high intelligence, but most computer chess programs do not play the game the way humans do. Human players, particularly masters, zero in on the strongest moves through a combination of recognizing positional patterns and combinations, and applying general principles, such as the desirability of gaining control of the central squares. Computer programs, however, take a more brute force approach, examining thousands of positions a second and scoring them according to various criteria such as whether they gain material or improve the position. Computer chess thus reflects more the triumph of processing power than robust artificial intelligence. (There have been attempts to use expert systems and other AI techniques to develop a computer chess player that "thinks" more like a human player, but these programs generally don't play as well as the brute force ones.)

EARLY SIMULATION GAMES

At the same time computer scientists were learning how to make computers play games, military planners had begun to focus on game theory, and the ability of the computer to simulate the effects of alternative strategies made it a powerful tool for research. By 1955, the military was running large-scale global cold war simulations pitting NATO against the USSR and the Eastern Bloc. Unlike deterministic board games, simulation games generally have complex rules and algorithms that attempt to capture the relationships between various factors (such as the morale, experience, and firepower of a military unit) and generate realistic results. In 1959, programmers at Carnegie Tech, Pittsburgh, unveiled their business simulation, "The Management Game."

Today, simulation games are used as training aids in the military and business, as well as to enrich the K-12 curriculum in social studies and other areas. Many games such as the kingdom-building simulator *Civilization,* while marketed primarily as entertainment, can hone planning and management skills.

ARCADE AND GRAPHIC GAMES

Starting in the 1960s, CRT displays gave the new minicomputers the means to display simple graphics. In 1962, an intrepid band of game hackers at MIT created Spacewar, the first interactive graphic game and the forerunner of the arcade game boom of the 1970s. When the first microcomputers such as the Apple II and Radio Shack TRS-80 hit the market in the late 1970s, they had rudimentary graphics capabilities and came with a simple BASIC programming language. Soon enthusiasts' magazines had hundreds of pages of game source code that could be typed in and run by hobbyists. Some popular examples included a lunar lander simulator and Trek, where players chased Klingon ships through a grid representing sectors of space. By the late 1970s, the home game cartridge machine had been introduced by Atari and other companies, while arcade games were in full swing, with Pac-Man the phenomenal success of 1980.

TODAY'S GAME MARKET

By the 1990s, the typical PC had a special video card capable of displaying from hundreds to millions of colors, together with onboard video memory that could hold complex images. Computer game graphics have become increasingly sophisticated, with 3D techniques, realistic textures, shading and light, and lifelike animation. While the dungeon monster or baseball player of 1980 had been a wireframe image or crude assemblage of polygons, the same character in 2000 is portrayed with near photographic and cinematic realism. (Compare, for example, the *Wizardry* of 1980, which used a wireframe dungeon, with the richly textured gothic world of 1999's *Diablo II.*) While action and role-playing games depend the most on graphics to be competitive, even strategy simulations typically have 3D objects on realistic terrain and animated combat (see COMPUTER GRAPHICS).

Next to graphical sophistication, the most important trend in the 1990s was toward the RTS (real-time simulation), where all players act simultaneously rather than taking turns. Increasingly, the players a gamer will encounter will be fellow gamers rather than an often unsatisfactory computer opponent. Many an hour of work time has been lost to office network games of *Quake* or *Warcraft.* Role-playing games have also gone on-line, where games such as *Everquest* and *Asheron's Call* are hosted on powerful servers and players invest hundreds of hours in building up their favorite characters.

A scene from the computer strategy game Civilization. *Some games specialize in realistic physical simulation while others (such as this one) embody sophisticated economic and strategic considerations.*

The emphasis on multimedia and multiplayer design has changed the way game development is done. The earliest microcomputer games were typically the product of a single programmer and reflected the personal style of developers such as Chris Crawford in *Balance of Power* and Richard Garriott ("Lord British") in the *Ultima* series. Today, however, commercial games are the product of teams that include graphics, animation and sound specialists, actors and voice talent, and specialist programmers in addition to the game designers. While earlier games had more the single authorship of books, many game developers now compare their industry to the movie industry and its dominant studios. And as with the movie industry, it has been argued that the high cost of development and of access to the market has led to much imitation of successful titles and less innovation. (However, there remains a place for shareware games developed by one or a few persons and distributed via the Internet.)

Further Reading

Bates, Bob. *Game Design: the Art & Business of Creating Games.* Roseville, Calif.: Prima Tech., 2001.

Crawford, Chris. *The Art of Computer Game Design.* Berkeley, Calif.: Osborne-McGraw Hill, 1982. (Electronic version available at http://www.vancouver.wsu.edu/fac/peabody/game-book/Coverpage.html)

Game Developer [magazine]. http://www.gdmag.com/

Poole, Steven. *Trigger Happy: Videogames and the Entertainment Revolution.* New York: Arcade, 2000.

Howland, Geoff. "Game Design: the Essence of Computer Games." http://www.lupinegames.com/articles/essgames.htm

von Neumann, J., and O. Morgenstern. *Theory of Games and Economic Behavior.* Princeton, N.J.: Princeton University Press, 1944.

computer graphics

Most early mainframe business computers produced output only in the form of punched cards, paper tape, or text printouts. However, system designers realized that some kinds of data were particularly amenable to a graphical representation. In the early 1950s, the first systems using the cathode ray tube (CRT) for graphics output found specialized application. For example, the MIT Whirlwind and the Air Force's SAGE air defense system used a CRT to display information such as the location and heading of radar targets. By 1960, the new relatively inexpensive minicomputers such as the DEC PDP series were being connected to CRTs by experimenters, who among other things created Spacewar, the first interactive video game.

By the late 1970s, the microcomputers from Apple, Radio Shack, Commodore, and others either included CRT monitors or had adapters that allowed them to be hooked up to regular television sets. These machines generally came with a version of the BASIC language that included commands for plotting lines and points and filling enclosed figures with color. While crude by modern standards, these graphics capabilities meant that spreadsheet programs could provide charts while games and simulations could show moving, interacting objects. Desktop computers that showed pictures on television-like screens seemed less forbidding than giant machines spitting out reams of printed paper.

Research at the Xerox PARC laboratory in the 1970s demonstrated the advantages of a graphical user interface based on visual objects, including menus, windows, dialog boxes, and icons (see USER INTERFACE). The Apple Macintosh, introduced in 1984, was the first commercially viable computer in which everything displayed on the screen (including text) consisted of bitmapped graphics. Microsoft's similar Windows operating environment became dominant on IBM architecture PCs during the 1990s. Today Apple, Microsoft, and UNIX-based operating systems include extensive graphics functions. Game and multimedia developers can call upon such facilities as Apple QuickDraw and Microsoft DirectX to create high resolution, realistic graphics.

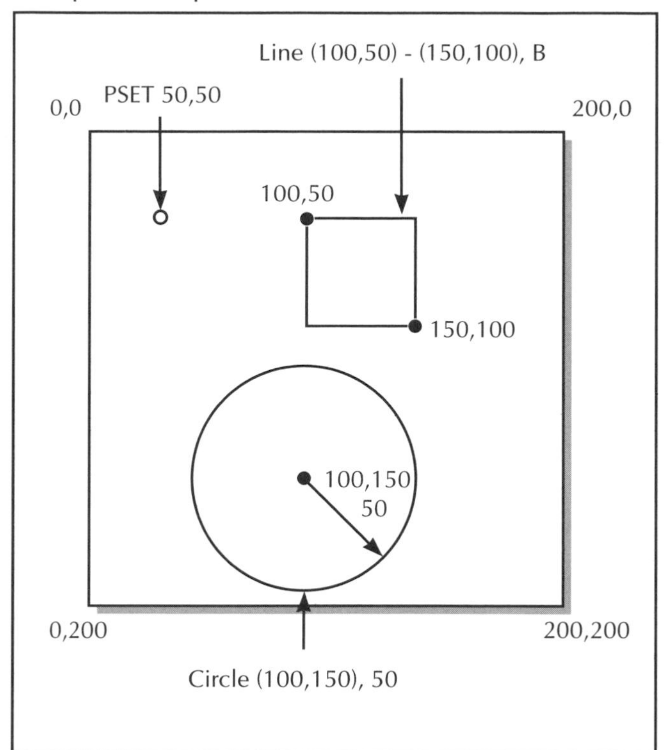

Some example figures plotted by BASIC graphics statements using screen coordinates.

BASIC GRAPHICS PRINCIPLES

The most basic capabilities needed for computer graphics are the ability to control the display of pixels (picture elements) on the screen and a way to specify the location of the spots to be displayed. A CRT screen is essentially a grid of pixels that correspond to phosphors (or groups of colored phosphors) that can be lit up by the electron beam(s). The first IBM PCs, for example, often displayed graphics on a 320 (horizontal) by 200 (vertical) grid, with 4 available colors.

A memory buffer is set up whose bytes correspond to the video display. (A simple monochrome display needs only one bit per pixel, but color displays must use additional space to store the color for each pixel.) A screen image is set up by writing the data bytes to the buffer, which then is sent to the video system. The video system uses the data to control the display device so the corresponding pixels are shown (in the case of a CRT, this means lighting up the "on" pixels with the electron gun[s]).

In most cases screen locations are defined in coordinates where point 0,0 is the upper left corner of the screen. The coordinates of the lower right corner depend on the screen resolution, At 320 by 200, the lower right corner would be 319,199.

For example, many versions of BASIC use statements such as the following:

```
PSET 50,50 ' draws a dot at X=50, Y=50
LINE (100,50)-(150,100), B ' draw square with UL
' corner at 100,50 and LR
' corner at 150,100
CIRCLE (100,150), 50, 4 ' draw a circle of
  radius 50
' with center at 200,200 and
' color 4 (red)
```

Languages such as C, C++ and JAVA don't have built-in graphics commands, but functions are provided in program libraries (see LIBRARY, PROGRAM). They would be used much like the BASIC commands given above.

GRAPHICS MODELS AND ENGINES

Modern applications (such as drawing programs and games) go well beyond simple two-dimensional objects. Indeed, multimedia developers typically use graphics engines designed to work with C++ or Java. A graphics engine provides a way to define and model 2D and 3D polygons. (Curves can be constructed by specifying "control points" for bicubic curves.)

Complex objects can be built up by specifying hierarchies (for example, a human figure might consist of a head, neck, upper torso, arms, hands, lower torso, legs, feet, and so on). By creating a hierarchy of arm, hand, fingers a transformation (scaling or rotation) of one object can be propagated to its dependent objects (see ANIMATION). In many cases graphics are created from real-world objects that have been digitally photographed or scanned, and then manipulated. (See IMAGE PROCESSING.)

In most scenes the relationships between graphical objects are also important. Modern graphics modeling programs use a virtual "camera" to indicate the position and angle from which the graphics are to be viewed. In rendering the scene, the Painter's Algorithm can be used to sort objects and draw closer surfaces on top of farther ones, as a painter might paint over the background. Alternatively, the Z-buffer algorithm stores depth information for each pixel to determine which ones are drawn. This technique requires less calculation (because surfaces don't need to be sorted), but more memory, since the depth of each pixel must be stored.

Within a scene, the effects of light (and its absence, shadows) must be realistically rendered. A simple technique can be used to calculate an overall light level for an object based on its angle in relation to the light source, plus a factor to account for ambient and diffuse light in the environment. The *Gouraud shading* technique can be used to smooth out the artifacts caused by the simple flat shading method. Another technique, *Phong shading*, can more realistically reproduce highlights (the sharp image of a light source being reflected within a surface). But the most realistic lighting effects are provided through ray tracing, which involves tracing how representative vectors (representing rays of light) reflect from or refract through various surfaces. However, ray tracing is also the most computationally intensive lighting technique.

Several techniques can be used to give objects more realistic surfaces. *Texture mapping* can be used to "paint" a realistic texture (perhaps scanned from a real-world object) onto a surface. For example, pieces in a chess game could be given a realistic wood grain or marble texture. This can be further refined through *bump mapping*, which calculates variations in the texture at each point based on light reflections.

APPLICATIONS AND TRADEOFFS

The most graphics-intensive applications today are games, multimedia programs, and scientific visualization or modeling applications. Because of the impact graphics have on users' perception of games and multimedia programs, developers spend a high proportion of their resources on graphics. Critics often complain that this is at the expense of core program functions. The software in turn places a high demand on user hardware: The contemporary "multimedia-ready" PC has a video card that includes special "video accelerator" hardware to speed up the display of graphics data and a video memory buffer of 32MB or more.

Complex 3D graphics with lighting, shading, and textures may have to be displayed at a relatively low resolution (such as 640 x 480) because of the limitations of the main processor (which performs necessary calculations) and the video card. However as processor speed and memory capacity continue to increase, many computer graphics now rival video and even film in realistic detail.

Further Reading

"Computer Graphics Research Centers." http://mambo.ucsc.edu/psl/cg.html

Foley, James D. [and others]. *Computer Graphics: Principles and Practice*. 2nd ed. Reading, Mass.: Addison-Wesley, 1995.

"Nan's Computer Graphics Page." http://www.cs.rit.edu/~ncs/graphics.html

Stevens, Roger. *Computer Graphics Dictionary: Including Animation, Game Development, and Photorealism*. Hingham, Mass.: Charles River Media, 2001.

computer industry

The U.S. computer industry began with the marketing of the Univac, designed by J. Presper Eckert and John Mauchly in the early 1950s. The first computers were made one at a time and only as ordered, and the market for the huge, expensive machines was thought to be limited to government agencies and the largest corporations. However, astute marketing by Sperry-Univac, Burroughs, and particularly, International Business Machines (IBM) convinced a growing number of companies that modern data processing facilities would be essential for managing their growing and increasingly complex business (see MAINFRAME).

The mainframe market was controlled by a handful of vendors who typically provided the complete computer system (including peripherals such as printers) and a long-term service contract. (Eventually, third-party vendors began to make compatible peripherals.) Companies that could not afford their own computers began to contract with service bureaus for their data processing needs, such as payroll processing.

By the 1960s, transistorized circuitry was replacing the vacuum tube, and somewhat smaller machines became practicable (see MINICOMPUTER). While these computers were the size of a desk, not a desktop, models such as Digital Equipment Corporation's PDP series and competition from companies such as Data General provided computing power for engineers and scientists to use in factories and laboratories. During the 1970s, the dedicated word processing machine marketed by the Wang Corporation began the digital transformation of the office. By the end of that decade, the first general-purpose desktop microcomputers were marketed. The Apple II made a modest inroad into business, fueled by VisiCalc, the first spreadsheet program.

This new market attracted the attention of IBM, viewed by many microcomputer enthusiasts as a dinosaurlike relic of the mainframe age. Uncharacteristically, IBM management gave the developers of their personal computer (PC) project free rein, and the result was the IBM PC introduced in 1981. The machine had two major advantages. One was the IBM name itself, which was comforting to executives contemplating a bewildering new technology. The other was that IBM (again, uncharacteristically) had followed Apple's lead in designing their PC with an "open architecture," meaning that third-party manufacturers could market a variety of expansion cards to increase the machine's capabilities. By 1990, about 10 million PCs worth about $80 billion were being sold annually (see IBM PC).

Although IBM tried to prevent other manufacturers from "cloning" the IBM chipset itself, it was unable to prevent companies such as Compaq from creating "IBM compatible" PCs that often surpassed the capabilities of the IBM models. (IBM introduced its microchannel architecture in the late 1980s in an unsuccessful attempt to regain proprietary advantage.) By the 1990s the IBM-compatible PCs (sometimes called "Wintel," for the Microsoft Windows operating system and Intel-compatible processor) had become an industry standard and a commodity manufactured and marketed by everything from the big name brands such as Dell and Gateway down to the corner computer store's backroom operation.

The announcement of Apple's Macintosh computer in 1984 made a vivid impression on the public (see MACINTOSH). With its fully graphical user interface, mouse, drawing program, and fonts, it seemed light-years ahead of the text-based IBM PCs. However, the Mac's slow speed, relatively high price, and closed architecture limited the Mac's penetration into the business market. The Mac did attract an enthusiastic minority of consumer users and achieved a lasting niche presence in education and among graphics and video professionals. Gradually, as Microsoft's graphical Windows operating system improved in the early 1990s, the Mac's advantages over the IBM-compatible machines diminished.

During the 1990s, desktop computers came with a series of increasingly powerful series of Pentium processors, matched by offerings from AMD and Cyrix. Multimedia (including high-end graphics and sound capabilities) became a standard feature, particularly on consumer PCs. Increasingly, the business PC was being connected to a local area network, and both business and consumer PCs included modems or broadband access to on-line services and the Internet. The need to manage network files and services (such as Web servers) led to the development of server PCs featuring high-capacity mass storage. At the same time, high-end PCs also challenged the graphics workstations made by companies such as Sun. The traditional minicomputer and high performance workstation category began to melt away. By 2002, an estimated 600 million personal computers were in use worldwide, with about half of them in homes.

The personal computer also grew smaller. The suitcase-sized "luggable" computers of the 1980s gave way to a range of laptop, notebook-sized, and palm-sized computers. Today wireless networking technology allows users of diminutive machines to access the full resources of the World Wide Web and local networks.

The idea of "appliance computing" has also been a recurrent theme among industry pundits. Proponents argue that there are still many people who feel intimidated by a standard computer interface but have become comfortable with other consumer electronic products such as televisions, CD players, or microwaves. If computer functions could be built into such devices, people might use them comfortably. For example, WebTV is a box that allows the user to surf the Web from the same armchair where he or she watches TV, using controls little more complicated than those found on a regular TV remote. Kitchen appliances might be transformed, with the microwave providing recipes and the refrigerator keeping an inventory and automatically ordering from the grocery store. However, as with the fully automated "wired home," featured in Sunday newspaper supplements, the appliance computer has remained difficult to market to consumers.

THE SOFTWARE INDUSTRY

Hardware is useless without software. Since the operating system (OS) is the software that enables all other software to access the computer, the OS market is a key part of the computer industry. Through a historical accident, a young programmer-entrepreneur named Bill Gates and his Microsoft Corporation received the contract to develop the operating system for the first IBM PC. Microsoft bought and adapted an existing operating system to create MS-DOS (also called PC-DOS). Until the end of the 1980s, DOS was the dominant operating system for IBM-compatible PCs (see MS-DOS). In the early 1990s, Microsoft introduced Windows 3.0, the first successful version of its graphical operating environment (see MICROSOFT WINDOWS). The dominance of Windows became so complete that it formed the basis for a federal antitrust case against Microsoft that as of 2002 had not been resolved.

The source of emerging challenges to Windows comes not from another desktop vendor but from the Internet, where Java offers the potential of delivering applications through the user's Web browser, regardless of whether that user is running Windows, the Macintosh OS, or Linux, a variant of UNIX that has been embraced by many enthusiasts. However, Java applications and Linux still represent only a tiny fraction of the market share held by Windows (see JAVA and UNIX).

The 1990s saw considerable consolidation in the office software arena. Microsoft's Office software suite has overwhelmed once formidable competitors such as WordPerfect and Corel. Packages such as Microsoft Office create their own mini-industries where developers create templates and add-ins. Today Java-based office suites delivered by Internet-based Application Service Providers (ASPs) "office clones" such as Star Office may be beginning to challenge Microsoft's near-monopoly in office applications.

Outside the office there is considerably more competition in the software industry. Today's consumers can choose from a wide variety of software that fills utility or other niche needs, including shareware ("try before you buy") offerings. In educational software and games some once-major innovators have been bought out or consolidated, but there is no one dominant company. Thousands of specialized software packages serve scientific, manufacturing, and business needs. While the general public is unaware of such programs, they make up much of the strength of the software industry.

OTHER PRODUCTS AND SERVICES

There are many other niches in computer hardware and software, and the landscape is constantly changing. Consumer electronics, particularly in the entertainment area, has become increasingly computer-like. Dedicated game machines and their software cartridges are major markets. Microsoft has become an increasing presence in both PC games and in dedicated game machines (introducing the XBox in 2001 to compete with Sony's PlayStation). Portable MP3 music players are becoming popular, and DVD is slowly gaining consumer acceptance as a superior way to deliver video and movies. Meanwhile, digital cameras are now competitive on a price-performance basis with "point-and-shoot" and even midrange SLR cameras.

The services sector of the computer industry lacks the visibility of new hardware products, but provides most of the industry's employment and much of its economic impact. In addition to the hundreds of thousands of programmers who provide business-related, consumer, and specialized software, there are the legions of help desk employees, computer and network technicians, creators of software development tools, writers of technical books and training products, industry investment analysts, reporters, and many others whose livelihood depends on the computer industry.

INDUSTRY TRENDS

As the new century begins, several trends evident during the 1990s are likely to continue. Delivery of applications via the Internet will grow in importance, possibly displacing the traditional CD package. The growing availability of broadband (high speed) Internet connections will facilitate the trend toward Web-based office computing and Web-based delivery of multimedia for entertainment and education. The growing use of mobile wireless Internet connections and home networks will make information more portable and communications closer to instantaneous. The merging of traditionally disparate media such as film, still photography, video, and sound (see DIGITAL CONVERGENCE) is likely to accelerate, with separate TV, VCR, and stereo sound systems being replaced by integrated systems. On the other hand, sales of PCs are relatively stagnant. Growth of sales in the United States was only 0.5 percent in 2001, and projected sales for 2003 are only 4.2 percent, according to an IDG research report.

The internationalization and globalization of the computer industry may be the most significant trend. By the 1960s, IBM was dominating the mainframe computer industry in Britain and Europe despite the efforts of indigenous companies and government initiatives. Japan was considerably more successful in developing a competitive electronics and computer industry under the long-term guidance of MITI (Ministry of International Trade and Industry). The Japanese became dominant in industrial robotics and strong in consumer electronics, including game machines (Sony), digital cameras (Sony and Fujitsu), and laptop computers (Toshiba). They have been less successful in desktop computers, Internet-related technology, and commercial software. Taiwan (and gradually, mainland China) have become important in the components and peripherals industry. The growing importance of Asia in the international computer industry is also underscored by the large number of programmers being trained in India, many of whom later immigrate to the United States.

Other trends are harder to assess, such as whether the dominance of Microsoft and its Windows family of operating systems will diminish during the first decade of the new century, perhaps precipitated by a court-ordered breakup of the company. As of 2001, high-flying "dot-com" stocks are crashing to earth, with the possibility of layoffs and slow growth leading to unemployment. But whatever happens, the central importance of the computer industry to the U.S. and world economy is unlikely to diminish.

Further Reading

Computer Industry Alamanac web site. http://www.c-i-a.com/
Computerworld. http://www.computerworld.com
Cortada, J. W. *Information Technology as Business History.* Westport, Conn.: Greenwood Press, 1996.
Datamation. http://www.datamation.com
International Data Corporation. http://www.idc.com
Petska, Karen. *Computer Industry Almanac.* 8th ed. Computer Industry Almanac, 1996. [Note: A ninth edition is tentatively scheduled for late 2002.]

computer literacy

As computers became integral to business, industry, trades, and professions, educators and parents became increasingly concerned that young people acquire a basic understanding of computers and master the related skills. The term *computer literacy* suggested that computer skills were now as important as the traditional skills of reading, writing, and arithmetic. However, there has been disagreement about the emphasis for a computer literacy curriculum. Some educators, such as Seymour Papert, computer scientist and inventor of the Logo language, believe that students can and should understand the concepts underlying computing, and be able to write and appreciate a variety of computer programs (see LOGO). By gaining an understanding of what computers can (and cannot) do, students will be able to think critically about how to appropriately use the machines, rather than simply mastering rote skills. Indeed, by gaining a good grasp of general principles, the student should be able to easily master specific skills.

An opposing view emphasizes the practical skills that most people (who will not become programmers) will need in everyday life and work. This sort of curriculum focuses on learning how to identify the parts of a computer and their functions, how to run popular applications such as word processors, spreadsheets, and databases, how to connect to the Internet and use its services, and so on. Computer literacy can also be broadened to include understanding the impact that computers are having on daily life and social issues that arise from computer use (such as security, privacy, and inequality).

Today computer literacy is an important part of every elementary and high school curriculum. Most students in middle-class or higher income brackets now have access to computers at home, and many thus gain considerable computer literacy outside of school. In addition, adult education and vocational schools often emphasize computer skills as a route to employment or career advancement. People also have the opportunity to learn on their own through books and videos.

Further Reading

Gookin, Dan. *PCs for Dummies.* 8th ed. New York: Hungry Minds, 2001.
Kershner, H. G. *Computer Literacy.* 3rd ed. Dubuque, Iowa: Kendall/Hunt, 1998.
White, Ron. *How Computers Work.* Millennium ed. Indianapolis, Ind.: Que, 1999.

computer science

Most generally, computer science is the study of methods for organizing and processing data in computers. The fundamental questions of concern to computer scientists range from foundations of theory to strategies for practical implementation.

FUNDAMENTAL THEORY

- What problems are susceptible to solving through an automated procedure? (See COMPUTABILITY AND COMPLEXITY.)
- Given that a problem is solvable, can it be solved without too much expenditure of time or computing resources?
- Can a step-by-step procedure be devised for solving a given problem? (See ALGORITHM.) How do different procedures (such as for sorting data) compare in efficiency and reliability? (See SORTING AND SEARCHING.)
- What methods of organizing data are most useful? (See DATA STRUCTURES.) What are the advantages and drawbacks of particular forms of organization? (See ARRAY, LIST PROCESSING, and QUEUE.)
- Which structures are best for representing the data needed for a given application? What is the best way to relate data to the procedures needed to manipulate it? (See ENCAPSULATION, CLASS, and PROCEDURES AND FUNCTIONS.)

THE TOOLS OF COMPUTING

- How can programs be structured so they are easier to read and maintain? (See STRUCTURED PROGRAMMING and OBJECT-ORIENTED PROGRAMMING.)
- Can programmers keep up with growth of operating systems and application programs that have millions of lines of code? (See SOFTWARE ENGINEERING and QUALITY ASSURANCE, SOFTWARE.)
- How can multiple simultaneous tasks (or even multiple processors) be coordinated to bring greater computing power to bear on problems? (See MULTITASKING and MULTIPROCESSING.)
- What is the best way to design an operating system, including the arrangement of different layers of the

operating system such as the hardware-specific drivers, kernel (essential functions), and interfaces (shells or visual environments)? (See OPERATING SYSTEM, KERNEL, DEVICE DRIVER, and SHELL.)

- What should be emphasized in designing a programming language? How does one specify the grammar of statements the declaration and handling of data types, and the mechanism for handling functions or procedures? (See BACKUS-NAUR FORM, DATA TYPES, PROCEDURES AND FUNCTIONS.)
- What considerations should be emphasized in designing a compiler for a given language? (See COMPILER.)
- How should a network be organized, and what protocols should be used for transferring data? (See NETWORK, INTERNET, DATA COMMUNICATIONS, TELECOMMUNICATIONS and TCP/IP.)

SPECIFIC APPLICATION AREAS

The general principles and tools must then be applied to a variety of application areas including:

- Text processing (See WORD PROCESSOR, TEXT EDITOR, and FONT.)
- Graphics (See COMPUTER GRAPHICS, IMAGE PROCESSING.)
- Database management, including file structures and file access (such as indexing and hashing), and database architecture (relational databases.) (See DATABASE MANAGEMENT SYSTEM, SQL, and XML).
- Business data processing issues, including the design of MIS (management information systems) and decision support systems.
- Scientific programming issues, including data acquisition, maintaining accuracy in calculations, and creating visualizations driven by the data. (See DATA ACQUISITION, NUMERIC DATA, and SCIENTIFIC COMPUTING APPLICATIONS.)
- User interface design (designing the interaction between human beings and the operating system or application). (See USER INTERFACE and WYSIWYG).
- The broad area of artificial intelligence, which affects ways of representing information and modeling reasoning processes. (See ARTIFICIAL INTELLIGENCE, NEURAL NETWORK, EXPERT SYSTEMS, and KNOWLEDGE REPRESENTATION.)
- Robotics and control systems (an older term, "cybernetics," has also been used for this field). (See ROBOTICS.)

Clearly the concerns of computer science overlap a number of related fields. The design of computer hardware is often considered to be computer engineering, but designers of hardware must be familiar with the algorithms that will be used to operate it (see also COMPUTER ENGINEERING). Both artificial intelligence and user interface design are affected by cognitive science (or psychology), the study of human thought processes. Biology both inspires and is illuminated by artificial life simulations, genetic algorithms, and neural networks. The most abstract questions of information processing touch on the field of information science (or information theory).

HISTORY OF THE FIELD

Not surprisingly, the field of computer science did not emerge until computers themselves became an established product. The early computer pioneers such as Alan Turing, J. Presper Eckert, and John Mauchly brought backgrounds in mathematics or engineering (see TURING, ALAN, ECKERT, J. PRESPER and MAUCHLY, JOHN). By the 1960s, however, a discipline and curriculum for computer science began to emerge. By the late 1990s more than 175 departments in American and Canadian universities offered a doctorate in computer science, with about a thousand new Ph.D.s being granted each year. (See EDUCATION IN THE COMPUTER FIELD for more details.)

The traditional computer science field emphasizes the theory of data representation, algorithms, and system architecture. In recent years a more practically oriented curriculum has emerged as an alternative. Under the titles of "Information Technology" or "Information Systems," this curriculum emphasizes application areas such as management information systems, database management, system administration, and Web development.

Further Reading

Biermann, Alan W. *Great Ideas in Computer Science: a Gentle Introduction.* 2nd ed. Cambridge, Mass.: MIT Press, 1997.

Brookshear, J. Glenn. *Computer Science: an Overview.* 6th ed. Reading, Mass.: Addison-Wesley, 2000.

"Computer Science."http://dir.yahoo.com/Science/Computer_Science/

Denning, P. *Talking to the Machine: Computers and Human Aspiration.* New York: Copernicus Books, 1999.

"Guide to Computer Science Internet Resources." http://www.library.ucsb.edu/istl/97-summer/internet2.html

Hillis, Daniel W. *The Pattern on the Stone: the Simple Ideas that Make Computers Work.* New York: Basic Books, 1998.

Knee, Michael. "Computer Science: A Guide to Selected Resources on the Internet." Association of College and Research Libraries. June 2001. http://www.ala.org/acrl/resjune01.html

computer virus

A computer virus is a program that is designed to copy itself into other programs. When the other programs are run, they carry out the virus's instructions, either instead of or in addition to their own. Since one of the primary tasks programmed into a virus is to reproduce itself, a virus program can spread rapidly. Viruses are generally programmed to seek out program files that are likely to be executed in the near future, such as those used by the operating system during the startup process. The result is a copy that can in turn generate an additional copy, and so on.

Appearing in the 1980s, the first computer viruses were generally spread by infecting programs on floppy disks, which were often passed between users. Today, viruses generally have instructions that enable them to gain access to network facilities (such as e-mail) to facilitate their spreading to other systems on a local network or on the Internet. The spread of viruses is complicated by the fact that operating systems (particularly Microsoft Windows) and applications (such as Microsoft Office) have the ability to run scripts or "macros" that are attached to documents. This facility can be useful for tasks such as sophisticated document formatting or form-handling, but it also means that viruses can attach themselves to scripts or macros and run whenever a document containing them is opened. Since modern e-mail programs have the ability to include documents as attachments to messages, this means that the unsuspecting recipient of a message can trigger a virus simply by opening a message attachment.

Viruses can be further disguised by programming them to remain dormant until a certain date, time, or other condition is reached. (Such a virus is sometimes called a *logic bomb*.) For example, a disgruntled programmer who is about to be dismissed might insert a virus that will wipe out payroll data at the beginning of the next month. A famous example of the time-triggered virus was the Michelangelo virus, so named because it was triggered to run on the artist's birthday, March 6, 1992. (See COMPUTER CRIME AND SECURITY.)

Viruses can be overtly destructive (such as by reformatting a computer's hard drive, wiping out its data). Other viruses can simply tie up system resources. The most infamous example of this was the "Internet Worm" introduced onto the network on November 2, 1988, by Robert Morris, Jr. This program was intended to reproduce slowly, planting its "segments" on networked computers by exploiting a flaw in the UNIX sendmail program. Unfortunately, Morris made an error that caused the worm to spread much more rapidly. Before the coordinated efforts of system administrators at affected sites came up with countermeasures, the worm had cost somewhere in the hundreds of thousands of dollars in lost computer and programmer time.

Viruses can also lurk quietly where they carry out tasks such as intercepting user passwords without the user's notice. (Such viruses are sometimes called *Trojan Horses* because they masquerade as innocuous system files.) Recently, viruses have also been used to spread and then launch simultaneous attacks on websites, bombarding them with spurious acknowledgment requests. (This is called a DDOS or "distributed denial of service attack.) In 1999 and 2000, some major e-commerce sites came under such attack.

COUNTERMEASURES

The only certain defense of a computer system from viruses would be through abstaining from contact between it and any other computers, either directly through a network or indirectly through exchange of programs on floppy disks or other removable media. In today's highly networked world, this is usually impractical. A more practical defense is to install antivirus software. Antivirus programs work by comparing the contents of files (either those already on the disk or entering via the Internet) with "signatures" or patterns of data found in known viruses. More sophisticated antivirus programs include the ability to recognize program code that is similar to that found in known viruses or that contains suspicious operations (such as attempts to reformat a disk or bypass the operating system and write directly to disk). If an antivirus program recognizes a virus, it warns the user and can be told to actually remove the virus. Because dozens of new viruses are identified each week, virus programs must be updated frequently with new virus signature files in order to remain effective. Some antivirus programs can update themselves by periodically linking to a website containing the update files.

The computer user is also an important line of defense against viruses. In addition to making sure antivirus scans are run regularly, users can minimize the risk of infection by not opening e-mail attachments from unknown sources (or even those from known correspondents that appear to be out of context). Scripting or macro facilities can also be disabled unless absolutely needed.

Further Reading

Computer Emergency Response Team (CERT).http://www.cert.org/

Cohen, F. B. *A Short Course on Computer Viruses*. 2nd ed. New York: John Wiley, 1994.

Denning, P. J., ed. *Computers Under Attack: Intruders, Worms, and Viruses*. New York: ACM Press, 1990.

McAfee.com. http://www.mcafee.com/

Symantec. http://www.symantec.com/

computer vision

In the biological world, vision is the process of receiving light signals from the environment through the eyes and optic nerves, from which the brain can extract patterns that contain useful information (such as recognizing food or a potential predator). Computer vision is the analogous process by which light is received by a sensor system (such as a digital camera). The light is then analyzed for meaningful patterns. Thus, a robot might be able to recognize the identity and positions of various parts on an assembly line.

Because computer vision involves pattern recognition, it is part of the discipline of artificial intelligence (see ARTIFICIAL INTELLIGENCE). The challenge is not in getting information about a visual scene from the camera and turning it into digital information (a grid of pixels). Rather, it is the ability to recognize meaningful patterns in fragmented images, something human infants learn to do almost from birth when they encounter human faces.

One way to approach the problem is to constrain the kinds of images the computer (or robot) has to deal with. If you can guarantee that a robot's field of vision will contain only a few fixed objects (a hopper, perhaps, or a conveyer belt) plus one or more distinctively shaped parts, it is relatively easy to program the dimensions of the possible objects into the vision system so that the robot can identify objects by comparing them with stored templates. However, if the robot encounters an object it isn't prepared for, such as a stray bit of packing material, it will be unable to identify (or properly deal) with the object.

Vision is also complicated by the problem of parsing three-dimensional objects in the visual field. Seen head-on, the side of a cube appears to be a two-dimensional square. Seen at an angle, it appears to be a three-dimensional assemblage with some faces visible and some not. To interpret these and more complicated objects, the robot might be programmed with rules that help it infer that an object is really a cube, that all cubes have six equal sides, and so on. Another strategy is to give the robot more than one "eye" so that images can be compared, much as humans do unconsciously with binocular vision. Finally, the robot can be given the ability to move its head and eyes in order to find a viewpoint that yields more information about an ambiguous object.

Human infants, of course, are not born with a fully developed understanding of the types of objects in their world. They are always learning new ways to distinguish, for example, a stuffed teddy bear from a live dog. Robot vision systems, too, can be programmed to learn (or at least, refine their ability to recognize objects). A statistical technique can be used to "sample" objects in the environment and find which characteristics most reliably "predict" the true nature of an object. Characteristics can be resampled from different viewpoints to see which ones remain invariant (unchanged). For example, a cube will always have four edges on each face. Another approach is to use a neural network, where the visual information is processed by a grid of nodes that are reinforced to the extent they are successful in identifying features (such as edges).

APPLICATIONS

Computer vision is a problem of great theoretical interest because it engages so many questions about perception, the ability to build models of the world, and the ability to learn. The field also has considerable practical potential. Currently, most robots are fixed to stations on factory floors where they work with a limited number of objects (parts) in a highly constrained, stable environment. However "service robots" have been gradually developed to work in a much less constrained environment (such as carrying supplies down hospital corridors or even serving as mobile assistants to astronauts in the weightless environment of the International Space Station). These robots would benefit greatly by having robust vision systems so that they can, for example, recognize individual human faces or detect potentially dangerous situations.

Further Reading

Davies, E. R. *Machine Vision.* San Diego, Calif.: Academic Press, 1997.

Fischler, M. and O. Firschein. *Readings in Computer Vision: Issues, Problems, Principles, and Paradigms.* San Francisco: Morgan Kaufmann, 1987.

Haralick, R. M. and L. G. Shapiro. *Computer and Robot Vision.* Reading, Mass.: Addison-Wesley, 1992.

"Machine Vision Research Links." http://www.machinevisiononline.org/links/

"Machine Vision Resources." http://www.eeng.dcu.ie/~whelanp/resources/resources.html

concurrent programming

Traditional computer programs do only one thing at a time. Execution begins at a specified point and proceeds according to decision statements or loops that control the processing. This means that a program generally cannot begin one step until a previous step ends.

Concurrent programming is the organization of programs so that two or more tasks can be executed at the same time. Each task is called a thread. Each thread is itself a traditional sequentially ordered program. One advantage of concurrent programming is that the processor can be used more efficiently. For example, instead of waiting for the user to enter some data, then performing calculations, then waiting for more data, a concurrent program can have a data-gathering thread and a data processing thread. The data-processing thread can work on previously gathered data while the data-gathering thread waits for the user to enter more data. The same principle is used in multitasking operating systems such as UNIX or Microsoft Windows. If the system has only a single processor, the programs are allocated "slices" of processor time according to some scheme of priorities. The result is that while the processor can be executing only one task (program) at a time, for practical purposes it appears that all the programs are running simultaneously (see MULTITASKING).

Multiprocessing involves the use of more than one processor. In such a system each task (or even each thread within a task) might be assigned its own processor. Multiprocessing is particularly useful for programs that involve intensive calculations, such as image processing or pattern recognition systems (see MULTIPROCESSING).

PROGRAMMING ISSUES

Regular programs written for operating systems such as Microsoft Windows generally require no special code to deal with the multitasking environment, because the operating system itself will handle the scheduling. (This is true with *preemptive multitasking,* which has generally supplanted an earlier scheme where programs were responsible for yielding control so the operating system could give another program a turn.)

Managing threads within a program, however, requires the use of programming languages that have special statements. Depending on the language, a thread might be started by a *fork* statement, or it might be coded in a way similar to a traditional subroutine or procedure. (The difference is that the main program continues to run while the procedure runs, rather than waiting for the procedure to return with the results of its processing.)

The coordination of threads is a key issue in concurrent programming. Most problems arise when two or more threads must use the same resource, such as a processor register (at the machine language level) or the contents of the same variable. Let's say two threads, A and B, have statements such as: Counter = Counter + 1. Thread A gets the value of Counter (let's say it's 10) and adds one to it. Meanwhile, thread B has also fetched the value 10 from Counter. Thread A now stores 11 back in counter. Thread B, now adds 1 and stores 11 back in Counter. The result is that Counter, which should be 12 after both threads have processed it, contains only 11. A situation where the result depends on which thread gets to execute first is called a *race condition.*

One way to prevent race conditions is to specify that code that deals with shared resources have the ability to "lock" the resource until it is finished. If thread A can lock the value of Counter, thread B cannot begin to work with it until thread A is finished and releases it. In hardware terms, this can be done on a single-processor system by disabling interrupts, which prevents any other thread from gaining access to the processor. In multiprocessor systems, an *interlock* mechanism allows one thread to lock a memory location so that it can't be accessed by any other thread. This coordination can be achieved in software through the use of a *semaphore,* a variable that can be used by two threads to signal when it is safe for the other to resume processing. In this scheme, of course, it is important that a thread not "forget" to release the sempahore, or execution of the blocked thread will halt indefinitely.

A more sophisticated method involves the use of message passing, where processes or threads can send a variety of messages to one another. A message can be used to pass data (when the two threads don't have access to a shared memory location). It can also be used to relinquish access to a resource that can only be used by one process at a time. Message-passing can be used to coordinate programs or threads running on a distributed system where different threads may not only be using different processors, but running on separate machines (a *cluster* computing facility).

Programming language support for concurrent programming has come through devising new dialects of existing languages (such as Concurrent Pascal), building facilities into new languages (such as Modula-2), or creating program libraries for languages such as C and C++.

Further Reading

Andrews, G. R. *Foundations of Multithreaded, Parallel, and Distributed Programming.* Reading, Mass.: Addison-Wesley, 2000.

Lea, D. *Concurrent Programming in Java.* 2nd ed. Reading, Mass.: Addison-Wesley, 2000.

conferencing systems

Conferencing systems are on-line communications facilities that allow users to log in and participate in discussions on a variety of topics. Although this is a rather amorphous category of software, some distinguishing characteristics can be identified. Conferencing is distinguished from chat or instant messaging systems because the messages are asynchronous (that is, one person at a time leaves a message, and there is no real-time interaction between participants). Unlike Netnews newsgroups, conferencing systems such as San Francisco Bay Area–based The Well tend to have users who are committed to long-term discussions in conferences (topical discussion areas) that tend to persist for weeks, months, or even years. Conferencing systems are often grouped under the umbrella term of Computer-Mediated Communications (CMC).

HISTORY

In the 1960s, researcher Murray Turoff at the Institute for Defense Analysis decided to adopt for computer use a discussion method called Delphi, developed at RAND corporation. This method was a collective process by which new ideas were discussed and voted on by a panel of experts. After he implemented Delphi as a system of messages passed via computer, he began to generalize his work into a more general method of facilitating on-line discussions. His Electronic Information Exchange System (EIES, pronounced "eyes") was designed to facilitate discussion within research communities of 10–50 members.

The emergence of topical on-line discussions can be seen in the development of the Usenet (or Netnews) newsgroups in the early 1980s, the development of communications or memo systems within large offices (particularly within the government), and the emergence of bulletin boards and on-line services for personal computer users. Most early news and bulletin board software had only rudimentary facilities for linking topics and responses. A more sophisticated approach to conferencing emerged within the PLATO educational computing network in the 1970s, in the form of Plato Notes. This system began as a simple way for users to leave messages or help requests in a text file, and evolved into a structure of "base notes" and linked response notes, a topic-and-response structure that became the general model for conferencing systems.

In the mid-1980s, the Well (Whole Earth 'Lectronic 'Link) began to provide on-line conferencing to anyone who subscribed. It used a text-based system called Picospan. With its improbable eclectic mix well salted with Grateful Dead fans and computer "nerds," the Well became a sort of petri dish for cultivating community (see VIRTUAL COMMUNITY). Long-term friendships (and feuds) and occasional romances have been nurtured by such conferencing systems.

TYPICAL STRUCTURE

A typical text-based conferencing system is divided into conferences, which are generally devoted to relatively broad subjects, such as UNIX, pop music, or politics. Each conference is further divided into topics, which usually reflect particular aspects of the general subject (such as a particular UNIX version, a pop music group, or a political issue). Most conferencing systems have a person or persons who act as a moderator (sometimes called a "host") who tries to encourage new users, keep discussions more or less on topic, and discourage personal attacks or vehement statements ("flames").

A user signs onto the system and "joins" one or more conferences. Each time the user visits a conference that he or she has joined, any topics (or responses in existing topics) that were posted since the last visit are presented. The user can read the postings and, if desired, enter a reply that becomes part of the thread of messages. (Users are also generally allowed to start new topics of their own.)

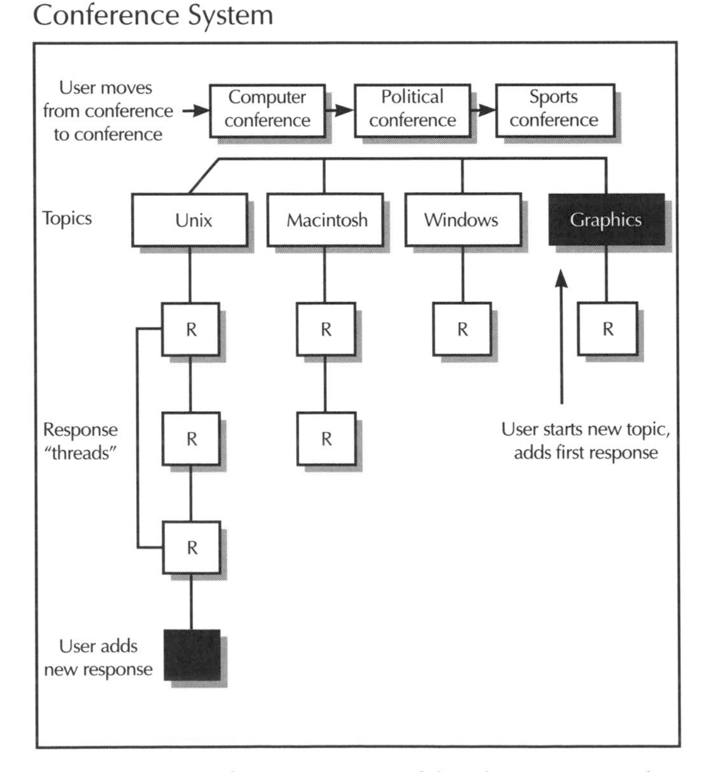

In the PicoSpan conferencing system each broad topic is assigned its own conference. Within the conference, users can start more specific topics. Once a user starts a topic other users can post responses, which are chained into "threads."

WEB-BASED CONFERENCING

Text-based systems such as Picospan are driven by the user entering command letters or words. While this paradigm is familiar to people who have experience with operating systems such as UNIX or MS-DOS, it can be more difficult for users who are used to the point-and-click approach of Windows programs and the World Wide Web. Many new conferencing systems use webpages to present conference topics and messages, with buttons replacing text commands. (The Well continues to offer both the text-based Picospan and the Web-based Engaged.)

Further Reading

"Conferencing Software for the Web." http://thinkofit.com/webconf/

Rheingold, Howard. *The Virtual Community: Homesteading on the Electronic Frontier.* Reading, Mass.: Addison-Wesley, 1993.

Woolley, David R. "PLATO: the Emergence of an On-Line Community." *Computer-Mediated Communication Magazine,* vol. 1, no. 3, July 1, 1994, 5. Available on-line at: http://www.december.com/cmc/mag/1994/jul/plato.html

constants and literals

Constants and literals are ways of describing data that does not change while a program runs. For example, a statement in C such as

```
const float pi = 3.14159;
```

expresses a value that will be used in calculations, but not changed. Constants can be of any data type, including character strings as well as numbers. String constants are usually enclosed in single or double quotes:

```
char * Greeting = "Hello, World";
```

Actual strings and numerals found in programs are sometimes called *literals*, meaning that they are to be accepted exactly as given (literally) rather than standing for some other value. Thus 3.14159 and Hello, World as given above can be considered to be numeric and string literals respectively.

Because many languages consider a value of 1 as representing a "true" result for a branch or loop test, and 0 as representing "false," programs in languages such as C often include declarations such as:

```
const True = 1;
const False = 0;
```

This lets you later have a loop construction such as

```
while (True) {
' body of program
} ;
```

which is a more readable way to code an endless loop than:

```
while (1) {
' body of program
} ;
```

However languages such as Pascal and C++ have a special boolean data type (*bool* in C++) that allows for constants or variables that will have one of two values, **true** or **false.**

Some languages provide a way to set up an ordered group of constant values. (See ENUMERATIONS AND SETS.)

CONSTANTS VS. VARIABLES

The difference between a constant and a variable is that a variable represents a quantity that can change (and is often expected to). For example, in the statement

```
int Counter = 0;
```

Counter is set to a starting value of zero, but will presumably be increased as whatever is to be counted is counted.

Most compilers will issue an error message if they detect an attempt to change the value of a constant. Thus the sequence of statements:

```
const float Tax_Rate = 8.25;
Tax_Rate = Tax_Rate + Surtax;
```

would be illegal, since Tax_Rate was declared as a constant rather than as a variable.

Many compilers, as part of code optimization, can discover values or expressions that will remain constant throughout the life of the program, even if they include variables. Such constants can be "propagated" or substituted for variables. This can speed up execution because unlike a variable, a constant does not need to be retrieved from memory (see COMPILER).

Further Reading

Sebesta, Robert W. *Concepts of Programming Languages.* Reading, Mass.: Addison-Wesley, 1999.

copy protection

Companies that produce software have had to cope with software that is expensive to develop, while the disks on which it is distributed are inexpensive to reproduce. The making and swapping of "pirated" copies of software is just about as old as the personal computer itself. Software piracy has taken a number of forms ranging from teenaged hackers making extra copies of games to virtual factories (often in Asia) that stamp out thousands of bogus copies of Windows operating systems and programs that would cost hundreds of dollars apiece if legitimate.

In order to prevent such copying, software producers in the 1980s often recorded the programs on floppy disks in a special format that made them hard to copy successfully. One way to do this is to record key information on disk tracks that are not normally read by the operating system and thus not reproduced by an ordinary copy command. When such a program runs, it can use a special device control routine to read the "hidden" track. If it doesn't find the identifying information there, it knows the disk is not a legitimate copy.

Another way to do copy protection is by having the program look for a small hardware device called a "dongle" connected to the computer, usually to the parallel printer port. Since the dongle is distributed only with the legitimate program, it can serve as an effective form of copy protection. (Encryption can also be used to render copies unusable without the key.)

DECLINE OF COPY PROTECTION

Copy protection has a number of drawbacks. Because disk-based copy protection writes on nonstandard tracks, even legitimate programs may not work with certain models of disk or CD drive. And because the legitimate user is unable to make a backup copy of the disk, if it is damaged, the user will be unable to use the program. Dongles, on the other hand, can interfere with the operation of other devices connected to the port, and a user might be required to use multiple dongles for multiple programs.

During the 1990s, copy protection was generally phased out, except for some games. A variety of other strategies are used against software piracy. The Software Publishers Association (SPA) maintains a program in which disgruntled users can report unauthorized copying of software at their workplace. Companies that allow unauthorized copying of software can be sued for violating the terms of their software license. International trade negotiations can include provisions for cracking down on the massive "cloning" of major software packages abroad.

Hackers and cyber-libertarians have often argued that the problem of software piracy has been overrated, and that allowing the copying of software would enable more people who would not otherwise buy programs to try them out. Once someone likes the program, they might buy it not only for legitimacy of ownership but in order to get access to the technical support and regular upgrades that are often required for complex business software packages. For less expensive software, the alternative of SHAREWARE allows for a "try before you buy" distribution of software.

Further Reading

Information Technology Association of America. "Intellectual Property Protection in Cyberspace: Toward a New Consensus." http://www.itaa.org/govt/pubs/ipp.htm

Smedinghoff, Thomas J. *The Software Publishers Association Legal Guide to Multimedia.* Reading, Mass.: Addison-Wesley, 1994.

CPU

The CPU, or central processing unit, is the heart of a computer, the place where data is brought in from input devices, processed, and sent to output devices. (This article describes the CPU from the point of view of desktop micromputers, where it is a single large silicon chip and supporting chips; see MAINFRAME for a discussion of that earlier architecture.)

The CPU consists of two major parts. The arithmetic-logic unit performs arithmetic or logical operations on pairs of numbers brought in from memory and stored in special locations called registers (see ARITHMETIC LOGIC UNIT). For example, the CPU can add a value from main memory to a value stored in a register and store the result back into memory. In addition to addition, subtraction, multiplication, and division, the CPU can logically compare the individual bits in two values, performing such operations as AND, where the result is 1 only if both bits are ones, or OR, where the result is 1 if either bit is one. The power of a CPU is measured either in the number of clock cycles that drive it each second (see CLOCK SPEED) or the number of standard instructions it can execute in a second. For modern PCs, clock speeds range into the billions of cycles per second (gigahertz) and millions of instructions per second (most instructions take more than one cycle to be completed).

The other key part of the CPU is the control unit, which determines when (and which) instructions will be executed. Operations to be performed are specified by instruction values that are the lowest level representation

CPU

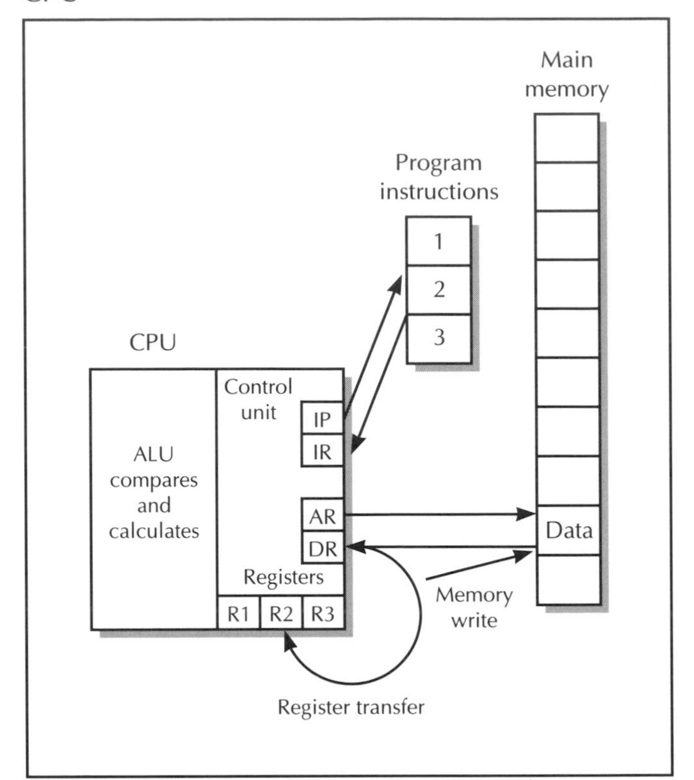

The CPU uses the Instruction Pointer (IP) to keep track of the address of the next instruction in memory, which is stored in the Instruction Register (IR). The Address Register (AR) and Data Register (DR) perform a similar function with program data. Data can also be moved between main memory and the CPU's registers, which are special fast-retrieval memory locations. Instructions are decoded by the control unit and passed to the Arithmetic Logic Unit (ALU) for execution.

of program code, sometimes called machine code. An index register is used to keep track of the current instruction. As instructions are processed, control signals can indicate special conditions, such as a result being negative. Based on the instructions and signals, the CPU can skip over some instructions, jumping to another location in the program.

The main memory or RAM (random access memory) contains both the program instructions and the data being used by the program, which in turn can be read from a disk or other medium or written back to storage. The effective speed of the system is derived not only from the clock speed but from the speed at which data travels over the system bus, a set of wires that each carry one data bit, as well as the operating speed of the memory chips themselves (see CLOCK SPEED and BUS).

The access of programs to the CPU is controlled in turn by the operating system. Modern operating systems share the CPU with several running programs, doling out execution time according to a scheduling algorithm that takes into account the possible special priority of some programs (see MULTITASKING).

Further Reading

Mano, M. Morris, and Charles R. Kime. *Logic and Computer Design Fundamentals.* 2nd ed. Upper Saddle River, N.J.: Prentice Hall, 2000.

White, Ron. *How Computers Work.* Millenium ed. Indianapolis, Ind.: Que, 1999.

Cray, Seymour

(1925–1996)
American
Computer Engineer, Inventor

Seymour Cray was a computer designer who pioneered the development of high-performance computers that came to be called supercomputers. Cray was born in Chippewa Falls, Wisconsin. After serving in World War II as an army electrical technician, Cray went to the University of Minnesota and earned a B.S. in electrical engineering and then an M.S. in applied mathematics. (This combination is a common background for many of the designers who would have to combine mathematics and engineering principles to create the first computers.)

Seymour Cray is considered by many people to be the father of the supercomputer. His innovative Cray computers looked—and performed—like something out of science fiction.

In 1951, he joined Engineering Research Associates (ERA), one of a handful of companies that sought to commercialize the digital computing technology that had been developed during and just after the war. Cray soon became known for his ability to grasp every aspect of computing from logic circuits to the infant discipline of software development. When ERA and its competitor, the Eckert-Mauchly Computer Company were bought by Remington Rand, Cray became the chief designer for the Univac, the first commercially successful computer. In 1957, however, Cray and two colleagues struck out on their own to form Control Data Corporation (CDC). Their CDC 1604 was one of the first computers to move from vacuum tubes to transistors. The CDC 6600 was considered by many to be technically superior to the IBM 360. However, by then IBM had become preeminent in the business computing market, while the CDC machines found favor with scientists.

By the late 1960s, Cray had persuaded CDC to provide him with production facilities within walking distance of his home in Chippewa Falls. There he designed the CDC 7600. This computer was hailed as the world's first supercomputer (see SUPERCOMPUTER). However CDC disagreed with Cray about the commercial feasibility of even more powerful computers. In 1972, Cray formed his own company, Cray Research, Inc. By then Cray's reputation as a computer architect was so great that investors flocked to buy stock in his company. His series of Cray supercomputers looked like sleek monoliths from a science fiction movie. The machines were the first supercomputers to use parallel processing, where tasks can be assigned to different processors to speed up throughput. While costing millions of dollars apiece, the Cray supercomputers made it possible to perform simulations in atomic physics, aerodynamics, and other fields that were far beyond the capabilities of earlier computers. However, the Cray Computer Corporation ran into financial problems and was bought by Silicon Graphics (SGI) in 1996.

Cray received many honors including the IEEE Computer Society Pioneer Award (1980) and the ACM/IEEE Eckert-Mauchly Award (1989). Cray died on October 5, 1996, in Colorado Springs, Colorado.

Further Reading

Bell, Gordon. "A Seymour Cray Perspective." http://research.microsoft.com/users/gbell/craytalk/

Breckenridge, Charles W. "A Tribute to Seymour Cray." http://www.cgl.ucsf.edu/home/tef/cray/tribute.html

Murray, C. J. *The Supermen: the Story of Seymour Cray and the Technical Wizards behind the Supercomputer.* New York: John Wiley, 1997.

Smithsonian Institute. National Museum of American History. "Seymour Cray Interview." http://americanhistory.si.edu/csr/comphist/cray.htm

cyberspace and cyber culture

The term *cyberspace* first came to prominence when it appeared in *Neuromancer,* a 1984 novel by science fiction writer William Gibson. The word is a combination of "cyber" (meaning related to computers) and "space." As another SF writer, Bruce Sterling, wrote in *The Hacker Crackdown* (1993), cyberspace is "the place between the phones. The indefinite place *out there,* where the two of you, human beings, actually meet and communicate."

While the elite telegraphers of the 19th century and later telephone users first experienced the sense of disembodied electronic communication, it took the development of widespread computer terminals, personal computers, and connecting networks to create a sense of an ongoing place in which people meet and interact. The first "villages" in cyberspace came into being during the 1970s as research networks (ARPA), and the Usenet newsgroups of UNIX users began to carry messages and news postings. During the 1980s, many more settlements began to light up the map of cyberspace, ranging from cities (large on-line services such as The Source, BIX, and CompuServe) to thousands of villages (tiny bulletin board systems running on personal computers). (See ON-LINE SERVICES and BULLETIN BOARD SYSTEMS.)

Wherever human beings build communities, they shape culture. The cyber culture that grew up in cyberspace has featured many diverse strands. Hackers (not originally a pejorative term) had their distinctive hangouts and lingo. Bulletin board cultures varied from the hacker hardcore to user groups that tried to assist beginners. On the nascent Internet multiplayer game worlds called MUDs (Multi-User Dungeons) and Muses used words to create richly detailed fantasy cyberspaces. Together with chat rooms and conferencing systems, they fostered virtual communities that, like physical communities, express a full range of human behavior (see CONFERENCING SYSTEMS, CHAT, and VIRTUAL COMMUNITY).

While cyber culture shares the characteristics of other human cultures, it also has unique characteristics that are dictated by the nature of the on-line, virtual medium. Since the on-line user reveals only what he or she chooses to reveal, identities can be fluid: playful or deceptive. While people are not physically vulnerable in cyberspace, they are certainly emotionally vulnerable. (Virtual eroticism, or "cyber sex" has even led to virtual rapes.) The issue of protecting privacy becomes important because sensitive personal information is constantly being exposed in order to carry on commerce.

THE FUTURE OF CYBERSPACE

By the end of the 1990s, the face of cyberspace was no longer that of text screens but that of the World Wide Web with its graphical pages. Multiplayer games now often feature graphics and even real-time voice communication is possible. With ubiquitous digital cameras, the boundary between cyberspace and physical space has become fluid, with people able to enter into each other's physical environments in realistic ways. Meanwhile, the development of virtual reality techniques has made computer-generated worlds much more vivid and realistic (see VIRTUAL REALITY). As more people are linked continually to the network by broadband and wireless connections, cyberspace may eventually disappear as a separate reality, having merged with physical space.

Further Reading

Dery, Mark. *Escape Velocity: Cyberculture at the End of the Century.* New York: Grove Press, 1997.

Gibson, William. *Neuromancer.* New York: Ace Books, 1984.

Hamman, Robin. "The Application of Ethnographic Methodology to the Study of Cybersex." http://www.socio.demon.co.uk/magazine/plummer.html

Lessig, Lawrence. *CODE and Other Laws of Cyberspace.* New York: Basic Books, 1999.

Negroponte, N. *Being Digital.* New York: Alfred E. Knopf, 1995.

Resource Center for Cyberculture Studies. http://otal.umd.edu/~rccs/

Turkle, Sherry. *Life on the Screen: Identity in the Age of the Internet.* New York: Simon & Schuster, 1995.

"Voice of the Shuttle: Cyberculture Page." http://vos.ucsb.edu/shuttle/cyber.html

D

data

Today the term *data* is associated in many peoples' minds mainly with computers. However, data (as in "given facts" or measurements) has been used as a term by scientists and scholars for centuries. Just as with a counting bead, a notch in a stick, or a handwritten tally, data as stored in a computer (or on digital media) is a *representation* of facts about the world. These facts might be temperature readings, customer addresses, dots in an image, the characteristics of a sound at a given instant, or any number of other things. But because computer data is not a fact but a representation of facts, its accuracy and usefulness depends not only on the accuracy of the original data, but on its *context* in the computer.

At bottom, computer data consists of binary states (represented numerically as ones or zeroes) stored using some physical characteristic such as an electrical or magnetic charge or a spot capable of absorbing or reflecting light. A string of ones and zeroes in a computer has no *inherent* meaning. Is the bit pattern 01000001 a number equivalent to 65 in the decimal system? Yes. Is it the capital letter "A"? It may be, if interpreted as an ASCII character code. Is it part of some larger number? Again, it may be, if the memory location containing this pattern is interpreted as part of a set of two, four, or more memory locations.

In order to be interpreted, data must be assigned a category such as integer, floating point (decimal), or character (see DATA TYPES). The programming language compiler uses the data type to determine how many memory locations make up that data item, and which bits in memory correspond to which bits in the actual number. Data items can be treated as a batch (see ARRAY) for convenience, or different kinds of data such as names, addresses, and Social Security numbers can be grouped together into records or structures that correspond to an entity of interest (such as a customer). In creating a structure within the program to represent the data, the programmer must be cognizant of its purpose and intended use.

The programming language and code statements define the context of data within the rules of the language. However, the *meaning* of data must ultimately be constructed by the human beings who use it. For example, whether a test score is good, bad, or indifferent is not a characteristic of the data itself, but is determined by the purposes of the test designer. This is why a distinction is often made between *data,* as raw numbers or characters, and *information* as data that has been placed in a meaningful context so that it can be useful and perhaps even enlightening to the user.

Further Reading

Bierman, Alan W. *Great Ideas in Computer Science: a Gentle Introduction.* 2nd ed. Cambridge, Mass.: MIT Press, 1997.

Hillis, Daniel W. *The Pattern on the Stone: the Simple Ideas that Make Computers Work.* New York: Basic Books, 1998.

data abstraction

Abstract data types are used to describe a "generic" type of data, specifying how the data is stored and what operations can be performed on it (see OBJECT-ORIENTED PROGRAMMING, LIST PROCESSING, STACK, and QUEUE).

For example, an abstract stack data type includes a structure for storing data (such as a list or array) and a set of operations, such as "pushing" an integer onto the stack and "popping" (removing) an integer from the stack. (For the process of combining data and operations into a single entity, see ENCAPSULATION.) Abstract data types can be implemented directly in object-oriented programming languages (see CLASS, C++, JAVA, and SMALLTALK).

One advantage of using abstract data types is that it separates a structure and functionality from its implementation. In designing the abstract stack type, for example, one can focus on what a stack does and its essential functions. One avoids becoming immediately bogged down with details, such as what sorts of data items can be placed on the stack, or exactly what mechanism will be used to keep track of the number of items currently stored. This approach also avoids "featuritis," the tendency to see how many possible functions or features one can add to the stack object. For example, while it might be useful to give a stack the ability to print out a list of its items, it is probably better to wait until one needs such a capability than to burden the basic stack idea with extra baggage that may make it more cumbersome or less efficient.

An abstract data type or its embodiment, a class, is not used directly by the program. Rather, it is used to create an entity (object) that is a particular instance of the abstract data type (for example, an actual stack that will be used to manipulate data). The data stored inside the object is not accessed directly, but through functions that the object receives from the abstract data type (such as the push and pop operations for a stack). (For more information about how such objects are used, see CLASS.)

Because the abstract data type is not directly used by the program, the implementation of how the data is stored or manipulated can be changed without affecting programs that use objects of that type. This *information hiding* is one of the chief benefits of object-oriented programming. Another advantage is inheritance, the ability to derive more specialized versions of the abstract data type or class. Thus, one can create a derived stack class that includes the printing function mentioned earlier.

Further Reading

Carrano, Frank M. and Janet L. Pritchard. *Data Abstraction and Problem Solving With Java: Walls and Mirrors.* Reading, Mass.: Addison-Wesley, 2001.

Cardelli, Luca, and Peter Wegner. "On Understanding Types, Data Abstraction, and Polymorphism." *Computing Surveys,* vol. 17, no. 4, December 1985, 471–522. Available on-line at http://research.microsoft.com/Users/luca/Papers/OnUnderstanding.pdf

"Towards the Clarification of the Object-Oriented Concepts." http://www.doc.mmu.ac.uk/STAFF/J.Gray/oopslang/aoblnts/intro/OOPCONCP.HTM

data acquisition

There are a variety of ways in which data (facts or measurements about the world) can be turned into a digital representation suitable for manipulation by a computer. For example, pressing a key on the keyboard sends a signal that is stored in a memory buffer using a value that represents the ASCII character code for the key pressed. Moving the mouse sends a stream of signals that are proportional to the rotation of the ball which in turn is calibrated into a series of coordinates and ultimately to a position on the screen where the cursor is to be moved. Digital cameras and scanners convert the varying light levels of what they "see" into a digital image.

Besides the devices that are familiar to most computer users, there are many specialized data acquisition devices (DAQs). Indeed, most instruments used in science and engineering to measure physical characteristics are now designed to convert their readings into digital form. (Sometimes the instrument includes a processor that provides a representation of the data, such as a waveform or graph. In other cases, the data is sent to a computer for processing and display.)

COMPONENTS OF A DATA ACQUISITION SYSTEM

The data acquisition system begins with a transducer, which is a device that converts a physical phenomenon (such as heat) into a proportional electrical signal. Transducers include devices such as thermistors, thermocouples, and pressure or strain gauges. The output of the transducer is then fed into a signal conditioning circuit. The purpose of signal conditioning is to make sure the signal fits into the range needed by the data processing device. Thus the signal may be amplified or its voltage may be adjusted or scaled to the required level. Another function of signal conditioning is to isolate the incoming signal from the computer to which the acquisition device is connected. This is necessary both to protect the delicate computer circuits from possible "spikes" in the incoming signal and to prevent "noise" (extraneous electromagnetic signals created by the computer itself) from distorting the signal, and thus the ultimate measurements. Various sorts of filters can be added for this purpose.

The conditioned signal is fed as an analog input into the data acquisition device, which is often a board inserted into a personal computer. The purpose of the board is to sample the signal and turn it into a stream of digital data. The digital data is stored in a buffer (either on the board or in the computer's main memory). Software then takes over, analyzing the data and creating

Data Acquisition

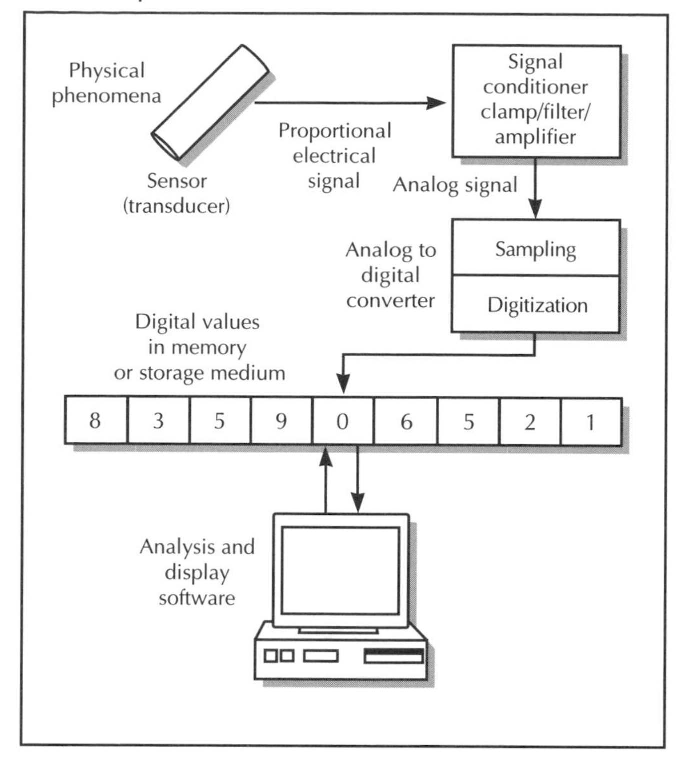

Data acquisition is the process of gathering real-time data from scientific instruments and making it available in digital form. Sensor signals are "conditioned" by filtering out extraneous values, and are then sampled and digitized. Software can now provide elaborate graphic displays as well as alert scientists to unusual readings.

appropriate displays (such as digital readings, graphs, or warning signals) as configured by the user. If the data is being displayed in real time, the speed of the software, the operating system, and the computer's clock speed may become significant (see CLOCK SPEED).

PERFORMANCE CONSIDERATIONS

The sampling rate, or the number of times the signal is measured per second, is of fundamental importance. A higher sampling rate usually means a more accurate representation of the physical data (thus audio sampled at higher rates sounds more "natural"). The faster the sampling rate, the larger the amount of data to be processed and the greater the amount of computer resources needed. Thus, picking a sampling rate usually involves a tradeoff between accuracy and speed (for a real-time application, data must be processed fast enough so that whoever is using it can respond to it as it comes in).

Three internal factors determine the performance of a DAQ. The *resolution* is the number of bits available to quantify each measurement. Clearly the ability to measure thousands of voltage levels is useless if the resolution of a system is only 8 bits (256 possible values.) The *range* is the distance between the minimum and maximum voltage levels the DAQ can recognize. If a signal must be "squeezed" into too narrow a range, a corresponding amount of resolution will be lost. Finally, there is the *gain* or the ratio between changes in the measured quantity and changes in the signal strength.

APPLICATIONS

Data acquisition systems are essential to gathering and processing the detailed data required by scientific and engineering applications. The automated control of chemical or biochemical processes requires the ability of the control software to assess real-time physical data in order to make timely adjustments to such factors as temperature, pressure, and the presence of catalysts, inhibitors, or other components of the process. The highly automated systems used in modern aviation and increasingly, even in ground vehicles, depend on real-time data acquisition. It is not surprising, then, that data acquisition is one of the fastest-growing fields in computing.

Further Reading

Beyon, Jeffrey Y. *LabVIEW Programming, Data Acquisition and Analysis.* Upper Saddle River, N.J.: Prentice Hall, 2000.

"Data Acquisition (DAQ) Fundamentals." http://www.ni.com/pdf/instrupd/appnotes/an007.pdf

James, Kevin. *PC Interfacing and Data Acquisition: Techniques for Measurement, Instrumentation and Control.* Boston: Newnes, 2000.

database administration

Database administration is the management of database systems (see DATABASE MANAGEMENT SYSTEM). Database administration can be divided into four broad areas: data security, data integrity, data accessibility, and system development.

DATA SECURITY

With regard to databases, ensuring data security includes the assignment and control of users' level of access to sensitive data and the use of monitoring tools to detect compromise, diversion, or unauthorized changes to database files (see DATA SECURITY). When data is proprietary, licensing agreements with both database vendors and content providers may also need to be enforced.

DATA INTEGRITY

Data integrity is related to data security, since the completeness and accuracy of data that has been compromised can no longer be guaranteed. However, data integrity also requires the development and testing of procedures for the entry and verification of data (input) as well as verifying the accuracy of reports (output).

Database administrators may do some programming, but generally work with the programming staff in maintaining data integrity. Since most data in computers ultimately comes from human beings, the training of operators is also important.

Within the database structure itself, the links between data fields must be maintained (referential integrity) and a locking system must be employed to ensure that a new update is not processed while a pending one is incomplete (see TRANSACTION PROCESSING).

Internal procedures and external regulations may require that a database be periodically audited for accuracy. While this may be the province of a specially trained information processing auditor, it is often added to the duties of the database administrator. (See also AUDITING IN DATA PROCESSING.)

DATA ACCESSIBILITY

Accessibility has two aspects. First, the system must be reliable. Data must be available whenever needed by the organization, and in many applications such as e-commerce, this means 24 hours a day, 7 days a week (24/7). Reliability requires making the system as robust as possible, such as by "mirroring" the database on multiple servers (which in turn requires making sure updates are stored concurrently). Failure must also be planned for, which means the implementation of onsite and offsite backups and procedures for restoring data (see BACKUP AND ARCHIVE SYSTEMS).

SYSTEM DEVELOPMENT

An enterprise database is not a static entity. The demand for new views or applications of data requires the development and testing of new queries and reports. While this is normally done by the database programmers, the administrator may need to consider its impact on the operation of the system. The administrator also helps plan for the needs of a growing, changing, organization by designing or evaluating proposals for expanding the system, possibly moving it to new hardware or a new operating system or migrating the database applications to a new database management system (DBMS).

Because of the importance of database management to corporations, government, and other organizations, database administration became a "hot" employment area in the 1990s. Most database administrators specialize in a particular database platform, such as Oracle or Microsoft Access. The growing need to make databases accessible via the Internet has added a new range of challenges to the database administrator, including the management of servers, remote authentication of users, and the mastery of Java, Common Gateway Interface (CGI), and scripting languages in order to tie the database to the server and user (see JAVA, CGI, PERL, and XML).

Further Reading

Hitchcock, Brian. *Sybase Database Administrator's Handbook.* Upper Saddle River, N.J.: Prentice Hall, 1993.

Kreines, David, Brian Laskey, and Deborah Russel, eds. *Oracle Database Administration: the Essential Reference.* Sebastopol, Calif.: O'Reilly, 1999.

Newton, Judith J., and Frankie E. Spielman, eds. *Data Administration: Standards and Techniques.* Collingdale, Penn.: DIANE Publishing, 1990.

database management system (DBMS)

A database management system consists of a database (a collection of information, usually organized into records with component fields) and facilities for adding, updating, retrieving, manipulating, and reporting on data.

DATABASE STRUCTURE

In the early days of computing, a database generally consisted of a single file that was divided into data blocks that in turn consisted of records and fields within records. The COBOL language was (and is) particularly suited to reading, processing, and writing data in such files. This *flat file* database model is still used for many simple applications including "home data managers." However, for more complex applications where there are many files containing interrelated data, the flat file model proves inadequate.

In 1970, computer scientist E. F. Codd proposed a *relational* model for data organization. In the relational model, data is not viewed as files containing records, but as a set of tables, where the columns represent fields

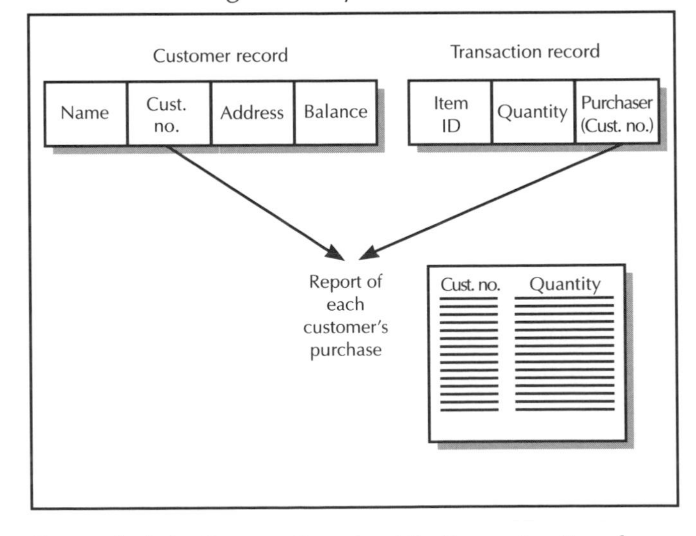

Because both the Customer Record and the Transaction Record include the Customer Number field, it is easy to pull information from both databases into a single report, such as a summary of purchases for each customer.

and the rows individual entities (such as customers or transactions).

A field (column) that two tables have in common (called the *key*) can be used to link the two. For example, consider a table of customer information (name, customer number, address, current balance, and so on) and a table of transaction information (product number, quantity, customer number of purchaser, and so on).

To find all the items purchased by a particular customer, the relational database uses the common field (the customer account number) to *join* the two tables. A query can then select all records in the transaction file whose customer number field matches the current customer in the customer file. (Notice that the validity of a key field depends on its being unique: If each customer doesn't have one [and only one] customer number, any report of purchases will not be dependable.)

A procedure called *normalization* is often used to create a set of tables from a set of data files and records, such that no fields contain duplicate information. This is necessary in order to ensure that a piece of information can be updated and the update "propagated" to the entire database without missing any instances.

Relational databases usually also enforce *referential integrity*. This means preventing changes to the database from causing inconsistencies. For example, if table A and table B are linked and a record is deleted from table A, any links to that record from records in table B must be removed. Similarly, if a change is made in a linked field in a table, records in a linked table must be updated to reflect the change.

During the 1980s, the dBase relational database program became the most popular DBMS on personal computers. Microsoft Access is now popular on Windows systems, and Oracle is prominent in the UNIX world.

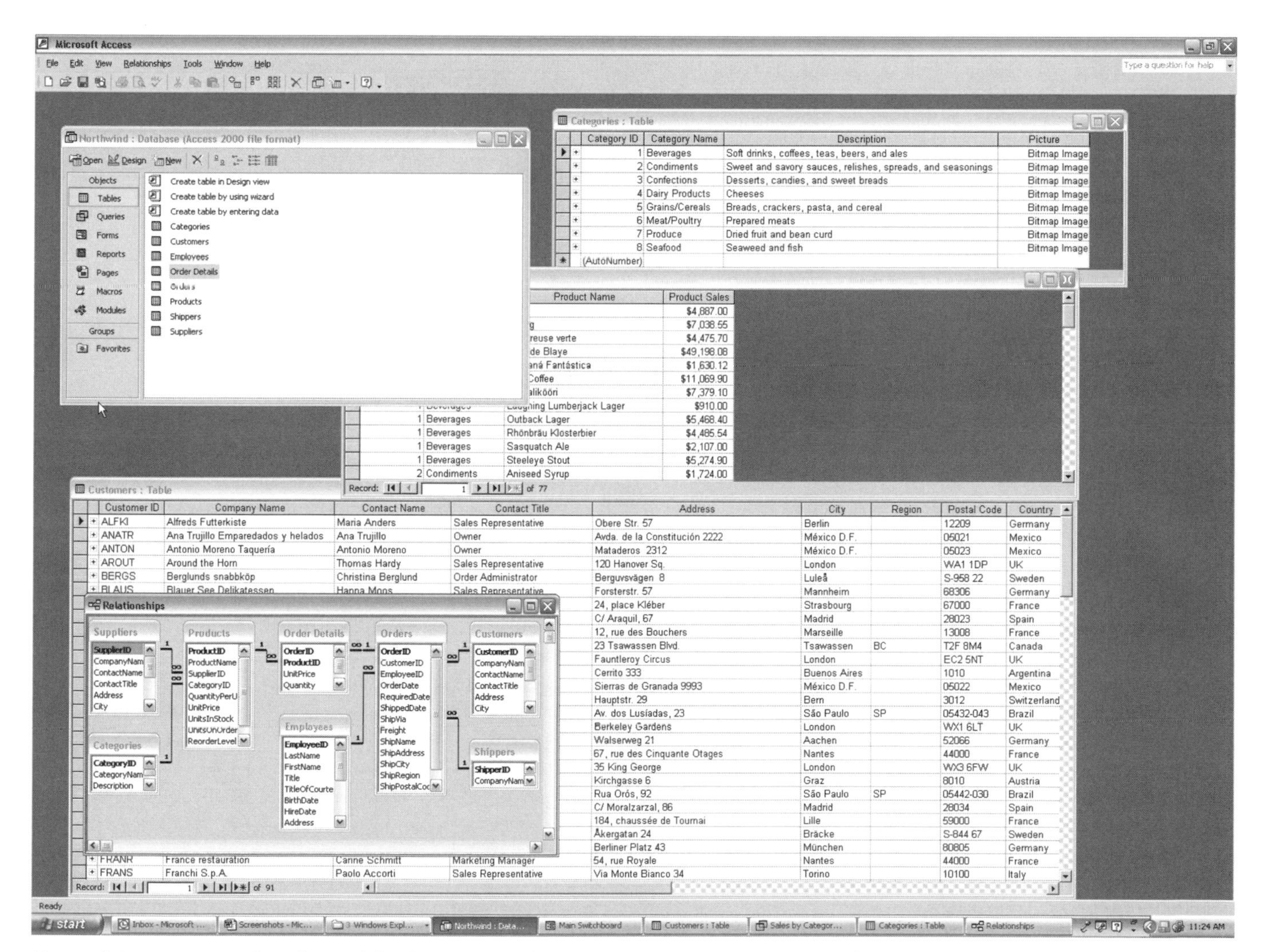

Microsoft Access is a popular relational database program for personal computers. It can be used for both simple ("flat file") databases and for complex databases with many interrelated files.

Beginning in the 1980s, SQL (Structured Query Language) became a widely used standard for querying and manipulating data tables, and most DBMS implement SQL (see SQL).

TRENDS

The embracing of object-oriented programming principles starting in the 1980s has led to development of object-oriented database structures (see OBJECT-ORIENTED PROGRAMMING). In this approach tables, queries, views, and other components of the DBMS are treated as objects that present their functionality through interfaces (much in the way a class in an object-oriented program does). This approach can improve data integrity, flexibility (such as through the ability to define new operations), and the development of new capabilities derived from predecessor objects. Object models are also helpful in dealing with a networked world in which data tables are often stored on separate computers.

Further Reading

Blaha, Michael. *Object-Oriented Modeling and Design for Database Applications.* Upper Saddle River, N.J.: Prentice-Hall, 1998.

Buyens, J. *Web Database Development: Step by Step.* Redmond, Wash.: Microsoft Press, 2000.

Harrington, Jan L. *Relational Database Design Clearly Explained.* 2nd ed. San Francisco: Morgan Kaufmann, 2002.

Hernandez, Michael J. *Database Design for Mere Mortals: A Hands-On Guide to Relational Database Design.* Reading, Mass.: Addison-Wesley, 1997.

Data Communications

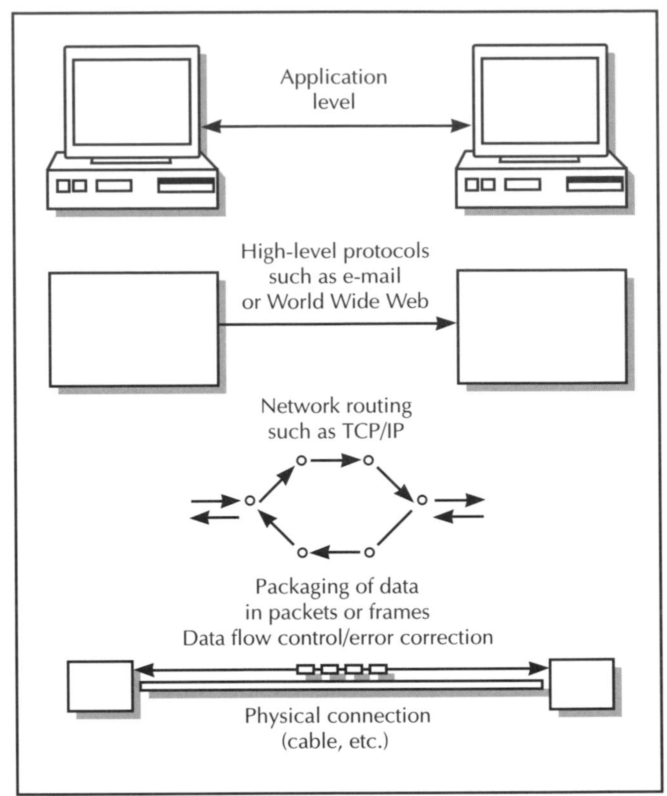

Modern data communications can be thought of as a series of layers, from the actual physical connection (such as a cable) at the "bottom" to the operations of software such as Web browsers or e-mail programs at the highest level.

data communications

Broadly speaking, data communications is the transfer of data between computers and their users. At its most abstract level, data communications requires two or more computers, a device to turn data into electronic signals (and back again), and a transmission medium. While telephone lines remain the most common medium for carrying data transmissions, fiber optic cable, network (Ethernet) cable, video cable, radio (wireless), or other kinds of links can be used. Finally, there must be software that can manage the flow of data.

Until recently, the modem was the main device used to connect personal computers to information services or networks (see MODEM). In general, data being sent over a communications link must be sent one bit at a time (this is called *serial* transmission, and is why an external modem is connected to a computer's serial port). However most phone cables and other links are *multiplexed,* meaning that they carry many channels (with many streams of data bits) at the same time.

To properly recognize data in a bit stream coming over a link, the transmission system must use some method of flow control and have some way to detect errors (see ERROR CORRECTION). Typically, the data is sent as groups or "frames" of bits. The frame includes a checksum that is verified by the receiver. If the expected and actual sums don't match, the recipient sends a "negative acknowledgment" message to the sender, which will retransmit the data. In the original system, the sender waited until the recipient acknowledged each frame before sending the next, but modern protocols allow the sender to keep sending while the frames being received are waiting to be checked.

The actual transmission of data over a line can be considered to be the lowest level of the data communications scheme. Above that is packaging of data as used and interpreted by software. Unless two computers are directly connected, the data is sent over a network, either a local area network (LAN) or a wide-area network such as the global Internet. A network consists of interconnected *nodes* that include switches or routers that direct data to its destination (see NETWORK). Networks such as the Internet use packet-switching: Data is sent as individual packets that contain a "chunk" of data, an address, and an indication of where the data fits within the message as a whole. The packets are routed at the routers using software that tries to find the fastest link to the destination. When the packets arrive at the destination, they are reassembled into the original message.

APPLICATIONS

Data communications are the basis both for networks and for the proper functioning of servers that provide services such as World Wide Web pages, electronic mail, on-line databases, and multimedia content (such as audio and streaming video). While webpage design and e-commerce are the "bright lights" that give cyberspace its character, data communications are like the plumbing without which computers cannot work together. The growing demand for data communications, particularly broadband services such as DSL and cable modems, translates into a steady demand for engineers and technicians specializing in the maintenance and growth of this infrastructure (see BROADBAND).

Besides keeping up with the exploding demand for more and faster data communications, the biggest challenge for data communications in the early 21st century is the integration of so many disparate methods of communications. A user may be using an ordinary phone line (19th-century technology) to connect to the Internet, while the phone company switches might be a mixture of 1970s or later technology. The same user might go to the workplace and use fast Ethernet cables over a local network, or connect to the Internet through DSL, an enhanced phone line. Traveling home, the user might use a personal digital assistant (PDA) with a wireless link to make a restaurant reservation (see WIRELESS COMPUTING). The user wants all these services to be seamless and essentially interchangeable, but today data communications is more like roads in the early days of the automobile—a few fast paved roads here and there, but many bumpy dirt paths.

Further Reading

Brown, Brian. "Introduction to Data Communications." [on-line interactive course] http://www.cit.ac.nz/smac/dc100www/dc_000.htm

Forouzan, Behrouz. *Data Communications and Networking.* New York: McGraw Hill, 2001.

Freeman, Roger R. *Practical Data Communications.* 2nd ed. New York: Wiley, 2001.

data compression

The process of removing redundant information from data so that it takes up less space is called data compression. Besides saving disk space, compressing data such as e-mail attachments can make data communications faster.

Compression methods generally begin with the realization that not all characters are found in equal numbers in text. For example, in English, letters such as *e* and *s* are found much more frequently than letters such as *j* or *x*. By assigning the shortest bit codes to the most common characters and the longer codes to the least common characters, the number of bits needed to encode the text can be minimized.

Huffman coding, first developed in 1952, is an ALGORITHM that uses a tree in which the pairs of the least probable (that is, least common) characters are linked, the next least probable linked, and so on until the tree is complete.

Another coding method, arithmetic coding, matches characters' probabilities to bits in such a way that the same bit can represent parts of more than one encoded character. This is even more efficient than Huffman coding, but the necessary calculations make the method somewhat slower to use.

Another approach to compression is to look for words (or more generally, character strings) that match those found in a dictionary file. The matching strings are replaced by numbers. Since a number is much shorter than a whole word or phrase, this compression method can greatly reduce the size of most text files. (It would not be suitable for files that contain numerical rather than text data, since such data, when interpreted as characters, would look like a random jumble.)

The Lempel-Ziv (LZ) compression method does not use an external dictionary. Instead, it scans the file itself for text strings. Whenever it finds a string that occurred earlier in the text, it replaces the later occurrences with an offset, or count of the number of bytes separating the occurrences. This means that not only common words

Data Compression

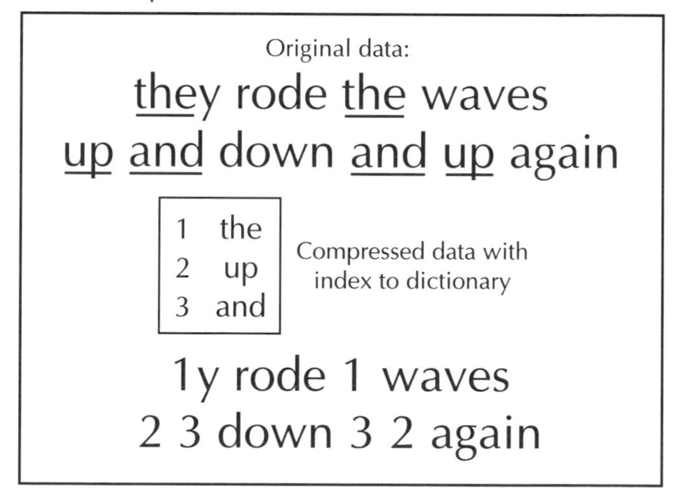

A basic approach to data compression is to look for recurring patterns and store them in a "dictionary." Each occurrence of the pattern can then be replaced by a brief reference to the dictionary entry. The resulting file may then be considerably smaller than the original.

but common prefixes and suffixes can be replaced by numbers. A variant of this scheme does not use offsets to the file itself, but compiles repeated strings into a dictionary and replaces them in the text with an index to their position in the dictionary.

Graphics files can often be greatly compressed by replacing large areas that represent the same color (such as a blue sky) with a number indicating the count of pixels with that value. However, some graphics file formats such as GIF are already compressed, so further compression will not shrink them much.

More exotic compression schemes for graphics can use fractals or other iterative mathematical functions to encode patterns in the data. Most such schemes are "lossy" in that some of the information (and thus image texture) is lost, but the loss may be acceptable for a given application. Lossy compression schemes are not used for binary (numeric data or program code) files because errors introduced in a program file are likely to affect the program's performance (if not "break" it completely). Though they may have less serious consequences, errors in text are also generally considered unacceptable.

TRENDS

There are a variety of compression programs used on UNIX systems, but variants of the Zip program are now the overwhelming favorite on Windows-based systems. Zip combines compression and archiving. Archiving, or the bundling together of many files into a single file, contributes a further reduction in file size. This is because files in most file systems must use a whole number of disk sectors, even if that means wasting most of a sector. Combining files into one file means that at most a bit less than one sector will be wasted.

Further Reading

"Compression Pointers." http://www.internz.com/compression-pointers.html

"Compression Reference Center." http://www.rasip.fer.hr/research/compress/index.html

Saloman, David. *Data Compression: the Complete Reference.* New York: Springer-Verlag, 2000.

Sayood, Khalid. *Introduction to Data Compression.* 2nd ed. San Francisco: Morgan Kaufmann, 2000.

data conversion

The developer of each application program that writes data files must define a format for the data. The format must be able to preserve all the features that are supported by the program. For example, a word processing program will include special codes for font selection, typestyles (such as bold or italic), margin settings, and so on.

In most markets there are more than one vendor, so there is the potential for users to encounter the need to convert files such as word processing documents from one vendor's format to another. For example, a Microsoft Word user needing to send a document to a user who has WordPerfect, or the user may encounter another user who also has Microsoft Word, but a later version.

There are some ways in which vendors can relieve some of their users' file conversion issues (and thus potential customer dissatisfaction). Vendors often include facilities to read files created by their major rivals' products, and to save files back into those formats. This enables users to exchange files. Sometimes the converted document will look exactly like the original, but in some cases there is no equivalence between a feature (and thus a code) in one application and a feature in the other application. In that case the formatting or other feature may not carry over into the converted version, or may be only partially successful.

Vendors generally make a new version of an application *downwardly compatible* with previous versions (see also COMPATABILITY AND PORTABILITY). This means that the new version can read files created with the earlier versions. (After all, users would not be happy if none of their existing documents were accessible to their new software!) Similarly, there is usually a way to save a file from the later version in the format of an earlier version, though features added in the later version will not be available in the earlier format.

Another strategy for exchanging otherwise incompatible files is to find some third format that both applications can read. Thus Rich Text Format (RTF), a format that includes most generic document features, is supported by most modern word processors. A user can thus export a file as RTF and the user of a different program will be able to read it. Similarly, many database and other programs can export files as a series of data values separated by commas (comma-delimited files), and the files can be then read by a different program and converted to its "native" format.

A variety of format conversion utilities are available as either commercial software or shareware. There are also businesses that specialize in data conversion. While their services can be expensive, using them may be the best way to convert large numbers of files, rather than having to individually load and save them. Data conversion services can also handle many "ancient" data files from the 1970s or even early 1980s whose formats are no longer supported by current software.

Further Reading

"DATACONV: Data Conversion Tools." http://home.snafu.de/webo/data_conv.html

data dictionary

A modern enterprise database system can contain hundreds of separate data items, each with important characteristics such as field types and lengths, rules for validating the data, and links to various databases that use that item (see DATABASE MANAGEMENT SYSTEM). There can also be many different *views* or ways of organizing subsets of the data, and stored procedures (program code modules) used to perform various data processing functions. A developer who is creating or modifying applications that deal with such a vast database will often need to check on the relationships between data elements, views, procedures, and other aspects of the system.

One fortunate characteristic of computer science is that many tools can be applied to themselves, often because the contents of a program is itself a collection of data. Thus, it is possible to create a database that keeps track of the elements of another database. Such a database is sometimes called a data dictionary. A data dictionary system can be developed in the same way as any other database, but many database development systems now contain built-in facilities for generating data dictionary entries as new data items are defined, and updating definitions as items are linked together and new views or stored procedures are defined. (A similar approach can be seen in some software development systems that create a database of objects defined within programs, in order to preserve information that can be useful during debugging.)

Data dictionaries are particularly important for creating data warehouses (see DATA WAREHOUSE), which are large collections of data items that are stored together with the procedures for manipulating and analyzing them.

Further Reading

Herbst, Maida Reavis. *12 Weeks to a Successful Data Dictionary.* Marblehead, Mass.: Opus Communications, 1997.

Wertz, Charles J. *The Data Dictionary: Concepts and Uses.* 2nd ed. New York: Wiley, 1993.

data mining

The process of analyzing existing databases in order to find useful information is called data mining. Generally, a database, whether scientific or commercial, is designed for a particular purpose, such as recording scientific observations or keeping track of customers' account histories. However, data often has potential applications beyond those conceived by its collector.

Conceptually, data mining involves a process of refining data to extract meaningful patterns—usually with some new purpose in mind. First, a promising set or subset of the data is selected or sampled. Particular fields (variables) of interest are identified. Patterns are found using techniques such as regression analysis to find variables that are highly correlated to (or predicted by) other variables, or through clustering (finding the data records that are the most similar along the selected dimensions). Once the "refined" data is extracted, a representation or visualization (such as a report or graph) is used to express newly discovered information in a usable form.

Similar (if simpler) techniques are being used to target or personalize marketing, particularly to on-line customers. For example, on-line bookstores such as Amazon.com can find what other books have been most commonly bought by people buying a particular title. (In other words, identify a sort of reader profile.) If a new customer searches for that title, the list of correlated titles can be displayed, with an increased likelihood of triggering additional purchases. Businesses can also create customer profiles based on their longer-term purchasing patterns, and then either use them for targeted mailings or sell them to other businesses (see E-COMMERCE). In scientific applications, observations can be "mined" for clues to phenomena not directly related to the original observation. For example, changes in remote sensor data might be used to track the effects of climate or weather changes.

TRENDS

Data mining of consumer-related information has emerged as an important application as the volume of e-commerce continues to grow, the amount of data generated by large systems (such as on-line bookstores and auction sites) increases, and the value of such information to marketers becomes established. However, as with many aspects of e-commerce, the long-term viability of business models based on data mining operations remains uncertain. Also, the use of consumer data for purposes unrelated to the original purchase, often by companies that have no pre-existing business relationship to the consumer, can raise privacy issues. (Data is often rendered anonymous by removing personal identification information before it is mined, but regulations or other ways to assure privacy remain incomplete and uncertain.)

Further Reading

Han, Jiawei, and Micheline Kamber. *Data Mining: Concepts and Techniques.* San Francisco: Morgan Kaufmann, 2000.

"KD Nuggets: Data Mining, Web Mining, Knowledge Discovery, and CRM Guide." http://www.kdnuggets.com/

Stilwell, Markus. "A Report on Data Mining." http://www.cs.usask.ca/homepages/grads/mgs310/Cmpt826S3/node5.html

data security

In most institutional computing environments, access to program and data files is restricted to authorized persons. There are several mechanisms for restricting file access in a multiuser or networked system.

USER STATUS

Because of their differing responsibilities, users are often given differing restrictions on access. For example, there might be status levels ranging from root to administrator to "ordinary." A user with root status on a UNIX system is able to access any file or resource. Any program run by such a user inherits that status, and thus can access any resource. Generally, only the user(s) with ultimate responsibility for the technical functioning of the system should be given such access, because commands used by root users have the potential to wipe out all data on the system. A person with administrator status may be able to access the files of other users and to access certain system files (in order to change configurations), but will not be able to access certain core system files. Ordinary users typically have access only to the files they create themselves and to files designated as "public" by other users.

FILE PERMISSIONS

Files themselves can have permission status. In UNIX, there are separate statuses for the user, any group to which the user belongs, and "others." There are also three different activities that can be allowed or disallowed: reading, writing, and executing. For example, if a file's permissions are

User	Group	Other
rwx	rw-	r—

the user can read or write the file or (if it is a directory or program), execute it. Members of the same group can read or write, but not execute, while others can only read the file without being able to change it in any way. Operating systems such as Windows NT use a somewhat different structure and terminology, but also provide for varying user status and access to objects.

RECORD-LEVEL SECURITY

Security on the basis of whole directories or even files may be too "coarse" for many applications. In a particular database file, different users may be given access to different data fields. For example, a clerk may have read-only access to an employee's basic identification information, but not to the results of performance evaluations. An administrator may have both read and write access to the latter. Using some combination of database management and operating system level capabilities, the system will maintain lists of user accounts together with the objects (such as record types or fields) they can access, and the types of access (read only or read/write) that are permitted. Rather than assigning access capabilities separately for each user, they may be defined for a group of similar users, and then individual users can be assigned to the group.

OTHER SECURITY MEASURES

Security is also important at the program level. Because a badly written (or malicious) program might destroy important data or system files, most modern operating systems restrict programs in a number of ways. Generally, each program is allowed to access only such memory as it allocates itself, and is not able to change data in memory belonging to other running programs. Access to hardware devices can also be restricted: an operating system component may have the ability to access the innermost core of the operating system (where drivers interact directly with devices), while an ordinary applications program may be able to access devices only through facilities provided by the operating system.

There are a number of techniques that unauthorized intruders can use to try to compromise operating systems (see COMPUTER CRIME AND SECURITY). Access capabilities that are tied to user status are vulnerable if the user can get the login ID and password for the account. If the account has a high (administrator or root) status, then the intruder may be able to give viruses, Trojan horses, or other malicious programs the status they need in order to be able to penetrate the defenses of the operating system (see also COMPUTER VIRUS).

Files that have intrinsically sensitive or valuable data are often further protected by encoding them (see ENCRYPTION). Encryption means that even intruders who gain

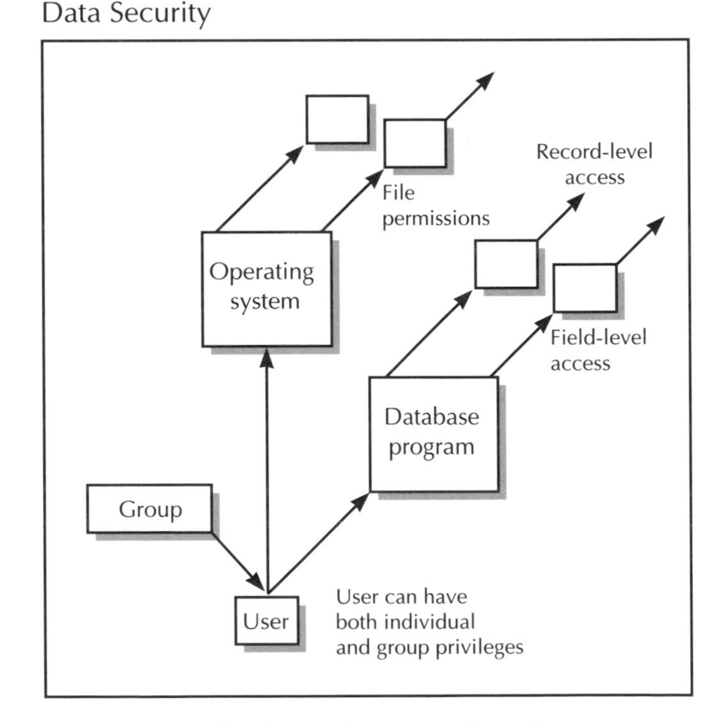

Data security involves layers of protection. Generally, users must provide an ID and password to access their accounts. The account itself may be assigned different levels of access to the system (such as user, supervisor, or superuser). At the operating system level, files and directories can have their own access permissions. A program such as a database can also control access to files, records, or even individual fields in a record.

read access to the file will need either to crack the encryption (very difficult without considerable time and computer resources) or somehow obtain the key. Encryption does not prevent the deletion or copying of a file, however, just the understanding of its contents.

There is a continuing tradeoff between security and ease of use. From the security standpoint, it might be assumed that the more barriers or checkpoints that can be set up for verifying authorization, the safer the system will be. However, as security systems become more complex, it becomes more difficult to ensure that authorized users are not unduly inconvenienced. If users are sufficiently frustrated, they will be tempted to try to bypass security, such as by sharing IDs and passwords or making files they create "public."

Further Reading

Frost, Jim. "Windows NT Security." http://world.std.com/~jimf/papers/nt-security/nt-security.html

Garfinkel, Simpson, and Gene Spafford. *Practical UNIX and Internet Security.* Sebastopol, Calif.: O'Reilly, 1996.

data structures

A data structure is a way of organizing data for use in a computer program. There are three basic components to a data structure: a set of suitable basic data types, a way to organize or relate these data items to one another, and a set of operations, or ways to manipulate the data.

For example, the ARRAY is a data structure that can consist of just about any of the basic data types, although all data must be of the same type. The way the data is organized is by storing it in sequentially addressable locations. The operations include storing a data item (element) in the array and retrieving a data item from the array.

TYPES OF DATA STRUCTURES

The data structures commonly used in computer science include arrays (as discussed above) and various types of lists. The primary difference between an array and a list is that an array has no internal links between its elements, while a list has one or more pointers that link the elements. There are several types of specialized list. A tree is a list that has a root (an element with no predecessor), and each other element has a unique predecessor. The guarantee of a unique path to each tree node can make the operations of inserting or deleting an item faster. A STACK is a list that is accessible only at the top (or front). Any new item is inserted ("pushed") on top of the last item, and removing ("popping") an item always removes the item that was last inserted. This order of access is called LIFO (last in, first out). A list can also be organized in a first in, first out (FIFO) order. This type of list is called a QUEUE, and is useful in a situation where tasks must "wait their turn" for attention.

Data Structure

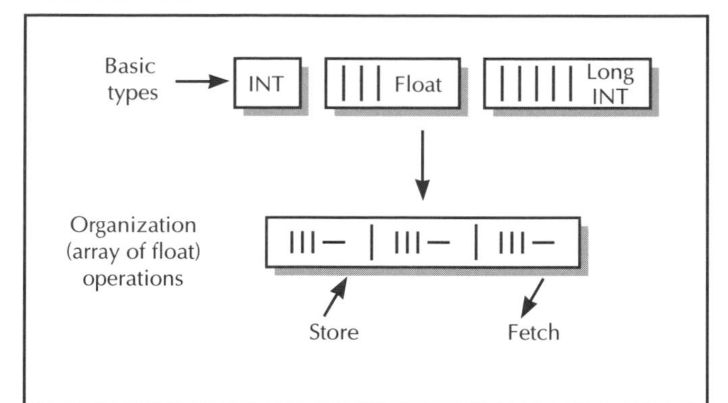

Int (integer), float (floating point), and long int (long integer) are all basic numeric data types. These types can be organized into more complex types, such as arrays. Operations can then be defined, such as storing or fetching elements from the array.

IMPLEMENTATION ISSUES

The implementation of any data structure depends on the syntax of the programming language to be used, the data types and features available in the language, and the algorithms chosen for the data operations that manipulate the structure. In traditional procedural languages such as C, the data storage part of a data structure is often specified in one part of the program, and the functions that operate on that structure are defined separately. (There is no mechanism in the language to link them.) In object-oriented languages such as C++, however, both the data storage declarations and the function declarations are part of the same entity, a CLASS. This means that the designer of the data structure has complete control over its implementation and use.

Together with algorithms, data structures make up the heart of computer science. While there can be numerous variations on the fundamental data structures, understanding the basic forms and being able to decide which one to use to implement a given algorithm is the best way to assure effective program design.

Further Reading

Horowitz, E., and S. Sahni. *Fundamentals of Data Structures in Pasical.* 2nd ed. Rockville, Md.: Computer Science Press, 1987.

Weiss, M. A. *Algorithms, Data Structures, and Problem Solving with C++.* Reading, Mass.: Addison-Wesley, 1997.

Wirth, N. *Algorithms + Data Structures = Programs.* Upper Saddle River, N.J.: Prentice Hall, 1976.

data types

As far as the circuitry of a computer is concerned, there's only one kind of data—a series of bits (binary digits) filling a series of memory locations. How those bits are to be interpreted by the people using the computer is entirely arbitrary. The purpose of data types is to define useful concepts such as integer, floating-point number, or character in terms of how they are stored in computer memory.

Thus, most computer languages have a data type called integer, which represents a whole number that can be stored in 16 bits (two bytes) of memory. When a programmer writes a declaration such as:

```
int Counter;
```

in the C language, the compiler will create machine instructions that set aside two bytes of memory to hold the contents of the variable Counter. If a later statement says:

```
Counter = Counter + 1;
```

(or its equivalent, Counter++) the program's instructions are set up to fetch two bytes of memory to the processor's accumulator, add 1, and store the result back into the two memory bytes.

Similarly, the data type long represents four bytes (32 bits) worth of binary digits, while the data type float stores a floating-point number that can have a whole part and a decimal fraction part (see NUMERIC DATA). The char (character) type typically uses only a single byte (8 bits), which is enough to hold the basic ASCII character codes up to 255 (see CHARACTERS AND STRINGS).

The Bool (Boolean) data type represents a simple true or false (usually 1 or 0) value (see BOOLEAN OPERATORS).

STRUCTURED DATA TYPES

The preceding data types all hold single values. However, most modern languages allow for the construction of data types that can hold more than one piece of data. The ARRAY is the most basic structured data type; it represents a series of memory locations that hold data of one of the basic types. Thus, in Pascal an array of integer holds integers, each taking up two bytes of memory.

Many languages have composite data types that can hold data of several different basic types. For example, the struct in C or the record in Pascal can hold data such as a person's first and last name, three lines of address (all arrays of characters, or strings), an employee number (perhaps an integer or double), a Boolean field representing the presence or absence of some status, and so on. This kind of data type is also called a user-defined data type because programmers can define and use these types in almost the same ways as they use the language's built-in basic types.

What is the difference between data types and data structures? There is no hard-and-fast distinction. Generally, data structures such as lists, stacks, queues, and trees are more complex than simple data types, because they include data relationships and special functions (such as pushing or popping data on a stack). However, a list is the fundamental data type in list-processing languages such as Lisp, and string operators are built into languages such as Snobol. (See LIST PROCESSING, STACK, QUEUE, and TREE.)

Further Reading

Aho, Alfred V., John E. Hopcroft, and Jeffrey Ullman. *Data Structures and Algorithms.* Reading, Mass.: Addison Wesley, 1983.

Carrano, Frank, and Janet J. Pritchard. *Data Abstraction and Problem Solving with C++: Walls and Mirrors.* 3rd ed. Reading, Mass.: Addison-Wesley, 2001.

Waite, Mitchell, and Robert Lafore. *Data Structures & Algorithms in Java.* Corte Madera, Calif.: Waite Group Press, 1998.

Watt, David A., and Deryck F. Brown. *Java Collections: an Introduction to Abstract Data Types, Data Structures, and Algorithms.* New York: Wiley, 2001.

Wirth, Niklaus. *Data Structures + Algorithms = Programs.* Englewood Cliffs, N.J.: Prentice Hall, 1976.

data warehouse

Modern business organizations create and store a tremendous amount of data in the form of transactions that become database records. Increasingly, however, businesses are relying on their ability to use data that was collected for one purpose (such as sales, customer service, and inventory) for purposes of research, planning, or decision support. For example, transaction data might be revisited with a view to identifying the common characteristics of the firm's best customers or determining the best way to market a particular type of product. In order to conduct such research or analysis, the data collected in the course of business must be stored in such a way that it is both accurate and flexible in terms of the number of different ways in which it can be queried. The idea of the data warehouse is to provide such a repository for data.

When data is used for particular purposes such as sales or inventory control, it is usually structured in records where certain fields (such as stock number or quantity) are routinely processed. It is not so easy to ask a different question such as "which customers who bought this product from us also bought this other product within six months of their first purchase?" One way to make it easier to query data in new ways is to store the data not in records but in arrays where, for example, one dimension might be product numbers and another categories of customers. This approach, called On-line Analytical Processing (OLAP) makes it possible to extract a large variety of relationships without being limited by the original record structure.

IMPLEMENTATION

The key in designing a data warehouse is to provide a way that researchers using analytical tools (such as statis-

Data Warehouse

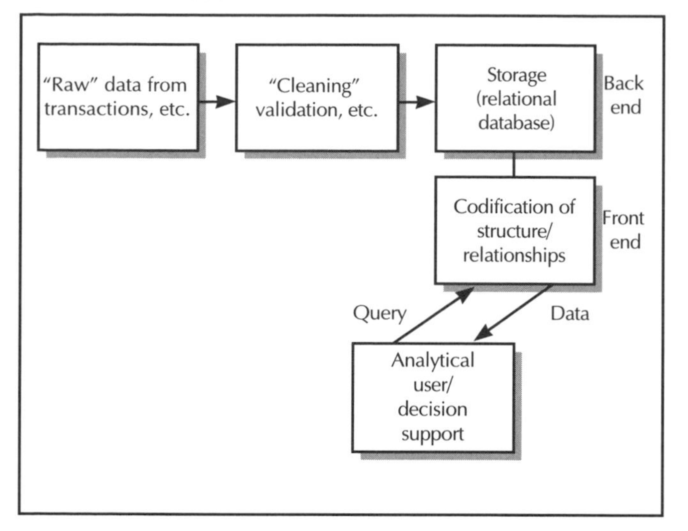

The general process of warehousing data. The data warehouse adds value to the data by further structuring it so relationships can be explored by analysts.

tics programs) can access the raw data in the underlying database. Software using query languages such as SQL can serve as such a link. Thus, the researcher can define a query using the many dimensions of the data array, and the OLAP software (also called *middleware*) translates this query into the appropriate combination of queries against the underlying relational database.

The data warehouse is closely related to the concept of data mining. In fact, data mining can be viewed as the exploitation of the collection of views, queries, and other elements that can be generated using the data warehouse as the infrastructure (see DATA MINING).

Further Reading

Kimball, R. *The Data Warehouse Lifecycle Toolkit: Expert Methods for Designing, Developing, and Deploying Data Warehouses.* New York: John Wiley, 1998.

Orr, Ken. "Data Warehousing Technology." Ken Orr Institute, 2000. http://www.kenorrinst.com/dwpaper.html

demon

The unusual computing term *demon* (sometimes spelled daemon) refers to a process (program) that runs in the background, checking for and responding to certain events. The utility of this concept is that it allows for automation of information processing without requiring that an operator initiate or manage the process.

For example, a print spooler demon looks for jobs that are queued for printing, and deals with the negotiations necessary to maintain the flow of data to that device. Another demon (called chron in UNIX systems) reads a file describing processes that are designated to run at particular dates or times. For example, it may launch a backup utility every morning at 1:00 A.M. E-mail also depends on the periodic operation of "mailer demons."

While the term *demon* originated in the UNIX culture, similar facilities exist in many operating systems. Even in the relatively primitive MS-DOS for IBM personal computers of the 1980s, the ability to load and retain small utility programs that could share the main memory with the currently running application allowed for a sort of demon that could spool output or await a special keypress. Microsoft Windows systems have many demon-like operating system components that can be glimpsed by pressing the Ctrl-Alt-Delete key combination.

The sense of autonomy implied in the term *demon* is in some ways similar to that found in *bots* or *software agents* that can automatically retrieve information on the Internet, or in the *Web crawler,* which relentlessly pursues, records, and indexes Web links for search engines.

Further Reading

Brock, Dean, and Bob Benites. *Mastering Tools, Taming Daemons: UNIX for the Wizard Apprentice.* Greenwich, Conn.: Manning Publications, 1995.

Stevens, W. Richard. *Advanced Programming in the UNIX Environment.* Reading, Mass.: Addison-Wesley, 1992.

"UNIX Daemons in Perl." http://www.webreference.com/perl/tutorial/9/

Dertouzos, Michael L.

(1936–2001)
Greek-American
Computer Scientist, Futurist

Born in Athens, Greece, on November 5, 1936, Michael Dertouzos spent adventurous boyhood years accompanying his father (an admiral) in the Greek navy's destroyers and submarines. He became interested in Morse Code, shipboard machinery, and mathematics. At the age of 16 he read an article about Claude Shannon's work in information theory and a project at the Massachusetts Institute of Technology that sought to build a mechanical robot "mouse." He quickly decided that he wanted to come to America to study at MIT.

After the hardships of the World War II years intervened, Dertouzos received a Fulbright scholarship that placed him in the University of Arkansas, where he earned his bachelor's and master's degrees while working on acoustic-mechanical devices for the Baldwin Piano Company. He was then able to fulfill his boyhood dream by receiving his Ph.D. from MIT, then promptly joined the faculty. He was director of MIT's Laboratory for Computer Science (LCS) starting in 1974. The lab has been a hotbed of new ideas in computing, including computer time-sharing, Ethernet networking, and public-key cryptography. Dertouzos also embraced the growing Internet and serves as coordinator of the World Wide Web consortium, a group that seeks to create standards and plans for the growth of the network.

Combining theoretical interest with an entrepreneur's eye on market trends, Dertouzos started a small company called Computek in 1968. It made some of the first "smart terminals" that included their own processors.

In the 1980s, Dertouzos began to explore the relationship between developments and infrastructure in information processing and the emerging "information marketplace." However, the spectacular growth of the information industry has taken place against a backdrop of the decline of American manufacturing. Dertouzos's 1989 book, *Made In America,* suggested ways to revitalize American industry.

During the 1990s, Dertouzos brought MIT into closer relationship with the visionary designers who were creating and expanding the World Wide Web. When Tim Berners-Lee and other Web pioneers were struggling to create the World Wide Web consortium to guide the future of the new technology, Dertouzos provided extensive guidance to help them set their agenda and structure. (See WORLD WIDE WEB and BERNERS-LEE, TIM.)

Dertouzos was dissatisfied with operating systems such as Microsoft Windows and with popular applications programs. He believed that their designers made it unnecessarily difficult for users to perform tasks, and spent more time on adding fancy features than on improving the basic usability of their products. In 1999, Dertouzos and the MIT LCS announced a new project called Oxygen. Working in collaboration with the MIT Artificial Intelligence Laboratory, Oxygen was intended to make computers "as natural a part of our environment as the air we breathe."

As a futurist, Dertouzos tried to paint vivid pictures of possible future uses of computers in order to engage the general public in thinking about the potential of emerging technologies. His 1995 book, *What Will Be,* paints a vivid portrait of a near-future pervasively digital environment. His imaginative future is based on actual MIT research, such as the design of a "body net," a kind of wearable computer and sensor system that would allow people to not only keep in touch with information but also to communicate detailed information with other people similarly equipped. This digital world will also include "smart rooms" and a variety of robot assistants, particularly in the area of health care. However, this and his 2001 publication, *The Unfinished Revolution,* are not unalloyed celebrations of technological wizardry. Dertouzos has pointed out that there is a disconnect between technological visionaries who lack understanding of the daily realities of most peoples' lives, and humanists who do not understand the intricate interconnectedness (and thus social impact) of new technologies.

Dertouzos was given an IEEE Fellowship and awarded membership in the National Academy of Engineering, He died on August 27, 2001, after a long bout with heart disease. He was buried in Athens near the finish line for the Olympic marathon.

Further Reading

Dertouzos, Michael L. *What Will Be: How the New World of Information Will Change Our Lives.* San Francisco: Harper, 1997.

———. *The Unfinished Revolution: Human-Centered Computers and What They Can Do For Us.* New York, N.Y.: HarperBusiness, 2001.

Schwartz, John. "Michael L. Dertouzos, 64, Computer Visionary, Dies." *New York Times,* August 30, 2001, B9.

desktop publishing

Traditionally documents such as advertisements, brochures, and reports were prepared by combining typed or printed text with pasted-in illustrations (such as photographs and diagrams). This painstaking layout process was necessary in order to produce "camera-ready copy" from which a printing company could produce the final product.

Starting in the late 1980s, desktop computers became powerful enough to run software that could be used to create page layouts. In addition, display hardware gained a high enough resolution to allow for pages to be shown on the screen in much the same form as they would appear on the printed page. (This is known by the acronym WYSIWYG, or "what you see is what you get."—see WYSIWYG.) The final ingredient for the creation of desktop publishing was the advent of affordable laser or inkjet printers that could print near print quality text and high-resolution graphics (see PRINTERS).

This combination of technologies made it feasible for trained office personnel to create, design, and produce many documents in-house rather than having to send copy to a printing company. Adobe's PageMaker program soon became a standard for the desktop publishing industry, appearing first on the Apple Macintosh and later on systems running Microsoft Windows. (The Macintosh's support for fonts and WYSIWYG displays gave it a head start over the Windows PC in the DTP industry, and to this day many professionals prefer it.)

There is no hard-and-fast line between desktop publishing and the creation of text itself. Modern word processing software such as Microsoft Word includes a variety of features for selection and sizing of fonts, and the ability to define styles for creating headings, types of paragraphs, and so on (see WORD PROCESSING). Word and other programs also allow for the insertion and placement of graphics and tables, the division of text into columns, and other layout features. In general, however, word processing emphasizes the creation of text (often for long documents), while desktop publishing software emphasizes layout considerations and the fine-tuning of a document's appearance. Thus, while a word processor might allow the selection of a font in a given point size, a desktop publishing program allows for the exact specification of leading (space between lines) and kerning (the adjustment of space between characters). Most desktop publishing programs can import text that was originally created in a word processor. This is helpful because using desktop publishing software to create the original text can be tedious.

Desktop publishing is generally used for short documents such as ads, brochures, and reports. Material to be published as a book or magazine article is normally submitted by the author as a word processing document. The publisher's production staff then creates a print-ready version. Books and other long documents are generally produced using in-house computer typesetting facilities.

Today desktop publishing is part of a range of technologies used for the production of documents and presentations. Document designers also use drawing programs (such as Corel Draw) and photo manipulation programs (such as Adobe Photoshop) in preparing illustrations. Further, the growing use of the Web means that many documents must be displayable on webpages as well as in print. Adobe's Portable Document Format (PDF) is one popular way of creating files that exactly portray printed text.

Further Reading

Blattner, D., and N. Davis, eds. *The QuarkXPress Book: For Macintosh and Windows.* Berkeley, Calif.: Peachpit Press, 1998.

"Desktop Publishing News." http://desktoppublishing.com/news1.html

Parker, Roger C. *Web Design and Desktop Publishing for Dummies.* 2nd ed. New York: Hungry Minds, 1997.

"Resources for Desktop Publishers." http://www.nlightning.com/dtpsbiblio.html

Shushan, R. and D. Wright, with L. Lewis. *Desktop Publishing by Design: Everyone's Guide to PageMaker6.* 4th ed. Redmond, Wash.: Microsoft Press, 1996.

device driver

A fundamental problem in computer design is the control of devices such as disk drives and printers. Each device is designed to respond to a particular set of control commands sent as patterns of binary values through the port to which the device is connected. For example, a printer will respond to a "new page" command by skipping lines to the end of the current page and moving the print head to the start of the next page, taking margin settings into account. The problem is this: When an applications program such as a word processor needs to print a document, how should the necessary commands be provided to the printer? If every application program has to include the appropriate set of commands for each device that might be in use, programs will be bloated and much development effort will be required for supporting devices rather than extending the functionality of the product itself. Instead, the manufacturers of printers and other devices such as scanners and graphics tablets typically provide a program called a driver. (A version of the driver is created for each major operating system in use.) The driver serves as the intermediary between the application, the operating system and the low-level device control system. It is sometimes useful to have drivers in the form of continually running programs that monitor the status of a device and wait for commands (see DEMON).

Modern operating systems such as Microsoft Windows typically take responsibility for services such as printing documents. When a printer is installed, its driver program is also installed in Windows. When the application program requests to print a document, Windows's print system accesses the driver. The driver turns the operating system's "generic" commands into the specific hardware control commands needed for the device.

While the use of drivers simplifies things for both program developers and users, there remains the need for

Device Driver

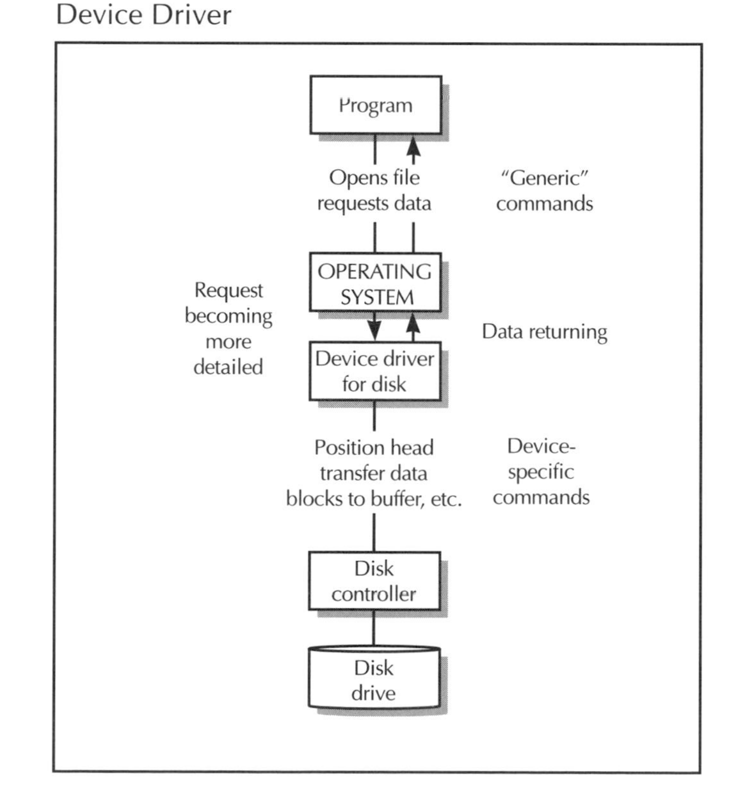

The device driver is the link between the operating system and the hardware that controls a specific device. Program requests are passed by the operating system to the device driver, which issues the detailed instructions needed by the device controller.

users to occasionally update drivers because of an upgrade either in the operating system or in the support for device capabilities. Both Windows and the Macintosh operating system implement a feature called plug and play. This allows for a newly installed device to be automatically detected by the system and the appropriate driver loaded into the operating system (see PLUG AND PLAY). Other device management components enable the OS to keep track of the driver version associated with each device. Some of the newest operating systems include auto-update features that can search on the Web for the latest driver versions and download them.

The need to provide drivers for popular devices creates something of a barrier to the development of new operating systems. In a catch-22, device manufacturers are unlikely to support a new OS that lacks significant market share, while the lack of device support in turn will discourage users from adopting the new OS. (Users of the Linux operating system faced this problem. However, that system's open source and cooperative development system made it easier for enthusiasts to write and distribute drivers without waiting for manufacturers to do so.)

Further Reading

"Device Driver Development for Microsoft Windows" http://www.chsw.com/ddk/

Egan, Janet and Thomas J. Teixeira. *Writing a UNIX Device Driver.* New York: Wiley, 1992.

"Mister Driver." http://www.mrdriver.com/

Oney, Walter, and Forrest Foltz. *Programming the Microsoft Windows Driver Model.* Redmond, Wash.: Microsoft Press, 1999.

Viscarola, Peter, and W. Anthony Mason. *Windows NT Device Driver Development.* Indianapolis, Ind.: New Rider Publishing, 1998.

digital cash

Also called digital money or e-cash, digital cash represents the attempt to create a method of payment for on-line transactions that is as easy to use as the familiar bills and coins in daily commerce (see E-COMMERCE). At present, credit cards are the principal means of making on-line payments. While using credit cards takes advantage of a well-established infrastructure, it has some disadvantages. From a security standpoint, each payment potentially exposes the payer to the possibility that the credit card number and possibly other identifying information will be diverted and used for fraudulent transactions and identity theft. While the use of secure (encrypted) on-line sites has reduced this risk, it cannot be eliminated entirely (see COMPUTER CRIME AND SECURITY). Credit cards are also impracticable for very small payments from cents to a few dollars (such as for access to magazine articles) because the fees charged by the credit card companies would be too high in relation to the value of the transaction.

One way to reduce security concerns is to make transactions that are anonymous (like cash) but guaranteed. Products such as DigiCash and CyberCash allow users to purchase increments of a cash equivalent using their credit cards or bank transfers, creating a "digital wallet." The user can then go to any website that accepts the digital cash and make a payment, which is deducted from the wallet. The merchant can verify the authenticity of the cash through its issuer. Since no credit card information is exchanged between consumer and merchant, there is no possibility of compromising it. The lack of wide acceptance and standards has thus far limited the usefulness of digital cash.

The need to pay for small transactions can be handled through micropayments systems. For example, users of a variety of on-line publications can establish accounts through a company called Qpass. When the user wants to read an article from the *New York Times,* for example, the fee for the article (typically $2–3) is charged against the user's Qpass account. The user receives one monthly credit card billing from Qpass, which settles accounts with the publications. Qpass, eCharge, and similar companies have had modest success. A similar (and quite successful) service is offered by companies such as PayPal and Billpoint, which allow winning auction bidders to send money from their credit card or bank account to the seller, who would not otherwise be equipped to accept credit cards. True micropayments would extend down to just a few cents.

As of 2001, there are many competing systems for anonymous payments and micropayments, with no clear market leader. However, the successful digital cash system is likely to have the following characteristics:

- Protects the anonymity of the purchaser (no credit card information transmitted to the seller)
- Verifiable by the seller, perhaps by using one-time encryption keys
- The purchaser can create digital cash freely from credit cards or bank accounts
- Micropayments can be aggregated at a very low transaction cost

As use of digital cash becomes more widespread, it is likely that tax and law enforcement agencies will press for the inclusion of some way to penetrate the anonymity of transactions for audit or investigation purposes. They will be opposed by civil libertarians and privacy advocates. One likely compromise may be requiring that transaction information or encryption keys be deposited in some sort of escrow agency, subject to being divulged upon court order.

Further Reading

Grabbe, Orlin J. "A Brief Guide to Digital Cash."http://www.aci.net/kalliste/dcguide.htm

McCullagh, Declan. "Feds: Digital Cash Can Thwart Us." *Wired News [on-line].* http://www.wired.com/news/politics/0,1283,38955,00.html

Wayner, Peter. *Digital Cash: Commerce on the Net.* 2nd ed. Boston: AP Professional, 1997.

digital convergence

Since the late 20th century, many forms of communication and information storage have been transformed from analog to digital representations (see ANALOG AND DIGITAL). For example, the phonograph record (an electromechanical analog format) gave way during the 1980s to a wholly digital format (see CD-ROM). Video, too, is now increasingly being stored in digital form (DVD or laser disks) rather than in the analog form of videotape. Voice telephony, which originally involved the conversion of sound to analogous electrical signals, is increasingly being digitized (as with many cell phones) and transmitted in packet form over the communications network.

The concept of digital convergence is an attempt to explore the implications of so many formerly disparate analog media now being available in digital form. All forms of digital media have key features in common. First, they are essentially pure information (computer data). This means that regardless of whether the data originally represented still images from a camera, video, or film, the sound of a human voice, music, or some other form of expression, that data can be stored, manipulated, and retrieved under the control of computer algorithms. This makes it easier to create seamless multimedia presentations (see also MULTIMEDIA and HYPERTEXT AND HYPERMEDIA). Services or products previously considered to be separate can be combined in new ways. For example, many radio stations now provide their programming in the form of "streaming audio" that can be played by such utilities as RealPlayer or Microsoft Windows Media Player (see STREAMING). Similarly, television news services such as CNN can offer selected excerpts of their coverage in the form of streaming video files. As more users gain access to broadband Internet connections (such as cable or DSL), it will eventually become feasible to deliver TV programs and even full-length feature films in digital format. (In a related development, digital TV recording systems now arriving on the market allow viewers to record more than one program at a time. These systems store digitized video on a hard disk rather than the tape used by the traditional VCR.)

Digital Convergence

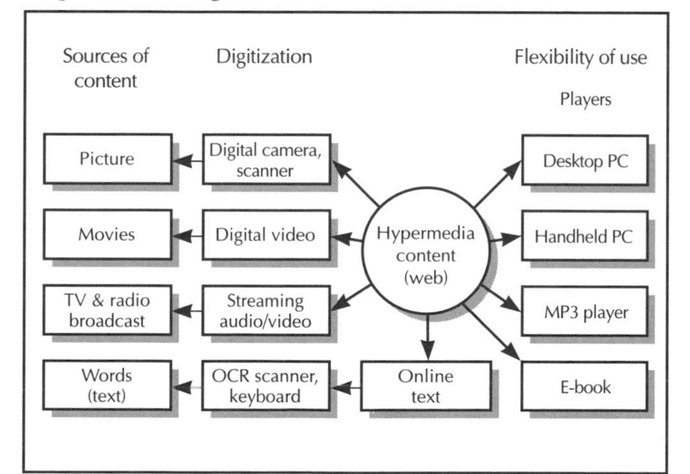

Digital convergence results when many formerly analog media (such as sound, film, and video) are acquired and processed digitally. Once in digital form, the content can be processed and played by a variety of software and used on many different platforms ranging from desktop computers to electronic books (e-books) and portable MP3 music players. Content can also be linked and organized using hypertext or hypermedia techniques, as on the Web.

EMERGING ISSUES

The merging of traditional media into a growing stream of digital content has created a number of difficult legal and social issues. Digital images or sounds from various sources can easily be combined, filtered, edited, or otherwise altered for a variety of purposes. As a result, the value of photographs as evidence may be gradually compromised. The ownership and control of the intellectual property represented by music, video, and film has also been complicated by the combination of digitization and the pervasive Internet. For example, during 2000–2001 the legal battles involving Napster, a program that allows users to share music files (many of them representing copyrighted works) pitted the rights of music producers and artists to control the distribution of their product against the technological capability of users to freely copy and distribute the material. While a variety of copy protection systems (both software and hardware-based) have been developed in an attempt to prevent unauthorized copying, historically such measures have had only limited effectiveness (see COPY PROTECTION, INTELLECTUAL PROPERTY AND COMPUTING).

Digital convergence also raises deeper philosophical issues. Musicians, artists, and scholars have frequently suggested that the process of digitization fails to capture subtleties of performance that might have been accessible in the original media. At the same time, the richness and immersive qualities of the new multimedia may be drawing people further away from the direct experience of the "real" analog world around them. Ultimately, the embodiment of digital convergence in the form of virtual reality likely to emerge in the early 21st century will pose questions as profound as those provoked by the invention of printing and the development of mass broadcast media (see VIRTUAL REALITY).

Further Reading

Covell, Andy. *Digital Convergence: How the Merging of Computers, Communications and Multimedia is Transforming Our Lives.* Newport, R.I.: Aegis Publishing, 2000.

Dertouzos, Michael L. *What Will Be: How the New World of Information will Change Our Lives.* San Francisco: HarperEdge, 1997.

digital divide

This term was coined in the late 1990s to describe the lag in access to personal computers (and particularly, to the Internet) among certain groups in the population. In general, according to figures from the National Telecommunications and Information Administration, the percentage of the population having regular access to the Internet grew rapidly toward the end of the decade, rising from 26.2 percent in December 1998 to 41.5 percent in August 2000. By the latter date more than 50 percent of American households had home computers, and 116.5 million Americans had some form of on-line access at home, work, school, or local libraries.

However, not all social groups have been gaining access at the same rate. There are a number of factors that correlate with the likelihood that a person or community will have access to the Web. People in lower-income brackets are less likely to own PCs. Phone service may be less reliable (particularly in rural areas), and Internet access may require expensive toll charges. While schools and public libraries can offer an alternative venue for Internet access, inner city schools have tended to lag behind in connecting to the Internet and in the ratio of networked computers to students. (The Net Day activities in the mid-1990s first publicized and sought to ameliorate this problem.)

Internet access also correlates to education. While persons lacking a college education are likely to be poorer than college graduates, they are also less likely to be working in jobs that include regular computer access. A deficiency in basic reading and keyboard skills can also serve as a barrier to participation in the on-line world (see also COMPUTER LITERACY). People over age 50 are also less likely to be on-line. They are more likely to have spent their career in noncomputerized jobs and may feel that they cannot master the new technology.

In general, however, statistics at the end of 2000 showed that many of the separate gaps that make up the digital divide have closed somewhat. Gaps based on age, gender (with women and men close to parity), and education seem to be closing the most swiftly. Minorities, too, have made considerable gains, with blacks and Hispanics reaching about 25 percent, representing a doubling in less than two years. Unfortunately, their participation rate is still only about half of that of whites.

Targeted attempts to close the digital divide through providing more Internet access through schools and libraries are likely to continue to be successful. The marketplace itself is perhaps making the biggest contribution, since the price of an Internet-capable PC with a basic dial-up connection is now around $500 plus about $10/month. (On the other hand, the growing use of broadband (cable and DSL) connections, which tend to be around $50/month, may be causing a new gap to emerge, between people with basic Internet access and those who have access to the rich media content that is practicable only over a high-speed connection.)

Improvement in the teaching of general literacy as well as technical skills in the K-12 schools is necessary if the next generation is to be able to participate fully and equally in the on-line world.

Further Reading

"Closing the Digital Divide." http://www.digitaldivide.gov/

U.S. Dept. of Commerce. National Telecommunications and Information Administration. "Falling Through the Net: Defining the Digital Divide." [revised November 1999] http://www.ntia.doc.gov/ntiahome/fttn99/contents.html

"Falling Through the Net: Toward Digital Inclusion." August 2000. http://www.ntia.doc.gov/ntiahome/digitaldivide/execsumfttn00.htm

Dijkstra, Edsger W.

(1930–)
Dutch
Computer Scientist

Edsger W. Dijkstra was born in Rotterdam, Netherlands, in 1930 into a scientific family (his mother was a mathematician and his father was a chemist). He received an intensive and diverse intellectual training, studying Greek, Latin, several modern languages, biology, mathematics, and chemistry. While majoring in physics at the University of Leiden in 1951, he attended a summer school at Cambridge that kindled what soon became a major interest in programming. He continued this pursuit at the Mathematical Center in Amsterdam in 1952 while finishing studies for his physics degree. At the time there were no degrees in computer science; indeed, programming did not yet exist as an academic discipline. Like most other computers of the time, the Mathematical Center's ARMAC was custom-built. With no high-level languages yet in use, programming required intimate familiarity with the machine's architecture and low-level instructions. Dijkstra soon found that he thrived in such an environment.

By 1956, Dijkstra had discovered an algorithm for finding the shortest path between two points. He applied

Edsger Dijkstra's ideas about structured programming helped develop the field of software engineering, enabling programmers to organize and manage increasingly complex software projects. (PHOTO COURTESY OF THE DEPARTMENT OF COMPUTER SCIENCES, UT AUSTIN)

the algorithm to the practical problem of designing electrical circuits that used as little wire as possible, and generalized it into a procedure for traversing treelike data structures.

During the 1960s, Dijkstra began to explore the problem of communication and resource-sharing within computers. He developed the idea of a semaphore. Like the railroad signaling device that allows only one train at a time to pass through a single section of track, the programming semaphore provides *mutual exclusion*, ensuring that two processes don't try to access the same memory or other resource at the same time.

Another problem Dijkstra tackled involved the sequencing of several processes that are accessing the same resources. He found ways to avoid a deadlock situation where one process had part of what it needed but was stuck because the process holding the other needed resource was in turn waiting for the first process to finish. His algorithms for allowing multiple processes (or processors) to take turns gaining access to memory or other resources would become fundamental for the design of new computing architectures.

During the 1970s, Dijkstra immigrated to the United States, where he became a research fellow at Burroughs, one of the major manufacturers of mainframe computers. During this time he helped launch the "structured programming" movement. His paper "GO TO Considered Harmful" criticized the use of that unconditional "jump" instruction because it made programs hard to read and verify. The newer structured languages such as Pascal and C affirmed Dijkstra's belief in avoiding or discouraging such haphazard program flow.

Since the 1980s, Dijkstra has been a professor of mathematics at the University of Texas at Austin and holds the Schlumberger Centennial Chair in Computer Science. In 1972, he won the Association for Computing Machinery's Turing Award, one of the highest honors in the field.

Further Reading

Dijkstra, Edsger Wybe. *A Discipline of Programming.* Upper Saddle River, N.J.: Prentice Hall, 1976.

Shasha, Dennis, and Cathy Lazere. *Out of their Minds: The Lives and Discoveries of 15 Great Computer Scientists.* New York: Springer-Verlag, 1995.

disabled persons and computing

The impact of the personal computer upon persons having disabilities involving sight, hearing, or movement has been significant but mixed. Computers can help disabled people communicate and interact with their environment, better enabling them to work and live in the mainstream of society. At the same time, changes in computer technology can, if not ameliorated, exclude some disabled persons from fuller participation in a society where computer access and skills are increasingly taken for granted.

COMPUTERS AS ENABLERS

Computers can be very helpful to disabled persons. With the use of text-to-speech software, blind people can have on-line documents read to them. (With the aid of a scanner, printed materials can also be input and read aloud.) Persons with low vision can benefit from software that can present text in large fonts or magnify the contents of the screen. Text can also be printed (embossed) in Braille. Deaf or hearing-impaired persons can now use e-mail or instant messaging software for much of their communication needs, largely replacing the older and more cumbersome teletype (TTY and TTD) systems. As people who have seen presentations by physicist Stephen Hawking know, even quadriplegics who have only the use of head or finger movements can input text and have it spoken by a voice synthesizer. Further, advances in coupling eye movements (and even brain wave patterns) to computer systems and robotic extensions offer hope that even profoundly disabled persons will be able to be more self-sufficient.

CHALLENGES

Unfortunately, changes in computer technology can also cause problems for disabled persons. The most pervasive problem arose when text-based operating systems such as MS-DOS were replaced by systems such as Microsoft Windows and the Macintosh that are based on graphic icons and the manipulation of objects on the screen. While text commands and output on the older system could be easily turned into speech for the visually impaired, everything, even text, is actually graphics on a Windows system. While it is possible to have software "hook into" the operating system to read text within Windows out loud, it is much more difficult to provide an alternative way for a blind person to find, click on, drag, or otherwise manipulate screen objects. Thus far, while Microsoft and other operating system developers have built some "accessibility" features such as screen magnification into recent versions of their products, there is no systematic, integrated facility that would allow a blind person to have the same facility as a sighted person.

The growth of the World Wide Web also poses problems for the visually impaired, since many webpages rely on graphical buttons for navigation. Software plug-ins can provide audio cues to help with screen navigation. While Web browsers usually have some flexibility in setting the size of displayed fonts, some newer features (such as Cascading Style Sheets) can remove control over font size from the user.

Because most computer systems today use graphical users interfaces, the failure to provide effective access may be depriving blind and visually impaired persons of employment opportunities. Meanwhile, the computer industry, educational institutions, and workplaces face potential challenges under the Americans with Disabilities Act (ADA), which requires that public and workplace facilities be made accessible to the disabled. Some funding through the Technology-Related Assistance Act has been provided to states for promoting the use of adaptive technology to improve accessibility.

Further Reading

"Adaptive Computer Products." http://www.makoa.org/computers.htm

Coombs, N., and C. Cunningham. *Information Access and Adaptive Technology.* Phoenix, Ariz.: Oryx Press, 1997.

Mates, Barbara T., Doug Wakefield, and Judith M. Dixon. *Adaptive Technology for the Internet: Making Electronic Resources Accessible to All.* Chicago, Ill.: American Library Association, 1999.

Microsoft Corporation. "Microsoft Windows Guidelines for Accessible Software Design." http://www.microsoft.com/enable/dev/guidelines/software.htm

disk array

Called RAID, or Redundant Array of Inexpensive Disks, disk array is a way of arranging a group of disk drives so that they work together to form a fast, reliable storage system. Greater speed is achieved by "striping" data, where a file is broken into pieces with each piece stored on a sector on a different drive. This means that instead of the head of a single drive having to jump around to read different parts of the file, the heads on all the drives can simultaneously read many parts of the file (which are then assembled into the proper order). The use of striping is required for the simplest level of disk array, called RAID-0.

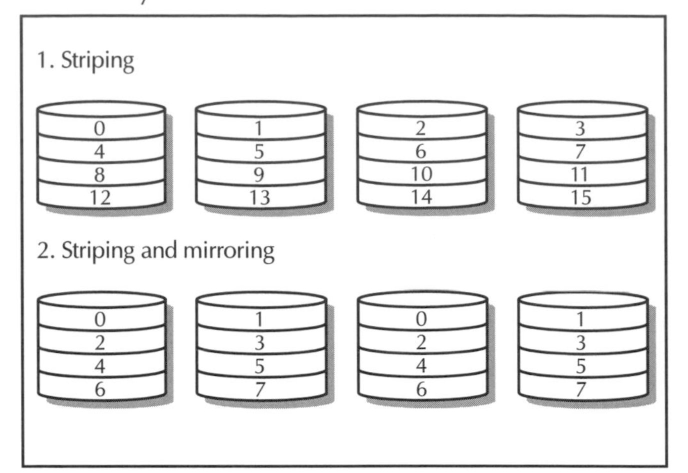

Striping spreads data across several disk drives so that a single head movement on each drive can fetch a large amount of data. Mirroring duplicates each sector of data on a second disk drive, ensuring that if one drive fails the data can still be retrieved. By combining striping and mirroring, both greater speed and greater safety can be achieved.

MIRRORING FOR REDUNDANCY

While striping can significantly improve the speed of data access for properly organized files, the system is if anything even more vulnerable to failure than one using a single drive. If any drive in a striped system fails, files throughout the system will become inaccessible. One way to improve reliability is to use two disk drives (or sets of drives) and copy all data to both. This is called mirroring. Since all data is stored on both disks, if one disk fails, the data is still available on the other. Since twice as many disks must be purchased to store the data, mirroring can be expensive (although the rapid increase in capacity and decrease in cost of disk drives in recent years may make this less of a consideration). While mirroring does not affect reading speed, data writing is slower since the data must be written to two sets of disks. Systems with mirroring are designated RAID-1.

ADVANCED TECHNOLOGIES

Higher levels of RAID offer additional features to improve reliability and/or speed. In RAID-2, data is striped (distributed across the disks) bit by bit or byte by byte, and synchronized actuators are used to control the disk

SUMMARY OF RAID LEVELS

LEVEL	STRIPING	ERROR-CORRECTION	COMMENTS
RAID-0	Basic striping by sector	None	
RAID-1	As in RAID-0	None	All data copied to two sets of disks.
RAID-2	Striping by bit or byte	Hamming error-correction code on additional disks.	All disk heads synchronized by actuator.
RAID-3	As in RAID-2	Single parity bit	Data from a faulty disk can be reconstructed.
RAID-4	Striping by block or sector	As in RAID-3	Independent actuators allow separate transactions to run concurrently.
RAID-5	As in RAID-4	Parity information spread across disks.	Distributing parity information improves concurrent processing.
RAID-6	As in RAID-4	Parity information determined separately for each layer.	Disks arranged in two dimensions (i.e. layers).
RAID-7	As in RAID-4	As in RAID-6	Dynamic mapping where data is not always stored in the same physical location.

heads. In addition, an error-correcting code is stored on a separate disk or disks, allowing data to be checked for corruption as it is read or written. Other versions of RAID are similar, but vary either in the type of error correction used (RAID-3 and RAID-5) or the way the data is striped across the disks (RAID-4).

In determining which form of RAID to use, system designers are guided not only by overall speed and reliability but also by the type of data typically being stored on the system. For example, RAID-2 works best for transferring large files, where the process of seeking and synchronization does not have to be performed very often. The above table summarizes the different levels of RAID.

Further Reading

Gibson, Garth A. *Redundant Disk Arrays: Reliable, Parallel Secondary Storage.* Cambridge, Mass.: MIT Press, 1992.

RAID Advisory Board. http://www.raid-advisory.com/

distributed computing

This concept involves the creation of a software system that runs programs and stores data across a number of different computers, an idea pervasive today. A simple form is the central computer (such as in a bank or credit card company) with which thousands of terminals communicate to submit transactions. While this system is in some sense distributed, it is not really decentralized. Most of the work is done by the central computer, which is not dependent on the terminals for its own functioning. However, responsibilities can be more evenly apportioned between computers (see CLIENT-SERVER COMPUTING).

Today the World Wide Web is in a sense the world's largest distributed computing system. Millions of documents stored on hundreds of thousands of servers can be accessed by millions of users' Web browsers running on a variety of personal computers. While there are rules for specifying addresses and creating and routing data packets (see INTERNET and TCP/IP), no one agency or computer complex controls access to information or communication (such as e-mail).

ELEMENTS OF A DISTRIBUTED COMPUTING SYSTEM

The term *distributed computer system* today generally refers to a more specific and coherent system such as a database where data objects (such as records or views) can reside on any computer within the system. Distributed computer systems generally have the following characteristics:

- The system consists of a number of computers (sometimes called *nodes*). The computers need not necessarily use the same type of hardware, though they generally use the same (or similar) operating systems.
- Data consists of logical objects (such as database records) that can be stored on disks connected to any computer in the system. The ability to move data around allows the system to reduce bottlenecks in data flow or optimize speed by storing the most frequently used data in places from which it can be retrieved the most quickly.
- A system of unique *names* specifies the location of each object. A familiar example is the DNS (Domain Naming System) that directs requests to webpages.
- Typically, there are many processes running concurrently (at the same time). Like data objects, processes can be allocated to particular processors to balance the load. Processes can be further broken down into *threads* (see CONCURRENT PROGRAMMING). Thus, the system can adjust to changing conditions (for example, processing larger numbers of incoming transactions during the day versus performing batches of "housekeeping" tasks at night).

- A remote procedure call facility enables processes on one computer to communicate with processes running on a different computer.
- In inter-process communication protocols specify the processing of "messages" that processes use to report status or ask for resources. Message-passing can be asynchronous (not time-dependent, and analogous to mailing letters) or synchronous (with interactive responses, as in a conversation).
- The capabilities of each object (and thus the messages it can respond to or send) are defined in terms of an interface and an implementation. The interface is like the declaration in a conventional program: It defines the types of data that can be received and the types of data that will be returned to the calling process. The implementation is the code that specifies how the actual processing will be done. The hiding of implementation details within the object is characteristic of object-oriented programming (see also CLASS).
- A distributed computing environment includes facilities for managing objects dynamically. This includes lower-level functions such as copying, deleting, or moving objects and system-wide capabilities to distribute objects in such as way as to distribute the load on the system's processors more evenly, to make backup copies of objects (replication), and to reclaim and reorganize resources (such as memory or disk space) that are no longer allocated to objects.

Three widely used systems for distributed computing are Microsoft's DCOM (Distributed Component Object Model), OMG's Common Object Request Broker Architecture (CORBA), and Sun's Java/Remote Method Invocation (Java/RMI). While these implementations are quite different in details, they provide most of the elements and facilities summarized above.

APPLICATIONS

Distributed computing is particularly suited to applications that require extensive computing resources and that may need to be scaled (smoothly enlarged) to accommodate increasing needs. Examples might include large databases, intensive scientific computing, and cryptography. A particularly interesting example is SETI@home, which invites computer users to install a special screen saver that runs a distributed process during the computer's idle time. The process analyzes radio telescope data for correlations that might indicate receipt of signals from an extraterrestrial intelligence.

Besides being able to marshal very large amounts of computing power, distributed systems offer improved fault tolerance. Because the system is decentralized, if a particular computer fails, its processes can be replaced by ones running on other machines. Replication (copying) of data across a widely dispersed network can also provide improved data recovery in the event of a disaster.

Further Reading

Farley, Jim, and Mike Loukides. *Java Distributed Computing.* Sebastopol, Calif.: O'Reilly, 1998.

Obasanjo, Dare, and Sanjay Bhatia. "An Introduction to Distributed Object Technologies." http://www.25hoursaday.com/IntroductionToDistributedComputing.html

Shan, Yen-Ping, Ralph H. Earle, and Marie A. Lenzi. *Enterprise Computing with Objects: from Client-Server Environments to the Internet.* Reading, Mass.: Addison-Wesley, 1997.

SETI@Home http://setiathome.ssl.berkeley.edu/

Distributed Computing

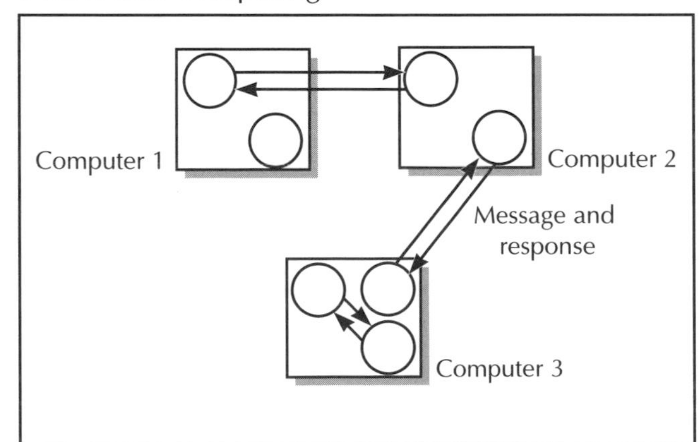

In a distributed computing system objects (representing processes in a database, for example) can use messages to coordinate processing. Objects can communicate with other objects on the same computer, or access objects running on other computers.

documentation of program code

Computer system documentation can be divided into two main categories based upon the intended audience. Manuals and training materials for users focus on explaining how to use the program's features to meet the user's needs (see DOCUMENTATION, USER). This entry, however, focuses on the creation of documentation for programmers and others involved in software development and maintenance (see also TECHNICAL WRITING).

Software documentation can consist of *comments* describing the operation of a line or section of code. Early programming with its reliance on punched cards had only minimal facilities for incorporating comments. (Some of the proponents of COBOL thought that the language's English-like syntax would make additional documentation unnecessary. Like the similar claim that trained programmers would no longer be needed, the reality proved otherwise.)

After the switch from punchcard input to the use of keyboards, adding comments became easier. For example, a comment in C looks like this:

```
printf("Hello, world\n"); /* Display the traditional message */
```

while C++ uses comments in this form:

```
cout << "Hello, World"; // This is also a comment
```

Each language provides a particular symbol or set of symbols for separating comments from executable code. The compiler ignores comments when compiling the program.

While proper commenting can help people understand a program's functions, the coding style should also be one that promotes clarity. This includes the use of descriptive and consistent names for variables and functions. This can also be influenced by the conventions of the operating system: For example, Windows has many special data structures that should be used consistently.

In addition to the commented source code, external documentation is usually provided. Design documents can range from simple flowcharts or outlines to detailed specifications of the program's purpose, structure, and operations. Rather than being considered an afterthought, documentation has been increasingly integrated into the practice of software engineering and the software development process. This practice became more prevalent during the 1960s and 1970s when it became clear that programs were not only becoming larger and more complex, but also that significant programs such as business accounting and inventory applications were likely to have to be maintained or revised for perhaps decades to come. (The lack of adequate documentation of date-related code in programs of this vintage became an acute problem in the late 1990s. See Y2K PROBLEM.)

DOCUMENTATION TOOLS

As programmers began to look toward developing their craft into a more comprehensive discipline, advocates of structured programming placed an increased emphasis not only on proper commenting of code but on the development of tools that could automatically create certain kinds of documentation from the source code. For example, there are utilities for C, C++, and Java (javadoc) that will extract information about class declarations or interfaces and format them into tables. Most software development environments now include features that cross-reference "symbols" (named variables and other objects). The combination of comments and automatically generated documentation can help with maintaining the program as well as being helpful for creating developer and user manuals.

While programmers retain considerable responsibility for coding standards and documentation, larger programming staffs typically have specialists who devote their full time to maintaining documentation. This includes the logging of all program change requests and the resulting new distributions or "patches," the record of testing and retesting of program functions, the maintenance of a "version history," and coordinating with technical writers in the production of revised manuals.

Further Reading

Denton, L., and J. Kelly. *Designing, Writing and Producing Computer Documentation.* New York: McGraw-Hill, 1993.

Kovitz, Benjamin L. *Practical Software Requirements: a Manual of Content and Style.* Greenwich, Conn.: Manning Publications, 1998.

documentation, user

As computing moved into the mainstream of offices and schools beginning in the 1980s and accelerating through the 1990s, the need to train millions of new computer users spawned the technical publishing industry. In addition to the manual that accompanied the software, third-party publishers produced full-length books for beginners and advanced users as well as "dictionaries" and reference manuals (see also TECHNICAL WRITING). A popular program such as WordPerfect or (today) Adobe Photoshop can easily fill several shelves in the computer section of a large bookstore.

A number of publishers targeted particular audiences and adopted distinctive styles. Perhaps the best known is the IDG "Dummies" series, which eventually diversified its offerings from computer-related titles to everything from home remodeling to investing. Berkeley, California, publisher Peachpit Press created particularly accessible introductions for Windows and Macintosh users. At the other end of the spectrum, publishers Sams, Osborne, Waite Group, and Coriolis targeted the developer and "power user" community and the eclectic, erudite volumes from O'Reilly grace the bookshelves of many UNIX users.

ON-LINE DOCUMENTATION

During the 1980s, the lack of a multitasking, window-based operating system limited the ability of programs to offer built-in (or "on-line") documentation. Traditionally, users could press the F1 key to see a screen listing key commands and other rudimentary help. However, both the Macintosh and Windows-based systems of the 1990s included the ability to incorporate a standardized, hypertext-based help system in any program. Users could now search for help on various topics and scroll through it while keeping their main document in view. Another facility, the "wizard," offered the ability to guide users step by step through a procedure.

The growth of the use of the Web has provided a new avenue for on-line help. Today many programs link users to their website for additional help. Even help files stored on the user's own hard drive are increasingly formatted in HTML for display through a Web browser. Additional sources of help for some programs include training videos and animated presentations using programs such as PowerPoint.

By the late 1990s, printed user manuals were becoming a less common component in software packages. (Instead, the manual was often provided as a file in the Adobe Acrobat format, which reproduces the exact appearance of printed material on the screen.) The computer trade book industry has also declined somewhat, but the bookstore still offers plenty of alternatives for users who are more comfortable with printed documentation.

Further Reading

Barker, T. T., and S. Dragga. *Writing Software Documentation: A Task-Oriented Approach.* New York: Allyn & Bacon, 1997.

Hargis, Gretchen, ed. *Developing Quality Technical Information: a Handbook for Writers and Editors.* Upper Saddle River, N.J.: Prentice Hall, 1997.

Kukulska-Hulme, Agnes. *Language and Communication: Essential Concepts for User Interface and Documentation Design.* New York: Oxford University Press, 1999.

Price, Jonathan, and Henry Korman. *How to Communicate Technical Information: A Handbook of Hardware and Software Documentation.* Reading, Mass.: Addison-Wesley, 1993.

Weiss, Edmond H. *How to Write Usable User Documentation.* 2nd ed. Phoenix, Ariz.: Oryx Press, 1991.

document model

Most early developers and users of desktop computing systems thought in terms of application programs rather than focusing on the documents or other products being created with them. From the application point of view, files are opened or created, content (text or graphics) is created, and the file is then saved. There is no connection between the files except in the mind of the user. The dominant word processors of the 1980s (such as WordStar and WordPerfect) were designed as replacements for the typewriter and emphasized the efficient creation of text (see WORD PROCESSING). Users who wanted to work with other types of information had to run completely separate applications, such as dBase for databases or Lotus 1-2-3 for spreadsheets. Working with graphics images (to the extent it was possible with early PCs) required still other programs.

This "application-centric" way of thinking suited program developers at a time when most computer systems (such as those running MS-DOS) could run only one program at a time. But increasing processor power, memory, and graphics display capabilities during the late 1980s made it possible to create an operating system such as Microsoft Windows that could display text fonts and formatting, graphics and other content in the same window, and run several different program windows at the same time (see MULTITASKING). In turn, this made it possible to present a model that was more in keeping with the way people had worked in the precomputer era.

In the new "document model," instead of thinking in terms of individual application programs working with files, users could think in terms of creating documents. A document (such as a brochure or report) could contain formatted text, graphics, and data brought in from database or spreadsheet programs. This meant that in the course of working with a document users would actually be invoking the services of several programs: a word processor, graphics editor, database, spreadsheet, and perhaps others. To the user, however, the focus would be on a screen "desktop" on which would be arranged documents (or projects), not on the process of running individual programs and loading files.

IMPLEMENTING THE DOCUMENT MODEL

There are two basic approaches to maintaining documents. One is to create large programs that provide all of the features needed, including word processing, graphics, and data management (see APPLICATION SUITE). While such tight integration can (ideally at least) create a seamless working environment with a consistent user interface, it lacks flexibility. If a user needs capabilities not included in the suite (such as, perhaps the ability to create an HTML version of the document for the Web), one of two cumbersome procedures would have to be followed. Either the operating system's "cut and paste" facilities might be used to copy data from another application into the document (possibly with formatting or other information lost in the process), or possibly the document could be saved in a file format that could be read by the program that was to provide the additional functionality (again with the possibility of losing something in the translation).

LINKING AND EMBEDDING

A more sophisticated approach is to create a protocol that applications could use to call upon one another's services. The Windows COM (Component Object Model) uses a technology formerly called OLE (Object Linking and Embedding). Using this facility, someone working on a document in Microsoft Word could "embed" another object such as an Excel spreadsheet or an Access database into the current document (which becomes the *container*). When the user double-clicks on the embedded object, the appropriate application is launched automatically, and the user sees the screen menus and controls from that application instead of those in Word. (One can also think of Word in this example being the client and Excel or Access as the server—see CLIENT-SERVER COMPUTING). All work done with the embedded object is automatically updated by the server application and everything is stored in the same document file. Alterna-

Document Model

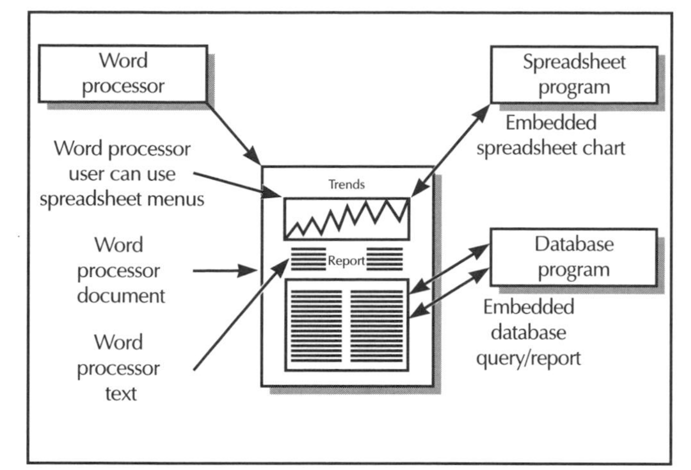

The document model focuses computing not on what programs are being used, but upon what the user is trying to create—a document. Some operating systems such as Microsoft Windows have facilities that allow a document from one application (such as a spreadsheet) to be embedded (inserted) into another document (such as in a word processor). In this case the word processor user can access the menus and other features of the embedded application.

tively, an application may be *linked* rather than embedded. In that case, the container document simply contains a pointer to the file in the other application. Whenever that file is changed, all documents that are linked to it are updated. Object embedding thus preserves a document-centric approach but works with any applications that support that facility, regardless of vendor. The Macintosh operating system offers a similar facility. A competing standard, OpenDoc, has been promoted by IBM and Apple, while documentation models are also being applied to the management of HTML documents on the Web.

Further Reading

Iseminger, David. *COM+ Developer's Reference Library.* Redmond, Wash.: Microsoft Press, 2000.

"OpenDoc vs. OLE." http://www.mactech.com/articles/mactech/Vol.10/10.08/OpenDocvsOLE/

Pritchard, Jason. *COM and CORBA Side-by-Side: Architectures, Strategies and Implementations.* Reading, Mass.: Addison-Wesley, 1999.

World Wide Web Consortium. "Document Object Model." http://www.w3.org/DOM/

Domain Name System

The operation of the Internet requires that each participating computer have a unique address to which data packets can be routed (see INTERNET and TCP/IP). The Domain Name System (DNS) provides alphabetical equivalents to the numeric IP addresses, giving the now familiar-looking Web addresses (URLs), e-mail addresses, and so on.

The system uses a set of "top-level" domains to categorize these names. One set of domains is based on the nature of the sites involved, including: .com (commercial, corporate), .edu (educational institutions), .gov (government), .mil (military), .org (nonprofit organizations), .int (international organizations), .net (network service providers, and so on).

The other set of top-level domains is based on the geographical location of the site. For example, .au (Australia), .fr (France), and .ca (Canada). (While the United States has the .us domain, it is generally omitted in practice, because the Internet was developed in the United States).

INTERNET COUNTRY CODES

AD	Andorra
AE	United Arab Emirates
AF	Afghanistan
AG	Antigua and Barbuda
AI	Anguilla
AL	Albania
AM	Armenia
AN	Netherlands Antilles
AO	Angola
AQ	Antarctica
AR	Argentina
AS	American Samoa
AT	Austria
AU	Australia
AW	Aruba
AZ	Azerbaijan
BA	Bosnia and Herzegovina
BB	Barbados
BD	Bangladesh
BE	Belgium
BF	Burkina Faso
BG	Bulgaria
BH	Bahrain
BI	Burundi
BJ	Benin
BM	Bermuda
BN	Brunei Darussalam
BO	Bolivia
BR	Brazil
BS	Bahamas
BT	Bhutan
BV	Bouvet Island
BW	Botswana
BY	Belarus
BZ	Belize
CA	Canada
CC	Cocos (Keeling) Islands
CF	Central African Republic
CG	Congo

CH Switzerland
CI Côte d'Ivoire (Ivory Coast)
CK Cook Islands
CL Chile
CM Cameroon
CN China
CO Colombia
CR Costa Rica
CS Czechoslovakia (former)
CU Cuba
CV Cape Verde
CX Christmas Island
CY Cyprus
CZ Czech Republic
DE Germany
DJ Djibouti
DK Denmark
DM Dominica
DO Dominican Republic
DZ Algeria
EC Ecuador
EE Estonia
EG Egypt
EH Western Sahara
ER Eritrea
ES Spain
ET Ethiopia
FI Finland
FJ Fiji
FK Falkland Islands (Malvinas)
FM Micronesia
FO Faroe Islands
FR France
FX France, Metropolitan
GA Gabon
GB Great Britain (UK)
GD Grenada
GE Georgia
GF French Guiana
GH Ghana
GI Gibraltar
GL Greenland
GM Gambia
GN Guinea
GP Guadeloupe
GQ Equatorial Guinea
GR Greece
GS S. Georgia and S. Sandwich Isls.
GT Guatemala
GU Guam
GW Guinea-Bissau
GY Guyana
HK Hong Kong
HM Heard and McDonald Islands
HN Honduras
HR Croatia (Hrvatska)
HT Haiti
HU Hungary
ID Indonesia
IE Ireland
IL Israel
IN India
IO British Indian Ocean Territory
IQ Iraq
IR Iran
IS Iceland
IT Italy
JM Jamaica
JO Jordan
JP Japan
KE Kenya
KG Kyrgyzstan
KH Cambodia
KI Kiribati
KM Comoros
KN Saint Kitts and Nevis
KP Korea (North)
KR Korea (South)
KW Kuwait
KY Cayman Islands
KZ Kazakhstan
LA Laos
LB Lebanon
LC Saint Lucia
LI Liechtenstein
LK Sri Lanka
LR Liberia
LS Lesotho
LT Lithuania
LU Luxembourg
LV Latvia
LY Libya
MA Morocco
MC Monaco
MD Moldova
MG Madagascar
MH Marshall Islands
MK Macedonia
ML Mali
MM Myanmar
MN Mongolia
MO Macau
MP Northern Mariana Islands
MQ Martinique
MR Mauritania
MS Montserrat
MT Malta
MU Mauritius
MV Maldives
MW Malawi
MX Mexico
MY Malaysia
MZ Mozambique
NA Namibia

NC	New Caledonia
NE	Niger
NF	Norfolk Island
NG	Nigeria
NI	Nicaragua
NL	Netherlands
NO	Norway
NP	Nepal
NR	Nauru
NT	Neutral Zone
NU	Niue
NZ	New Zealand (Aotearoa)
OM	Oman
PA	Panama
PE	Peru
PF	French Polynesia
PG	Papua New Guinea
PH	Philippines
PK	Pakistan
PL	Poland
PM	St. Pierre and Miquelon
PN	Pitcairn
PR	Puerto Rico
PT	Portugal
PW	Palau
PY	Paraguay
QA	Qatar
RE	Reunion
RO	Romania
RU	Russian Federation
RW	Rwanda
SA	Saudi Arabia
SB	Solomon Islands
SC	Seychelles
SD	Sudan
SE	Sweden
SG	Singapore
SH	St. Helena
SI	Slovenia
SJ	Svalbard and Jan Mayen Islands
SK	Slovak Republic
SL	Sierra Leone
SM	San Marino
SN	Senegal
SO	Somalia
SR	Suriname
ST	Sao Tome and Principe
SU	USSR (former)
SV	El Salvador
SY	Syria
SZ	Swaziland
TC	Turks and Caicos Islands
TD	Chad
TF	French Southern Territories
TG	Togo
TH	Thailand
TJ	Tajikistan
TK	Tokelau
TM	Turkmenistan
TN	Tunisia
TO	Tonga
TP	East Timor
TR	Turkey
TT	Trinidad and Tobago
TV	Tuvalu
TW	Taiwan
TZ	Tanzania
UA	Ukraine
UG	Uganda
UK	United Kingdom
UM	US Minor Outlying Islands
US	United States
UY	Uruguay
UZ	Uzbekistan
VA	Vatican City State (Holy See)
VC	Saint Vincent and the Grenadines
VE	Venezuela
VG	Virgin Islands (British)
VI	Virgin Islands (U.S.)
VN	Viet Nam
VU	Vanuatu
WF	Wallis and Futuna Islands
WS	Samoa
YE	Yemen
YT	Mayotte
YU	Yugoslavia
ZA	South Africa
ZM	Zambia
ZR	Zaire
ZW	Zimbabwe

DOMAINS AND ADDRESSES

A complete Internet address generally consists of a word representing the name of the organization or company, possibly followed by the name of a department or division. This is followed by the top-level domain. Here are some examples:

well.com The Well conferencing system, a business in the U.S.

acm.org The Association for Computing Machinery, a nonprofit professional organization

state.gov United States Department of State

berkeley.edu University of California, Berkeley

www2.physics.ox.ac.uk Department of Physics, Oxford University, Oxfordshire, United Kingdom.

To access a service at a given site, the host address is prefixed to indicate the server or service. Most commonly, this is www for World Wide Web. Thus www.well.com indicates the Web server at the well.com host, while ftp.well.com would indicate the ftp (file transfer protocol) server. (In some cases, if there is no prefix, www will be assumed.)

A complete Web address or URL (Uniform Resource Locator) also includes a prefix for the protocol to be used (see WORLD WIDE WEB). Most commonly this is http:// (for hypertext transfer protocol), though most Web browsers will treat this as the default and not require that it be typed. ftp:// can be used to access ftp servers via the Web. Finally, a URL must include the path to the directory that actually contains the HTML document or other resource, as well as its filename. Thus a complete address for a hypothetical user's home page might be:

http://www.BigUniversity.edu/users/tomr/index.html

INTERNAL ADDRESSING

When a Web user types such an address, the Web browser connects to a nearby *name server.* This program translates the name into an IP (Internet Protocol) address. The address consists of four 8-bit numbers called *tuples,* separated by periods. For example, the domain name www.well.com currently translates to 208.178.101.2. The first number represents one of five classes of networks, with the first three classes (A-C) organized according to the number and size of networks and D and E being reserved for one-to-many "broadcast" transmissions and experimentation respectively.

To obtain a domain name, a person or organization contacts one of several registration services accredited by ICANN (the nonprofit Internet Corporation for Names and Numbers), which took over the service originally provided by a single company, Network Solutions. Each name must be unique. Considerable legal disputation has occurred when someone not connected with a company has registered a domain containing that company's name. The tremendous growth of e-commerce has made distinctive or easy-to-remember domain names a scarce and valuable commodity. Foreseeing this, some speculators bought up attractive domains in the hope (sometimes realized) of selling them to corporations at a huge profit. Anti–"domain squatting" laws were passed in reaction. In other cases, disgruntled employees or consumers have registered domains for websites critical of major corporations such as airlines and telephone companies. In the courts, this pits the right of free speech against the right of a company to control the use of its name.

EXPANDING THE SYSTEM

The expansion of the Internet has strained the capacity of the existing DNS. The shortage of "name space" is being addressed by the release of IP Version 6, which replaces the 32-bit addresses with 128-bit ones. In addition, in November 2000 ICANN announced the creation of seven new top-level domains: .aero (air transport), .biz (business), .coop (cooperatives), .info (general-purpose), .museum (museums), .name (personal sites), and .pro (professionals such as lawyers, accountants, and physicians). However, the situation is muddled by the existence of competing proposals and the use of unofficial DNS systems that provide their own domains (but require special software for access, since they are not recognized by regular DNS servers).

Further Reading

Albitz, Paul, and Cricket Liu. *DNS and BIND,* 4th ed. Sebastopol, Calif.: O'Reilly, 2001.

ICANN (Internet Corporation for Assigned Names and Numbers), http://www.icann.org/

InterNIC. http://www.internic.net/

InterNIC Registry WhoIs [domain look-up]. http://www.internic.net/whois.html

Dreyfus, Hubert

(1929–)

American

Philosopher, Cognitive Psychologist

As the possibilities for computers going beyond "number crunching" to sophisticated information processing became clear starting in the 1950s, the quest to achieve artificial intelligence (AI) was eagerly embraced by a number of innovative researchers. For example, Allen Newell, Herbert Simon, and Cliff Shaw at the RAND Corporation, attempted to write programs that could "understand" and intelligently manipulate symbols rather than just literal numbers or characters. Similarly, MIT's Marvin Minsky (see MINSKY, MARVIN) was attempting to build a robot that could not only perceive its environment, but in some sense understand and manipulate it. (See ARTIFICIAL INTELLIGENCE and ROBOTICS.)

Into this milieu came Hubert Dreyfus, who had earned his Ph.D. in philosophy at Harvard. Dreyfus had specialized in the philosophy of perception (how meaning can be derived from a person's environment) and phenomenology (the understanding of processes). When Dreyfus began to teach a survey course on these areas of philosophy, some of his students asked him what he thought of the artificial intelligence researchers who were taking an experimental and engineering approach to the same topics the philosophers had discussed abstractly.

Philosophy had attempted to explain the process of perception and understanding (see also COGNITIVE SCIENCE). One tradition, the rationalism represented by such thinkers as Descartes, Kant, and Husserl took the approach of formalism and attempted to elucidate rules governing the process. They argued that in effect the human mind was a machine (albeit a wonderfully complex and versatile one). The opposing tradition, represented by the phenomenologists Wittgenstein, Heidegger, and Merleau-Ponty, took a holistic approach in which physical states, emotions, and experience were inextricably intertwined in creating the world that people perceive and relate to.

If computers, which at that time had only the most rudimentary "senses" and no emotions could perceive and understand in the way humans did, then the rules-based approach of the rationalist philosophers would be vindicated. But when Dreyfus had examined the AI efforts, he wrote a paper titled "Alchemy and Artificial Intelligence." His comparison of AI to alchemy was provocative in that it suggested that like the alchemists, the modern AI researchers had met with only limited success in manipulating their materials (such as by teaching computers to perform such intellectual tasks as playing checkers and even proving mathematical theorems). However, Dreyfus concluded that the kind of flexible, intuitive, and ultimately robust intelligence that characterizes the human mind couldn't be matched by any programmed system. Each time AI researchers demonstrated the performance of some complex task, Dreyfus examined the performance and concluded that it lacked the essential characteristics of human intelligence. Dreyfus expanded his paper into the book *What Computers Can't Do*. Meanwhile, critics complained that Dreyfus was moving the goal posts after each play, on the assumption that "if a computer did it, it must not be true intelligence."

Two decades later, Dreyfus reaffirmed his conclusions in *What Computers Still Can't Do,* while acknowledging that the AI field had become considerably more sophisticated in creating systems of emergent behavior (such as neural networks).

Currently a professor in the Graduate School of Philosophy at the University of California, Berkeley, Dreyfus continues his work in pure philosophy (including a commentary on phenomenologist philosopher Martin Heidegger's *Being and Time*) while still keeping an eye on the computer world in his latest publication, *On the Internet*.

Further Reading

Dreyfus, Hubert. *What Computers Can't Do: a Critique of Artificial Reason*. New York: Harper and Row, 1972.

———. *What Computers Still Can't Do*. Cambridge, Mass.: MIT Press, 1992.

Dreyfus, Hubert, and Stuart Dreyfus. *Mind Over Machine: the Power of Human Intuitive Expertise in the Era of the Computer.* rev. ed. New York: Free Press, 1988.

E

Eckert, J. Presper

(1919–1995)
American
Computer Engineer

J. Presper Eckert played a key role in the design of what is often considered to be the first general-purpose electronic digital computer, then went on to pioneer the commercial computer industry. An only child, Eckert grew up in a prosperous Philadelphia family that traveled widely and had many connections with Hollywood celebrities such as Douglas Fairbanks and Charlie Chaplin. He was a star student in his private high school and also did well at the University of Pennsylvania, where he graduated in 1941 with a degree in electrical engineering and a strong mathematics background.

Continuing at the university as a graduate student and researcher, Eckert met an older researcher, John Mauchly. They found they shared a deep interest in the possibilities of electronic computing, a technology that was being spurred by the needs of war research. After earning his master's degree in electrical engineering, in 1942 Eckert joined Mauchly in submitting a proposal to the Ballistic Research Laboratory of the Army Ordnance Department for a computer that could be used to calculate urgently needed firing tables for guns, bombs, and missiles. The Army granted the contract, and they organized a team that grew to 50 people. Begun in April 1943, their ENIAC (Electronic Numerical Integrator and Computer) was finished in 1946. While it was too late to aid the war effort, the room-size machine filled with 18,000 vacuum tubes demonstrated the practicability of electronic computing. Its computation rate of 5,000 additions per second far exceeded other calculators of the time.

With some input from mathematician John von Neumann, Eckert and Mauchly began to develop a new machine, EDVAC, for the University of Pennsylvania (see VON NEUMANN, JOHN). While this effort was still under way, they formed their own business, the Eckert-Mauchly Computer Corporation and began to develop the BINAC (BINary Automatic Computer), which was intended to be a (relatively) compact and lower-cost version of ENIAC. This machine demonstrated a key principle of modern computers—the storage of program instructions along with data. The ability to store, manipulate, and edit instructions vastly increased the flexibility and ease of use of computing machines (see also HISTORY OF COMPUTING).

By the late 1940s, Eckert and Mauchly began to develop Univac I, the first commercial implementation of the new computing technology. When financial difficulties threatened to sink their company in 1950, it was acquired by Remington Rand. Working as a division within that company, the Eckert-Mauchly team completed Univac I in time for the computer to make a remarkably accurate forecast of the 1952 presidential election results.

Eckert continued with the Sperry-Rand Corporation (later called Univac and then Unisys Corporation) and

became a vice president and senior technical adviser. He retired in 1989. He received an honorary doctorate from the University of Pennsylvania in 1964. In 1969, he was awarded the National Medal of Science, the nation's highest award for achievement in science and engineering.

Further Reading

Eckstein, P. "Presper Eckert." *IEEE Annals of the History of Computing* 18, vol. 1, Spring 1996, 25–44.

McCartney, Scott. *Eniac: the Triumphs and Tragedies of the World's First Computer.* New York: Berkley Books, 1999.

Smithsonian Institution. National Museum of American History. "Presper Eckert Interview." http://americanhistory.si.edu/csr/comphist/eckert.htm

e-commerce

Electronic commerce is the use of electronic computers and networks to carry out business transactions. The transactions can be business to business (B2B), where, for example, a manufacturing company orders parts or a retail store orders finished goods. Transactions can also be retail (consumer), ranging from banking to the buying of books, music CDs, or even groceries and prescription drugs on-line.

DEVELOPMENT OF E-COMMERCE

Starting in the 1960s, electronic networks became a key part of the infrastructure for processing transactions (see BANKING AND COMPUTERS). However, the use of computers for transferring and reconciling payments between banks as well as processing consumer credit card transactions did not truly change the nature of the transactions. Essentially, existing forms and systems of payment were put into computerized form for speed and efficiency. Without the benefits of this automation the economy probably could not have sustained its overall high rate of growth over the past four decades. Further, making transaction information machine-readable meant that it could be integrated with other functions (such as inventory and accounts receivable) and that information (such as changes in consumer preferences) could be analyzed and responded to more quickly, a necessity in a highly competitive market. Thus information about current transactions could be quickly fed to the marketing department and used to adjust marketing programs or to justify the creation of new ones.

E-COMMERCE APPLICATIONS TODAY

The advent of on-line services such as America On-line and CompuServe during the 1980s and particularly, the ubiquity of the Internet's World Wide Web starting in the mid-1990s have led to an e-commerce that goes beyond transaction and payment processing to a reshaping of the way consumers buy goods and services. For example, today millions of consumers buy many of their books through on-line booksellers such as Amazon.com and Barnes and Noble rather than at a local chain or independent bookstore. This has many ramifications. On the one hand, e-commerce sites can track a consumer's activity through the use of *cookies* (small identifying files stored on the user's hard drive). Earlier, businesses could track aggregate transactions and discern patterns and trends, but today's software can create what amounts to a customized, individualized marketing program for each consumer. Thus Amazon's book recommendations will reflect not only books that the consumer has previously purchased, but books typically bought by *other* people who bought the current purchase will also be suggested.

E-commerce

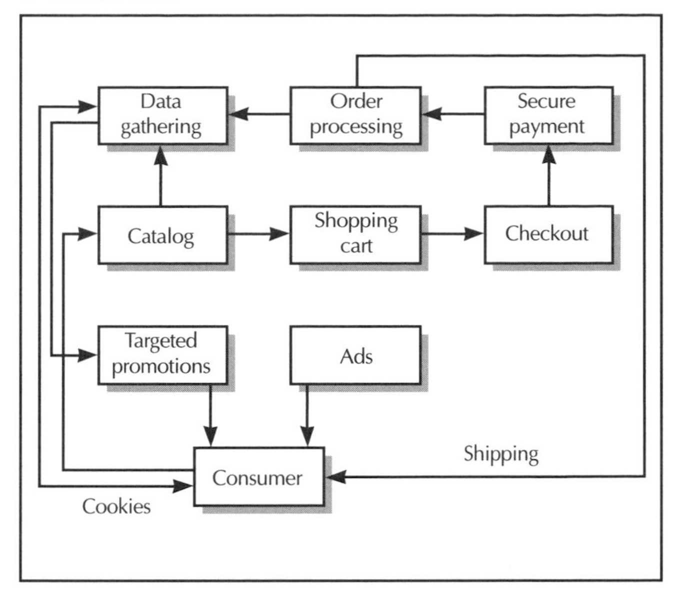

E-commerce involves far more than just advertising and selling goods and services. In a typical e-commerce system a "shopping cart" records a consumer's selections. The items ordered must be processed against inventory and prepared for shipping. Meanwhile, information about the user's selections and viewing is fed into a database from which patterns of consumer behavior can be extracted. Some techniques of information gathering raise privacy concerns, however.

Of course modern e-commerce involves more than the selling of goods such as books, music CDs, and software. The earlier assumption was that bulky or perishable goods (such as groceries) were not suitable for on-line sales. However, companies such as Webvan and regular supermarkets such as Safeway have been moderately successful at selling groceries on-line for delivery from local distribution centers. On-line pharmacies are now offering to serve consumer's prescription needs.

Another seemingly natural application for e-commerce is the selling of *content* such as news, entertainment, and reference resources. On-line portals such as

Yahoo! and About.com seek to have consumers look to their site as the starting point for their Web research (see PORTAL). An increasing number of newspapers put all or some of their stories and features on their website. (Usually current and recent material is offered for free, while older articles are priced at a few dollars each.) Thus, even traditional businesses such as newspapers and television networks, while not yet deriving significant revenue from e-commerce, are feeling obligated to have a "presence" on the Web, seeing it as increasingly important in maintaining consumer awareness.

Critics of modern e-commerce tend to focus on issues of security, privacy, and what might be called "quality of experience." The now-standard use of secure (encrypted) transactions has gradually made consumers more confident that credit card details and other personal information will not be intercepted by criminal hackers. (Nevertheless, once such information is stored on the vendor's server it can still be vulnerable to compromise.) Privacy is a more complex issue. The same software that can tailor the buying experience to each individual can also be used to create a "profile" of that individual that can be sold to other marketers (see PRIVACY IN THE DIGITAL AGE). Thus, many people feel their privacy is being invaded and the information they provide is being misused, and there has been pressure to pass legislation that might, for example, allow use of information only from consumers who have "opted in."

The broadest issue is whether e-commerce might have unfortunate social consequences. The convenience and efficiency of on-line consumer e-commerce may force many traditional "brick and mortar" stores out of business. Thus, small specialty bookstores, which had already been suffering from the assault of the giant chains, are now facing competition from on-line stores, including those owned by the chains. Critics suggest that the large amount of information about books and other products available on-line is not a substitute for the personal relationship that consumers can form with a knowledgeable bookseller. Finally, the growth of e-commerce may mean that people without on-line access may have to pay more for goods and have fewer choices (see DIGITAL DIVIDE).

FUTURE TRENDS

The stock market downturn starting in early 2001 particularly affected "dot-coms," or corporations devoted to a purely Internet-based business model. According to the e-commerce research company Webmergers, 222 major Internet companies closed in 2000, and by mid-2001 another 269 had folded. Of the failures, 47 percent involved e-commerce companies, with content-related firms accounting for 27 percent and infrastructure firms (such as ISPs) 10 percent. Among the casualties was Webran.

Many imaginative ideas for selling goods or services on the Web have failed to be viable. Even such giants as Amazon.com and Yahoo! have had their long-term viability questioned as revenues have failed to meet expectations. However, both business-to-business and consumer-oriented e-commerce applications are now sufficiently well established to suggest that this way of selling and buying is here to stay, even if the specific forms and applications change.

Further Reading

Computer Industry Almanac. http://www.c-i-a.com

"E-Commerce Guide." http://ecommerce.internet.com/

Kalakota, R., and A. Whinston. *E-Business: Roadmap for Success.* Reading, Mass.: Addison-Wesley, 1999.

Organization for Economic Cooperation and Development. "The Economic and Social Impacts of Electronic Commerce: Preliminary Report and Research Agenda." http://www.oecd.org/subject/e_commerce/summary.htm

"Tutorial: Finding Your Way Around E-Commerce." http://idm.internet.com/features/Ecommercetut.html

education and computers

Computers are increasingly being used in educational institutions from elementary school to college. While computers have had as yet little impact on the structure or organization of schools, educational software and the use of the Internet has had a growing impact on how education is delivered.

HISTORY

During the 1950s and early 1960s, computer resources were generally too scarce, expensive, and cumbersome to be used for teaching, although universities aspired to have computers to aid their graduate and faculty researchers. However, during the 1960s computer engineers and educators at the Computer-based Education Research Laboratory at the University of Illinois, Urbana, formed a unique collaboration and designed a computer system called PLATO. The PLATO system used mainframe computers to deliver instructional content to up to 1,000 simultaneous users at terminals throughout the University of Illinois and other educational institutions in the state. PLATO pioneered the interactive approach to instruction and the use of graphics in addition to text. The PLATO system was later marketed by Control Data Corporation (CDC) for use elsewhere. During this time Stanford University also set up a system for delivering computer-assisted instruction (CAI) to users connected to terminals throughout the nation. (See COMPUTER-AIDED INSTRUCTION.)

By the early 1980s, microcomputers had become relatively affordable and capable of running significant educational software including graphics. Apple Computer's Apple II became an early leader in the school market, and the introduction of the Macintosh in 1984 with the Hypercard scripting language inspired many teachers and other enthusiasts to create their own educational software. By the early 1990s, IBM compatible PCs with Windows were catching up. Commercially available computer games (such as *Civilization* or *Railroad Tycoon*) also offered ways to enrich social studies and other classes (see COMPUTER GAMES).

The advent of the World Wide Web and graphical Web browsing in the mid-1990s spurred schools to connect to the Internet. The Web offered the opportunity for educators to create resources that could be accessed by colleagues and students anywhere in the world. The use of Web portals such as Yahoo!, library catalogs, and online encyclopedias gave teachers and students potential access to a far greater variety of information than could possibly be found in textbooks. The Web also offered the opportunity for students at different schools to participate in collaborative projects, such as community surveys or environmental studies.

APPLICATIONS

Educational applications of computing can be divided into several categories, as summarized in the following table.

While small compared to the business market, the educational software industry is a significant market, targeting both schools and parents seeking to improve their children's academic performance. However, the educational use of computers extends far beyond specialized software. Schools are in effect a major industry in themselves, requiring much of the same support software as large businesses.

TRENDS

The growth of the World Wide Web has led to some shift of emphasis away from stand-alone, CD-ROM based applications running on local PCs or networks. Educators are excited about the possibilities for on-line collaboration. Public concern about children achieving an adequate level of technical skill (see COMPUTER LITERACY) has fueled an increasing commitment of funds for computer hardware, software, and networking for schools.

Some visionaries speak of a 21st-century "virtual school" that has no classroom in the conventional sense, but uses the Internet and conferencing software to bring teachers and students together. While there has been only limited experimentation in creating virtual secondary schools, thousands of university courses are now offered on-line, and many degree programs are now available. Some institutions such as the University of Phoenix have made such "distance learning" a core part of their growth strategy.

Several factors have caused other observers to have misgivings about the rush to get schools onto the "information superhighway." Many schools lack adequate physical facilities and teacher training. Under those circumstances other priorities might deserve precedence

Application Area	Users	Examples
Computer-Aided Instruction (CAI)	Generally high school and up	Course modules for science, social studies, etc. Students evaluated and materials presented on the basis of student performance.
Drill-and-practice	Elementary school students	Sets of math problems, geography quizzes, etc. Sounds or graphics used for reward for correct answers.
Online collaborative learning	Elementary and high school students	Students from different schools use e-mail or chat to coordinate a project, such as creating a website about local environmental issues.
Educational Simulations	Junior high and older students	Gamelike programs that simulate real-world problems, such as managing a city or investing in the stock market. Often commercially available games can be used.
Reference and Resources	Elementary and older students	CD-ROM or Internet based general encyclopedias or resources in particular areas such as biology. Often includes extensive multimedia (graphics, video clips, sounds).
General Software Applications	Students and teachers	Use of general-purpose software such as word processors, publishing, or presentation programs for creating class projects and reports. Also use of e-mail, chat, and conferencing systems for collaboration and after-hours communication between students and teachers.
Administrative Applications	Teachers and administrators	Use of specialized or general-purpose software to maintain attendance, grades, and other class and school administration functions.

over the installation of technology that may not be effectively utilized. At the same time, the lagging in access to technology by minorities and the poor may suggest that schools must play a significant role in providing such access and enabling the coming generation to catch up (see DIGITAL DIVIDE).

On a philosophical level, some critics of information technology such as Clifford Stoll have suggested that much educational use of computing and the Internet is "over-hyped" and may be destructive to the nurturing of a direct, intimate relationship between teacher and learner. Computers do not appear to be a panacea for dealing with the problems of low performance, poor motivation, and lack of accountability that many see as endemic to the American educational system. However, the pervasiveness of information technology and the promising capabilities that it offers ensure that computing will continue to play a central role in education.

Further Reading

"The Global Schoolhouse." http://www.globalschoolhouse.org/

Gooden, Andrea R. *Computers in the Classroom: How Teachers and Students are Using Computers to Transform Learning.* San Francisco: Jossey-Bass, 1996.

Stoll, Clifford. *Silicon Snake Oil: Second Thoughts on the Information Superhighway.* New York: Doubleday, 1995.

education in the computer field

Education and training in computer-related fields runs the gamut from courses in basic computer concepts in adult education or junior college programs to postgraduate programs in computer science and engineering. Curricula can be roughly divided into the following areas

- Computer literacy and applications
- Computer science
- Information systems

COMPUTER LITERACY AND APPLICATIONS

There is a general consensus that basic knowledge of computer terminology and mastery of widely used types of software will be essential for a growing number of occupations (see COMPUTER LITERACY). The elementary and junior high school curriculum now generally includes computer classes or "labs" where students learn the basics of word processing, spreadsheets, databases, graphics software, and use of the World Wide Web. There may also be introductory courses in programming, usually featuring easy-to-use programming languages such as Logo or BASIC.

Some high schools offer a track geared toward preparation for college studies in computer science. This track may include courses in more advanced languages such as C++ or Java. Because of public interest and marketability, courses in graphics (such as use of Adobe Photoshop), multimedia, and Web design are also increasingly popular. Adult education and community college programs feature a similar range of courses. Many of today's adult workers went to school at a time when personal computers were not readily available and computer literacy was not generally emphasized. The career prospects of many older workers are thus increasingly limited if they don't receive training in basic computer skills.

Technical or vocational schools offer tightly focused programs that are geared toward providing a set of marketable skills, often in conjunction with gaining industry certifications (see CERTIFICATION OF COMPUTER PROFESSIONALS).

COMPUTER SCIENCE

In the early 1950s, knowledge of computing tended to have an ad hoc nature. On the practical level, computing staffs tended to train newcomers in the specific hardware and machine-level programming languages in use at a particular site. On the theoretical level, programmers in scientific fields were likely to come from a background in electronics, electrical engineering, or similar disciplines.

As it became clear that computers were going to play an increasingly important role, courses specific to computing were added to curricula in mathematics and engineering. By the late 1950s, however, leading people in the computing field had become convinced that a formal curriculum in computer science was necessary for further advance in an increasingly sophisticated computing arena (see COMPUTER SCIENCE). By the early 1960s, efforts at the University of Michigan, University of Houston, Stanford, and other institutions had resulted in the creation of separate graduate departments of computer science. By the mid-1960s, the National Academy of Sciences and the President's Science Advisory Committee had both called for a major expansion of efforts in computer science education to be aided by federal funding. During the 1970s and 1980s, mathematical and engineering societies (in particular the Association for Computing Machinery (ACM) and Institute for Electrical and Electronic Engineering (IEEE) worked to established detailed computer science curricula that extended to undergraduate study. By 2000, there were 155 accredited programs in computer science in the United States.

INFORMATION SYSTEMS

The traditional computer science curriculum emphasizes theoretical matters such as algorithm and program design and computer architecture. Hiring managers in corporate information systems departments have observed that computer science graduates often have little experience in such practical considerations as systems analysis, or the designing of computer systems to meet business requirements. There has also been an increasing need for systems administrators, database administrators, and networking professionals who are well versed in the management and maintenance of particular systems.

In response to demand from industry, many universities have instituted degree programs in information systems (sometimes called MIS or Management Information Systems) as an alternative to computer science. While these programs include some study of theory, they focus on practical considerations and often include internships or other practical work experience. Some programs offer more ambitious students a dual track leading to an MBA.

Further Reading

Association for Computing Machinery. *ACM Model High School Computer Science Curriculum.* http://www.acm.org/education/hscur/index.html

Association for Computing Machinery. "ACM Student Web Site." http://www.acm.org/membership/student/

"CRA Taulbee Survey [Survey of employment of Ph.D.'s in Computer Science]." www.cra.org/statistics

Davis, G. B., and others. *IS '97: Model Curriculum and Guidelines for Undergraduate Degree Programs in Information Systems. Report of the Joint (ACM/AIS/AITP) Curriculum Task Force.* New York: ACM Press, 1997.

E-mail

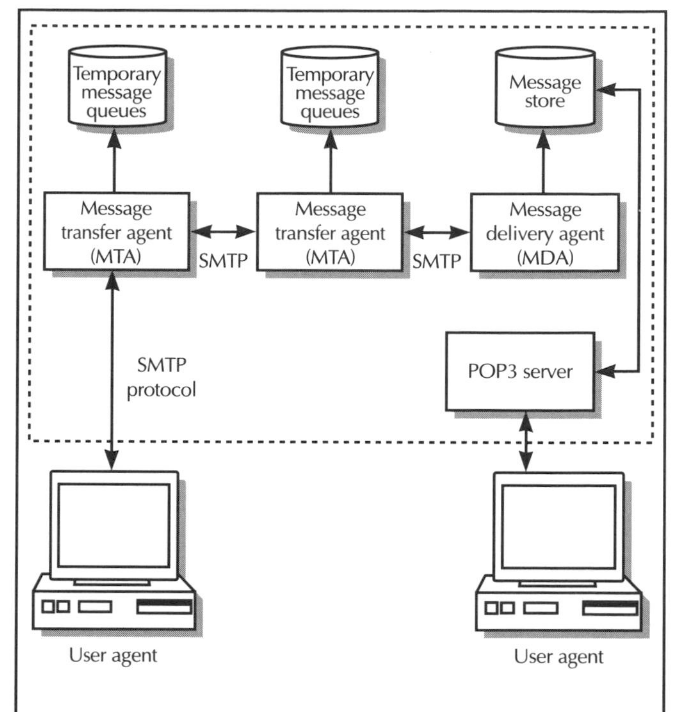

Transmission of an e-mail message depends on widely used protocols such as SMTP, which controls message format and processing, and POP3, which handles interaction between mail servers and client programs. As long as the formats are properly followed, users can employ a wide variety of mail programs (agents) and service providers can use a variety of mail server programs.

e-mail

Electronic mail is perhaps the most ubiquitous computer application in use today. E-mail can be defined as the sending of a message to one or more individuals via a computer connection.

DEVELOPMENT AND ARCHITECTURE

The simplest form of e-mail began in the 1960s as a way that users on a time-sharing computer system could post and read messages. The messages consisted of text in a file that was accessible to all users. A user could simply log into the system, open the file, and look for messages. In 1971, however, the ARPANET (ancestor of the Internet—see INTERNET) was used by researchers at Bolt Beranek and Newman (BBN) to send messages from a user at one computer to a user at another. The availability of e-mail helped fuel the growth of the ARPANET through the 1970s and beyond.

As e-mail use increased and new features were developed, the question of a standardized protocol for messages became more important. By the mid-1980s, the world of e-mail was rather fragmented, much like the situation in the early history of the telephone, where users often had to choose between two or more incompatible systems. Apranet (or Internet) users used SMTP (Simple Mail Transport Protocol) while a competing standard (OSI MHS, or Message Handling System) also had its supporters. Meanwhile, the development of consumer-oriented on-line services such as CompuServe and America On-line threatened a further balkanization of e-mail access, though systems called *gateways* were developed to transport messages from one system to another.

By the mid-1990s, however, the nearly universal adoption of the Internet and its TCP/IP protocol had established SMTP and the ubiquitous Sendmail mail transport program had established a uniform infrastructure for e-mail. The extension of the Internet protocol to the creation of intranets has largely eliminated the use of proprietary corporate e-mail systems. Instead, companies such as Microsoft and Netscape compete to offer full-featured e-mail programs that include group-oriented features such as task lists and scheduling (see also PERSONAL INFORMATION MANAGER).

E-MAIL TRENDS

The integration of e-mail with HTML for Web-style formatting and MIME (for attaching graphics and multimedia files) has greatly increased the richness and utility of the e-mail experience. E-mail is now routinely used within organizations to distribute documents and other resources. However, the addition of capabilities has also opened security vulnerabilities. For example, Microsoft Windows and the popular Microsoft Outlook e-mail client together provide the ability to run programs (scripts) directly from attachments (files associated with e-mail messages). This means that it is easy to create a virus program that will run when an enticing-looking attachment is opened. The virus can then find the user's mailbox and mail copies of itself to the people found there. E-mail has thus replaced the floppy disk as the preferred medium for such mischief. (See COMPUTER VIRUS.)

Beyond security issues, e-mail is having considerable social and economic impact. E-mail has largely replaced postal mail (and even long-distance phone calls) as a way for friends and relatives to keep in touch. As more companies begin to use e-mail for providing routine bills and statements, government-run postal systems are seeing their first-class mail revenue drop considerably. Despite the risk of viruses and the annoyance of electronic junk mail ("spam"), e-mail has become as much a part of our way of life as the automobile and the telephone.

Further Reading

Costales, Bryan, and Eric Allman. *Sendmail.* Sebastopol, Calif.: O'Reilly, 1997.

Hafner, K., and M. Lyon. *Where Wizards Stay Up Late.* New York: Simon & Schuster, 1996.

Schneier, B. *Email Security: How to Keep Your Electronic Messages Private.* New York: John Wiley, 1995.

embedded system

When people think of a computer, they generally think of a general-purpose computing system housed in a separate box, for use on the desk or as a laptop or hand-held device. However, the personal computer and its cousins are only the surface of a hidden web of computing capability that reaches deep into numerous devices used in our daily lives. Modern cars, for example, often contain several specialized computer systems that monitor fuel injection or enhance the car's grip on the road under changing conditions. Many kitchen appliances such as microwaves, dishwashers, and even toasters contain their own computer chips. Communications systems ranging from cell phones to TV satellite dishes include embedded computers. Most important, embedded systems are now essential to the operation of critical infrastructure such as medical monitoring systems and power transmission networks. (The potential vulnerability of embedded systems to the Y2K date-related problems was a major concern in the months leading up to 2000, especially because many embedded systems might have to be replaced rather than just reprogrammed. In the event, it turned out that there were relatively few date-dependent systems and only minor disruptions were experienced. See Y2K PROBLEM.)

CHARACTERISTICS OF EMBEDDED SYSTEMS

What most distinguishes an embedded system from a desktop computer is not that it is hidden inside some other device, but that it runs a single, permanent program whose job it is to monitor and respond to the environment in some way. For example, an oven controller would accept a user input (the desired temperature), monitor a sensor or thermostat, and control the heat to ensure that the correct temperature is being maintained. Embedded systems are thus similar to robots in that they sense and manipulate their environment.

Architecturally, an embedded system typically consists of a microprocessor, some *nonvolatile memory* (memory that can maintain its contents indefinitely), sensors (to receive readings from the environment), signal processors (to convert inputs into usable information), and "effectuators" (switches or other controls that the embedded system can use to change its environment). In practice, an embedded system may not have its own sensors or effectors, but instead interface with other systems (such as avionics or steering).

Programmers of embedded systems often use special compilers or languages that are particularly suited for creating embedded software (see ADA and FORTH). Because available memory is limited, embedded program code tends to be compact. Since embedded systems are often responsible for critical infrastructure, their operating programs must be carefully debugged. Designers try to make programs "robust" so they can respond sensibly to unex-

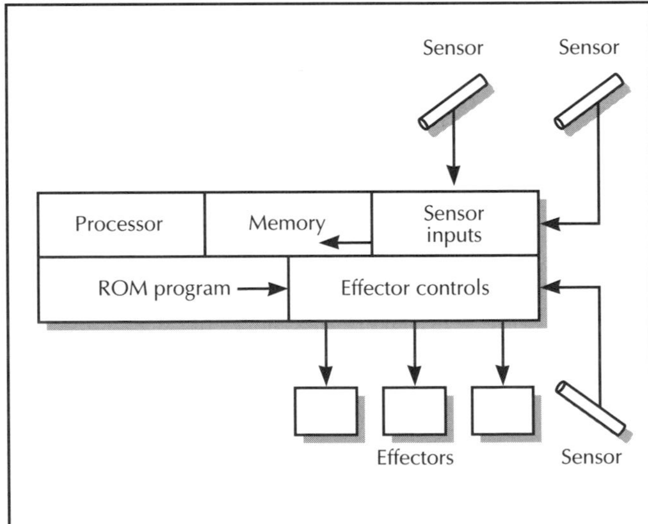

An embedded system is a computer processor that is part of a "real-world" device that must interact with its environment. Sensor inputs (such as torque or pressure sensors) provide real-time data about conditions faced by the device (such as a vehicle). This data is processed by the onboard processor under the control of a permanent (ROM) program, and commands are issued to the effector controls, which might, for example, apply braking pressure.

pected conditions or at least "fail gracefully" in a way least likely to cause damage. Other strategies to improve the reliability of embedded systems include the use of overdesigned, fault-tolerant components (as in the military "milspec") and the use of separate, redundant systems so that a failing system can be "locked out" and processing can continue elsewhere.

Further Reading

Baron, C., J. C. Geffory, and G. Motet, eds. *Embedded System Applications.* New York: Kluwer Academic Press, 1997.

Elbert, T. F. *Embedded Programming in Ada.* New York: Van Nostrand Reinhold, 1986.

Ganssle, J. G. *The Art of Programming Embedded Systems.* New York: Academic Press, 1992.

Heath, S. *Embedded Systems Design.* Oxford, U.K.: Butterworth-Heinemann, 1997.

employment in the computer field

The number of computer-related positions has grown rapidly over the past few decades. According to the U.S. Bureau of Labor Statistics, by the mid-1990s the fastest-growing professions in the United States were systems analysts, computer scientists, and computer engineers.

Computer-related employment can be broken down into the following general categories:

- Hardware design and manufacturing, including computer systems, peripherals, communications and network hardware, and other devices
- The software industry, ranging from business applications to consumer software, games, and entertainment
- The administrative sector (systems administration, network administration, database administration, computer security, and so on)
- The growing Web sector, including ISPs, Web hosts and page developers, and e-commerce applications
- The support sector, including training and education, computer book publishing, technical support, and systems repair and maintenance

In addition to these "pure" computer-related jobs, there are many other positions that involve working with PCs. These include word processing/desktop publishing, statistics, scientific research, accounting and billing, shipping, retail sales and inventory, and manufacturing. (See also PROGRAMMING AS A PROFESSION.)

JOB MARKET CONSIDERATIONS

In the late 1990s, a number of sources forecast a growing gap between the number of positions opening in computer-related fields and the number of new people entering the job market (estimates of the gap's size ranged into the hundreds of thousands nationally). Particularly in the Internet sector, demand for programmers and system administrators meant that new college graduates with basic skills could earn unprecedented salaries, while experienced professionals could often become highly paid consultants. Despite the growing emphasis on computing in secondary and higher education, computer science and engineering candidates were in particularly short supply. As a result, many companies received permission to hire larger numbers of immigrants from countries such as India.

By early 2001, the Internet "bubble" seemed on the verge of bursting, with many "dot-coms" going out of business or downsizing. Nor was the decline limited to the previously hot Internet sector. For example, chipmaker Intel began in April 2001 to offer bonuses to previously recruited students in exchanging for not holding the company to its original job offers.

To the extent that retrenchment is occurring, employers will be able to be more selective in requiring specific skills and experience. This is likely to accelerate the growing use of industry certification (see CERTIFICATION OF COMPUTER PROFESSIONALS). However, computers are now so pervasively integrated into every aspect of the economy that qualified jobseekers will continue to find opportunity and the number of openings will increase as the economy recovers. The challenge to tomorrow's jobseekers is to obtain solid qualifications while maintaining the flexibility to respond to changes in technology. For example, the growing sophistication of Web-authoring software and the automation of website management tools may reduce the need for (or status of) webmasters while some emerging sectors (such as computer-assisted genetic engineering) may experience explosive growth.

Socially, the key challenges that must be met to ensure a healthy computer-related job market are the improvement of education at all levels (see EDUCATION AND COMPUTERS) and the increasing of ethnic and gender diversity in the field (which is related to the fostering of more equal educational opportunity).

Further Reading

Computing Research Association. "Recruitment and Retention of Underrepresented Minority Graduate Students in Computer Science." http://www.cra.org/reports/r&rminorities.pdf

Freeman, P., and W. Aspray. *The Supply of Information Technology Workers in the United States.* Washington, D.C.: Computing Research Association, 1999. Available on-line at http://www.cra.org/reports/wits/cra.wits.html

Henderson, Harry. *Career Opportunities in Computers and Cyberspace.* New York: Facts On File, 1999.

Menn, Joseph. "Once High-Flying Dot-Com Crowd Comes Back to Earth Rapidly." *Los Angeles Times,* April 23, 2001, C1.

U.S. Bureau of Labor Statistics. http://www.bls.gov/home.htm

———. "Computer and Mathematical Occupations." http://www.bls.gov/oes/2000/oes150000.htm

emulation

One consequence of the universal computer concept (see VON NEUMANN, JOHN) is that in principle any computer can be programmed to imitate the operation of any other. An emulator is a program that runs on one computer but accurately processes instructions written for another (see also MICROPROCESSOR and ASSEMBLER). For example, fans of older computer games can now download emulation programs that allow modern PCs to run games originally intended for an Apple II microcomputer or an Atari game machine. Emulators allowing Macintosh and Linux users to run Windows programs have also achieved some success.

In order to work properly, the emulator must set up a sort of virtual model of the target microprocessor, including appropriate registers to hold data and instructions and a suitably organized segment of memory. While carrying out instructions in software rather than in hardware imposes a considerable speed penalty, if the processor of the emulating PC is much faster than the one being emulated, the emulator can actually run faster than the original machine.

In the past, emulation was sometimes used to allow programmers to develop software for large, expensive mainframes while using smaller machines. Emulators can also consist of a combination of specially-designed chips and software, as in the case of the "IBM 360 on a chip" that became available for the IBM PCs.

The term *emulation* is also sometimes used to refer to a program that accurately simulates the operation of a hardware device. For example, when printers that included hardware for processing the PostScript typographical language were expensive, programs were developed that could process the PostScript instructions in the PC itself and then send the output as graphics to a less expensive printer.

Further Reading

Habib, S., ed. *Microprogramming and Firmware Engineering.* New York: Van Nostrand Reinhold, 1988.

Somogyi, Stephen. "PC Emulators." *MacWorld.* http://www.macworld.com/2000/01/reviews/pcemulators.html

encapsulation

In the earliest programming languages, any part of a program could access any other part simply by executing an instruction such as "jump" or "goto." Later, the concept of the subroutine helped impose some order by creating relatively self-contained routines that could be "called" from the main program. At the time the subroutine is called, it is provided with necessary data in the form of global variables or (preferably) parameters, which are variable references or values passed explicitly when the subroutine is called. When the subroutine finishes processing, it may return values by changing global variables or changing the values of variables that were passed as parameters (see PROCEDURES AND FUNCTIONS).

While an improvement over the totally unstructured program, the subroutine mechanism has several drawbacks. If it is maintained as part of the main program code, one programmer may change the subroutine while another programmer is still expecting it to behave as previously defined. If not properly restricted, variables within the subroutine might be accessed directly from outside, leading to unpredictable results. To minimize these risks, languages such as C and Pascal allow variables to be defined so that they are "local"—that is, accessible only from code within the function or procedure. This is a basic form of encapsulation.

The class mechanism in C++ and other object-oriented languages provides a more complete form of encapsulation (see OBJECT-ORIENTED PROGRAMMING, CLASS, and C++). A class generally includes both *private* data and procedures or methods (accessible only from within the class) and *public* methods that make up the interface. Code in the main program uses the class interface to create and manipulate new objects of that class.

Encapsulation thus both protects code from uncontrolled modification or access and hides information (details) that programmers who simply want to use functionality don't need to know about. Thus, high-quality classes can be designed by experts and marketed to other developers who can take advantage of their functionality without having to "reinvent the wheel."

Further Reading

Booch, G. *Object-Oriented Analysis and Design with Applications.* 2nd ed. Reading, Mass.: Addison-Wesley, 1994.

Müller, Peter. "Introduction to Object-Oriented Programming Using C++." http://www.gnacademy.org/uu-gna/text/cc/Tutorial/tutorial.html

encryption

The use of encryption to disguise the meanings of messages goes back thousands of years (the Romans, for example, used *substitution ciphers,* where each letter in a message was replaced with a different letter). Mechanical cipher machines first came into general use in the 1930s. During World War II the German Enigma cipher machine used multiple rotors and a configurable plugboard to create a continuously varying cipher that was thought to be unbreakable. However, Allied codebreakers built electromechanical and electronic devices that succeeded in exploiting flaws in the German machine (while incidentally advancing computing technology). During the cold war Western and Soviet cryptographers vied to create increasingly complex cryptosystems while deploying more powerful computers to decrypt their opponent's messages.

In the business world, the growing amount of valuable and sensitive data being stored and transmitted on computers by the 1960s led to a need for high-quality commercial encryption systems. In 1976, the U.S. National Bureau of Standards approved the Data Encryption Standard (DES), which originally used a 56-bit key to turn each 64-bit chunk of message into a 64-bit encrypted ciphertext. DES relies upon the use of a complicated mathematical function to create complex permutations within blocks and characters of text. DES has been implemented on special-purpose chips that can encrypt millions of bytes of message per second.

PUBLIC-KEY CRYPTOGRAPHY

Traditional cryptosystems such as DES use the same key to encrypt and decrypt the message. This means that the key must be somehow transmitted to the recipient before the latter can decode the message. As a result, security may be compromised. However, the same year DES was officially adopted, Whitfield Diffie and Martin Hellman proposed a very different approach, which became known as public-key cryptography. In this scheme each user has two keys, a private key and a public key. The user publishes his or her public key, which enables any interested person to send the user an encrypted message that can be decrypted only by using the user's private key, which is kept secret. The system is more secure because the private key is never transmitted. Further, a user can distribute a message encrypted with his or her private key that can be decrypted only with the corresponding public key. This provides a sort of signature for authenticating that a message was in fact created by its putative author.

In 1978, Ron Rivest, Adi Shamir, and Leonard Adelman announced the first practical implementation of public-key cryptography. This algorithm, called RSA, became the prevailing standard in the 1980s. While keys may need to be lengthened as computer power increases, RSA is likely to remain secure for the foreseeable future.

Encryption

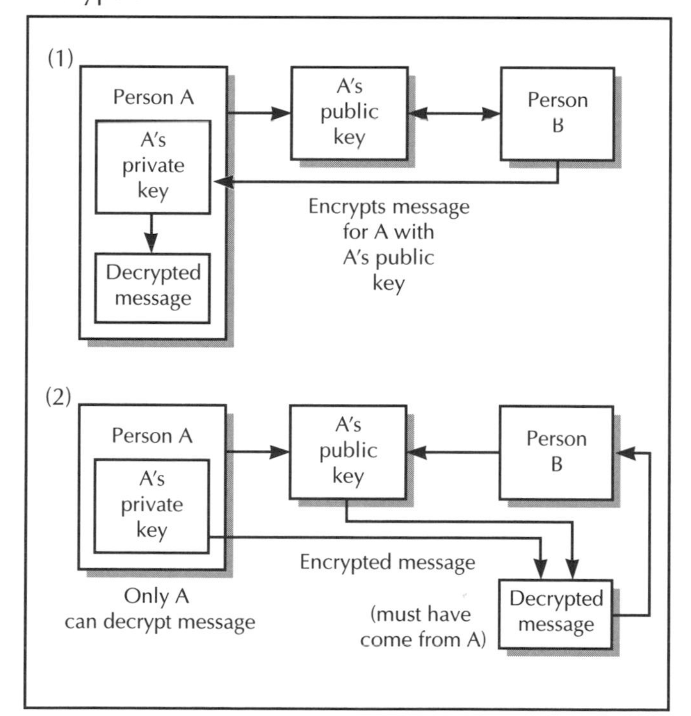

Public-key encryption allows users to communicate securely without having to exchange their private keys. In part 1, person A publishes a public key, which can be used by anyone else (such as person B) to encrypt a message that only person A can read. In part 2, person A encrypts a message with his or her private key. Since this message can be decrypted only by using person A's public key, person B can use the published public key to verify that the message is indeed from person A.

LEGAL CHALLENGES

Until the 1990s, the computer power required for routine use of encryption was generally beyond the reach of most small business and consumer users, and there was little interest in a version of the RSA algorithm for microcomputers. Meanwhile, the U.S. federal government tried to maintain tight controls over encryption technology, including prohibitions on the export of encryption software to many foreign countries.

However, the growing use of electronic mail and the hosting of commerce on the Internet greatly increased concern about security and the need to implement an easy-to-use form of encryption. In 1990, Philip Zimmermann wrote an RSA-based email encryption program that he called Pretty Good Privacy (PGP). However, RSA, Inc. refused to grant him the necessary license for its distribution. Further, FBI officials and sympathetic members of Congress seemed poised to outlaw the use of any form of encryption that did not include a provision for government agencies to decode messages.

Believing that people's liberty and privacy were at stake, Zimmermann gave copies of PGP to some friends. The program soon found its way onto computer bulletin boards, and then spread worldwide via Internet newsgroups and ftp sites. Zimmermann then developed PGP 2.0, which offered stronger encryption and a modular design that made it easy to create versions in other languages. The U.S. Customs Department investigated the distribution of PGP but dropped the investigation in 1996 without bringing charges. (At about the same time a federal judge ruled that mathematician Daniel Bernstein had the right to publish the source code for an encryption algorithm without government censorship.)

Government agencies eventually realized that they could not halt the spread of PGP and similar programs. In the early 1990s, the National Security Agency (NSA), the nation's most secret cryptographic agency, proposed that standard encryption be provided to all PC users in the form of hardware that became known as the Clipper Chip. However, the hardware was to include a "back door" that would allow government agencies and law enforcement (presumably upon fulfilling legal requirements) to decrypt any message. Civil libertarians believed that there was far too much potential for abuse in giving the government such power, and a vigorous campaign by privacy groups resulted in the mandatory Clipper Chip proposal being dropped by the mid-1990s in favor of a system called "key escrow." This system would require that a copy of each encryption key be deposited with one or more trusted third-party agencies. The agencies would be required to divulge the key if presented with a court order. However, this proposal has been met with much the same objections that had been made against the Clipper Chip.

In the early 21st century, the balance is likely to continue to favor the code-makers over the code-breakers. While it is rumored that the NSA can use arrays of supercomputers to crack any encrypted message given enough time, and a massive eavesdropping system called Echelon for analyzing message traffic has been partially revealed, as a practical matter most of the world now has access to high-quality cryptography. Only radically new technology (see QUANTUM COMPUTING) is likely to reverse this trend.

Further Reading

Henderson, Harry. *Privacy in the Information Age* (Library in a Book) New York: Facts On File, 1999.

"The International PGP Home Page." http://www.pgpi.org/

Levy, Stephen. *Crypto*. New York: Viking, 2001.

Singh, Simon. *The Code Book: the Science of Secrecy from Ancient Egypt to Quantum Cryptography*. New York: Anchor Books, 2000.

Engelbart, Douglas

(1925–)
American
Computer Engineer

Douglas Engelbart invented key elements of today's graphical user interface, including the use of windows, hypertext links, and the ubiquitous mouse. Engelbart grew up on a small farm near Portland, Oregon, and acquired a keen interest in electronics. His electrical engineering studies at Oregon State University were interrupted by wartime service in the Philippines as a radar technician. During that time he read a seminal article by Vannevar Bush entitled "As We May Think." Bush presented a wide-ranging vision of an automated, interlinked text system not unlike the development that would become hypertext and the World Wide Web (see BUSH, VANNEVAR).

After returning to college for his Ph.D. (awarded in 1955), Engelbart worked for NACA (the predecessor of NASA) at the Ames Laboratory. Continuing to be inspired by Bush's vision, Engelbart conceived of a computer display that would allow the user to visually navigate through information displays. Engelbart received his doctorate in electrical engineering in 1955 at the University of California, Berkeley, taught there a few years, and then went to the Stanford Research Institute (SRI), a hotbed of futuristic ideas. In 1962, Engelbart wrote a seminal paper of his own, titled "Augmenting Human Intellect: a Conceptual Framework." In this paper Engelbart emphasized the computer not as a mere aid to calculation, but as a tool that would enable people to better visualize and organize complex information to meet the increasing challenges of the modern world. The hallmark of Engelbart's approach to computing would continue to be his focus on the central role played by the user.

In 1963, Engelbart left SRI and formed his own research lab, the Augmentation Research Center. During the 1960s and 1970s, he worked on implementing linked text systems (see HYPERTEXT AND HYPERMEDIA). In order to help users interact with the computer display, he came up with the idea of a device that could be moved to control a pointer on the screen. Soon called the "mouse," the device would become ubiquitous in the 1980s.

Engelbart also took a key interest in the development of the ARPANET (ancestor of the Internet) and adapted his NLS hypertext system to help coordinate network development. (However, the dominant form of hypertext on the Internet would be Tim Berners-Lee's World Wide Web. [See BERNERS-LEE, TIM.]) In 1989, Engelbart founded the Bootstrap Institute, an organization dedicated to improving the collaboration within organizations, and thus their performance. During the 1990s, this nurturing of new businesses and other organizations would become his primary focus.

Engelbart received the MIT-Lemuelson Award in 1997 and the National Medal of Technology in 2000.

Further Reading

Bardini, Thierry. *Bootstrapping: Douglas Engelbart, Coevolution, and the Origins of Personal Computing*. Stanford, Calif.: Stanford University Press, 2000.

Engelbart, Christina. "The Stage is Set: from a Biographical Sketch of Douglas Engelbart." http://www.bootstrap.org/chronicle/index.html#2

"Internet Pioneers: Doug Engelbart." http://www.ibiblio.org/pioneers/Engelbart.html

"The Mouse." http://www0.mercurycenter.com/svtech/news/special/engelbart/

enterprise computing

This concept refers to the organization of data processing and communications across an entire corporation or other organization. Historically, computing technology and infrastructure often developed at different rates in the various departments of a corporation. For example, by the 1970s, departments such as payroll and accounting were making heavy use of electronic data processing (EDP) using mainframe computers. The introduction of the desktop computer in the 1980s often resulted in operations such as marketing, corporate communications, and planning being conducted using a disparate assortment of software, databases, and document repositories. Even the growing use of networking often meant that an enterprise had several different networks with at best rudimentary intercommunication.

The movement toward enterprise computing, while often functioning as a buzzword for the selling of new networking and knowledge management technology, conveys a real need both to manage and leverage the growing information resources used by a large-scale enterprise. The infrastructure for enterprise computing is the network, which today is increasingly built using Internet protocol (see TCP/IP and INTRANET) although legacy networks must often still be supported. Enterprise-oriented software uses the client-server model, with an important decision being which operating systems to support (see CLIENT-SERVER COMPUTING).

The need for flexibility in making data available across the organization is leading to a gradual shift from the older relational database (RDBMs) to object-oriented databases (OODBMs). One advantage of object-oriented databases is that it is more scalable (able to be expanded without running into bottlenecks) and data can be distributed dynamically to take advantage of available computing resources. (An alternative is the central depository. See DATA WAREHOUSE.) The dynamic use of storage resources is also important (see DISK ARRAY).

The payoffs for a well-integrated enterprise information system go beyond efficiency in resource utilization and information delivery. If, for example, the marketing department has full access to data about sales, the data can be analyzed to identify key features of consumer behavior (see DATA MINING).

Further Reading

Blanding, Stephen F. *Enterprise Operations Management Handbook, Second Edition.* Grand Rapids, Mich.: CRC Press, 1999.

Goodyear, Mark, ed. *Enterprise System Architectures.* Grand Rapids, Mich.: CRC Press, 1999.

entrepreneurs in computing

Much publicity has been given to figures such as Microsoft founder and multibillionaire Bill Gates, who turned a vest-pocket company selling BASIC language tapes into the dominant seller of operating systems and office software for PCs. Historically, however, the role of key entrepreneurs in the establishment of information technology sectors repeats the achievements of such 19th- and early 20th-century technology pioneers as Thomas Edison and Henry Ford. There appear to be certain times when scientific insight and technological capability can be translated into businesses that have the potential to transform society while making the pioneers wealthy.

Like their counterparts in earlier industrial revolutions, the entrepreneurs who created the modern computer industry tend to share certain common features. In positive terms one can highlight imagination and vision such as that which enabled J. Presber Eckert and John Mauchly to conceive that the general-purpose electronic computer could find an essential place in the business and scientific world (see ECKERT, J. PRESPER and MAUCHLY, JOHN). In the software world, observers point to Bill Gates's intense focus and ability to create and market not just an operating system but also an approach to computing that would transform the office (see GATES, WILLIAM, III). The Internet revolution, too, was sparked by both an "intellectual entrepreneur" such as Tim Berners-Lee, inventor of the World Wide Web (see BERNERS-LEE, TIM) and by Netscape founders Mark Andreessen and Jim Clark, who turned the Web browser into an essential tool for interacting with information both within and outside of organizations.

While technological innovation is important, the ability to create a "social invention"—such as a new vehicle or plan for doing business, can be equally telling. At the beginning of the 21st century, the World Wide Web, effectively less than a decade old, is seeing the struggle of entrepreneurs such as Amazon.com's Jeff Bezos, eBay's Pierre Omidyar, and Yahoo!'s Jerry Yang to expand significant toeholds in the marketing of products and information into sustainable businesses.

Historically, as industries mature, the pure entrepreneur tends to give way to the merely effective CEO. In the computer field, however, it is very hard to sort out the waves of innovation that seem to follow close upon one another. Some sectors, such as the selling of computer systems (a sector dominated by entrepreneurs such as Michael Dell [Dell Computers] and Compaq's Rod Canion) seem to have little remaining scope for innovation. In other sectors, such as operating systems (an area generally dominated by Microsoft), an innovator such as Linus Torvalds (developer of Linux) can suddenly emerge as a viable challenger. And as for the Internet and e-commerce, it is too early to tell whether the pace of innovation has slowed and the shakeout now under way will lead to a relatively stable landscape.

Further Reading

Cringely, R. X. *Accidental Empires.* New York: Harper, 1997.

Henderson, Harry. *A to Z of Computer Scientists.* New York: Facts On File, 2003.

———. *Communications and Broadcasting.* (Milestones in Science and Invention). New York: Facts On File, 1996.

Jager, Rama Dev and Rafael Ortiz. *In the Company of Giants.* New York: McGraw-Hill, 1997.
Reid, R. H. *Architects of the Web.* New York: John Wiley, 1997.
Spector, Robert. *amazon.com: Get Big Fast.* New York: HarperBusiness, 2000.

enumerations and sets

It is sometimes useful to have a data structure that holds specific, related data values. For example, if a program is to perform a particular action for data pertaining to each day of the week, the following Pascal code might be used:

```
type Day is (Monday, Tuesday, Wednesday, Thurs-
  day, Friday, Saturday, Sunday)
```

Such a data type (which is also available in Ada, C, and C++) is called an *enumeration* because it enumerates, or "spells out" each and every value that the type can hold.

Once the enumeration is defined, a looping structure can be used to process all of its values, as in:

```
var Today: Day;
for Today: = Monday to Sunday do (some state-
ments)
```

Pascal, C, and C++ do not allow the same item to be used in more than one enumeration in the same name space (area of reference). Ada, however, allows for "overloading" with multiple uses of the same name. In that case, however, the name must be qualified by specifying the enumeration to which it belongs, as in:

```
If Day = Days ('Monday') . . .
```

As far as the compiler is concerned, an enumeration value is actually a sequential integer. That is, Monday = 0, Tuesday = 1, and so on. Indeed, built-in data types such as Boolean are equivalent to enumerations (false = 0, true = 1) and in a sense the integer type itself is an enumeration consisting of 0, 1, 2, 3, . . . and their negative counterparts. Pascal also includes built-in functions to retrieve the preceding value in the enumeration (pred), the following element (succ), or the numeric position of the current element (ord).

The main advantage of using explicit enumerations is that a constant such as "Monday" is more understandable to the program's reader than the value 0. Enumerations are frequently used in C and C++ to specify a limited group of items such as flags indicating the state of device or file operation.

Unlike most other languages Pascal and Ada also allow for the definition of a *subrange*, which is a sequential portion of a previously defined enumeration. For example, once the Day type has been defined, an Ada program can define subranges such as:

```
subtype Weekdays is Days range Monday . . Fri-
day;
subtype Weekend is Days range Saturday . . Sun-
day;
```

SETS

The set type (found only in Pascal and Ada) is similar to an enumeration except the order of the items is not significant. It is useful for checking to see whether the item being considered belongs to a defined group. For example, instead of a program checking whether a character is a vowel as follows:

```
if (char = 'a') or (char = 'e') or (char = 'i')
or (char = 'o') or (char = 'u') . . .
```

the program can define:

```
type Vowels = (a, e, i, o, u);
if char in Vowels . . .
```

Further Reading

Sebesta, Robert W. *Concepts of Programming Languages.* Reading, Mass.: Addison-Wesley, 1999.

ergonomics of computing

Ergonomics is the study of the "fit" between people and their working environment. Because computers are such a significant part of the working life of so many people, finding ways for people to maximize efficiency and reduce health risks associated with computer use is increasingly important.

Since the user will be looking at the computer monitor for hours on end, it is important that the display be large enough to be comfortably readable and that there be enough contrast. Glare on the monitor surface should be avoided. It is recommended that the monitor be placed so that the top line of text is slightly below eye level. A distance of about 18 inches to two feet (roughly arm's length) is recommended. There has been concern about the health effects of electromagnetic radiation generated by monitors. Most new monitors are designed to have lower emissions.

While the "standard" keyboard has changed little in 20 years of desktop computing, there have been attempts at innovation. One, the Dvorak keyboard, uses an alternative arrangement of letters to the standard "QWERTY." Although it is a more logical arrangement from the point of view of character frequency, studies have generally failed to show sufficient advantage that would compensate for the effort of retraining millions of typists. There have also been specially shaped "ergonomic" keyboards that attempt to bring the keys into a more natural relationship with the hand (see KEYBOARD).

The use of a padded wrist rest remains controversial. While some experts believe it may reduce strain on the arm and neck, others believe it can contribute to Carpal Tunnel Syndrome. This injury, one of the most serious repetitive stress injuries (RSIs), is caused by compression of a nerve within the wrist and hand.

Because of reliance on the mouse in many applications, experts suggest selecting a mouse that comfortably fits the hand, with the buttons falling "naturally" under the fingers. When moving the mouse, the forearm, wrist, and fingers should be kept straight (that is, in line with the mouse). Some people may prefer the use of an alternative pointing device (such as trackball or "stub" within the keyboard itself, often found in laptop computers).

A variety of so-called ergonomic chairs of varying quality are available. Such a chair can be a good investment in worker safety and productivity, but for best results the chair must be selected and adjusted after a careful analysis of the individual's body proportions, the configuration of the workstation, and the type of applications being used. In general, a good ergonomic chair should have an adjustable seat and backrest and feel stable rather than rickety.

The operating system and software in use are also important. Providing clear, legible text, icons or other controls and a consistent interface will contribute to the user's overall sense of comfort, as well as reducing eyestrain. It is also important to try to eliminate unnecessary repetitive motion. For example, it is helpful to provide shortcut key combinations that can be used instead of a series of mouse movements. Beyond specific devices, the development of an integrated design that reduces stress and improves usability is part of what is sometimes called human factors research.

In March 2001, President Bush cancelled new OSHA standards that would have further emphasized reporting and mitigating repetitive stress and musculo-skeletal disorders (MSDs). However, the legal and regulatory climate is likely to continue to place pressure on employers to take ergonomic considerations into account.

Further Reading

Coe, Marlana. *Human Factors for Technical Communicators*. New York: Wiley, 1996.

"UCLA Ergonomics." http://ergonomics.ucla.edu/Ergowebv2.0/office_ergonomics.htm

U.S. Department of Labor. Occupational Safety and Health Administration. "Ergonomics." http://www.osha-slc.gov/SLTC/ergonomics/

Vredenberg, Karel, Scott Isensee, and Carol Righi. *User-Centered Design: an Integrated Approach*. Upper Saddle River, N.J.: Prentice Hall, 2001.

error correction

Transmitting data involves the sending of bits (ones and zeros) as signaled by some alternation in physical characteristics (such as voltage or frequency). There are a number of ways in which errors can be introduced into the data stream. For example, electrical "noise" in the line might be interpreted as spurious bits, or a bit might be "flipped" from one to zero or vice versa. Generally speaking, the faster the rates at which bits are being sent, the more sensitive the transmission is to effects that can cause errors.

While a few wrong characters might be tolerated in some text messages or graphics files, binary files representing executable programs must generally be received perfectly, since random changes can make programs fail or produce incorrect results. Data communications engineers have devised a number of methods for checking the accuracy of data transmissions.

The simplest scheme is called *parity*. A single bit is added to each eight-bit byte of data. In even parity, the extra (parity) bit is set to one when the number of ones in the byte is odd. In odd parity, a one is added if the data byte has an even number of ones. This means that the receiver of the data can expect it to be even or odd respectively. When the byte arrives at its destination, the receiving program checks the parity bit and then counts the number of ones in the rest of the byte. If, for example, the parity is even but the data as received has an odd number of ones, then at least one of the bits must have been changed in error. Parity is a fast, easy way to check for errors, but it has some unreliability. For example, if there were *two* errors in transmission such that a one became a zero and a different zero became a one, the parity would be unchanged and the error would not be detected.

The checksum method offers greater reliability. The binary value of each block of data is added and the sum is sent along with the block. At the destination, the bits in the block are again added to see if they still match the sum. A variation, the *cyclical redundancy check* or CRC, breaks the data into blocks and divides them by a fixed number. The remainder for the division for each block is

Error Correction

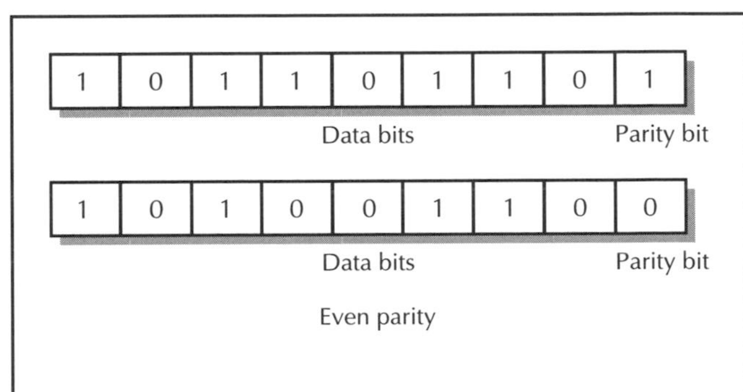

For even parity, if the number of ones in the byte is odd, the parity bit is set to one to make the total number of ones even. Odd parity would work the same way, except the parity bit would be set when necessary to ensure an odd number of ones.

appended to the block and the calculation is repeated and checked at the destination. Today most modem control software implements parity or CRC checking.

A more sophisticated method called the *Hamming Code* offers not only high reliability but also the ability to automatically correct errors. In this scheme the data and check bits are encrypted together to create a code word. If the word received is not a valid code word, the receiver can use a series of parity checks to find the original error. Increasing the ratio of redundant check bits to message bits improves the reliability of the code, but at the expense of having to do more processing to encrypt that data and requiring more time to transmit it.

Further Reading

Wicker, S. B. *Error Control Systems for Digital Communication and Storage.* Upper Saddle River, N.J.: Prentice Hall, 1995.

error handling

An important characteristic of quality software is its ability to handle errors that arise in processing (also called *run-time errors*). Before it is released for general use, a program should be thoroughly tested with a variety of input (see QUALITY ASSURANCE, SOFTWARE). When errors are found, the soundness of the algorithm and its implementation must be checked, as well as the program logic (see ALGORITHM). Interaction between the program and other programs (including the operating system) as well as with hardware must also be considered. (See BUGS AND DEBUGGING.)

However, even well-tested software is likely to encounter errors. Therefore a program intended for widespread use must include instructions for dealing with errors, anticipated or otherwise. The process of error handling can be divided into four stages: validation, detection, communication, and amelioration.

Data validation is the first line of defense. At the "front end" of the program, data being entered by a user (or read from a disk file or communications link) is checked to see whether it falls within the prescribed parameters. (In the case of a program such as a data management system, the user interface plays an important role. Data input fields can be designed so that they accept only valid characters. On-line help and error messages can explain to users why a particular input is invalid.)

However, data validation can ensure only that data falls within the generally acceptable parameters. Some particular combination or context of data might still be erroneous, and calculations performed within the program can also produce errors. Some examples include a divisor becoming zero (not allowable mathematically) or a number overflowing or underflowing (becoming too large or too small for register or memory space allotted for it).

Error communication is generally handled by a set of error codes (special numeric values) returned to the main program by the function used to perform the calculation. In addition, errors that arise in file processing (such as "file not found") also return error codes. For example, suppose there is a division function in C++

```
double Quotient(double dividend, double divisor)
throw(ZERODIV)

{
   if (0.0 == divisor)
      throw ZERODIV();
   return dividend / divisor;
}
```

In C++ "throw" means to post an error that can be "caught" by the appropriate error-handling routine. Thus, the corresponding "catch" code might have:

```
catch( ZERODIV )
{
   cout << "Division by zero error!" << endl;
}
```

Once an error has been detected and communicated, decision statements (branches or loops) can check for the presence of error codes and execute appropriate instructions based on what is encountered. (In object-oriented languages such as C++ special classes and objects are often used to handle errors.)

Many simple utility programs respond to errors by issuing an error message and then quitting. However, many real-world applications must be able to respond to errors and continue processing (for example, a program reading data from a scientific instrument may have to deal with the occasional "outlier" or a strange value caused by a burst of interference). Depending on circumstances, the error amelioration code might simply reject the erroneous data or result, ask for the data to be resent, or keep a log or statistics of the number and kind of errors encountered. More sophisticated approaches based on mathematical error analysis are also possible.

Further Reading

Lichten, W. *Data and Error Analysis.* Upper Saddle River, N.J.: Prentice Hall, 1998.

Taylor, J. R. *An Introduction to Error Analysis.* Mill Valley, Calif.: University Science Books, 1997.

expert systems

An expert system is a computer program that uses encoded knowledge and rules of reasoning to draw conclusions or solve problems. Since reasoning (as opposed to mechanical calculation) is a form of intelligent behavior, the field of expert systems (also called knowledge representation or knowledge engineering) is part of the broader field of AI (see ARTIFICIAL INTELLIGENCE).

HISTORY AND APPLICATIONS

By the end of the 1950s, early research in artificial intelligence was producing encouraging results. A number of tasks associated with human reasoning seemed to be well within the capabilities of computers. Early checkers and chess programs, while far from expert level, were steadily improving. Computer programs were proving geometry theorems. One of the most important AI pioneers, John McCarthy, declared that in principle all human knowledge could be encoded in such a way that programs could "understand" and reason from that knowledge to new conclusions.

Two disparate approaches to achieving AI gradually emerged. In the early 1960s, many researchers tried to generalize the automated reasoning process so that a program could analyze and solve a wide variety of problems, much in the way a human being can. The resulting programs were indeed flexible, but it was difficult to work with anything other than simplified problems. (The SHRDLU program, for example, worked in an abstract world of blocks on a table.)

The other approach was to try to provide exhaustively specified rules for dealing with a more narrowly defined realm of knowledge. The DENDRAL program, developed in the mid-1960s by Edward A. Feigenbaum and associates, was designed to analyze the mass spectra of organic molecules according to theories employed by chemists (see FEIGENBAUM, EDWARD). It eventually became clear that the key to the success for such program lay more in the "capturing" and encoding of expert knowledge than in the development of more flexible methods of reasoning. The methods for encoding and working with the knowledge were refined and further developed into a variety of expert systems during the 1970s.

In the 1980s, expert system technology became mature enough to leave the laboratories and play a role in industry. Two early applications were Digital Equipment Corporation's XCON, which automatically configured minicomputers from component parts at a rate and accuracy far surpassing that of human engineers. Another, Dipmeter Advisor, used real-time data to predict the dip (tilt) of rock layers in a drill bore. (This information was crucial for determining the feasibility of an oil or gas well.)

Today expert systems are a mature technology (and indeed, the most tangible success of AI research in practical applications). Expert systems are used in applications as diverse as engine troubleshooting, diagnosis of rare diseases, and investment analysis.

ANATOMY OF AN EXPERT SYSTEM

An expert system has two main components, a *knowledge base* and an *inference engine*. The knowledge base consists of a set of assertions (facts) or of rules expressed as if . . . then statements that specify conditions that, if true, allow a particular inference to be drawn (see PROLOG). The inference engine accepts new assertions or queries and tests them against the stored rules. Because satisfying one rule can create a condition that is to be tested by a subsequent rule, chains of reasoning can be built up. If the reasoning is from initial facts to an ultimate conclusion, it is called forward chaining. If a conclusion is given and the goal is to prove that conclusion, there can be backward chaining from the conclusion to the assertions (similar to axioms in mathematical proofs).

While some rules are ironclad (for example, if a closed straight figure has three sides, it's a triangle) in many real-world applications it is necessary to take a probabilistic approach. For example, experience might suggest that if a customer buys reference books there is a 40 percent chance the customer will also buy a related CD-ROM product. Thus, rules can be given weights or *confidence factors* and as the rules are chained, a cumulative probability for the conclusion can be generated and some threshold probability for asserting a conclusion can be specified. (See also FUZZY LOGIC.)

While rules-based inference systems are relatively easy to traverse automatically, they may lack the flexibility to codify the knowledge needed for complex activities (such as automatic analysis of news stories). An alternative approach involves the construction of a knowledge base consisting of *frames*. A frame (also called a *schema*) is an

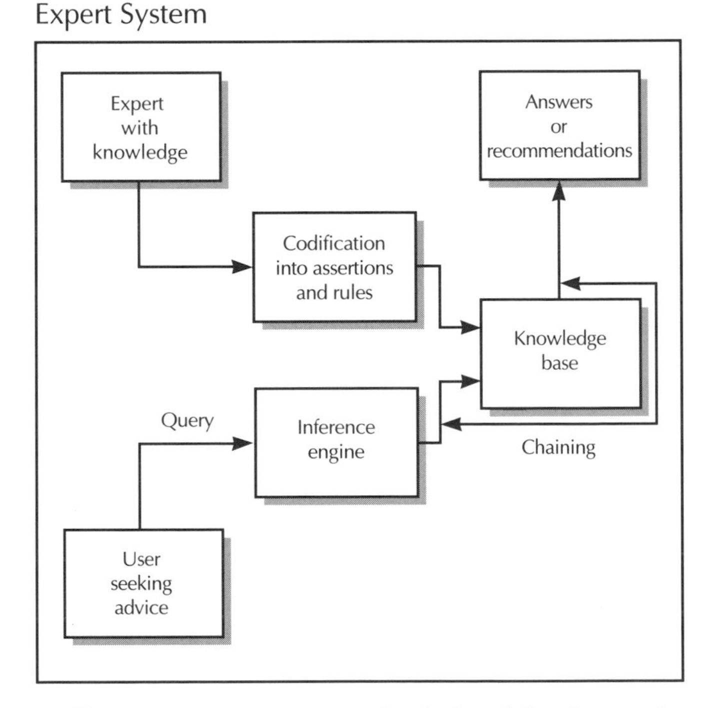

Building an expert system requires that the knowledge of experts be "captured" in the form of a series of assertions and rules called a knowledge base. Once the knowledge base is established, users seeking advice can use an inference engine to examine the knowledge base for valid conclusions that can be expressed as recommendations, often with varying degrees of confidence.

encoded description of the characteristics and relationships of entities. For example, an expert system designed to analyze court cases might have frames that describe the roles and interests of the defendant, defense counsel, prosecutor, and so on, and other frames describing the trial and sentencing process. Using this knowledge, the system might be able to predict what sort of plea agreement a particular defendant might reach with the state. While potentially more robust than a rules-based system, a frames-based system faces the twin challenges of building and maintaining a complex and open-ended knowledge base and of developing methods of reasoning more akin to generalized artificial intelligence (see ARTIFICIAL INTELLIGENCE).

TRENDS

Expert systems (particularly of the rules-based variety) now have an established place in business, industry, and science. The field of genomics and genetic engineering, widely seen as the "technology of the 21st century" may be a particularly fruitful applications area for analytical expert systems. Another promising area is the use of expert systems for e-commerce marketing analysis (see DATA MINING). An emerging emphasis in expert system development is the use of object-oriented concepts (see OBJECT-ORIENTED PROGRAMMING) and distributed database and knowledge sharing technology to build and maintain large knowledge bases more efficiently.

Further Reading

[Expert Systems Resources] http://www.sonic.net/~foggy/Leha/html/expert.html

Feigenbaum, E., P. McCorduck, and H. P. Nii. *The Rise of the Expert Company.* New York: Times Books, 1988.

Stefik, M. *Introduction to Knowledge Systems.* San Francisco: Morgan Kaufmann, 1995.

F

Feigenbaum, Edward

(1936–)
American
Computer Scientist

Edward Feigenbaum was a pioneer artificial intelligence researcher, best known for his development of expert systems (see ARTIFICIAL INTELLIGENCE). Feigenbaum was born in Weehawken, New Jersey. His father, a Polish immigrant, died before Feigenbaum's first birthday. His stepfather, an accountant and bakery manager, was fascinated by science and regularly brought young Edward to the Hayden Planetarium's shows and to every department of the vast Museum of Natural History. The electromechanical calculator his father used to keep accounts at the bakery particularly fascinated Edward. His interest in science gradually turned to a perhaps more practical interest in electrical engineering.

While at the Carnegie Institute of Technology (now Carnegie Mellon University), Feigenbaum was encouraged to venture beyond the more mundane curriculum to the emerging field of computation. He became interested in John Von Neumann's work in game theory and decision making and also met Herbert Simon, who was conducting pioneering research into how organizations made decisions (see VON NEUMANN, JOHN). This in turn brought Feigenbaum into the early ferment of artificial intelligence research in the mid-1950s. Simon and Alan Newell had just developed Logic Theorist, a program that simulated the process by which mathematicians proved theorems through the application of heuristics, or strategies for breaking problems down into simpler components from which a chain of assertions could be assembled leading to a proof.

Feigenbaum quickly learned to program IBM mainframes and then began writing AI programs. For his doctoral thesis, he explored the relation of artificial problem solving to the operation of the human mind. He wrote a computer program that could simulate the human process of perceiving, memorizing, and organizing data for retrieval. Feigenbaum's program, the Elementary Perceiver and Memorizer (EPAM), was a seminal contribution to AI. Its "discrimination net," which attempted to distinguish between different stimuli by retaining key bits of information, would eventually evolve into the *neural network* (see NEURAL NETWORK). Together with Julian Feldman, Feigenbaum edited the 1962 book *Computers and Thought,* which summarized both the remarkable progress and perplexing difficulties encountered during the field's first decade.

During the 1960s, Feigenbaum worked to develop systems that could perform induction (that is, derive general principles based on the accumulation of data about specific cases). Working on a project to develop a mass spectrometer for a Mars probe, Feigenbaum and his fellow researchers became frustrated at the computer's lack of knowledge about basic rules of chemistry. Feigenbaum then decided that such rules (or knowledge) might be encoded in such a way that the program could apply it to the data being gathered from chemical samples. The result in 1965 was Dendral, the first of what would become a host of successful and productive expert systems (see EXPERT SYSTEM). A further advance came in 1970 with Meta-Dendral, a program that could not only apply existing rules to determine the structure of a compound, it could also compare known structures with the existing database of rules and infer new rules, thus improving its own performance.

During the 1980s, Feigenbaum coedited the four-volume *Handbook of Artificial Intelligence.* He also introduced expert systems to a lay audience in two books, *The Fifth Generation* (co-authored with Pamela McCorduck) and *The Rise of the Expert Company.*

Feigenbaum combined scientific creativity with entrepreneurship in founding a company called IntelliGenetics and serving as a director of Teknowledge and IntelliCorp. These companies pioneered the commercialization of expert systems. In doing so, Feigenbaum and his colleagues publicized the discipline of "knowledge engineering"—the capturing and encoding of professional knowledge in medicine, chemistry, engineering, and other fields so that it can be used by an expert system. In what he calls the "knowledge principle" he asserts that the quality of knowledge in a system is more important than the algorithms used for reasoning. Thus, Feigenbaum has tried to develop knowledge bases that might be maintained and shared as easily as conventional databases.

Remaining active in the 1990s, Feigenbaum was second president of the American Association for Artificial Intelligence and (from 1994 to 1997) chief scientist of the U.S. Air Force. In 1995, Feigenbaum received the Association for Computing Machinery's prestigious A. M. Turing Award.

Further Reading

Feigenbaum, Edward, Julian Feldman, and Paul Armer, eds. *Computers and Thought.* Cambridge, Mass.: MIT Press, 1995.

Feigenbaum, Edward, Pamela McCorduck, and H. Penny Nii. *The Rise of the Expert Company: How Visionary Companies are Using Artificial Intelligence to Achieve Higher Productivity and Profits.* New York: Vintage Books, 1989.

Shasha, Dennis, and Cathy Lazere. *Out of Their Minds: The Lives and Discoveries of 15 Great Computer Scientists.* New York: Copernicus/Springer-Verlag, 1995.

file

At bottom, information in a computer is stored as a series of bits, which can be grouped into larger units such as bytes or "words" that represent particular numbers or characters. In order to be stored and retrieved, a collection of such binary data must be given a name and certain attributes that describe how the information can be accessed. This named entity is the file.

FILES AND THE OPERATING SYSTEM

Files can be discussed at three levels, the physical layout, the operating system, and the application program. At the physical level, a file is stored on a particular medium. (See FLOPPY DISK, HARD DISK, CD-ROM and TAPE DRIVES.) On disk devices a file takes up a certain number of sectors, which are portions of concentric tracks. (On tape, files are usually stored as contiguous segments or "blocks" of data.)

The file system is the facility of the operating system that organizes files (see OPERATING SYSTEM). For example, on DOS and Windows PCs, there is a file allocation table (FAT) that consists of a linked list of clusters (each cluster consists of a fixed number of sectors, varying with the overall size of the disk). When the operating system is asked to access a file, it can go through the table and find the clusters belonging to that file, read the data and send it to the requesting application. Modern file systems further organize files into groups called folders or directories, which can be nested several layers deep. Such a hierarchical file system makes it easier for users to organize the dozens of applications and thousands of files found on today's PCs. For example, a folder called Book might have a subfolder for each chapter, which in turn contains folders for text and illustrations relating to that chapter.

Besides storing and retrieving files, the modern file system sets characteristics or attributes for each file. Typical attributes include write (the file can be changed), read (the file can be accessed but not changed), and archive (which determines whether the file needs to be included in the next backup). In multi-user operating systems such as UNIX there are also attributes that indicate ownership (that is, who has certain rights with regard to the file). Thus a file may be executable (run as a program) by anyone, but writeable (changeable) only by someone who has "superuser" status (see also DATA SECURITY).

FILES AND APPLICATIONS

The ultimate organization of data in a file depends on the application. A typical approach is to define a data record with various fields. The program might have a loop that repeatedly requests a record from the file, processes it in some way, and repeats until the operating system tells it that it has reached the end of the file. This would be a *sequential* access; a program can also be set up for *random* access, which means that an arbitrary record can be requested and that request will be translated into the correct physical location in the file. The two approaches can be combined in ISAM (Indexed Sequential Access Method), where the records are stored sequentially but fields are indexed so a particular record can be retrieved.

Since files such as graphics (images), sound, and formatted word processing documents can only be read and used by particular applications, files are often given names with extensions that describe their format. When a Windows user sees, for example, a Microsoft Word document, the filename will have a .DOC extension (as in chapter.doc) and will be shown with an icon registered by the application for such files. Further, a file association will be registered so that when a user opens such a file the Word program will run and load it.

From a user interface point of view, the use of the file as the main unit of data has been criticized as not corresponding to the actual flow of most kinds of work. While from the computer's point of view, the user is opening, modifying, and saving a succession of separate files, the user often thinks in terms of working with documents (which may have components stored in a number of separate files.) Thus, many office software applications offer a document-oriented or project-oriented view of data that hides or minimizes the details of individual files (see DOCUMENT MODEL).

Further Reading

Bach, Marice J. *The Design of the UNIX Operating System*. Englewood Cliffs, N.J.: Prentice Hall, 1986.

Nagar, Rajeev. *Windows NT File System Internals: a Developer's Guide*. Sebastopol, Calif.: O'Reilly, 1997.

file server

The growth in desktop computing since the 1980s has resulted in much data being moved from mainframe computers to desktop PCs, which are now usually linked by networks. While a network enables users to exchange files, there remains the problem of storing large files or collections of files (such as databases) that are too large for a typical PC hard drive or that need to be accessed and updated by many users.

The common solution is to obtain a computer with large, fast disk drives (see also DISK ARRAY). This computer, the file server, is equipped with software (often included with the networking package) that serves (provides) files as requested by users or applications on the other PCs on the network. (See also CLIENT-SERVER COMPUTING.) The specifics of configuring the server for optimum efficiency, providing adequate security, and arranging for backup or archiving varies with the particular network operating system in use (the most popular environments are Windows NT and its successor, XP, and the various versions of UNIX and Linux).

The file server has many advantages over storing the files needed by each user on his or her own PC. By storing the files on a central server, ordinary users' PCs do not need to have larger, more expensive disk drives. Central storage also makes it easier to ensure that backups are run regularly (see BACKUP AND ARCHIVE SYSTEMS).

There are some potential problems with this approach. With central storage, a failure of the file server could bring work throughout the network to a halt. (The use of RAID (see DISK ARRAY) with its redundant "mirror" disks is designed to prevent the failure of a single drive from making data inaccessible). As the network and/or size of the data store gets larger, multiple servers are usually used. The performance of a file server is also greatly affected by the efficiency of the caching mechanism used (see CACHE).

As the amount of data that must be accessible increases, organizations will consider storage area network (SAN) and network attached storage (NAS) technologies. SAN makes it easier for numerous users to share a resource such as an automated tape library or disk RAID, while NAS is an efficient way to allow files to be centrally stored but readily shared.

Further Reading

Eckstein, Robert, and David Collier-Brown. *Using Samba*. Sebastopol, Calif.: O'Reilly, 1999.

Hill, Jeff and Judy Wong. "Selecting the Right NAS File Server for your Workgroup LAN." http://www.dmreview.com/portal.cfm?NavID=91&EdID=2155&PortalID=14&Topic=15

Minasi, Mark. *Mastering Windows NT Server 4*. 7th ed. San Francisco: Sybex, 2000.

Preston, W. Curtis. *Using SANs and NAS*. Sebastopol, Calif.: O'Reilly, 2002.

file transfer protocols

With today's networked PCs and the use of e-mail attachments it is easy to send a copy of a file or files from one computer to another, because networks already include all the facilities for doing so. Earlier, many PCs were not networked but could be connected via a dial-up modem. To established the connection, a terminal program running on one PC had to negotiate with its counterpart on the other machine, agreeing on whether data would be sent in 7- or 8-bit chunks, and the number of parity bits that would be included for error-checking (see ERROR CORRECTION). The sending program would inform the receiving program as to the name and basic type of the file. For binary files (files intended to be interpreted as literal binary codes, as with executable programs, images, and so on) the contents would be sent unchanged. For text files, there might be the issue of which character set (7- bit or 8-bit ASCII) was being used, and whether the ends of lines were to be marked with a CR (carriage return) character, an LF (linefeed), or both (see CHARACTERS AND STRINGS).

IMPLEMENTATIONS

Once the programs agree on the basic parameters for a file transfer, the transfer has to be managed to ensure that it completes correctly. Typically, files are divided into blocks of data (such as 1K, or 1024 bytes each). During the 1970s, Ward Christensen developed Xmodem, the first widely used file transfer program for PCs running CP/M (and later, MS-DOS and other operating systems). Xmodem was quite reliable because it incorporated a checksum (and later, a more advanced CRC) to check the integrity of each data block. If an error is detected, the receiving program requests a retransmission.

The Ymodem program adds the capability of specifying and sending a batch of files. Zmodem, the latest in this line of evolution, automatically adjusts for the amount of errors caused by line conditions by changing the size of the data blocks used and also includes the ability to resume after an interrupted file transfer. Another widely used file transfer protocol is Kermit, which has been implemented for virtually every platform and operating system. Besides file transfer, Kermit software offers terminal emulation and scripting capabilities. However, despite their robustness and capability, Zmodem and Kermit have been largely supplanted by the ubiquitous Web download link.

In the UNIX world, the ftp (file transfer protocol) program has been a reliable workhorse for almost 30 years. With ftp, the user at the PC or terminal connects to an ftp server on the machine that has the desired files. A variety of commands are available for specifying the directory, listing the files in the directory, specifying binary or text mode, and so on. While the traditional implementation uses typed text commands, there are now many ftp clients available for PCs that use a graphical interface with menus and buttons and allow files to be selected and dragged between the local and remote machines.

Even though many files can now be downloaded through HTML links on Web pages, ftp is still the most efficient way to transfer batches of files, such as for uploading content to a Web server.

Further Reading

"Beginner's Guide to Using ftp." http://www.tldp.org/HOWTO/mini/FTP-3.html

"The Kermit Project." http://www.columbia.edu/kermit/

Veljkov, Mark D., and George Hartnell. *Pocket Guides to the Internet: Transferring with File Transfer Protocol.* Westport, Conn.: Mecklermedia, 1994.

film industry and computing

Anyone who compares a science fiction film of the 1960s or 1970s with a recent offering will be struck by the realism with which today's movie robots, monsters, or aliens move against vistas of giant starships and planetary surfaces. The computer has both enhanced the management of cinematic production processes and made possible new and startling effects.

The role of the computer in film begins well before the first camera rolls. Writers can use computers to write scripts, while specialized programs can be used to lay out storyboards. Using 3D programs somewhat like CAD (drafting) programs, set designers can experiment with the positioning of objects before deciding on a final design and obtaining or creating the physical props. For mattes (backgrounds against which the characters will be shot in a scene), a computer-generated scene can now be inserted directly into the film without the need for an expensive, hand-painted backdrop.

Similarly, animation and special effects can now be rendered in computer animation form and integrated into the storyboard so that the issues of timing and combining of effects can be dealt with in the design stage. The actual effects can then be created (such as by using extremely realistic computer-controlled puppets and models together with computer generated imagery, or CGI) with the assurance that they will properly fit into the overall sequence. The ability to combine physical modeling, precise control, and added textures and effects can now create a remarkably seamless visual result in which the confrontation between a beleaguered scientist and a vicious velociraptor seems quite believable.

Just as the physical and virtual worlds are frequently blended in modern moviemaking, the traditional categories of visual media have also merged. Disney's fully animated films such as *The Lion King* benefit from the same computer-generated lighting and textures as the filming of live actors. Using 3D graphics engines, computer game scenes are now rendered with almost cinematic quality (see COMPUTER GAMES). Even characters from old movies can be digitally combined (composited) with new footage. (Of course, the artistic value of such efforts may be controversial.)

Computer technology, while still relatively expensive, can also give the generally lower budget world of television access to higher-quality effects. As computers continue to become more powerful yet cheaper, amateur or independent filmmakers will gradually gain abilities previously reserved to big Hollywood studios.

The delivery of film and video has also been greatly affected by digitization. Classic movies can be digitized to rescue them from deteriorating film stock, while videos can be delivered digitally over cable TV systems or over the Internet. The ability to easily copy digital content does raise issues of piracy or theft of intellectual property (see INTELLECTUAL PROPERTY AND COMPUTING).

Further Reading

De Leeuw, B. *Digital Cinematography.* New York: Academic Press, 1997.

Fielding, R. A. *Technological History of Motion Pictures.* Berkeley: University of California Press, 1983.

Rogers, Pauline B. *The Art of Visual Effects: Interviews on the Tools of the Trade.* Boston: Focal Press, 1999.

"Special Effects: *Titanic* and Beyond." http://www.pbs.org/wgbh/nova/specialfx2/

Yahoo! "Computer Generated Visual Arts." http://dir.yahoo.com/Arts/Visual_Arts/Computer_Generated/

financial software

Large businesses use complex database systems, spreadsheets, and other applications for activities such as accounting, planning/forecasting, and market research (see BUSINESS APPLICATIONS OF COMPUTERS). Here we will consider the variety of consumer and small business software applications that are available to help with the planning and management of financial activities, such as

- Home budgeting and money management
- Investment and retirement planning
- College financing
- Tax planning and filing
- Home buying or selling
- Basic accounting, inventory, and other activities for small business

Basic home money management programs (such as the popular Quicken) handle the budgeting and recording of daily and monthly expenses. The program can usually also interface with on-line banking services (see BANKING AND COMPUTERS) as well as exporting data to tax filing software.

For small or home-based businesses, programs such as QuickBooks can provide basic management of inventory, sales, taxes, expenses, and other functions. There are also niche programs for applications such as managing on-line auctions or Web-based sales.

For financial planning, there are a variety of programs (ranging from small free or shareware utilities available on-line to full commercial packages) that offer special calculators, graphs, and other aids for planning for the future. For example, the future value of a savings account at various points can be calculated given the interest rate, or the full cost of a loan or mortgage similarly calculated. Full-featured programs usually include helpful explanations of the various types of financial instruments. Some programs conduct an "interview" where the program asks the user about his or her objectives, priorities, or tolerance for risk, and then recommends a course of action. Such programs can be helpful even though they lack the experience and breadth of knowledge available to a human financial planner.

Tax preparation software is perhaps the fastest-growing consumer financial application. Programs normally must be purchased each year to incorporate the latest changes in tax law. An important incentive has been created by the Internal Revenue Service encouraging electronic filing of tax returns by promising speedier refunds to "e-filers."

TRENDS

Publishers of respected guidebooks (such as for college admissions and financial aid) are creating electronic versions that can be easier to use and more up to date than the printed counterpart. Meanwhile, many websites are offering utilities such as financial calculators, implemented in Java and run on-line without any software having to be downloaded by the user. The services can be offered to attract users for paid services or simply to acquire e-mail addresses for solicitation. Users should be cautious about revealing sensitive identification or financial data to unknown on-line sites.

The growth in small and home-based businesses is likely to continue in an economy that continues to offer new opportunities while reducing job security. While starting a small business is always an uncertain enterprise, easy-to-use accounting software offers the budding entrepreneur a better chance of being able to stay on top of expenses during the crucial first months of business.

The growing complexity of financial choices available to average consumers and the need for more people to take responsibility for their retirement planning is likely to increase the range and capability of financial planning applications in the future.

Further Reading

Nelson, Stephen L. *Quicken 2001 for Dummies.* New York: Hungry Minds, 2000.

Heady, Christy, Robert K. Heady, and Dennis Fertig. *The Complete Idiot's Guide to Managing Your Money.* 3rd. ed. Indianapolis, Ind.: Alpha Books, 2001.

Ivens, Kathy. *Quickbooks 2001: the Official Guide.* Berkeley, Calif.: Osborne McGraw Hill, 2001.

finite state machine

There are many calculations or other processes that can be described using a specific series of states or conditions. For example, the state of a combination lock depends not only on what numeral is being dialed or punched at the moment, but on the numbers that have been previously entered. An even simpler example is a counter (such as a car odometer), whose next output is equal to one increment plus its current setting. In other words, a state-based device has an inherent "memory" of previous steps.

In computing, a program can be set up so that each possible input, when combined with the current state, will result in a specified output. That output becomes the new state of the machine. (Alternatively, the machine can be set so that only the current state determines the output, without regard to the previous state.) This is supported by the underlying structure of the logic switching within computer circuits as well as the "statefulness" of all calculations. (Given n, $n+1$ is defined, and so on.) Alan Turing showed that combining the state mechanism with an infinite memory (conceptualized as an endless roll of tape) amounted to a universal computer—that is, a

Finite State Machine

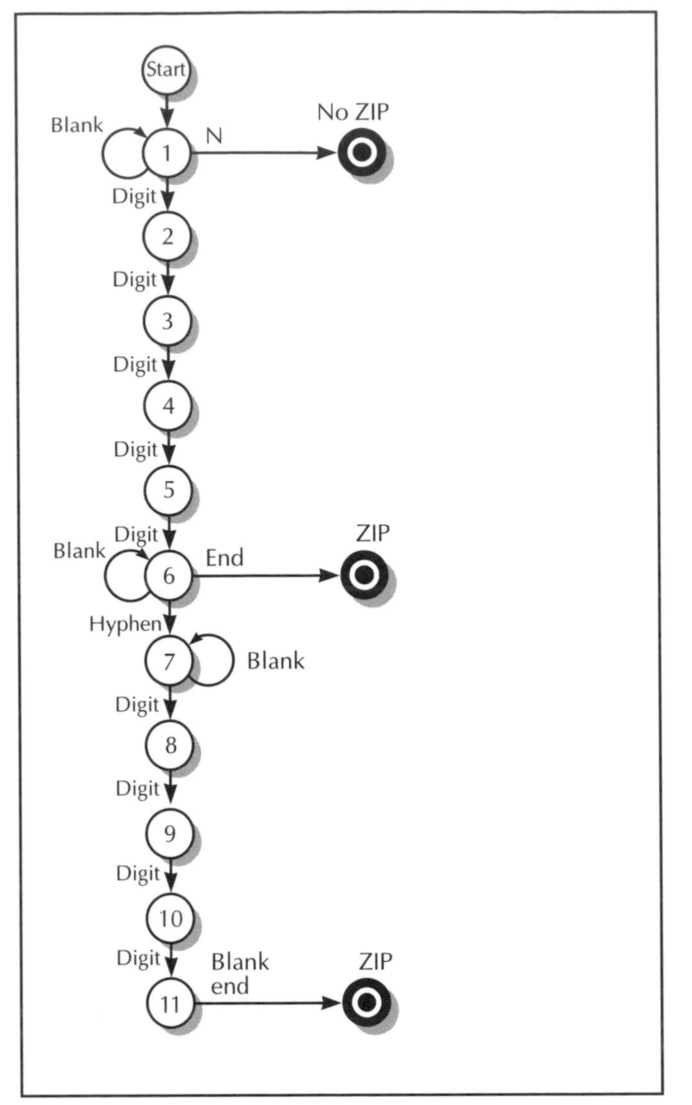

This diagram shows a finite state representation of a ZIP code. The arrows link each state (within a circle) to its possible successor. In this simple example, each digit must be followed by another digit until the fifth digit, which can either be followed by a blank (indicating a five-digit ZIP code) or four more digits for a 9-digit ZIP.

mechanism that could perform any valid calculation, given enough time (see TURING, ALAN).

The idea of the sequential (or state) machine is closely related to *automata,* which are entities whose behavior is controlled by a state table. The interaction of such automata can produce astonishingly complex patterns (see CELLULAR AUTOMATA).

APPLICATIONS

Many programs and operating systems are structured as an endless loop where an input (or command) is processed, the results returned, the next input is processed, and so on, until an exit command is received. A *mode* or state can be used to determine the system's activity. For example, a program might be in different modes such as waiting for input, processing input, displaying results, and so on. The program logic will refer to the current state to determine what to do next and at some point the logic will *transition* the system to the next state in the sequence. The validity of some kinds of programs, protocols, or circuits can therefore be proven by showing that there is an equivalent finite state machine—and thus that all possible combinations of inputs have been accounted for.

Further Reading

"Finite State Machine." http://www.c3.lanl.gov/mega-math/gloss/pattern/dfa.html

Hennie, F. C. *Finite-State Models for Logical Machines.* New York: John Wiley, 1968.

Holzmann, G. J. *Design and Validation of Computer Protocols.* Upper Saddle River, N.J.: Prentice Hall, 1991.

firewall

The vulnerability of computer systems to malicious or criminal attack has been greatly increased by the growing number of connections between computers (and local networks) and the worldwide Internet (see COMPUTER CRIME AND SECURITY, INTERNET, and TCP/IP). The growing use of permanent broadband connections by consumers (such as DSL and cable modem links) has increased the risk to home users. Intruders can use "port scanning" programs to determine what connections a given system or network has open, and can use other programs to snoop and steal or destroy sensitive data.

A firewall is a program (or combination of software and hardware) that sits between a computer (or local network) and the Internet. Typical firewall functions include:

- Examining incoming data packets and blocking those that include commands to examine or use unauthorized ports or IP addresses
- Blocking data packets that are associated with common hacking techniques such as "trojans" or "back-door" exploitations
- Hiding all the internal network addresses on a local network, presenting only a single address to the outside world (this is also called NAT, or Network Address Translation)
- Monitoring particular applications such as ftp (file transfer protocol) and telnet (remote login), restricting them to certain addresses. Often a special address called a proxy is established rather than allowing direct connections between the outside and the local network.

Firewalls are usually configured by providing a rule that specifies what is to be done based on the origin address or other characteristics of an incoming packet. Because connections made by local programs to the outside can also compromise the system, rules are also created for such applications. The firewall package may come with a set of default rules for common applications and situations. When something not covered by the rules happens, the user will be prompted and guided to establish a new rule.

Internet security packages for home users often combine a firewall with other services such as virus protection, parental control, and blocking of objectionable content or advertising.

Further Reading

"Firewalls Update." http://techupdate.cnet.com/enterprise/0-6133457-724-6949731.html?tag=st.cn.sr1.ssr.

Northcutt, Stephen. *Network Intrusion Detection: an Analyst's Handbook*. Indianapolis, Ind.: New Rider Publishing, 1999.

Zwicky, Elizabeth [and others] *Building Internet Firewalls*. 2nd ed. Sebastopol, Calif.: O'Reilly, 2000.

flag

A flag is a variable that is used to specify a particular condition or status (see VARIABLE). Usually a flag is either true or false. For example, a flag Valid_Form could be set to true before the input form is processed. If the validation check for any data field fails, the flag would be set to false. After the input procedure has ended, the main program would check the Valid_Form flag. If it's true, the data on the form is processed (for example, continuing on to the payment process). If the flag is false, the input form might be redisplayed with errors or omissions highlighted.

Flags can be combined to check multiple conditions. For example, suppose the input form routine also looked up the customer's account and checked to make sure the customer was approved for purchasing. The test for this might read:

```
If Valid_Form and Valid_Customer then
// continue processing else
// display error messages
```

In such cases, the flags are combined using the appropriate *and* or *or* operators (see BOOLEAN OPERATORS).

While flags are often used inside a routine to keep track of processing, modern programming practice discourages the use of "global" flags at the top level of the program. As with other global variables, such flags are vulnerable to being unpredictably changed or to having two parts of the program check the same flag without being able to rely on its state. (Thus a routine relies on a global flag being true but calls another routine that sets the flag to false without the original routine checking it again.) If several routines (or even programs) are being run at the same time, the situation gets even more complicated and a semaphore that can be controlled by one process at a time is more appropriate (see CONCURRENT PROGRAMMING). However, a main program that sets a flag to indicate the program mode and does not allow the flag to be changed by routines within the program is relatively safe.

Flags can also have more than two valid conditions, such as for specifying a number of possible states for a file or device. This usage is found mostly in operating systems.

Further Reading

Myers, Gene. "Becoming Bit Wise." C-Scene Issue 09. http://cscene.org/CS9/CS9-02.html

Vincent, Alan. "Flag Variables, Validation and Function Control." http://wsabstract.com/javatutors/valid1.shtml

flat-panel display

The traditional computer display uses a cathode ray tube (CRT) like that in a television set (see MONITOR). The flat-panel display is an alternative used in most laptop computers and some higher-end desktop systems. The most common type uses a liquid crystal display (LCD). The display consists of a grid of cells with one cell for each of the three colors (red, green, and blue) for each pixel.

The LCD cells are sandwiched between two polarizing filter layers that consist of many fine parallel grooves. The two filters are set so that the grooves on the second are rotated 90 degrees with respect to the first. By default, the light is polarized by the first filter, twisted by the liquid crystals so it is parallel to the grooves of the second filter, and thus passes through to be seen by the viewer. (For color displays, the light is first passed through one of three color filters to make it red, green, or blue as set for that pixel.) However, if current is applied to a crystal cell, the crystals realign so that the light passes through them without twisting. This means that the second polarizing filter now blocks the light and the cell appears opaque (or dark) to the viewer.

Color LCD displays can use two different mechanisms for sending the current through the crystals. In passive matrix displays, the current is timed so that it briefly charges the correct crystal cells. The charges fade quickly, making the image look dim, and the display cannot be refreshed quickly because of the persistence of ghost images. This means that such displays do not work well with games or other programs with rapidly changing displays.

In an active matrix display, each display cell is controlled by its own thin film transistor (TFT). These displays are sharper, brighter, and can be refreshed more frequently, allowing better displays for animations and games. However, fabrication costs for TFT displays are

Flat-Panel Screen Display

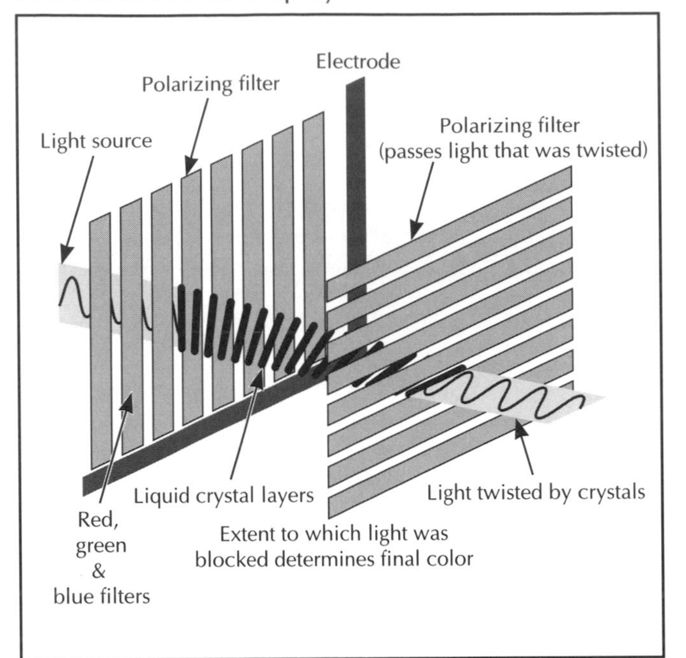

When the current is off, the liquid crystals remain twisted so the light passes through both polarizing panels and illuminates the display. However, when current is applied, the crystals straighten out, causing the light to be blocked by the second polarizing panel.

higher, and the displays are also vulnerable to having a few transistors fail, leading to permanent dark spots on the display. Active matrix displays also use more power, reducing battery life on laptop PCs. A general disadvantage of flat panel displays is that their pixel dimensions are fixed, so setting the display to a resolution smaller than its full dimensions usually results in an unsatisfactory image.

As newer technologies bring down the cost of flat-panel LCD displays they are increasingly being seen on desktop PCs, where they have the advantage of taking up much less space than conventional monitors while drawing less power. However, flat panel displays still cost about three times that of standard CRT monitors with equivalent display size and resolution.

Further Reading

"The PC Technology Guide." http://www.pctechguide.com/07panels.htm

White, Ron. *How Computers Work*. Millennium Ed. Indianapolis, Ind.: Que, 1999.

floppy disk

The floppy disk or diskette has until recent years been the primary method for distributing software and providing removable data storage for personal computers. Diskettes first appeared in the late 1960s on IBM minicomputers, and became more widespread on a variety of minicomputers and early microcomputers during the 1970s.

The now obsolete 8-inch and 5-1/4 inch disks were made from Mylar with a metal oxide coating, the assembly being housed in a flexible cardboard jacket (hence the term "floppy disk"). The more compact 3.5-inch diskettes first widely introduced with the Apple Macintosh in 1984 became the standard type for all PCs by the 1990s. These diskettes are no longer truly "floppy" and come in a rigid plastic case.

A typical floppy disk drive has a controller with two magnetic heads so that both sides of the diskette can be used to hold data. The surface is divided into concentric tracks that are in turn divided into sectors. (For more on disk organization, see HARD DRIVE.) The heads are precisely positioned to the required track/sector location using stepper motors under control of the disk driver. The data capacity of a disk depends on how densely tracks can be written on it. Today's 3.5-inch diskettes typically hold 1.44MB of data.

In recent years, drive technology has advanced so that many more tracks can be precisely written in the same amount of surface. The result is found in products such as the popular Zip disks, which can hold 100MB or even 250MB, making them comparable in capacity and speed with older, smaller hard drives.

Since the late 1990s, the traditional floppy disk has become less relevant for most users. With more computers connected to networks, the use of network copying commands or e-mail attachments has made it less necessary to exchange files via floppy, a practice dubbed "sneaker-net." When data needs to be backed up or archived, the high-capacity Zip-type drive, tape, or writable CD is a more practical alternative to low-capacity floppies. (See BACKUP AND ARCHIVE SYSTEMS.) With its iMac line, Apple actually discontinued including a floppy drive as standard equipment. In PC-compatible laptops, a floppy drive is often available as a plug-in module that can be alternated with other devices. Desktop systems still typically come with a single 3.5-inch drive.

Further Reading

White, Ron. *How Computers Work*. Millennium Ed. Indianapolis, Ind.: Que, 1999.

flowchart

A flowchart is a diagram showing the "flow" or progress of operations in a computer program. Flowcharting was one of the earliest aids to program design and documentation, and a plastic template with standard flowcharting symbols was a common programming accessory. Today CASE (computer-aided software engineering) systems often include utilities that can automatically generate flowcharts based on the control structures and procedure calls found in the program code (see CASE).

The standard flowchart symbols include blocks of various shapes that represent input/output, data processing, sorting and collating, and so on. Lines with arrows indicate the flow of data from one stage or process to the next. A diamond-shaped symbol indicates a decision to be made by the program. If the decision is an "if" (see BRANCHING STATEMENTS) separate lines branch off to the alternatives. If the decision involves repeated testing (see LOOP), the line returns back to the decision point while another line indicates the continuation of processing after the loop exits. Devices such as printers and disk drives have their own symbols with lines indicating the flow of data to or from the device.

Complex software systems can employ several levels of flowcharts. For example, a particular routine within a program might have its own flowchart. The routine as a whole would then appear as a symbol in a higher-level flowchart representing the program as a whole. Finally, a *system chart* might show each program that is run as part of an overall data processing system.

While still useful, flowcharting is often supplemented by other techniques for program representation (see PSEUDOCODE). Also, modern program design tends to shift the emphasis from charting the flow of processing to elucidating the properties and relationships of objects (see OBJECT-ORIENTED PROGRAMMING).

Further Reading

Boillot, M. H., G. M. Gleason, and L. W. Horn. *Essentials of Flowcharting*. New York: WCB/McGraw-Hill, 1995.

Flowchart

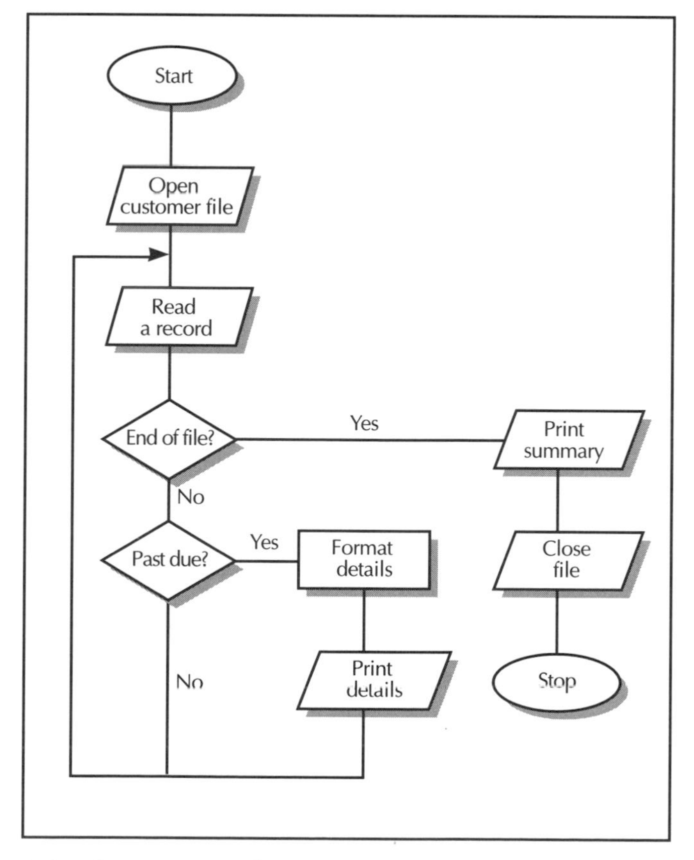

A flowchart uses a set of simple symbols to describe the steps involved in a data-processing operation. The parallelograms indicate an input/output operation (such as reading or writing a file). The "decision diamonds" have yes and no branches depending on the result of a test or comparison.

font

In computing, a font refers to a typeface that has a distinctive appearance and style. In most word processing, desktop publishing, and other programs the user can select the point size at which the font is to be displayed and printed (in traditional typography each point size would be considered to be a separate font). Operating systems such as Windows and Macintosh usually come with an assortment of fonts, and applications can register additional fonts to make them available to the system.

Fonts are often presented as a "family" that includes the same type design with different *attributes* such as boldface and italic. The spacing of letters could be uniform (monospace) as in the Courier font often used for printing computer program code or proportional (as with most text fonts). For proportional fonts the design can include *kerning*, or the precise fitting together of adjacent letters for a more attractive appearance. Fonts are also described as serif if they have small crossbars on the ends of letters such as at the end of the crossbar on a T in the Times Roman font. Other fonts such as Arial lack the tiny bars and are called sans serif (without serif).

There are two basic ways to store font data in the computer system. Bitmapped fonts store the actual pattern of tiny dots that make up the letters in the font. This has the advantage of allowing each letter in each point size to be precisely designed. The primary disadvantage is the amount of memory and system resources required to store a font in many point sizes. In practice, this consideration results in only a relatively few fonts and sizes being available.

The alternative, an *outline or vector* font uses a "page description language" such as Adobe PostScript or TrueType to provide graphics commands that specify the drawing of each letter in a font. When the user specifies a font, the text is rendered by processing the graphics commands in an interpreter. Since the actual bitmap doesn't need to be stored and all point sizes of a font can be generated from one description, outline fonts save memory and disk space (although they require additional

Font Characteristics

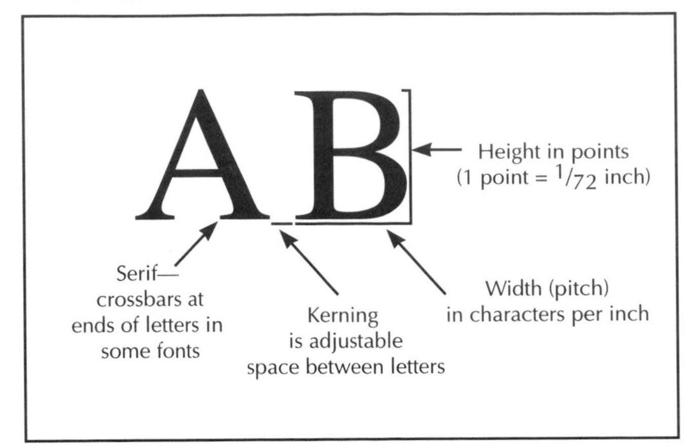

Strictly speaking, a particular type design is called a typeface, and a font is a rendering of a typeface with specified characteristics such as height in points and possibly width or pitch in characters per inch. Thus, there are usually many fonts for each typeface.

processor resources for rendering). While sophisticated scaling techniques are used to maintain a pleasing appearance as the font size changes, outline fonts will not look as polished as bitmapped fonts that are hand-designed at each point size.

A Windows Font Selector dialog box. Notice that the user can set the typeface name, size in points, and style (such as bold or italic). Various kinds of special effects can also be specified.

Further Reading

Aaron, B. *TrueType Display Fonts.* San Francisco: Sybex, 1993.

King, Jean Callan, and Tony Esposito. *The Designer's Guide to Postscript Text Type.* revised ed. New York: Wiley, 1997.

Young, Doyald. *Fonts & Logos: Font Analysis, Logotype Design, Typography, Type Comparison.* Sherman Oaks, Calif.: Delphi Press, 1999.

Forth

The unusual Forth programming language was designed by Charles H. Moore in 1970. An astronomer, Moore was interested in developing a compact language for controlling motors to drive radio telescopes and other equipment.

LANGUAGE STRUCTURE

Forth has a very simple structure. The Forth system consists of a collection of *words*. Each word is a sequence of operations (which can include other existing words). For example, the DUP word makes a copy of a data value. Data is held by a stack. For example, the arithmetic expression written as 2 + 3 in most languages would be written in Forth as + 2 3. When the + operator (which in Forth is a pre-defined word) executes, it adds the next two numbers it encounters (2 and 3) together, and puts the sum on the stack (where in turn it might be fetched for further processing by the next word in the program (see STACK). This representation is also called *postfix notation* and is familiar to many users of scientific calculators.

The words in the dictionary are "threaded" or linked so that each word contains the starting address of the next one. The Forth interpreter runs a simple loop where it fetches the next *token* (one or more characters delimited by spaces) and scans the dictionary to see if it matches a defined word (including variables). If a word is found, the code in the word is executed. If no word is found, the interpreter interprets the token as a numeric constant, loads it on the stack, and proceeds to the next word.

A key feature of Forth is its extensibility. Once you have defined a word, the new word can be used in exactly the same way as the predefined words. The various forms of *defining words* allow for great control over what happens when a new word is created and when the word is later executed. (In many ways Forth anticipated the principles of object-oriented programming, with words as objects with implicit constructors and methods. A well-organized Forth program builds up from "primitive" operations to the higher-level words, with the program itself being the highest-level word.)

Forth has always attracted an enthusiastic following of programmers who appreciate a close communion with the flow of data in the machine and the ability to precisely tailor programs. The language is completely interactive, since any word can be typed at the keyboard to execute it and display the results. Forth was also attractive in the early days of microcomputing because the lack of need for a sophisticated interpreter or compiler meant that Forth systems could run comfortably on systems that had perhaps 16K or 64K of available RAM.

Forth never caught on with the mainstream of programmers, however. Its very uniqueness and the unusual mindset it required probably limited the number of people willing to learn it. While Forth programs can be clearly organized, badly written Forth programs can be virtually impossible to read. However, Forth is sometimes found "under the hood" in surprising places (for example, the PostScript page description language is similar to Forth) and the language still has a considerable following in designing hardware control devices (see EMBEDDED SYSTEMS).

Further Reading

Brodie, L. *Starting FORTH.* 2nd ed. Upper Saddle River, N.J.: Prentice Hall, 1987.

———. *Thinking FORTH.* 2nd ed. Upper Saddle River, N.J.: Prentice Hall, 1994.

Forth Interest Group. http://www.forth.org

FORTRAN

As computing became established throughout the 1950s, the need for a language that could express operations in a more "human-readable" language began to be acutely felt. In a high-level language, programmers define variables and write statements and expressions to manipulate them. The programmer is no longer concerned with specifying the detailed storage and retrieval of binary data in the computer, and is freed to think about program structure and the proper implementation of algorithms.

FORTRAN (FORmula TRANslator) was the first widely used high-level programming language. It was developed by a project begun in 1954 by a team under the leadership of IBM researcher John Backus. The goal of the project was to create a language that would allow mathematicians, scientists, and engineers to express calculations in something close to the traditional notation. At the same time, a compiler would have to be carefully designed so that it would produce executable machine code that would be nearly as efficient as the code that would have been created through the more tedious process of using assembly languages. (See COMPILER and ASSEMBLER.)

The first version of the language, FORTRAN I, became available as a compiler for IBM mainframes in 1957. An improved (and further debugged version) soon followed. FORTRAN IV (1963) expanded the number of supported data types, added "common" data storage, and included the DATA statement, which made it easier to load literal numeric values into variables. This mature version of FORTRAN was widely embraced by scientists and engineers, who created immense libraries of code for dealing with calculations commonly needed for their work.

By the 1970s, the structured programming movement was well under way. This school of programming emphasized dividing programs into self-contained procedures into which data would be passed, processed, and returned. The use of unconditional branches (GOTO statements) as was common in FORTRAN was now discouraged. A new version of the language, FORTRAN 77 (or F77), incorporated many of the new structural features. The next version, FORTRAN 90 (F90), added support for recursion, an important technique for coding certain kinds of problems (see RECURSION). Mathematics libraries were also modernized.

SAMPLE PROGRAM

The following simple example illustrates some features of a traditional FORTRAN program:

```
INTEGER INTARRAY(10)
INTEGER ITEMS, COUNTER, SUM, AVG
SUM = 0
READ *, ITEMS
DO 10 COUNTER = 1, ITEMS
   READ *, INTARRAY(COUNTER)
   SUM = SUM + INTARRAY(COUNTER)
10 CONTINUE
   AVG = SUM / ITEMS
   PRINT  'SUM OF ITEMS IS: ', SUM
   PRINT 'AVERAGE IS: ', AVG
STOP
END
```

The program creates an array holding up to ten integers (see ARRAY). The first number it reads is the number of items to be added up. It stores this in the variable ITEMS. A DO loop statement then repeats the following two statements once for each number from 1 to the total number of items. Each time the two statements are executed, COUNTER is increased by 1. The statements read the next number from the array and add it to the running total in SUM. Finally, the average is calculated and the sum and average are printed.

Like its contemporary, COBOL, FORTRAN is viewed by many modern programmers as a rather clumsy and anachronistic language (because of its use of line number references, for example). However, there is a tremendous legacy of tested, reliable FORTRAN code and powerful math libraries. (For example, a FORTRAN program can call library routines to quickly get the sum or cross-product of any array or matrix.) These features ensure that FORTRAN has continuing appeal and utility to users who are more concerned with getting fast and accurate results than with the niceties of programming style.

Further Reading

"Fortran Resources." http://www.lahey.com/other.htm

Sleighthome, Jane, and I. D. Chivers. *Introducing Fortran 95.* New York: Springer-Verlag, 2000.

Vowels, Robin A. *Algorithms and Data Structures in F and Fortran.* Tucson, Ariz.: Unicomp, 1999.

fractals in computing

Fractals and the related idea of chaos have profoundly changed the way scientists think about and model the world. Around 1960, Benoit Mandelbrot noticed that supposedly random economic fluctuations were not distributed evenly but tended to form "clumps." As he investigated other sources of data, he found that many other things exhibited this odd behavior. He also discovered that the patterns of distribution were "self-similar"—that is, if you magnified a portion of the pattern it looked like a miniature copy of the whole. Mandelbrot coined the term *fractal* (meaning fractured, or broken up) to describe such patterns. Eventually, a number of simple mathematical functions were found to exhibit such behavior in generating values.

Fractals offered a way to model many phenomena in nature that could not be handled by more conventional geometry. For example, a coastline that might be measured as 1,600 miles on a map might be many thousands of miles when measured on local maps, as the tiny inlets at every bay and beach are measured. Fractal functions could replicate this sort of endless generation of detail in nature.

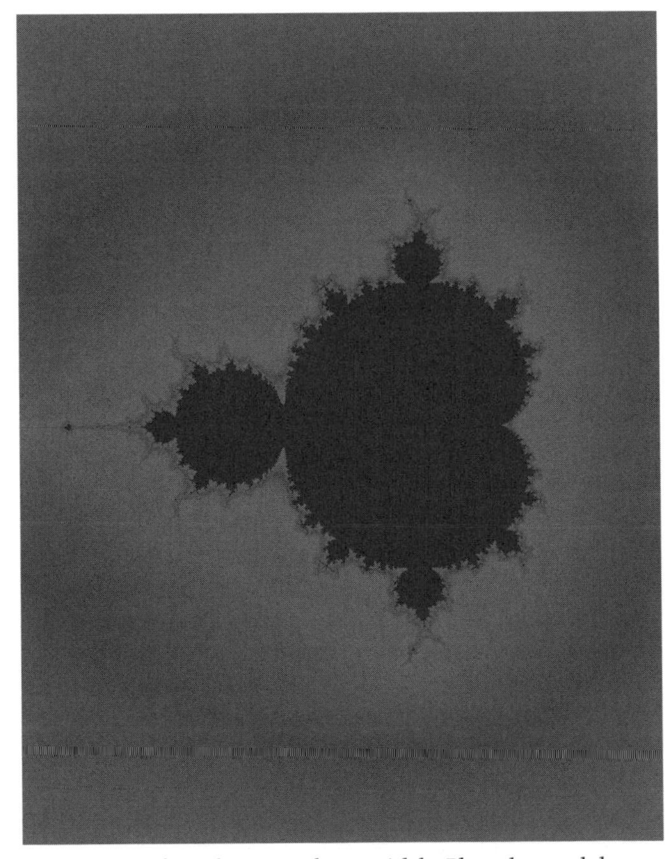

A Mandelbrot fractal generated using Adobe Photoshop and the KPT (Kai's Power Tools) Fraxplorer filter.
(COURTESY OF LISA YOUNT)

Fractals showed that seemingly random or chaotic data could form a web of patterns. At the same time, Mandelbrot and others had discovered that the pattern radically depended on the precise starting conditions: A very slight difference at the start could generate completely different patterns. This "sensitive dependence on initial conditions" helped explain why many phenomena such as weather (as opposed to overall climate) resisted predictability.

COMPUTING APPLICATIONS

Many computer users are familiar with the colorful fractal patterns generated by some screen savers. There are hundreds of "families" of fractals (beginning with the famous Mandelbrot set) that can be color-coded and displayed in endless detail. But there are a number of more significant applications. Because of their ability to generate realistic textures at every level of detail, many computer games and simulations use fractals to generate terrain interactively. Fractals can also be used to compress large digital images into a much smaller equivalent by creating a mathematical transformation that preserves (and can be used to re-create) the essential characteristics of the image. Military experts can use fractal analysis either to distinguish artificial objects from surrounding terrain or camouflage, or to generate more realistic camouflage. Fractals and chaos theory are likely to produce many surprising discoveries in the future, in areas ranging from signal analysis and encryption to economic forecasting.

Further Reading

"Fractal Resources." http://www.fignations.com/resources/home.html?page=/resources/frl.html

Gleick, J. *Chaos: The Making of a New Science.* New York: Viking, 1987.

Mandelbrot, Benoit. *The Fractal Geometry of Nature.* New York: W.H. Freeman, 1982.

Peitgen, H.-O., and P. H. Richter. *The Beauty of Fractals.* New York: Springer-Verlag, 1986.

functional languages

Most commonly used computer languages such as C++ and FORTRAN are *imperative* languages. This means that a statement is like a "sentence" in which the value of an expression or the result of a function is used in some way, such as assigning it to another variable or printing it. For example:

```
A = cube(3)
```

passes the parameter 3 to the cube function, which returns the value 27, which is then assigned to the variable A.

In a functional language, the values of functions are not assigned to variables (or stored in intermediate locations as functions are evaluated). Instead, the functions are manipulated directly, together with data items (atoms) arranged in lists. The earliest (and still best-known) functional language is LISP (see LISP). Programming is accomplished by defining and arranging functions until the desired processing is accomplished. (The decision making accomplished by branching statements in imperative languages is accomplished by incorporating conditionals in function definitions.)

Many functional languages (including LISP) for convenience incorporate some features of imperative languages. The ML language, for example, includes data type declarations. A similar language, Haskell, however, eschews all such imperative features.

APPLICATIONS

Functional languages have generally been used for specialized purposes, although they can in principle perform any task that an imperative language can. APL, which is basically a functional language, has devotees who appreciate its compact and powerful syntax for performing calculations (see APL). LISP and its variants have long been favored for many artificial intelligence applications, particularly natural language processing, where its representation of data as lists and the facility of its list-processing functions seems a natural fit.

Proponents of functional languages argue that they free the programmer from having to be concerned with explicitly setting up and using variables. In a functional language, problems can often be stated in a more purely mathematical way. Further, because functional programs are not organized as sequentially executed tasks, it may be easier to implement parallel processing systems using functional languages.

However, critics point out that imperative languages are much closer to how computers actually work (employing actual storage locations and sequential operation) and thus produce code likely to be much faster and more efficient than that produced by functional languages.

Further Reading

Bird, R. *Introduction to Functional Programming Using Haskell.* 2nd ed. London: Prentice Hall, 1998.

Peyton Jones, S. L. *The Implementation of Functional Programming Languages.* New York: Prentice Hall, 1987.

Szymanski, B. K., ed. *Parallel Functional Languages and Compilers.* New York: ACM Press, 1991.

Thompson, S. Haskell. *The Craft of Functional Programming.* 2nd ed. Reading, Mass.: Addison-Wesley, 1996.

fuzzy logic

At bottom, a data bit in a computer is "all or nothing" (1 or 0). Most decisions in computer code are also all or nothing: Either a condition is satisfied, and execution takes one specified path, or the condition is not satisfied and it goes elsewhere. In real life, of course, many situations fall between the cracks. For example, a business might want to treat a credit applicant who *almost* qualifies for "A" status different from one who barely made "B." While a program could be refined to include many gradations between B and A, another approach is to express the degree of "closeness" (or certainty) using fuzzy logic.

In 1965, mathematician L. A. Zadeh introduced the concept of the *fuzzy set.* In a fuzzy set, a given item is not simply either a member or not a member of a specified set. Rather, there is a degree of membership or "suitability" somewhere between 0 (definitely not a member) and 1 (definitely a member). A program using fuzzy logic must include a variety of rules for determining how much certainty to assign in a given case. One way to create rules is to ask experts in a given field (such as credit analysis) to articulate the degree of certainty or confidence they would feel in a given set of circumstances. For physical systems, data can also be correlated (such as the relationship of temperature to the likelihood of failure of a component) and used to create a rule to be followed by, for example, a chemical process control system.

Fuzzy logic is particularly applicable to the creation of EXPERT SYSTEMS that are better able to cope with uncertainty and the need to weigh competing factors in coming to a decision. It can also be used in engineering to allow designers to specify which factors they want to tightly constrain (such as for safety reasons) and which can be allowed more leeway. The system can then come up with optimized design specifications. Fuzzy logic has also been applied to areas such as pattern recognition and image analysis where a number of uncertain observations must often be accumulated and a conclusion drawn about the overall object.

Further Reading

Dubois, D., H. Prade, and R. R. Yager, eds. *Readings in Fuzzy Sets for Intelligent Systems.* San Francisco: Morgan Kaufmann, 1993.

"Fuzzy Logic Tutorial." http://www.seattlerobotics.org/encoder/mar98/fuz/flindex.html

"Introduction to Fuzzy Logic." http://www.cs.tamu.edu/research/CFL/fuzzy.html

Zadeh, L. A., and others. *Fuzzy Sets and Applications.* New York: John Wiley, 1987.

G

Gates, William, III (Bill)

(1955–)
American
Entrepreneur, Programmer

Bill Gates built Microsoft, the dominant company in the computer software field and in doing so, became the world's wealthiest individual, with a net worth measured in the tens of billions. Born on October 28, 1955, to a successful professional couple in Seattle, Gates's teenage years coincided with the first microprocessors becoming available to electronics hobbyists.

Gates showed both technical and business talent as early as age 15, when he developed a computerized traffic-control system. He sold his invention for $20,000, then dropped out of high school to work as a programmer for TRW for the very respectable salary of $30,000. By age 20, Gates had returned to his schooling and become a freshman at Harvard, but then he saw a cover article in *Popular Electronics*. The story introduced the Altair, the first commercially available microcomputer kit.

Gates believed that microcomputing would soon become a significant industry. To be useful, however, the new machines would need software, and Gates and his friend Paul Allen began by creating an interpreter for the BASIC language that could run in only 4 KB of memory, making it possible for people to write useful applications without having to use assembly language. This first prod-

Bill Gates is the multibillionaire CEO of the Microsoft Corporation, the leader in operating systems and software for personal computers. The company has faced antitrust actions since the late 1990s. (COURTESY OF MICROSOFT CORPORATION)

uct was quite successful, although to Gates's annoyance it was illicitly copied and distributed for free.

In 1975, Gates and Allen formed the Microsoft Corporation. Most of the existing microcomputer companies, including Apple, Commodore, and Tandy (Radio Shack) signed agreements to include Microsoft software with their machines. However, the big breakthrough came in 1980, when IBM decided to market its own microcomputer. When negotiations for a version of CP/M (then the dominant operating system) broke down, Gates agreed to supply IBM with a new operating system. Buying one from a small Seattle company, Microsoft polished it a bit and sold it as MS-DOS 1.0. Sales of MS-DOS exploded as many other companies rushed to create "clones" of IBM's hardware, each of which needed a copy of the Microsoft product.

In the early 1980s, Microsoft was only one of many thriving competitors in the office software market. Word processing was dominated by such names as WordStar and WordPerfect, Lotus 1-2-3 ruled the spreadsheet roost, and dBase II dominated databases (see WORD PROCESSING, SPREADSHEET, and DATABASE MANAGEMENT SYSTEM). But Gates and Microsoft used the steady revenues from MS-DOS to undertake the creation of Windows, a much larger operating system that offered a graphical user interface (see USER INTERFACE). While the first versions of Windows were clumsy and sold poorly, by 1990 Windows (with versions 3.1 and later, 95 and 98) had become the new dominant OS and Microsoft's annual revenues exceeded $1 billion (see MICROSOFT WINDOWS). Gates relentlessly leveraged both the company's technical knowledge of its own OS and its near monopoly in the OS sector to gain a dominant market share for the Microsoft word processing, spreadsheet, and database programs.

By the end of the decade, however, Gates and Microsoft faced formidable challenges. The growth of the Internet and the use of the Java language with Web browsers offered a new way to develop and deliver software, potentially getting around Microsoft's operating system dominance (see JAVA). That dominance, itself, was being challenged by Linux, a version of UNIX created by Finnish programmer Linus Torvalds (see UNIX). Gates responded that Microsoft, too, would embrace the networked world and make all its software fully integrated with the Internet and distributable in new ways.

However, antitrust lawyers for the U.S. Department of Justice and a number of states began legal action in the late 1990s, accusing Microsoft of abusing its monopoly status by virtually forcing vendors to include its software with their systems. In 2000, a federal judge agreed with the government. In November 2002, an appeals court accepted a proposed settlement that would not break up Microsoft but would instead restrain a number of its unfair business practices.

Gates's personality often seemed to be in the center of the ongoing controversy about Microsoft's behavior. Positively, he has been characterized as having incredible energy, drive, and focus in revolutionizing the development and marketing of software. But that same personality is viewed by critics as showing arrogance and an inability to understand or acknowledge the effects of its actions. Gates often appears awkward and even petulant in his appearance in public forums. However, his achievements and resources guarantee that he will be a major factor in the computer industry and the broader American economy for years to come.

Further Reading

Gates, Bill, Nathan Myhrvold, and Peter M. Rinearson. *The Road Ahead.* rev ed. New York: Penguin Books, 1996.

Lowe, Janet C. *Bill Gates Speaks: Insight from the World's Greatest Entrepreneur.* New York: Wiley, 1998.

Wallace, James, and Jim Erickson. *Hard Drive: Bill Gates and the Making of the Microsoft Empire.* New York: HarperBusiness, 1993.

genetic algorithms

The normal method for getting a computer to perform a task is to specify the task clearly, choose the appropriate approach (see ALGORITHM), and then implement and test the code. However, this approach requires that the programmer first know the appropriate approach, and even when there are many potentially suitable algorithms, it isn't always clear which will prove optimal.

Starting in the 1960s, however, researchers began to explore the idea that an evolutionary approach might be adaptable to programming. Biologists today know that nature did not begin with a set of highly optimized algorithms. Rather, it addressed the problems of survival through a proliferation of alternatives (through mutation and recombination) that are then subjected to natural selection, with the fittest (most successful) organisms surviving to reproduce. Researchers began to develop computer programs that emulated this process.

A genetic program consists of a number of copies of a routine that contain encoded "genes" that represent elements of algorithms. The routines are given a task (such as sorting data or recognizing patterns) and the most successful routines are allowed to "reproduce" by exchanging genetic material. (Often, further "mutation" or variation is introduced at this stage, to increase the range of available solutions.) The new "generation" is then allowed to tackle the problem, and the process is repeated. As a result, the routines become increasingly efficient at solving the given problem, just as organisms in nature become more perfectly adapted to a given environment.

APPLICATIONS

Variations of genetic algorithms or "evolutionary programming" have been used for many applications. In engineering development, a virtual environment can be set up in which a simulated device such as a robot arm can be allowed to evolve until it is able to perform to acceptable specifications. Different versions of an expert system program can be allowed to compete at performing tasks such as predicting the behavior of financial markets. Finally, a genetic program is a natural way to simulate actual biological evolution and behavior in fields such as epidemiology.

Further Reading

Back, T. *Evolutionary Algorithms in Theory and Practice: Evolution Strategies, Evolutionary Programming, Genetic Algorithms.* Oxford, U.K.: Oxford University Press, 1996.

"Genetic Algorithms Archive." http://www.aic.nrl.navy.mil/galist/

"Genetic Algorithms FAQ." http://www-2.cs.cmu.edu/afs/cs.cmu.edu/project/airepository/ai/html/faqs/ai/genetic/top.html

Mitchell, M. *An Introduction to Genetic Algorithms.* Cambridge, Mass.: MIT Press, 1996.

graphics card

Prior to the late 1970s, most computer applications (other than some scientific and experimental ones) did not use graphics. However, the early microcomputer systems such as the Apple II, Radio Shack TRS-80, and Commodore PET could all display graphics, either on a monitor or (with the aid of a video modulator) on an ordinary TV set. While primitive (low resolution; monochrome or just a handful of colors) this graphics capability allowed for a thriving market in games and educational software.

The earliest video displays for mainstream PCs provided basic text display capabilities (such as the MDA, or monochrome display adapter, with 25 lines of text up to 80 characters per line) plus the ability to create graphics by setting the color of individual pixels. The typical low-end graphics card of the early 1980s was the CGA (Color Graphics Adapter), which offered various modes such as 320 by 200 pixels with four colors. Computers marketed for professional use offered the EGA (Enhanced Graphics Adapter), which could show 640 by 350 pixels at 16 colors.

The ultimate video display standard during the time of IBM dominance was the VGA (Video Graphics Array), which offered a somewhat improved high resolution of 640 by 480 pixels at 16 colors, with an alternative of a lower 320 by 280 pixels but with 256 colors. Because of its use of a color palette containing index values, the 256 colors can actually be drawn from a range of 262,144 possible choices. VGA also marked a break from earlier standards because in order to accommodate such a range of colors it had to convert digital information to analog signals to drive the monitor, rather than using the digital circuitry found in earlier monitors.

Graphics Card

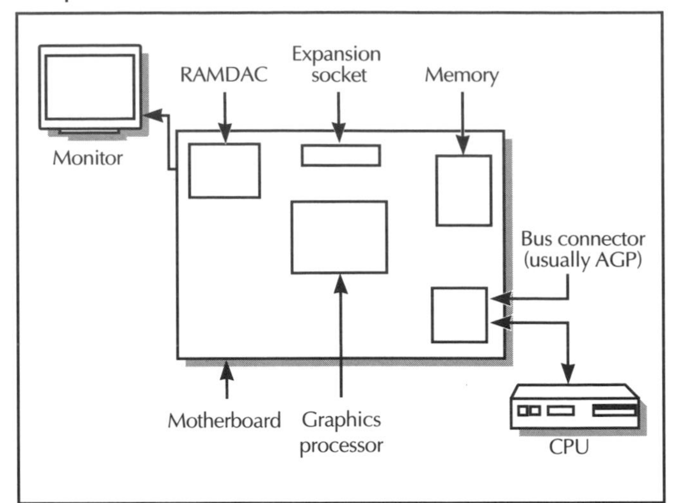

The basic parts of a graphics card. The card is connected to the CPU by the bus (often a special bus called the AGP, or Accelerated Graphics Port). Graphics data can be generated by the CPU and transferred directly to the graphics card's memory, but most cards today perform a lot of the graphics processing using the card's own onboard processor. Once data is stored, it is transferred from the card's memory to the monitor by way of a digital-to-analog converter called a RAMDAC, which is necessary because the monitor is an analog device.

Modern video cards can be loosely described as implementing SVGA (Super VGA), but there are no longer discrete standards. Typical display resolutions for desktop PCs today are 1024 by 768 or 1280 by 1024 pixels. (Laptops traditionally have had a lower-resolution 800 by 600 display, but many are now comparable to desktop displays.) The range of colors is vast, with up to 16,777,216 possible colors stored as 32 bits per pixel.

Storing 32 bits (4 bytes) for each of the pixels on a 1024 by 768 screen requires more than 3 megabytes. However, this is just for static images. Games, simulations, and other applications use moving 3D graphics. Since a computer screen actually has only two dimensions, mathematical algorithms must be used to transform the representation of objects so they look as if they have three dimensions, appearing in proper perspective, with regard to what objects are behind other objects, and with realistic lighting and shading (see COMPUTER GRAPHICS).

Traditionally, all of the work of producing the actual screen data was undertaken by the PC's main processor, executing instructions from the application program and display driver. By putting a separate processor on the video card (called a video accelerator), together with its own supply of memory (now up to 64MB), the main system was freed from this burden. A new high-bandwidth connection between the PC motherboard and the graphics card became available with the development of the

AGP (Accelerated Graphics Port). (See BUS.) Memory used on video cards is also optimized for video operations, such as by using types of memory such as Video RAM (VRAM) that do not need to be refreshed as frequently.

Increasingly, the algorithms for creating realistic images (such as lighting, shading, and texture mapping) are now supported by the software built into the video card. Of course, the applications program needs a way to tell the graphics routines what to draw and how to draw it. In systems running Microsoft Windows, a program function library called Direct3D (part of a suite called DirectX) has become the standard interface between applications and graphics hardware. Video card manufacturers in turn have optimized their cards to carry out the kinds of operations implemented in DirectX. (A nonproprietary standard called OpenGL has also achieved some acceptance, particularly on non-Windows systems.)

In evaluating video cards, the tradeoff is between the extent to which advanced graphic features are supported and the number of frames per second that can be calculated and sent to the display. If the processing becomes too complicated, the frame rate will slow down and the display will appear to be jerky instead of smooth.

Further Reading

Engel, Wolfgang F., Amir Geva, and Andre Lamothe, eds. *Beginning Direct3D Game Programming*. Roseville, Calif., PrimaTech, 2001.

Ferraro, Richard F. *Programmer's Guide to the EGA, VGA, and SuperVGA cards*. 3rd ed. Reading, Mass.. Addison-Wesley, 1994.

Pabst, Tom [and others]. *Tom's Hardware Guide: High Performance PC Secrets*. Indianapolis, Ind.: Que, 1998.

"PC 3D Graphics Accelerators FAQ." http://www.faqs.org/faqs/pc-hardware-faq/3dgraphics-cards/part1/

PC Guide. "Video Cards." http://www.pcguide.com/ref/video/

PC Tech. Guide. "Multimedia/Graphics Cards." http://www.pctechguide.com/05graphics.htm

Woo, Mason [and others]. *OpenGL Programming Guide: The Official Guide to Learning OpenGL, Version 1.2*. 3rd ed. Reading, Mass.: Addison-Wesley, 1999.

graphics formats

Broadly speaking, a graphics file consists of data that specifies the color of each pixel (dot) in an image. Since there are many ways this information can be organized, there are a variety of graphics file formats. The most important and widely used ones are summarized below.

BMP (WINDOWS BITMAP)

In a bitmap format there is a group of bits (i.e. a binary value) that specifies the color of each pixel. Windows provides standard bitmap (BMP) formats for 1-bit (2 colors or monochrome), 4-bit (16 colors), 8-bit (256 colors), or 24-bit (16 million colors). The Windows bitmap format is also called a DIB (device-independent bitmap) because the stored colors are independent of the output device to be used (such as a monitor or printer). The relevant device driver is responsible for translating the color to one actually used by the device. Because it is "native" to Windows, BMP is widely used, especially for program graphics resources.

Bitmap formats have the advantage of storing the exact color of every pixel without losing any information. However, this means that the files can be very large (from hundreds of thousands of bytes to several megabytes for Windows screen graphics). BMP and other bitmap formats do support a simple method of compression called run-length encoding (RLE), where a series of identical pixels is replaced by a single pixel and a count. Bitmap files can be further compressed through the use of utilities such as the popular Zip program (see DATA COMPRESSION).

EPS

EPS (Encapsulated PostScript) is a vector-based rather than bitmap (raster) format. This means that an EPS file consists not of the actual pixel values of an image, but the instructions for drawing the image (including coordinates, colors, and so on). The instructions are specified as a text file in the versatile PostScript page description language. This format is usually used for printing, and requires a printer that supports PostScript (there are also PostScript renderers that run entirely in software, but they tend to be slow and somewhat unreliable).

GIF

GIF, or Graphics Interchange Format, is a bitmapped format promulgated by CompuServe. Instead of reserving enough space to store a large number of colors in each pixel, this format uses a color table that can hold up to 256 colors. Each pixel contains a reference (index into) the color table. This means that GIF works best with images that have relatively few colors and for applications (such as webpages) where compactness is important. GIF also uses compression to achieve compactness, but unlike the case with JPEG it is a lossless compression called LZW. There is also a GIF format that stores simple animations.

JPEG

JPEG, which stands for Joint Photographic Experts Group, is widely used for digital cameras because of its ability to highly compress the data in a color graphics image, allowing a reasonable number of high-resolution pictures to be stored in the camera's onboard memory. The compression is "lossy," meaning that information is lost during compression (see DATA COMPRESSION). At relatively low compression ratios (such as 10:1, or 10 percent of the original image size) changes in the image due to data loss are unlikely to be perceived by the human eye. At higher ratios (approaching 100:1) the image becomes seriously degraded. JPEG's ability to store thousands of colors (unlike GIF's limit of 256) makes the format particularly suitable for the subtleties of photography.

PCX

PCX is a compressed bitmap format originally used by the popular PC Paintbrush program. In recent years it has been largely supplanted by BMP and TIFF.

TIFF

TIFF, or Tagged Image File Format, is also a compressed bitmap format. There are several variations by different vendors, which can lead to compatibility problems. Implementations can use various compression methods, generally leading to ratios of 1.5 to 1 to about 2 to 1.

Further Reading

Brown, C. Wayne, and Barry J. Shepherd. *Graphics File Formats Reference and Guide.* Greenwich, Conn.: Manning Publications, 1995.

"GIF Animation on the WWW." http://members.aol.com/royalef/gifanim.htm

Graphics File Formats page. http://www.dcs.ed.ac.uk/home/mxr/gfx/index-hi.html

JPEG FAQ. http://www.faqs.org/faqs/jpeg-faq/

Murray, James D., William vanRyper. *Encyclopedia of Graphics File Formats.* 2nd ed. (on CD-ROM). Sebastopol, Calif.: O'Reilly, 1996.

"Wotsit's Format: the Programmer's Resource." http://www.wotsit.org/

graphics tablet

While conventional pointing devices (see MOUSE) are quite satisfactory for making selections and even manipulating objects, many artists prefer the control available only through a pen or pencil, which allows the angle and pressure of the stylus tip to be varied, creating precise lines and shading. A graphics tablet (also called a *digitizing tablet*) is a device that uses a specially wired pen or pencil with a flat surface (tablet). Besides tracking the location of the pen and translating it into X/Y screen coordinates, the tablet also has pressure sensors (depending on sensitivity, the tablet can recognize 256, 512, or 1024 levels of pressure). In combination with buttons on the pen, the pressure level can be used to control the line thickness, transparency, or color. In addition, the driver software for some graphics tablets includes additional functions such as the ability to program the pen to control features of such applications as Adobe Photoshop.

The tablet is connected to the PC (usually through a USB port). The pen may be connected to the tablet by a tether, or it may be wireless. If the pen has an onboard battery, it can provide additional features at the expense of weight and the need to replace batteries occasionally.

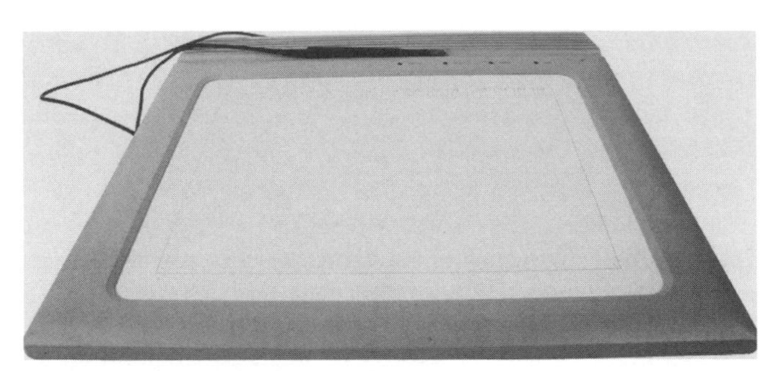

Many graphics tablets use a stylus or pen. The system can track the pen's position and often, the amount of pressure being exerted, and draw the line accordingly.

A variant implementation uses a small "puck" instead of a pen. The puck, which can be moved smoothly over the tablet surface, often has a window with crosshairs in the center. This makes it particularly useful for tracing detailed drawings such as in engineering applications.

Many artists find that wielding a pen with a graphics tablet offers not only finer control, but also more natural and less fatiguing method of input than with the mouse.

Further Reading

About.com. "Is a Graphics Tablet Right for You?" http://graphicssoft.about.com/compute/graphicssoft/library/weekly/aa001213a.htm

Kolle, Iril C. *Graphics Tablet Solutions.* Cincinnati, Ohio: Muska & Lipman, 2001.

green PC

This is a general term for features that reduce the growing environmental impact of the manufacture or use of computers. This impact has several aspects: energy consumption, resource consumption, pollution, and waste disposal or recycling.

ENERGY CONSUMPTION

The greatest part of a typical computer system's power consumption is from the monitor, followed by the hard drive and CPU. It follows that considerable energy can be saved if these components are powered down when not in use. On the other hand, most users do not want to go through the whole computer startup process several times a day. One solution is to design a computer system so that it turns off many components when not in use but is still able to restore full function in a few seconds.

When applied to a personal computer, the federally adopted Energy Star designation indicates a computer system that includes an energy saving mode that can power down the monitor, hard drive, or CPU after a specified period elapses without user activity, such that the inactive system consumes no more than 30 watts. In the ultimate energy-saving feature a suspend mode saves the current state of the computer's memory (and thus of program operation) to a disk file. When the user presses a key (or moves the mouse), the computer "wakes up" and reloads its memory contents from the disk, resuming operation where it left off. By 2000, virtually all new PCs were Energy Star compliant, though many users fail to actually enable the power-saving features.

RESOURCE CONSUMPTION

Computers consume a variety of resources, starting with their manufacturing and packaging. Resource consumption can be reduced by building more compact units and by designing components so they can be more readily stripped and recycled or reused. Adopting reusable storage media (such as rewritable CDs), recycling printer toner cartridges, and changing office procedures to minimize the generation of paper documents are also ways to reduce resource consumption.

POLLUTION

While computer use doesn't create significant pollution, the same can't be said of the manufacture of computers and other electronic devices. Fabrication of computer chips in more than 200 large plants around the world employs a variety of toxic chemicals and waste products. (California's Silicon Valley has 29 toxic sites under the EPA's Superfund program.) There has been some progress in reducing the use of toxic chemicals in electronics plants. For example, the use of ozone-depleting chlorofluorocarbons (CFCs) has been largely replaced by much less harmful cleaning agents. However, as of 2000 industry and regulators had barely made a dent in the overall production of toxic waste. (Since many forms of electrical generation involve pollution, the energy-saving mode described earlier can also be viewed as pollution reduction.)

Further Reading

Anzovin, Stephen. *The Green PC: Practical Choices that Make a Difference.* 2nd ed. New York: Windcrest/McGraw Hill, 1994.

Matthews, James. "How Green is Your PC?" *Computer User,* Dec. 20, 2000. Available on-line at http://www.computeruser.com/articles/daily/8,6,1,1220,00.html

groupware

When PCs were first introduced into the business world, they tended to be used in isolation. Individual workers would prepare documents such as spreadsheets and database reports and then print them out and distribute them as memos, much in the way of traditional paper documents. However, as computers began to be tied together into local area networks (see LOCAL AREA NETWORK) in the 1980s, focus began to shift toward the use of software to facilitate communication, coordination, and collaboration among workers. This loosely defined genre of software was dubbed groupware.

Popular groupware software suites such as Lotus Notes and Microsoft Exchange generally offer at least some of the following features

- E-mail coordination, including the creation of group or task-oriented mail lists
- Shared calendar, giving each participant information about all upcoming events
- Meeting management, including scheduling (ensuring compatibility with everyone's existing schedule) and facilities booking
- Scheduling tasks with listing of persons responsible for each task, progress (milestones met), and checking off completed tasks
- Real-time "chat" or instant message capabilities
- Documentation systems that allow a number of people to make comments on the same document and see and respond to each other's comments
- "Whiteboard" systems that allow multiple users to draw a diagram or chart in real time, with everyone able to see and possibly modify it

With many networks now implemented as intranets, groupware is increasingly integrated with the Internet, with documents and shared resources (calendars, schedules, and so on) implemented in HTML as webpages or Web-linked databases. (See also INTRANET and PERSONAL INFORMATION MANAGER.)

Groupware is likely to be an increasingly important aspect of institutional information processing in a global, mobile economy. With workgroups often geographically distributed (as well as including telecommuters), traditional face-to-face meetings become increasingly impractical as well as often being considered wasteful and inefficient. (Concerns about air travel following the September 11, 2001 attacks may also play a role.) Groupware offers the ability to create as many "virtual meetings" or conferences as needed. On the other hand, this very facility might lead to inefficiency if people spend too much time with "chitchat" or unfocused meetings.

Further Reading

Chaffey, Dave. *Groupware, Workflow and Intranets: Reengineering the Enterprise with Collaborative Software.* Boston: Digital Press, 1998.

Coleman, David, and Raman Khanna, eds. *Groupware: Technologies and Applications.* Upper Saddle River, N.J.: Prentice Hall, 1995.

"Groupware Servers." http://serverwatch.internet.com/groupservers.html
Hills, Mellanie. *Intranet as Groupware.* New York: Wiley, 1996.
Jessup, Leonard M., and Joseph S. Valacich, eds. *Group Support Systems: New Perspectives.* New York: Macmillan, 1993.

Grove, Andrew S.

(1936–)
Hungarian-American
Entrepreneur

Andrew Grove is a pioneer in the semiconductor industry and builder of Intel, the corporation whose processors now power the majority of personal computers. Grove was born András Gróf on September 2, 1936, in Budapest to a Jewish family. Grove's family was disrupted by the Nazi conquest of Hungary early in World War II. Andrew's father was conscripted into a work brigade and then into a Hungarian formation of the German army. Andrew and his mother, Maria, had to hide from the Nazi roundup in which many Hungarian Jews were sent to death in concentration camps.

Although the family survived and was reunited after the war, Hungary had come under Soviet control. Andrew, now 20, believed his freedom and opportunity would be very limited, so he and a friend made a dangerous border crossing into Austria. Grove came to the United States, where he lived with his uncle in New York and studied chemical engineering. He then earned his Ph.D. at the University of California at Berkeley and became a researcher at Fairchild Semiconductor in 1963 and then assistant director of development in 1967. He soon became familiar with the early work toward what would become the integrated circuit, key to the microcomputer revolution that began in the 1970s and wrote a standard textbook (*Physics and Technology of Semiconductor Devices*).

In 1968, however, he joined colleagues Robert Noyce and Gordon Moore in leaving Fairchild and starting a new company, Intel. Grove switched from research to management, becoming Intel's director of operations. He established a management style that featured what he called "constructive confrontation"—a vigorous, objective discussion where opposing views could be aired without fear of reprisal. Critics, however, sometimes characterized the confrontations as more harsh than constructive.

Grove became a formidable competitor. In the late 1970s, it was unclear whether Intel (maker of the 8008, 8080, and subsequent processors) or Motorola (with its 68000 processor) would dominate the market for microprocessors to run the new desktop computers. Grove emphasized the training and deployment of a large sales force, and by the time the IBM PC debuted in 1982, it and its imitators would all be powered by Intel chips.

During the 1980s, Grove would be challenged to be adaptable when Japanese companies eroded Intel's share of the DRAM (memory) chip market, often "dumping" product below their cost. Grove decided to get Intel out of the memory market, even though it meant downsizing the company until the growing microprocessor market made up for the lost revenues. In 1987, Grove had weathered the storm and become Intel's CEO. He summarized his experience of the rapidly changing market with the slogan "only the paranoid survive."

During the 1990s, Intel introduced the popular Pentium line, having to overcome mathematical flaws in the first version of the chip and growing competition from Advanced Micro Devices (AMD) and other companies that made chips compatible with Intel's. Grove also had to fight prostate cancer, apparently successfully, and relinquished his CEO title in 1998, remaining chairman of the board.

Through several books and numerous articles, Grove has had considerable influence on the management of modern electronics manufacturing. He has received many industry awards, including the IEEE Engineering Leadership Recognition award (1987), and the AEA Medal of Achievement award (1993). In 1997, he was CEO of the Year (*CEO* magazine) and *Time* magazine's Man of the Year.

Further Reading

"Andy's Biography." http://www.andygrove.com/intel/people/asg/biography/biography.htm
Grove, Andrew S. *High Output Management.* 2nd ed. New York: Vintage Books, 1995.
———. *One-on-One with Andy Grove.* New York: Putnam, 1987.
———. *Only the Paranoid Survive.* New York: Currency Doubleday, 1996.

H

hackers and hacking

Starting in the late 1950s, in computer facilities at MIT, Stanford, and other research universities people began to encounter persons who had both unusual programming skill and an obsession with the inner workings of the machine. While ordinary users viewed the computer simply as a tool for solving particular problems, this peculiar breed of programmers reveled in extending the capabilities of the system and creating tools such as program editors that would make it easier to create even more powerful programs. The movement from mainframes that could run only one program at a time to machines that could simultaneously serve many users created a kind of environmental niche in which these self-described *hackers* could flourish. Indeed, while administrators sometimes complained that hackers took up too much of the available computer time, they often depended on them to fix the bugs that infested the first versions of time-sharing operating systems. Hackers also tended to work in the wee hours of the night while normal users slept.

Early hackers had a number of distinctive characteristics and tended to share a common philosophy, even if it was not always well articulated:

- Computers should be freely accessible, without arbitrary limits on their use (the "hands-on imperative").
- "Information wants to be free" so that it can reach its full potential. Conversely, government or corporate authorities that want to restrict information access should be resisted or circumvented.
- The only thing that matters is the quality of the "hack"—the cleverness and utility of the code and what it lets computers do that they couldn't do before.
- As a corollary to the above, the reputation of a hacker depends on his (it was nearly always a male) work—not on age, experience, academic attainment, or anything else.
- Ultimately, programming was a search for truth and beauty and even a redemptive quality—coupled with the belief that technology can change the world.

Hackers were relatively tolerated by universities and sometimes prized for their skills by computer companies needing to develop sophisticated software. However, as the computer industry grew, it became more concerned with staking out, protecting, and exploiting intellectual property. To the hacker, however, intellectual property was a barrier to the unfettered exploration and exploitation of the computer. Hackers tended to freely copy and distribute not only their own work but also commercial systems software and utilities.

During the late 1970s and 1980s, the microcomputer created a mass consumer software market, and a new generation of hackers struggled to get the most out of machines that had a tiny amount of memory and only rudimentary graphics and sound capabilities. Some became successful game programmers. At the same time a new term entered the lexicon, *software piracy.* Pirate hackers cracked the copy protection on games and other commercial software so the disks could be copied freely and exchanged at computer fairs, club meetings, and on illicit bulletin boards (where they were known as "warez"). (See COPY PROTECTION and INTELLECTUAL PROPERTY AND COMPUTING.)

The growing use of on-line services and networks in the 1980s and 1990s brought new opportunities to exploit computer skills to vandalize systems or steal valuable information such as credit card numbers. The popular media used the term *hacker* indiscriminately to refer to clever programmers, software pirates, and people who stole information or spread viruses across the Internet. People who are aware of the original hacker tradition continue to fight to distinguish true hackers from modern criminal "crackers." Indeed, the wide availability of scripts for password cracking, website attacks, and virus creation means that destructive crackers often have little real knowledge of computer systems and do not share the attitudes and philosophy of the true hackers who sought to exploit systems rather than destroy them.

During the 1980s, a new genre of science fiction called *cyberpunk* became popular. It portrayed a fractured, dystopian future where elite hackers could "jack into" computers, experiencing cyberspace directly in their mind, as in William Gibson's *Neuromancer* and *Count Zero.* In such tales the hacker became the high-tech analog of the cowboy or samurai, a virtual gunslinger who fought for high stakes on the newest frontier. Meanwhile, lurid stories about such notorious real-world hackers as Kevin Mitnick and Kevin Poulsen and their pursuit by security experts such as Tsutomo Shimomura brought the dark side of hacking into popular consciousness.

In response to public fears about hackers' capabilities, federal and local law enforcement agencies stepped up their efforts to find and prosecute people who crack or vandalize systems or websites. Antiterrorism experts now worry that well-financed, orchestrated hacker attacks could be used by rogue nations or terrorist groups to paralyze the American economy and perhaps even disrupt vital infrastructure such as power distribution and air traffic control. In this atmosphere the older, more positive image of the hacker seems to be fading—although the free-wheeling creativity of hacking at its best continues to be manifested in such efforts as the development of the Linux operating system.

Further Reading

2600 magazine. http://www.2600.com

Gibson, William. *Neuromancer.* West Bloomfield, Mich.: Phantasia Press, 1986.

Hafner, Katie, John Markoff. *Cyberpunk: Outlaws and Hackers on the Computer Frontier.* New York: Simon & Schuster, 1991.

Levy, Stephen. *Hackers: Heroes of the Computer Revolution.* New York: Doubleday, 1984.

Littman, J. *The Fugitive Game: On-line with Kevin Mitnick.* Boston: Little, Brown, 1996.

Raymond, Eric. *The New Hacker's Dictionary.* 3rd ed. Cambridge, Mass.: MIT Press, 1996.

Sterling, Bruce. *The Hacker Crackdown: Law and Disorder on the Electronic Frontier.* New York: Bantam Books, 1992.

Turkle, S. *The Second Self: Computers and the Human Spirit.* New York: Simon & Schuster, 1984.

handwriting recognition

While the keyboard is the traditional means for entering text into a computer system, both designers and users have long acknowledged the potential benefits of a system where people could enter text using ordinary script or printed handwriting and have it converted to standard computer character codes (see CHARACTERS AND STRINGS). With such a system people would not need to master a typewriter-style keyboard. Further, users could write commands or take notes on handheld or "palm" computers the size of a small note pad that are too small to have a keyboard (see PORTABLE COMPUTERS). Indeed, such facilities are available to a limited extent today.

A handwriting recognition system begins by building a representation of the user's writing. With a pen or stylus system, this representation is not simply a graphical image but includes the recorded "strokes" or discrete movements that make up the letters. The software must then create a representation of features of the handwriting that can be used to match it to the appropriate character templates. Handwriting recognition is actually an application of the larger problem of identifying the significance of features in a pattern.

One approach (often used on systems that work from previously written documents rather than stylus strokes) is to identify patterns of pixels that have a high statistical correlation to the presence of a particular letter in the rectangular "frame" under consideration. Another approach, used in the Apple Newton handheld computer, is to try to identify groups of strokes or segments that can be associated with particular letters. In evaluating such tentative recognitions, programs can also incorporate a network of "recognizers" that receive feedback on the basis of their accuracy (see NEURAL NETWORK). Finally, where the identity of a letter remains ambiguous, lexical analysis can be used to determine the most probable letter in a given context, using a dictionary or a table of letter group frequencies.

IMPLEMENTATION AND APPLICATIONS

A number of handheld computers beginning with Apple's Newton in the mid-1990s and the now popular PalmPilot

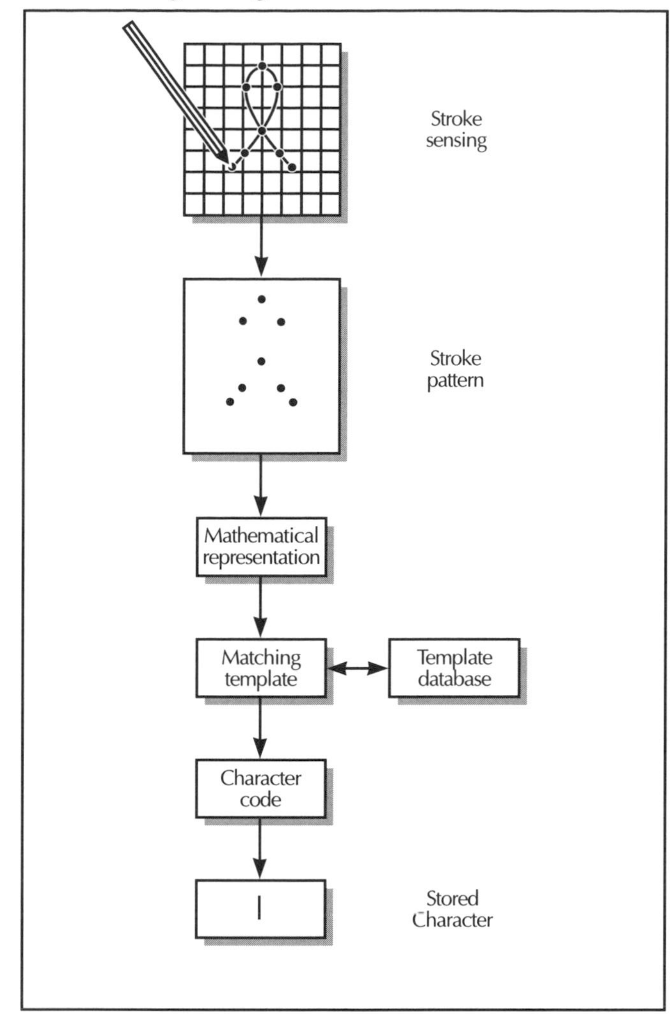

One approach to handwriting recognition involves the extraction of a stroke pattern and its comparison to a database of templates representing various letters and symbols. Ultimately, the corresponding ASCII character is determined and stored.

have some ability to recognize handwriting. However, current systems can be frustrating to use because accuracy often requires that users write very carefully and consistently or (as in the case of the PalmPilot) even replace their usual letter strokes with simplified alternatives that the computer can more easily recognize. If the user is allowed to use normal strokes, the system must be gradually "trained" by the user giving writing samples and confirming the system's guess about the letters. As the software becomes more adaptable and processing power increases (allowing more sophisticated algorithms or larger neural networks to be practical) users will be able to write more naturally and systems will gain more consumer acceptance.

Currently, handwriting recognition is used mainly in niche applications, such as collecting signatures for delivery services or filling out "electronic forms" in applications where the user must be mobile and relatively hands-free (such as law enforcement).

Further Reading

Center for Pattern Recognition and Machine Intelligence (CENPARMI). http://www.cenparmi.concordia.ca/

Center of Excellence for Document Analysis and Recognition (CEDAR). http://www.cedar.buffalo.edu/

Dao, Jeff. "Handwriting Recognition Comes of Age." *Windows CE Journal* 1, no. 1, March 1998. http://www.cetj.com/archives/9803/9803hand.shtml

"Handwriting Recognition with Neural Networks." http://www.willamette.edu/~nhorton/Newton.htm

Impedovo, Sebastiano. *Fundamentals in Handwriting Recognition.* New York: Springer-Verlag, 1994.

[Pattern, Handwriting Recognition Links] http://www.wspc.com.sg/books/compsci/phr.html

hard disk

Even after decades of evolution in computing, the hard disk drive remains the primary means of fast data storage and retrieval in computer systems of all sizes. The disk itself consists of a rigid aluminum alloy platter coated with a magnetic oxide material. The platter can be rotated at speeds of more than 10,000 rpm. A typical drive consists of a stack of such platters mounted on a rotating spindle, with a read/write head mounted above each platter.

Early hard drive heads were controlled by a stepper motor, which positioned the head in response to a series of electrical pulses. (This system is still used for floppy drives.) Today's hard drives, however, are controlled by a voice-coil actuator, similar in structure to an audio speaker. The coil surrounds a magnet. When a current enters the coil, it generates a magnetic field that interacts with that of the permanent magnet, moving the coil and thus the disk head. Unlike the stepper motor, the voice coil is continuously variable and its greater precision allows data tracks to be packed more tightly on the platter surface, increasing disk capacity.

The storage capacity of a drive is determined by the number of platters and the spacing (and thus number) of tracks that can be laid down on each platter. Capacities have steadily increased while prices have plummeted: In 1980, for example, a hard drive for an Apple II microcomputer cost more than $1,000 and held only 5MB of data. By 2001, hard drives with a capacity of 20–40GB (billions of bytes) sold for a couple hundred dollars.

Data is organized on the disk by dividing the tracks into segments called *sectors.* When the disk is prepared to receive data (a process called *formatting*), each sector is tested by writing and reading sample data. If an error occurs, the operating system marks the sector as unusable (virtually any hard disk will have at least a few such bad sectors).

Hard Disk

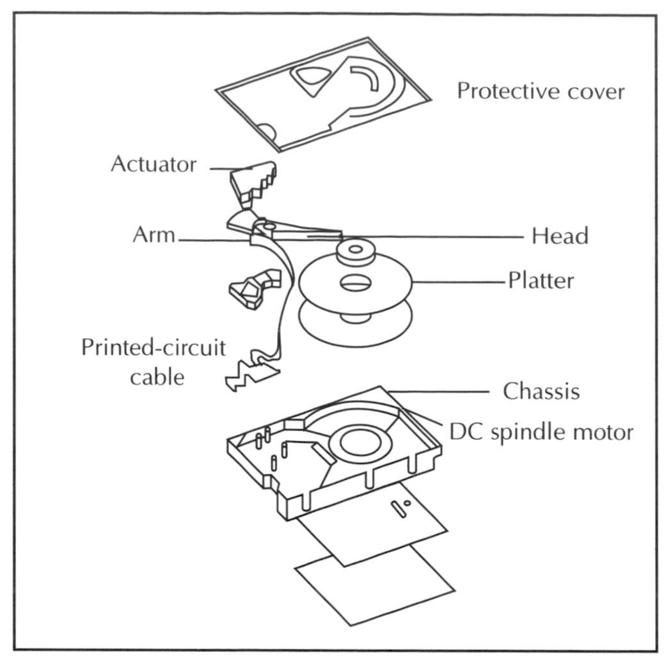

Parts of a typical hard disk drive. Many hard drives have multiple heads and platters to allow for storage of larger amounts of data.

The set of vertical corresponding tracks on the stack of platters that make up the drive is called a *cylinder.* Since the drive heads are connected vertically, if a head is currently reading or writing for example sector 89 on one platter, it is positioned over that same sector on all the others. Therefore, the operating system normally stores files by filling the full cylinder before going to a new sector number.

Another way to improve data flow is to use *sector interleaving.* Because many disk drives can read data faster than the operating system can read it from the disk's memory buffer, data is often stored by skipping over adjacent sectors. Thus, instead of storing a file on sectors 1, 2, and 3, it might be stored on sectors 1, 3, and 5 (this is called a 2:1 interleave). Moving the head from sector 1 to sector 3 gives the system enough time to process the data. (Otherwise, by the time the system was ready to read sector 2, the disk would have rotated past it and the system would have to wait through a complete rotation of the disk.) Newer CPUs are often fast enough to keep up with contiguous sectors, avoiding the need for interleaving.

Data throughput tends to decrease as a hard drive is used. This is due to *fragmentation.* The operating system runs out of sufficient contiguous space to store new files and has to write new files to many sectors widely scattered on the disk. This means the head has to be moved more often, slowing data access. Using an operating system (or third party) defragmentation utility, users can periodically reorganize their hard drive so that files are again stored in contiguous sectors.

Files can also be reorganized to optimize space rather than access time. Because the operating system has a minimum cluster size (4K with the FAT32 system used in Windows 98) a single file with only 32 bytes of data will still consume 4,096 bytes. However, if all the files are written together as one huge file (with an index that specifies where each file begins) that waste of space would be avoided. This is the principle of *disk compression.* Disk compression does slow access somewhat (due to the need to look up and position to the actual data location for a file) and the system becomes more fragile (since garbling the giant file would prevent access to the data in perhaps thousands of originally separate files). The low cost of high capacity drives today has made compression less necessary.

INTERFACING HARD DRIVES

When the operating system wants to read or write data to the disk (either for its own purposes or to carry out a request from an application program) it must send commands to the *driver,* a program that translates high-level commands to the instructions needed to operate the *disk controller,* which in turn operates the motors controlling the disk heads. The two most common standards for interfacing and controlling hard drives are EIDE (Enhanced Integrated Drive Electronics) and SCSI (Small Computer System Interface). EIDE (successor to the earlier IDE) is the most common interface for personal computers. SCSI is more expensive but has several advantages: It has the ability to organize incoming commands for greater efficiency and also features greater flexibility (an EIDE controller can connect only two hard drives, while SCSI can "daisy chain" a large number of disk drives or other peripherals). In practice, the two interfaces perform about equally well. USB (Universal Serial Bus) can also be used to interface with external hard drive units (see USB).

The capacity of hard drives is likely to continue to increase, with data able to be written more densely or perhaps in multiple layers on the same disk surface. Denser storage also offers the ability to make drives more compact. Already hard drives with a diameter of about an inch have been built by IBM and others for use in digital cameras. At the same time removable storage technologies (such as DVD-ROM and cartridge drives) are also growing in speed and capacity so that users have a variety of ways to supplement the hard drive's capacity.

Further Reading

Pabst, Tom. *Tom's Hardware Guide: High Performance PC Secrets.* Indianapolis, Ind.: Que, 1998.

Simmons, Curt. *Windows 2000 Hardware and Disk Management.* Upper Saddle River, N.J.: Prentice Hall, 2000.

White, Ron. *How Computers Work.* Millennium Edition. Indianapolis, Ind.: Que, 1999.

hashing

A *hash* is a numeric value generated by applying a mathematical formula to the numeric values of the characters in a string of text (see CHARACTERS AND STRINGS). The formula is chosen so that the values it produces are always the same length (regardless of the length of the original text) and are very likely to be unique. (Two different strings should not produce the same hash value. Such an event is called a *collision.*)

APPLICATIONS

The two major application areas for hashing are information retrieval and cryptographic certification. In databases, an index table can be built that contains the hash values for the key fields and the corresponding record number for each field, with the entries in hash value order. To search the database, an input key is hashed and the value is compared with the index table (which can be done using a very fast binary search). If the hash value is found, the corresponding record number is used to look up the record. This tends to be much faster than searching an index file directly.

Alternatively, a "coarser" but faster hashing function can be used that will give the same hash value to small groups (called *bins*) of similar records. In this case the hash from the search key is matched to a bin and then the records within the bin are searched for an exact match.

In cryptography an encrypted message can be hashed, producing a unique fixed-length value. (The fixed length prevents attackers from using mathematical relationships that might be discoverable from the field lengths.) The hashed message can then be encrypted again to create an electronic signature (see CERTIFICATE, DIGITAL). For long messages this is more efficient than having to apply the signature function to each block of the encrypted message, yet the unique relationship between the original message and the hash maintains a high degree of security.

Finally, hashing can be used for error detection. If a message and its hash are sent together, the recipient can hash the received text. If the hash value generated matches the one received, it is highly likely the message was received intact (see also ERROR CORRECTION).

Hashing

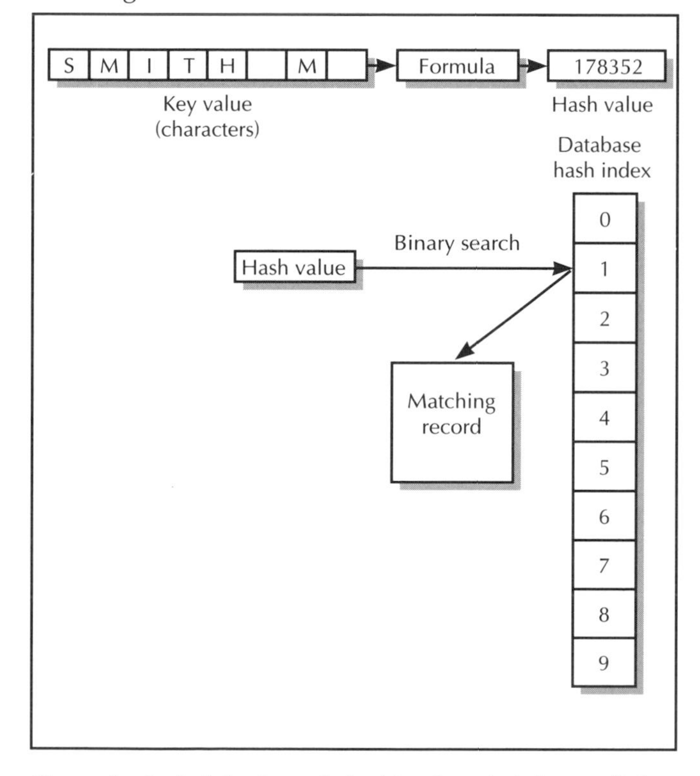

To search a hashed database, the hashing formula is first applied to the search key, yielding a hash value. That value can then be used in a binary search to quickly zero in on the matching record, if any.

Further Reading

Pieprzyk, Josef, and Babak Sadeghiyan. *Design of Hashing Algorithms.* New York: Springer-Verlag, 1993.

Uzgalis, Robert. "Hashing Concepts and the Java Programming Language." http://www.serve.net/buz/hash.adt/java.000.html

heap

In operating systems and certain programming languages (such as LISP), a *heap* is a pool of memory resources available for allocation by programs. The memory segments (sometimes called *cells*) can be the same size or of variable size. If the same size, they are linked together by pointers (see LIST PROCESSING). Memory is then allocated for a variable by traversing the list and setting the required number of cells to be "owned" by that variable. (While some languages such as Pascal and C use explicit memory allocation or deallocation functions, other languages such as LISP use a separate runtime module that is not the responsibility of the programmer.)

Deallocation (the freeing up of memory no longer needed by a variable so it can be used elsewhere) is more complicated. In many languages several different pointers can be used to refer to the same memory location. It is therefore necessary not only to disconnect a given pointer from the cell, but to track the total number of pointers connected to the cell so that the cell itself is deallocated only when the last pointer to it has been disconnected. One way to accomplish this is by setting up an internal variable called a *reference counter* and incrementing or decrementing it as pointers are connected or disconnected. The disadvantages of this approach include the memory overhead needed to store the counters and the execution overhead of having to continually check and update the counters.

An alternative approach is *garbage collection*. Here the runtime system simply connects or disconnects pointers as required by the program's declarations, without making an attempt to reclaim the disconnected ("dead") cells. If and when the supply of free cells is exhausted, the runtime system takes over and begins a three-stage process. First, it provisionally sets the status indicator bit for each cell to show that it is "garbage." Each pointer in the program is then traced (that is, its links are followed) into the heap, and if a valid cell is found that cell's indicator is reset to "not garbage." Finally, the garbage cells that remain are linked back to the pool of free cells available for future allocation. The chief drawback of garbage collection is that the more cells actually being used by the program, the longer the garbage-collecting process will take (since all of these cells have to be traced and verified). Yet it is precisely when most cells are in use that garbage collection is most likely to be required.

The need for garbage collection has diminished in many programming environments because modern computers not only have large amounts of memory, most operating systems also implement virtual memory, which allows a disk or other storage device to be treated as an extension of main memory.

Note: the term *heap* is also used to describe a particular type of binary tree. (See TREE.)

Further Reading

Heap (Data Structures and Algorithms). http://swww.ee.uwa.edu.au/~plsd210/ds/heaps.html

Sebesta, Robert W. *Concepts of Programming Languages*. Reading, Mass.: Addison-Wesley, 1999.

help systems

In the early days of computing, the programmers of a system tended to also be its users and were thus intimately familiar with the program's operation and command set. If not a programmer, the user of a mainframe program was probably at least a well-trained operator who could work with the aid of a brief summary or notes provided by the programmer. However, with the beginnings of office automation in the 1970s and the growing use of desktop computers in office, home, and school in the 1980s, increasingly complex programs were being put in the hands of users who often had only minimal computer training (see COMPUTER LITERACY).

While programs often came with one or more tutorial or reference manuals, designers realized that offering help through the program itself would have some clear advantages. First, the user would not have to switch attention from the computer screen to look things up in a manual. Second, the help system could be programmed to not only provide information, but also to help the user find the information needed in a given situation. For example, related topics could be linked together and a searchable index provided.

IMPLEMENTATION

Programs running under the text-based MS-DOS of the 1980s tended to have only rudimentary help screens (often invoked by pressing the F1 key). Generally, these were limited to brief summaries of commands and associated key combinations. However, with the growing use of Microsoft Windows (and the similar Macintosh interface), a more complete and versatile help system was possible. Since these systems allowed multiple windows to be displayed on the screen, the user could consult help information while still seeing the program's main screen. This allowed for trying a recommended procedure and observing the results.

Windows and Macintosh help systems also featured highlighted links in the text that could be used to jump to related topics (see HYPERTEXT AND HYPERMEDIA). A topic word can also be typed into an index box, bringing up any matching topics. If all else fails, the entire help file could be indexed so that any word could be used to find matching topics.

More recent Windows programs also include *wizards*. A wizard is a step-by-step procedure for accomplishing a particular task. For example, if a Microsoft Word user want to learn how to format text into multiple columns, the help system can offer a wizard that takes the user through the procedure of specifying the number of columns, column size, and so on. The steps can even be applied directly to the document with the wizard "driving" the program accordingly.

Recently, many programs have implemented their help in the form of webpages, stored either on the user's computer or at the vendor's website (see HTML). HTML has the advantage that it is now a nearly universal format that can be used on a variety of platforms and (if hosted on a website) the help can be continually improved and updated. However, HTML help systems tend to lack the thorough indexing and wizards found in Windows.

A variety of shareware and commercial help authoring systems such as RoboHelp are available to help developers create help in Windows or HTML format. UNIX systems, which have always included an on-line manual, now typically offer HTML-based help as well.

With printed documentation being increasingly eschewed for cost-cutting reasons, users of many programs today must depend on the help system as well as on on-line documents (such as PDF files) and Web-based support.

Further Reading

Duffy, Thomas M., James E. Palmer, and Brad Mehlenbacher. *Online Help: Design and Evaluation*. Norwood, N.J.: Ablex, 1992.

"Help Authoring Resource Center."http://www.helpauthoring.com/
James-Tanny, Char. *Sams Teach Yourself RoboHELP 2000.* Carmel, Ind.: Sams, 1999.

hexadecimal system

The base 16 or hexadecimal system is a natural way to represent the binary data stored in a computer. It is more compact than binary because four binary digits can be replaced by a single "hex" digit.

The following table gives the corresponding decimal, binary, and hex values from 0 to 15:

DECIMAL	BINARY	HEX
0	0	0
1	0001	1
2	0010	2
3	0011	3
4	0100	4
5	0101	5
6	0110	6
7	0111	7
8	1000	8
9	1001	9
10	1010	A
11	1011	B
12	1100	C
13	1101	D
14	1110	E
15	1111	F

Note that decimal and hex digits are the same from 0 to 9, but hex uses the letters A–F to represent the digits corresponding to decimal 10–15. The system extends to higher numbers using increasing powers of 16, just as decimal uses powers of 10: For example, hex FF represents binary 11111111 or decimal 255. Many of the apparently arbitrary numbers encountered in programming can be better understood if one realizes that they correspond to convenient groupings of bits: FF is eight bits, sufficient to hold a single character (see CHARACTERS AND STRINGS). In low-level programming memory addresses are also usually given in hex (see ASSEMBLER).

Further Reading

Matz, Kevin. "Introduction to Binary and Hexadecimal." http://www.comprenica.com/atrevida/atrtut01.html

history of computing

With the digital computer now more than 50 years old, there has been growing interest in its history and development. Although it would take a library of books to do the subject justice, providing a summary of the main themes and trends of each decade of computing will give readers of this book some helpful context for understanding the other entries.

PREHISTORY

In a sense, the idea of mechanical computation emerged in prehistory when early humans discovered that they could use physical objects such as piles of stones, notches, or marks as a counting aid. The ability to perform computation beyond simple counting extends back to the ancient world: For example, the abacus developed in ancient China could still beat the best mechanical calculators as late as the 1940s (see CALCULATOR). The mechanical calculator began in the West in the 17th century, most notably with the machines created by philosopher-scientist Blaise Pascal. Other devices such as "Napier's bones" (ancestor of the slide rule) depended on proportional logarithmic relationships (see also ANALOG COMPUTER).

While the distinction between a calculator and true computer is subtle, Charles Babbage's work in the 1830s delineated the key concepts. His "analytical engine," conceived but never built, would have incorporated punched cards for data input (an idea taken over from the weaving industry), a central calculating mechanism (the "mill"), a memory ("store"), and an output device (printer). The ability to input both program instructions and data would enable such a device to solve a wide variety of problems (see BABBAGE, CHARLES).

Babbage's thought represented the logical extension of the worldview of the industrial revolution to the problem of calculation. The computer was a "loom" that wove mathematical patterns. While Babbage's advanced ideas became largely dormant after his death, the importance of statistics and information management would continue to grow with the development of the modern industrial state in Europe and the United States throughout the 19th century. The punch card as data store and the creation of automatic tabulation systems would re-emerge near the end of the century (see HOLLERITH, HERMAN).

During the early 20th century, mechanical calculators and card tabulation and sorting machines made up the data processing systems for business, while researchers built special-purpose analog computers for exploring problems in physics, electronics, and engineering. By the late 1930s, the idea of a programmable digital computer emerged in the work of theoreticians (see TURING, ALAN and VON NEUMANN, JOHN).

1940s

The highly industrialized warfare of World War II required the rapid production of a large volume of accurate calculations for such applications as aircraft design, gunnery control, and cryptography. Fortunately, the field was now ripe for the development of programmable digital computers. Many reliable components were available to the computer designer including switches and relays from the telephone industry and card readers and punches (manufactured by Hollerith's descendant, IBM), and vacuum tubes used in radio and other electronics.

Early computing machines included the Mark I (see AIKEN, HOWARD), a huge calculator driven by electrical relays and controlled by punched paper tape. Another machine, the prewar Atanasoff-Berry Computer (see ATANASOFF, JOHN) was never completed, but demonstrated the use of electronic (vacuum tube) components, which were much faster than electromechanical relays. Meanwhile, a German inventor built a programmable binary computer that combined a mechanical number storage mechanism with telephone relays (see ZUSE, KONRAD). Zuse also proposed building an electronic (vacuum tube) computer, but the German government decided not to support the project.

During the war, British and American code breakers built a specialized electronic computer called Colossus, which read encoded transmissions from tape and broke the code of the supposedly impregnable German Enigma machines.

John Mauchly shown with a portion of the ENIAC. Completed in 1946, the ENIAC is often considered to be the first large-scale electronic digital computer. (PHOTO COURTESY OF COMPUTER MUSEUM HISTORY CENTER)

The most viable general-purpose computers were developed by J. Presper Eckert and John Mauchly starting in 1943 (see ECKERT, J. PRESPER and MAUCHLY, JOHN). The first, ENIAC, was completed in 1946 and had been intended to perform ballistic calculations. While its programming facilities were primitive (programs had to be set up via a plugboard), ENIAC could perform 5,000 arithmetic operations per second, about a thousand times faster than the electromechanical Mark I. ENIAC had about 19,000 vacuum tubes and consumed as much power as perhaps a thousand modern desktop PCs.

1950s

The 1950s saw the establishment of a small but viable commercial computer industry in the United States and parts of Europe. Eckert and Mauchly formed a company to design and market the UNIVAC, based partly on work on the experimental EDVAC. This new generation of computers would incorporate the key concept of the *stored program:* Rather than the program being set up by wiring or simply read sequentially from tape or cards, the program instructions would be stored in memory just like any other data. Besides allowing a computer to fetch instructions at electronic rather than mechanical speeds, storing programs in memory meant that one part of a program could refer to another part during operation, allowing for such mechanisms as branching, looping, the running of subroutines, and even the ability of a program to modify its own instructions.

The UNIVAC became a hit with the public when it was used to correctly predict the outcome of the 1952 presidential election. Government offices and large corporations began to look toward the computer as a way to solve their increasingly complex data processing needs. Forty UNIVACs were eventually built and sold to such customers as the U.S. Census Bureau, the U.S. Army and Air Force, and insurance companies. Sperry (having bought the Mauchly-Eckert company), Bendix, and other companies had some success in selling computers (often for specialized applications), but it was IBM that eventually captured the broad business market for mainframe computers.

The IBM 701 (marketed to the government and defense industry) and 702 (for the business market) incorporated several emerging technologies including a fast electronic (tube) memory that could store 4,096 36-bit data words, a rotating magnetic drum that could store data that is not immediately needed, and magnetic tape for backup. The IBM 650, marketed starting in 1954, became the (relatively) inexpensive workhorse computer for businesses (see also MAINFRAME). The IBM 704, introduced in 1955, incorporated magnetic core memory and also featured floating-point calculations.

1960s

The 1960s saw the advent of a "solid state" computer design featuring transistors in place of vacuum tubes and the use of ferrite magnetic core memory (introduced commercially in 1955). These innovations made computers both more compact (although they were still large by modern standards), more reliable, and less expensive to operate (due to lower power consumption.) The IBM 1401 was a typical example of this new technology: It was compact, relatively simple to operate, and came with a fast printer that made it easier to generate data.

There was a natural tendency to increase the capacity of computers by adding more transistors, but the hand-wiring of thousands of individual transistors was difficult and expensive. As the decade progressed, however, the concept of the integrated circuit began to be implemented in computing. The first step in that direction was to attach a number of transistors and other components to a ceramic substrate, creating modules that could be handled and wired more easily during the assembly process.

IBM applied this technology to create what would become one of the most versatile and successful lines in the history of computing, the IBM System/360 computer. This was actually a series of 14 models that offered successively greater memory capacity and processing speed while maintaining compatibility so that programs developed on a smaller, cheaper model would also run on the more expensive machines. Compatibility was ensured by devising a single 360 instruction set that was implemented at the machine level by *microcode* stored in ROM (read-only memory) and optimized for each model. By 1970 IBM had sold more than 18,000 360 systems worldwide.

By the mid-1960s, however, a new market segment had come into being: the minicomputer. Pioneered by Digital Equipment Corporation (DEC) with its PDP line, the minicomputer was made possible by rugged, compact solid-state (and increasingly integrated) circuits. Architecturally, the mini usually had a shorter data word length than the mainframe, and used indirect addressing (see ADDRESSING) for flexibility in accessing memory. Minis were practical for uses in offices and research labs that could not afford (or house) a mainframe (see MINICOMPUTER). They were also a boon to the emerging use of computers in automating manufacturing, data collection, and other activities, because a mini could fit into a rack with other equipment (see also EMBEDDED SYSTEMS). In addition to DEC, Control Data Corporation (CDC) produced both minis and large high-performance machines (the Cyber series), the first truly commercially viable supercomputers (see SUPERCOMPUTER).

In programming, the main innovation of the 1960s was the promulgation of the first widely-used, high-level programming languages, COBOL (for business) and FORTRAN (for scientific and engineering calculations), the result of research in the late 1950s. While some progress had been made earlier in the decade in using symbolic names for quantities and memory locations (see ASSEMBLER), the new higher-level languages made it easier for professionals outside the computer field to learn to program and made the programs themselves more readable, and thus easier to maintain. The invention of the COMPILER (a program that could read other programs and translate them into low-level machine instructions) was yet another fruit of the stored program concept.

1970s

The 1970s saw minis becoming more powerful and versatile. The DEC VAX ("Virtual Address Extension") series allowed larger amounts of memory to be addressed and increased flexibility. Meanwhile, at the high end, Seymour Cray left CDC to form Cray Research, a company that would produce the world's fastest supercomputer, the compact, freon-cooled Cray-1. In the mainframe mainstream, IBM's 370 series maintained that company's dominant market share in business computing.

The most striking innovation of the decade, however, was the microcomputer. The microcomputer (now often called the "computer chip") combined three basic ideas: an integrated circuit so compact that it could be laid on a single silicon chip, the design of that circuit to perform the essential addressing and arithmetic functions required for a computer, and the use of microcode to embody the fundamental instructions. Intel's 4004 introduced in late 1971 was originally designed to sell to a calculator company. When that deal fell through, Intel started distribut-

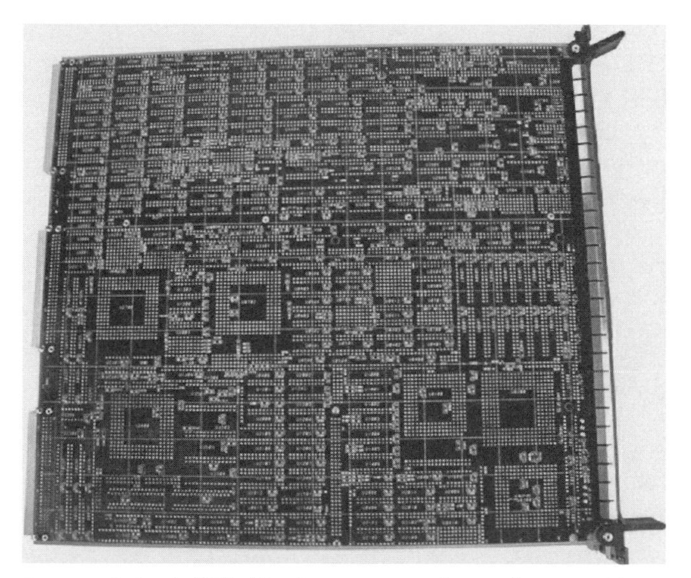

Integrated circuit (IC) chips for memory and control were making for increasingly powerful, compact, and reliable computer components. The microprocessor supplied the remaining ingredient needed for a true desktop personal computer.

ing the microprocessors in developer's kits to encourage innovators to design computers around them. Soon Intel's upgraded 8008 and 8080 microprocessors were available, along with offerings by Rockwell, Texas Instruments, and other companies.

Word of the microprocessor spread through the electronic hobbyist community, being given a boost by the January 1975 issue of *Popular Electronics* that featured the Altair computer kit, available from an Albuquerque company called MITS for about $400. Designed around the Intel 8080, the Altair featured an expansion BUS (an idea borrowed from minis).

The Altair was hard to build and had very limited memory, but it was soon joined by companies that designed and marketed ready-to-use microcomputer systems, which soon became known as personal computers (PCs). By 1980, entries in the field included Apple (Apple II), Commodore (Pet), and Radio Shack (TRS-80). These computers shared certain common features: a microprocessor, memory in the form of plug-in chips, read-only memory chips containing a rudimentary operating system and a version of the BASIC language, and an expansion bus to which users could connect peripherals such as disk drives or printers.

The spread of microcomputing was considerably aided by the emergence of a technical culture where hobbyists and early adopters wrote and shared software, snatched up a variety of specialized magazines, talked computers in user groups, and evangelized for the cause of widespread personal computing.

Meanwhile, programming and the art of software development did not stand still. Innovations of the 1970s included the philosophy of structured programming (featuring well-defined control structures and methods for passing data to and from subroutines and procedures). New languages such as Pascal and C, building on the earlier Algol, supported structured programming design to varying degrees (see STRUCTURED PROGRAMMING). Programmers on college campuses also had access to UNIX, a powerful operating system containing a relatively simple kernel, a shell for interaction with users, and a growing variety of utility programs that could be connected together to solve data processing problems (see UNIX). It was in this environment that the government-funded ARPANET developed protocols for communicating between computers and allowing remote operation of programs. Along with this came e-mail, the sharing of information in newsgroups (Usenet), and a growing web of links between networks that would eventually become the Internet (see INTERNET).

1980s

In the 1980s, the personal computer came of age. IBM broke from its methodical corporate culture and allowed a design team to come up with a PC that featured an open, expandable architecture. Other companies such as Compaq legally created compatible systems (called "clones"), and "PC-compatible" machines became the industry standard. Under the leadership of Bill Gates, Microsoft gained control of the operating system market and also became the dominant competitor in applications software (particularly office software suites).

Although unable to gain market share comparable to the PC and its clones, Apple's innovative Macintosh, introduced in 1984, adapted research from the Xerox PARC laboratory in user interface design. At a time when PC compatibles were still using Microsoft's text-based MS-DOS, the Mac sported a graphical user interface featuring icons, menus, and buttons, controlled by a mouse (see USER INTERFACE). Microsoft responded by developing the broadly similar Windows operating environment, which started out slowly but had become competitive with Apple's by the end of the decade.

The 1980s also saw great growth in networking. University computers running UNIX were increasingly linked through what was becoming the Internet, while office computers increasingly used local area networks (LANs) such as those based on Novell's Netware system. Meanwhile, PCs were also being equipped with modems, enabling users to dial up a growing number of on-line services ranging from giants such as CompuServe to a diversity of individually run bulletin board systems (see BULLETIN BOARD SYSTEMS).

In the programming field a new paradigm, object-oriented programming (OOP) was offered by languages such as SMALLTALK and C++, a variant of the popular C language. The new style of programming focused on programs as embodying relationships between objects that are responsible for both private data and a public interface represented by methods, or capabilities offered to users of the object. Both structured and object-oriented methods attempted to keep up with the growing complexity of large software systems that might incorporate millions of lines of code. The federal government adopted the Ada language with its ability to precisely manage program structure and data operations. (See OBJECT-ORIENTED PROGRAMMING and ADA.)

1990s

By the 1990s, the PC was a mature technology dominated by Microsoft's Windows operating system. UNIX, too, had matured and become the system of choice for university computing and the worldwide Internet. Although the potential of the Internet for education and commerce was beginning to be explored, at the beginning of the decade the network was far from friendly for the average consumer user.

This changed when Tim Berners-Lee, a researcher at Geneva's CERN physics lab, adapted hypertext (a way to link documents together) with the Internet protocol to implement the World Wide Web. By 1994, Web browsing software that could display graphics and play sounds was available for Windows-based and other computers (see WORLD WIDE WEB and WEB BROWSER). The remainder of the decade became a frenzied rush to identify and exploit business plans based on e-commerce, the buying and selling of goods and services on-line (see E-COMMERCE). Meanwhile, educators demanded Internet access for schools.

In the office, the Intranet (a LAN based on the Internet TCP/IP protocol) began to supplant earlier networking schemes (see INTRANET). Belatedly recognizing the threat and potential posed by the Internet, Bill Gates plunged Microsoft into the Web server market, included the free Internet Explorer browser with Windows, and vowed that all Microsoft programs would work seamlessly with the Internet.

Moore's Law, the dictum that computer power roughly doubles every 18 months, continued to hold true as PCs went from clock rates of a few tens of MHz to more than 1GHz. RAM and hard disk capacity kept pace, while low-cost color printers, scanners, digital cameras, and video systems made it easier than ever to bring rich media content into the PC and the on-line world.

BEYOND 2000

The new decade began with great hopes, particularly for the Web and multimedia "dot-coms," but their stocks, inflated by unsustainable expectations, took a significant dip in 2000 2001 and continued to suffer in the general economic malaise of 2002. While it is not possible to predict exactly how the industry will shake out, the computer, particularly the PC and the Internet, is now so integral a part of daily life and of the economy that investment will continue to pour into promising technical developments. Some of the most promising areas for the 2000–2010 period appear to be:

- Genetic and biomedical research
- Data mining / automatic data analysis
- Image processing
- Robotics
- Military applications (such as providing individual soldiers with real-time battlefield information)
- Security systems (in the wake of the September 11, 2001 attacks)

Thus far, each decade has brought new technologies and methods to the fore, and few observers doubt that this will be true in the future.

Note: for a more detailed chronology of significant events in computing, see Appendix 1: "Chronology of Computing."

Further Reading

Bashe, C. J. [and others]. *Early Computers*. Cambridge, Mass.: MIT Press, 1986.

Greenia, Mark W. *History of Computing: An Encyclopedia of the People and Machines that Made Computer History*. revised CD ed. Lexikon Services, 2001.

Kidder, Tracy. *The Soul of a New Machine*. New York: Modern Library, 1997.

Pugh, E. W., L. R. Johnson, and J. J. H. Palmer. *IBM's 360 and Early 370 Systems*. Cambridge, Mass.: MIT Press, 1991.

Spencer, Donald D. *The Timetable of Computing*. Ormond Beach, Fla.: Camelot Publishing, 1999.

Hollerith, Herman

(1860–1929)
American
Inventor

Herman Hollerith invented the automatic tabulating machine, a device that could read the data on punched cards and display running totals. His invention would become the basis for the data tabulating and processing industry. Hollerith was born in Buffalo, New York, and graduated from the Columbia School of Mines. After graduation, he went to work for the U.S. Census as a statistician. Among other tasks he compiled vital statistics for Dr. John Shaw Billings, who suggested to Hollerith that using punched cards and some sort of tabulator would help the Census Department keep up with the growing volume of demographic statistics.

Hollerith studied the problem and decided that he could build a suitable machine. He went to MIT, where he taught mechanical engineering while working on the machine, which was partly inspired by an earlier device that had used a piano-type roll rather than punched cards as input. The peripatetic Hollerith soon got a job with the U.S. Patent Office, partly to learn the procedures he would need to follow to patent his tabulator. He applied for several patents, including one for the punched-card tabulator. He tested the device with vital statistics in Baltimore, New York, and the state of New Jersey.

Hollerith's mature system included a punch device that a clerk could use to record variable data in many categories on the same card (a stack of cards could also be prepunched with constant data, such as the number of the census district). The cards were then fed into a device something like a small printing press. The top part of the press had an array of spring-loaded pins that connected to tiny pots of mercury (an electrical conductor) in the bottom. The pins were electrified. Where a pin encountered a punched hole in the card, it penetrated through to the mercury, allowing current to flow. The current created a magnetic field that moved the corresponding counter dial forward one position. The dials could be read after a batch of cards was finished, giving totals for each cate-

A portrait of Herman Hollerith next to one of his punch card tabulator machines. Such machines made it possible to complete the 1890 U.S. census in a fraction of the time needed by earlier methods. (PHOTO COURTESY COMPUTER MUSEUM HISTORY CENTER)

gory, such as an ethnicity or occupation. The dials could also be connected to count multiple conditions (for example, the total number of foreign-born citizens who worked in the clothing trade).

Aided by Hollerith's machines, a census unit was able to process 7,000 records a day for the 1890 census, about ten times the rate in the 1880 count. Starting around 1900, Hollerith brought out improved models of his machines that included such features as an automatic (rather than hand-fed) card input mechanism, automatic sorters, and tabulators that boasted a much higher speed and capacity. Hollerith machines soon found their way into government agencies involved with vital statistics, agricultural statistics, and other data-intensive matters, as well as insurance companies and other businesses.

Facing vigorous competition and in declining health, Hollerith sold his patent rights to the company that eventually evolved into IBM, the company that would come to dominate the market for tabulators, calculators, and other office machines. The punched card, often called the Hollerith card, would become a natural choice for computer designers and would remain the principal means of data and program input for mainframe computers until the 1970s.

Further Reading

Austrian, G. D. *Herman Hollerith: Forgotten Giant of Information Processing.* New York: Columbia University Press, 1982.

Kistermann, F. W. "The Invention and Development of the Hollerith Punched Card." *Annals of the History of Computing,* 13, 245–259.

home office

The widespread use of the personal computer and associated peripherals such as printers has made it more practical for many people to do at least part of their work from their homes. In addition to traditional freelance occupations such as writing and editing, many other businesses including consulting, design, and sales can now be conducted from a home office. Computer hardware and software makers began to target a distinctive market niche that is sometimes referred to as SOHO (Small Office / Home Office), thus including both actual home offices and small commercial offices.

As a market, the SOHO has somewhat different requirements than the large offices traditionally served by major computer vendors:

- Relatively modest PCs as compared to heavy-duty file servers or workstations
- Peripherals shared by two or more PCs (although the plummeting price of printers made it common to provide each PC with its own printer)
- The need for a small "footprint"—that is, minimizing the space taken up by the equipment. Multifunction peripherals (typically incorporating printer, scanner, copier, and perhaps a fax machine) are a popular solution to this requirement.
- A simple local network (see LOCAL AREA NETWORK and INTRANET) with shared Internet access
- Low-end or midrange software (such as Microsoft Works or Office Small Business edition as opposed to the full-blown Office suite)
- Available installation and support (since many home users lack technical hardware or system administration skills)

Although the home or small office remains a significant market segment, specific targeting to the segment has become more difficult. With falling PC prices and increasing capabilities, there is little difference today between a mid-level "consumer" computer system and the kinds of systems previously marketed for home office use.

Further Reading

Bredin, Alice, and Kirsten M. Lagatree. *The Home Office Solution: How to Balance Your Professional and Personal Lives While Working at Home.* New York: Wiley, 1998.

Phillips, Barty. *The Home Office Planner.* San Francisco: Chronicle Books, 2000.

Hopper, Grace Murray

(1906–1992)
American
Computer Scientist

Grace Brewster Murray Hopper was an innovator in the development of high-level computer languages in the 1950s and 1960s. She is best known for her role in the development of COBOL, which became the premier language for business data processing.

Hopper was born in New York City. She graduated with honors with a B.A. in mathematics and physics from Vassar College in 1928, and went on to receive her M.A. and Ph.D. in mathematics at Yale University. She taught at Vassar from 1931 to 1943, when she joined the U.S. Naval Reserve at the height of World War II. As a lieutenant (J.G.), she was assigned to the Bureau of Ordnance, where she worked in the Computation Project at Harvard under pioneer computer designer Howard Aiken (see AIKEN, HOWARD). She became one of the first "coders" (that is, programmers) for the Mark I. After the war, Hopper worked for a few years in Harvard's newly established Computation Laboratory. In 1949, however, she became senior mathematician at the Eckert-Mauchly Corporation, the world's first commercial computer company, where she helped with program design for the famous UNIVAC. She stayed with what became the UNIVAC division under Remington Rand (later Sperry Rand) until 1971.

While working with UNIVAC, Hopper's main focus was on the development of programming languages that could allow people to use symbolic names and descriptive statements instead of binary codes or the more cryptic forms of assembly language (see ASSEMBLER). In 1952, she developed A-0, the first COMPILER (that is, a program that could translate language statements to the corresponding low-level machine instructions). She then developed A-2 (a compiler that could handle mathematical expressions), and then in 1957 she developed Flow-Matic. This was the first compiler that worked with English-like statements and was designed for a business data processing environment.

In 1959, Hopper joined with five other computer scientists to plan a conference that would eventually result in the development of specifications for a "Common Business Language." Her earlier work with Flow-Matic and her design input played a key role in the development of what would become the COBOL language.

Hopper retained her Navy commission and even after her retirement in 1966 she was recalled to active duty to work on the Navy's data processing needs. She finally retired in 1986 with the rank of rear admiral. Hopper spoke widely about data processing issues, especially the need for standards in computer language and architecture, the lack of which she said cost the government bil-

Grace Murray Hopper created the first computer program compiler and was instrumental in the design and adoption of COBOL. When she retired, she was the first woman admiral in U.S. Navy history. (PHOTO COURTESY OF UNISYS CORPORATION)

lions of dollars in wasted resources. Admiral Hopper died on January 1, 1992, in Arlington, Virginia.

Hopper received numerous awards and honorary degrees, including the National Medal of Technology. (The navy named a suitably high-tech Aegis destroyer after her in 1996.) The Association for Computing Machines (ACM) created the Grace Murray Hopper Award to honor distinguished young computer professionals. Hopper has become a role model for many girls and young women considering careers in computing.

Further Reading

Billings, C. W. *Grace Hopper: Navy Admiral and Computer Pioneer.* Hillfield, N.J.: Enslow Publishers, 1989.

"Grace Murray Hopper (1906–1992)." http://www.unh.edu/womens-commission/hopper.html

Spencer, Donald D. *Great Men & Women of Computing.* Ormond Beach, Fla.: Camelot Publishing, 1999.

HTML

In developing the World Wide Web, Tim Berners-Lee (see BERNERS-LEE, TIM) had to provide several basic facilities. One was a protocol, HTTP, for requesting documents over the network (see WORLD WIDE WEB). Another was a system of links between documents (see HYPERTEXT AND HYPERMEDIA). The third was a way to embed instructions in the pages so that the Web browser could properly display the text and graphics. Berners-Lee created HTML (Hypertext Markup Language) for this purpose. It is based on the more elaborate SGML (Standard Generalized Markup Language).

The basic "statement" in HTML is the *tag*. Tags are delimited by angle brackets (<>). Tags that affect a document or section of a document come in pairs, with the second member of the pair preceded by a slash. For example, the tags

```
<HTML>

</HTML>
```

indicate the beginning and end of an HTML document, while <BOLD> and </BOLD> delimit text that should be rendered in boldface.

Besides specifying such things as headings, font, font size, and typestyles, HTML includes tags for Web-related functions. One of the most useful is the A, or "anchor" tag. As with some other HTML tags, the A tag is used with *attributes* that further specify what it so be done. The A tag is usually used with the <HREF> or Hypertext Reference attribute, which specifies a document that is to be linked to the current document so that the user can click on a highlight to go there. For example:

```
<A HREF="http://www.MySite.Pages/Glossary">Glossary of Computer Terms</A>
```

specifies a link to a particular page at a particular site. The link will appear in the browser as the highlighted text **Glossary of Computer Terms.** If clicked, the browser will load the HTML page titled Glossary.

IMPLEMENTATION AND EXTENSIONS

Inserting HTML tags by hand is a tedious and error-prone process (for example, it's easy to omit a bracket or a slash or add "illegal" spaces within tags). Fortunately, there are now many HTML editor programs that let users insert the appropriate elements much in the way word processors make it easy to specify fonts and formatting. (Indeed, programs such as Microsoft Word allow users to convert and save documents in HTML format.)

HTML is being extended in a number of ways. First, new features have been added to later versions of the language, including better support for frames, columns, tables, and other formats. Browser developers have also adopted a system called Cascading Style Sheets, which allows document authors to define general styles to ensure consistent document appearance. Style sheets can

HTML

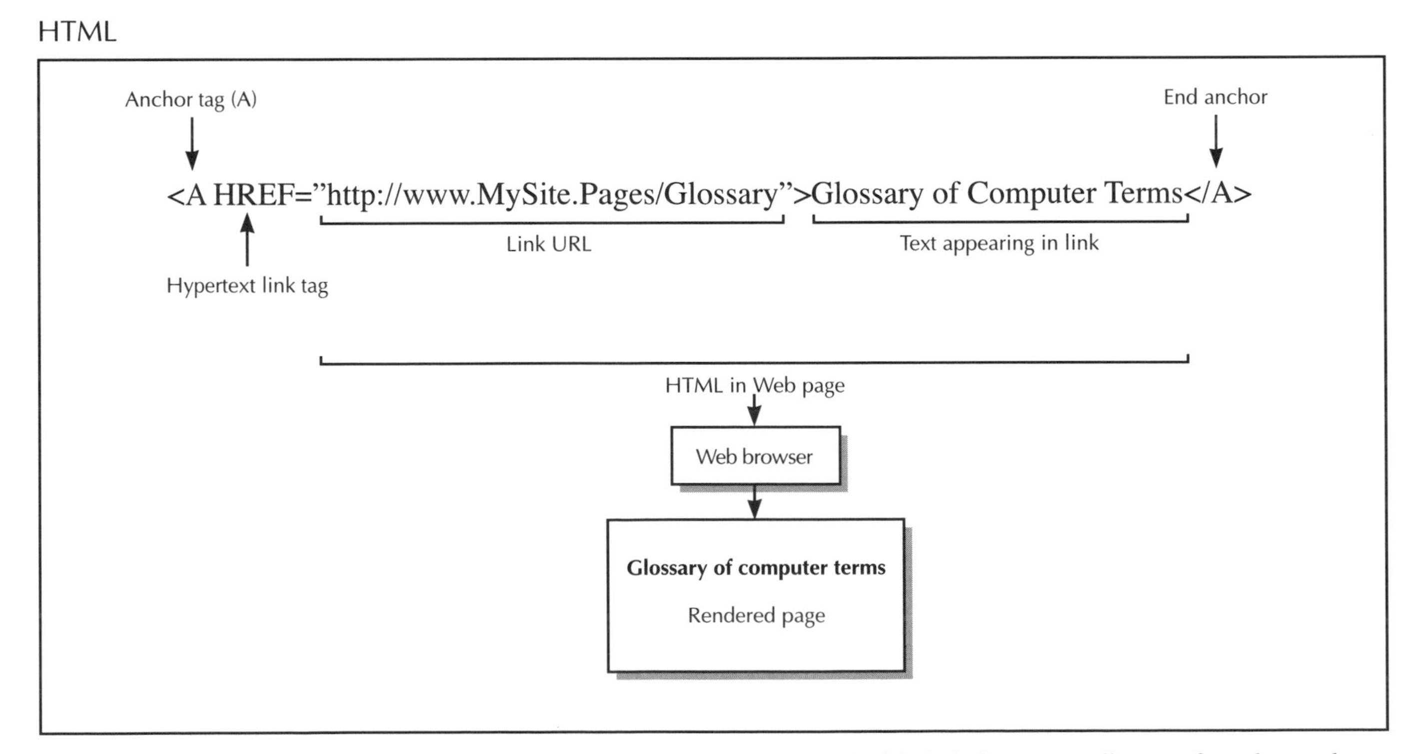

An HTML hyperlink embedded in a webpage. The anchor link gives the address (URL) of the linked page, as well as specifying the text that will appear in the link, which will rendered by the Web browser in a special color or font.

inherit styles from other style sheets, allowing an organization to create general style sheets that can then be refined to create specialized styles for particular types of documents. Another HTML offshoot, Dynamic HTML (DHTML) adds capabilities that allow all elements on an HTML page to respond interactively to user actions. For example, this allows the creation and processing of forms using instructions in JavaScript.

Another elaboration of HTML is called the Extensible Markup Language (see XML). XML allows document authors to create additional tags to identify data structures and relationships within a document, which can in turn be used by document management, database, or other programs.

Further Reading

Carey, Patrick. *New Perspectives on Creating Web Pages with HTML and Dynamic HTML Comprehensive.* Course Technology, 2000.

"Dynamic HTML Resources." http://developer.netscape.com/tech/dynhtml/resources.html

"Hypertext Markup Language" http://www.w3.org/MarkUp/

Lemay, Laura. *Sams Teach Yourself Web Publishing with HTML in 21 Days.* 2nd ed. Indianapolis, Ind.: Sams, 2000.

Meyer, Eric A. *Cascading Style Sheets: the Definitive Guide.* Sebastopol, Calif.: O'Reilly, 2000.

Powell, Thomas A. *HTML: the Complete Reference.* 3rd ed. New York: McGraw-Hill, 2000.

hypertext and hypermedia

Most computer users today are familiar with the concept of hypertext, even if they don't often use the term itself. Each time a Web user clicks on a link on a webpage, he or she is using hypertext. Most on-line help systems also use hypertext to take the reader from one topic to another, related topic. The term *hypermedia* acknowledges modern systems' use of many kinds of resources other than plain text, including still images, videos, and sound recordings.

In a traditional document, the reader is generally assumed to proceed sequentially from the beginning to the end. (Although there may well be footnotes or cross-references within the document, these are generally experienced as temporary divergences from the primary, sequential narrative.) Generally speaking, each reader might be expected to acquire roughly the same set of facts from the document.

In a hypertext document, however, the links between topics create multiple potential paths for readers. To the extent the author has provided links between all related topics, the reader is free to pursue his or her particular interests rather than being bound by a sequential structure imposed by the author. For example, in a document that discusses various organisms in an ecology and the effects of climate and vegetation, one reader might choose to explore one organism in depth, following links from it to other resources devoted to that organism (including outside webpages, images, videos, and so on). Another reader might be interested specifically in the effects of rainfall on the ecology as a whole and follow a completely different set of links to sites having climatological data.

HISTORY AND DEVELOPMENT

In 1945, a time when the very first digital computers were coming on-line, Vannevar Bush, a pioneer designer of analog computers, proposed a mechanism he called the Memex (see BUSH, VANNEVAR). This system would link portions of documents to allow retrieval of related information. The proposal was impracticable in terms of the very limited capacity of computers of the time. By the 1960s, when computers had become more powerful (and the minicomputer was beginning to be a feasible purchase for libraries and schools), another visionary, Theodore Nelson, coined the terms *hypertext* and *hypermedia.* He suggested that networking (a technology then in its infancy) could allow for what would eventually amount to a worldwide database of interconnected information. Nelson developed his specifications for a system he called Xanadu, but he was unable to create a working version of the system until the late 1990s. However, in 1968 Douglas Engelbart (also known as the inventor of the computer mouse) demonstrated a more limited but workable hypertext system called NLS/Augment.

During the 1970s and 1980s, a variety of hypertext systems were created for various platforms, including Guide and Toolbook for MS-DOS and Windows PCs. Perhaps the most influential system was Hypercard, developed for Apple's Macintosh. While Hypercard did not have a complete set of facilities for creating hypertext, the flexible, programmable, linkable "cards" could be used to implement hypertext documents. Many encyclopedias and other reference products on CD-ROM began to implement some form of hypertext links.

The true explosion of hypertext came with the development and growth of the World Wide Web throughout the 1990s. Hypertext on the Web is implemented through the use of HTTP (HyperText Transport Protocol) over the Internet's TCP/IP protocol and by coding documents in HTML (Hypertext Markup Language). (See HTML, INTERNET, TCP/IP, and WORLD WIDE WEB.)

IMPLEMENTATION

A hypertext document consists of *nodes.* A node can be a part of a document that conveys a logical "chunk" of information, such as the text that would be under a particular heading in a traditional document. In some systems nodes can be grouped together as a *composite*—for example, the second-level headings under a first-level heading might be considered nodes making up a single composite.

The text contains *links*. A link specifies an *anchor* or specific location to which it points. The user normally doesn't see the anchor, but rather the *marker,* which is some form of highlighting (such as a different color) that indicates that an area is a link that can be clicked on. (In systems such as the Web, link markers need not be textual. Small pictures are often used as visual link markers.) Web browsers and other hypertext programs often supplement the use of links with various navigation aids. These can include buttons for traversing back or forward through a list of recently visited links, a *history* list from which previous links can be selected, and *bookmarks* that allow the user to save and descriptively label important links for easier future access.

Hypertext is becoming the dominant paradigm for presenting technical or other reference information. With less-structured text, hypertext links are usually considered to be supplemental to the traditional structure. Despite suggestions to the contrary, hypertext seems to be problematic with regard to fiction, unless a work is constructed as an explicit hypertext. If hypertext literature becomes popular, it will require that both authors and readers radically change their role and expectations with regard to the text.

Further Reading

Bush, Vannevar. "As We May Think." *Atlantic Monthly* 176, 101–108. Available on-line at http://www.theatlantic.com/unbound/flashbks/computer/bushf.htm

Nelson, Theodore. *Computer Lib/Dream Machines.* rev. ed. Chicago: Hugo's Books, 1987.

World Wide Web Consortium. "Open Hypertext Protocols." http://www.w3.org

Snyder, I. *Hypertext: the Electronic Labyrinth.* New York: New York University Press, 1997.

I

IBM PC

By 1981, a small but vigorous personal computer (PC) industry was offering complete desktop computer systems. Apple's Apple II offered color graphics and expandability through an "open architecture"—slots into which cards designed by third-party vendors could be plugged. While the Apple II had its own DOS (disk operating system) as did Radio Shack's TRS-80, most microcomputers sold in the business market used CP/M, an operating system developed by Gary Kildall and his company Digital Research.

Meanwhile, IBM, the world's largest computer company, had quietly created a special team headed by Phillip ("Don") Estridge and tasked with designing a personal computer. Unlike the case with the company's mainframe development, the team was given considerable freedom in choosing architecture and components—but they were told they would have to have a machine ready for the market in one year.

Because of the short time frame, the team chose third-party components already well established in the market, including the monitor, floppy disk drive, and a printer. Unlike Apple and most other companies, IBM created two separate video display systems, one monochrome (MDA) for sharp text for business applications and the three-color CGA system for the game and education markets (see GRAPHICS CARD).

The IBM team also adopted standards from the emerging microcomputer industry instead of trying to use existing mainframe standards. For example, they used the ASCII code to represent characters, not the EBCDIC code used on IBM mainframes. They also chose the Intel 8086 and 8088 microprocessors, which had an instruction set similar to that of the Intel 8080 used in many CP/M systems (see MICROPROCESSOR). This would make it easy for software developers to create IBM PC versions of their software quickly so that the new machine would have a repertoire of business software.

One might have expected that IBM would also adopt a version of CP/M as the PC's operating system, taking advantage of the closest thing to an existing industry standard. However, CP/M was relatively expensive, and negotiations with Digital Research stumbled, leaving an opening for a much smaller company, Microsoft, to sell a DOS based on software it had licensed from Seattle Computer Products. While IBM did offer CP/M and another operating system based on the UC San Diego Pascal development system, Microsoft DOS, which became known as PC-DOS (and later MS-DOS), was cheapest and effectively became the default offering (see MS-DOS).

When IBM officially announced its PC in April 1981, Apple took out full-page ads "welcoming" the new competitor to what it considered to already be a mature industry. But by the end of 1983, a million IBM PCs had been sold, dwarfing Apple and other brands. From then on, while Apple would go on to announce its distinctive Macintosh in 1984, the IBM machine would set the industry standard. To most people, "PC" would mean "IBM PC."

OPEN STANDARDS AND EXPANSION

As more businesses bought IBM PCs, the company steadily expanded the machine's capabilities to meet the demands of the business environment. The next model, the PC-XT, introduced in 1982, included a hard disk drive and more system memory. As software became more demanding, the need for a faster and more capable processor also became apparent. In 1984, IBM responded with the PC-AT, which used the Intel 80286 processor, combining the faster processor with a wider (16-bit) and faster data bus (see BUS).

However, IBM would not have the market to itself. A consequence of the use of an open, expandable architecture and "off the shelf" processor and other components is that other companies could market PCs that were compatible with IBM's (that is, they could run the same operating system and applications software). Although competitors could not legally make a simple copy of the read-only memory (ROM) BIOS, the code that enabled the components to communicate, they could reverse-engineer a functional equivalent. The first major competitor in what became known as the "PC Clone" market was Compaq, which also offered an improved video display and a transportable model. Zenith, Tandy (Radio Shack), and HP also offered "name-brand" PC clones.

In 1987, IBM tried to establish a proprietary standard by introducing the PS/2 line, which featured a 3.5-inch floppy drive (standard PC compatibles used 5.25-inch drives), a new high-resolution graphics standard (VGA), a new system bus (MCA or Microchannel Architecture), and a new operating system (OS/2). Despite some technical advantages, the PS/2 achieved only modest success. Since the card slots were incompatible with the previous standard, existing expansion products could not be used. Microsoft soon came out with a new operating environment, Windows, which while inferior in multitasking capabilities to OS/2 was easier to use (see USER INTERFACE and MICROSOFT WINDOWS).

By the 1990s, it was clear that IBM no longer controlled the standards for PCs. (Indeed, IBM soon abandoned the PS/2 MCA architecture and returned to the earlier standard, which competitors had never left.) Instead, the industry incrementally built upon what had become known as the ISA (Industry Standard Architecture), supplementing it with a new kind of expansion card connector called PCI. Currently, IBM is in the second tier in PC sales behind industry leaders Dell and Compaq, having a market share comparable to Hewlett-Packard and Gateway. IBM has also done relatively well in the laptop computer sector with its Thinkpad series.

Today's industry standards are effectively determined by two companies: the chip-maker Intel and the software giant Microsoft. Indeed, "standard" PCs are now often called "Wintel" machines. The direct-order giant Dell and its competitors Gateway and Compaq dominate the "commodity PC" market. However, by creating a standard that was flexible enough for two decades of PC development, IBM made a lasting contribution to computing comparable to its innovations in the mainframe arena.

Further Reading

Dell, Deborah A,. and J. Jerry Purdy. *Thinkpad: a Different Shade of Blue.* Indianapolis, Ind.: Sams, 1999.

Dell, Michael. *Direct from Dell: Strategies that Revolutionized an Industry.* New York: HarperBusiness, 2000.

Gilster, Ron. *PC Hardware: a Beginner's Guide.* New York: McGraw-Hill, 2001.

Hoskins, Jim, and Bill Wilson. *Exploring IBM Personal Computers.* 10th ed. Gulf Breeze, Fla.: Maximum Press, 1999.

image processing

Image processing is a general term for the manipulation of a digitized image to produce an enhanced or more convenient version. Some of the earliest applications were in the military (aerial and, later, satellite reconnaissance) and in the space program. The military and space programs had a great need for extracting as much useful information as possible from images that were often gathered under extreme or marginal conditions. They also needed to make cameras and other hardware components simultaneously more compact and more efficient, and generally had the funds to pay for such specialized developments.

Once developed, higher-quality image processing systems found their way into other applications such as domestic surveillance and medical imaging. The development of cameras that could directly turn light into digitized images (see PHOTOGRAPHY, DIGITAL) made image processing seamless by avoiding the necessity of scanning images from traditional film.

Image processing applications can be divided into three general categories: enhancement, interpretation, and maintenance.

ENCHANCEMENT

Enhancement includes bringing out objects of interest (such as enemy vehicles or a particular rock formation on Mars) from the surrounding background by enhancing contrast or applying appropriate *filters* to block out the background. More sophisticated filters can also be used to compensate for defects in the original image, such as "red-eye," blur, and loss of focus. Today's image processing programs, such as the popular Adobe Photoshop, make relatively sophisticated image manipulation techniques available to interested amateurs as well as professionals. More sophisticated image enhancement techniques include the creation of 3D images based upon the differences calculated from a number of photos shot from slightly different angles.

INTERPRETATION
Interpretation refers to manipulation designed to help human observers obtain more and better information from the image. For example, "false color" can be used to heighten otherwise imperceptible color differences in the original image, or to translate nonvisual information (such as heat or radio emission levels) into visual terms.

Artificial intelligence algorithms can also be employed to automatically analyze images for features of interest (see also COMPUTER VISION). In fields such as military reconnaissance this might allow a high volume of imagery to be prescreened, with images meeting certain criteria "flagged" for the attention of human interpreters.

MAINTENANCE
Maintenance includes archiving of images, often with the aid of compression to reduce the amount of storage space required (see DATA COMPRESSION). It can also include the restoration of images that may have been degraded (as from chemical decomposition of stored film.) This can be done either by creating a reversible mathematical model of the degradation process (thus, for example, restoring colors that have changed through oxidation or other processes) or by creating a model of how the image was formed in the first place and comparing its output to the existing image.

Further Reading

NIH Public Domain Image Processing Software. http://rsb.info.nih.gov/nih-image/

Russ, J. C. *The Image Processing Handbook.* 3rd ed. Boca Raton, Fla.: CRC Press, 1999.

Sanz, J. L. C. *Image Technology: Advances in Image Processing, Multimedia, and Machine Vision.* New York: Springer-Verlag, 1996.

information retrieval

While much attention is paid by system designers to the representation, storage and manipulation of information in the computer, the ultimate value of information processing software is determined by how well it provides for the effective retrieval of that information. The quality of retrieval is dependent on several factors: hardware, data organization, search algorithms, and user interface.

At the hardware level, retrieval can be affected by the inherent seek time of the device upon which the data is stored (such as a hard disk), the speed of the central processor, and the use of temporary memory to store data that is likely to be requested (see CACHE). Generally, the larger the database and the amount of data that must be retrieved to satisfy a request, the greater is the relative importance of hardware and related system considerations.

Data organization includes the size of data records and the use of indexes on one or more fields. An index is a separate file that contains field values (usually sorted alphabetically) and the numbers of the corresponding records. With indexing, a fast binary search can be used to match the user's request to a particular field value and then the appropriate record can be read (see also HASHING).

There is a tradeoff between storage space and ease of retrieval. If all data records are the same length, random access can be used; that is, the location of any record can be calculated essentially by multiplying the record's sequence number by the fixed record length. However, having a fixed record size means that records with shorter data fields must be "padded," wasting disk space. Given the low cost of disk storage today, space is generally less of a consideration.

The search algorithms used by the program can also have a major impact on retrieval speed (see SORTING AND SEARCHING). As noted, if a binary search can be done against a sorted list of fields or records, the desired record can be found in only a few comparisons. At the opposite extreme, if a program has to move sequentially through a whole database to find a matching record, the average number of comparisons needed will be half the number of records in the file. (Compare looking up something in a book's index to reading through the book until you find it.)

Real-world searching is considerably more complex, since search requests can often specify conditions such as "find e-commerce but not amazon.com" (see BOOLEAN OPERATORS). Searches can also use wildcards to find a word stem that might have several different possible endings, proximity requirements (find a given word within so many words of another), and other criteria. Providing a robust set of search options enables skilled searchers to more precisely focus their searches, bringing the number of results down to a more manageable level. The drawback is that complex search languages result in more processing (often several intermediate result sets must be built and internally compared to one another). There is also more likelihood that searchers will either make syntax errors in their requests or create requests that do not have the intended effect.

While database systems can control the organization of data, the pathways for retrieval and the command set or interface, the World Wide Web is a different matter. It amounts to the world's largest database—or perhaps a "metabase" that includes not only text pages but file resources and links to many traditional database systems. While the flexibility of linkage is one of the Web's strengths, it makes the construction of search engines difficult. With millions of new pages being created each week, the "webcrawler" software that automatically traverses links and records and indexes site information is hard pressed to capture more than a diminishing fraction of the available content. Even so, the number of "hits" is often unwieldy (see SEARCH ENGINE).

Information Retrieval

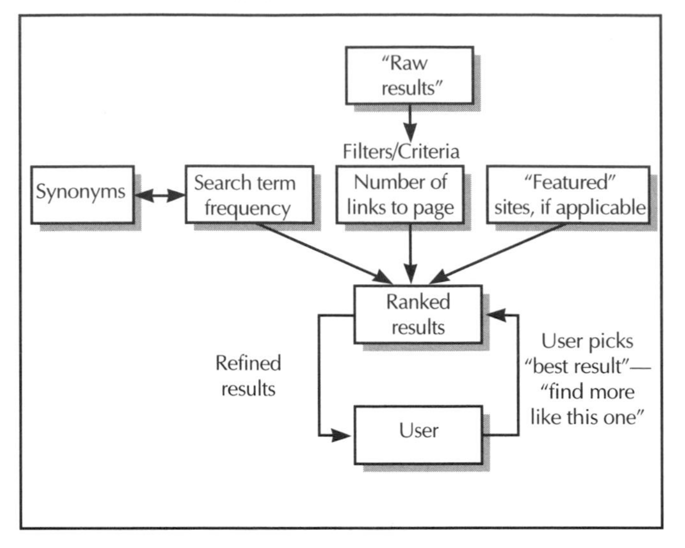

A number of criteria can be used by Web search engines to determine the likely relevance of search results. Perhaps the most important tool, however, is feedback from the user.

A number of strategies can be used to provide more focused search results. The title or full text of a given page can be checked for synonyms or other ideas often associated with the keyword or phrase used in the search. The more such matches are found, the higher the degree of *relevance* assigned to the document. Results can then be presented in declining order of relevance score. The user can also be asked to indicate a result document that he or she believes to be particularly relevant. The contents of this document can then be compared to the other result documents to find the most similar ones, which are presented as likely to be of interest to the researcher.

Information retrieval from either stand-alone databases or the Web can also be improved by making it unnecessary for users to employ structured query languages or even carefully selected keywords. Users can simply type in their request in the form of a question, using ordinary language: For example, "What country in Europe has the largest population?" The search engine can then translate the question into the structured queries most likely to elicit documents containing the answer. Ask Jeeves and similar search services have thus far been only modestly successful with this approach.

Finally, encoding more information about content and structure within the document itself can provide more accurate and useful retrieval. The use of XML (an extension of HTML) and work toward a "semantic Web" offers hope in that direction (see XML and BERNERS-LEE, TIM).

Further Reading

Baeza-Yates, R., and Berthier Ribeiro-Neto. *Modern Information Retrieval*. Reading, Mass.: Addison-Wesley, 1999.

Chang, George, ed. *Mining the World Wide Web: An Information Search Approach*. Boston: Kluwer Academic, 2001.

Minsky, Marvin. *Semantic Information Processing*. Cambridge, Mass.: MIT Press, 1968.

information superhighway (or Information Highway)

This metaphor, widely touted in the mid-1990s by both Democratic vice president Al Gore and Republican house speaker Newt Gingrich, attempts to convey both the desirability of speedy access to large amounts of information over the Internet and the importance of such access to daily life (see INTERNET). Today's information needs are implicitly compared to the transportation needs of the booming postwar economy that led to the massive federal investment in the interstate highway system in the 1950s and 1960s. The interstate highways led to tremendous changes in the location of centers of employment and business. Similarly, it is argued that widespread Internet access will change the nature of business and employment in the 21st century.

Government involvement in the Internet dates back to its beginnings, arising from Defense Department research funding. Proponents of large-scale federal investment today argue that only such involvement can guarantee that access to the Internet's information resources is available to citizens in all economic groups and that the system is robust and capacious enough to keep up with rapidly growing data traffic and the growing use of media (such as streaming sound and video) that requires higher bandwidth connections (see BROADBAND and STREAMING). Under this scenario, the National Information Infrastructure (NII) would make Internet access as ubiquitous as the telephone is today.

Critics of the information superhighway metaphor believe that government control of the overall development of the Internet it is likely to raise issues of privacy and other civil liberties. For example, government initiatives to expand Internet capabilities might include such features as government-accessible "backdoors" to encryption (like the ill-fated V-Chip), requirements for content filtering, and so on (see ENCRYPTION). Federal money usually comes with politically inspired strings attached. Further, critics point out that centralized, top-down initiatives are at odds with the very nature of the Internet as a decentralized system where standards come from technical competition and emerging consensus rather than by fiat.

While there have been some federal initiatives (such as a telephone tax used to fund school Internet connections) and continuing policy-making and planning activity, no comprehensive initiative for universal broadband access has been passed by Congress, and it does not seem to be a major priority for the Bush administration as of

2002. However, about 180 research universities and corporations continue to work on the collaborative Internet 2 (I2) project, which deploys high-capacity "backbone" Internet connections and advanced technology for an infrastructure designed to accommodate the Internet's future growth.

Further Reading

"Internet 2." [home page] http://www.internet2.edu/

"National Information Infrastructure Agenda for Action." http://www.ibiblio.org/nii/toc.html

information theory

Information theory is the study of the fundamental characteristics of information and its transmission and reception. As a discipline, information theory took its impetus from the ideas of Claude Shannon (see SHANNON, CLAUDE).

In his seminal paper "A Mathematical Theory of Communication" published in the *Bell System Technical Journal* in 1948, Shannon analyzed the redundancy inherent in any form of communication other than a series of purely random numbers. Because of this redundancy, the amount of information (expressed in binary bits) needed to convey a message will be less than the number in the original message. It is because of redundancy that data compression algorithms can be applied to text, graphics, and other types of files to be stored on disk or transmitted over a network (see DATA COMPRESSION).

Shannon also analyzed the unpredictability or uncertainty of information as it is received—that is, the number of possibilities for the next bit or character. This is related to the number of possible symbols, but since all symbols are usually not equally likely, it is actually a sum of probabilities. Shannon used the physics term *entropy* to refer to this measure. It is important because it makes it possible to analyze the probability of error (caused by such things as "line noise") in a communications circuit. Shannon's basic formula is:

$$C = B\log_2(1 + P / N)$$

where the channel capacity C is in bits per second, B is the bandwidth, P the signal power, and N the Gaussian noise power.

Shannon found that if as long as the actual data transmission rate is less than the channel capacity C, an error-correcting code can be devised to ensure that any desired accuracy rate is achieved (see ERROR CORRECTION). A related formula can also be used to find the lowest transmission power needed given a specified amount of noise.

The influence of Shannon and his disciples on computing has been pervasive. Information theory provides the fundamental understanding needed for applications in data compression, signal analysis, data communication, and cryptography—as well as problems in other fields such as the analysis of genetic mutation or variation.

Further Reading

Hillman, Chris. "Entropy in Information and Coding Theory." http://www.math.psu.edu/gunesch/Entropy/infcode.html

Mackay, David J. C. "A Short Course in Information Theory." Cavendish Laboratory, Cambridge, 1995. http://131.111.48.24/pub/mackay/info-theory/course.html

McEliece, R. J. *The Theory of Information and Coding*. Reading, Mass.: Addison-Wesley, 1977.

Shannon, Claude, and Warren Weaver. "A Mathematical Theory of Communication." Urbana: University of Illinois Press, 1998.

Input/Output (I/O)

While the heart of a computer is its central processing unit or CPU (the part that actually "computes"), a computer must also have a "circulatory system" through which data moves between the CPU, the main memory, input devices (such as a keyboard or mouse), output devices (such as a printer), and mass storage devices (such as a hard or floppy disk drive). Input/Output or I/O processing is the general term for the management of this data flow (see also BUS, PARALLEL PORT, SERIAL PORT, and USB).

I/O processing can be categorized according to how a request for data is initiated, what component controls the process, and how the data flows between devices. In most early computers the CPU was responsible for all I/O activities (see CPU). Under program control, the CPU initiated a data transfer, checked the status of the device (or area of memory) that would be sending or receiving the data, and monitored the flow of data until it was complete. While this arrangement simplified computer architecture and reduced the cost of memory units or peripheral devices (at a time when computer hardware was hand-built and relatively costly), it also meant that the CPU could perform no other processing until I/O was complete.

In most modern computers, responsibility for I/O has largely been removed from the CPU, freeing it to concentrate on computation. There are several ways to implement such architecture. One method that has been used on microcomputers since their earliest day is *interrupt-driven I/O*. This means that the CPU has separate circuits on which a device requesting I/O service can "post" a request. The CPU periodically checks the circuits for an interrupt request (IRQ). If one is found, it can send a query to each device on a list until the correct one is found (the latter is called *polling*). Alternatively, the overhead involved in polling can be eliminated by having the IRQ include either a device identification number or a memory address that contains an interrupt service routine (this is called *vectored* interrupts). While interrupts

Input/Output

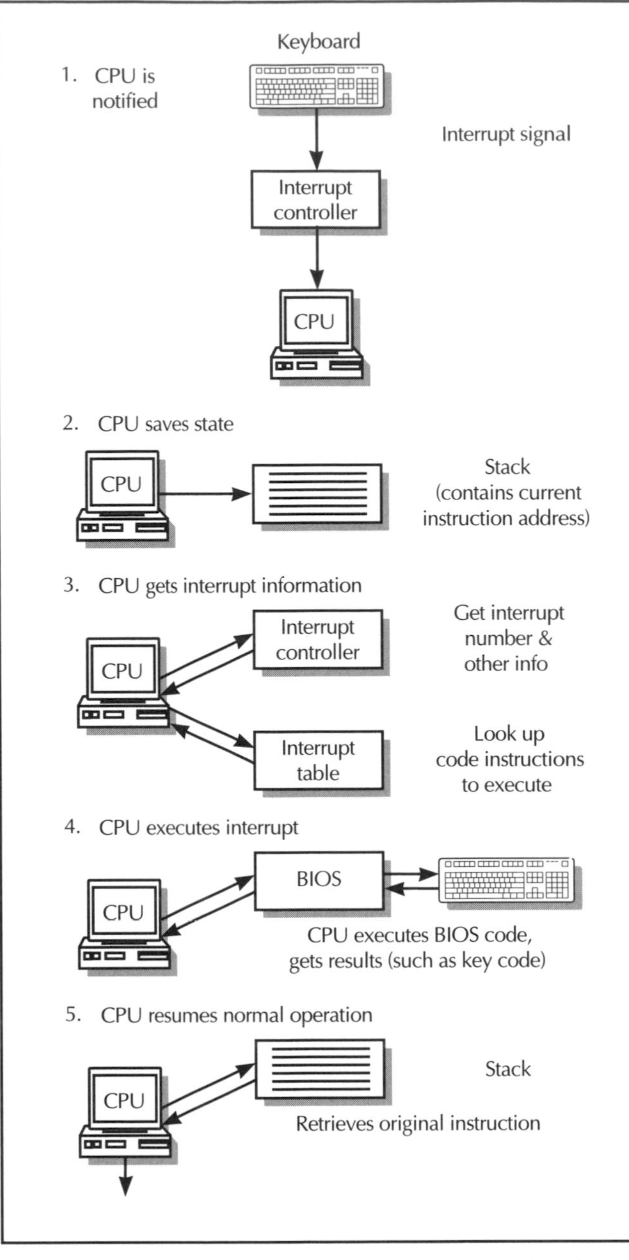

Steps in processing an interrupt request (IRQ) in a PC. (1) The device requesting attention signals the Interrupt Controller, which in turn sends a special signal to the CPU. (2) The CPU saves its state (including internal data and the address of the current instruction) to a stack. (3) The CPU gets the interrupt number and other information from the Interrupt Controller, then looks up a set of instructions for processing that particular interrupt. (4) The CPU executes the interrupt processing code, which generally links to BIOS code for handling a device such as the keyboard. (5) The CPU reloads its state information from the stack and resumes the interrupted processing.

alone do not free the CPU of the need to manage the I/O, they do remove the overhead of having to frequently check all devices for I/O.

The actual I/O process can also be moved out of the CPU through the use of direct memory access (DMA). Here a separate control device takes over control of the system from the CPU when I/O is requested. It then transfers data directly between a device (such as a hard disk drive) and a buffer in main memory. Although the CPU is idle during this process, the transfer is accomplished much more quickly because the full capacity of the bus can be used to move data rather than having to be shared with the flow of program instructions in the CPU.

A more sophisticated I/O control device is called a *channel*. A channel controller can operate completely independently of the CPU without requiring that the CPU become idle during a transfer. Channels can also act as a sort of specialized CPU or *coprocessor*, running program instructions to monitor the data transfer. There are also channels capable of monitoring and controlling several devices simultaneously (this is called *multiplexing*). The use of channels in mainframes such as the IBM 360 and its descendants is one reason why mainframes still perform a workhorse role in high-volume data processing.

In microcomputers the trend has also been toward offloading I/O from the CPU and the main bus to separate controllers or channels. For example, the AGP (Accelerated Graphics Port) found on most modern PCs acts as a channel between main memory and the graphics controllers (see GRAPHICS CARD). This means that as a program generates graphics data it can be automatically transferred from memory to the graphics controllers without any load on the CPU, and over a bus that is faster than the main system bus.

Further Reading

Hayes, J. P. *Computer Architecture and Organization.* 3rd ed. New York: WCB/McGraw-Hill, 1998.

White, Ron. *How Computers Work.* Millennium Edition. Indianapolis, Ind.: Que, 1999.

installation of software

While not often covered in computer science or software engineering courses, the process of getting a program to work on a given computer is often nontrivial. In the early days of PCs, installation generally involved simply copying the main program file and any needed settings files to a disk directory and possibly setting up the appropriate driver for the user's printer. (A cryptic user interface sometimes made the latter procedure a frequent occasion for technical support calls.) Users generally did not have to make many choices about what components to install or where to put them. On the other hand, installation programs sometimes made changes to a user's system without notification or the ability to "back out."

The ascension of Microsoft Windows to dominance as a PC operating system improved the installation process in several ways. Since the operating system and device drivers written by hardware vendors took over responsibility for installing and configuring printers and other devices, users generally didn't have to worry about configuring programs to work with specific hardware. Particularly with Windows 95 and later versions, a standard "installation kit" allows software developers to provide a familiar, step-by-step installation procedure to guide users. Generally, installation consists of an introductory screen, legal agreement, and the opportunity to choose a hard drive folder for the program. A moving "progress bar" then shows the files being copied from the installation CD to the hard drive. A "readme" file giving important considerations for using the program is usually provided. Increasingly, software registration is done by launching the user's Web browser and directing it to the vendor's website where a form is presented.

The installation of drivers accompanying new hardware such as a printer or scanner has been simplified even more through the "Plug and Play" feature in modern versions of Windows. This allows the system to automatically detect the presence of a new device and either install the driver automatically or prompt the user to insert a disk or CD (see PLUG AND PLAY).

Installation becomes a much more complicated matter when an enterprise has to install from tens to hundreds or thousands of copies of a program on employees' PCs. While small businesses may simply buy consumer-packaged software and install one copy on each PC, large businesses generally obtain a site license allowing a certain number of installations (or in some cases, unlimited on-site installations). Organizations must monitor the number of installations of a particular program package to ensure that licensing agreements are not violated while trying to use available software assets as efficiently as possible. (This is sometimes called *software asset management* or SAM.)

An automated installation script can be used to install a copy of the same software on each PC on the company's network—or a utility can be used to copy an exact hard disk image, including fully configured operating system and applications, to each PC. Alternatively, it is possible to buy networked versions of some programs. In this case the application actually runs on a server and is accessed from (but not copied to) each user's PC. This technique has also been adopted to provide consumers with an alternative to stand-alone installation (see APPLICATION SERVICE PROVIDER).

Linux and UNIX systems have also evolved more sophisticated installation systems in order to keep up with today's more complex applications and distributions. One common solution used by Red Hat and other Linux vendors is a "package" system where the user selects programs and features and the system identifies the components (packages) that must be installed to enable them.

Further Reading

Baker, Robert S. *The Official InstallShield for Windows Installer Developer's Guide (with CD-ROM)*. New York: Hungry Minds, 2001.

"SAM Solutions." http://www.corpsoft.com/Centers/samdirectory.htm

"SEPP—Software Installation and Sharing System." [for UNIX] http://people.ee.ethz.ch/~oetiker/sepp/

"Windows 2000 Software Installation and Maintenance." http://www.microsoft.com/windows2000/techinfo/administration/management/siamwp.asp

"WISE Home Page." http://www.wise.com/index.asp?bhcp=1

intellectual property and computing

Intellectual property can be defined as the rights the creator of an original work (such as an invention or a book) has to control its reproduction or use. Developers of new computer hardware, software, and media content must be able to realize a return on their time and effort. This return is threatened by the ease with which programs and data on disks can be illicitly copied and redistributed. Several legal mechanisms can be used to deter such behavior.

LEGAL PROTECTION MECHANISMS

Intellectual property represented by the design of new hardware can be protected through the patent system. A patent gives the inventor the exclusive right to sell or license the invention for 20 years after the date of filing. The basic requirements for a device to be patentable are that it represents an actual physical device or process and that it be sufficiently original and useful. A mere idea for a device, a mathematical formula, or a law of nature is not patentable in itself. In computing, a patent can be given for an actual physical device that meets the originality and usefulness requirements. Software that works with that device to control a physical process can be part of the patent, but an algorithm is not patentable by itself.

Because of these restrictions, most software is protected by copyright rather than by patent. A computer program is considered to be a written work akin to a book. (After all, a computer program can be thought of as a special type of narrative description of a process. When compiled into executable code and run on a suitable computer, a program has the ability to physically carry out the process it describes.)

Like other written works, a program has to be sufficiently original. Once copyrighted, protection lasts for the life of the author (programmer) plus 70 years. (Works made for hire are covered for 95 years from first publication or 120 years from creation.) Given the pace of change in computing, such terms are close to "forever." While not strictly necessary, registration of the work with the U.S. Copyright Office and the inclusion of a copyright statement serve as effective legal notice and prevent infringers from claiming that they did not know the work was copyrighted.

Content (that is, text or multimedia materials) presented in a computer medium can be copyrighted in the same way as its traditional printed counterpart. However, in 1996 the U.S. Supreme Court declared that a program's user interface as such could not be copyrighted (see *Lotus Development Corp. v. Borland International,* U.S. 94-2003).

Computer programs have also received protection as trade secrets. Under the Uniform Trade Secrets Act, as adopted in many states, a program can be considered a trade secret if gaining economic value from it depends upon it not being generally known to competitors, and that "reasonable effort" is undertaken to maintain its secrecy. The familiar confidentiality and non-disclosure agreements signed by many employees of technical firms are used to enforce such secrecy.

FIRST AMENDMENT ISSUES

In a few cases the government itself has sought to limit access to software, citing national security. In the 1996 case of *Bernstein v. U.S.*, however, the courts ultimately ruled that computer program code was a form of writing protected by the First Amendment, so government agencies seeking to prevent the spread of strong encryption software could not prevent its publication.

However, First Amendment arguments have been less effective in challenging private software protection mechanisms. In 2001 a U.S. District judge ruled that Princeton University computer scientist Edward Felten and his colleagues had no legal basis to challenge provisions of the Digital Millennium Copyright Act (DMCA). The scientists had claimed that a letter from the Recording Industry Association of America (RIAA) had cast a "chilling effect" on their research into DVD-protection software by threatening them with legal action if they published academic papers about copy protection software used by online music services. The RIAA had withdrawn its letter, and the courts ruled there was no longer anything to sue about. Critics of the decision claim that it still leaves the academics in a sort of legal limbo since there is no guarantee that they would not be sued if they published something.

In another widely watched case the U.S. Court of Appeals in New York affirmed a ruling that Eric Corley, editor of the hacker magazine *2600* could not publish the code for DeCSS, a program that would allow users to read encrypted DVD disks, bypassing publisher's restrictions. The Court said that the DMCA did not infringe upon First Amendment rights. This decision would appear to conflict with *Bernstein,* although the latter has to do with government censorship, not copyright. The Supreme Court is likely to hear one or more computer-related copyright cases in the years to come.

FAIR USE AND COPY PROTECTION

Although the purchase of software may look like a simple transfer of ownership, most software is accompanied by a license that actually grants only the right to use the program under certain conditions. For example, users are typically not allowed to make copies of the program and run the program on more than one computer (unless the license is specifically for multiple uses). However, as part of "fair use" users are allowed to make an archival or backup copy to guard against damage to the physical media.

Until the 1990s, it was typical for many programs (particularly games) to be physically protected against copying (see COPY PROTECTION). Talented hackers or "software pirates" are usually able to defeat such measures, and "bootleg" copies of programs outnumber legitimate copies in some Asian markets, for example. Copy protection and/or encryption is also typically used for some multimedia products such as DVD movies.

In recent years the Motion Picture Association of America (MPAA) has been fighting a legal battle against Linux programmers who created an open source (freely available) player program for movie DVDs. They did so because the MPAA deemed it uneconomical to create a licensed program for the Linux system.

In another well-known case the music-sharing service Napster was sued by the Recording Industry Association of America (RIAA). The trade group argued that Napster's only purpose is to facilitate the illegal copying and distribution of copyrighted music. The courts ultimately agreed and Napster, deprived of much of its access to "free" music, essentially disappeared in 2002. (However, it should be noted that the legal victory of the recording industry may yet be thwarted by the popularity of decentralized music-sharing services such as Gnutella that cannot be easily suppressed.)

Many recording companies, movie studios and other content providers believe that only the mandatory inclusion of anticopying features in the hardware of PCs and player devices would truly be effective in dealing with the piracy problem. A proposed bill, the Consumer Broadband and Digital Television Promotion Act of 2002, would give the entertainment and electronics industries up to 18 months to agree on a standard system to be incorporated in all devices. The bill was introduced by Senator Ernest Hollings and is expected to be also introduced in the House later in the year. However, the bill faces formidable obstacles, and is opposed by computer industry giants such as IBM, Hewlett-Packard, and Dell. Consumer advocates also argue that the forced incorporation of anticopying devices would prevent traditional fair use and force consumers to buy new hardware—devices that might be less flexible and capable than existing ones.

CHALLENGES OF NEW MEDIA

The convergence of software with written, visual and audio content also complicates copyright issues. Publishers of such material rely upon digital rights management (DRM) software, which uses electronic watermarking and encryption to try to prevent unauthorized copying. However, a Russian programmer, Dmitri Sklyarov, was arrested at a Las Vegas computer security conference in 2001. He and his company were providing software that allowed users to bypass the DRM for Adobe's popular PDF electronic publication format.

While legal in most of the world, Sklyarov's program was held to violate the U.S. DMCA. Programmers quickly rallied to his defense, arguing that the DMCA was far too broad in making people liable for the creation of software tools that *might* be used to violate copyright. Further, activists argued that the DRM software prevented the exercise of fair use by the user, such as by preventing content from being able to lent or used on a different computer platform. In late 2001 charges against Sklyarov were dismissed in return for his agreeing to testify against the company making the software.

The owners of intellectual property tend to win in the courts, but the fundamental nature of the computer medium makes enforcement of such property rights an endless battle. All forms of creative expression are increasingly being carried in digital form (see DIGITAL CONVERGENCE). It is much easier to copy data than to prevent its being copied; even encryption can be broken or bypassed. The economic repercussions of the collision between legal protection, technological capability, and new forms of distribution are difficult to predict.

Further Reading

Boyle, James. *Shamans, Software and Spleens: Law and the Construction of the Information Society.* Cambridge, Mass.: Harvard University Press, 1996.

Cornell University Law School. Legal Information Institute. "Copyright: an Overview." http://www.law.cornell.edu/topics/copyright.html

Electronic Frontier Foundation. http://www.eff.org

Fishman, Stephen. *Copyright Your Software.* 3rd. ed. Berkeley, Calif.: Nolo Press, 2001.

Glazier, Stephen C. *e-Patent Strategies for Software, e-Commerce, the Internet, Telecom Services, Financial Services, and Business Methods (with Case Studies and Forecasts).* Washington, D.C.: LBI Institute, 2000.

Litman, Jessica. *Digital Copyright: Protecting Intellectual Property on the Internet.* Amherst, N.Y.: Prometheus Books, 2001.

National Research Council. *The Digital Dilemma: Intellectual Property in the Information Age.* Washington, D.C.: National Academy Press, 2000.

U.S. Patent and Trademark Office. "Intellectual Property and the National Information Infrastructure." http://www.uspto.gov/web/offices/com/doc/ipnii/

Internet

The Internet is the worldwide network of all computers (or networks of computers) that communicate using a particular protocol for routing data from one computer to another (see TCP/IP). As long as the programs they run follow the rules of the protocol, the computers can be connected by a variety of physical means including ordinary and special phone lines, cable, fiber optics, and even wireless or satellite transmission.

HISTORY AND DEVELOPMENT

The Internet's origins can be traced to a project sponsored by the U.S. Defense Department. Its purpose was to find a way to connect key military computers (such as those controlling air defense radar and interceptor systems). Such a system would require a great deal of redundancy, routing communications around installations that had been destroyed by enemy nuclear weapons. The solution was to break data up into individually addressed packets that could be dispatched by routing software that could find whatever route to the destination was viable or most efficient. At the destination, packets would be reassembled into messages or data files.

By the early 1970s, a number of research institutions including the pioneer networking firm Bolt Beranek and Newman (BBN), Stanford Research Institute (SRI), Carnegie Mellon University, and the University of California at Berkeley were connected to the government-funded and administered ARPANET (named for the Defense Department's Advanced Research Projects Agency). Gradually, as use of the ARPANET's protocol spread, gateways were created to connect it to other networks such as the National Science Foundation's NSFnet. The growth of the network was also spurred by the creation of useful applications including e-mail and Usenet, a sort of bulletin-board service (see the Applications section below).

Meanwhile, a completely different world of on-line networking arose during the 1980s in the form of local bulletin boards, often connected using a store-and-forward system called FidoNet, and proprietary on-line services such as CompuServe and America On-line. At first there were few connections between these networks and the ARPANET, which had evolved into a general-purpose network for the academic community under the rubric of NSFnet. (It was possible to send e-mail between some networks using special gateways, but a number of different kinds of address syntax had to be used.)

In the 1990s, the NSFnet was essentially privatized, passing from government administration to a corporation that assigned domain names (see DOMAIN NAME SYSTEM). However, the impetus that brought the Internet into the daily consciousness of more and more people was the development of the World Wide Web by Tim Berners-Lee at the European particle research laboratory CERN (see BERNERS-LEE, TIM and WORLD WIDE WEB). With a standard way to display and link text (and the addition of graphics and multimedia by the mid-1990s), the Web *is* the Internet as far as most users are concerned (see WEB BROWSER). What had been a network for academics and adventurous professionals became a mainstream medium by the end of the decade (see also E-COMMERCE).

APPLICATIONS

A number of applications are (or have been) important contributors to the utility and popularity of the Internet.

- E-mail was one of the earliest applications on the ancestral ARPANET and remains the single most popular Internet application. Standard e-mail using SMTP (Simple Mail Transport Protocol) has been implemented for virtually every platform and operating system. In most cases once a user has entered a person's e-mail address into the "address book," e-mail can be sent with a few clicks of the mouse. While failure of the outgoing or destination mail server can still block transmission of a message, e-mail today has a high degree of reliability (see E-MAIL).
- Netnews (also called Usenet, for UNIX User Network) is in effect the world's largest computer bulletin board. It began in 1979, when Duke University and the University of North Carolina set up a simple mechanism for "posting" text files that could be read by other users. Today there are tens of thousands of topical "newsgroups" and millions of messages (called articles). Although still impressive in its quantity of content, many Web users now rely more on discussion forums based on webpages (see NETNEWS AND NEWSGROUPS).
- Ftp (File Transport Protocol) enables the transfer of one or more files between any two machines connected to the Internet. This method of file transfer has been largely supplanted by the use of download links on webpages, except for high-volume applications (where an ftp server is often operated "behind the scenes" of a Web link). FTP is also used by Web developers to upload files to a website (see FILE TRANSFER PROTOCOLS).
- Telnet is another fundamental service that brought the Internet much of its early utility. Telnet allows a user at one computer to log into another machine and run a program there. This provided an early means for users at PCs or workstations to, for example, access the Library of Congress catalog online. However, if program and file permissions are not set properly on the "host" system, telnet can cause security vulnerabilities. The telnet user is also vulnerable to having IDs and passwords stolen, since these are transmitted as clear (unencrypted) text. As a result, some on-line sites that once supported telnet access now limit access to Web-based forms. (Another alternative is to use a program called "secure shell" or ssh, or to use a telnet client that supports encryption.)
- Gopher was developed at the University of Minnesota and named for its mascot. Gopher is a system of servers that organize documents or other files through a hierarchy of menus that can be browsed by the remote user. Gopher became very popular in the late 1980s, only to be almost completely supplanted by the more versatile World Wide Web.
- WAIS (Wide Area Information Service) is a gateway that allows databases to be searched over the Internet. WAIS provided a relatively easy way to bring large data resources on-line. It, too, has largely been replaced by Web-based database services.
- The World Wide Web as mentioned above, is now the main means for displaying and transferring information of all kinds over the Internet. Its flexibility, relative ease of use, and ubiquity (with Web browsers available for virtually all platforms) has caused it to subsume most earlier services. The utility of the Web has been further enhanced by the development of many search engines that vary in thoroughness and sophistication (see WORLD WIDE WEB and SEARCH ENGINE).
- Streaming Media protocols allow for a flow of video and/or audio content to users. Player applications for Windows and other operating systems, and growing use of high-speed consumer Internet connections (see BROADBAND) have made it possible to present "live" TV and radio shows over the Internet.
- E-commerce or the carrying out of business transactions on the Web was the most highly touted application of the late 1990s. While a few companies such as Amazon.com and eBay have managed to create large revenue streams with expectations of eventual profitability, other entrepreneurs have concentrated on business-to-business (B2B) commerce applications, such as the replacement of traditional forms of wholesale distribution with on-line-brokered systems. The stock market declines of the early 2000s have cast doubt on the projected rate of growth of e-commerce.

CURRENT AND FUTURE TRENDS

After the rapid growth of the late 1990s, Internet access in the United States seems to be reaching saturation levels with a declining rate of growth. Internationally, Internet growth has been strong through the developed world, particularly in Europe and Asia. Unless it can be heavily subsidized (including the provision or upgrading of basic infrastructure), it is unlikely that Internet access will become a mass phenomenon in the developing world for the foreseeable future. (See also DIGITAL DIVIDE and INFORMATION SUPERHIGHWAY.)

Increasingly, proprietary local area network (LAN) systems have been replaced by networks using the same protocols as the Internet (see INTRANET). This means that the same software (web servers, browsers, search engines, and so on) used to connect to the Internet is used to manage an organization's internal network. This seamless connection raises security concerns and requires the deployment of effective measures to protect the local network from intruders (see FIREWALL).

In the United States, the transition from relatively slow dial-up access to broadband (mainly cable or DSL) is likely to be steady. In the early 2000s, the continuing growth in bandwidth-hungry applications such as streaming video and audio is likely to increase congestion as its demands increase faster than the provision of new high-speed "backbone" connections. Initiatives based on the Internet 2 proposal are likely to relieve congestion in the long run.

Security remains a pressing concern in the wake of well-publicized "denial of service" attacks that have occasionally brought down major sites such as Yahoo! as well as the continuing spread of viruses, particularly through exploiting vulnerabilities in the dominant Microsoft server and client products (see COMPUTER CRIME AND SECURITY and COMPUTER VIRUS). Together with concerns about terrorism in the post-9/11 world and threats to privacy, security issues are likely to keep the growing Internet under critical scrutiny and invite further government regulation.

In the longer term, however, what we call the Internet today is likely to become so ubiquitous that people will no longer think of it as a separate system or entity. Household appliances, cars, cell phones, televisions, and virtually every other device used in daily life will communicate with other devices and with control systems using Internet protocols. In effect, people may eventually live "inside" a World Wide Web.

Further Reading

"CNET Internet Guide." http://www.cnet.com/internet/0-3761.html?tag=stbc.gp

CyberAtlas. http://cyberatlas.internet.com/ http://www.c-i-a.com/iia_info.htm

Hafner, K., and M. Lyon. *Where Wizards Stay Up Late: The Origins of the Internet.* New York: Simon & Schuster, 1996.

Hobbes' Internet Timeline. http://www.zakon.org/robert/internet/timeline/

Internet Society [website]. http://www.isoc.org

Internet Statistics Reference Desk. http://www.refdesk.com/netsnap.html

Juliussen, Egil, and Karen Petska Juliussen. *Internet Industry Almanac.* Glenbrook, Nev.: Computer Industry Almanac, 1998. [Note: a new edition may appear in 2002] See publisher's website at http://www.c-i-a.com/iia_info.htm

Segaller, S. *Nerds 2.0.1: A Brief History of the Internet.* New York: TV Books, 1998.

Internet appliance

An Internet appliance (sometimes called a Web appliance) is a simplified computer system designed to serve as a low-cost, user-friendly alternative to the standard personal computer as a means of Internet access. Generally, such systems (such as Gateway's Touch Pad and Compaq's iPaq introduced in late 2000) feature a compact case with a separate or built-in display and keyboard. The user simply plugs the system into power and phone lines and turns it on. Systems are generally preset to use a particular ISP such as America On-line or MSN. This means that the user can browse right away without having to install any software or go through a sign-up or configuration process. (The drawback is that the user is usually "locked in" to a particular service, which the user may later find to be inadequate or worse still, unavailable.)

WebTV represents an alternative approach to the Internet appliance that uses the home TV as the display device. The appliance is connected to the TV in much the same way as would be done with a video gaming console (some of which now also have Internet access options). Backed by Microsoft and a number of manufacturers, WebTV has gained more than 800,000 subscribers who pay somewhere between $100 and $200 for the box and between $10 and $25/month for the Internet connection. WebTV offers the ability to watch TV and surf the Web at the same time, to send and receive e-mail and instant messages, and even to "talk back" to special interactive TV broadcasts. In its cheapest configuration WebTV lacks a keyboard, which requires that users click on letters on an onscreen "virtual keyboard" in order to compose e-mail or other text. However, an external keyboard is available. Devices that use home TVs for display also suffer from limited resolution, with text even on large-screen TVs often being fuzzier and harder to read than on regular PC monitors.

As of the early 2000s, it's unclear whether the Internet appliance will survive as a discrete market niche. One advantage, the low cost that arises from discarding the hard drives and other components of traditional PCs, has been largely overtaken by the fall of low-end consumer PCs below $500. Another advantage, compactness and portability, is now shared by the ubiquitous palmtop computer, although these devices tend to have limited display space and lack a usable keyboard. Finally, the trend toward keeping applications on the server (see APPLICATION SERVICE PROVIDER) and the use of diskless workstations or "network computers" in some offices may also provide many of the same functions associated with an Internet appliance.

On the other hand, it is likely that Internet access will be increasingly built into conventional appliances and communications devices, ranging from in-car consoles to enhanced cell phones and portable music players.

Further Reading

Cole, Bernard Conrad. *The Emergence of Net-Centric Computing: Network Computers, Internet Appliances, and Connected PCs.* Upper Saddle River, N.J.: Prentice Hall PTR, 1999.

Rischpater, Ray. *Internet Appliances: a Wiley Tech. Brief.* New York: Wiley, 2001.

USA Today Tech. Reviews. "Net appliances on parade." http://www.usatoday.com/life/cyber/tech/review/crh629.htm.

Internet applications programming

The growth of the Internet and its centrality in business, education, and other fields has led many programmers to specialize in Internet-related applications. These can include the following:

- Web servers, including specialized servers such as secure e-commerce servers
- Other e-commerce applications including inventory, stock display, and user shopping cart features (see also E-COMMERCE)
- Interactive Web-based forms, such as forms used to handle technical support queries or to search databases
- Gateways or "front ends" between Web forms and databases
- Web browsers and browser "plug-ins" (modules that extend the browser's ability to display certain kinds of content) (See WEB BROWSER and PLUG-IN)
- Streaming video and audio servers and players (see STREAMING)
- Search engines, both general-purpose and specialized (see SEARCH ENGINE)
- "Bots" or programs designed to automatically search the Web and analyze information, such as comparative price data (see SOFTWARE AGENT)

Internet applications programmers use a variety of languages and other programming tools (often in combination) to implement these applications. Some of the most common are:

- C++ is generally used for fundamental applications, particularly those that must work at the system level and for which speed and efficiency are prerequisites. Examples would include Web servers and browsers and some browser plug-ins (see C++).
- Java has largely supplanted C++ as a general-purpose language for programming small applications ("applets") that are hosted by websites and run on the user's browser. With a syntax that differs in only a few respects from C++, Java can also be used to write standalone applications (see JAVA).
- HTML is not really a full-fledged programming language, but it defines the layout and formatting of webpages, as well as providing for hyperlinks and the embedding of applications. In many cases, HTML no longer has to be coded directly but can be generated from word processor-like page design programs (see HTML).
- Scripting languages are an important tool for Internet and Web development. CGI (Common Gateway Interface) is a facility that allows scripts to control the interaction between HTML forms on a webpage and other programs such as databases (see CGI). CGI scripts are written in scripting languages (see PERL, PYTHON, and SCRIPTING LANGUAGES). Use of CGI is being gradually supplanted by applets written in Java as well as other scripting languages such as JavaScript and VBScript.
- Active Server Pages (ASP) is a facility that uses Windows ActiveX components to process scripts created in Visual Basic, which in turn create HTML pages "on the fly" and send them to the user's Web browser.
- Microsoft's recent .NET initiative represents an attempt to integrate Internet connectivity and distributed operation into the programming framework for all major languages.

TRENDS

Experienced programmers will continue to be needed for creating and extending the infrastructure for the Internet and Web and for providing increasingly powerful and easy-to-use tools for developing websites. However, the wide variety of tools now available means that people with less experience will be able to design and implement attractive and effective webpages, plugging in functionality such as on-line shopping, conferencing, and site-specific search engines. If web development follows the same course as traditional programming, predictions that specialized programmers will no longer be needed will prove premature. At the same time, generalist web developers will be able to do more.

Further Reading

Bates, Chris. *Web Programming: Building Internet Applications.* New York: Wiley, 2000.

"Developing Internet Applications for Education." http://www.dc.peachnet.edu/~shale/education/programming.html

"Microsoft Active Server Pages Roadmap." http://www.aspdeveloper.net/iasdocs/aspdocs/roadmap.asp

World Wide Web Consortium. "CGI: Common Gateway Interface." http://www.w3.org/CGI/

Internet service provider (ISP)

An Internet service provider is any organization that provides access to the Internet. While nonprofit organizations such as universities and government agencies can be considered to be ISPs, the term is generally applied to a commercial, fee-based service.

Typically, a user is given an account that is accessed by logging in through the operating system's Internet connection facility by supplying a user ID and password. Once connected, the user can run Web browsers, e-mail clients, and other programs that are designed to work with an Internet connection. Most ISPs now charge flat monthly fees ranging from $20 or so for dial-up access to around $40–$60 for high-speed cable or DSL connections. Some services such as America On-line and CompuServe include ISP service as part of a package that also includes such features as software libraries, discussion forums, and instant messaging. On-line services tend to be more expensive than "no frills" ISP services.

Most personal ISP accounts include a small allotment of server space that users can use to host their personal webpages. There are generally extra charges for larger allotments of space, for sites that generate high traffic, and for commercial sites. Business-oriented ISPs typically provide a more generous starting allotment along with more extensive technical support and more reliable and higher-capacity servers that are managed 24 hours a day.

The rapid growth in Internet use in the mid-1990s encouraged many would-be entrepreneurs to start ISPs. However, with so many providers entering the field and with the price for basic Internet connections falling, it soon became apparent that the survival prospects for "generic" ISPs would be poor. People entering the business today strive to provide added-value services such as superior webpage hosting facilities, or to focus on specialized services for particularly industries (such as real estate). The big on-line services (particularly America On-line and to a lesser extent Microsoft Network [MSN]) continue to be successful by providing ISP service that is integrated with other services.

A number of providers decided that there was little money to be made in selling basic ISP service, so instead they provide the connection free, relying on the accompanying stream of advertising to provide revenue. This approach has generally failed, as Internet business models that rely on advertising revenue are increasingly questioned. Finally, some companies have bundled low-cost or even "free" PCs with long-term contracts for ISP service.

Further Reading

Burris, Anne M. *Service Provider Strategy: Proven Secrets for xSPs*. Upper Saddle River, N.J.: Prentice Hall PTR, 2002.

Yahoo! "Internet Service Providers." Access through Home > Business and Economy > Business to Business > Communications and Networking > Internet and World Wide Web > Network Service Providers > Internet Service Providers (ISPs).

Internet telephony

Many people who have ordinary voice telephone service now have Internet-connected personal computers using either those same phone lines or higher-speed DSL or cable connections (see also TELECOMMUNICATIONS). Since most PCs have microphone inputs and simple voice-recording software, all the ingredients are in place for carrying voice telephone calls as a stream of Internet Protocol (IP) data packets (see TCP/IP). The result is Internet telephony (also called IP telephony or voice over IP).

With suitable software installed on users' PCs, conversations can be carried either over a local network (see INTRANET), replacing intercoms or local phone calls, or between users anywhere on the worldwide network (replacing long-distance phone service). The obvious advantage is cost: Since most Internet users have local, toll-free connections to their Internet service providers and pay a flat rate for Internet service (see INTERNET SERVICE PROVIDER), an expensive long-distance or metered toll call can be replaced by a virtually free conversation over the Internet.

Internet telephony has several disadvantages compared to standard voice telephony, however. First, poor-quality microphones can mean that the fidelity of the voice transmission going out over the net is also poor. Congestion on the net can result in the voice transmission becoming choppy or interrupted. Also, two parties wishing to have an Internet phone conversation must generally prearrange (through e-mail or instant messaging) a time when both of them will be on-line and at their PC. Finally, Internet telephony can generally be used only when both parties have a PC, an Internet connection, and suitable software.

However, a hybrid service can allow someone to make an Internet call from a PC to someone having only an ordinary phone (or vice versa). For example, a user who subscribes to an Internet telephony service provider (ITSP) can transmit voice over the Internet to the service provider, which then carries over the Internet to the node nearest to the other party in the conversation. At the node, the call is patched into the regular phone system. Since the long-distance transmission is over the Internet rather than over the phone system, there are no long-distance charges. (This is actually similar in principle to a free service offered for many years by amateur radio operators.)

Internet Telephony

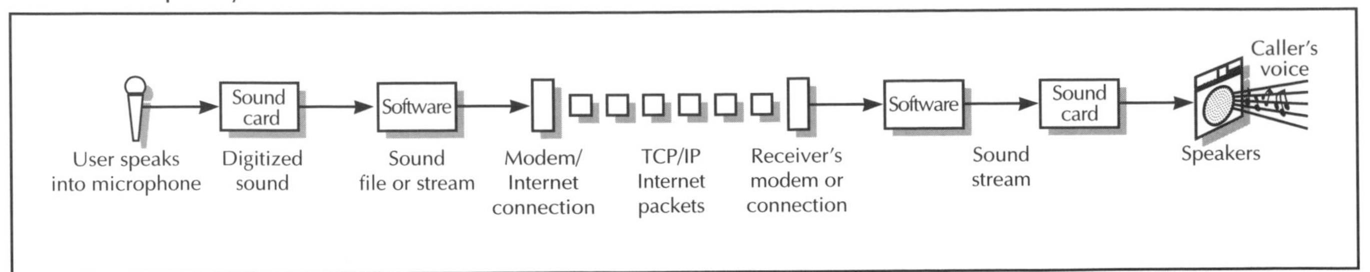

A regular telephone carries the voice as an analog signal over the phone line. For Internet (IP) telephony, however, the user's voice from the microphone is converted to a digital signal that is carried by standard Internet packets. At the destination, the packets are reassembled into a stream of digital data that is then sent to the sound card to be turned back into voice sounds to be played through the system speaker.

TRENDS

The ability to leverage the Internet to provide inexpensive telephone service continues to be attractive, and such services continue to grow as companies such as the Internet Telephony Exchange Carrier Corporation (ITXC) build an infrastructure of service providers, increasing the areas covered by Internet telephony services. However, telephone companies have sometimes objected to the uncompensated burden potentially placed on their local service by ITSPs and to the potential loss of revenue from traditional long-distance service. Thus far, federal regulators have been reluctant to create a tariff mechanism that would require ISPs to pay phone companies for the use of their lines for Internet telephony. The outcome of this controversy is hard to predict because it is part of the larger question of whether Internet users and service providers should pay more for their use of the public telephone network. Additionally, since Internet Service Providers must provide sufficient bandwidth to carry the data associated with Internet telephony, they, too, need to find a way to identify and recoup the costs associated with it.

The convergence of telephone and Internet technology may eventually result in traditional voice telephone service being phased out in favor of Internet telephony. While the investment in new phone equipment would be substantial, the Internet packet-switching model allows for far more efficient use of infrastructure since traffic can be flexibly and automatically directed in accordance with changing conditions, rather than each call occupying a particular physical circuit (circuit-switching).

Further Reading

Internet Telephony Exchange Carrier Corporation (ITXC). http://www.itxc.com/intro.html

McKnight, Lee W. "Internet Telephony and Open Communications Policy." [Harvard University, Kennedy School of Government Information Infrastructure Project]http://www.ksg.harvard.edu/iip/iicompol/Papers/McKnight.html

McKnight, Lee W., William Lehr, and David D. Clark,.eds. *Internet Telephony.* Cambridge, Mass.: MIT Press, 2001.

MIT Internet Telephony and Telecoms Convergence Consortium. http://itel.mit.edu/

Wright, David J. *Voice Over Packet Networks.* New York: Wiley, 2001.

interpreter

An interpreter is a program that analyzes (parses) programming commands or statements in a high-level language (see PROGRAMMING LANGUAGES), creates equivalent executable instructions in machine code (see ASSEMBLER) and executes them. An interpreter differs from a compiler in that the latter converts the entire program to an executable file rather than processing and executing it a statement at a time (see COMPILER).

Many earlier versions of the BASIC programming language were implemented as interpreters. Since an interpreter only has to hold one program statement at a time in memory, it could run on early microcomputers that had only a few tens of thousands of bytes of system memory. However, interpreters run programs considerably more slowly than a compiled program would run. One reason is that an interpreter "throws away" each source code statement after it interprets it. This means that if a statement runs repeatedly (see LOOP), it must be re-interpreted each time it runs. A compiler, on the other hand, would create only one set of machine code instructions for the loop and then move on. Also, because a compiler keeps the entire program in memory, it can analyze the relationship between multiple statements and recognize ways to rearrange or substitute them for greater efficiency.

Interpretation can also be used to bridge differences in hardware platforms. For example, in the UCSD Pascal system developed in the 1970s, an interpreter first translates the Pascal source code into a standardized "P-code" (pseudocode) for a generic processor called a P-machine. To run the program on a particular actual machine, a sec-

Entry Interpreter

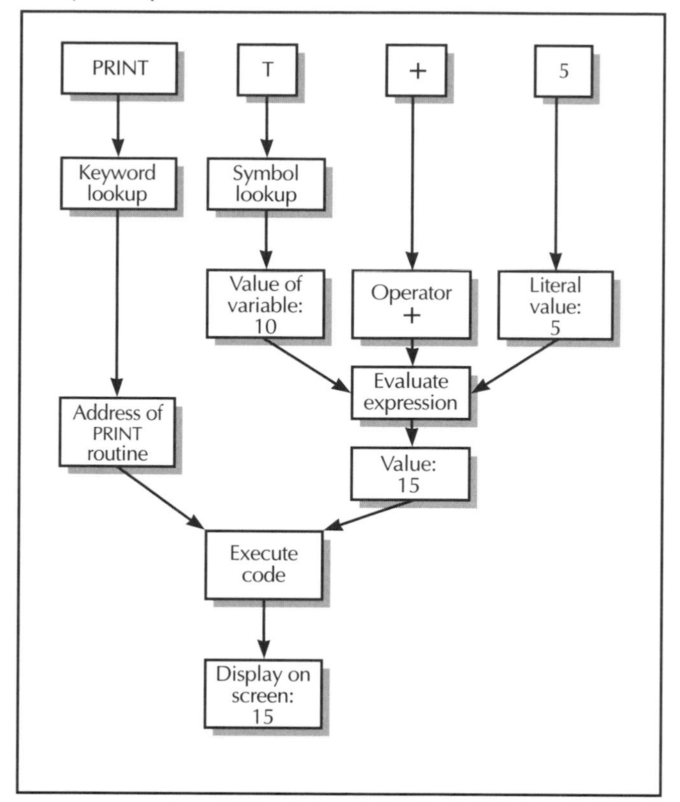

An interpreter scans a program code or command statement to determine what each token (word or symbol) represents. Keywords such as PRINT are looked up in a dispatch table that contains instructions for dealing with that function. Variables are looked up in a symbol table that gives their current value. Values and operators make up expressions that are interpreted to yield their final value. In this case the final value of 15 is given as data to the PRINT routine, which is executed to put the number 15 on the screen.

ond interpreter translates the P-code into specific executable machine instructions for that machine. Today Java uses a similar idea. A Java programming system translates source code into an intermediate "bytecode," which is interpreted by a Java Virtual Machine, usually running with a Web browser.

In practice, with today's high-speed computers and graphical operating environments, interpretative and compilation functions are often seamlessly integrated into a programming environment where code is checked for syntax as it is entered, incrementally compiled (such that only changed code is recompiled), and the programmer receives the same kind of rapid feedback that was the hallmark of the early BASIC interpreters (see PROGRAMMING ENVIRONMENT). Purely interpretive systems survive mainly in the form of text command processors for operating systems (see SHELL).

Further Reading

Kamin. S. N. *Programming Languages: An Interpreter-Based Approach.* Reading, Mass.: Addison-Wesley, 1990.

Lindholm, T., and F. Yellin. *The Java Virtual Machine Specification.* Reading, Mass.: Addison-Wesley, 1996.

intranet

During the 1980s, many organizations developed internal networks of PCs using network operating systems such as Novell's Netware and Microsoft's Windows NT (see LOCAL AREA NETWORK). During the 1990s, however, many of these same computers were also connected to an external network, the Internet, using a different protocol (see TCP/IP). As more business databases, documents, and e-mail began to be accessed through Web servers and other Internet-enabled programs, it became clear to many network administrators and developers that Internet tools such as Web servers and browsers, scripting languages, search engines, e-mail, and instant messaging could form the basis for an in-house information management and communication system. Instead of using proprietary networking software, the local network could be run under TCP/IP as a restricted, localized version of the Internet—an *intranet.*

In a typical intranet, the network server runs a Web server for the in-host website (or sites) as well as databases, document stores, e-mail, and other resources. (Microsoft software is used for these purposes in most corporate intranets, although a significant minority runs UNIX-based versions of such programs under the Linux operating system.) Users access these resources using the same browser and other tools that they would use to access external websites. An organization's intranets at different locations can also be connected through special secure Internet links, forming an *extranet.* A combination of hardware and software is used to protect the intranet from outside attack or intrusion (see FIREWALL).

The intranet has become an attractive alternative to proprietary local area networks for several reasons:

- The organization is not locked into a proprietary network operating system. Any system that supports TCP/IP networking can be used.
- Training costs are reduced because users can use the same Web browsers and e-mail programs they use on the Internet.
- Many tools are now available to make it easy to organize and search information on websites, and many word processors, publishing programs, presentation software, spreadsheets, and databases can now produce output ready for use on webpages.

Intranets do pose challenges, however. The very flexibility inherent in a nonproprietary system also means that an intranet may grow haphazardly and not scale up smoothly as hardware, users, and data resources are added. Converting in-house information to Web-ready formats can be costly. Firewalls and other security measures must be maintained and kept up to date. Nevertheless, intranets are gradually supplanting proprietary networks.

Further Reading

Bernard, Ryan. *The Corporate Intranet.* 2nd ed. New York: Wiley, 1997.

"Complete Intranet Resource site." http://www.intrack.com/intranet/

Intranet Journal. http://www.intranetjournal.com/

Linthicum, David S. *David Linthicum's Guide to Client/Server and Intranet Development.* New York: Wiley, 1997.

J

Java

Java is a computer language similar in structure to C++. Although Java is a general-purpose programming language, it is most often used for creating applications to run on the Internet, such as Web servers. A special type of Java program called an *applet* can be linked into webpages and run on the user's Web browser.

As an object-oriented language, Java uses classes that provide commonly needed functions including the creation of user interface objects such as windows and buttons. A variety of sets of classes ("class frameworks") are available, such as the AWT (Abstract Windowing Toolkit).

PROGRAM STRUCTURE

A Java program begins by importing or defining classes and using them to create the objects needed for the program's functions. Code statements then create the desired output or interaction from the objects, such as drawing a picture or putting text in a window. Here is a simple Java applet program:

```
import java.applet.Applet;
import java.awt.Graphics;
public class HelloWorld extends Applet {
  public void paint(Graphics g) {
    g.drawString("Hello world!", 50, 25);
  }
}
```

The first two lines import (bring in) standard classes. The applet class is the foundation on which applet programs are built. The AWT (Abstract Windowing Toolkit) is a set of classes that provide a graphical user interface.

The program then declares a new class called HelloWorld and specifies that it is built on (extends) the applet class.

Java

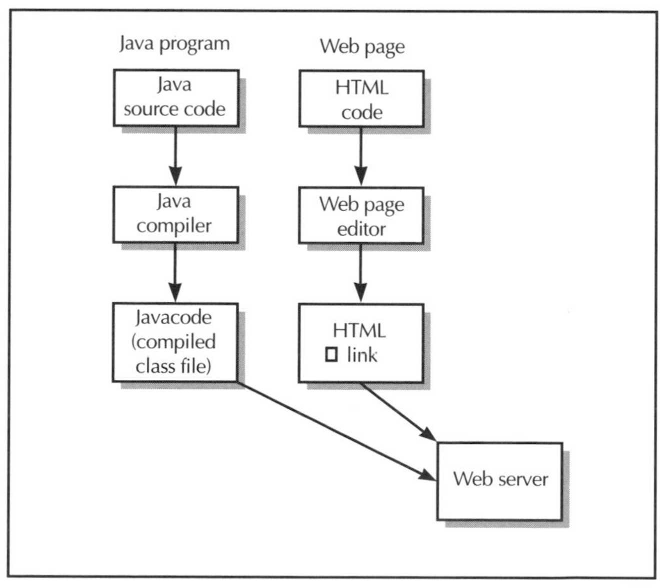

After an embedded Java program (called an applet) is compiled, its executable file (Javacode) is stored on the Web server, together with the HTML file for the webpage to which the program is linked.

Next is a declaration for a method (procedure for doing something) called paint. This method uses a graphics object g that includes various capabilities for drawing things on the screen. Finally, the program uses the graphic object's predefined drawstring method to draw a string of text.

To develop this program, the programmer compiles it with the Java compiler. He or she then creates an HTML page that includes a tag that specifies that this code is to be run when the link is activated (see HTML).

DEVELOPMENT OF JAVA

Java was created by James Gosling (1955–). It began as an in-house project at Sun Microsystems to design a language that could be used to program "smart" consumer devices such as an interactive television. When this project was abandoned, Gosling, Bill Joy, and other developers realized that the language could be adapted to the rapidly growing Internet. Developers of webpages needed an easier way to create programs that could run when the page was accessed by a user. By implementing user controls on webpages, the designers could give Web users the ability to interact on-line in much the same way they interact with objects on the screen on a Macintosh or Windows PC.

ADVANTAGES

Java has largely fulfilled this promise for Web developers. C++ programmers have an easy learning curve to Java, since the two languages have very similar syntax and a similar use of classes and other object-oriented features. On the other hand, programmers who don't know C++ benefit from Java being more streamlined than C++. For example, Java avoids the necessity to use pointers (see POINTERS AND INDIRECTION) and uses classes as the consistent building block of program structure. Software powerhouses such as Microsoft (until recently) and IBM have joined Sun in promoting Java.

Another much-touted feature of Java is its platform independence. The language itself is separate from the various operating system platforms. For each platform, a Java Virtual Machine (JVM) is created, which interprets or compiles the code generated by the Java compiler so it can run on that platform.

Theoretically, this means that a program written to standard Java specifications will run on any computer and any operating system that has a Java Virtual Machine. In practice, however, Microsoft and other developers have created Java programming environments that include classes that are specific to certain platforms (such as Windows). While such implementations can be more efficient and take better advantages of a particular platform, programs written using such Java "dialects" may not run on other platforms without modification.

Java Applet and JVM

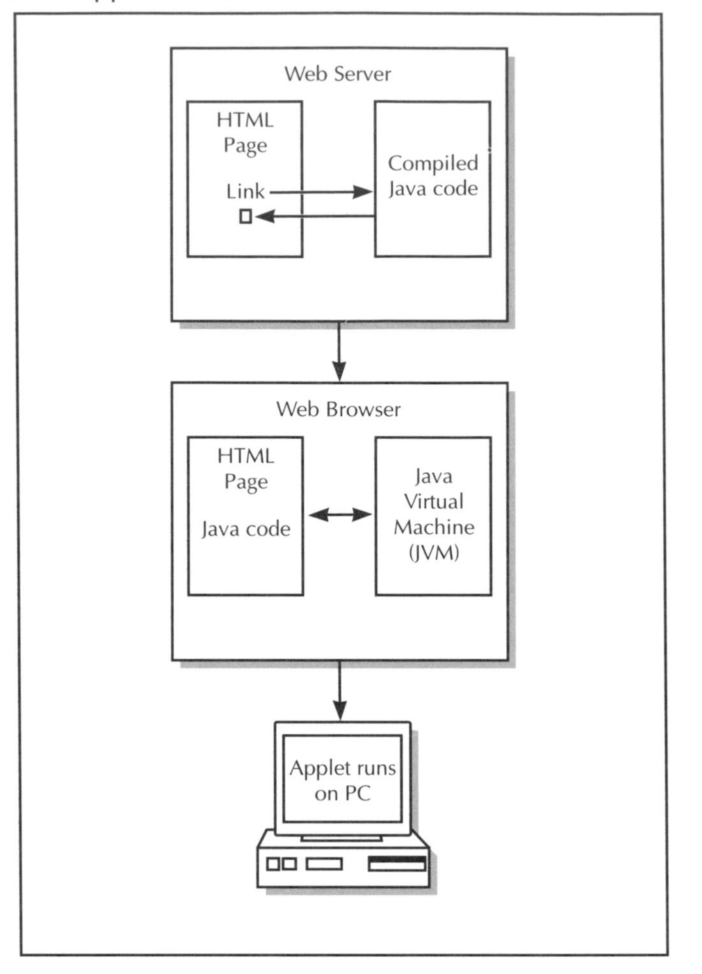

To run the Java applet, the user loads the linked page in the Web browser. The applet may then run automatically, or it may be connected to a particular link or a control such as a button. Once activated, the applet is downloaded by the Web browser, which then runs its Javacode using a module called a Java Virtual Machine (JVM). There is a separate JVM for each type of computer system.

For security, Java applets run within a "sandbox" or restricted environment so the user is protected from malicious Java programs. (For example, programs are not allowed to access the user's disk or to connect the user's machine to another website.) Web browsers can also be set to disable the running of Java applets.

A NEW WAY TO DELIVER SOFTWARE?

Java has been touted as a way to bring useful software directly to Web users, regardless of platform and operating system. Because of the need to interpret Java code, Java programs generally run slower than C or C++ programs compiled for a particular hardware platform. However, the development of true native code compilers and the prevalence of faster Internet connections may result in software such as an office suite (word processor, database, spreadsheet, contact manager, and so on) that users could rent on the Internet rather than buying from Microsoft or another traditional vendor. Such users could also run free on lost-cost operating systems such as Linux that are Java-compatible rather than more expensive products such as Windows (see APPLICATIONS SERVICE PROVIDER).

However, by summer 2001 Microsoft had announced that it would not be including the Java Virtual Machine with its upcoming Windows XP (although users would be able to obtain it from third parties and install it easily). While the company said it was seeking to avoid charges of "bundling" in the wake of the decision in its antitrust suit, some observers believe that Microsoft is seeking to relegate Java to secondary status in favor of its new ".net" initiative for Web development. The effects of these developments remain to be seen.

Further Reading

Arnold, Ken, and James Gosling. *The Java Programming Language*. 2nd ed. Reading, Mass.: Addison-Wesley, 1997.

Flanagan, David. *Java in a Nutshell*. 4th ed. Sebastopol, Calif. O'Reilly, 2002.

IBM Java Developer's Resources. http://www.ibm.com/developer/java/

Java Tutorial with many code examples. http://java.sun.com/docs/books/tutorial/index.html

JavaWorld magazine. www.javaworld.com

job control language

In the early days of computing, data processing generally had to be done in batches. By modern standards the memory capacity of the computer was very limited (see MAINFRAME). Typically, programs had to be loaded one at a time from punch cards or tape. The data to be processed by each program also had to be made available by being mounted on a tape drive or inserted as a stack of cards into the card reader (see PUNCHED CARDS AND PAPER TAPE). After the program ran, its output would consist of more data cards or tape, which might in turn be used as input for the next program.

For example, a series of programs might be used to read employee time cards and calculate the payments due after various items of withholding. That data might in turn be input into a program to print the payroll checks and another program to print a summary report.

In order for all this to work, the computer's operating system must be told which files (on which devices) are to be used by the program, the memory partition in which the program is to be run, the device to which output will be written or saved, and so on. This is done by giving the computer instructions in job control language (JCL). (In the punch card days, the JCL cards were put at the top of the deck before the cards with the instructions for the program itself.)

For a simple example, we will use some elements of IBM MVS JCL. In this version of job control language the general form for all statements is

```
//name operation operands comment
```

where *name* is a label that can be used to reference the statement from elsewhere, *operation* indicates one of a set of defined JCL language commands, *operand* is a series of values to be passed to the system, and *comment* is optional explanatory text.

The three basic types of statement found in most job control languages are JOB, EXEC, and DD. The JOB statement identifies the job and the user running it and sets up some parameters to specify the handling of the job.

```
//JOB,CLASSPROJ1,GROUP=J999996,USER=P999995,
//PASSWORD=?
```

This statement passes information to the system that identifies the job name, group as assigned by the facility, and user ID. The PASSWORD parameter is given a question mark to indicate that it will be prompted for at the terminal. Other parameters can be used to specify such matters as the amount of computer time to be allocated to the job and the way in which any error messages will be displayed.

The EXEC statement identifies the program to be run. Some systems can also have a library of stored JCL procedures that can also be specified in the EXEC statement. This means that frequently run jobs can be run without having to specify all the details each time. An example EXEC statement is:

```
//Datasort EXEC BINSORT,BUFFER=256K
```

Here the statement is labeled Datasort so it can be referenced from another part of the program. The procedure to be executed is named BINSORT, and it is passed a parameter called BUFFER with a value of 256K (presumably this is the amount of memory to be used to hold data to be sorted).

One or more DD (Data Definition) statements are used to specify sets (sources) of data to be used by the program. This includes a specification of the type (such as disk or tape) and format of the data. It also includes instructions specifying what is to happen to the data set. For example, the data set might be old (existing) or newly created by the program. It may also be temporary (possibly to be passed on to the next program) or permanent ("cataloged").

Since interactive, multitasking operating systems such as Windows and UNIX are now the norm in most computing, JCL is used less frequently today. However, it is still needed in large computer installations running operating systems such as IBM MVS (see MAINFRAME) and for some batch processing of scientific or statistical programs (such as in FORTRAN or SAS).

Further Reading

Brown, Gary Deward. *System 390 Job Control Language*. 4th ed. New York: John Wiley, 1998.

Malaga, Ernie, and Ted Holt. *Complete CL: The Definitive Control Language Programming Guide*. 3rd ed. Carlsbad, Calif.: Midrange Computing, 1999.

Jobs, Steven Paul

(1955–)
American
Entrepreneur

Steve Jobs was co-founder of Apple Computer and shaped the development and marketing of its distinctive Macintosh personal computer. Jobs showed an enthusiastic interest in electronics starting in his high school years and gained experience through summer work at Hewlett-Packard, one of the dominant companies of the early Silicon Valley. In 1974, he began to work for pioneer video game designer Nolan Bushnell at Atari. He also became a key member of the Homebrew Computer Club, a group of hobbyists who designed their own microcomputer systems using early microprocessors.

Meanwhile, Jobs's friend Steve Wozniak had developed plans for a complete microcomputer system that could be built using a single-board design and relatively simple circuits (see WOZNIAK, STEVEN). In it Jobs saw the potential for a standardized, commercially viable microcomputer system. They formed a company called Apple Computer (named apparently for the vanished orchards of Silicon Valley) and built a prototype they called the Apple I. Although they could only afford to build a few dozen of the machines, they made a favorable impression on the computer enthusiast community. By 1977, they were marketing a more complete and refined version, the Apple II.

Unlike kits that could be assembled only by experienced hobbyists, the Apple II was ready to use "out of the box." It included a cassette tape recorder for storing programs. When connected to a monitor or an ordinary TV, the machine could create color graphics that were dazzling compared to the monochrome text displays of most computers. Users could buy additional memory (the first model came with only 4K of RAM) as well as cards that could drive devices such as printers or add other capabilities.

The ability to run a program called VisiCalc (see SPREADSHEET) propelled the Apple II into the business world, and about 2 million of the machines were eventually sold. In 1982, when *Time* magazine featured the personal computer as its "man of the year," Jobs's picture appeared on the cover. As he relentlessly pushed Apple forward, supporters pointed to Jobs's charismatic leadership, while detractors said that he could be ruthless when anyone disagreed with his vision of the company's future.

However, 1982 also brought industry giant IBM into the market. Its 16-bit computer was more powerful than the Apple II, and IBM's existing access to corporate purchasing departments resulted in the IBM PC and its "clones" quickly dominating the business market (see IBM PC).

Jobs responded to this competition by designing a PC with a radically different user interface, based largely on work during the 1970s and the Xerox PARC laboratory. The first version, called the Lisa, featured a mouse-driven graphical user interface that was much easier to use than the typed-in commands required by the Microsoft/IBM DOS. While the Lisa's price tag of $10,000 kept it out of the mainstream market, its successor, the Macintosh, attracted millions of users, particularly in schools, although the IBM PC and its progeny continued to dominate the business market (see MACINTOSH). Meanwhile, Jobs had recruited John Sculley, former CEO of PepsiCo, to serve as Apple's CEO.

After a growing divergence with Sculley over management style and Apple's future priorities, Jobs left the company in 1985. Using the money from selling his Apple stock, Jobs bought a controlling interests in Pixar, a graphics studio that had been spun off from LucasFilm. He also founded a company called NextStep. The company focused on high-end graphics workstations that used a sophisticated object-oriented operating system. However, while its software (particularly its development tools) was innovative, the company was unable to sell enough of its hardware and closed that part of the business in 1993.

In 1998, Jobs returned as CEO of Apple. By then the company was struggling to maintain market share for its Macintosh line in a world that was firmly in the "Wintel" (Windows on Intel-based processors) camp. He had some success in revitalizing Apple's consumer product line with the iMac, a colorful, slim version of the Macintosh. He also focused on development of the new Mac OS X, a blending of the power of UNIX with the ease-of-use of the traditional Macintosh interface.

Although the technical innovations were carried out by others, Jobs's vision and entrepreneurial spirit played a key role in bringing personal computing to a larger audience. The Macintosh interface that he championed would become the norm for personal computing, albeit in the form of Microsoft Windows on IBM PC compatible systems.

Further Reading

Angelelli, Lee. "Steve Jobs. Exiled Silicon Valley Prophet Returns to His People," http://members.nbci.com/Steve_Jobs/

Deutschman, Alan. *The Second Coming of Steve Jobs*. New York: Broadway Books, 2000.

Levy, Steven. *Insanely Great: The Life and Times of the Macintosh, the Computer that Changed Everything*. New York: Penguin Books, 2000.

journalism and computers

The pervasive use of computers and the Internet has changed the practice of journalism in many ways. This entry will focus on the general impact of technology on the creation and dissemination of news content. For discussion of software used in the production of publications, see DESKTOP PUBLISHING, and WORD PROCESSING. For the role that journalism plays in the computer industry, see JOURNALISM AND THE COMPUTER INDUSTRY.

RESEARCH AND NEWSGATHERING

The gathering of on-scene information at newsworthy events began to change in the 1980s, when notebook-sized portable computers became available. Instead of having to "file" stories with the newspaper by telegraph or phone, the reporter could write the piece and send it to the newspaper's computer using a phone connection (see MODEM) or later, Internet-based e-mail.

The ability of reporters (particularly investigative reporters) to do in-depth research has been greatly enhanced by the Internet. Traditionally, reporters looking for background material for an assignment could consult printed reference works, their publications' archives of printed articles (the "morgue"), and various public records, usually in paper form. This process was necessarily slow, and it was difficult to widen research to include a greater variety of sources while still remaining timely.

Today most publications produce and store their material electronically and make it available through various commercial database services such as LexisNexis. Reporters thus have virtually instant access to articles written by their colleagues around the world. Instead of having to rely on a few press releases, position papers, or wire stories, reporters can search the Internet to delve more deeply into the underlying source material, such as original documents or statistics. An increasing number of public records are also available on-line.

CHANGING STANDARDS AND NEW CHALLENGES

After being submitted electronically, reporters' stories can be edited, revised as necessary, and submitted to the computer-controlled typesetting systems that have now become standard in most publications. Besides saving production costs, computer-based newspaper production also makes it easier to make last-minute changes as well as to create special editions that include regional news.

However, at the same time the greater use of information technology has made print journalists more productive, it has also contributed to trends that continue to challenge the viability of print journalism itself. The nature of the Internet poses new challenges to reporter-researchers. The accuracy of traditionally published books or articles is backed implicitly by the reputation of the publisher as well as that of the author. By offering a wide variety of materials produced outside the mainstream publishing process, often by unknown authors, the Internet can provide a much greater diversity of viewpoints. The downside is that the reporter-researcher has little assurance of the veracity or accuracy of facts given on unknown websites. This creates a greater burden of fact checking in responsible journalism or, alternatively, a relaxation of the traditional standards. (The most famous example of the latter is Matt Drudge, a self-made Internet-based journalist who sometimes dramatically "scooped" his more plodding colleagues but did not adhere to the old journalistic standard of finding two independent sources for each key fact.)

The use of the Internet as both a research tool and a medium of publication is also bound up with the ever-accelerating pace of the "news cycle," or the time it takes for a story to be disseminated and responded to. Broadcast journalism with the advent of 24-hour news networks such as CNN has steadily increased the pace of the broadcast news cycle. Many newspapers and magazines have found having websites to be a competitive necessity. The Internet potentially combines the immediacy of broadcast journalism with the ability to use text to convey information in depth. The organization of webpages (see HYPERTEXT AND HYPERMEDIA) avoids the physical limitations of the printed medium.

In addition to websites that mirror and expand the contents of printed newspapers, a number of distinctive Internet-only sites emerged in the mid to late 1990s. Examples include salon.com, an "on-line newsmagazine" that also includes regular featured columnists and discussion forums. However, the downturn in the Internet-based economy that became pronounced in 2001 has made the original idea of having free access supported by advertising less viable. Such sites are now trying to convert to a subscription-based model similar to that of print-based publications, but it is unclear whether they will be able to attract enough paying subscribers to bring in revenue sufficient to maintain rich and distinctive content.

At the same time, in an era when a stream of both images and the printed word is on tap 24 hours a day, print journalism faces a shrinking market and the need to justify itself to consumers. The industry has responded since the 1970s by an increasing number of mergers of metropolitan daily newspapers as well as the merging of newspapers into broader-based media companies. Many people have grown up with the daily routine of a newspaper at the breakfast table, and there is still a cachet for prestigious publications such as the *New York Times* and the *Wall Street Journal*. Futurists have predicted that newspapers might eventually be delivered to "electronic book" devices, perhaps through a wireless connection. This might combine the immediacy of the Internet with the physical convenience and portability of a newspaper.

Further Reading

De Volk, Roland. *Introduction to On-line Journalism: Publishing News and Information.* Boston, Mass.: Allyn and Bacon, 2001.

Hall, Jim. *On-line Journalism: A Critical Primer.* Sterling, Va.: Pluto Press, 2001.

Lamble, Stephen. "Computer-Assisted Reporting." http://members.optusnet.com.au/~slamble/index.html

New York Times. www.nytimes.com

Reddick, Randy, and Elliot King. *The On-line Journalist: Using the Internet and Other Resources.* 3rd ed. Belmont, Calif.: Wadsworth, 2000.

Salon.com. www.salon.com

University of Southern California. Annenberg School for Communication. "On-line Journalism Review." http://ojr.usc.edu/

journalism and the computer industry

Developments in the computer industry and user community have been chronicled by a great variety of printed and on-line publications. As computer science began to emerge as a discipline in the late 1950s and 1960s, academically oriented groups such as the Association for Computing Machinery (ACM) and Institute for Electrical and Electronics Engineers (IEEE) began to issue both general and special-interest journals. Meanwhile, the computer industry developed both computer science-oriented publications (such as the *IBM Systems Journal*) and independent industry periodicals such as *Datamation*.

The development of microcomputer systems in the mid- to late-1970s was accompanied by a proliferation of varied and often feisty publications. *Byte* magazine, which coined the term *PC* in 1976, became a respected trade publication that introduced new technologies while showcasing what programmers could do with the early systems. The weekly newspaper *InfoWorld* provided more immediate and detailed coverage of industry developments, and was joined by similar publications such as *Information Week* and *Computerworld*. Meanwhile, technically savvy programmers and do-it-yourself engineers turned to such publications as the exotically named *Dr. Dobbs' Journal of Computer Calisthenics and Orthodontia* (eventually shortened to *Dr. Dobbs' Journal*). Many groups of people who owned particular systems (see USER GROUPS) also published their own newsletters with technical tips.

The success of the IBM PC family of computers established a broad-based consumer computing market. It was accompanied by the success of *PC Magazine*, which addresses a wide spectrum of both general consumers and "power users." As the revenue for the PC industry grew in the 1990s, the trade publications grew fatter with advertising. The popularity of the Internet and particularly the World Wide Web in the latter part of the decade provided niches for a spate of new publications including *Internet World* and *Yahoo! Internet Life*. At the same time, many traditional publications began to offer expanded content via websites. For example, Ziff Davis, publisher of *PC Magazine* and other computer magazines created ZDNet, which offered a large amount of content from the magazines plus expanded news and extensive shareware and utility libraries.

Like earlier technological developments, the PC and the Internet have also spawned cultural expressions. The culture growing around the Internet and a generation of young programmers, artists, and writers saw expression in another genre of publications, ranging from small, eclectic printed or Web "zines" to the slick *Wired* magazine.

The downturn in the computer industry that began in 2000 has had a significant impact on computer-related publications. With computer hardware becoming a commodity with lower profit margins and the failure of many Internet-based businesses or "dot-coms," advertising revenue has shrunk, and along with it, the size of some publications. The venerable *Byte* became an Internet-only publication in 1998, for example. However, ZDNet and other Web-based sources of industry news and resources have continued, with some weaker services being merged with stronger ones.

Further Reading

Byte Magazine [on-line]. www.byte.com

"Computer Magazines and E-zines." http://www.compinfo-center.com/itmags.htm

"Elsop's Directory of Computer Publications on the Web." http://www.elsop.com/wrc/pubindex.htm

ZDNet. www.zdnet.com

Joy, Bill

(1955–)
American
Software Engineer, Entrepreneur

Bill Joy developed many of the key utilities used by users and programmers on UNIX systems (see UNIX). He then became one of the industry's leading entrepreneurs and later, a critic of some aspects of computer technology.

As a graduate student in computer science and electrical engineering at the University of California at Berkeley in the 1970s, Joy worked with UNIX designer Ken Thompson (1943–) to add features such as virtual memory (paging) and TCP/IP networking support to the operating system (the latter work was sponsored by DARPA, the Defense Advanced Research Projects Agency). These development eventually led to the distribution of a distinctive version of UNIX called Berkeley Software Distribution (BSD), which rivaled the original version developed at AT&T's Bell Laboratories. The BSD system also popularized features such as the C shell (a command processor) and the text editors "ex" and "vi." (See SHELL.)

As opposed to the tightly controlled AT&T version, BSD UNIX development relied upon what would become known as the open source model of software development (see OPEN SOURCE MOVEMENT). This encouraged programmers at many installations to create new utilities for the operating system, which would then be reviewed and integrated by Joy and his colleagues. BSD UNIX gained industry acceptance and was adopted by the Digital Equipment Corporation (DEC), makers of the popular VAX series of minicomputers.

In 1982, Joy left UC Berkeley and co-founded Sun Microsystems, a company that became a leader in the manufacture of high-performance UNIX-based workstations for scientists, engineers, and other demanding users. Even while becoming a corporate leader, he continued to refine UNIX operating system facilities, developing the Network File System (NFS), which was then licensed for use not only on UNIX systems but on VMS, PC-DOS, and Macintosh systems. Joy's versatility also extended to hardware design, where he helped create the Sun SPARC reduced instruction set (RISC) microprocessor that gave Sun workstations much of their power.

Bill Joy made key contributions to the Berkeley Software Distribution (BSD) version of UNIX, including developing its Network File System (NFS). As a co-founder of Sun Microsystems, Joy then helped develop innovative workstations and promoted Java as a major language for developing Web applications.

In the early 1990s, Joy turned to the growing world of Internet applications and embraced Java, a programming language created by James Gosling (see JAVA). He developed specifications, processor instruction sets, and marketing plans. Java became a very successful platform for building applications to run on Web servers and browsers and to support the needs of e-commerce. As Sun's chief scientist since 1998, Joy has led the development of Jini, a facility that would allow not just PCs but many other "Java-enabled" devices such as appliances and cell phones to communicate with one another.

Recently, however, Joy has expressed serious misgivings about the future impact of artificial intelligence and related developments on the future of humanity. Joy remains proud of the achievements of a field to which he has contributed much. However, while rejecting the violent approach of extremists such as Unabomber Theodore Kaczynski, Joy points to the potentially devastating unforeseen consequences of the rapidly developing capabilities of computers. Unlike his colleague Ray Kurzweil's optimistic views about the coexistence of humans and sentient machines, Joy points to the history of biological evolution and suggests that superior artificial life forms will displace humans who will be unable to compete with them. He believes that given the ability to reproduce themselves, intelligent robots or even "nanobots" (see NANOTECHNOLOGY) might soon be uncontrollable.

Joy also expresses misgivings about biotechnology and genetic engineering, seen by many as the dominant scientific and technical advance of the early 21st century. He has proposed that governments develop institutions and mechanisms to control the development of such dangerous technologies, drawing on the model of the agencies that have more or less successfully controlled the development of nuclear energy and the proliferation of nuclear weapons for the past 50 years.

Joy received the ACM Grace Murray Hopper Award for his contributions to BSD UNIX before the age of 30. In 1993, he was given the Lifetime Achievement Award of the USENIX Association, "For profound intellectual achievement and unparalleled services to the UNIX community."

Further Reading

Joy, Bill. "Why the Future Doesn't Need Us." *Wired,* 8.04, April 2000, available on-line at http://www.wired.com/wired/archive/8.04/joy.html.

Joy, Bill, ed. *The Java Language Specification, Second Edition.* Reading, Mass.: Addison-Wesley, 2000.

O'Reilly, Tim. *A Conversation with Bill Joy.* Available on-line at http://www.openp2p.com/pub/a/p2p/2001/02/13/joy.html

Williams, Sam. "Bill Joy Warns of Tech's Dangerous Evolution." *Upside.com,* January 18, 2001, available on-line at http://www.upside.com/Open_Season/3a648a96b.html

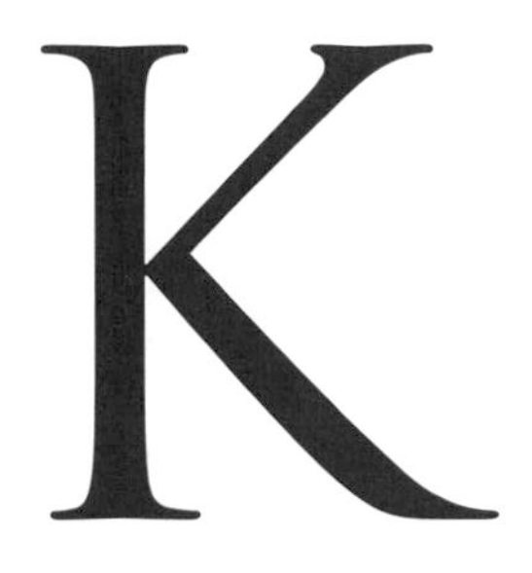

Kay, Alan

(1940–)
American
Computer Scientist

Alan Kay developed a variety of innovative concepts that changed the way people use computers. Because he devised ways to have computers accommodate users' perceptions and needs, Kay is thought by many to be the person most responsible for putting the "personal" in personal computers. Kay also made important contributions to object-oriented programming, changing the way programmers organized data and procedures in their work.

Kay's father developed prostheses (artificial limbs) and his mother was an artist and musician. These varied perspectives contributed to Kay's interest in interaction with and perception of the environment. In the late 1960s, while completing work for his Ph.D. at the University of Utah, Kay developed his first innovations in both areas. He helped Ivan Sutherland with the development of a program called Sketchpad that enabled users to define and control onscreen objects, while also working on the development of Simula, a language that helped introduce new programming concepts (see SIMULA and OBJECT-ORIENTED PROGRAMMING). Indeed, Kay coined the term *object-oriented* in the late 1960s. He viewed programs as consisting of objects that contained appropriate data that could be manipulated in response to "messages" sent from other objects. Rather than being rigid, top-down procedural structures, such programs were more like teams of cooperating workers. Kay also worked on parallel programming, where programs carried out several tasks simultaneously (see CONCURRENT PROGRAMMING). He likened this structure to musical polyphony, where several melodies are sounded simultaneously.

Kay participated in the Defense Advanced Research Projects Agency (DARPA)—funded research that was leading to the development of the Internet. One of these DARPA projects was FLEX, an attempt to build a computer that could be used by nonprogrammers through interacting with onscreen controls. While the bulky technology of the late 1960s made such machines impracticable, FLEX incorporated some ideas that would be used in later PCs, including multiple onscreen windows.

During the 1970s, Kay worked at the innovative Xerox Palo Alto Research Center (PARC). Kay designed a laptop computer called the Dynabook, which featured high-resolution graphics and a graphical user interface. While the Dynabook was only a prototype, similar ideas would be used in the Alto, a desktop personal computer that could be controlled with a new pointing device, the mouse (see also ENGELBART, DOUGLAS). A combination of high price and Xerox's less than aggressive marketing kept the machine from being successful commercially, but Steven Jobs (see JOBS, STEVEN) would later use its interface concepts to design what would become the Macintosh.

On the programming side Kay developed Smalltalk, a language that was built from the ground up to be truly object-oriented (see SMALLTALK). Kay's work showed that there was a natural fit between object-oriented programming and an object-oriented user interface. For example, a button in a screen window could be represented by a button object in the program, and clicking on the screen button could send a message to the button program object, which would be programmed to respond in specific ways.

After leaving Xerox PARC in 1983, Kay briefly served as chief scientist at Atari and then moved to Apple, where he worked on Macintosh and other advanced projects. In 1996, Kay became a Disney Fellow and Vice President of Research and Development at Walt Disney Imagineering. Kay has won numerous awards including the Association for Computing Machinery Software Systems award.

Further Reading

Gasch, Scott. "Alan Kay." http://ei.cs.vt.edu/~history/GASCH.KAY.HTML

Kay, Alan. "The Early History of Smalltalk." in Thomas J. Bergin, Jr., and Richard G. Gibson, Jr., eds. *History of Programming Languages II.* New York: ACM; Reading, Mass.: Addison-Wesley, 1996.

"Alan C. Kay: A Clear Romantic Vision," in Shasha, Dennis and Cathy Lazere, eds. *Out of their Minds: The Lives and Discoveries of 15 Great Computer Scientists.* New York: Copernicus, 1995.

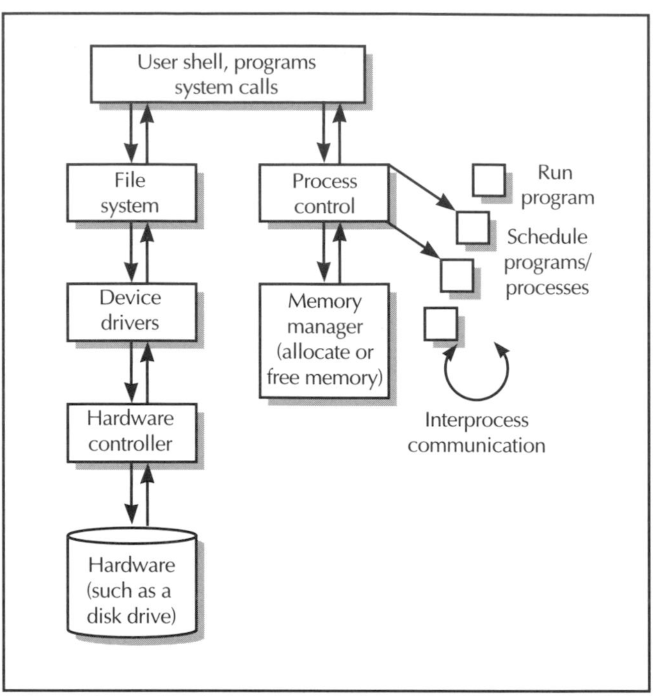

The kernel is an intermediary between users and programs and the hardware system. It provides the functions necessary for allocating and controlling processes and system resources.

kernel

The idea behind an operating system kernel is that there is a relatively small core set of "primitive" functions that are necessary for the operation of system services (see also OPERATING SYSTEM). These functions can be provided in a single component that can be adapted and updated as desirable. The fundamental services include:

- Process control—scheduling how the processes (programs or threads of execution within programs) share the CPU, switching execution between processes, creating new processes, and terminating existing ones (see MULTITASKING).
- Interprocess communication—sending "messages" between processes enabling them to share data or coordinate their data processing.
- Memory management—allocating and freeing up memory as requested by processes as well as implementing virtual memory, where physical storage is treated as an extension of main (RAM) memory. (See MEMORY MANAGEMENT.)
- File system services—creating, opening, reading from, writing to, closing, and deleting files. This includes maintaining a structure (such as a list of nodes) that specifies the relationship between directories and files. (See FILE.)

In addition to these most basic services, some operating systems may have larger kernels that include security functions (such as maintaining different classes of users with different privileges), low-level support for peripheral devices, and networking (such as TCP/IP).

The decision about what functions to include in the kernel and which to provide through device drivers or system extensions is an important part of the design of operating systems. Many early systems responded to the very limited supply of RAM by designing a "microkernel" that could fit entirely in a small amount of memory reserved permanently for it. Today, with memory a relatively cheap resource, kernels tend to be larger and include functions that are paged dynamically into and out of memory.

In the UNIX world (and particularly with Linux) the kernel is constantly being improved through informal collaborative efforts. Many Linux enthusiasts regularly install new versions of the kernel in order to stay on the "leading edge," while more conservative users can opt for waiting until the next stable version of the kernel is released.

Further Reading

Bach, Maurice J. *The Design of the UNIX Operating System.* Englewood Cliffs, N.J.: Prentice Hall, 1986.

Bovet, Daniel P., and Marco Cesati. *Understanding the Linux Kernel.* Sebastopol, Calif.: O'Reilly, 2000.

Silberschatz, Abraham, Peter Baer Galvin, and Greg Gagne. *Operating System Concepts, Sixth Edition.* New York: Wiley, 2001.

Torvalds, Linus. "The Story of the Linux Kernel." http://www.linuxworld.com/linuxworld/lw-1999-03/lw-03-opensources.html

keyboard

Although most of today's personal computers feature a point-and-click graphical interface (see USER INTERFACE and MOUSE) the keyboard remains the main means for entering text and other data into computer applications. The modern computer keyboard traces its ancestry to the typewriter, and the layout of its alphabetic and punctuation keys remains that devised by typewriter pioneer Christopher Latham Sholes in the late 1860s.

The principal difference in operation is that while a typewriter needs only to transfer the impression of a key through a ribbon onto a piece of paper, the computer keyboard must generate an electrical signal that uniquely identifies each key. This technology dates back to the 1920s with the adoption of the teletypewriter (often known by the brand name Teletype), which allowed operators to type text at a keyboard and send it over telephone lines to be printed. The transmissions used the Baudot character code, which used five binary (off or on) positions to encode letters and characters. This gave way to the ASCII code in the 1960s (see CHARACTERS AND STRINGS) at about the time that remote time-sharing services allowed users to interact with computers through a Teletype connection.

The modern personal computer keyboard was standardized in the mid-1980s when IBM released the PC AT. This expanded keyboard now has 101 or 102 keys. It supplements the standard typewriter keys with cursor-control (arrow) keys, scroll control keys (such as Page Up and Page Down), a dozen function keys that can be assigned to commands by software, and a separate calculator-style keyboard for numeric data entry. During the 1990s, Microsoft introduced a few extra keys for Windows-specific functions.

The advent of laptop (or notebook) computers required some compromises. The keys are generally

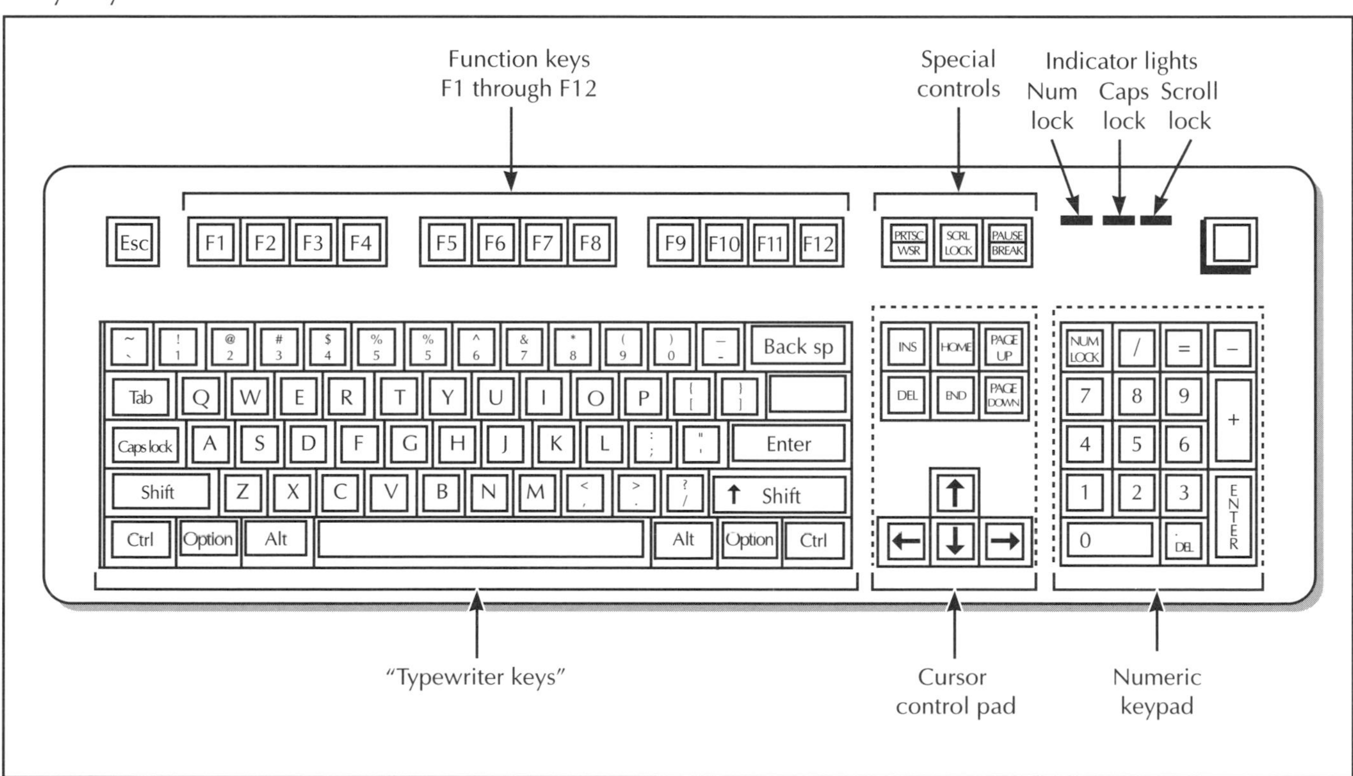

A standard 101-key PC desktop keyboard. Keyboards for laptop computers must generally compromise, for example, by eliminating the separate cursor keypad or using a special shift key to "double up" the function keys with other keys.

smaller, although on the better units they are still far enough apart to allow for comfortable touch-typing. Laptops often combine the function keys and cursor control keys with the regular keys, using a special "Fn" key to shift between them.

In recent years, there has been some interest in adopting an alternative key layout devised by August Dvorak in the 1950s. The theory behind this layout was that arranging the keyboard so the most commonly used keys were directly under the fingers would be more efficient than the Sholes layout, which legend claims was devised primarily to slow down typists to a speed that early typewriters could handle without jamming. However, researchers have generally been unable to find a significant improvement in either performance or ergonomics between use of the standard and Dvorak layouts, and the latter has not caught on commercially.

Concern with repetitive strain injury (RSI) has led to experiments in designing a keyboard more suited to the human wrist and hand (see ERGONOMICS OF COMPUTING). Some designs such as the Microsoft Natural Keyboard divide the layout into left and right banks of keys and angle them toward one another to reduce strain on the wrists. An extreme form of the design actually breaks the keyboard into two pieces. Such extreme designs have not found wide acceptance.

The Microsoft Natural Keyboard features an ergonomic layout designed to help reduce typing stress. (PHOTO COURTESY OF MICROSOFT CORPORATION)

It is possible that the further development of voice recognition software might allow spoken dictation to supplant the keyboard for data entry. Currently, however, such technology is limited in speed and accuracy (see SPEECH RECOGNITION AND SYNTHESIS).

Further Reading

"Alternative Keyboard Designs FAQ." http://www.tifaq.com/keyboards.html

Baber, Christopher. *Beyond the Desktop: Designing and Using Interaction Devices.* San Diego, Calif.: Academic Press, 1996.

knowledge representation

The earliest concern of computer science was the representation of "raw" data such as numbers in programs (see DATA TYPES). Such data can be used in calculations, and actions taken based on tests of data values, using branching (IF) or looping structures.

However, facts are more than data. A fact is an *assertion,* for example about a relationship, as in "Joe is a son of Mike," often expressed in a form such as son (Joe, Mike). Implications can also be defined as proceeding from facts, such as

```
son (Joe, Mike) implies father (Mike, Joe) or
son (Joe, Mike) and son (Mike, Phil) implies
grandson (Joe, Phil)
```

While it can be expressed in a variety of different forms of notation, this *predicate calculus* forms the basis for many automated reasoning systems that can operate on a "knowledge base" of assertions, prove the validity of a given assertion, and even generate new conclusions based upon existing knowledge (see also EXPERT SYSTEMS).

An alternative form of knowledge representation used in artificial intelligence programs is based on the idea of frames. A frame is a structure that lists various characteristics or relationships that apply to a given individual or class. For example, the individual "cat" might have a frame that includes characteristics such as "warm-blooded" and "bears live young." In turn, these characteristics are also assigned to the class "mammal" such that any individual having those characteristics belongs to that class. A program can then follow the linkages and conclude that a cat is a mammal. Linkages can also be diagrammed as a "semantic network" in a structure called a directed graph, with the lines between nodes labeled to show relationships.

Knowledge representation systems have different considerations depending on their intended purpose. A KR system in an academic research setting might be intended to demonstrate *completeness:* that is, it can generate all possible conclusions from the facts given. However, expert systems designed for practical use usually do not attempt to generate all possible conclusions (which might be computationally impracticable) but to generate *useful* conclusions that are likely to serve the needs of the knowledge consumer.

It is also important to note that epistemology (the theory of knowledge) plays an important role in understanding and evaluating KR systems. As an example, the assertion "Mary believes she is 600 years old" might be a fact (Mary is observed to hold such a belief), but the contents of the belief are presumably not factual. The context of this belief might also be different if Mary is an adult as opposed to being a five-year-old child. Similarly, ontological (state of being) considerations can also complicate the evaluation of assertions. For example, should a fire be treated as an object in itself, a process, or an attribute of a burning object? Knowledge representation thus intertwines philosophy and computer science.

Further Reading

Hayes, P. J., and K. Ford, eds. *Reasoning Agents in a Dynamic World: the Frame Problem.* Boston, JAI Press, 1992.

"Introduction to Frame Based Knowledge Representation." http://www.cs.utexas.edu/users/qr/algy/algy-expsys/node2.html

"The Knowledge Museum: Representation Exhibits." http://www.akri.org/museum/represen/index.htm

"Knowledge Representation." http://ijgj229.infj.ulst.ac.uk/BillsWeb/PGCert/InfoSys/4.knowrep.html

Lenat, D. B., and R. V. Guha. *Building Large Knowledge-Based Systems.* Reading, Mass.: Addison-Wesley, 1990.

Knuth, Donald

(1938–)
American
Computer Scientist

Donald Knuth has contributed to many aspects of computer science, but his most lasting contribution is his monumental work, *The Art of Computer Programming,* which is still in progress.

Born in Milwaukee on January 10, 1938, Knuth's initial background was in mathematics. He received his master's degree at the Case Institute of Technology in 1960 and his Ph. D. from the California Institute of Technology (Caltech) in 1963. As a member of the Caltech mathematics faculty Knuth became involved with programming and software engineering, serving both as a consultant to the Burroughs Corporation and as editor of the Association for Computing Machinery (ACM) publication *Programming Languages.* In 1968, Knuth confirmed his change of career direction by becoming professor of computer science at Stanford University.

In 1971, Knuth published the first volume of *The Art of Computer Programming* and received the ACM Grace Murray Hopper Award. His broad contributions to the field as well as specific work in the analysis of algorithms and computer languages garnered him the ACM Turing Award, the most prestigious honor in the field. Knuth also did important work in areas such as LR (left-to-right, rightmost) parsing, a context-free parsing approach used in many program language interpreters and compilers (see PARSING).

However, Knuth then turned away from writing for an extended period. His primary interest became the development of a sophisticated software system for computer-generated typography. He developed both the TeX document preparation system and METAFONT, a system for typeface design that was completed during the 1980s. TeX found a solid niche in the preparation of scientific papers, particularly in the fields of mathematics, physics, and computer science where it can accommodate specialized symbols and notation.

Knuth did return to *The Art of Computer Programming* and by the late 1990s he had completed two more of a projected seven volumes. With his broad interests and contributions and "big picture" approach to the evaluation of programming languages, algorithms, and software engineering methodologies, Knuth can fairly be described as one of the "Renaissance persons" of the computer science field. His numerous awards include the ACM Turing Award (1974), IEEE Computer Pioneer Award (1982), American Mathematical Society's Steele Prize (1986), and the IEEE's John von Neumann Medal (1995).

Further Reading

Frenkel, Karen A. "Donald E. Knuth: Scholar with a Passion for the Particular." *Profiles in Computing, Communications of the ACM,* vol. 30, no. 10, October 1987.

Knuth, Donald E. *Literate Programming.* Stanford, Calif.: Center for the Study of Language and Information, 1992.

———. *The Art of Computer Programming.* 3rd ed. vols. 1–3. Reading, Mass.: Addison-Wesley, 1998.

Slater, Robert. *Portraits in Silicon.* Cambridge, Mass.: MIT Press, 1987.

L

law enforcement and computers

Besides his superb reasoning skills, perhaps Sherlock Holmes's most important asset was his extensive collection of notes that provided a cross-referenced index to London's criminal underworld. Today computer applications have given law enforcers investigative, forensic, communication, and management tools that Holmes and his rivals in the old Scotland Yard could not have imagined.

For the officer on the street, the ability to obtain auto license, stolen vehicle, or outstanding warrant information in near real-time provides a much better picture of the potential risk in making stops or arrests. If a criminal case is opened, a variety of software applications come into play. These include case management programs for keeping track of evidence and witness interviews. Evidence must be properly logged at all times to maintain a legally defensible chain of custody against accusations of tampering.

The investigation of a crime involves many computerized forensic aids. Besides automated matching of fingerprints and, increasingly other physical data (see BIOMETRICS), records can also be searched to detect patterns such as crimes with related modus operandi (MOs). The ability to access information from other jurisdictions and to interface federal, state, and local agencies is also very important, particularly for cases involving organized crime, interstate fugitives, and terrorism.

Since data stored on computers is an increasingly prevalent form of evidence, law enforcement specialists must also employ tools to recover data that may have been partially erased or encrypted by suspects. Computers can be more active instruments of crime (see COMPUTER CRIME AND SECURITY). Such traditional tools as wiretapping must be adapted to new forms of communication such as e-mail while addressing concerns about civil liberties and privacy (see PRIVACY IN THE DIGITAL AGE).

High-level planning for law enforcement budgets and priorities requires access to detailed crime statistics. At the national level, the Justice Department's Bureau of Justice Statistics is a definitive information source. Law enforcers, like other professionals, increasingly use websites, chat areas, and e-mail lists to discuss computer-related law enforcement issues with colleagues.

Law enforcement agencies also use the same "bread and butter" software needed by any substantial organization, including word processing, spreadsheet, payroll, and other accounting programs.

Further Reading

Anderson, Michael. "Computer Evidence Processing." http://www.forensics-intl.com/art3.html

Casey, Eoghan. *Digital Evidence and Computer Crime: Forensic Science, Computers, and the Internet.* San Diego, Calif.: Academic Press, 2000.

"Law Enforcement Products and Services Directory." http://www.officer.com/products.htm

Sammes, Tony, Brian Jenkinson, and A. J. Sammes. *Forensic Computing: A Practitioner's Guide.* New York: Springer-Verlag, 2000.

Woodward, Mark. "Computers, the Internet and Criminal Investigation." http://www.co-asn-rob.org/Articles/computers.htm

libraries and computing

The library is the institution traditionally charged with the collection and distribution of humanity's collective heritage of written information. It is thus not surprising that the development of modern information technology has meant that libraries have had to undergo pervasive changes in their practices and responsibilities.

One of the earliest applications for automation in libraries was cataloging. By the 1960s, the ever-increasing volume of books and serials (periodicals) published each year was placing a growing burden on the manual cataloging system. Under this system, catalogers at large libraries (and particularly the Library of Congress) prepared a catalog record for each new publication. These records were distributed by the Library of Congress in the form of catalog card proof slips. These, as well as compiled card images from other libraries, could be used by each library to prepare catalog records for its own holdings.

As mid-size computers became more affordable, it became practicable for at least large library systems to put their catalog records on-line. In 1968, the MARC (Machine Readable Cataloging) standard was first promulgated. A MARC record uses specific, numbered fields to describe the elements of a book, such as its catalog card number, main entry, title, imprint, collation (pagination), and subject headings.

At first, MARC records were distributed mainly on magnetic tape in place of card slips. However, by the late 1970s large on-line cataloging systems such as OCLC (On-line College Library Center) and RLIN (Research Library Information Network) were enabling libraries to search for and download cataloging information in real time, and in turn upload their own original catalog records to the shared database. This greatly reduced redundant cataloging effort. If a library receives a new book, a library assistant can search for a preexisting catalog record. The record can then be easily modified for local use, such as by adding a call number and holdings information. The problem of authorities (standardized entries for names) is also made more manageable by being able to check entries on-line.

By the 1980s, the next logical step was under way: The card catalog began to be replaced by a wholly electronic catalog, enabling library patrons to search the catalog at a terminal. Besides saving money, the on-line catalog also offers researchers many more ways to search for materials: for example, they can use keywords and not rely only on titles and subject headings.

Along with cataloging, libraries began to automate their circulation and acquisitions systems as well. As these systems become integrated, libraries can both monitor the demand (finding materials that are in heavy use and need additional copies) and speed up the supply, by integrating the acquisitions system with ordering systems maintained by book distributors.

However, while most librarians consider the computer to be a boon to their profession, there are criticisms and further challenges. Nicholas Baker, for example, has decried the abandonment of information in card catalogs that was not carried over into electronic form. Baker has also criticized the replacement of bound archives of periodicals with microfilm, which is often of poor quality and prone to deterioration. The storage of publications on computer media has also met with concerns that the physical durability of the media has not been sufficiently investigated, and that in a rapidly changing technological world data formats can become obsolete, no longer supported, and potentially unreadable (see also BACKUP AND ARCHIVE SYSTEMS).

The growth of the World Wide Web has also presented libraries with both opportunities and challenges. Catalogers and reference librarians are struggling to find new ways to categorize and retrieve the always-changing and ephemeral content of webpages. Meanwhile, librarians have faced not only funding and training issues in providing expanded public Web access in libraries, but have also had to deal with demands that Web content be filtered to protect children from objectionable content. (The American Library Association opposes such filtering as a form of censorship.)

Further Reading

American Library Association. Office for Information Technology Policy. http://www.ala.org/oitp/

Burke, John J. *Neal-Schuman Library Technology Companion: a Basic Guide for the Library Staff.* New York: Neal Schuman, 2001.

Library and Information Technology Association (LITA). http://www.lita.org/

"Library Automation: Current Trends." http://mingo.info-science.uiowa.edu/~asis/libaut.html

Pitkin, Gary M. *The Impact of Emerging Technologies on Reference Service and Bibliographic Instruction.* Westport, Conn.: Greenwood Press, 1995.

library, program

Programming is a labor-intensive activity, especially when the time required to test, debug, and verify the operation of the program code is included. It is not surprising, then, that even the earliest programmers sought ways to reuse the code for commonly needed operations such as data input, sorting, calculation, and formatting rather than writing it from scratch. If a well-organized collection or library of program routines is available, developers of new applications can concentrate on the aspects particular to the current problem and use the library code for routine operations.

In the mainframe world, the use of program libraries was also mandated by the limited amount of main memory available. A data processing task was often accomplished by retrieving a series of card decks or tapes from the library and mounting them in turn. Intermediate results could be passed between programs under the control of a special script (see JOB CONTROL LANGUAGE).

Some programming languages, notably C and its descendants C++ and Java, are designed to provide a small core of essential features (such as control structures, data types, and operators). Other functions, such as math routines, data I/O (input/output), and formatting are provided in library files that are invoked by programs that need particular features. There are several advantages to this approach. The core language is kept simple because it doesn't have to deal with issues such as the actual storage of data in memory that are dependent on the particular architecture of each type of machine. To "port" the language to a new machine, specialists in its architecture can implement the standard library functions. In addition to the standard libraries included with the compiler, programmers are also free to create additional libraries to support particular applications such as graphics.

Library, Program 1

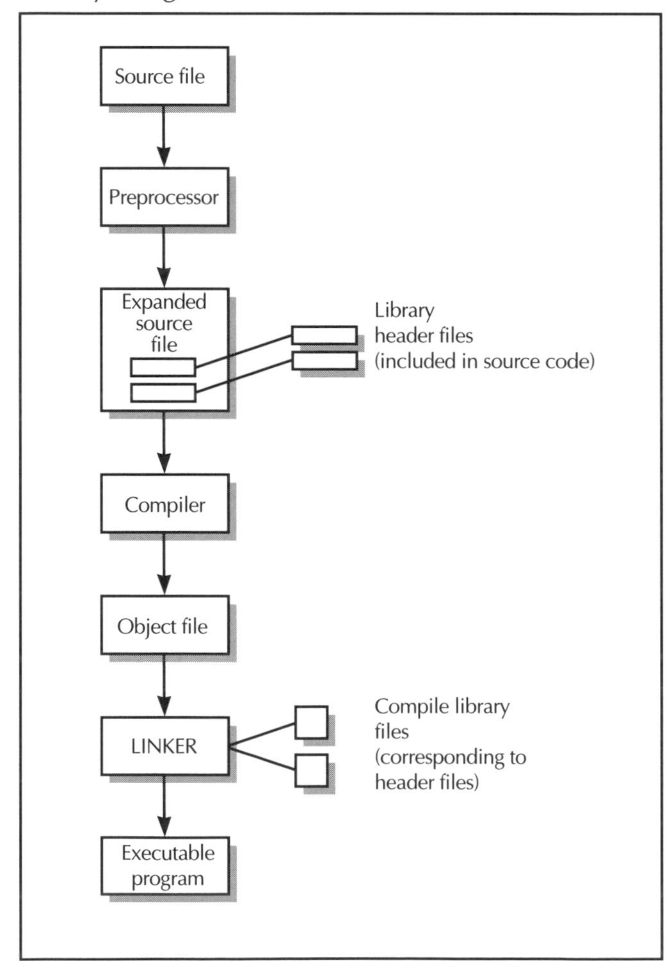

To use a program library, the programmer includes the appropriate header file in the source code. After the source code is compiled, the linker links it to the compiled object code file corresponding to the header file, creating a single executable file.

Library, Program 2

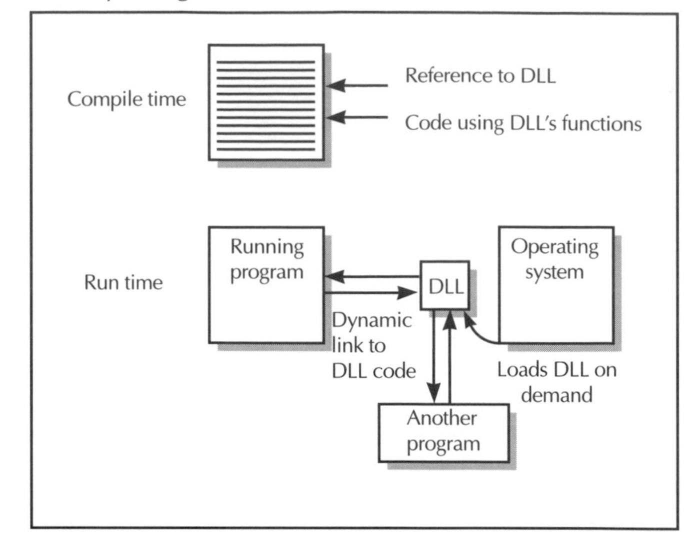

Dynamic linking is an alternative approach to library use. The program is compiled with a reference to the library, but it is not linked to the library code until the program is actually running. Since several different running programs can link to the same dynamic link library (DLL), memory is saved.

With a traditional library the library routines invoked in the source code are included in the final executable file. With most modern operating systems, however, many programs are active in memory at the same time (see MULTITASKING). Storing the same commonly used routines (such as standard I/O) with each program wastes memory. Therefore, operating systems such as Microsoft Windows use *dynamic linking*. This means that instead of compiling the library code into the program to create the executable file, the program links to the library at execution time. If another program is using the library, the new program links to the same copy in memory rather than having to store another copy. (Dynamically linked libraries [DLLs] include special code to keep track of the invocation of the library functions by each separate program.)

Licklider, Joseph Carl Robnett

(1915–1990)
American
Computer Scientist, Psychologist

Most of the early computer pioneers came from backgrounds in mathematics or engineering. This naturally led them to focus on the computer as a tool for computation and information processing. Joseph Licklider, however, brought an extensive background in psychology to the problem of designing interactive computer systems that could provide better communication and access to information for users.

Licklider was born on March 11, 1915, in St. Louis, Missouri. During the 1930s, he attended Washington University in St. Louis, earning B.A. degrees in psychology, mathematics, and physics. He then concentrated on psychology for his graduate studies, earning an M.A. at Washington University and then receiving his Ph.D. from the University of Rochester in 1942.

While at Rochester, Licklider participated in a study group led by Norbert Wiener, pioneer in the new field of cybernetics, in the late 1940s. This brought him into contact with emerging computer technology and its exciting prospects for the future. In turn, Licklider's psychology background allowed him a perspective quite different from the mathematical and engineering background shared by most early computer pioneers.

Cybernetics emphasized the computer as a system that could interact in complex ways with the environment. Licklider added an interest in human-computer interaction and communication. He began to see the computer as a sort of "amplifier" for the human mind. He believed that humans and computers could work together to solve problems that neither could successfully tackle alone. The human could supply imagination and intuition, while the computer provided computational "muscle." Ultimately, according to the title of his influential paper, it might be possible to achieve a true "Man-Computer Symbiosis."

During the 1950s, Licklider taught psychology at the Massachusetts Institute of Technology, hoping eventually to establish a full-fledged psychology department that would elevate the concern for what engineers call "human factors." From 1957 to 1962 he also served in the private sector as a vice president for engineering psychology at Bolt Beranek and Newman, the company that would become famous for pioneering networking technology.

In 1962, the federal Advanced Research Projects Agency (ARPA) appointed Licklider to head a new office focusing on leading-edge development in computer science. Licklider soon brought together research groups that included in their leadership three of the leading pioneers in artificial intelligence: John McCarthy, Marvin Minsky, and Allen Newell (see ARTIFICIAL INTELLIGENCE; MCCARTHY, JOHN; MINSKY, MARVIN). By promoting university access to government funding, Licklider also fueled the growth of computer science graduate programs at major universities such as Carnegie Mellon University, University of California at Berkeley, Stanford University, and the Massachusetts Institute of Technology.

In his research activities, Licklider focused his efforts not so much on AI as on the development of interactive computer systems that could promote his vision of human-computer symbiosis. This included time-sharing systems, where many users could share a large computer system, and networks that would allow users on different computers to communicate with one another. He believed that the cooperative efforts of researchers and programmers could develop complex programs more quickly than teams limited to a single agency or corporation (see also OPEN SOURCE MOVEMENT).

Licklider's efforts to focus ARPA's resources on networking and human-computer interaction would provide the resources and training that would, in the late 1960s, begin the development of what would become the Internet. Licklider spent the last two decades of his career teaching at MIT. Before his death in 1990, he presciently predicted that by 2000 people around the world would be linked in a global computer network.

Further Reading

J. C. R. Licklider (1915–1990) [Biographical Timeline]. http://www.columbia.edu/~jrh29/years.html

Licklider, J. C. R. "The Computer as Communication Device." *Science and Technology,* April 1968. Available on-line in http://www.memex.org/licklider.pdf

———. "Man-Computer Symbiosis." *IRE Transactions on Human Factors in Electronics,* vol. HFE-1, 4–11, March 1960. Available on-line in http://www.memex.org/licklider.pdf

Waldrop, M. Mitchell. *The Dream Machine: J. C. Licklider and the Revolution that Made Computing Personal.* New York: Viking, 2001.

linguistics and computing

The study of human language and advances in computer science have been closely intertwined. The field of computational linguistics uses computer systems to investigate the structure of natural language. In turn, the area of natural language processing involves the creation of software that can apply linguistic principles to process written or spoken human language.

As simple low-level instruction codes began to evolve into complex high-level programming language, language designers had to struggle to give precise, complete, and unambiguous definitions for the language's structure. This is essential for language users to be confident that their programs will yield the desired results. It is also important that developers trying to implement a language on different hardware platforms and operating systems have rigorous language specifications so the compiler on the new system will produce programs equivalent to those on the system where the language was first developed.

When computer scientists turned to linguistics for help in defining programming languages, they found the work of

Noam Chomsky, perhaps the 20th century's preeminent linguist, to be particularly helpful. Chomsky developed a concept of formal language in which grammar could be specified as a series of rules built up a level at a time. For example, at the lowest level, there is an alphabet from which recognized words are generated. Next there are rules for generating phrases (such as a noun phrase consisting of a noun with optional adjectives and a verb phrase consisting of a verb with optional adverbs). In turn, phrases can be combined to form sentences.

Because grammatical structures are created by applying rules to strings of symbols (words), the result is called a *generative grammar.* Chomsky sought to apply this concept of a "transformational generative grammar" as a universal structure applicable to all human languages. Meanwhile, computer scientists could use formal grammar rules to define the valid statements in programming languages (see also BACKUS-NAUR FORM). This in turn allows a compiler *parser* to break down high-level language statements and convert them into low-level instruction codes that can actually be executed by the CPU (see ASSEMBLER; PARSING).

As new languages and more powerful hardware gave computers increased power to deal with complex systems, computer scientists (and artificial intelligence researchers in particular) applied themselves to the problem of computer processing of human languages. Success in this field might lead not only to computer systems that humans could communicate with far more naturally, but also to automatic machine translation that could, for example, allow an English speaker and a Chinese speaker to communicate via e-mail.

However, developers of natural language systems face formidable challenges. Most fundamentally, while computers process symbols using a restrictive, deterministic procedure that Chomsky classifies as *finite state* (see FINITE STATE MACHINE), human languages must be understood using the more complex transformational grammar. The language processing system must therefore have rules that can cope with the often ambiguous structure of actual human speech. (For example, does the word *fly* in a given sentence mean an insect, a baseball batted high in the air, or perhaps a zippered opening in one's trousers?)

One way to limit the problem is to deal with a restricted realm of discourse. For example, a natural language "front end" to a database might assume that all input nouns refer to entities that exist in the database, such as employees, positions, salaries, and so on. It then becomes a matter of translating a query such as "How many employees in the human resources department make more than $50,000 a year" into something like:

```
find quantity (employee.department = "human
resources") and (employee.salary > 50,000)
```

Understanding unrestricted text such as that found in newspaper stories is much more complex, since fewer assumptions can be made about the subject of the discourse. Here the AI concept of frames can prove useful. A frame is a sort of script that describes the elements of life's common events or transactions. For example, suppose a news story begins "Joe X was arrested yesterday for the murder of Sarah Y. He was arraigned today and bail was denied." A system reading the story might see "arrested" and see that it links to an internal frame called "crime." The crime frame might have slots for "accused person," "charge," "victim," and "custodial status." The system could then interpret the story as indicating that Joe is the accused person, murder is the charge, Sarah is the victim. For the custodial status the system might look to another frame called "arraignment" that includes the rule that if bail is allowed and paid, the person's status is "released until trial" while if the bail is either not allowed or not paid, the status is "in custody."

Machine translation from one language to another has been a cherished goal since the early 1950s. There are two basic approaches. The first involves trying to find words or phrases that match between the languages as well as rules for turning grammatical structures in one language into equivalents in the other. This is roughly comparable to trying to read German by looking up each word in a German-English dictionary while keeping in mind, for example, the different subject-verb orders in

Linguistics and Computing

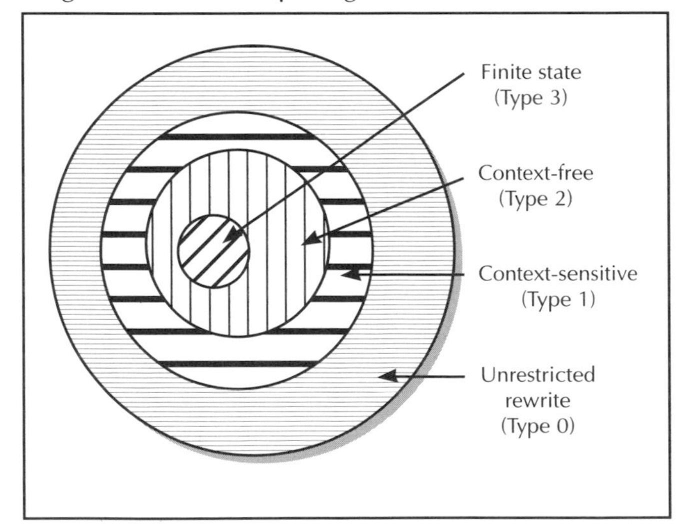

Noam Chomsky's hierarchy of language grammars. Type 0 corresponds to the Turing Machine. Some human languages tend to be context-sensitive (Type 1) in that the position of a word in the sentence matters. Some other natural languages (such as Latin) as well as push-down automata are Type 2, context-free, where position does not affect meaning. Type 3 grammar (also called regular grammar) applies to finite state automata as well as to many aspects of natural language. Note that the concentric circles indicate that the higher types are subsets of the lower ones.

the two languages. Unfortunately, such an approach produces results of limited value. It is hard to allow for idioms or ambiguous sentence structure that would pose no problem to a native speaker of the language. Machine translation of this type is now offered on both fee-based services and free via websites such as Babelfish. While ensuring correct and complete understanding still requires a review by a human translator, rough machine translation is usable for such things as understanding a foreign language item posted on a Web auction site.

More advanced translation systems take the approach of translating the text first to an intermediate language (an interlingua), usually with a simplified, restricted grammar. This representation could then be translated into any target language by applying a set of rules tailored to that language. (This is similar to an approach used in languages such as Java where programs are first translated into a PSEUDOCODE ("J-code") which in turn can be interpreted by an interpreter designed for each type of computer. While ultimately more promising than simple matching systems, this approach is far more complex and progress has been slower. Gradually, however, grammars of major world languages are being represented in a form suitable for computer processing.

Computational linguistics and natural language processing are likely to be of increasing interest in years to come. With the World Wide Web bringing the world's languages into more pervasive contact, the ability to translate or automatically summarize webpages and e-mail will be very marketable. It is also likely that advanced, secret research in the field is also being carried out by organizations such as the National Security Agency (NSA), which monitor worldwide communications.

Further Reading

Allen, J. *Natural Language Understanding.* 2nd ed. Reading, Mass.: Addison-Wesley, 1995.

Association for Computational Linguistics [home page]. http://www.cs.columbia.edu/~acl/home.html

Babelfish. http://www.babelfish.com.

[Computational Linguistics Resources]. http://www.ai.mit.edu/projects/iiip/nlp.html

Grosz, B., K. Sparck-Jones, and B. Webber. *Readings in Natural Language Processing.* San Francisco: Morgan Kauffmann, 1986.

Lawler, John M., and Helen Aristar Dry, eds. *Using Computers in Linguistics: a Practical Guide.* New York: Routledge, 1998.

Suereth, R. *Developing Natural Language Interfaces: Processing Human Conversations.* New York: McGraw-Hill, 1997.

LISP

As interest in AI (see ARTIFICIAL INTELLIGENCE) developed in the early 1950s, researchers soon became frustrated by the low-level computer languages of the day, which emphasized computation and other manipulation of numbers rather than the processing of symbolic data.

At the 1956 Dartmouth Summer Research Project on Artificial Intelligence, a gathering that brought together the key early pioneers in the field, John McCarthy presented his concepts for a different kind of computer language. Such a language, he believed, should be able to deal with mathematical functions in their own terms—by manipulating symbols, not just calculating numbers.

Together with Marvin Minsky, McCarthy began to implement a language called LISP (for "list processor"). (See MCCARTHY, JOHN and MINSKY, MARVIN.) As the name suggests, the language uses lists to store data (see LIST PROCESSING) and features many functions for manipulating list elements. List can consist of single elements (called "atoms") as in

```
(A B C D)
```

but lists can also include other lists, as in

```
(A (B C) D)
```

Each list item is stored as a "node" containing both a pointer to its data value and a pointer to the next item in the linked list. The LISP system typically includes housekeeping functions such as "garbage collection," where the memory from discarded list items is returned to the free memory pool for later allocation.

LISP programs look forbidding at first sight because they tend to have many nested parentheses. However, expressions and functions are actually constructed in a much simpler way than in most other languages. Without the need for complicated parsing, the LISP interpreter (called "eval" because it evaluates its input) looks at the stream of data and first asks whether the next item is a constant (such as a number, quoted symbol, string, quoted list, or keyword). If so, its value is returned. Otherwise, the interpreter checks to see if the item is a defined variable and, if so, returns its value. Finally, the interpreter checks to see if there is a list. If so, the list is considered to be a function call followed by its arguments. The function is called, given the data, and the result is returned.

The following table shows some items in a list program and how the interpreter will evaluate them:

EXAMPLES OF LISP ITEMS

Type of item	Example	Evaluation
integer	24	24
float	5.5	5.5
ratio	3/4	0.75
keyword	defun	defines function
quoted integer	'24	24
quoted list	'(3 1 4 1 5)	(3 1 4 1 5)
boolean	nil	false
function call	+24	6
variable	a	its value
quoted variable	'a	a

Languages such as Algol, Pascal or C emphasize statements and procedures. LISP, on the other hand, was the first FUNCTIONAL language (see FUNCTIONAL LANGUAGES). The heart of a LISP program is functions that are evaluated together with their arguments. LISP includes many built-in, or *primitive* functions. Besides the usual mathematical operations, there are primitives for basic list-processing functions. For example, the list function creates a list from its arguments: (list 1 2 3) returns the list (1 2 3), while the cons function inserts an atom into the beginning of a list, and the append function tacks it onto the end. Programmers define their own functions using the defun keyword.

LISP has two other features that make it a powerful and flexible language for manipulating symbols and data. LISP allows for recursive functions (see RECURSION). For example, the following function raises a variable x to the power y:

```
(defun power (x y)
  (if (= y 0) 1
    (* x (power x (1- y)))))
```

Here the if expression checks to see whether y is 0. If not, the second expression invokes the function (power) itself, which performs the same test. The result is that the function keeps calling itself, storing temporary values, until y gets down to 0. It then "winds itself back up," multiplying x by itself y times.

But perhaps the most interesting feature of LISP is that it makes no distinction between programs (functions) and data. Since a function call and its arguments themselves constitute a list, a function can be fed as data to other functions. This makes it easy to write programs that modify their own operation.

LANGUAGE DEVELOPMENT

LISP quickly caught on with artificial intelligence researchers, and the version called LISP 1.5 was considered robust enough for writing large-scale applications. While "mainstream" computer scientists often used Algol and its descendants as a universal language for expressing algorithms, LISP became the lingua franca for AI people.

However, a number of dialects such as Mac-LISP diverged as versions were written to support new hardware or were promoted by companies such as LMI and Symbolics. While researchers liked the interactive nature of interpreted LISP (where functions could be defined and immediately tried out at the keyboard), practical applications required compilation into machine language to achieve adequate speed.

A widely used LISP variant is Scheme, developed by MIT researchers in the mid-1970s. Scheme simplifies LISP syntax (while still preserving the spirit), and at the same time generalizes further by allowing functions to have all the capabilities of data entities. That is, functions can be passed as parameters to other functions, returned as values, and assigned to variables and lists.

By the 1980s, personal computers became powerful enough to run LISP, and the proliferation of LISP variants running on different platforms led to a standardization movement that resulted in Common LISP in 1984. Common LISP combines the features of many existing dialects, includes a rich variety of data types, and also makes greater allowance for the imperative, sequential programming approach of languages such as C. It thus accommodates varying styles of programming. It is widely available today in both commercial and shareware versions.

Seymour Papert created a LISP-like language called Logo, which has been used to teach sophisticated computer science ideas to young students (see LOGO).

Further Reading

"Common LISP Implementations." [Including downloadable free versions] http://www.lisp.org/table/systems.htm

"Common LISP Primer." http://grimpeur.tamu.edu/~colin/lp/

Friedman, Daniel P. *The Little LISPer.* 3rd ed. Chicago: Science Research Associates, 1989.

———. *The Little Schemer.* 4th ed. Cambridge, Mass.: MIT Press, 1996.

Steele, G., Jr. *Common Lisp: The Language.* 2nd ed. Bedford, Mass.: Digital Press, 1990.

Steele, G. L., Jr., and R. P. Gabriel. "The Evolution of LISP," in *History of Programming Languages II,* eds. Bergin, T. J. and R. J. Gibson, Jr. New York: Addison-Wesley, 1996.

list processing

A *list* is a series of data items that can be accessed sequentially by following links from one item to the next. Lists can be very useful for ordering or sorting data items and for storing them on a stack or queue.

There are two general approaches to constructing lists. In a data list used with procedural programming languages such as C, each list item consists of a structure consisting of a data member and a pointer. The pointer, called "next," contains the address of the next item. A program can easily "step through" a list by starting with the first item, processing its data, then using the pointer to move to the next item, continuing until some condition is met or the end of the list is reached.

In LISP-type languages, however, a more general structure is used, since essentially all data is part of a list. Here each item is a *node* that can contains a pointer to any valid object and a pointer to the next node. One advantage of this scheme is that since fixed-length data fields are not used, the list can be "hooked up" to objects of varying sizes and types. This can also use memory more efficiently, though at the cost of additional processing being needed to periodically reclaim memory ("garbage collection").

List Processing (1)

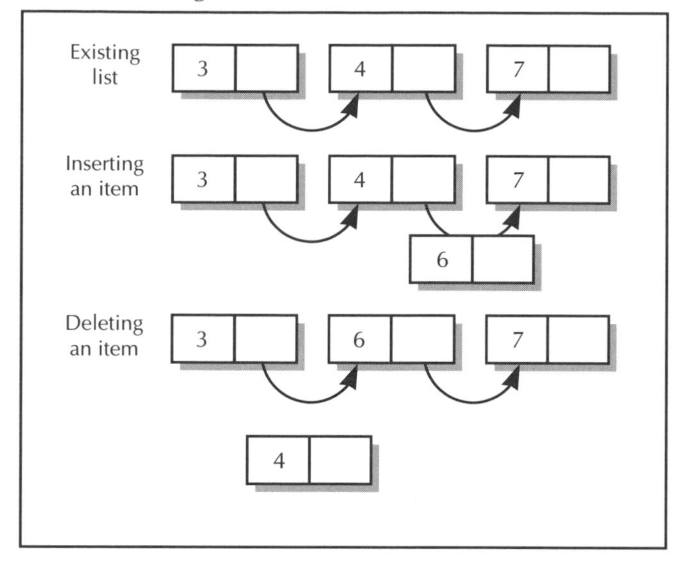

A singly linked list. Each node (item) includes a value and a pointer to the next node. Inserting a new node is simply a matter of adjusting the pointer of an existing node to point to the new node, with the new node's pointer in turn pointing to the next item (or the end of the list). A node is removed by disconnecting its pointer.

Besides traversing (stepping through) a list by following its "next" pointer, the basic list-processing operations are insertion and deletion. It is easy to insert a new element into a list: You first move to the item after which the new item is to be inserted. Next, you connect that item's "next" pointer (link) to the new item. You then connect the new item's next link to the item that originally followed the insertion point. Deleting an item is even simpler: You "snip out" the item by connecting the item that originally linked to it to the item that was originally after it.

List Processing (2)

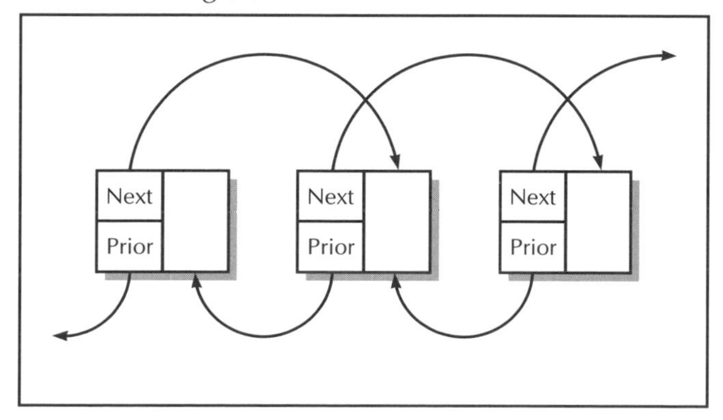

A doubly linked list. Each node has two pointers, one to the next item and one to the prior (preceding) item. While doubly linked lists use more memory, they can be processed more quickly because they can be traversed in either direction.

Sometimes lists are set up so that each item has two pointers: one to the next item and one to the previous one. Such *doubly linked lists* can be traversed in either direction, making retrieval faster in some situations, though at the cost of storing the extra pointers. Lists are also used to implement some specialized data structures (see STACK and QUEUE).

APPLICATIONS

Lists are generally used to provide convenient access to relatively small amounts of data where flexibility is required. Unlike an array, a list need use only as much memory as it needs to accommodate the current number of items (including their associated pointers). A LISP-style node list can be even more flexible in that items with varying sizes and types of data can be included in the same list. Lists are thus a more flexible way to implement such things as look-up tables. (See also ARRAY.)

Further Reading

"List Processing" [in LISP]. http://www.gnu.org/manual/emacs-lisp-intro/html_node/List-Processing.html

McGriff, Duane. *A Beginner's Guide to List Processing: The Other Side of LOGO.* Englewood Cliffs, N.J.: Prentice Hall, 1986.

local area network (LAN)

Starting in the 1980s, many organizations sought to connect their employees' desktop computers so they could share central databases, share or back up files, communicate via e-mail, and collaborate on projects. A system that links computers within a single office or home, or a larger area such as a building or campus, is called a local area network (LAN). (Larger networks linking branches of an organization throughout the country or world are called wide area networks, or WANs. See also NETWORK.)

HARDWARE ARCHITECTURE

There are two basic ways to connect computers in a LAN. The first, called Ethernet, was developed by a project at the Xerox Palo Alto Research Center (PARC) led by Robert Metcalfe. Ethernet uses a single cable line called a *bus* to which all participating computers are connected. Each data packet is received by all computers, but processed only by the one it is addressed to. Before sending a packet, a computer first checks to make sure the line is free. Sometimes, due to the time delay before a packet is received by all computers, another computer may think the line is free and start transmitting. The resulting *collision* is resolved by having both computers stop and wait varying times before resending.

Because connecting all computers to a single bus line is impractical in larger installations, Ethernet networks

Local Area Network (1)

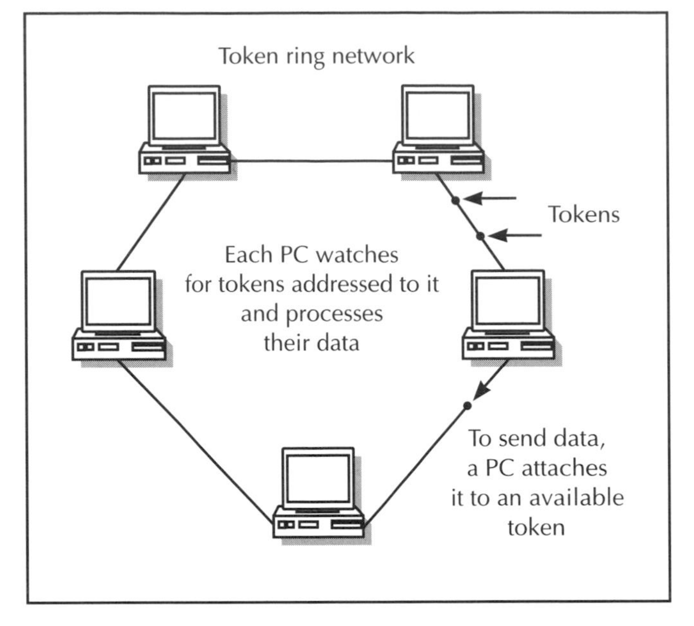

A Token Ring network connects the machines in a "chain" around which messages called tokens travel. Any PC can "grab" a passing token and attach data and the address of another PC to it. Each PC in turn watches for tokens that are addressed to it.

Local Area Network (2)

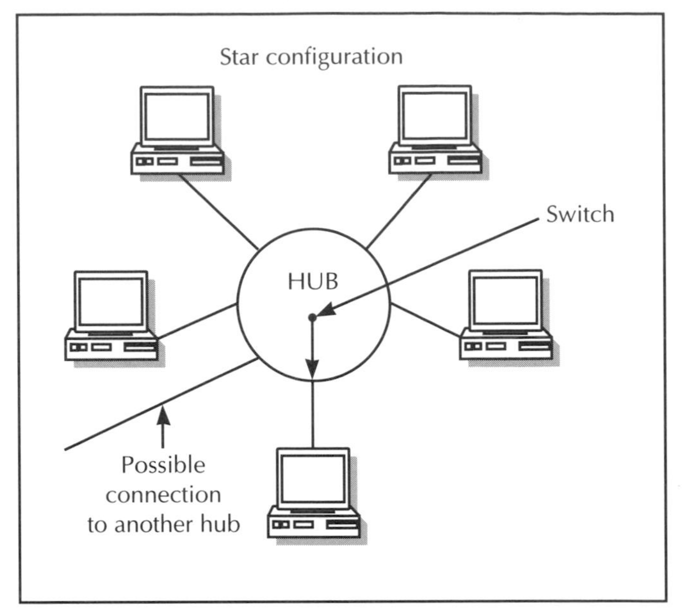

The Star network configuration uses a central hub to which each PC is attached. To extend the network (such as into other offices), the hubs can be connected to one another so they function as switches. When a token arrives that is addressed to one of its PCs, the hub will route it to the appropriate machine.

are frequently extended to multiple offices by connecting a bus in each office to a switch, creating a subnetwork or segment (this is sometimes called a *star topology*). The switches are then connected to a main bus. Packets are first routed to the switch for the segment containing the destination computer. The switch then dispatches the packet to the destination computer. Another advantage of this *switched Ethernet* system is that more-expensive, high-bandwidth cable can be used to connect the switches to move the packets more quickly over greater distances, while less-expensive cabling can be used to connect each computer to its local switch.

An alternative way to arrange a LAN is called *token ring*. Instead of the computers being connect to a bus that ends in a terminator, they are connected in a circle where the last computer is connected to the first. Interference is prevented by using a special packet called the token. Like the use of a "talking stick" in a tribal council, only the computer holding the token can transmit at a given time. After transmitting, the computer puts the token back into circulation so it can be grabbed by the next computer that wants to send data.

LAN SOFTWARE

Naturally there must be software to manage the transmission and reception of data packets. The structure of a packet (sometimes called a *frame*) has been standardized with a preamble, source and destination addresses, the data itself, a checksum, and two special layers that interface with the differing ways that Ethernet and token ring networks physically handle the packets.

The low-level processing of data packets must also be interfaced with the overall operating system so that, for example, a user on a desktop PC can "see" folders and files on the file server and whole files can be transferred between server and desktop PC. From the 1980s to the mid-1990s the most common LAN operating system for DOS and later Windows-based PCs was Novell Netware, while Macintosh users used AppleTalk. Later versions of Windows (notably Windows NT) then incorporated their own networking support, and Netware use declined somewhat. Linux and other UNIX variants are also used, particularly in academic settings.

The latest trend in LANs is to put TCP/IP, the Internet protocol "on top of" whatever network operating system is being used (see TCP/IP). This means that the software uses IP packets that are carried within, for example, Ethernet packets. The universality of the Internet and the wide variety of tools available makes TCP/IP the most flexible protocol for LANs (see INTRANET).

The desire of many families to share a broadband Internet connection using computers in different rooms has led to an increase in home LANs. Because traditional twisted-pair cable can be awkward to install in homes, many users have opted instead for networks that send data over existing phone lines, power lines, or even wireless transceivers. These technologies generally transfer data at rates of 1–10MB/s, about 10–100 times slower than cabled Ethernet. However, this speed is generally adequate for shared Internet access and playing on-line games. Another alternative for connecting nearby computers is to use network modules that connect to the computer's USB ports (see USB).

Further Reading

About.com "Computer Networking." [Resources] http://compnetworking.about.com/library/weekly/mpreviss.htm

Hallberg, Bruce A. *Networking: a Beginner's Guide.* Berkeley, Calif.: Osborne McGraw-Hill, 2001.

"Home Networking." [Overview] http://www.broadbandcompass.com/search/jsp/learnmore/homenetworking_overview.jsp?partnerID=microsoft

Spurgeon, Charles E. *Ethernet: The Definitive Guide.* Sebastopol, Calif.: O'Reilly, 2000.

Logo

Logo is a derivative of LISP (see LISP) that preserves much of that language's list processing and symbolic manipulation power while offering simpler syntax, easier interactivity, and graphics capabilities likely to appeal to young people. Logo has often been used as a first computer language for students in elementary and junior high school grades. As Harold Abelson noted in his *Apple Logo* primer in 1982, "Logo is the name for a philosophy of education and a continually evolving family of programming languages that aid in its realization."

Logo was developed starting in 1967 by educator Seymour Papert and his colleagues at Bolt, Beranek and Newman, Inc. Papert, a mathematician and AI pioneer, had became interested in devising an education-oriented computer language after working with developmental psychologist Jean Piaget. Papert focused particularly on Piaget's emphasis on "constructivism"—the idea that people learn mainly by fitting new concepts into an existing framework built from the experience of daily life. Papert came to believe that abstract computer languages such as FORTRAN or even BASIC were hard for children to assimilate because their algebraic formulas and syntax had little in common with daily activities such as walking, playing, drawing, or making things.

For example, most computer languages implement graphics using statements that specify screen points using Cartesian coordinates (X, Y). A square, for example, might be drawn by statements such as:

```
PLOT 100, 100
LINETO 150, 100
LINETO 150, 150
LINETO 100, 150
LINETO 100, 100
```

While familiarity with the coordinate system eventually allows one to visualize this operation, it is far from intuitive.

Papert, however, includes a "turtle" in his Logo language. The turtle was originally an actual robot that could be programmed to move around; in most systems today it is represented by a cursor on the screen. As the turtle moves, it uses a "pen" to leave a "trail" that draws the graphic.

With turtle commands, a square can be drawn by:

```
FD 50 (that is, forward 50)
RT 90 (turn right 90 degrees)
FD 50
RT 90
FD 50
RT 90
FD 50
RT 90
```

Here, the student programmer can easily visualize walking and turning until he or she arrives back at the starting point. In keeping with Piaget's theories, the learning is congruent with the physical world and daily activities.

Logo includes control structures similar to those in other languages, so the above program can be rewritten as simply:

```
REPEAT 4 [FD 50 RT 90]
```

Logo is much more than a set of simple drawing commands, however. Students can also be encouraged

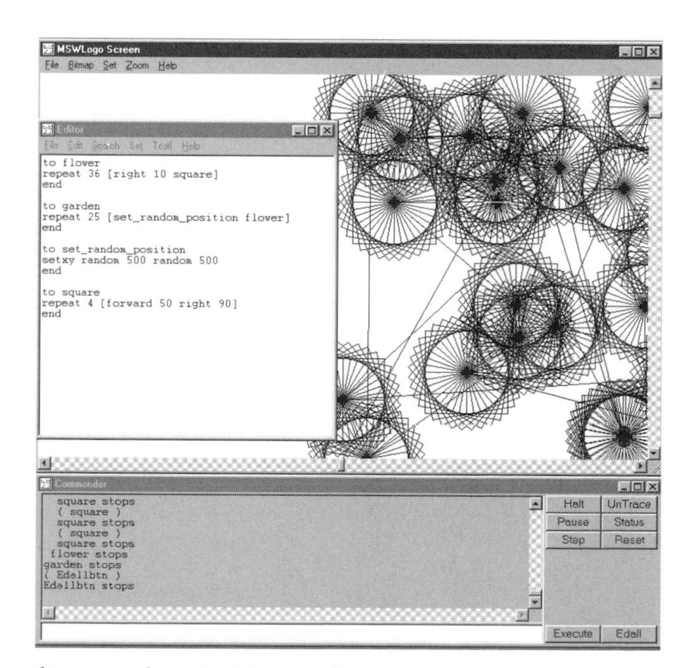

This screen shows MSWLogo, a free version of the Logo language. Notice how a program with intricate output can be built up from a series of simple procedures.

to use the list-processing commands to create everything from computer-generated poetry to adventure games. Unlike LISP's obscurely named commands such as car and cdr, Logo's list commands are readily understandable. For example, first returns the first item in a list, while butfirst returns all of the list except the first item.

Logo procedures are introduced by the **to** keyword, implying that the programmer is "teaching the computer" how to do something. For example, a procedure to draw a square with a variable size and starting position might look like this:

```
to square :X :Y :Size
setxy :X :Y
repeat 4 [fd :Size rt 90]
end
```

Logo has been steadily enhanced over the years, and includes not only a full set of math functions, but also many versions include special sound, graphics, and multimedia functions for Windows or Macintosh systems. By the mid-1980s, Logo had been combined with the popular LEGO building toy to create LEGO Logo. This popular kit enables students to build and control a variety of robots and other gadgets.

By the 1990s, Logo had to some extent become a casualty to the pressure on educators to provide "real world" programming skills using languages such as C++ or Java. However, Logo using educators have continued to flourish in parts of Europe, Japan, and Latin America. Logo has also been energized by the development of two recent versions. MicroWorlds Logo took advantage of the Macintosh interface to provide a full-featured multimedia environment, and it was later adapted for Windows systems. Another version, StarLogo, emphasizes parallel processing concepts, and is able to control thousands of separate turtles that can be programmed to simulate behaviors such as bird flocks or traffic flows. As Brian Harvey's books show, Logo's accessible, interactive nature continues to make it a good choice for teaching computer science to adults as well.

Further Reading

Harvey, Brian. *Computer Science Logo Style.* Vols. 1–3, 2nd ed. Cambridge, Mass.: MIT Press, 1997.

"Introduction to StarLogo." http://el.www.media.mit.edu/groups/el/projects/starlogo/

jFuller Educational. "An Introduction to MSW Logo." http://www.southwest.com.au/~jfuller/logotut/menu.htm

LOGO Foundation [home page]. http://el.www.media.mit.edu/groups/logo-foundation/logo/

Papert, Seymour. *Mindstorms: Children, Computers, and Powerful Ideas.* New York: Basic Books, 1993.

"Welcome to MSW Logo." [Free downloadable Logo for Windows] http://www.softronix.com/logo.html

loop

If computers were merely fast sequential calculators, they would still be of some use. However, much of the power of the computer comes from its ability to carry out repetitive tasks without supervision. The *loop* is the programming language structure that controls such activities. Virtually every language has some form of loop construct, with variations in syntax ranging from the relatively English-like COBOL and Pascal to the more cryptic C. We will use BASIC for our examples, since its syntax is easy to read.

The standard while loop performs the specified actions *as long as* the specified condition is true. For example:

```
While NOT EOF (Input_File)
   Read_Record
   Process_Record
Wend
Print "Done!"
```

This loop first checks to see whether the end of the input file (opened earlier) has been reached. If not, it reads and processes a record (using procedures defined elsewhere). The "Wend" marks the end of the statements controlled by the loop. When the end of the file is reached, the test fails (returns false) and control skips to the statement following Wend. See the accompanying flowchart for a visual depiction of the operation of this loop.

Flowchart for Loop

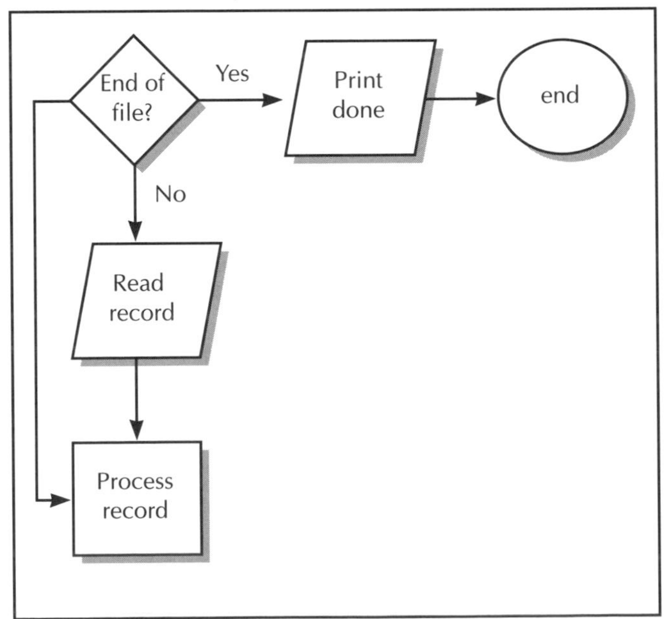

Flowchart for a loop that reads and processes records until it reaches the end of the file. Programmers must make sure that the end condition of a loop is properly defined, or the loop may run endlessly, "hanging" the program.

A variant form of while loop performs the test *after* executing the enclosed instructions. For example:

```
Do
   Print "Enter a number: "
   Input Number
   Print "You entered: ";Number
While (Number <> 0)
Print "I'm Done!"
```

This loop will display each number the user enters, then test it for zero. After a zero is encountered, control will skip to the final print statement.

Note that because this second form of while loop does not perform the test until it has performed the specified actions at least once, it would not be appropriate for the first example. In that case, the loop would attempt to get a record before discovering it had reached the end of the file, and an error would result.

The for loop is useful when an action is to be repeated for each of a limited series of cases. For example, this loop would print out the ASCII characters corresponding to the codes from 32 through 65:

```
For CharVal = 32 to 65 Step 1
   Print Char$(CharVal)
Next CharVal
```

Here Char$ is a function that output the character corresponding to the supplied ASCII character code. The step clause specifies the interval over which the variable within the loop is to be incremented. Here it's not strictly necessary, since it defaults to 1.

Loops of all sorts can be "nested" so that an inner loop executes completely for each step of the outer loop. For example:

```
For Vertical = 0 to 767
   For Horizontal = 0 to 1023
      Print_Pixel (Vertical, Horizontal)
   Next Horizontal
Next Vertical
```

Here the program will move across each line of the screen, printing the contents of each pixel. Each time the inner loop finishes, the outer loop increments, moving the scanning down to the next line. Indention is used to make the relation between the outer and inner loops clear.

In programming loops it's important to frame the test conditions correctly so that they terminate appropriately. An "endless loop" can cause a program to "hang" indefinitely. However, some programs do code an endless outer loop to indicate the program is to run indefinitely unless closed by the operating system. For example, the loop

```
While (1)
   ' Instructions go here
Wend
```

will execute indefinitely, since the value one is equivalent to "true."

Since many programs spend most of their time repeatedly executing loops, programmers seeking to improve the performance of their code pay especial attention to the code within the body of a loop. Any code such as a variable assignment, conversion, or calculation that needs to be done only once should be moved outside the loop.

Further Reading

Sebesta, Robert W. *Concepts of Programming Languages.* Reading, Mass.: Addison-Wesley, 1999.

Stroustrup, Bjarne. *The C++ Programming Language.* special ed. Reading, Mass.: Addison-Wesley, 2000.

M

Macintosh

Since its inception in 1984, Apple's Macintosh line of personal computers has offered a distinctive, innovative alternative to the more mainstream IBM-compatible PCs. When the Macintosh came out, it was billed as the computer "for the rest of us." Unlike the text-based, command-driven DOS-based IBM PC and its "clones," the "Mac" offered an interface that consisted of menus, folders, and icons that could be manipulated by clicking and dragging (see USER INTERFACE and MOUSE). The system came out of the box with a paint/draw program and a word processor that could show documents using the actual font sizes and styles that would appear in printed text. This "WYSIWYG" (What You See Is What You Get) feature quickly made the Mac the machine of choice for desktop publishers and graphic artists (see WYSIWYG). The Mac also met with some success in the educational market, where the way had been paved by the earlier Apple II.

However, there were factors would limit the Mac to a minority market share. The first models ran slowly. Although its Motorola 68000 processor was comparable to the Intel 80286 used by the IBM XT and AT series, the need to draw extensive graphics placed a heavier burden on the Mac's CPU.

Marketing decisions also proved to be problematic. The IBM PC had an "open architecture." Clone makers were able to legally produce machines that were functionally equivalent, and Microsoft was able to license to clone manufacturers essentially the same DOS operating system that IBM used. This created a robust market as manufacturers competed with added features or lower prices.

Apple, on the other hand, jealously guarded the Apple's hardware and the ROM (read only memory) that held the key operating system code. Apple made only a brief and half-hearted attempt to license the Mac OS to third parties in 1995, and by then it was probably too late. Apple CEO Steve Jobs kept prices relatively high, betting that the Mac's unique operating system and interface would entice people to buy the more expensive machine.

But something of a vicious circle set in. Since the Mac used a unique operating system, developing new applications (or porting existing ones) to the Mac was expensive. And since the Mac market represented only a small fraction of the PC-compatible market, developers were reluctant to create such software. Some flagship products such as Aldus PageMaker and Adobe Photoshop did cater to the Mac's graphic strengths. In general, however, the PC-compatible owner had a far wider range of software to choose from, and businesses were traditionally more comfortable with IBM equipment, even if IBM didn't make it.

Microsoft helped develop some successful Mac software, including versions of its Word and Office programs. But Microsoft CEO Bill Gates responded to the Mac's interface advantages over MS-DOS by developing a new

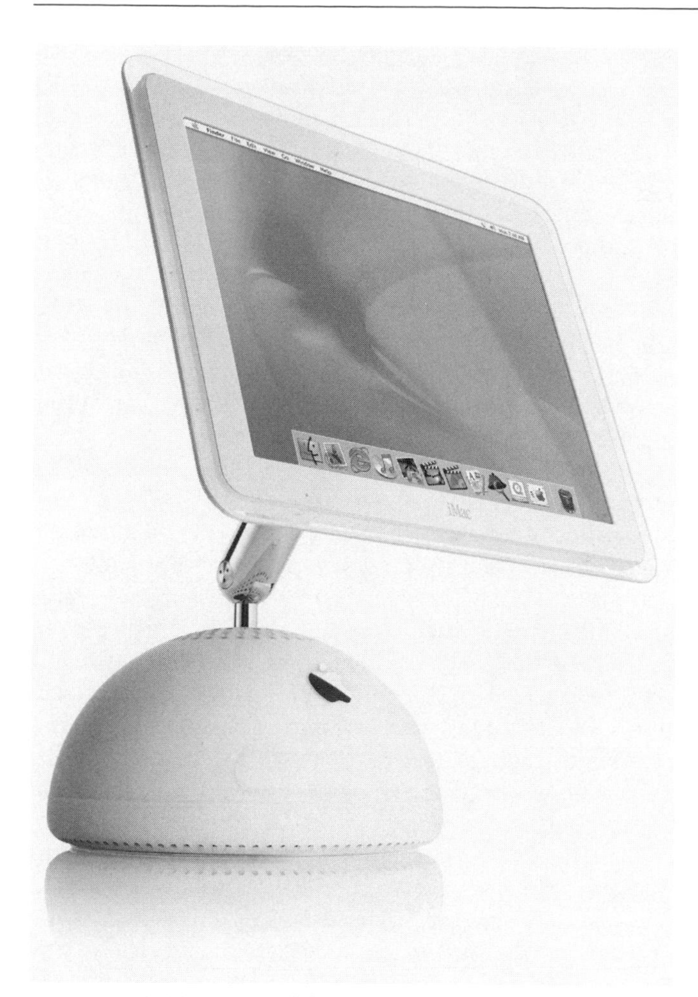

The latest (2001) version of the Apple iMac features distinctive styling with its swiveled flat panel display. The entire "works" of the computer are built into the base. (PHOTO COURTESY OF APPLE COMPUTER, INC.)

operating environment, Windows. Apple sued, claiming that Microsoft had gone beyond the license it had negotiated with Apple for use of elements of the Mac interface. By the early 1990s, however, Apple had lost the lawsuit. While the early versions of Windows were clumsy and met with little success, version 3.0 and, later, Windows 95 succeeded in providing a user experience that was increasingly close to that achieved by the Mac.

Apple kept trying to innovate and carve out a larger market share, designing both Power Macs that used the PowerPC RISC (reduced instruction set) microprocessor and PowerBook laptops. Toward the end of the 1990s, Apple tried to address the low end of the market with the iMac, a colorful, sleek machine packed with features such as home video editing, and achieved modest success in attracting new customers.

However, starting in 2000 Apple seemed to be changing course. It released an entirely new operating system, OS X, featured a sophisticated UNIX operating system underneath the traditional Mac interface. While Apple has not abandoned the consumer market (and offered a new version of the iMac with a round lamplike base and tiltable flat screen), the latest leading-edge Macs seem to be targeting the graphics, scientific, and engineering workstation market rather than the consumer PC arena.

Further Reading

Colby, Clifford [and others], eds. *The Macintosh Bible.* 8th Ed. Berkeley, Calif.: Peachpit Press, 2001.

Levy, Stephen. *Insanely Great: The Life and Times of Macintosh, the Computer That Changed Everything.* New York: Penguin, 1995.

"The Mac Observer." http://www.macobserver.com/

"Our Favorite Macintosh On-line Resources." http://www.quillserv.com/www/macres/macon-line.html

macro

For both programmers and ordinary users, the ability to "package" a group of instructions so that it can be invoked with a single command can save a lot of effort. The term *macro* is used for such instruction packages in a variety of contexts.

In the early days, programmers had to work with low-level machine instructions (see ASSEMBLER). Developers soon realized that a program could be used to write other programs. This program, called a macro assembler, lets the programmer write a group of instructions such as:

```
COMPARE macro

LOAD %1 ' load first data item
STOREX ' store in register X
LOAD %2 ' load second data item
STOREY ' store in register Y
CMPXY ' compare X and Y registers

endm
```

Now, if the programmer wants to compare the contents of two memory locations (say COUNTER and LIMIT), he or she can write simply:

```
COMPARE COUNT LIMIT
```

The assembler replaces COMPARE with the sequence of instructions above, substituting COUNT and LIMIT for %1 and %2.

Macros are also used in some higher-level languages, notably C. A module called the macro processor performs the required substitutions into the source code before the code is parsed and compiled. For example, a C programmer might include this macro in a program file:

```
#define IS_LOWERCASE(x) (( (x)>='a') && ( (x)
<='z') )
```

Somewhere in the program there might appear a statement such as:

```
if IS_LOWERCASE (Letter)
```

The macro processor will replace this with:

```
if (( (Letter)>='a') && ((Letter) <='z') )
```

This saves typing as well as reducing the chance of a typo creating a hard-to-find bug.

Macros are similar to procedures and functions (see PROCEDURES AND FUNCTIONS) in that they let the programmer treat a group of instructions as a single unit, simplifying coding. However, a given procedure appears in the code only once, although it may be called upon from many different parts of the program. A macro, on the other hand, is not "called." Each time it is mentioned, the macro is *replaced* by the corresponding instructions. Thus macros increase the size of the source code.

Many programmers today prefer using functions with the appropriate code rather than macros. Using functions saves space, since each function's code need only appear once. Although there is some processing overhead at runtime in calling the function, the function approach also ensures that the data sent to the function will be checked to make sure it is of the proper type. The macro, on the other hand, usually leaves it up to the programmer to make sure the data type being used is appropriate.

APPLICATION MACROS

The term *macro* is also used with applications software. Here is means a series of commands (such as cursor positioning or text formatting) that are recorded and assigned to a certain key combination. For example, a word processor user might define a macro called Letter and record the keystrokes and/or mouse movements needed to open a new document, insert a letterhead from a file, update the date, insert a salutation, and position the cursor to continue writing the letter. The recorded keystrokes might be assigned to the key combination Control + L.

Further Reading

Brown, P. J. *Macro Processors and Techniques for Portable Software.* New York: John Wiley, 1974.

Kernighan, Brian, and Dennis M. Ritchie. *The C Programming Language.* 2nd ed. Upper Saddle River, N.J.: Prentice Hall, 1988.

mainframe

In the era of vacuum tube technology, all computers were large, room-filling machines. By the 1960s, the use of transistors (and later, integrated circuits), enabled the production of smaller (roughly, refrigerator-sized) systems (see MINICOMPUTER). By the late 1970s, desktop computers were being designed around newly available computer chips (see MICROPROCESSOR). Although they, too, now use integrated circuits and microprocessors, the largest scale machines are still called mainframes.

The first commercial computer, the UNIVAC I (see ECKERT, J. PRESPER and MAUCHLY, JOHN) entered service in 1951. These machines consisted of a number of large cabinets. The cabinet that held the main processor and main memory was originally referred to as the "mainframe" before the name was given to the whole class of machines.

Although the UNIVAC (eventually taken over by Sperry Corp.) was quite successful, by the 1960s the quintessential mainframes were those built by IBM, which controlled about two-thirds of the market. The IBM 360 (and in the 1970s, the 370) offered a range of upwardly compatible systems and peripherals, providing an integrated solution for large businesses.

Traditionally, mainframes were affordable mainly by large businesses and government agencies. Their main application was large-scale data processing, such as the census, Social Security, large company payrolls, and other applications that required the processing of large amounts of data, which were stored on punched cards or transferred to magnetic tape. Programmers typically punched their COBOL or other commands onto decks of punched cards that were submitted together with processing instructions (see JOB CONTROL LANGUAGE) to operators who mounted the required data tapes or cards and then submitted the program cards to the computer.

By the late 1960s, however, *time-sharing* systems allowed large computers to be partitioned into separate areas so that they can be used by several persons at the same time. The punched cards began to be replaced by Teletypes or video terminals at which programs or other commands could be entered and their results displayed or printed. At about the same time, smaller computers were being developed by Digital Equipment Corporation (DEC) with its PDP series (see MINICOMPUTER).

With its range of compatible systems the IBM 360 series and its successor, the 370, would dominate business mainframe computing for decades. (IMAGE COURTESY OF THE IBM CORPORATE ARCHIVES)

With increasingly powerful minicomputers and later, desktop computers, the distinction between mainframe, minicomputer, and microcomputer became much less pronounced. To the extent it remains, the distinction today is more about the bandwidth or amount of data that can be processed in a given time than about raw processor performance. Powerful desktop computers combined into networks have taken over many of the tasks formerly assigned to the largest mainframe computers. With a network, even a large database can be stored on dedicated computers (see FILE SERVER) and integrated with software running on the individual desktops.

Nevertheless, mainframes such as the IBM System/390 are still used for applications that involve processing large numbers of transactions in near real-time. Indeed, many of the largest e-commerce organizations have a mainframe at the heart of their site. The reason is that while the raw processing power of high-end desktop systems today rivals that of many mainframes, the latter also have high-capacity *channels* for moving large amounts of data into and out of the processor.

Early desktop PCs relied upon their single processor to handle most of the burden of input/output (I/O). Although PCs now have I/O channels with separate processors (see BUS), mainframes still have a much higher data throughput. The mainframe can also be easier to maintain than a network, since software upgrades and data backups can be handled from a central location. On the other hand, a system depending on a single mainframe also has a single point of vulnerability, while a network with multiple mirrored file servers can work around the failure of an individual server.

Further Reading

Butler, Janet G. *Mainframe to Client-Server Migration: Strategic Planning Issues and Techniques.* Charleston, S.C.: Computer Technology Research Corporation, 1996.

Pugh, Emerson W., Lyle R. Johnson, and John H. Palmer. *IBM's 360 and Early 370 Systems.* Cambridge, Mass.: MIT Press, 1991.

Yahoo Commercial Directory. "Mainframes." http://dir.yahoo.com/Business_and_Economy/Business_to_Business/Computers/Hardware/Systems/Mainframes/

management information system

The first large-scale use of computers in business in the late 1950s and 1960s focused on fundamental data processing. Companies saw computers primarily as a way to automate such functions as payroll, inventory, orders, and accounts payable, hoping to keep up with the growing volume of data in the expanding economy while saving labor costs associated with manual methods. The separate data files and programs used for basic business functions were generally not well integrated and could not be easily used to obtain crucial information about the performance of the business.

By the 1970s, the growing capabilities of computers encouraged executives to look for ways that their information systems could be used to competitive advantage. Clearly, one possibility was that reporting and analysis software could be used to help them make faster and better decisions, such as about what products or markets to emphasize. To achieve this, however, the "data processing department" had to be transformed into a "management information system" (MIS) that could allow analysis of business operations at a variety of levels.

THE MIS PYRAMID

If one thinks of the information infrastructure of an enterprise as being shaped like a pyramid, the bottom of the pyramid consists of the transactions themselves, where products and services are delivered, and the supporting point of sale, inventory, and distribution systems that keep track of the flow of product.

The next layer up begins the process of integration and operational control. For example, previously separate sales and inventory system (perhaps updated through a daily batch process) now become part of an integrated system where a sale is immediately reflected in reduced inventory, and the inventory system is in turn interfaced with the order system so more of a product is ordered when it goes out of stock.

Management Information Systems

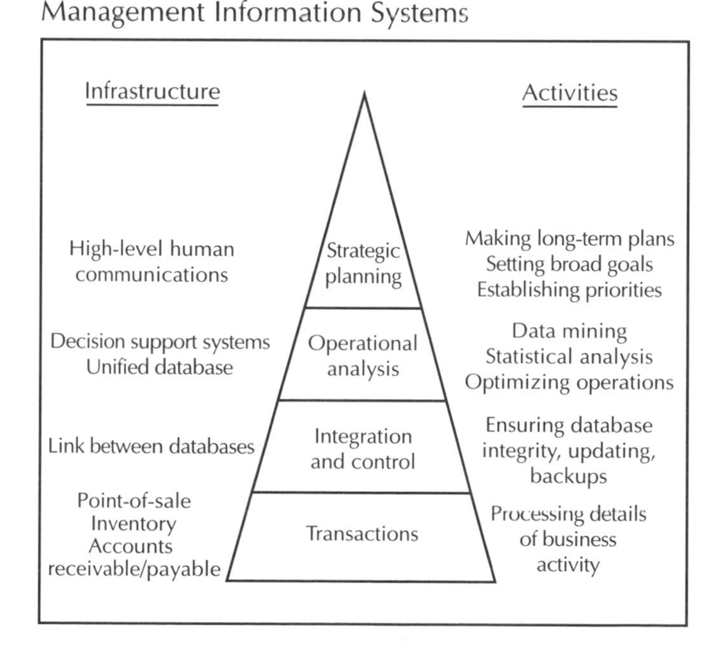

The activities involved in managing an enterprise's information infrastructure can be drawn as a pyramid. The raw material of transactions at the bottom are stored in databases. Moving up the pyramid, these data sources are integrated and refined to provide better information about business operations as well as material for operational analysis and strategic planning.

The next layer can be called the operational analysis layer. Here such functions as sales, inventory, and ordering aren't simply connected; they are part of the same system of databases. This means that both simple and complex queries and analysis can be run against a database containing every type of transaction that the business engages in. In addition to routine reports such as sales by region or product line, market researchers or strategic planners can receive the data they need to answer questions such as:

- What products are staying on the shelf the longest?
- What is the ratio between profitability and shelf space for particular items?
- What is the relationship between price reductions, sales, and profits for a certain category of items?

The goal of this layer is to help managers identify the variables that affect the performance of their store or other business division and to determine how to optimize that performance.

The very top level can be called the strategic planning layer. Here top-level executives are interested in the overall direction of the business: determining which divisions of a company should receive the greatest long-term investment, and which perhaps should be phased out. For example:

- Which kind of sales are growing the fastest: in-store, mail-order catalog, or Internet on-line store?
- How is our market share trending compared to various classes of competitors?
- How are sales trending with regard to various types (demographics) of customers?

SOFTWARE SUPPORT

There are many considerations to choosing appropriate software to support the users who are trying to answer questions at the various levels of management. At minimum, to create a true management information system, the information from daily transactions must be made accessible to a variety of query or analysis programs.

In the past three decades many established businesses have had to go through a painful process of converting a variety of separate databases and "legacy software" (often written in COBOL in the 1960s or 1970s) into a modern relational database such as Oracle or Microsoft Access. Sometimes a company has decided that the cost of rewriting software and converting data is simply too high, and instead, opts for a patchwork of utility programs to convert data from one program to another.

The growth of networking in the 1980s and Web-based intranets in the 1990s required that the old model of a large, centralized data repository accessed directly by only a few users be replaced by a less centralized model, sometimes going as far as using a distributed database system where data "objects" can reside throughout the network yet be accessed quickly by any user (see DATABASE MANAGEMENT SYSTEM). An alternative is the data repository that includes queries and other tools (see DATA WAREHOUSE).

FUTURE OF MIS

With the prominence of the Internet and e-commerce today, MIS has had to cope with an even more complex and fast-moving world. On the one hand, widespread e-commerce enables the capturing of more detailed data about transactions and consumer behavior in general. New tools for analyzing large repositories of data (see DATA MINING) make it possible to continually derive new insights from the recent past. It is thus clear that information is not just a tool but also a corporate asset in itself. On the other hand, fierce competition and often shrinking profit margins in e-commerce have placed increasing pressure on MIS departments to find the greatest competitive advantage in the shortest possible time.

The importance of MIS has also been reflected in its place in the corporate hierarchy. The top-level executive post of Chief Information Officer (CIO) has perhaps not yet achieved parity with the Chief Financial Officer (CFO), but healthy budgets for MIS even in constrained economic times testify to its continuing importance.

Further Reading

About.com Guide to MIS Software. http://businesssoft.about.com/cs/vendorssoftware/

Cook, Melissa A. *Building Enterprise Information Architecture: Reengineering Information Systems.* Upper Saddle River, N.J.: Prentice Hall PTR, 1996.

Laudon, K. C., and J. P. Laudon. *Management Information Systems: New Approaches to Organization and Technology.* Upper Saddle River, N.J.: Prentice Hall, 1999.

Lucas, Henry C. *Information Technology for Management.* Boston: Irwin/McGraw-Hill, 2000.

"Management Science Programs. Management Information Systems." http://dir.yahoo.com/Education/By_Subject/Business_and_Economy/College_and_University/Business_Schools__Departments__and_Programs/Management_Science/Management_Information_Systems/

Turban, Efraim, Ephraim McClean, and James Wetherbe. *Information Technology for Management: Improving Quality and Productivity.* New York: John Wiley, 1999.

marketing of software

The way software has been produced and marketed has changed considerably in the past five decades. In the nascent computer industry of the 1950s, commercial software was developed and marketed by the manufacturers of computer systems—firms such as Univac (later Sperry-Univac), Burroughs, and of course, IBM (see MAINFRAME).

However, a separate (third-party) software industry emerged as early as 1955 with the founding of Computer Usage Corporation (CUC) by two former IBM employees. Nevertheless, the primary competition was between hardware manufacturers, with software seen as part of the overall package.

By the early 1960s, larger software companies emerged such as Computer Science Corporation (CSC) and Electronic Data Systems (EDS), which became an empire under the energetic, albeit often controversial leadership of H. Ross Perot, as well as the European giant SAP. These companies specialized in providing customized software solutions for users who could not meet their needs with the software library offered by the maker of their computer system (see also BUSINESS APPLICATIONS OF COMPUTERS).

By the 1970s, however, vendor-supplied and contracted custom software alternatives were being increasingly accompanied by "off the shelf" software packages. By 1976, 100 software products from 64 software companies had reached the $1 million mark in sales.

The 1980s saw the emergence of a completely new sector: desktop computer (PC) users. Traditionally, software had been marketed to programmers or managers, but now individual users or office managers could buy and install word processing programs, spreadsheets, database, and other programs. At the same time, a market for software for use in the home and schools, particularly education, personal creativity, and game programs required new methods of marketing. For the first time ads for software began to appear on TV and in general-interest magazines.

While large businesses still required custom-made software, most small to medium businesses looked for powerful and integrated office software solutions (see APPLICATION SUITE, OFFICE AUTOMATION, and GATES, WILLIAM). By the mid-1990s, Microsoft's Office suite had dominated this market, although that dominance is now threatened to some extent by software written in Java for Windows or hosted on Linux systems (see also OPEN SOURCE MOVEMENT).

The growth in the Internet (see E-COMMERCE) has also offered new venues for the marketing and distribution of software. Sites such as ZDNet and CNet have to some extent displaced computer magazines as sources for product reviews. These sites also offer extensive libraries of "try before you buy" software (see SHAREWARE), some of which is trial versions of full-blown commercial products. The local "mom and pop" PC software store has largely vanished, with software now marketed mainly by chain stores such as Electronics Boutique or CompUSA, and increasingly, through Web-based stores, often established by the chains, as well as the giant on-line bookstore Amazon.com. The movement of much marketing activity on-line has also reduced somewhat the importance of the annual pilgrimage to the COMDEX computer trade show.

Further Reading

"An Overview of the History of the Software Industry." http://www.softwarehistory.org/overview.htm

Fisher, Franklin M., James W. McKie, and Richard B. Mancke. *IBM and the U.S. Data Processing Industry.* New York: Praeger, 1983.

Hoch, Detley J. [and others]. *Secrets of Software Success: Management Insights from 100 Software Firms Around the World.* Cambridge, Mass.: Harvard Business School Press, 1999.

Petska-Juliussen, Karen, and Egil Juliussen. *Computer Industry Almanac.* 8th ed. Buffalo Grove, Ill.: Computer Industry Almanac Inc., 1996.

Platner, Hasso, and William McKone (translator). *Anticipating Change: Secrets Behind the SAP Empire.* Roseville, Calif.: PrimaTech, 2000.

Reid, Robert H. *Architects of the Web: 1,000 Days that Built the Future of Business.* New York: Wiley, 1997.

Software History Center. http://www.softwarehistory.org/index.html

mathematics of computing

The roots of modern computer science lie in an interest in rapid computation. Simple mechanical calculators (see CALCULATOR) may date back to ancient times; however, it is the work of mathematicians Blaise Pascal (1623–1662) and Gottfried Leibniz (1646–1716) that gave rise to the first practical mechanical calculators. By the mid-19th century, Charles Babbage (1791–1871) had conceptualized and designed mechanical computers that included the essential features (programs, processor, memory, input/output) of the modern digital computer (see BABBAGE, CHARLES). His motivation was the need for rapid, accurate calculation of statistical tables made necessary by the manufacturing economy of the Industrial Revolution. By the end of the century, the volume of such data had increased to the point where mechanical calculators and tabulators (see HOLLERITH, HERMAN) had become the only practical way to keep up.

Mathematically, a computer can be seen as a way to rapidly and automatically execute procedures that have been proven to lead to reliable solutions to a problem (see ALGORITHM). Once computers came on the scene, mathematical principles for verifying or proving algorithms would acquire new practical importance.

By the early 20th century, however, mathematicians were beginning to examine the problem of determining what propositions were provable, and in 1931 Kurt Godel published a proof that any mathematical system necessarily allowed for the formation of propositions that could not be proven using the axioms of that system. An analogous question was determining what problems were computable. Working independently, two researchers (see CHURCH, ALONZO and TURING, ALAN) formulated models that could be used to test for computability. Turing's model, in particular, provided a theoretical construct (the Turing Machine) that could, using combinations of a few simple operations, calculate anything that was computable.

By the 1940s, electromechanical (relays) or electronic (tube) switching elements made it possible to build practical high-speed computers. Computer circuit designers could draw upon the advances in symbolic logic in the 19th century (see BOOLEAN OPERATORS). Boolean logic, with its true/false values, would prove ideal for operating computers constructed from on/off switched elements.

The mathematical tools of the previous 150 years could now be used to design systems that could not only calculate but also manipulate symbols and achieve results in higher mathematics (see the next article, MATHEMATICS SOFTWARE).

MATHEMATICS AND MODERN COMPUTERS

A variety of mathematical disciplines bear upon the design and use of modern computers. Simple or complex algebra using variables in formulas is at the heart of many programs ranging from financial software to flight simulators. Indeed, one of the most enduring scientific and engineering languages takes its name from the process of translating formulas into computer instructions (see FORTRAN).

Geometry, particularly the analytical geometry based upon the coordinate system devised by Rene Descartes (1596–1650) is fundamental to computer graphics displays, where the screen is divided into X (vertical) and Y (horizontal) axes. Modern graphics systems have added 3D depiction and sophisticated algorithms to allow the rapid display of complex objects. Beyond graphics, the Cartesian insight that converted geometry into algebra makes a variety of geometrical problems accessible to computation, including the finding of optimum paths for circuit design. Design of computer and network architectures also involves the related field of topology. The fascinating field of fractal geometry has found use in computer graphics and data storage techniques (see FRACTALS IN COMPUTING).

Aspects of number theory, often considered the most abstract branch of mathematics, have found surprising relevance in computer applications. These include randomization (random number generation) and the factoring of large numbers, which is crucial for cryptography.

Mathematics also bears on computer networking with regard to communications theory (see BANDWIDTH and SHANNON, CLAUDE) and techniques for error correction.

THE COMPUTER'S CONTRIBUTION TO MATHEMATICS

Mathematics as a discipline is thus essential to its younger sibling, computer science. In turn, however, computer science and technology have enriched the pursuit of mathematical truth in surprising ways. As early as 1956, a program called *Logic Theorist,* written by Herbert Simon (1916–) and Allen Newell (1927–1992) demonstrated how a program (that is, a collection of algorithms) could prove mathematical propositions given axioms and rules. While these early programs worked on a somewhat hit-or-miss basis, later theorem-solving programs produced solutions different from the standard ones known to mathematicians, and sometimes more elegant. Thus the computer, which began as an aid to calculation, became an aid to symbol manipulation and to some extent an independent creative source.

Further Reading

Bell Labs Computing Sciences Research Center. http://cm.bell-labs.com/cm/cs/

Chudnovsky, David V., and Richard D. Jenks, eds. *Computers in Mathematics.* New York: M. Dekker, 1990.

Feinerman, Robert P. *Using Computers in Mathematics.* Needham Heights, Mass.: Ginn Press, 1991.

Lipschutz, Seymour. *Schaum's Outline of Essential Computer Mathematics.* New York: McGraw Hill, 1982.

Logic and Foundations of Computing Group. http://www.math-stat.uottawa.ca/lfc/

Took, D. James, and Norma Henderson, eds. *Using Information Technology in Mathematics Education.* New York: Haworth Press, 2001.

mathematics software

As explained in the preceding article, computer science looked to mathematics to create and verify its algorithms. In turn, computer software has greatly aided many levels of mathematical work, ranging from simple calculations to manipulation of symbols and abstract forms.

At the simplest level, computers overlap the functions of simple electronic calculators. Indeed, operating systems such as Microsoft Windows and UNIX systems include calculator utilities that can be used to solve problems requiring a basic four function or more elaborate scientific calculator.

The true power of the computer became more evident to ordinary users when spreadsheet software was introduced commercially in 1979 with VisiCalc (see SPREADSHEET). Spreadsheets make it easy to maintain and update summaries and other reports generated by formulas. Later versions of spreadsheet programs such as Lotus 1-2-3 and Microsoft Excel have the ability to create a wide variety of plots and charts to show relationships between variables in visual terms.

Moving from simple formulas to the manipulation of symbolic quantities (as in algebra), the Association for Computing Machinery (ACM) classification system describes several broad areas of computer-aided mathematics. These include *numerical analysis* (techniques for solving, linear, non-linear, and differential equations), *discrete mathematics* (combinatorial and graph theory), and probability and statistics.

There are two general approaches to mathematical software. One is the creation of libraries of routines or procedures that address particular kinds of problems. A programmer who is creating software that must deal with particular mathematical problems can link these routines to the program, call the procedures with appropriate variables or data, and return the results to the main program for further processing (see PROCEDURES AND FUNCTIONS). The language FORTRAN is still widely used for developing mathematics libraries, and there is a legacy of tens of thousands of routines available. Modern systems have the ability to link these procedures to programs written in more recent languages such as C.

The advantage of using program libraries is that they don't require learning new programming techniques. Each routine can be treated as a "black box." However, it is often desirable to work with traditional mathematical notation (what one might see on a blackboard in a calculus class, rather than typed into computer code). A stand-alone software package such as Macsyma, Mathcad, Maple6, Matlab, or Mathematica can automatically simplify or solve algebraic expressions or perform hundreds of traditional mathematical procedures. For statistical analysis, programs such as SPSS can apply all of the standard statistical tests to data and provide a large variety of graphics.

Further Reading

Association for Computing Machinery. *Transactions on Mathematical Software.* Information at: http://math.nist.gov/toms/

Green, Samuel B. [and others]. *Using SPSS for Windows: Analyzing and Understanding Data.* Upper Saddle River, N.J.: Prentice Hall, 1999.

Martinez, Wendy L., and Angel R. Martinez. *Computational Statistics Handbook with Matlab.* Grand Rapids, Mich.: CRC Press, 2001.

"Mathematics Software." [Reference Links] http://www.math.psu.edu/MathLists/Software.html

Netlib Repository: a Collection of Mathematical Software, Papers, and Databases. http://www.netlib.org/

National Institute of Standards and Technology. "NIST Guide to Available Mathematical Software." http://gams.nist.gov/

Wolfram, Stephen. *The Mathematica Book.* 4th ed. New York: Cambridge University Press, 1999.

Mauchly, John William

(1907–1980)
American
Inventor, Computer Scientist

John Mauchly was codesigner of the earliest full-scale digital computer, ENIAC, and its first commercial successor, Univac (see also ECKERT, J. PRESPER). His and Eckert's work went a long way toward establishing the viability of the computer industry in the early 1950s.

Mauchly was born on August 30, 1907, in Cincinnati, Ohio. He attended the McKinley Technical High School in Washington, D.C., and then began his college studies at Johns Hopkins University, eventually changing his major from engineering to physics. The spectral analysis problems he tackled for his Ph.D. (awarded in 1932) and in postgraduate work required a large amount of painstaking calculation. So, too, did his later interest in weather prediction, which led him to design a mechanical computer for harmonic analysis of weather data (see also ANALOG COMPUTER). He also learned about binary switching circuits ("flip-flops") and experimented with building electronic counters, which used vacuum tubes and were much faster than counters using electromagnetic relays.

Mauchly taught physics at Ursinus College in Philadelphia from 1933 to 1941. On the eve of World War II, however, he went to the University of Pennsylvania's Moore School of Engineering and took a course in military applications of electronics. He then joined the staff and began working on contracts to prepare artillery firing tables for the military. Realizing how intensive the calculations would be, in 1942 he wrote a memo proposing that an electronic calculator be built to tackle the problem. The proposal was rejected at first, but by 1943 table calculation by mechanical methods was falling even further behind. Herman Goldstine, who had been assigned by the Aberdeen Proving Ground to break the bottleneck, approved the calculator project.

With Mauchly providing theoretical design work and J. Presper Eckert heading the engineering effort, the Electronic Numerical Integrator and Computer, better known as ENIAC, was completed too late to influence the outcome of the war. However, when the machine was demonstrated in February, 1946, it showed that a programmable electronic computer was not only about a thousand times faster than an electromechanical calculator, it could be used as a general-purpose problem-solver that could do much more than existing calculators.

Mauchly and Eckert left the Moore School after a dispute about who owned the patent for the computer work. They jointly founded what became known as the Eckert-Mauchly Computer Corporation, betting on Mauchly's confidence that there was sufficient demand for computers not only for scientific or military use, but for business applications as well. By 1950, however, they were struggling to sell and build their improved computer, Univac, while fulfilling existing government contracts for a scaled-down version called BINAC. In 1950, they sold their company to Remington Rand, while continuing to work on Univac. In 1952, Univac stunned the world by correctly predicting the presidential election results on election night long before most of the votes had come in.

Early on, Mauchly saw the need for a better way to write computer programs. Univac and other early computers had been programmed through a mixture of rewiring, setting of switches, and entering numbers into registers. This made programming difficult, tedious, and error-prone. Mauchly wanted a way that variables could be represented symbolically: for example, Total rather than a register number such as 101. Under Mauchly's supervision William Schmitt wrote what became known as Brief Code. It allowed two-letter combinations to stand for both variables and operations such as multiplication or exponentiation. A special program read these instructions and converted them to the necessary register and machine operation commands (see INTERPRETER). While primitive compared to later languages (see ASSEMBLER and PROGRAMMING LANGUAGES), Brief Code represented an important leap forward in making computers more usable.

Mauchly stayed with Remington Rand and its successor Sperry Rand until 1959, but then left over a dispute about the marketing of the Univac. He continued his career as a consultant and lecturer. Mauchly and Eckert also became embroiled in a patent dispute arising from their original work with ENIAC. Accused of infringing Sperry Rand's ENIAC patents, Honeywell claimed that the ENIAC patent was invalid, with another computer pioneer, John Atanasoff, claiming that Mauchly and Eckert had obtained crucial ideas after visiting his laboratory in 1940 (see ATANASOFF, JOHN VINCENT).

In 1973, Judge Earl Richard Larson ruled in favor of Atanasoff and Honeywell. However, many historians of the field give Mauchly and Eckert the lion's share of the credit because it was they who had built full-scale, practical machines.

Mauchly played a key role in founding the Association for Computing Machinery (ACM), one of the field's premier professional organizations. He served as its first vice president and second president. He received many tokens of recognition from his peers, including the Howard Potts Medal of the Franklin Institute. In turn, the ACM established an Eckert-Mauchly award for excellence in computer design. John Mauchly died on January 8, 1980.

Further Reading

McCartney, Scott. *ENIAC: The Triumphs and Tragedies of the World's First Computer.* New York: Berkeley Books, 1999.

Stern, N. "John William Mauchly: 1907–1980," *Annals of the History of Computing* 2, 2, 100–103.

McCarthy, John

(1927–)
American
Computer Scientist, AI Pioneer

Starting in the 1950s, John McCarthy played a key role in the development of artificial intelligence as a discipline, as well as developing LISP, the most popular language in AI research.

John McCarthy was born on September 4, 1927, in Boston, Massachusetts. He completed his B.S. in mathematics at the California Institute of Technology, then earned his Ph.D. at Princeton University in 1951. During the 1950s, he held teaching posts at Stanford University, Dartmouth College, and the Massachusetts Institute of Technology.

Although he seemed destined for a prominent career in pure mathematics, he encountered computers while working during the summer of 1955 at an IBM laboratory. He was intrigued with the potential of the machines for higher-level reasoning and intelligent behavior (see ARTIFICIAL INTELLIGENCE). The following year he put together a conference that brought together people who would become key AI researchers, including Marvin Minsky (see MINSKY, MARVIN). He proposed that "the study is to proceed on the basis of the conjecture that every aspect of learning or any other feature of intelligence can in principle be so precisely described that a machine can be made to simulate it. An attempt will be made to find how to make machines use language, form abstractions and concepts, solve kinds of problems now reserved for humans, and improve themselves."

Mathematics had well-developed symbolic systems for expressing its ideas. McCarthy decided that if AI researchers were to meet their ambitious goals, they would need a programming language that was equally capable of expressing and manipulating symbols. Starting in 1958, he developed LISP, a language based on lists that could flexibly represent data of many kinds and even allowed programs to be fed as data to other programs (see LISP). LISP would be used in the coming decades to code most AI research projects, and McCarthy continued to play an important role in refining the language, while moving to Stanford in 1962, where he would spend the rest of his career.

McCarthy also contributed to the development of Algol, a language that would in turn greatly influence modern procedural languages such as C. He also helped develop new ways for people to use computers. Consulting with Bolt, Beranek and Newman (the company that would later build the beginnings of the Internet), he helped design time-sharing, a system that allowed many users to share the same computer, bringing down the cost of computing and making it accessible to more people. He also sought to make computers more interactive, designing a system called THOR, which used video display terminals. Indeed, he pointed the way to the personal computer in a 1972 paper on "The Home Information Terminal."

In 1971, McCarthy received the prestigious A. M. Turing award from the Association for Computing Machinery. In the 1970s and 1980s, he taught at Stanford and remained a prominent spokesperson for AI, arguing against critics such as philosopher Hubert Dreyfus (see

DREYFUS, HUBERT), who claimed that machines could never achieve true intelligence. As of 2001, McCarthy is professor emeritus of computer science at Stanford University.

Further Reading

"John McCarthy's Home Page." http://www-formal.stanford.edu/jmc/

McCarthy, John. "The Home Information Terminal." *Man and Computer: Proceedings of the International Conference, Bordeaux, France, 1970.* Basel: S. Karger, 1972, 48–57.

———. "Philosophical and Scientific Presuppositions of Logical AI," in McCarthy, H. J. and Vladimir Lifschitz, eds., *Formalizing Common Sense: Papers by John McCarthy.* Norwood, N.J.: Ablex, 1990.

McCorduck, Pamela. *Machines Who Think.* New York: W. H. Freeman, 1979.

measurement units used in computing

Newcomers to the computing world often have difficulty mastering the variety of ways in which computer capacity and performance are measured. A good first step is to look at the most common metric prefixes that indicate the magnitude of various units (see table).

COMMON METRIC PREFIXES USED IN COMPUTING

Prefix	Magnitude
kilo	10^3 (1 thousand)
mega	10^6 (1 million)
giga	10^9 (1 billion)
tera	10^{12} (1 trillion)
milli	10^{-3} (1 thousandth)
micro	10^{-6} (1 millionth)
nano	10^{-9} (1 billionth)
pico	10^{-12} (1 trillionth)

Strictly speaking, most computer measurements are based on the binary system, using powers of two. Thus *kilo* actually means 2^{10}, which is actually 1,024, and *mega* is actually 2^{20}, or 1,048,576. However, this distinction is generally not important for gaining a sense of the magnitudes involved. In 1998, the International Electrotechnical Commission promulgated a new set of prefixes for these base two computer-related magnitudes, such that for example, mebi- is supposed to be used instead of mega-. There is little evidence thus far that this scheme is being widely adopted.

We will now consider some of the main areas in which computer capacity or performance is measured.

STORAGE CAPACITY

The smallest unit of information, and thus of data storage, is a bit (*bi*nary digi*t*). A bit can be either 1 or 0 and is physically represented in different ways according to the memory or storage device being used. On most computers the most-used storage unit is the *byte,* which contains eight bits. Since this represents eight binary digits, or 2^8, a byte can hold values from 0 to 255 (decimal). The following table gives some typical units of storage.

DATA STORAGE UNITS

Unit	Typical Use
bit	Processor data handling capacity. Most processors today can handle 32 or 64 bits at a time.
byte (8 bits)	Holds an ASCII character value or a small number, 0-255.
kilobyte	Used to measure RAM (random access memory) and floppy disk capacity for early PCs.
megabyte	RAM capacity in modern PCs; hard drive capacity in older PCs.
gigabyte	Hard drive capacity in modern PCs.
terabyte	Disk arrays used for larger servers and mainframes (see DISK ARRAY).

GRAPHICS

Printed output is generally measured in dots per inch (dpi). Screen images and images used in digital photography are measured in pixels or megapixels. However, the amount of data needed to specify (and thus store) a pixel in an image depends on the number of colors and other information to be stored. (See GRAPHICS FORMATS.)

PROCESSOR SPEED

Processor speed is measured in millions of cycles per second (megahertz or MHz). The earliest microprocessors had speeds measured in 1-2 MHz or so. PCs of the 1980s ranged from about 8 to 50 MHz. In the 1990s, speeds ramped up to the hundreds of MHz, and in 2002 PC speeds are often measured in gigahertz (GHz).

CALCULATION SPEED

The speed at which a computer can perform calculations depends on more than raw processor speed. For example, a processor that can store or fetch 32-bit numbers can perform many calculations faster than one with only a 16-bit capacity even if the two processors have the same clock speed in cycles per second.

Calculation speed is often measured in "flops" or floating-point operations per second (see NUMERIC DATA), or for modern processors, megaflops. While this measurement is often touted in product literature, savvy users look to more reliable benchmarks that re-create actual conditions of use, including calculation-intensive, data transfer intensive, or graphics-intensive operations.

DATA COMMUNICATIONS AND NETWORKING

The speed at which data can be transferred over a modem or network connection is measured in bits per second (BPS). A related term, *baud,* was used (somewhat inaccurately) with the earlier modems (see BANDWIDTH and MODEM).

TYPICAL DATA TRANSFER SPEEDS IN BITS PER SECOND (BPS)

Range	Examples
100 Mbps	Fast Ethernet
44.7 Mbps	T-3 data line
10–16 Mbps	Token ring and Ethernet LANs (see LOCAL AREA NETWORK), newer home phone line networking; 802.11b wireless networks.
1.5 Mbps	T-1 data line, many home wireless networks.
1.0–2.0 Mbps	Older home phone line networking, 802.11 standard wireless data transmission.
0.75–1.5 Mbps	DSL and cable modem (depending on load) (see BROADBAND).
128 Kbps	ISDN
28–56 Kbps	PC modems, depending on type, standard, and line quality.

Further Reading

Rowlett, Russ. "How Many? A Dictionary of Units of Measurement." http://www.unc.edu/~rowlett/units/dictM.html

United States Department of Commerce. National Institute of Standards and Technology. "The International System of Units (SI)." http://physics.nist.gov/Pubs/SP330/sp330.pdf

Ziff Davis Corp. "ZD Webopedia." http://www.webopedia.com/

medical applications of computers

Since health care delivery is a business (indeed, one of the largest sectors of the economy), any hospital, health plan, or independent medical practice involves much the same software as any other large business. This includes accounts receivable and payable, payroll, and supplies inventory. Both general and customized industry software can be used for these functions; however, this article focuses on applications specific to medicine.

MEDICAL INFORMATION SYSTEMS

The management of information specifically related to medical care is sometimes called *medical informatics.* The type of information gathered depends on many factors including the type of institution, ranging from a small doctor's office to a large clinic to a full hospital and the nature and scope of the treatment provided. However, one can make some generalizations.

For outpatients, the required information includes an extensive medical record for each patient, including records of medical tests and their results, prescriptions and their status, and so on. For hospital patients, there are also admissions records, an extensive list of itemized charges, and records that must be maintained for public health or other governmental purposes. Hospitals increasingly use customized, integrated hospital information systems (HIS) that integrate billing, medical records, and pharmacy.

Additional record keeping needs arise from the mechanisms used to pay for health care. Each health payment system, whether government-run (such as Medicare or the Veterans' Administration) or a private health maintenance organization (HMO), has extensive rules and procedures about how each surgery, treatment, test, or medication can be submitted for payment. The software must be able to use recognized classifications systems such as the DSM-IV (Diagnostic and Statistical Manual of Mental Disorders).

CLINICAL INFORMATION MANAGEMENT

The modern hospital generates extensive real-time data about the condition of patients, particularly those in critical or intensive care or undergoing surgery. Many hospitals have bedside or operating room terminals where physicians or nurses can review summaries of data such as vital statistics (blood pressure, heart function, and so on). Data can also be entered or reviewed using handheld computers (see PORTABLE COMPUTERS). The ultimate goal of such systems is to provide as much useful information as possible without overwhelming medical personnel with data entry and related tasks that might detract from patient care.

In 2001, a new group, the Patient Safety Institute, was formed in an attempt to create a nationwide standardized format for electronic patient records. This would make it possible for emergency personnel to download a patient's record into a handheld computer and access potentially life-saving information such as medications and allergies.

There has also been some progress in medical decision support systems. Going beyond data summarization, such systems can analyze changes or trends in medical data and highlight those of clinical significance. Such systems can also aid in the compilation of medical charts or possibly compile portions of the chart automatically for later review.

DIAGNOSTIC AND TREATMENT SYSTEMS

The diagnosis and treatment of many conditions has been profoundly enhanced by the use of computer-assisted medical instruments. At the beginning of the last century the use of X rays revolutionized the imaging of the anatomy of living things. X rays, however, were limited in detail and depth of imaging. Techniques of tomography, involving synchronized movement of the X-ray tube and film, were then developed to create a sharp focus deeper within the target structure. The development of computerized tomography (CT or CAT) scanning in the 1970s used a different and more effective approach: A beam of X rays is swept through the target area while computerized radiation detectors precisely calculate the absorption of radiation, and thus the density of the tissue or other structure at each point. This results in a highly detailed image that can be viewed as a series of layers or combined into a three-dimensional holographic display.

Another widely used imaging technique is positron emission tomography (PET) scanning, which tracks the radiation emission from a short-lived radioisotope injected into the patient. It is particularly helpful for studying the flow of blood or gas and other physiological or metabolic changes. Magnetic resonance imaging (MRI) uses the absorption and re-emission of radio waves in a strong magnetic field to identify the characteristic signature of the hydrogen nucleus (i.e. a proton) in water within the body, and thus delineate the surrounding structures.

Besides controlling the scanning process (especially in CAT scanning), the computer is essential for the creation and manipulation of the resulting images. A typical image processing (IP) system is actually an array of many individual processors that perform calculations and comparisons on parts of the image to enhance contrast and extract information that can lead to a more precise depiction of the area of interest. The resulting images (consisting of an array of pixels and associated information) can be further enhanced in a variety of ways using video processing software. Other software using pattern recognition techniques can be programmed to look for tumors or other anomalous structures (see IMAGE PROCESSING).

TRENDS

Medical informatics is likely to be a strong growth area in coming decades. As the population ages, demand for medical care will increase. At the same time, there will be growing pressure to control costs. Although technology is expensive, there is a general belief that information can be leveraged to provide more cost-effective treatment and management of health care delivery.

Medical systems are likely to become more integrated. There have been proposals to create permanent, extensive electronic medical records that patients might even "wear" in the form of a small implanted chip. However, concern about the consequences of violation of privacy and misuse of medical information (such as by employers or insurers) raises significant challenges (see also PRIVACY IN THE DIGITAL AGE).

There are many exciting possibilities for computer-assisted medical treatment. It may eventually be possible to provide all the detail of a CAT scan or MRI while a medical procedure is being performed. At any rate, surgeons will be able to see ever more clearly what they are doing, and robot-controlled surgical instruments (such as lasers) are already operating with a precision that cannot be matched by human hands. Such instrumentation also allows for the possibility that skilled surgeons might be able to operate through telepresence, bringing lifesaving surgery to remote areas (see TELEPRESENCE).

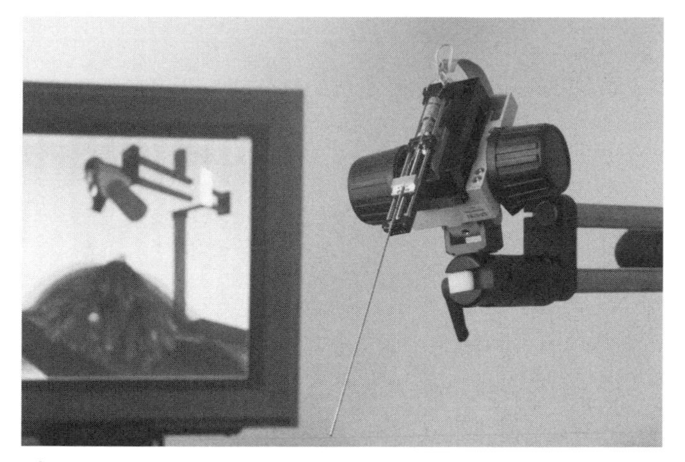

This NASA project is developing a "smart" probe that could provide instant analysis of breast tumors in order to guide surgeons in their work. Such instruments could make surgery more accurate and effective, and reduce unnecessary operations. (NASA PHOTO)

Further Reading

American Medical Informatics Association. "Journal of the American Medical Informatics Association."http://www.jamia.org/

La Tella, Ralph R. "The 2000 Guide to Handheld and Palmtop Computing Resources for Health Care Professionals." http://www.medicalsoftwareforpdas.com/

"Medical Computing Today." http://www.medicalcomputingtoday.com/

Shortliffe, Edward H., ed. [and others]. *Medical Informatics: Computer Applications in Health Care and Biomedicine.* 2nd ed. New York: Springer-Verlag, 2000.

memory

Generally speaking, memory is a facility for temporarily storing program instructions or data during the course of processing. In modern computers the main memory is random access memory (RAM) consisting of silicon chips. Today's personal computers typically have from between 64MB (megabytes) and 512MB of main memory.

DEVELOPMENT OF THE TECHNOLOGY

In early calculators "memory" was stored as the positions of various dials. Charles Babbage conceived of a "store" of such dials that could hold constants or other values needed during processing by his Analytical Engine (see BABBAGE, CHARLES).

A number of forms of memory were used in early electronic digital computers. For example, a circuit with an inherent delay could be used to store a series of pulses that could be "refreshed" every fraction of a second to maintain the data values. The Univac I, for example, used a mercury delay line memory. Researchers also experimented with cathode ray tubes (CRTs) to store data patterns.

The most practical early form of memory was the ferrite core, which consisted of an array of tiny donut-shaped magnets, crisscrossed by electrical lines so that any element can be addressed by row and column number. By converting data into appropriate voltage levels, the magnetic state of the individual elements can be switched on and off to represent 1 or 0. In turn, a current can be passed through any element to read its current state—although the element must then be remagnetized. Ferrite cores were relatively fast but expensive, and "core" became programmers' shorthand for the amount of precious memory available.

By the 1960s, the use of transistors and integrated circuits made electronic solid-state memory systems possible. Since then, the MOSFET (Metal Oxide Semiconductor Field Effect Transistor) using CMOS (Complementary Metal Oxide) fabrication has been the dominant way to implement DRAM (dynamic random access memory). Here "dynamic" means that the memory must be "refreshed" by applying current after data is read in each cycle, and "random access" means that any desired memory location can be accessed directly rather than requiring locations to be read sequentially.

Thanks to VLSI (Very Large Scale Integration), millions of individual transistors can be fabricated onto a single chip. As of 2001, 64 and 128MB memory modules are available for about $100 or less, so today's computers now often have more main memory capacity than the previous generation had hard disk capacity.

Static RAM is used in some computer components where maximum memory speed is desirable. Static memory is faster because it does not need to be refreshed after each reading cycle. However, it is also considerably more expensive.

Memory performance is also dependent on how quickly locations in the memory can be addressed. The earliest forms of DRAM required that the row and column of the desired memory location be sent in separate cycles. EDO (Extended Data Out) and more recent technologies allow the row to be requested one time, and then just the column given for adjacent or nearby locations. Timing and pipelining techniques can also be used to start a new request while the previous one is still being processed.

Memory speed is limited by the inherent response time of the memory chip, but also by the number of clock cycles per second initiated by the data bus (see BUS). In the late 1990s, two new variants of dynamic RAM offered the ability to use faster bus speeds. Synchronous DRAM (SDRAM) allows bus speeds up to about 100 MHz, while Rambus DRAM (RDRAM) may be able to support speeds of up to 800 MHz.

In actual systems, a small amount of faster memory (see CACHE) is used to hold the data that is most likely to be immediately needed. A proper balance between primary and secondary cache and main memory in the system chipset makes it less necessary to use the fastest, most expensive form of main memory.

Many computers also have ROM (Read-Only Memory) or PROM (Programmable Read-Only Memory). This memory holds permanent system settings and data (see BIOS) that are needed during the startup process (see BOOT SEQUENCE).

Further Reading

Keeth, Brent, and R. J. Baker. *DRAM Circuit Design: A Tutorial.* New York: IEEE Press, 2000.

"PC Guide: DRAM Technologies." http://www.pcguide.com/ref/ram/tech_SDRAM.htm

Prince, Betty. *High Performance Memories: DRAMs and SRAMs: Evolution and Function.* rev. ed. New York: Wiley, 1999.

memory management

Whatever memory chips or other devices are installed in a computer, the operating system and application programs must have a way to allocate, use, and eventually release portions of memory. The goal of memory management is to use available memory most efficiently. This can be difficult in modern operating environments where dozens of programs may be competing for memory resources.

Early computers were generally able to run only one program at a time. These machines didn't have a true operating system, just a small loader program that loaded the application program, which essentially took over control of the machine and accessed and manipulated the memory. Later systems offered the ability to break main memory into several fixed partitions. While this allowed more than one program to run at the same time, it wasn't very flexible.

VIRTUAL MEMORY

From the very start, computer designers knew that main memory (RAM) is fast but relatively expensive, while secondary forms of storage (such as hard disks) are slower but relatively cheap. Virtual memory is a way to treat such auxiliary devices (usually hard drives) as though they were part of main memory. The operating system allocates some storage space (often called a swapfile) on the disk. When programs allocate more memory than is available in RAM, some of the space on the disk is used instead. Because RAM and disk are treated as part of the same address space (see ADDRESSING), the application requesting memory doesn't "know" that it is not getting "real" memory. Accessing the disk is much slower than accessing main memory, so programs using this secondary memory will run more slowly.

Virtual memory has been a practical solution since the 1960s, and it has been used extensively on PCs running operating systems such as Microsoft Windows. However, with prices of RAM falling drastically in the new century, there is likely to be enough main memory on the latest systems available to run most popular applications.

MEMORY ALLOCATION

Most programs request memory as needed rather than a fixed amount being allocated as part of program compilation. (After all, it would be inefficient for a program to try to guess how much memory it would need, and possibly tie up memory that could be used more efficiently by other programs.) The operating system is therefore faced with the task of matching the available memory with the amounts being requested as programs run.

One simple algorithm for memory allocation is called *first fit*. When a program requests memory, the operating system looks down its list of available memory blocks and allocates memory from the first one that's large enough to fulfill the request. (If there is memory left over in the block after allocation, it becomes a new block that is added to the list of free memory blocks.)

As a result of repeated allocations using this method, the memory space tends to become fragmented into many leftover small blocks of memory. As with fragmentation of files on a disk, memory fragmentation slows down access, since the hardware (see MEMORY) must issue repeated instructions to "jump" to different parts of the memory space.

Using alternative memory allocation algorithms can reduce fragmentation. For example, the operating system can look through the entire list (see also HEAP) and find the *smallest* block that is still large enough to fulfill the allocation request. This *best fit* algorithm can be efficient. While it still creates fragments from the small leftover pieces, the fragments usually don't amount to a significant portion of the overall memory.

The operating system can also enforce standard block sizes, keeping a "stockpile" of free blocks of each permitted size. When a request comes in, it is rounded to the nearest amount that can be made from a combination of the standard sizes (much like making change). This approach, sometimes called the *buddy system,* means that programs may receive somewhat more or less memory than they want, but this is usually not a problem.

Memory Management

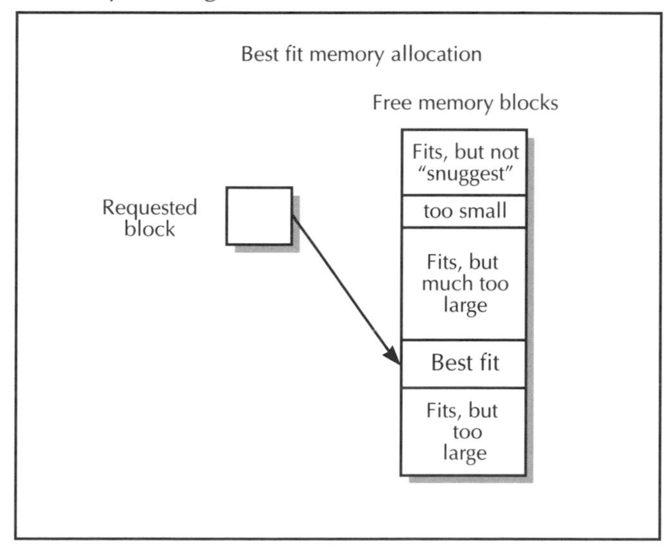

The "best fit" method of memory management tries to find the smallest available block that is large enough to contain the requested amount of memory.

RECYCLING MEMORY

In a multitasking operating system, programs should release memory when it is no longer needed. In some programming environments memory is released automatically when a data object is no longer valid—see VARIABLE) while in other cases memory may need to be explicitly freed by calling the appropriate function.

Recycling is the process of recovering these freed-up memory blocks so they are available for reallocation. To reduce fragmentation, some operating systems analyze the free memory list and combine adjacent blocks into a single, larger block (this is called *coalescence*). Operating systems that use fixed memory block sizes can do this more quickly because they can use constants to calculate where blocks begin and end.

Many more sophisticated algorithms can be used to improve the speed or efficiency of memory management. For example, the operating system may be able to receive information (explicit or implicit) that helps it determine whether the requested memory needs to be accessed extremely quickly. In turn, the memory management system may be designed to take advantage of particular processor architecture. Combining these sources of knowledge, the memory manager might decide that a particular requested memory block be allocated from special high speed memory (see CACHE).

While RAM is now cheap and available in relatively large quantities even on desktop PCs, the never-ending race between hardware resources and the demands of ever larger database and other applications guarantees that memory management will remain a concern of operating system designers. In particular, distributed database systems where data objects can reside on many different machines in the network require sophisticated algorithms that take not only memory speed but also network load and speed into consideration.

Further Reading

Jones, Richard. *Garbage Collection: Algorithms for Automatic Dynamic Memory Management.* New York: Wiley, 1996.

"The Memory Management Reference Beginner's Guide: Overview." http://www.memorymanagement.org/articles/begin.html

message passing

In the early days of computing, a single program usually executed sequentially, with interruptions for calls to various procedures or functions that would perform data processing tasks and then return control to the main program (see PROCEDURES AND FUNCTIONS). However, by the 1970s UNIX and other operating systems had introduced the capability of running several programs at the same time (see MULTITASKING). Additionally, it became common to create a large program that would manage data and smaller programs that could link users to that service (see CLIENT-SERVER COMPUTING). Further, programs themselves began to be organized in a new way (see OBJECT-ORIENTED PROGRAMMING). A program now consisted of a number of entities (objects) representing data and methods (things that can be done with the data).

Thus, both at the operating system and application level it became necessary to have various objects communicate with one another. For example, a client program requests a service from the server. The server performs the required service and reports its completion. The mechanism by which information can be sent from one program to another (or between objects in a program) is called *message passing.*

In one message-passing scheme, two objects (such as client and server) agree on a standard memory location called a *port*. Each program checks the port regularly to see if a message (containing instructions, data, or an address where data can be found) is pending. In turn, outgoing messages can be left at the port.

The client-server idea can be found within operating systems as well. For example, there can be a component devoted to providing file-related services, such as opening or reading a file (see FILE). An application that wants to open a file leaves an appropriate message to the operating system. The operating system has a message dispatcher that examines incoming messages and routes them to the correct component (the file system manager in this case).

Within an object-oriented program, an object is sent a message by invoking one of its *methods* (Smalltalk and other languages) or *member functions* (C++ or Java). For example, suppose there's an object call Speaker that represents the system's internal speaker. As part of a user alert procedure, there might be a call to

```
Speaker.Beep (500)
```

which might be defined to mean "sound a beep for 500 milliseconds."

There are a number of issues involved in setting up message-passing systems. For example, it is convenient for many programs or objects to use the same port or other facility for leaving and retrieving messages, but that means the operating system must spend additional time routing or dispatching the messages. On the other hand, if two objects create a *bound* port, no others can use it, so each can assume that any message left there is from the other object.

During the 1992–1994 period, a standard called MPI (Message Passing Interface) was established by a group of more than 40 industry organizations. It has since been superseded by MPI-2.

Further Reading

Gropp, William, Ewing Lusk, and Rajeev Thakur. *Using MPI: Portable Parallel Programming With the Message-Passing Interface.* 2nd ed. Cambridge, Mass.: MIT Press, 1999.

"The Message Passing Interface (MPI) Standard." http://www-unix.mcs.anl.gov/mpi/

Petzold, Charles. *Programming Windows: the Definitive Guide to the Win32 API.* 5th ed. Redmond, Wash.: Microsoft Press, 1998.

microprocessor

A microprocessor is an integrated circuit chip that contains all of the essential components for the central processing unit (CPU) of a microcomputer system such as a personal computer.

Microprocessor development began in the 1960s when a new company called Intel was given a contract to develop chips for programmable calculators for a Japanese firm. Marcian E. "Ted" Hoff headed the project. He decided that rather than hard-wiring most of the calculator logic into the chips, he would create a general-purpose chip that could read instructions and data, perform basic arithmetic and logical functions, and transfer data between memory and internal locations called registers.

The resulting *microprocessor,* when combined with some RAM (random access memory), some preprogrammed ROM (Read Only Memory), and an input/output (I/O) chip constituted a tiny but complete CPU, soon dubbed "a computer on a chip." This first microprocessor, the Intel 4004, had only a few thousand transistors, could handle data only 4 bits at a time, and ran at 740 KHz (about one three-thousandth the speed of the latest Pentium IV chips).

Microprocessor

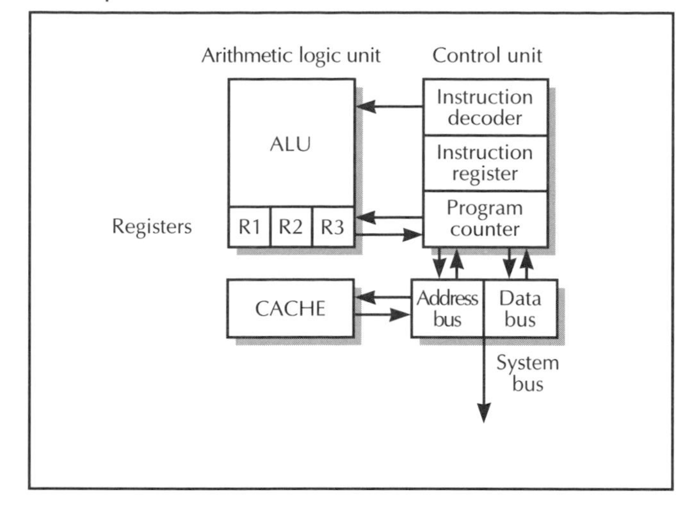

Schematic of a simple microprocessor. The control unit is responsible for fetching and decoding instructions, as well as fetching or writing data to memory. The Arithmetic Logic Unit (ALU) does the actual computing (including arithmetic and logical comparisons). The registers hold data being currently used by the ALU, while the cache contains instructions that have been prefetched because they are likely to be needed soon.

Intel gradually refined the chip, giving it the logic circuits to enable it to perform additional instructions, more internal stack and register space, and 8 KB of space to store programs. The 8008 could handle 8 bits of data at a time, while the 8080 became the first microprocessor that was capable of serving as the CPU for a practical microcomputer system. Its descendants, the 8088 and 8086 (16-bit) powered industry-standard IBM-compatible PCs. Meanwhile, other companies such as Motorola (68000), Zilog (Z-80), and MOS Technology (6502) powered competing PCs from Apple, Atari, Commodore, and others.

With the dominance of the IBM PC and its clones (see IBM PC), the Intel 80 x 86 series in turn dominated the microprocessor market. (The *x* refers to successive digits, as in 80286, 80386, and 80486.) At the next level this nomenclature was replaced by the Pentium series, which is up to the Pentium 4 as of 2002.

According to a famous dictum called Moore's Law, the density (number of transistors per cubic area) and speed (in terms of clock rate) of microprocessors has roughly doubled every 18 months to two years. Intel expects to be making microprocessors with 1 billion transistors by 2007.

MICROPROCESSOR AND MICROCOMPUTER

A microcomputer is a system consisting of a microprocessor and a number of auxiliary chips. The microprocessor chip serves as the central processing unit (CPU). It contains a clock that regulates the flow of data and instructions (each instruction takes a certain number of clock cycles to execute). There is also an index register that keeps track of the instruction being executed. A small number of locations called *registers* within the CPU allow for storing or retrieving the data being used by instructions much more quickly than retrieval from main memory (RAM).

Typically, the instruction register advances to the next instruction. The instruction is fetched, decoded, and sent to the CPU's ALU (arithmetic logic unit) for processing. Data needed to be processed by the instruction are either fetched from a register or, through an *address register,* fetched from RAM. (Some processors store one operand for an arithmetic operation in a special register called the *accumulator.*)

Floating-point operations (those involving numbers that can include decimal points) require special registers that can keep track of the decimal position. Until the mid-1990s, many systems used a separate microprocessor called a *coprocessor* to handle floating point operations. However, later chips such as the Pentium series integrate floating point operations into the main chip.

In order to function as the heart of a microcomputer, the CPU must communicate with a variety of other devices by interacting with special controller chips. For example, there is a bus interface chip (see BUS) that decodes memory addresses and routes requests to the appropriate devices on the motherboard. When data is requested from memory, a memory controller must physically fetch the data from RAM (see MEMORY). There is also

The MITS Altair (1975) was the first microcomputer available commercially. It was usually purchased in kit form. While the Altair didn't have much processing capacity, it aroused great interest and inspired other computer builders such as Apple's Steve Wozniak and Steve Jobs. (PHOTO COURTESY COMPUTER MUSEUM HISTORY CENTER)

a cache controller that interfaces with one or two levels of high-speed cache memory (see CACHE). The algorithms implemented in the cache controller aim to have the next instructions and the most-likely needed data already in the cache when the CPU requests them.

Other devices such as disk drives, modems, printers, and video cards are all connected to the CPU through input/output (I/O) interfaces that connect to the system bus. Most of the devices connected to the bus have their own microprocessors. Software (see DEVICE DRIVER) translates high-level programming instructions (such as to open a file) to the appropriate device commands.

The CPU and many other devices also contain ROM (read only memory) chips that have permanent basic instructions stored on them (see BIOS). This enables the CPU and other devices to perform the necessary actions to enter into communication when the system starts up (see BOOT SEQUENCE).

NEW FEATURES EMERGE

Improvements in microprocessors during the 1980s included wider data paths and the ability to address a larger amount of memory. For example, the Intel 80386 was the first 32-bit processor for PCs and could address 4GB of memory. (Earlier processors such as the 80286 had to divide memory into segments or use paging to swap memory in and out of a smaller space to make it look like a larger one.) Over the years microprocessors tended to add more built-in cache memory, enabling them to have more instructions or data ready for immediate use.

Another way to get more performance out of a microprocessor is to increase the speed with which instructions can be executed. One technique, called *pipelining*, breaks the processor into a series of segments, each of which can execute a particular operation. Instead of waiting until an instruction has been completely executed and then turning to the next one, a pipelined microprocessor moves the instruction from segment to segment as its operations are executed, with following instructions moving into the vacated segments. As a result, two or more instructions can be undergoing execution at the same time.

In addition to pipelining, the Pentium series and other recent chips can have instructions executing simultaneously using different arithmetic logic units (ALUs) or floating-point units (FPUs).

Another way to improve instruction processing is to use a simpler set of instructions. First introduced during the 1980s for minicomputers and high-end workstations (such as the Sun SPARC series), reduced instruction set computer (RISC) chips have smaller, more uniform instructions that can be more easily pipelined, as well as many registers for holding the results of the intermediate processing. During the 1990s, RISC concepts were also adopted in PC processor designs such as the 80486 and Pentium (see REDUCED INSTRUCTION SET COMPUTER).

In the new century, it is unclear when physical limitations will eventually slow down the tremendous rate of increase in microprocessor power. As the chips get denser and smaller, more heat is generated with less surface through which it can be removed. At still greater densities, quantum effects may also begin to be a problem. On the other hand, new technologies might take the elements of the processor down to a still smaller level (see MOLECULAR COMPUTING and NANOTECHNOLOGY).

While the stand-alone desktop, laptop, or handheld computer is the most visible manifestation of the microprocessing revolution, there are probably hundreds of "invisible" microprocessors in use for every visible computer. Today microprocessors help monitor and control everything from home appliances to cars to medical devices (see EMBEDDED SYSTEMS).

Further Reading

"Intel Microprocessor Quick Reference Guide." http://www.intc.com/pressroom/kits/quickref.htm

Jackson, Tim. *Inside Intel: Andy Grove and the Rise of the World's Most Powerful Chip Company.* New York: Dutton, 1997.

Shanley, Tom. *Pentium Pro and Pentium II System Architecture.* 2nd ed. Reading, Mass.: Addison-Wesley, 1998.

Shriver, Bruce D. *The Anatomy of a High-Performance Microprocessor: A Systems Perspective.* Los Alamitos, Calif.: IEEE Computer Society, 1998.

Microsoft Windows

Often simply called Windows, Microsoft Windows refers to a family of operating systems now used on the majority of personal computers. Windows PCs run Intel or Intel-compatible microprocessors and use IBM-compatible hardware architecture.

HISTORY AND DEVELOPMENT

By 1984, the IBM PC and its first "clones" from other manufacturers dominated the market for personal computers, quickly overtaking the previously successful Apple II and various machines running the CP/M operating system. Through a combination of initiative and luck, Microsoft CEO Bill Gates had licensed what became its MS-DOS operating system to IBM, while retaining the rights to license it also to the clone manufacturers (see also GATES, WILLIAM).

However, 1984 also brought Apple back into contention with the Macintosh. Using a graphical user interface (GUI) largely based on research done at Xerox's Palo Alto Research Center (PARC) in the 1970s, the Macintosh was strikingly more attractive and user friendly than the all-text, command-line driven MS-DOS. As third parties began to offer GUI alternative to DOS, Microsoft rushed to complete its own GUI, called Windows. Although it was actually announced well before the coming of the Macintosh, Windows 1.0 was not released until

1985. Its poor fonts, graphics, and window operation made it compare unfavorably to the Macintosh. Through the rest of the 1980s, Microsoft struggled to improve Windows. The acceptance of Windows was aided by several large software manufacturers such as Aldus (PageMaker) writing software for the new operating system as well as Microsoft's designing or porting its own software such as the Excel spreadsheet.

Windows 3.0, released in 1990, was considerably improved and began to attract significant numbers of users away from MS-DOS–based programs. Microsoft was also greatly aided by its ability to leverage its operating system dominance to make it economically imperative for PC manufacturers to "bundle" Windows with new PCs.

About the same time, Microsoft had been working with IBM on a system called OS/2. Unlike Windows, which was actually a program running "on top of" MS-DOS, OS/2 was a true operating system that had sophisticated capabilities such as multitasking, multithreading, and memory protection. Microsoft eventually broke off its relationship with IBM, abandoned OS/2, but incorporated some of the same features into a new version of Windows called NT (New Technology), first released in 1993. NT, which progressed through several versions, was targeted at the high-end server market, while the consumer version of Windows continued to evolve incrementally as Windows 95 and Windows 98 (released in those respective years). These versions included improved support for networking (including TCP/IP, the Internet standard) and a feature called "PLUG AND PLAY" that allowed automatic installation of drivers for new hardware.

Toward the end of the century, Microsoft began to merge the consumer and server versions of Windows. Windows 2000 incorporated some NT features and provided somewhat greater security and stability for

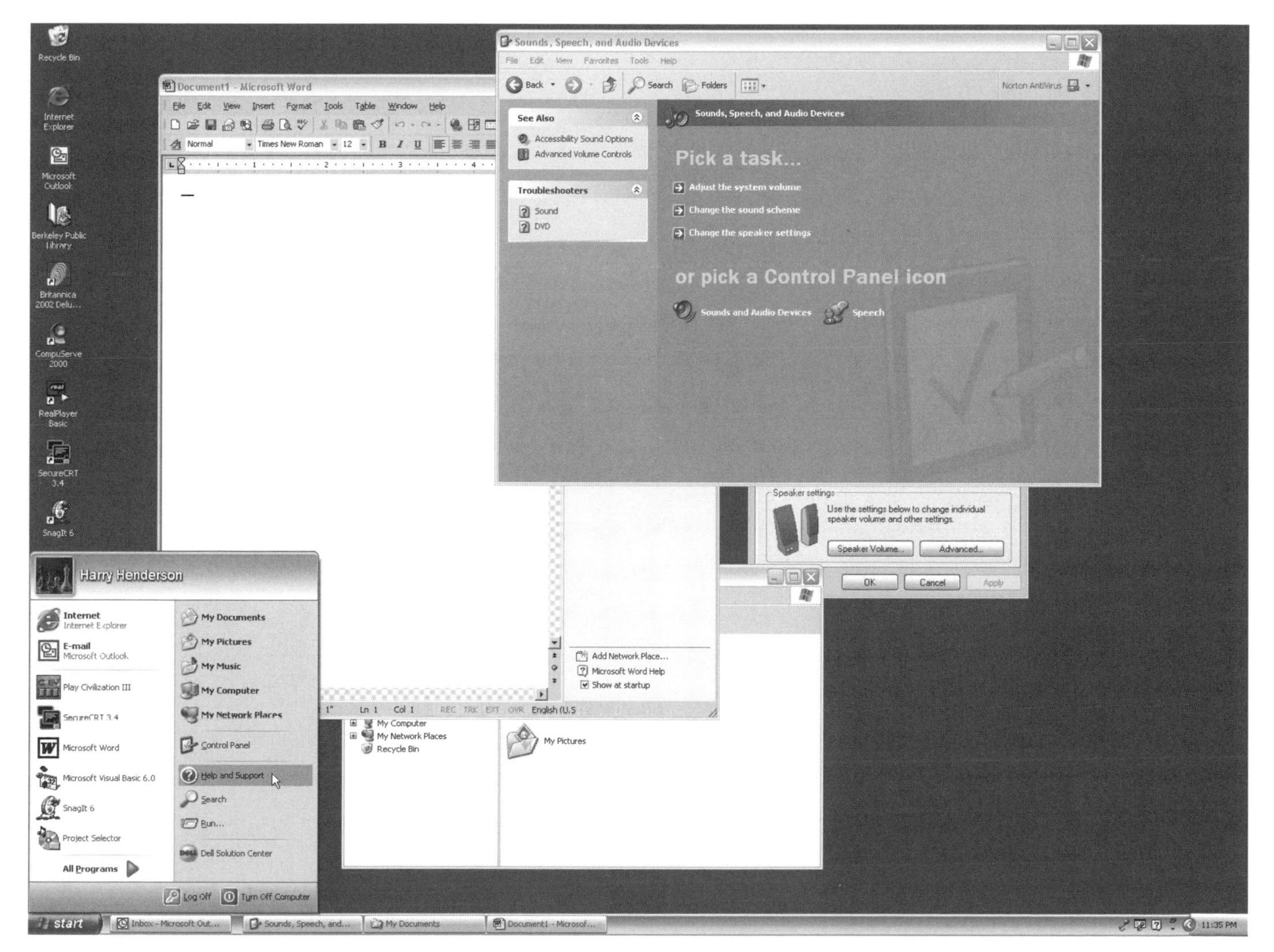

Microsoft Windows XP, the 2001 incarnation of the operating system, represents a unification of the consumer and "professional" Windows versions. The screen shows the Start menu, Microsoft Word, and one of the Control Panel windows used to make system settings.

consumers. Finally, with Windows XP, released in 2001, the separate consumer and NT versions of Windows have disappeared entirely, to be replaced by home and "professional" versions of XP.

USER'S PERSPECTIVE

From the user's point of view, Windows is a way to control and view what is going on with the computer. The user interface consists of a standard set of objects (windows, menus, buttons, sliders, and so on) that behave in generally consistent ways. This consistency, while not absolute, reduces the learning curve for mastering a new application. Programs can be run by double-clicking on their icon on the underlying screen (called the desktop), or by means of a set of menus.

Windows users generally manage their files through a component called Windows Explorer or My Computer. Explorer presents a treelike view of folders on the disk. Each folder can contain either files or more folders, which in turn can contain files, perhaps nested several layers deep. Folders and files can be moved from place to place simply by clicking on them with the mouse, moving the mouse pointer to the destination window or folder, and releasing the button (this operation is called dragging). Another useful feature is called a context menu. Accessed by clicking with the right-hand mouse button, the menu brings up a list of operations that can be done with the currently selected object. For example, a file can be renamed, deleted, or sent to a particular destination.

Windows includes a number of features designed to make it easier for users to control their PC. Most settings can be specified through windows called *dialog boxes*, which include buttons, check boxes, or other controls. Most programs also use Windows's Help facility to present help pages using a standard format where related topics can be clicked. Most programs are installed or uninstalled using a standard "wizard" (step-by-step procedure), and wizards are also used by many programs to help beginners carry out more complex tasks (see HELP SYSTEMS).

MULTITASKING

From the programmer's point of view, Windows is a multitasked, event-driven environment (see MULTITASKING). Programmers must take multitasking into account in recognizing that certain activities (such as I/O) and resources (such as memory) may vary with the overall load on the system. Responsible programs allocate no more memory than they need, and release memory as soon as it is no longer needed. If the pool of free memory becomes too low, Windows starts swapping the least recently used segments of memory to the hard drive. This scheme, called virtual memory, allows a PC to run more and larger programs than would otherwise be possible, but since accessing the hard drive takes considerably longer than accessing RAM, the system as a whole starts slowing down.

Windows also has a rather small amount of memory reserved for its GDI (Graphics Device Interface), a system used for displaying graphical interface objects such as icons. If this resource pool (which has been made somewhat more flexible in later versions of Windows) runs out, the system can grind to a halt.

PROGRAMMING PERSPECTIVE

Programmers moving to Windows from more traditional systems (such as MS-DOS) must also deal with a new paradigm called event-driven programming. Most traditional programs are driven by an explicit line of execution through the code—do this, make this decision, and depending on it, do that—and so on. Windows programs, however, typically display a variety of menus, buttons, check boxes, and other user controls. They then wait for the user to do something. The user thus has considerable freedom to move about in the program, performing tasks in different orders.

A Windows program, therefore, is driven by *events*. An event is generally some form of user interaction such as clicking on a menu or button, moving a slider, or typing into a text box. The event is conveyed by a message (see MESSAGE PASSING) that Windows dispatches to the affected object. For example, if the user presses down (clicks) the left mouse button while the mouse pointer is over a window, a WM_BUTTONDOWN message is sent to that window.

Each of these interface objects (collectively called *controls*) has a message-handling procedure that identifies the message. The object must then have appropriate program code that responds to each possible type of event. For example, if the user clicks on the File menu and then clicks on Open, the code will display a standard dialog box that allows for selecting the file to be opened.

Fortunately for the programmer, Windows provides developers with a large collection of types of windows, dialog boxes, and controls that can be displayed using a function call. For example, this code (after some preliminary declarations), displays a type of window called a list box:

```
HWND MyWindow;
hMyWindow = CreateWindow("LISTBOX","Available
Services",
  WS_CHILD|WS_VISIBLE,
  0,0,100,200
  hwndParent,NULL,hINst,NULL);
```

Here the various parameters passed to the CreateWindow function specify the type of window, window title, characteristics, and location. The function returns a "window handle," which is a pointer that holds the window's address and allows it to be accessed later.

Most Windows programming environments, including C++ and particularly, Visual Basic, now let program designers avoid having to specify code such as the above to create windows and other objects. Instead, the programmer can click and drag various objects onto a design screen to establish the interface that will be seen by the program's user. The programmer can then use Properties settings to specify many characteristics of the screen objects without having to explicitly program them.

Microsoft and third-party developers also provide ready-made programming code in *dynamic link libraries* (DLLs). These resources (see LIBRARY, PROGRAM) can be called by any application, which can then use any object or function defined in the library. Windows also provides a facility called OLE (Object Linking and Embedding). This lets an application such as a word processor "host" another application such as a spreadsheet. Thus, the Microsoft Word, for example, can embed a Microsoft Excel spreadsheet into a document, and the spreadsheet can be worked with using all the usual Excel commands. In other words, OLE lets applications make their features, controls, and functionality accessible to other applications. Indeed, collections of controls are often packaged as OCX (OLE controls) and sold to developers.

Despite all this available help, Windows presents a steep learning curve for many programmers. There are hundreds of functions for handling interface objects, drawing graphics, managing files, controlling devices, and other tasks. With the growing use of object-oriented programming languages (see OBJECT-ORIENTED PROGRAMMING and C++) in the late 1980s and 1990s, Microsoft devised the Microsoft Foundation Classes (MFC). This framework defines all of the interface objects and other entities (such as data structures) as C++ classes.

Using MFC, a programmer, instead of calling a function to create a window, creates an object of a particular Window class. To customize a window, the programmer can use inheritance to derive a new window class. The various functions for controlling windows are then defined as member functions of the window class. This use of object-oriented, class-based design organizes much of the great hodgepodge of Windows functions into a logical hierarchy of objects and makes it easier to master and to use.

For example, using the traditional Windows API (see APPLICATIONS PROGRAMMING INTERFACE) one puts a text string into a list box using this code:

```
LRESULT LRes;
LRes =
SendMessage(hMyListBox,LB_ADDSTRING,0,"Network
Services");
```

(LRes is a number that will hold a code that says whether the item was successfully added)

Using MFC, this code can be rewritten as:

```
CListBox * pListBox;
int nRes;
nRes = pListBox->AddString ("Network Services");
```

Here a pointer is declared to an object of the ListBox class, and a member function of that class, AddString, is then called. While this code may not look simpler, it uses a consistent object-oriented approach.

TRENDS

By just about any standard Microsoft Windows has achieved remarkable success, capturing and largely holding the lion's share of the PC operating system market. However, Windows has been persistently criticized on grounds of reliability and security. Although later versions of Windows such as NT, 2000, and particularly XP are considerably more stable than early ones, most Windows users who run a variety of taxing applications periodically experience the dreaded "Blue Screen of Death," signifying that the system has crashed and must be rebooted, possibly resulting in loss of data.

Microsoft has included powerful facilities that allow Windows applications to be controlled by other applications or remotely (see also SCRIPTING LANGUAGES). Unfortunately, these facilities have proven to be quite vulnerable to computer viruses that can use them to damage systems connected to the Internet. There seems to be a never-ending race between developers of program "patches" designed to plug security holes and inventive, albeit malicious virus writers.

The Internet has also helped create some alternatives to Windows for PC users. The Java programming language (see JAVA) offers a way to create programs that can run on Windows, Macintosh, UNIX, and other systems via Web browsers. If a significant amount of key software applications are available in multiplatform Java versions, PC users might eventually find it less compelling to use Windows. The popular UNIX variant called Linux (see UNIX) also offers a robust platform that has become especially favored for running Web servers. While Linux has traditionally suffered from a dearth of business-oriented software, a number of office software suites are now available (see also OPEN SOURCE MOVEMENT).

While Windows still remains the dominant PC operating system with tens of thousands of applications and at least several hundred million users around the world, it is likely that the PC operating systems of 2020 will be as different from Windows 2000 as the latter is from the MS-DOS of the early 1980s.

Further Reading

Microsoft Developer Network (MSDN). http://msdn.microsoft.com/

Petzold, Charles. *Programming Windows: The Definitive Guide to the Win32 API.* 5th ed. Redmond, Wash.: Microsoft Press, 1998.

Prosise, Jeff. *Programming Windows with MFC.* 2nd ed. Redmond, Wash.: Microsoft Press, 1999.

Simpson, Alan, and John Grimes. *The Little Windows 98 Book.* Berkeley, Calif.: Peachpit Press, 1998.

"Windows XP Developer Resources." http://windowsxp.devx.com/default.asp

middleware

Often two applications that were originally created for different purposes must later be linked together in order to accomplish a new purpose. For example, a company selling scientific instruments may have a large database of product specifications, perhaps written in COBOL some years ago. The company has now started selling its products on the Internet, using its Web server and e-commerce applications (see E-COMMERCE). Prospective customers of the website need to be able to access detailed information about the products. Unfortunately, the Web software (perhaps written in Java) has no easy way to get information from the company's old product database. Rather than trying to convert the old database to a more modern format (which might take too long or be prohibitively expensive), the company may choose to create a *middleware* application that can mediate between the old and new applications.

There are a variety of types of middleware applications. The simplest and most general type of facility is the RPC (Remote Procedure Call), which allows a program running on a client computer to execute a program running on the server. DCE (Distributed Computing Environment) is a more robust and secure implementation of the RPC concept that provides file-related other operating system services as well as executing remote programs.

More elaborate architectures are used to link complex applications such as databases where a program running on one computer on the network must get data from a server. For example, an Object Request Broker (ORB) is used in a CORBA (Common Object Request Broker Architecture) system to take a data request generated by a user and find servers on the network that are capable of fulfilling the request.

Middleware is often inserted into a program to allow for better monitoring or control of distributed processing. For example a TP (transaction processing) monitor is a middleware program that keeps track of a transaction that may have to go through several stages (such as point of sale entry, credit card processing, and inventory update). The TP monitor can report whether any stage of the transaction processing failed (see TRANSACTION PROCESSING).

Middleware can also be put in charge of *load balancing.* This means distributing transactions so that they are evenly apportioned among the servers on the network, in order to avoid creating delays or bottlenecks.

While use of middleware may not be as "clean" a solution as designing an integrated system from the bottom up, the economic realities of a fast-changing information environment (particularly with regard to deployment on the Web) often makes middleware an adequate second-best choice.

Further Reading

Britton, Chris. *IT Architectures and Middleware: Strategies for Building Large, Integrated Systems.* Reading, Mass.: Addison-Wesley, 2000.

"Middleware." http://www.doit.wisc.edu/services/middleware/

military applications of computers

War has always been one of the most complex of human enterprises. Even leaving actual combat aside, the U.S. military and defense establishment constitute a huge employer, research and training agency, and transportation network. Managing all these activities require sophisticated database, inventory, tracking, and communications systems. When thousands of private defense contractors of varying sizes are considered as part of the system, the complexity and scope of the enterprise become even larger.

Specifically, military information technology applications can be divided into the following broad areas: logistics, training, operations, and battle management.

LOGISTICS

It is often said that colonels worry about tactics while generals preoccupy themselves largely with logistics. Logistics is the management of the warehousing, distribution, and transportation systems that supply military establishments and forces in the field with the equipment and fuel they need to train and to fight. Logistics within the United States is analogous to similar problems for very large corporations. The same bar codes, point of use terminals, and other tracking, inventory, and distribution systems that Amazon.com uses to get books quickly to customers while avoiding excessive inventory are, in principle, applicable to modernizing military logistic systems.

An added dimension emerges when logistical support must be supplied to forces operating in remote countries, possibly in the face of efforts by an enemy to disrupt supply. Such considerations as efficient loading procedures to accommodate limited air transport capacity, prioritization of shipping to provide the most urgently needed items, and transportation security can all come into play. (The military has pioneered the use of retinal scanners and other systems for controlling access to sensitive areas. See BIOMETRICS.)

The need for mobility and compactness makes laptops and even palmtops the form factors of choice. Military or "milspec" versions of computer hardware are generally built with more rugged components and greater resistance to heat, moisture, or dust.

TRAINING

The use of automated systems to provide training goes back at least as far as the World War II era Link trainer, which used automatic controls and hydraulics to place trainee pilots inside a moveable cockpit that could respond to their control inputs. Today computer simulations with sophisticated graphics and control systems can provide highly realistic depictions of flying a helicopter or jet fighter or driving a battle tank. The military has even adapted commercial flight simulators for training purposes. Simulations can also cover Special Forces operations and tactical decision making. Indeed, many real-time simulations (RTS) sold as popular commercial games and avidly played by young people already contain enough realistic detail to be adopted by the military as is. For example, the game Rainbow Six, based on operations in Tom Clancy novels, simulates tactical counterterrorism operations. In turn, the U.S. Army has recently begun to use a simulation game to give young gamers a taste of the military life.

OPERATIONS

Aircraft, ships, and land vehicles used by the military have been fitted with a variety of computerized systems. The "glass cockpit" in aircraft is replacing the increasingly unmanageable maze of dials and switches with information displays that can keep the pilot focused on the most crucial information while making other information readily available. Traditional keyboards and joystick-type controllers can be replaced by touch screens and even by systems that can understand a variety of voice commands (see SPEECH RECOGNITION AND SYNTHESIS). Similar control interfaces can be used in tanks or ships.

Robotics offers a variety of intriguing possibilities for extending the reach of military forces while minimizing casualties. Remote-control robots can be used to clear minefields or perform reconnaissance. (The Predator armed reconnaissance drone was first used successfully in anti-terrorist operations in Afghanistan in 2002.) Armed robots could assault enemy strong points without risking soldiers. The development of autonomous robots that can plan their own missions, select targets, and make other decisions is a longer-term prospect that depends on the application of artificial intelligence in the extremely challenging and chaotic battlefield environment.

BATTLE MANAGEMENT

Battle management is the ability to gather, synthesize, and present crucial information about the environment around the military unit and enable military personnel to make rapid, accurate decisions about threats and the best way to neutralize them.

The earliest example, the SAGE (Strategic Air Ground Environment) computer system, resulted from a massive development effort in the 1950s that strained the capacity of early vacuum tube-based computers to its limit. The purpose of SAGE was to provide an integrated tracking and display system that could give the Strategic Air Command (SAC) complete real-time information about any Soviet nuclear bomber strikes in progress against the continental United States Descendents of this system were able to track ballistic missiles.

The Aegis system first deployed aboard selected navy ships in the 1970s is a good example of a tactical battle management system on a somewhat smaller scale. Aegis is a computerized system that can integrate information from sophisticated shipboard radar and sonar arrays as well as receiving and merging data from other ships and reconnaissance assets (such as helicopters). The captain of an Aegis cruiser or destroyer therefore has a real-time picture showing the locations, headings, and speeds of friendly and enemy ships, aircraft, and missiles. The system can also automatically distribute the available munitions to most effectively engage the most threatening targets.

Ultimately, the military hopes to give each unit in the field and even individual soldiers a battle management display that would pinpoint enemy vehicles and other activity. Unpiloted drone aircraft such as the Predator can loiter over the battlefield and feed video and other data into the battle management system.

While the ability to transmit and process large amounts of information can lead to strategic or tactical advantage, it also demands increased attention to security. If an enemy can jam the information processing system, its advantages could be lost at a crucial moment. Worse, if an enemy can "spoof" the system or introduce deceptive data, the military's information system could become a weapon in the enemy's hands (see COMPUTER CRIME AND SECURITY and ENCRYPTION).

Further Reading

Campen, Alan D., and Douglas H. Dearth. *Cyberwar 2.0: Myths, Mysteries & Reality.* Fairfax, Va.: Armed Force Communications and Electronics Association, 1998.

Denning, Dorothy D. *Information Warfare and Security.* Reading, Mass.: Addison-Wesley, 1998.

Jackson, Nancy Beth. "Ready for Battle, Wielding a Stylus. Palmtop Makers and Software Writers are Taking Aim at the Military Market." *New York Times,* November 8, 2001, D1.

Tiron, Roxana. "Computers in Combat: Double-Edged Swords." *National Defense,* vol. 85, issue 571, June 2001, 14.

minicomputer

The earliest general-purpose electronic digital computers were necessarily large, room-size devices. In the 1960s, however, the replacement of tubes with transistors (and gradually, integrated circuits) gave designers the choice of either keeping computers large and packing more processing and memory capacity into them, or making smaller computers that still had considerable power. The latter option led to the *minicomputer* as contrasted with the larger mainframe (see MAINFRAME).

Compared to mainframes, minicomputers often handled data in smaller "chunks" (such as 16 bits as compared to 32 or 64) and had a smaller memory capacity. Minicomputers also tended to have more limited input/output (I/O) capacity. However, while large businesses still needed mainframes to handle their large databases and volume of transactions, the minicomputer offered a relatively low cost (tens of thousands of dollars rather than hundreds of thousands), computing facility

Minicomputers such as this DEC PDP-8 brought computing power to many academic and scientific institutions for the first time. They also encouraged a culture of cooperative software development that led to such innovations as the UNIX operating system. (PHOTO COURTESY OF THE PAUL PIERCE COMPUTER COLLECTION)

for scientific laboratories, university computing centers, industrial control, and various specialized needs.

The pioneering and most successful minicomputer company was the Digital Equipment Corporation (DEC). In 1960, DEC introduced its PDP-1, which was followed in 1965 by the quite successful PDP-8, which sold for only $18,000. By the early 1970s, DEC had been joined by competitors such as Data General and the availability of integrated memory circuits (RAM) and microprocessors packed more speed and capacity into each succeeding model.

The minicomputer had several important effects on the development of computer science and the "computer culture" as a whole (see HACKERS AND HACKING). Minicomputers gave university students direct, interactive access to computers through time-sharing, Teletype terminals, or CRT display terminals. Because minicomputers usually lacked the extensive (and expensive) software packages that came with mainframes, university users developed and eagerly swapped software such as program editors and debuggers. This cooperative effort achieved its most striking result in the development of the UNIX operating system.

The reader has probably noticed that this article refers to minicomputers in the past tense. The minicomputer didn't really disappear, but rather was transmogrified. By the late 1980s and certainly the 1990s, the personal desktop computer had taken advantage of more powerful microprocessors and ever more densely packed memory chips to create workstations that rivaled or exceeded the power of established minicomputers. Eventually, the minicomputer as a category virtually disappeared, its functions taken over by machines such as the powerful graphics workstations developed by companies such as Sun Microsystems and Silicon Graphics.

Further Reading

Eckhouse, R., and R. Morris. *Minicomputer Systems: Organization, Programming and Applications (PDP-11)*. Upper Saddle River, N.J.: Prentice Hall, 1979.

Kidder, Tracy. *The Soul of a New Machine*. New York: Modern Library, 1997.

Pearson, Jamie. *Digital at Work: Snapshots from the First Five Years*. Burlington, Mass.: Digital Press, 1992.

Minsky, Marvin Lee

(1927–)
American
Computer Scientist

Starting in the 1950s, Marvin Minsky played a key role in the establishment of ARTIFICIAL INTELLIGENCE (AI) as a discipline. Combining cognitive psychology and computer science, Minsky developed ways to make computers function in "brain-like" ways (see NEURAL NETWORK) and then developed provocative insights about how the human brain might be organized.

Marvin Minsky was born in New York City on August 9, 1927. His father was a medical doctor, and Marvin proved to be a brilliant science student at the Bronx High School of Science and the Phillips Academy. Although he majored in mathematics at Harvard, he also showed a strong interest in biology and psychology. In 1954, he received his Ph.D. in mathematics at Princeton. In 1956, he was a key participant in the seminal Dartmouth conference that established the goals of the new discipline of artificial intelligence.

One of the most important of those goals was to explore the relationship between thinking in the human brain and the operation of computers. Earlier in the century, research into the electrical activities of neurons (the brain's information-processing cells) had led to speculation that the brain functioned something like an intricate telephone switchboard, carrying information through millions of tiny connections. During the 1940s, researchers had begun to experiment with creating electronic circuits that mimicked the activity of neurons.

In 1957, Fran Rosenblatt built a device called a perceptron. It consisted of a network of electronic nodes that can transmit and respond to signals that function much like nerve stimuli in the brain (see NEURAL NETWORK). For example, a perceptron could "recognize" shapes by selectively reinforcing the stimuli from light hitting an array of photocells. In 1969, Minsky and Seymour Papert coauthored a very influential book on the significance and limitations of perceptron research. Their work not only spurred research into neural networks and their possible practical applications, but also proved a strong impetus for the new field of *cognitive psychology*, bridging the study of human mental processes and the insights of computer science (see COGNITIVE SCIENCE).

Meanwhile, Minsky had joined with John McCarthy (see MCCARTHY, JOHN) to found the Artificial Intelligence Laboratory at the Massachusetts Institute of Technology (MIT). In moving from basic perception to the higher order ways in which humans learn, Minsky developed the concept of frames. Frames are a way to categorize knowledge about the world, such as how to plan a trip. Frames can be broken into subframes. For example, the trip-planning frame might have subframes about air transportation, hotel reservations, and packing. Minsky's frames concept became a key to the construction of expert systems that today allow computers to advise on such topics as drilling for oil or medical diagnosis (see EXPERT SYSTEMS and KNOWLEDGE REPRESENTATION). In the 1970s, Minsky and his colleagues at MIT designed robotic systems to test the ability to use frames to accomplish simpler tasks, such as navigating around the furniture in a room.

Minsky believed that the results of research into simulating cognitive behavior had fruitful implications for human psychology. In 1986, Minsky published *The Society of Mind*. This book suggests that the human mind is not a single entity (as classical psychology suggests) or a system with a small number of often-warring subentities (as psychoanalysis asserted). It is more useful, Minsky suggests, to think of the mind as consisting of a multitude of independent agents that deal with different parts of the task of living and interact with one another in complex ways. What we call mind or consciousness, or a sense of self is, therefore, what emerges from this ongoing interaction.

Minsky continues his research at MIT. He has received numerous awards, including the ACM Turing Award (1969) and the International Joint Conference on Artificial Intelligence Research Excellence Award (1991).

Further Reading

Franklin, Stan. *Artificial Minds*. Cambridge, Mass.: MIT Press, 1995.

"Marvin Minsky." [Home page] http://www.media.mit.edu/people/minsky/

Minsky, Marvin. *The Society of Mind*. New York: Simon & Schuster, 1986.

Minsky, Marvin, and Seymour Papert. *Perceptrons: An Introduction to Computational Geometry*. 2nd ed. Cambridge, Mass.: MIT Press, 1988.

modem

As computers proliferated and users experienced an increasing need to exchange data and communicate, it became logical to tap into the telephone system, a communications technology that already linked millions of places around the world.

The problem is that the conventional telephone is an analog rather than digital device. It converts sound (such as speech) into continuously varying electrical signals. Computers, on the other hand, use discrete pulses of on/off (binary) data. However, it proved relatively easy to build a device that could "modulate" the data pulses, imposing them on a sort of carrier wave and thus converting them into electrical signals that could travel along telephone lines. At the other end of the line a corresponding device could "demodulate" that telephone signal, converting it back into data pulses. This "modulator-demodulator" device is known as a *modem* for short.

A modem contains both the modulator and demodulator circuit, with a connection to a cable and a phone jack on one side and a connection to the computer on the other. The computer connection can be provided by connecting to a standard port on the outside of the PC (see SERIAL PORT) or by mounting the modem on a card that slides into the PC's internal bus (see BUS) and connects to the outside phone line through a jack. The modem must also have a component that generates the dialing pulses needed to establish a phone connection.

The first modems for PCs appeared in the early 1980s and were very slow by modern standards, transmitting data at 300 bps (bits per second). However, speed steadily improved, reaching 1,200, 2,400, 9,600 and so on up to 56,000, which is about the maximum practical speed for this technology over ordinary phone lines.

Phone lines are far from hermetically sealed, and random fluctuations called "line noise" can sometimes be misinterpreted by the modem as part of the data signal, leading to errors. However, modern modems include sophisticated error-correcting protocols (see ERROR CORRECTION) and can automatically negotiate with each other to reduce data transmission speed over noisy lines. Data compression techniques also make it possible to have an effectively greater transmission speed by packing more information into less data. In the 1990s, there were some problems caused by competing standards, but today most modems meet the International Telecommunications Union (ITU) v.90 standard for 56 kbps transmission. The modem is now a reliable, stable commodity included as standard equipment in most new PCs.

Modems have met with increasing competition as a means to connect homes to the Internet. Data can be transmitted over video cable or special phone lines (such as DSL or ADSL) at 20–30 times faster than for a modem on an ordinary phone line (see BROADBAND). However, besides being two to three more times expensive than typical dial-up services, broadband technologies tend to be concentrated in urban areas. Nevertheless, the versatile modem is becoming a secondary means of data communication for many users.

Further Reading

Bingham, John. *The Theory and Practice of Modem Design.* New York: Wiley, 1988.
"Modems Rosenet." [Resources] http://modems.rosenet.net/or/

molecular computing

While the electronic digital computer is by far the most prevalent type of calculating device in use today, it is also possible to build computational devices that exploit natural laws and processes to solve problems (see ANALOG COMPUTER). One of the most intriguing approaches is based upon chemistry and biology rather than electronics.

Consider that all living things possess a detailed "database system" of coded information, namely, the DNA sequences that define their genetic code. DNA consists of strands composed of four bases: adenine (A), cytosine (C), guanine (G), and thymine (T). There are a variety of ways in which biologists can "sequence" a strand of DNA, that is, determine the order of bases in it. It is also relatively easy to make many copies of a given chain by using the polymerase chain reaction (PCR) technique.

Molecular Computing

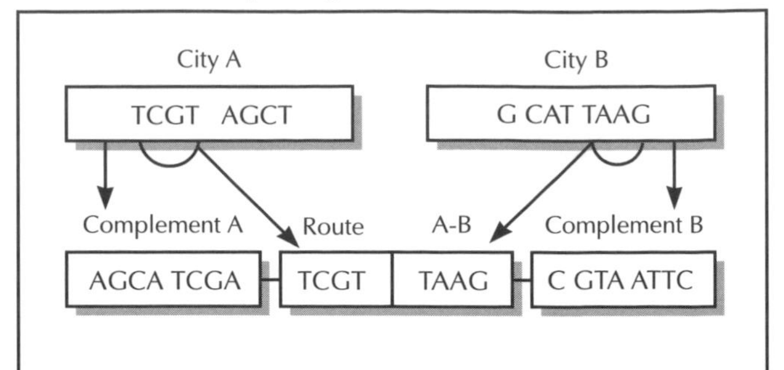

Molecular computing takes advantage of the properties of molecules such as DNA to create what is in effect a massive array of parallel processors. In this example, DNA strands can be coded to represent cities and possible routes between them so that they will chemically solve the Traveling Salesman Problem.

This stockpile of coded DNA strands can be used to solve combinatorial problems. This type of problem becomes exponentially harder to solve through "brute force" computation as the number of elements increases. An example is the famous "Traveling Salesman Problem." Here the goal is to determine a route that visits all of a list of cities while visiting each city only once.

As Leonard Edelman pointed out in his 1994 article in *Science,* a DNA-based approach to the traveling salesman problem begins by assigning two sets of four bases to each city. Next, a similar DNA combination is assigned to each available direct route between two cities, using half (four bases) of the sequence assigned to the respective cities. That is, if one city is coded TCGTAGCT and another city is coded GCATTAAG, then a route from the first city to the second would be coded TCGTTAAG.

When binding one DNA strand to another, T always binds with A, and C always binds with G. Therefore a "complement" can be defined that will bind with a given DNA string. For example, the complement of TCGTAGCT would be AGCATCGA.

Next, the strands representing the complements for the cities are mixed with the ones representing routes. If a city complement runs into a route containing that city, they bind together. The other end (representing the other end of the route) might then encounter another route strand, thus extending the route to a third city and so on, until there are strands representing potentially complete routes to all the cities. After the mixing and combining is completed, separation and sequencing techniques can be used to find the shortest strand that includes all the cities. This represents the solution to the problem.

The attractiveness of molecular computing lies in its being "massively parallel" (see MULTIPROCESSING). Although molecular operations are individually much slower than electronics, DNA strands can be replicated and assembled in great numbers, potentially allowing them to go through quintillions (10^{18}) of combinations at the same time. In 1996, Dan Boneh designed an approach using DNA combinations that could be used to break the Data Encryption Standard (DES) encryption scheme by testing huge numbers of keys simultaneously.

Although this application suggests the potential power in molecular computing, the approach has significant drawbacks. There are many ways that damage can occur to DNA strands during combination and processing, leading to errors. Even for the combinatorial problems that are molecular computing's strong suit, conventional electronic computers using large arrays of parallel processors are able to offer comparable power and a much easier interface. However, molecular computing illustrates the rich way in which information and information processing are embedded in nature and the potential for harnessing it for practical applications.

Further Reading

Adelman, Leonard. "Molecular Computation of Solutions to Combinatorial Problems." *Science,* 266, Nov. 11, 1994, 1021–1024.

Boneh, Dan, Richard Lipton, and Chris Dunworth. "Breaking DES Using a Molecular Computer." in *Proceedings of DIMACS Workshop on DNA Computing,* Providence, RI: American Mathematical Society, 1995.

Calude, Cristian S. *Computing with Cells and Atoms: an Introduction to Quantum, DNA, and Membrane Computing.* New York: Taylor & Francis, 2000.

"World Wide DNA Computing, Computational Biology, and Molecular Computing." http://www.cis.udel.edu/~dna3/DNA/dnacomp.html

monitor

As designers strove to make computers more interactive and user-friendly, the advantages of the cathode ray tube (CRT) already used in television became clear. Not only could text be displayed without wasting time and resources on printing but the individually addressable dots (pixels) could be used to create graphics. While such displays were used occasionally in defense and research systems in the 1950s, the first widespread use of CRT video monitors came with the new generation of smaller computers developed in the 1960s (see MINICOMPUTER). Since such computers were often used for scientific, engineering, industrial control, and other real-time applications, the combination of video display and keyboard (i.e. a Video Display Terminal, or VDT) was a much more practical way for users to oversee the activities of such systems. (This oversight function also led to the term *monitor.*)

A monitor can be thought of as a television set that receives a converted digital signal rather than regular TV programming. To send an image to the screen, the PC first assembles it in a memory area called a video buffer (modern video cards can store up to 64 MB of complex graphics data. See COMPUTER GRAPHICS). Ultimately, the graphics are stored as an array of memory locations that represent the colors of the individual screen dots, or pixels. The video card then sends this data through a digital to analog converter (DAC), which converts the data to a series of voltage levels that are fed to the monitor.

The monitor has electron "guns" that are aimed according to these voltages. (A monochrome monitor has only one gun, while a color monitor, like a color TV, has separate guns for red, blue, and green. The electrons from the guns pass through a lattice called a shadow mask, which keeps the beams properly separated and

Monitor

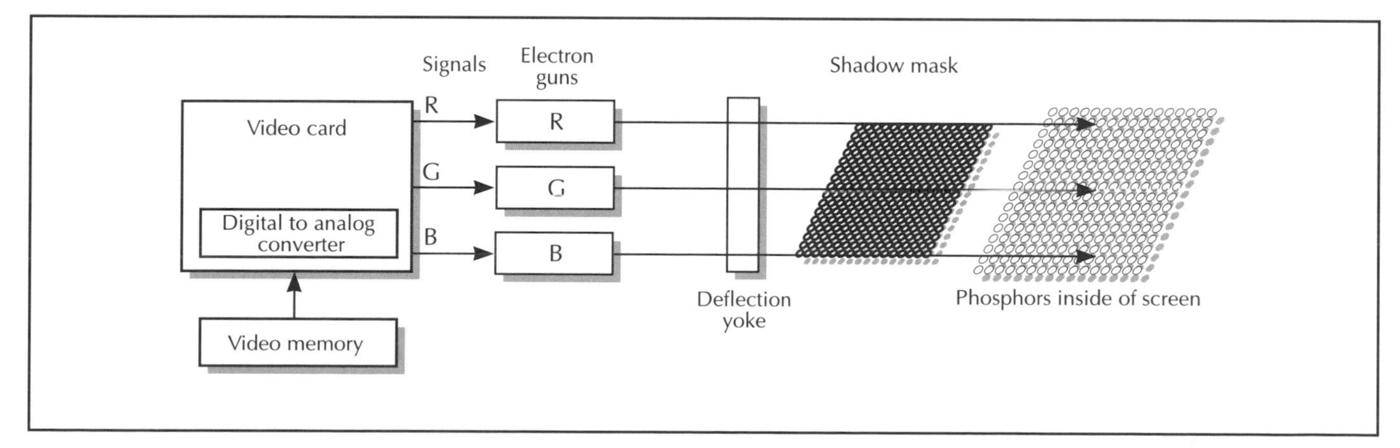

A standard computer monitor works much like an ordinary color TV set. The difference is that the signal is derived not from a broadcast program but from the contents of video memory as processed and converted by the computer's graphics card.

aligned. Each pixel location on the inner surface of the CRT is coated with phosphors, one that responds to each of the three colors.

The intensity of the beam hitting each color determines the brightness of the color, and the mixture of the red, blue, and green color levels determines the final color of the pixel. (Today's graphics systems can generate more than 16.7 million different colors, although the human eye cannot make such fine distinctions.)

The beam sweeps along a row of pixels and then turns off momentarily as it is refocused and set to the next row. The process of scanning the whole screen in this way is repeated 60 times a second, too fast to be noticed by the human eye. Less expensive monitors were sometimes designed to skip over alternate lines on each pass so that each line is refreshed only 30 times a second. This *interlaced* display can have noticeable flicker, and falling prices have resulted in virtually all current monitors being noninterlaced.

Another factor influencing the quality of a CRT monitor is the size of the screen area devoted to each pixel. The spacing in the shadow mask that defines the pixel areas is called the *dot pitch*. A smaller dot pitch allows for a sharper image.

During the 1980s, emerging video standards offered increasing screen resolution and number of colors, starting with the first IBM PC color displays at 320 x 200 pixels, 4 colors up to video graphics array (VGA) displays at 1024 x 768 pixels and at least 256 colors. The latter is considered the minimum standard today, with some displays going as high as 1600 x 1200 with millions of colors.

Meanwhile, the CRT monitor has become a commodity item with steadily falling prices. A 19-inch color monitor now costs only a few hundred dollars. Ergonomically, it is important for the combination of display size and resolution to be set to avoid eyestrain. There has been some concern about users receiving potentially damaging nonionizing radiation from CRT displays, but studies have generally been unable to confirm such effects. Modern monitors are generally designed to minimize this radiation.

The CRT monitor's bright, sharp image has made it the display of choice for desktop systems. The designers and users of laptop computers, however, cannot accept the bulk and power consumption of a CRT monitor (see PORTABLE COMPUTERS). Instead, they turn to the use of a liquid crystal display (LCD). (See FLAT-PANEL DISPLAY.)

Further Reading

Buccola, Chris. "How It Works: Monitors." http://firingsquad.gamers.com/guides/hiwmonitors/default.asp/

White, Ron. *How Computers Work*. Millennium Ed. Indianapolis, Ind.: Que, 1999.

"ZDNet Reviews: Monitors." http://www.zdnet.com/products/filter/guide/0,7267,1500121,00.html

motherboard

Large computers generally had separate large cabinets to hold the central processing unit (CPU) and memory (see MAINFRAME). Personal computers, built in an era of integrated electronics, use a single large circuit board to serve as the base into which chips and expansion boards are plugged. This base is called the motherboard.

The motherboard has a special slot for the CPU (see also MICROPROCESSOR). Data lines (see BUS) connect the CPU to RAM (see MEMORY) and various device controllers. Besides compactness, use of a motherboard minimizes the use of possibly fragile cable connections. It also provides expansion capability. Assuming its pins are compatible with the slot and it is operationally compatible, a PC user can plug a more powerful processor into the slot on the motherboard, upgrading performance. Memory expansion is also provided using a row of memory sockets. Memory, originally inserted as rows of separate chips plugged into individual sockets, is now provided in single modules called DIMMs that can be easily slid into place.

The motherboard also generally includes about six general-purpose expansion slots. These follow two different standards, ISA (industry standard architecture) and PCI (peripheral component interconnect) with PCI now predominating (see BUS). These slots allow users to mix and match such accessories as graphics (video) cards,

Motherboard

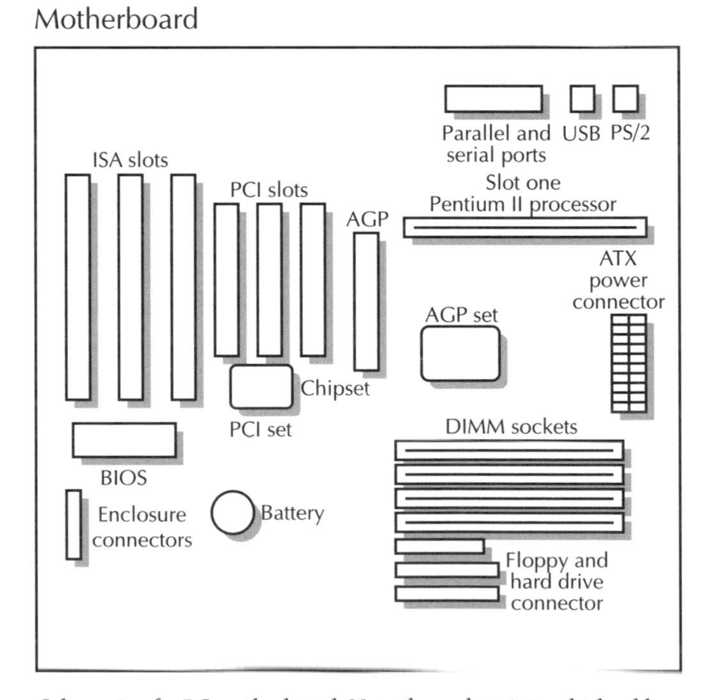

Schematic of a PC motherboard. Note the sockets into which additional RAM memory chips (DIMM) modules can be inserted, as well as the slots for ISA and PCI standard expansion cards. (Although a Pentium is assumed here, other types of processors can use different types of sockets.)

disk controllers, and network cards. Additionally, the motherboard includes a chip that stores permanent configuration settings and startup code (see BIOS), a battery, a system clock, and a power supply.

The most important factors in choosing a motherboard are the type and speeds of processor it can accommodate, the bus speed, the BIOS, system chipset, memory and device expansion capacity, and whether certain features (such as video) are integrated into the motherboard or provided through plug-in cards. Generally, users must work within the parameters of their system's motherboard, although knowledgeable people who like to tinker can buy a motherboard and build a system "from scratch" or keep their current peripheral components and upgrade the motherboard.

Further Reading

Chambers, Mark. *Building a PC for Dummies.* Foster City, Calif.: IDG Books, 1998.

Papst, Thomas. *Tom's Hardware Guide.* Indianapolis, Ind.: Sams, 1998.

White, Ron. *How Computers Work.* Millennium Ed. Indianapolis, Ind.: Que, 1999.

Mouse

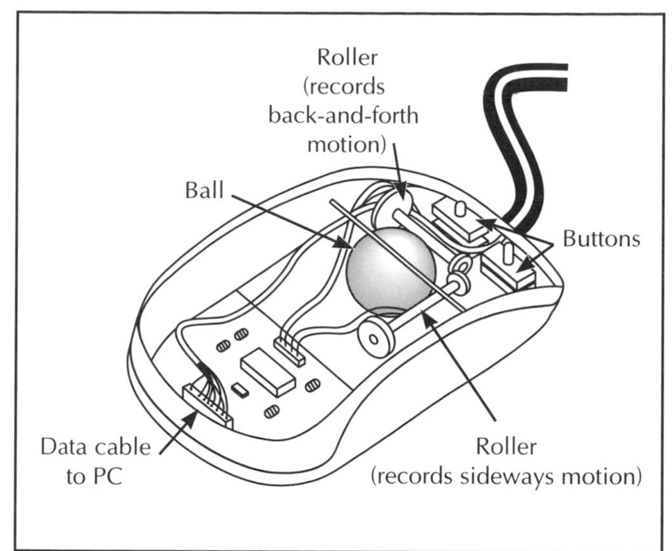

This type of mouse is mechanical, using the rolling of the ball to change the coordinates generated by the mouse. While mechanical mice are the most common, there are also optical mice that can sense changes in position without moving parts.

mouse

Traditionally, computers were controlled by typing in commands at the keyboard. However, as far back as the mid-1960s researchers had begun to experiment with providing users with more natural ways to interact with the machine. In 1965, Douglas Engelbart at the Stanford Research Institute (SRI) devised a small box that moved over the desk on wheels and was connected to the computer by a cable. As the user moved the box around, it sent signals representing its motion. These signals in turn were used to draw a pointer on the screen. Engelbart found that this system was less taxing on users than alternative such as light pens or joysticks (see ENGELBART, DOUGLAS).

This device, dubbed a "mouse," remained largely a laboratory novelty. In the 1970s, however, Xerox designed a mouse-driven graphical user interface for its Alto system, which saw only limited use. In 1984, however, Apple introduced the mouse to millions of users of its Macintosh. By the early 1990s, millions more users were switching their IBM-compatible PCs from text commands (see MS-DOS) to the mouse-driven Windows interface. (See MICROSOFT WINDOWS.) Today a desktop PC without a mouse would be as unthinkable as one without a keyboard.

Meanwhile, the mouse became smaller and sleeker. Instead of wheels, the contemporary mouse uses a rolling ball that turns two adjacent rollers inside the mouse. A mouse pad with a special surface is generally used to provide uniform traction. A new type of mouse uses optical sensors instead of rollers to sense its changing position, and does not require a mouse pad. Some mice are also cordless, using infrared or wireless data connections.

Optical and wireless mice (such as this Microsoft Wireless Intellimouse Explorer) have no cords to become tangled. (PHOTO COURTESY OF MICROSOFT CORPORATION)

Since mice are generally impracticable for laptop use (see PORTABLE COMPUTERS), designers have offered a variety of alternatives. These include a trackball (a rolling ball built into the keyboard), a touch-sensitive finger pad, or a small stub that can be moved like joystick by the fingertip.

Most mice now have at least two buttons. Generally, the left button is used for selecting objects, opening menus, or launching programs. The right button is used to bring up a menu of actions that can be done with the selected object. Activating a button is called *clicking*. It is the operating system that assigns significance to clicking or double-clicking (clicking twice in rapid succession) or *dragging* (holding a button down while moving the pointer). Some mice have a third button and/or a small wheel that can be used to scroll the display, but only certain software recognizes these functions.

Further Reading

Pang, Alex Soojung-Kim. *The Making of the Mouse. American Heritage of Invention & Technology* 17, no. 3, Winter 2002. Available on-line at http://www.americanheritage.com/it/2002/03/mouse.shtml

White, Ron. *How Computers Work*. Millennium Ed. Indianapolis, Ind.: Que, 1999.

MS-DOS

The MS-DOS operating system became standard for personal computers built by IBM and its imitators (see IBM PC) during the 1980s. Today it has been virtually displaced by various versions of Microsoft Windows (see MICROSOFT WINDOWS). However, MS-DOS is important as an expression of both the limitations of the first generation of personal computers and the remarkable patience and ingenuity of its developers and users.

DEVELOPMENT

By the end of the 1970s, there were a number of rudimentary operating systems for personal computers that used a variety of microprocessors. Generally, their capabilities were limited to loading and running programs and providing basic file organization and access.

The most sophisticated early PC operating system was CP/M, developed by Gary Kildall's Digital Research for machines based on the Intel 8008 microprocessor. CP/M offered more advanced capabilities such as the ability to use not only floppy but also hard disks, and included improved commands for listing file directories. CP/M even offered rudimentary programming tools, such as an editor and assembler, as well as an expandable architecture that allowed programmers to write utilities that could be in effect added to the operating system (see ASSEMBLER).

In one of computer history's greatest missed opportunities, Kildall and IBM failed to come to an agreement in 1980 for creating a version of CP/M for the IBM PC, which was being developed using the new 16-bit 8086 processor. IBM turned instead to Bill Gates and Microsoft, who had achieved something of a reputation for their widely used BASIC language package for personal computers (see GATES, WILLIAM). Gates agreed to provide IBM with an operating system, and did so by buying a program called QDOS ("quick and dirty operating system"), which had been developed by Tim Paterson of Seattle Computer Products. This program was released for the IBM PC as PC-DOS in 1981. However, Microsoft did not sell IBM an exclusive license, so when "clone" makers proved able to legally build IBM-compatible machines, Microsoft could sell them a generic version called MS-DOS. As the PC market boomed, this provided Microsoft with a large revenue stream, and the company never looked back.

FEATURES

MS-DOS offered a rather "clean" design that separates the operating system into three parts. There is a hardware-independent I/O system (stored as the file MSDOS.SYS), which processes requests from programs for access to disk files or to other devices such as the screen. The routines needed to actually communicate with the devices are stored in a separate file, IO.SYS, which is written by each computer manufacturer. (As users from the early 1980s remember, "PC-compatible" machines often had proprietary variations in areas such as video.) Finally, the command processor (COMMAND.COM) displays the once familiar C:\> prompt and waits for the user to type commands. For example, the DIR command followed by a path specification such as C:\TEMP lists the contents of that directory. Programs, too, can be run by typing their names at the prompt.

The MS-DOS file system, which remained largely unchanged until the most recent versions of Windows, uses a FAT (file allocation table) to indicate the disk allocation units or "clusters" assigned to each file. Starting with MS-DOS 2.0 in 1983, a hierarchical scheme of directories and subdirectories was introduced, allowing for better organization of the larger amount of space on hard disks.

One interesting feature of MS-DOS is the ability to load a program into memory and keep it available even while other programs are in use. This "terminate and stay resident" (TSR) function was soon used by enterprising developers to provide utilities such as notepads, calendars or shortcuts (see MACRO) that users could activate through special key combinations.

Users, however, had to struggle to keep enough memory free for their applications, resident programs, and device drivers. A combination of CPU addressing limitations and the high price of memory meant that early IBM PCs had a maximum of 640K of memory to hold the operating system and application programs. A trick called "expanded memory" was developed to allow data to be swapped back and forth between the 640K of usable memory and the 1–2 MB of additional memory that became available in the later 1980s.

By the early 1990s, MS-DOS (now up to version 6.0) was offering an alternative command processor (called DOSSHELL) that included some mouse operations, better support for larger amounts of memory, and the ability to switch between different application programs. However, by that time Windows 3.0 was proving increasingly successful, and by 1995 most new PCs were being shipped with Windows. Many new users scarcely used MS-DOS at all. Finally, with the advent of Windows NT, 2000 and XP, the MS-DOS program code that still lurked within the process of running Windows disappeared entirely.

Further Reading

"Information and Help with Microsoft DOS." http://www.computerhope.com/msdos.htm

Paterson, Tim. "An Inside Look at MS-DOS." *Byte,* vol. 8, no. 6, June 1983, 250–252. Also available on-line at http://www.patersontech.com/Dos/Byte/InsideDos.htm

multimedia

The earliest computers produced only numeric output or text (which itself actually consists of numbers—see CHARACTERS AND STRINGS). During the 1960s, CRT graphics (see MONITOR) came into limited use, mainly on computers used for scientific and engineering applications (see MINICOMPUTER). However, most business computer users continued to receive only textual output. A notable exception in the 1970s was PLATO, a system of networked educational computer terminals that combined text, graphics, and sound. It is this combination that became known as *multimedia.*

While much less powerful than mainframes or minicomputers, the hobbyist and early commercial PCs (see GRAPHICS CARD) of the late 1970s generally did have the capability of producing simple monochrome or color graphics on a monitor or TV screen. The Apple Macintosh, first released in 1984, was a considerable leap forward: Its user interface was inherently graphical, with even text being rendered as graphic bitmaps (see MACINTOSH).

The arrival of the PC greatly encouraged the development of entertainment software (see COMPUTER GAMES) as well as educational programs. As PCs became more powerful and gained hard drives and, by the late 1980s, CD-ROM drives (see CD-ROM), it became practical to put extensive multimedia content on systems in the home and school. One popular application has been encyclopedias, where the text from the printed version can be enhanced with graphics such as photographs, maps, and charts. Besides being more compelling and easier to use than the printed version, multimedia encyclopedias can be updated easily through annual upgrades, as well as allowing for linking to websites that can further amplify or update the content.

Encyclopedias and other educational programs also benefited from the use of links that the user can click with the mouse, bringing up additional or related information or illustrations (see HYPERTEXT AND HYPERMEDIA). Bill Atkinson's Hypercard, released for the Macintosh in 1987, provided a multimedia "construction set" that could be used by nonprogrammers to create simple hyperlinked presentations, educational programs, and even games. Hypertext and linking are the "glue" that binds multimedia into an integrated experience.

Multimedia business presentations are now routinely created using software such as Microsoft PowerPoint, then projected at meetings. While simple presentations can emulate the traditional "slide show," one-upmanship inevitably leads to more elaborate animations.

MULTIMEDIA AND DAILY LIFE

DVD-ROM drives, with about six times the storage capacity of CDs, now make it practical to include video or even feature-length movies as part of a PC multimedia package. Meanwhile, the video capabilities of PCs continue to grow, with many PCs as of 2002 having 64 MB or more of video memory. Combined with processors running at up to 2.5 GHz, this allows computer-generated graphics to rival the quality of live video.

However, the most important trend is probably the delivery of on-line multimedia content (see INTERNET, ONLINE SERVICES, and WORLD WIDE WEB). The widespread marketing of the Mosaic and Netscape browsers (see WEB BROWSER) in the mid-1990s changed the Internet from an arcane, text-driven experience to a multimedia platform. The ability to deliver a continuous "feed" of video and audio (see STREAMING) allows content such as TV news reports to be carried with full video and radio broadcasts carried "live" with good fidelity. Newspapers and broadcast outlets are increasingly investing in on-line versions of their content, viewing a Web presence as a business necessity. As more Internet users gain access to high-speed cable and DSL services (see BROADBAND), multimedia may become as pervasive a part of the computing experience as television is in daily life.

Many facets of that daily life are likely to be affected by multimedia technology in the coming decades. The ability to deliver real-time, high-quality multimedia content, as well as the use of cameras (see VIDEOCONFERENCING and WEB CAM) has made "virtual" meetings not only possible but also routine in some corporate settings. When applied to lectures, this technology can facilitate "distance learning" where teachers work with students without them occupying the same room (see EDUCATION AND COMPUTERS). Video "chat" services and immersive, pervasive on-line games may become important social outlets for many people, with the experience becoming ever more realistic (see VIRTUAL REALITY).

Further Reading

"MIT Media Lab." http://www.media.mit.edu/

Packer, Randal, and Ken Jordan, eds. *Multimedia: from Wagner to Virtual Reality.* New York: Norton, 2001.

Pavlik, John V. *New Media Technology: Cultural and Commercial Perspectives.* Needham Heights, Mass.: Allyn and Bacon, 1998.

multiprocessing

One way to increase the power of a computer is to use more than one processing unit. In early computers (see MAINFRAME) a single processor handled both program execution and input/output (I/O) operations. In the late 1950s, however, machines such as the IBM 709 introduced the concept of *channels,* or separate processing units for I/O operations. In such systems the central processor sends a set of I/O commands (such as to read a file into memory) to the channel, which has its own processor for carrying out the operation.

True multiprocessing, however, involves the use of more than one central processing unit (CPU). One successful design, Control Data Corporation's CDC 6600 (1964), contained both multiple arithmetic/logic units (the part of the CPU that does calculations) and multiple controllers for I/O and memory access control. IBM soon added multiprocessing capability to its 360 line of mainframes.

Multiprocessing can be either asymmetric or symmetric. Asymmetric multiprocessing essentially maintains a single main flow of execution with certain tasks being "handed over" by the CPU to auxiliary processors. (For example, the Intel 80386 processor could be purchased with an additional floating-point processor, allowing such calculations to be performed using more efficient hardware. When the Pentium line was developed, floating-point was integrated into the main CPU).

Symmetric multiprocessing (SMP) has multiple, full-fledged CPUs, each capable of the full range of operations. The processors share the same memory space, which requires that each processor that accesses a given memory location be able to retrieve the same value. This *coherence* of memory is threatened if one processor is in the midst of a memory access while another is trying to write data to that same memory location. This is usually handled by a "locking" mechanism (see also CONCURRENT PROGRAMMING) that prevents two processors from simultaneously accessing the same location.

A subtler problem occurs with the use by processors of separate internal memory for storing data that is likely to be needed (see CACHE). Suppose CPU "A" reads some

Multiprocessing

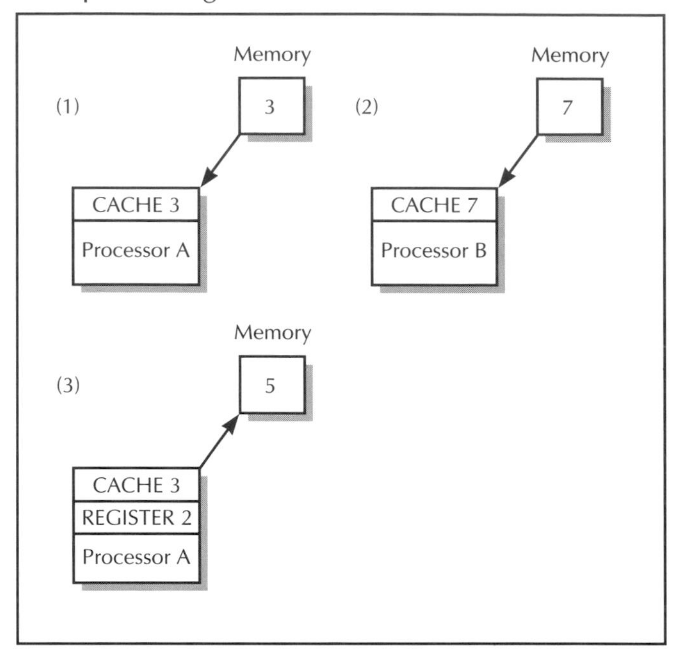

This example shows what can happen if processes don't properly manage a shared memory resource. At (1) processor A retrieves 3 from the memory location. At (2) processor B copies 7 from its cache to that same memory location. Finally, at (3) processor A adds the 3 it had retrieved to a 2 in its register, storing 5 back in a location where processor B probably expects there to still be a 7.

data from memory and stores it in its cache. A moment later, CPU "B" writes to that memory location, changing the data. At this point the data in "A's" cache no longer matches that in the actual memory. One way to deal with this problem is called *bus snooping*. Each CPU includes a controller that monitors the data line (see BUS) for memory locations being used by other CPUs. When it sees an address that refers to an area of memory currently being stored in the cache, the controller updates the memory from the cache. This write operation sends a signal that lets other CPUs know that any cached data they have for that location is no longer valid. This means the other CPUs will go back to memory and reread the current data.

Alternatively, all CPUs can be given a single shared cache. While less complicated, this approach limits the number of CPUs to the maximum data-handling capacity of the bus.

Larger-scale multiprocessing systems consist of latticelike arrays of hundreds or even thousands of CPUs, which are referred to as *nodes*. Indeed, small *clusters* of CPUs using the architecture given above can be connected together to form larger arrays. Each cluster can have its own shared memory cache. Because accessing memory at a remote node takes considerably longer than accessing "local" memory within the cluster, maintaining coherence through bus monitoring is impracticable. Instead, memory is usually organized into data objects that are *distributed* optimally to reduce the necessity for remote access, and the objects are shared by CPUs requesting them through a directory system.

MULTIPROGRAMMING

In order for a program to take advantage of the ability to run on multiple CPUs, the operating system must have facilities to support multiprocessing, and the program must be structured so its various tasks are most efficiently distributed among the CPUs. These separate tasks are generally called *threads*. A single program can have many threads, each executing separately, perhaps on a different CPU, although that is not required.

The operating system can use a number of approaches to scheduling the execution of processes or threads. It can simply assign the next idle (available) CPU to the thread. It can also give some threads higher priority for access to CPUs, or let a thread continue to "own" its CPU until it has been idle for some specified time.

The use of threads is particularly natural for applications where a number of activities must be carried on simultaneously. For example, a scientific or process control application may have a separate thread reading the data being returned from each instrument, another thread monitoring for alarm conditions, and other threads generating graphical output.

Threads also allow the user to continue interacting with a program while the program is busy carrying out earlier requests. For example, the user of a Web browser can continue to use menus or navigation buttons while the browser is still loading graphics needed for the currently displayed webpage. A search program can also launch separate threads to send requests to multiple search engines or to load multiple pages.

Support for multiprogramming and threads can now be found in versions of most popular programming languages, and some languages such as Java are explicitly designed to accommodate it.

Further Reading

Culler, D., J. P. Singh, and A. Gupta. *Parallel Computer Architecture: A Hardware/Software Approach*. San Francisco: Morgan Kaufmann, 1999.

Hwang, K., and Z. Xu. *Scalable Parallel Computing*. New York: McGraw-Hill, 1998.

McCormick, Bruce. "Symmetric Multiprocessing (SMP)." http://www.nswc.navy.mil/cosip/nov97/osa1197-1.shtml

multitasking

Users of modern operating systems such as Microsoft Windows are familiar with multitasking, or running several programs at the same time. For example, a user might be writing a document in a word processor, pause to check the e-mail program for incoming messages, type a page address into a Web browser, then return to writing. Meanwhile, the operating system may be running a number of other programs tucked unobtrusively into the background, such as a virus checker, task scheduler, or system resource monitor.

Each running program "takes turns" using the PC's central processor. In early versions of Windows, multitasking was *cooperative*, with each program expected to periodically yield the processor to Windows so it could be assigned to the next program in the queue. One weakness of this approach is that if a program crashes, the CPU might be "locked up" and the system would have to be rebooted. However, Windows NT, 2000, and XP (as well as operating systems such as UNIX) use preemptive multitasking. The operating system assigns a "slice" of processing (CPU) time to a program and then switches it to the next program regardless of what might be happening to the previous program. Thus, if a program "crashes," the CPU will still be switched to the next program, and the user can maintain control of the system and shut down the offending program.

Systems with preemptive multitasking often give programs or tasks different levels of priority that determine how big a slice of CPU time they will get. For example, the "active" program (in Windows, the one whose win-

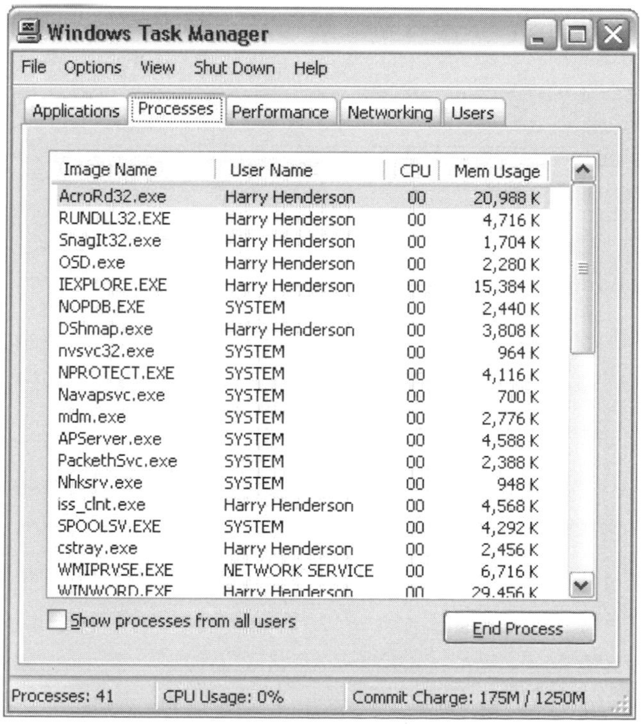

Windows Task Manager

File Options View Shut Down Help

Applications | Processes | Performance | Networking | Users

Image Name	User Name	CPU	Mem Usage
AcroRd32.exe	Harry Henderson	00	20,988 K
RUNDLL32.EXE	Harry Henderson	00	4,716 K
SnagIt32.exe	Harry Henderson	00	1,704 K
OSD.exe	Harry Henderson	00	2,280 K
IEXPLORE.EXE	Harry Henderson	00	15,384 K
NOPDB.EXE	SYSTEM	00	2,440 K
DShmap.exe	Harry Henderson	00	3,808 K
nvsvc32.exe	SYSTEM	00	964 K
NPROTECT.EXE	SYSTEM	00	4,116 K
Navapsvc.exe	SYSTEM	00	700 K
mdm.exe	SYSTEM	00	2,776 K
APServer.exe	SYSTEM	00	4,588 K
PackethSvc.exe	SYSTEM	00	2,388 K
Nhksrv.exe	SYSTEM	00	948 K
iss_clnt.exe	Harry Henderson	00	4,568 K
SPOOLSV.EXE	SYSTEM	00	4,292 K
cstray.exe	Harry Henderson	00	2,456 K
WMIPRVSE.EXE	NETWORK SERVICE	00	6,716 K
WINWORD.EXE	Harry Henderson	00	29,456 K

Show processes from all users | End Process

Processes: 41 | CPU Usage: 0% | Commit Charge: 175M / 1250M

Windows users can bring up a window listing all processes or tasks running on the system, and shut down any task that has stopped responding to input.

dow has been selected for interaction by the user) will be given preference over a background program such as a print spooler. Also, the operating system can more intelligently assign CPU time according to what a given program is doing. Thus, if a program is waiting for user input, it may be given only an occasional slice of CPU time so it can check to see whether input has been received. (The user, after all, is millions of times slower than the CPU.) When some input (such as a menu selection) is ready for processing, the program can be given higher priority.

Priority can be expressed in two different ways. One way is to move a program up in the list of running tasks (see QUEUE). This ensures it gets a turn before any lower-priority task. The other way is to have turns of varying length, with the higher-priority program getting a longer turn.

Even operating systems with preemptive multitasking can provide facilities that programs can use to communicate their own sense of their priority. In UNIX systems, this is referred to as "niceness." A "nice" program gives the operating system permission to interrupt lengthy calculations so other programs can have a turn, even if the program's priority would ordinarily entitle it to a greater share of the CPU.

Multitasking should be distinguished from two several similar-sounding terms. Multitasking refers to entirely separate programs taking turns executing on a single CPU. *Multithreading,* on the other hand, refers to separate pieces of code within a program executing simultaneously but sharing the program's common memory space. Finally, *multiprocessing* or parallel processing refers to the use of more than one CPU in a system, with each program or thread having its own CPU (see MULTIPROCESSING).

Further Reading

Bach, Maurice J. *Design of the UNIX Operating System.* Englewood Cliffs, N.J.: Prentice Hall, 1987.

Tannenbaum, Andrew, and Albert S. Woodhull. *Operating Systems: Design and Implementation.* 2nd ed. Upper Saddle River, N.J.: Prentice Hall, 1997.

music, computer

Computers have had a variety of effects on the performance, rendering, and composition of music. At the same time, the sound capabilities of standard personal computers have improved greatly, and music and other sounds have become an integral part of games and educational software (see MULTIMEDIA).

After the invention of the vacuum tube, a number of electronic instruments were devised. The best known is the theremin, invented by Lev Termin, a Russian physicist, in 1919. The instrument consists of a vacuum tube connected to two antennas. The player varies the pitch and volume of its eerie sound by moving his or her hands near the antennas.

Some composers became fascinated by electronic music, both for its sense of modernity and its promise of breaking the bonds of traditional form and instrumentation. In 1953, German composer Karlheinz Stockhausen (1928–) founded an Electronic Music Studio in Cologne and created electronic works.

Meanwhile, inventors experimented with electronic synthesizers such as the RCA MKI and MKII, which used vacuum tubes and could be programmed with punched paper tape. The advent of solid-state circuitry in the 1960s made synthesizers far more reliable and compact. The Moog synthesizer in particular became a staple of leading-edge rock and avant-garde music. It was now time for the computer to catch up to the potential of electronic sound.

In the 1970s, digital music synthesizers with keyboards and microprocessor-controlled sound generation became available to adventurous (and fairly well-to-do) musicians. Ray Kurzweil's digital music synthesis system, introduced in 1984, achieved a new level of sonic realism by using programming stored in read-only memory (ROM) to emulate subtle characteristics such as attack and timbre, realistically re-creating the sounds of many types of orchestral instruments.

Computer music synthesis enabled composers to experiment with algorithmic composition. That is, they could use programs to create new works by combining randomization with the permutation of patterns (serialism). Compositions have also been based on applying mathematical structures (such as fractals) and the concepts being discovered by computer scientists, including adaptive structures such as neural nets and genetic algorithms.

Like most avant-garde music, computer music composition remained largely unknown to most people. However, the technology of music synthesis was to become democratically available to everyday musicians as well. As the personal computer began to bring increasingly powerful microprocessors to consumers, it became practicable to in effect add a music synthesizer to the PC. The musical instrument digital interface, MIDI, provides a protocol for connecting traditional musical instruments such as pianos and guitars to a personal computer. MIDI specifies the pitch, volume, attack (how a note increases to maximum volume), and decay (how it dies away). The musician then uses the instrument as an input device, with the notes played being recorded as MIDI data. Different tracks can then be edited (such as to transpose to a different key), and combined in various ways to create complete compositions. Because MIDI stores instructions, not actual digitized sound, it is a quite compact way to store music. MIDI brought the synthesizer within reach of just about any serious musician—and many amateurs.

PC sound cards can play sound in two ways. *Wave Table Synthesis* uses a table of stored digital samples of notes played by various instruments, and algorithmically manipulates them to reproduce the MIDI-encoded music. *FM Synthesis* attempts to create waves that replicate the intended sounds, based on a model of what happens in a given instrument. It is less faithful to the original sound, since it does not capture the detailed "texture" of a digital sample.

Today's PCs have sound cards that can handle both playback of audio CDs and rendering of digitized and synthesized sounds. The cards have the capacity to support many simultaneous voices (polyphony) as well as rendering speech faithfully. While early PCs tended to have only tiny internal speakers, most PCs today come with speakers (often including subwoofers and even multiple speakers for "surround sound") comparable to midrange home stereo systems.

Further Reading

Computer Music Journal. Cambridge, Mass.: MIT Press Journals.

Dodge, Charles, and Jerse Thomas. *Computer Music: Synthesis, Composition, and Performance.* 2nd ed. New York: Schirmer, 1997.

European Telematic Network for Education. "Computers in Music Education." http://www.xtec.es/rtee/eng/

Moore, F. Richard. *Elements of Computer Music.* Upper Saddle River, N.J.: Prentice Hall, 1998.

N

nanotechnology

Ordinary refining and manufacturing involve the use of grinding, cutting, heating, application of chemicals, and other processes that affect large numbers of atoms or molecules at once. These processes are necessarily imprecise: Some atoms or molecules will end up unprocessed or somehow out of alignment. The resulting material will thus fall short of its maximum theoretical strength or other characteristics.

In a talk given in 1959, physicist Richard Feynman suggested that it might be possible to manipulate atoms individually, spacing them precisely. As Feynman also pointed out, the implications for computer technology are potentially very impressive. A current commercial DIMM memory module about the size of a person's little finger holds about 250 megabytes (MB) worth of data. Feynman calculated that if 100 precisely arranged atoms were used for each bit of information, the contents of all the books that have ever been written (about 10^{15} bits) could be stored in a cube about 1/200 of an inch wide, just about the smallest object the unaided human eye can see. Further, although the density of computer logic circuits in microprocessors is millions of times greater than it was with the computers of 1959, computers built at the atomic scale would be billions of times smaller still. Indeed, they would be the smallest (or densest) computers possible short of one that used quantum states within the atoms themselves to store information (see QUANTUM COMPUTING). "Nanocomputers" could also efficiently dissipate heat energy, overcoming a key problem with today's increasingly dense microprocessors.

Feynman offered some possible methods of manufacture, and discussed some of the obstacles that would have to be overcome to do engineering at a molecular or atomic scale. These include lubrication, the effects of heat, and electrical resistance. He invited adventurous high school students to develop science projects to explore this new technology.

The idea of atomic-level engineering lay largely dormant for about two decades. Starting with a 1981 paper, however, K. Eric Drexler began to flesh out proposed structures and methods for a branch of engineering he termed *nanotechnology.* (The "nano" refers to a nanometer, or one billionth of a meter.) Research in nanotechnology today focuses on two broad areas: assembly and replication. Assembly is the problem of building tools (called assemblers) that can deposit and position individual atoms. Since such tools would almost certainly be prohibitively expensive to manufacture individually, research has focused on the idea of making tools that can reproduce themselves. This area of research began with John von Neumann's 1940s concept of self-replicating computers (see VON NEUMANN, JOHN). If an assembler can assemble other assemblers from the available "feed stock" of atoms, then obtaining the number of assemblers necessary to manufacture the intended product would be no problem. (As science fiction writers have pointed out, the ultimate problem would be making sure the self-reproducing assemblers don't get out of control and start turning everything around them, potentially the whole Earth, into more of themselves.)

Researchers at IBM have already created some "nanoscale" devices, including a transistor with a 60 nanometer gate length. A structure called a carbon nanotube has also been used to create logic circuits on a molecular scale. Research on self-assembly has been more rudimentary, but Thomas N. Theis, Director of Physical Sciences at IBM's Thomas J. Watson Research Center believes there are several promising possibilities involving such things as self-assembling crystals and lattices.

Further Reading

Drexler, K. Eric. *Engines of Creation*. Garden City, N.Y.: Anchor Press/Doubleday, 1986.

———. "Molecular Engineering: An approach to the development of general capabilities for molecular manipulation." *Proceedings of the National Academy of Sciences, USA* 78, no. 9, 5275–5278, September 1981. Also available on-line at http://www.imm.org/PNAS.html

———. *Nanosystems: Molecular Machinery, Manufacturing, and Computation*. New York: Wiley, 1992.

Institute for Molecular Manufacturing. http://www.imm.org/

"Introduction to Nanotechnology and Nanocomputers." http://www.mitre.org/research/nanotech/intronano.html

Santarini, Michael. "IBM Scientist Sees Nanotechnology Supplanting Transistors." *EE Times*, November 6, 2001. Available on-line at http://www.eetimes.com/story/OEG20011106S0033

netiquette

As each new means of communication and social interaction is introduced, social customs and etiquette evolve in response. For example, it took time before the practice of saying "hello" and identifying oneself became the universal way to initiate a phone conversation.

By the 1980's, a system of topical news postings (see NETNEWS AND NEWSGROUPS) carried on the Internet was becoming widely used in universities, the computer industry, and scientific institutions. Many new users did not understand the system, and posted messages that were off topic. Others used their postings as to insult or attack ("flame") other users, particularly in newsgroups discussing perennially controversial topics such as abortion. When a significant number of postings in a newsgroup are devoted to flaming and counter-flaming, many users who had sought civilized, intelligent discussion leave in protest.

In 1984, Chuq von Rospach wrote a document entitled "A Primer on How to Work with the Usenet Community." It and later guides to net etiquette or "netiquette" offered useful guidelines to new users and to more experienced users who wanted to facilitate civil discourse. These suggestions include:

- Learn about the purpose of a newsgroup before you post to it. If a group is moderated, understand the moderator's guidelines so your postings won't be rejected.
- Before posting, follow some discussions to see what sort of language, tone, and attitude seems to be appropriate for this group.
- Don't post bulky graphics or other attachments unless the group is designed for them.
- Avoid "ad hominem" (to the person) attacks when discussing disagreements.
- Don't post in ALL CAPS, which is interpreted as "shouting."
- Check your postings for proper spelling and grammar. On the other hand, avoid "flaming" other users for their spelling or grammar errors.
- When replying to an existing message, include enough of the original message to provide context for your reply, but no more.
- If you know the answer to a question or problem raised by another user, send it to that user by e-mail. That way the newsgroup doesn't get cluttered up with dozens of versions of the same information.

In 1994, a firm of immigration attorneys enraged much of the on-line community by posting messages offering their services in each of the thousands of different newsgroups. "Spam" was born. Technically savvy users responded by creating "cancelbots" or programs that attempt to detect and automatically delete postings containing spam. Today, spam is mainly conveyed by e-mail, with mail servers and client programs offering various options for blocking it.

NETIQUETTE IN THE 21ST CENTURY

Netiquette concerns have evolved with the changing face of the Internet. While newsgroups are still an important source of information and social interaction, many users have changed their focus to live, real-time chat groups (see CHAT, ON-LINE). Although chat is more like conversation than writing news postings, the general advice about determining the purpose and tone of a given venue still applies, as does the stricture about flaming.

Further Reading

"Netiquette Guidelines." http://www.dtcc.edu/cs/rfc1855.html

"Netiquette Home Page." http://www.albion.com/netiquette/

Rinaldi, Arlene H. "The Net: User Guidelines and Netiquette." http://www.fau.edu/netiquette/net/

Also see the newsgroup www.news.announce.newusers.org, which includes the latest version of Chuq von Rospach's "A Primer on How to Work with the Usenet Community."

netnews and newsgroups

Originally called Usenet and originating in the UNIX user community in the late 1970s, netnews is distributed today over the Internet in the form of thousands of newsgroups devoted to just about every imaginable topic.

DEVELOPMENT

By the late 1970s, researchers at many major universities were using the UNIX operating system (see UNIX). In 1979, a suite of utilities called UUCP was distributed with the widely used UNIX Version 7. These utilities could be used to transfer files between UNIX computers that were linked by some form of telephone or network connection.

Two Duke University graduate students, Tom Truscott and Jim Ellis, decided to set up a way in which users on different computers could share a collection of files containing text messages on various topics. They wrote a simple set of shell scripts that could be used for distributing and viewing these message files. The first version of the news network linked computers at Duke and at the University of North Carolina. Soon these programs were revised and rewritten in the C language and distributed to other UNIX users as the "A" release of the News software.

During the 1980s, the news system was expanded and features such as moderated newsgroups were added. As the Internet and its TCP/IP protocol (see TCP/IP) became a more widespread standard for connecting computers, a version of News using the NNTP (Network News Transmission Protocol) over the Internet was released in 1986. Netnews is a mature system today, with news reading software available for virtually every type of computer.

STRUCTURE AND FEATURES

Netnews postings are simply text files that begin with a set of standard headers, similar to those used in e-mail. (Like e-mail, news postings can have binary graphics or program files attached, using a standard called MIME, for Multipurpose Internet Mail Extensions.)

The files are stored on news servers—machines that have the spare capacity to handle the hundreds of gigabytes of messages now posted each week. The files are stored in a typical hierarchical UNIX fashion, grouped into approximately 75,000 different newsgroups.

As shown in the following table, the newsgroups are broken down into 10 major categories. The names of individual groups begin with the major category and then specify subdivisions. For example, the newsgroup comp.sys.ibm.pc deals with IBM PC-compatible personal computers, while comp.os.linux deals with the Linux operating system.

MAIN DIVISIONS OF NETNEWS NEWSGROUPS

Category	Coverage
alt	An alternative system with its own complete selection of topics.
biz	Business-related discussion, products, etc.
comp	Computer hardware, software and operating systems.
humanities	Arts and literature, philosophy, etc.
misc.	Various topics that don't fit in another category.
news	Announcements and information relating to the news system itself.
rec	Sports, games, and hobbies.
sci	The sciences.
soc	Social and cultural issues.
talk	Current controversies and debates.

DISTRIBUTION AND READING

The servers are linked into a branching distribution system. Messages being posted by users are forwarded to the nearest major regional "node" site, which in turn distributes them to other major nodes. In turn, when messages arrive at a major node from another region, they are distributed to all the smaller sites that share the *newsfeed.* Due to the volume of groups and messages, many sites now choose to receive only a subset of the total newsfeed. Sites also determine when messages will expire (and thus be removed from the site).

There are dozens of different news reading programs that can be used to view the available newsgroups and postings. On UNIX systems, programs such as elm and tin are popular, while other newsreaders cater to Windows, Macintosh, and other systems. Major Web browsers such as Netscape and Internet Explorer offer simplified news reading features. To use these news readers, the user accesses a newsfeed at an address provided by the Internet Service Provider (ISP). There are also services that let users simply navigate through the news system by following the links on a webpage. The former service called DejaNews, now purchased by the Web search service Google, is the best known and most complete such site.

Further Reading

[Google News Service] http://groups.google.com/googlegroups/deja_announcement.htmlnews.announce.newusers Information for new users of the Netnews system.

Netnews and Newsgroups

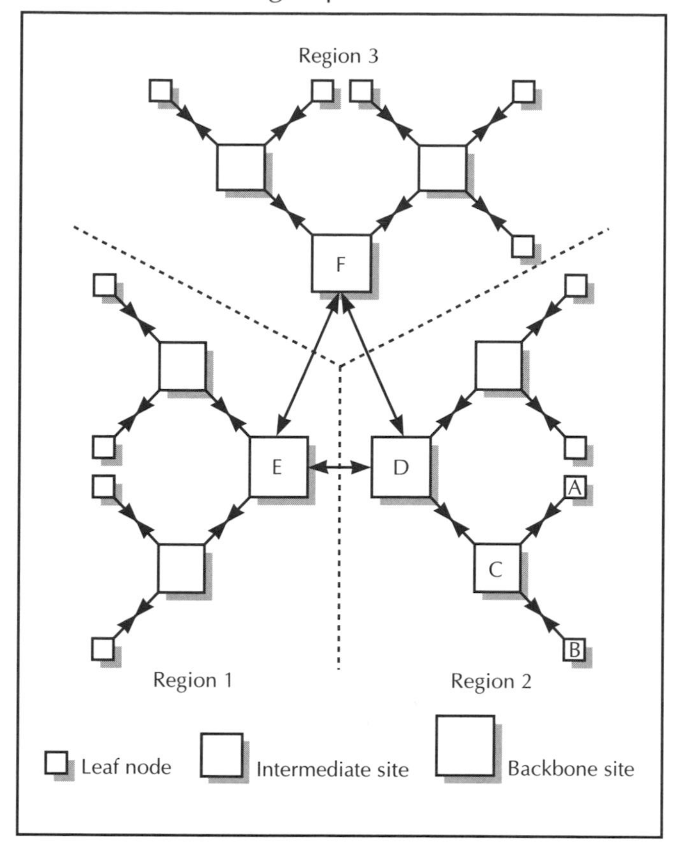

Netnews (or USENET) used an ingenious method for propagating news articles. Well-connected "backbone" nodes collected messages from nearby systems and forwarded them to other backbone nodes, which in turn distributed the messages (as a "newsfeed") to smaller nearby nodes. (Today most users read news with their Web browsers, getting the newsfeed from their Internet Service Provider).

Newsgroup Descriptions. http://livinginternet.com/u/uu_desc.htm

Pfaffenberger, Bryan. *The USENET Book: Finding, Using, and Surviving Newsgroups on the Internet.* Reading, Mass.: Addison-Wesley, 1995.

Spencer, Henry, and David Lawrence. *Managing Usenet.* Sebastopol, Calif.: O'Reilly, 1998.

network

In the 1940s, the main objective in developing the first digital computers was to speed up the process of calculation. In the 1950s, the machines began to be used for more general data-processing tasks by governments and business. By the 1960s, computers were in use in most major academic, government, and business organizations. The desire for users to share data and to communicate both within and outside their organization led to efforts to link computers together into networks.

Computer manufacturers began to develop proprietary networking software to link their computers, but they were limited to a particular kind of computer, such as a DEC PDP minicomputer, or an IBM mainframe. However, the U.S. Defense Department, seeing the need for a robust, decentralized network that could maintain links between their computers under wartime conditions, funded the development of a protocol that, given appropriate hardware to bridge the gap, could link these disparate networks (see INTERNET, LOCAL AREA NETWORK).

NETWORK ARCHITECTURE

Today's networks are usually defined by open (that is, nonproprietary) specifications. According to the OSI (open systems interconnection) model, a network can be considered to be a series of seven layers laid one atop another (see DATA COMMUNICATION).

The physical layer is at the bottom. It specifies the physical connections between the computers, which can be anything from ordinary phone lines to cable, fiber optic, or wireless. This layer specifies the required electrical characteristics (such as voltage changes and durations that constitute the physical signal that is recognized as either a 1 or 0 in the "bit stream."

The next layer, called the data link layer, specifies how data will be grouped into chunks of bits (frames or packets) and how transmission errors will be dealt with (see ERROR CORRECTION).

The network layer groups the data frames as parts of a properly formed data packet and routes that packet from the sending node to the specified destination node. A variety of routing algorithms can be used to determine the most efficient route given current traffic or line conditions.

The transport layer views the packets as part of a complete transmission of an object (such as a webpage) and ensures that all the packets belonging to that object are sorted into their original sequence at the destination. This is necessary because packets belonging to the same message may be sent via different routes in keeping with traffic or line conditions.

The session layer provides application programs communicating over the network with the ability to initiate, terminate, or restart an interrupted data transfer.

The presentation layer ensures that data formats are consistent so that all applications know what to expect. This layer can also provide special services (see ENCRYPTION or DATA COMPRESSION).

Finally, the application layer gives applications high-level commands for performing tasks over the network, such as FILE TRANSFER PROTOCOL (ftp).

Most modern operating systems support this model. The Internet protocol (see TCP/IP) has become the lingua franca for most networking, so modern versions of Microsoft Windows and the Macintosh Operating System as well as all versions of UNIX provide the services that applications need to make and manage TCP/IP connections.

Networks that link computers remotely (such as over phone lines) are sometimes called wide area networks, or WANs. Networks that link computers within an office, home, or campus, usually using cables, are called local area networks (LANs). See LOCAL AREA NETWORK for more details about LAN architecture and software and INTRANET for the use of TCP/IP in local networks.

TRENDS

It has become the norm for desktop and portable computers to have access to the Internet. A computer from which one cannot send or receive e-mail or view webpages almost gives the perception of being crippled, because so many applications now assume that they can access the network. For example, the latest antivirus programs regularly check their manufacturer's website and download the latest virus definitions and software patches. Recent versions of Windows, too, include a built-in update facility that can obtain security patches and newer versions of device drivers.

The flip side of the power of networking to keep every PC (and its user) up to date is the vulnerability to both intrusion attempts and viruses (see COMPUTER CRIME AND SECURITY). Virtually all networks include a layer of software whose job it is to attempt to block intrusions and protect sensitive information (see FIREWALL).

Besides attending to security, network administrators and engineers must continually monitor the traffic on the network, looking for bottlenecks, such as an often-requested database being stored on a file server with a relatively slow hard drive. Besides upgrading key hardware, another approach to relieve congestion is to adopt a distributed database (see DATABASE MANAGEMENT SYSTEM) that stores "data objects" throughout the network and can dynamically relocate them to improve access.

The growing appetite for data-rich applications such as high-fidelity audio and video (see STREAMING and MULTIMEDIA) tends to put a strain on the capacity of most networks. In response, institutional users look to optical fiber and other high capacity connections (see BANDWIDTH), while home users gradually switch from dial-up service on regular phone lines (see MODEM) to DSL phone lines and cable.

During the late 1990s, government agencies announced several initiatives for replacing the existing Internet infrastructure with high speed/high bandwidth alternatives. These proposals include the National Science Foundation's vBNS (very high Bandwidth Network Service), the Internet2 initiative, and NGI, or Next Generation Internet. It remains unclear whether or how these proposals will bear fruit for ordinary network users.

Further Reading

Berkowitz, Howard C. *Designing Routing and Switching Architectures for Enterprise Networks.* Indianapolis, Ind.: Macmillan Technical Publishing, 1999.

Comer, Douglas. *Internetworking with TCP/IP: vol. 1: Principles, Protocols and Architecture.* 4th ed. Upper Saddle River, N.J.: Prentice Hall, 2000.

Internet2 Home Page. http://www.internet2.edu

NGI Initiative. http://www.ngi.gov

Tannenbaum, A. S. *Computer Networks.* 3rd ed. Upper Saddle River, N.J.: Prentice Hall, 1996.

neural network

When digital computers first appeared in the late 1940s, the popular press often referred to them as "electronic brains." However, computers and living brains operate very differently. The human brain contains about 100 billion neurons, and each neuron can form connections to as many as a thousand neighboring ones. Neurons respond to electronic signals that jump across a gap (called a synapse) and into electrodelike dendrons. The incoming signals form combinations that in turn determine whether the neuron becomes "excited" and in turn emits a signal through its axon. Clumps of neurons, therefore, act as networks that in effect sum up incoming signals and develop a response to them. That is, they "learn."

Neural Network (1)

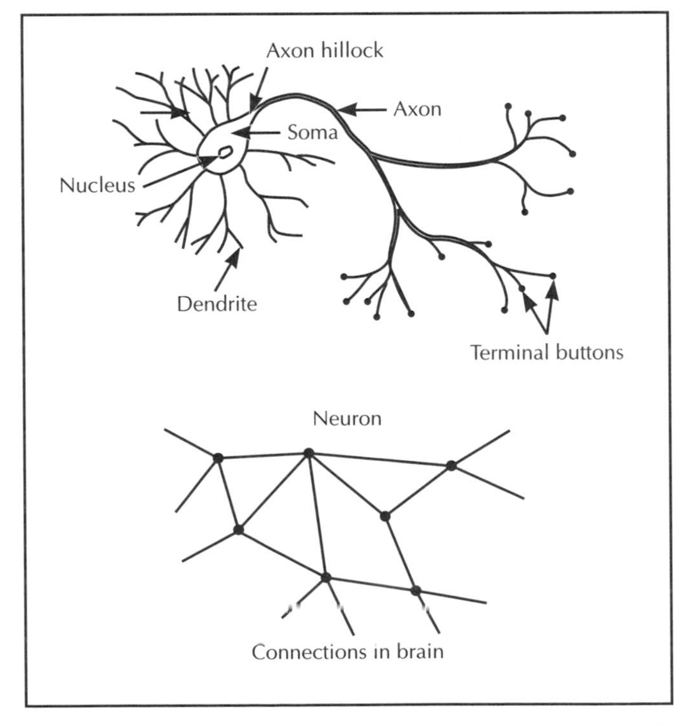

A neuron in the brain has a number of terminals or connectors and thus can connect to many neighboring neurons.

In a conventionally operated computer, the "neurons" (memory locations) are not inherently connected, and the central processing unit (CPU) uses arbitrary, interchangeable memory locations for storing data. Algorithms written by a programmer and implemented in instructions executed by the CPU impose cognition, to the extent one can speak of it in computers. In the brain, however, cognition seems to be something that emerges from the cooperating activities and connections of the neurons in response to sense stimuli, and possibly the creation of agentlike entities, as described in Marvin Minsky's book *The Society of Mind.*

Alan Turing and John von Neumann (see TURING, ALAN and VON NEUMANN, JOHN) had established the universality of the computer. That is, any calculation or logical operation that can be performed at all can be performed by an appropriate computer program. This means that the "brain" model of a network of interconnected neurons can also be implemented in a computer. During the 1940s, Warren S. McCulloch and Walter Pitts developed an electronic "neuron" in the form of a binary (on/off) switch that could be linked into networks and used to perform logical functions.

During 1950s, Marvin Minsky, working at the MIT Artificial Intelligence Laboratory (see MINSKY, MARVIN) further developed these concepts, and Frank Rosenblatt developed a classic form of neural network called a Perceptron. This consists of a network of processing elements (that is, functions), each of which are presented with weighted inputs (called *vectors*) from which it calculates an output value of either true (1) or false (0). The designer of the system knows what the correct output should be. If a given element (or node) produces the correct output, no changes are made. If it produces the wrong output, however, the weights given for each input are changed by some increment, plus a further adjustment or "bias" factor. This adjustment is repeated for all units as necessary until the output is correct. In other words, each neuron is constantly adapting the way it evaluates its inputs and thus its output, and that output is in turn being fed into the evaluation process of the neighboring neurons. (In practice, a neural network can have several layers of processing units, with one layer providing inputs to the next.)

For example, suppose a neural network is being trained to recognize objects based on the light being received from an array of sensors. The sensor readings are interpreted by a number of "neurons," which should output 1 if part of the desired object exists at the location scanned by its sensor. At first there will be many false readings—points at which part of the object is not recognized, or is falsely recognized. However, after many cycles of adjustment this "supervised learning" process results in a neural network that has a high probability of being able to identify all objects of a given general form. What is significant here is that a *generalized* ability has been achieved, and it has emerged without any specific programming being required!

Neural Network (2)

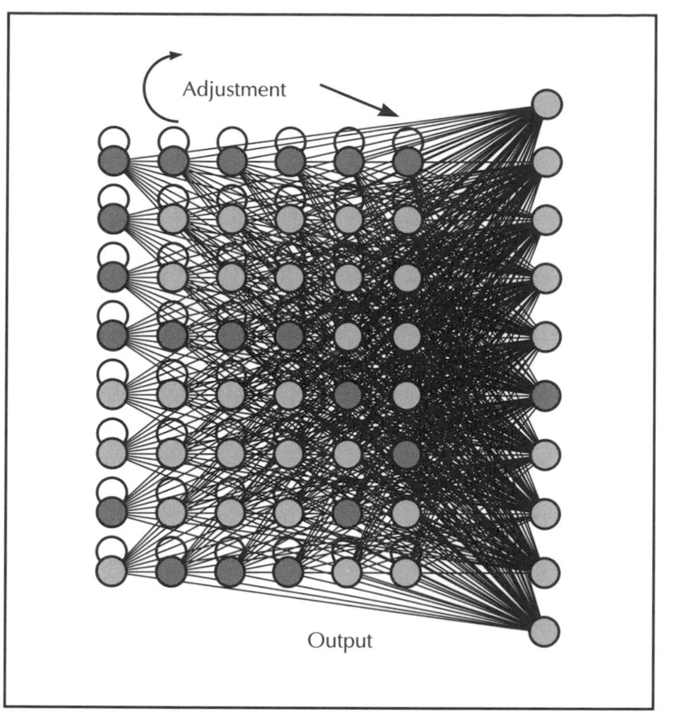

In a computer neural network the "neurons" or nodes are "trained" to detect a pattern by being adjusted until they successfully register it.

Neural networks are slowly making their way into commercial applications. They can be used to help robots recognize the key components of their environment (see ROBOTICS and COMPUTER VISION), for interpreting spoken language (see SPEECH RECOGNITION AND SYNTHESIS), and for problems in classification and statistical analysis (see also DATA MINING). In general, the neural network approach is most useful for applications where there is no clear algorithmic approach possible—in other words, applications that deal with the often "fuzzy" realities of daily life.

Further Reading

Bartlett, Peter L., and Martin M. Anthony. *Neural Network Learning: Theoretical Foundations.* New York: Cambridge University Press, 1999.

Brookshear, J. Glenn. *Computer Science: an Overview.* 6th ed. Reading, Mass.: Addison-Wesley, 2000.

"Introduction to Neural Networks." http://vv.carleton.ca/~neil/neural/neuron.html

Swingler, Kevin. *Applying Neural Networks: a Practical Guide.* San Diego, Calif.: Academic Press, 1996.

nonprocedural languages

Most computer languages are designed to facilitate the programmer declaring suitable variables and other data structures, then encoding one or more procedures for manipulating the data to achieve the desired result (see DATA TYPES and PROCEDURES AND FUNCTIONS). A further refinement is to join data and data manipulation procedures into *objects* (see OBJECT-ORIENTED PROGRAMMING).

However, since the earliest days of computing, programmers and language designers have tried to create higher-level, more abstract ways to specify what a program should do. Such higher-level specifications are, after all, easier for people to understand. And if the computer can do the job of translating a high-level specification such as "Find all the customers who haven't bought anything in 30 days and send them this e-mail message" into the appropriate procedural steps, people will be able to spend less time coding and debugging the program.

It is actually best to think of a continuum that has at one end highly detailed procedures (see ASSEMBLER) and at the other end an English-like syntax like that given above. Already in an early language like FORTRAN the emphasis is moving away from the details of how you multiply numbers and store the result to simply specifying the operation much like the way a mathematician would write it on a blackboard. such as T = I + M. COBOL can render such specifications even more readable, albeit verbose: ADD I TO M GIVING T, for example. However, these languages are still essentially procedural.

Some languages are less procedural in that they hide most of the details (or subprocedures) involved in carrying out the desired operation. For example, in modern database languages such as SQL what would be a procedure (or a set of procedures) in some languages is treated as a *query* at a high level (see SQL). For example:

```
select customer where (today - customer.lastpur-
chasedate) > 30
```

Programming packages such as Mathematica are also nonprocedural in that they allow for problems to be stated using the same symbolic notation that mathematicians employ, and many standard procedures for solving or transforming equations are then carried out automatically.

Other examples of relatively nonprocedural languages include logic-programming languages (see PROLOG and EXPERT SYSTEMS) and languages where the desired results are built up from defining functions rather than through a series of procedural steps (see LISP and FUNCTIONAL LANGUAGES).

Further Reading

Clocksin, W. F., and C. S. Mellish. *Programming in Prolog.* 4th ed. New York: Springer-Verlag, 1994.

Truitt, Thomas D., Stuart B. Mindlin, and Tarralyn A. Truitt. *An Introduction to Nonprocedural Languages: Using NPL.* New York: McGraw Hill, 1983.

numeric data

Text characters and strings can be stored rather simply in computer memory, such as by devoting 8 bits (one byte) or 16 bits to each character. The storage of numbers is more complex because there are both different formats and different sizes of numbers recognized by most programming languages.

Integers (whole numbers) have the simplest representation, but there are two important considerations: the total number of bits available and whether one bit is used to hold the sign.

Since all numbers are stored as binary digits, an *unsigned* integer has a range from 0 to 2^{bits} where "bits" is

Numeric Data (1)

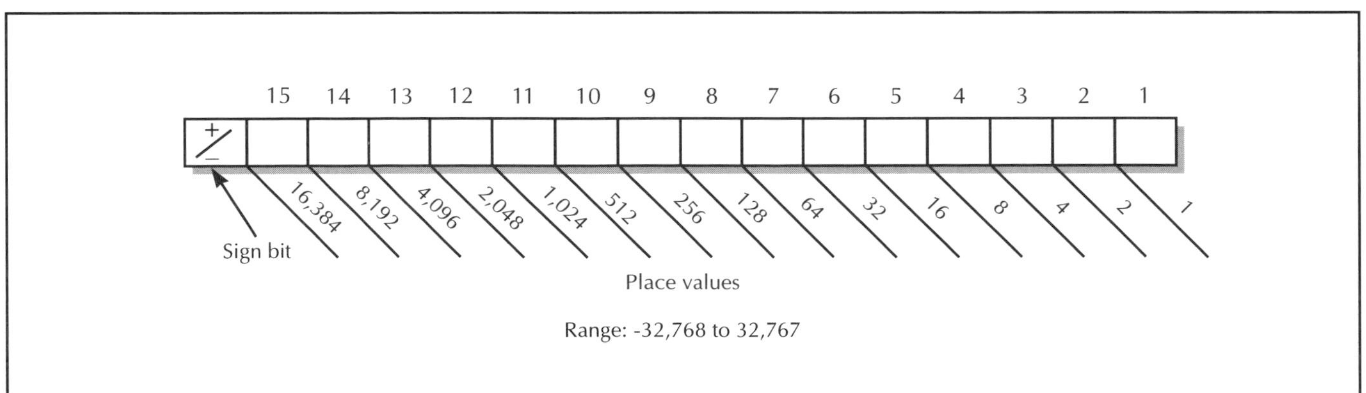

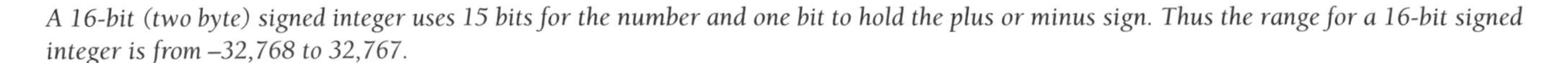

A 16-bit (two byte) signed integer uses 15 bits for the number and one bit to hold the plus or minus sign. Thus the range for a 16-bit signed integer is from –32,768 to 32,767.

Numeric Data (2)

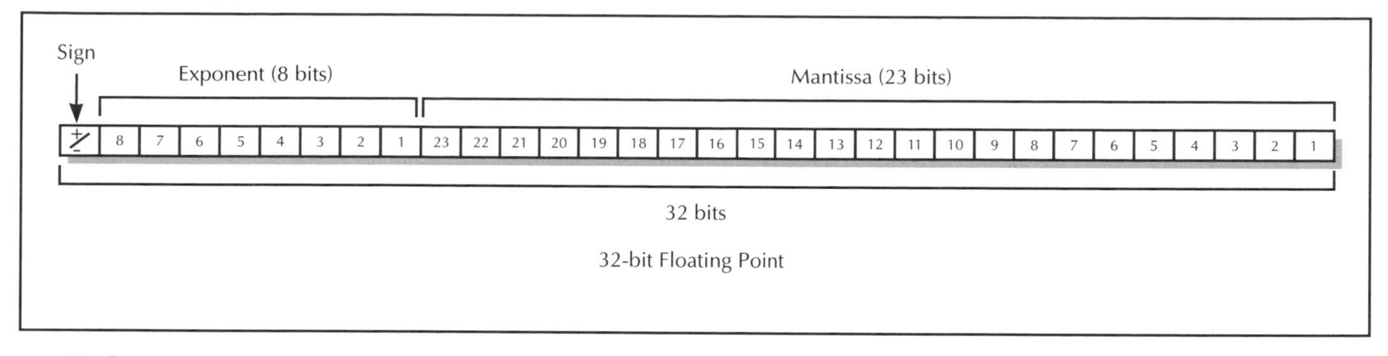

A 32-bit floating-point decimal number uses one bit for the sign, 8 bits for the exponent, and 23 bits for the significant digits (the mantissa).

the total number of bits available. Thus if there are 16 bits available, the maximum value for an integer is 65535. If negative numbers are to be handled, a signed integer must be used (in most languages such as C, C++, and Java, an integer is signed unless unsigned is specified). Since one bit is used to hold the sign and each bit doubles the maximum size, it follows that a signed integer can have only half the range above or below zero. Thus, a 16-bit signed integer can range from –32,768 to 32,767.

One complication is that the available sizes of integers depend on whether the computer system's native data size is 16, 32, or 64 bits. In most cases the native size is 32 bits, so the declaration "int" in a C program on such a machine implies a signed 32-bit integer that can range from -2^{31} or –2,147,483,647 to $2^{31}-1$, or 2,147,483,647. However, if one is using large numbers in a program, it is important to check that the chosen type is large enough. The *long* specifier is often used to indicate an integer twice the normal size, or 64 bits in this case.

FLOATING POINT NUMBERS

Numbers with a fractional (decimal) part are usually stored in a format called floating point. The "floating" means that the location of the decimal point can be moved as necessary to fit the number within the specified digit range. A floating point number is actually stored in four separate parts. First comes the *sign,* indicating whether the number is negative or positive. Next comes the *mantissa,* which contains the actual digits of the number, both before and after the decimal point. The *radix* is the "base" for the number system used. Finally, the *exponent* determines where the decimal point will be placed.

For example, the base 10 number 247.35 could be represented as 24735×10^{-2}. The –2 moves the decimal point at the end two places to the left. However, floating-point numbers are *normalized* to a form in which there is just one digit to the left of the decimal point. Thus, 247.35 would actually be written 2.4735×10^{2}. This system is also known as scientific notation.

As noted earlier, actual data storage in modern computers is always in binary, but the same principle applies. According to IEEE Standard 754, 32-bit floating-point numbers use 1 bit for the sign, 8 bits for the exponent, and 23 bits for the mantissa (also called the *significand,* since it expressed the digits that are significant—that is, guaranteed not to be "lost" through overflow or underflow in processing). The *double precision* float, declared as a "double" in C programs, uses 1, 11, and 52 bits respectively.

Programmers who use relatively small numbers (such as currency amounts) generally don't need to worry about loss of precision. However, if two numbers being multiplied are large enough, even though both numbers fit within the 32-bit size, their product may well generate more digits than can be held within the 23 bits available for the mantissa. This means that some precision will be lost. This can be avoided to some extent by using the "double" size.

Since floating-point calculations use more processor cycles (see MICROPROCESSOR) than integer calculations, processor designers have paid particular attention to improving floating-point performance. Indeed, processors are often rated in terms of "megaflops" (millions of floating-point operations per second) or even "gigaflops" (billions of flops).

Further Reading

Hollasch, Steve. "IEEE Standard 754 Floating Point Numbers." http://research.microsoft.com/~hollasch/cgindex/coding/ieeefloat.html

Sebesta, Robert W. *Concepts of Programming Languages.* Reading, Mass.: Addison-Wesley, 1999.

object-oriented programming

During the last two decades the way in which programmers view the data structures and functions that make up programs has significantly changed. In simplified form the earliest approach to programming was roughly the following:

- Determine what results (or output) the user needs
- Choose or devise an algorithm (procedure) for getting that result
- Declare the variables needed to hold the input data
- Get the data from the file or user input
- Assign the data to the variables
- Execute the algorithm using those variables
- Output the result

While this type of approach often works well for small "quick and dirty" programs, it becomes problematic as the complexity of the program increases. In real-world applications data structures (such as for a customer record or inventory file) are accessed and updated by many different routines, such as billing, inventory, auditing, summary report generation, and so on. It is easy for a programmer working on one part of the program to make a change in a data field specification (such as changing its size or underlying data type) without other programmers finding out. Suddenly, other parts of the program that relied on the original definitions start to "break," giving errors, or worse, silently produce incorrect results.

During the 1970s, computer scientists advocated a variety of reforms in programming practices (see STRUCTURED PROGRAMMING) in an attempt to make code both more readable and safer from unwanted side effects. For example, the "goto" or arbitrary jump from one part of the program to another was discouraged in favor of strictly controlled iterative structures (see LOOP). Also encouraged was the declaration of local variables that could not be changed from outside the procedure in which they were defined.

DEVELOPMENT OF OBJECT-ORIENTED LANGUAGES

However, a more radical programming paradigm was also in the making. In existing languages, there is no inherent connection between data and the procedures that operate upon that data. For example, the employee record may be declared somewhere near the beginning of the program, while procedures to update fields in the record, copy the record, print the record, and so on may well be found many pages deeper into the program.

A new approach, object-oriented programming is based on the fact that in daily life we interact with thousands of objects. An object, such as a ball, has properties (such as size and color) and capabilities (such as bouncing). In interacting with an object, we use its capabilities. It is much more natural to think of an object as a whole than to have its properties and capabilities jumbled together with those of other objects.

Simula 67, developed in the late 1960s, was the first object-oriented language (see SIMULA). It was followed in the 1970s by Smalltalk, a language developed at the Xerox PARC laboratory, home of innovative research in graphical user interfaces. Smalltalk, like Windows today, treats each window, menu, and other control on the screen as an object (see SMALLTALK). Finally, during the 1980s C++ came into prominence, adding the essential features of object-oriented programming to the already very popular C language. Today most popular mainstream languages, including C++, Java, and Visual Basic, are object-oriented (see C++ and JAVA). Many specialized database languages are also object-oriented.

ELEMENTS OF OBJECT-ORIENTED PROGRAMMING

The various object-oriented languages differ somewhat in capabilities, and of course in syntax. However, being object-oriented generally implies that the language has the following features.

CLASSES AND OBJECTS

An object is defined using a template called a class. A class contains both the data needed to characterize the object and the procedures (sometimes called methods or member functions) needed to work with the object (see CLASS). Thus, there could be a class for circles to be drawn on a graphics display. The class might include as its data the x and y coordinates for the center of the circle, the size of the radius, whether the circle is filled, the color to be used for filling, and so on. (See C++ for more examples.)

When the program needs to use an object of the class, it declares it in the same way it would an ordinary built-in data type such as an integer. Languages such as C++ provide for a special function called a constructor that can be used to define the processing needed when a new object is created—for example, memory allocation and setting initial values for variables.

To access data or functions within a class, the name of an object of that class is used, followed by a variable or function. Thus, if there's a class called *circle*, a program might specify the following:

```
MyCircle Circle; // Declare an object of the
Circle class
MyCircle.X = 100; // X coordinate on screen
MyCircle.Y = 50; // Y coordinate on screen
MyCircle.Radius = 25; // Radius in pixels
MyCircle.Filled = True; // A Boolean constant
equal to 1
MyCircle.FillColor = Blue; // a previously
defined color constant
```

Object-Oriented Programming (1)

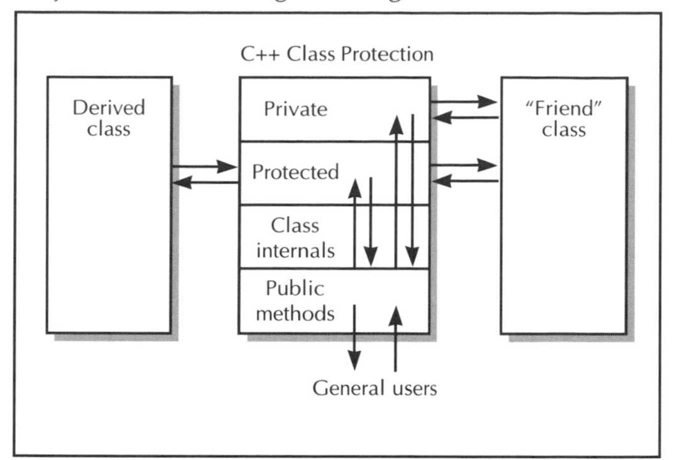

In the object-oriented C++ programming language data within a class can be restricted in several ways. Private data can be accessed only from within the class itself, or from another class declared to be a "friend" of the containing class. Protected data has these forms of access plus it can also be accessed from any class derived from the containing class. Finally, public data or functions (methods) can be accessed from anywhere in the program, and provide the interface by which the class is used.

Once these specifications have been made, the circle can be drawn by calling upon its "draw" method or member function:

```
MyCircle.Draw;
```

The designer of a class can choose to restrict access to certain data items or functions, using a keyword such as private or protected. For example, instead of having the part of the program that uses the class directly set the x and y coordinates, it could keep those variables private and instead provide a method called SetPos. The class might then take the coordinates specified by the user and adjust them to fit the screen dimensions. The Draw method would then use the adjusted internal coordinates rather than those supplied originally by the user.

INHERITANCE

Many objects are more elaborate or specialized variations of more basic objects. For example, in Microsoft Windows the various kinds of dialog boxes are specialized versions of the general Window class. Therefore, the specialized version is created by declaring it to be derived from a "base class." Put another way, the specialized class *inherits* the basic data and functions available in the base (parent) class. The programmer can then add new data or functions or modify the inherited ones to create the necessary behavior for the specialized class.

Languages such as C++ allow for a class to be derived from more than one base class. This is called multiple

Object-Oriented Programming (2)

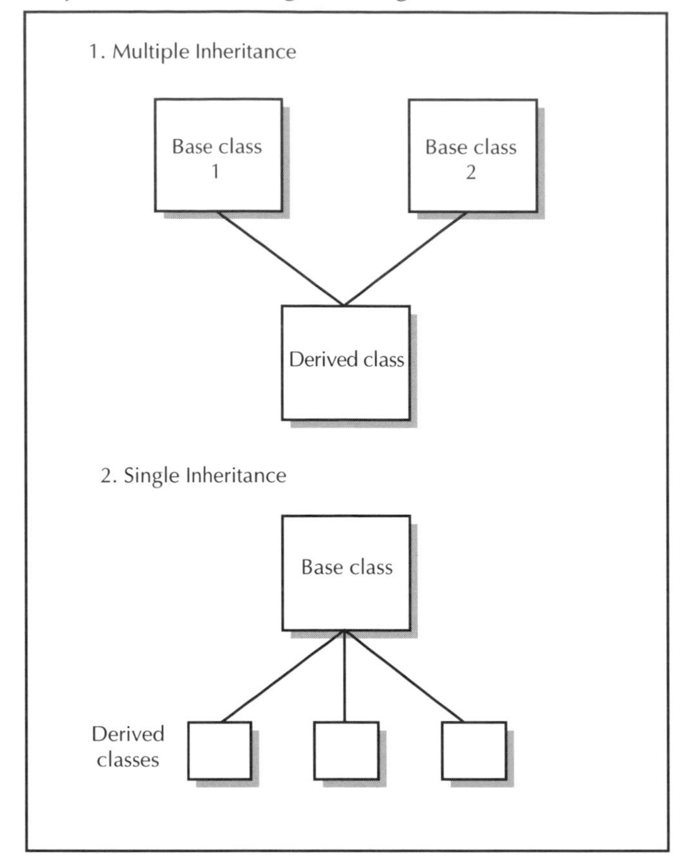

C++ and some other object-oriented languages support multiple inheritance, where a class can be derived from (and obtain functionality from) two or more parent or "base" classes. However, some languages such as Java support only single inheritance, where a derived class can have only one base class.

inheritance. For example, a Message Window class might inherit its overall structure from the Window class and its text-display capabilities from the Message class. However, it can sometimes be difficult to keep the relationships between multiple classes clear. The Java language takes the alternative approach of being limited to only single inheritance of classes, but allowing interfaces (specifications of how a class interacts with the program) to be multiply inherited.

POLYMORPHISM AND OVERLOADING

Different kinds of objects often have analogous methods. For example, suppose there is a series of classes that represent various polygons: square, triangle, hexagon, and so forth. Each class has a method called "perimeter" that returns the total distance around the edges of the object. If each of these classes is derived from a base polygon class, each class inherits the base class's perimeter method and adapts it for its own use. Thus, a square might calculate its perimeter simply by multiplying the length of a side by four, while the rectangle would have to add up different-sized pairs of sides, and so on.

Similarly, the same operator in a language can have different meanings depending on what data types it is being applied to. The plus (+) operator, for example, is defined in most languages so that various types of integers or floating-point values can be added (see NUMERIC DATA).

Object-oriented languages such as C++ allow operators to be given additional definitions so they can handle additional data types, including classes defined by the user. For example, what might adding the string "object" and the string "oriented" yield? The most sensible answer is a new string that contains both of the original strings: "object oriented." If one defines a String class, then one can also define the + operator as a member function of that class, such that when something like String1 + String2 is encountered, the expression will be evaluated as the combination (concatenation) of the two strings. The + operator is said to have been *overloaded* for use with the String class.

ENCAPSULATION

The ability to keep the detailed workings of a class private promotes program reliability (see ENCAPSULATION). Software developers can create well-organized libraries of classes that other programmers can use simply by referring to the interface specifications (see LIBRARY, PROGRAM). Encapsulation also makes programs more readable. Once one understands the capabilities of the objects, it is relatively easy to understand the overall operation of the program without getting bogged down in details. Object-oriented programming takes the encapsulation achieved through the earlier structured programming movement and makes it more integral to the language structure.

TRENDS

Object-oriented programming was initially decried as a fad by some critics. The initial learning curve for traditionally trained programmers and the overhead that made early implementations of languages such as Smalltalk run slowly inhibited acceptance of the new paradigm at first. However, the introduction of C++ by Bjarne Stroustrup provided a fairly easy path for C programmers into the object-oriented world. For example, the class was syntactically similar to the familiar struct.

The movement toward object-oriented programming and design was also spurred by the more or less coincidental popularity of graphical user interfaces such as Microsoft Windows. Since these systems are built upon event-driven programming using a variety of coexisting objects, the object-oriented class approach fit such operating systems much more naturally. Thus, during the late

1980s and 1990s, many Windows programmers began to use the Microsoft Foundation Classes (MFC) as their way to structure their access to the operating system.

An object-based approach also fits more naturally into environments where programs and data may be running on many interconnected computers (see NETWORK and MULTIPROCESSING). Treating the client and server programs as interacting objects thus makes sense, as does treating databases as collections of data objects (see DATABASE MANAGEMENT SYSTEM). For all of these reasons the object paradigm is likely to remain dominant for the foreseeable future.

Further Reading

Ambler, Scott W. *The Object Primer.* 2nd ed. New York: Cambridge University Press, 2001.

"Cetus Links: Object-Orientation." http://www.sente.ch/cetus/software.html

Müller, Peter. "An Introduction to Object-Oriented Programming Using C++." http://www.desy.de/user/projects/C++/courses/cc/Tutorial/tutorial.html

Shalloway, Alan, and James R. Trott. *Design Patterns Explained: A New Perspective on Object-Oriented Design.* Reading, Mass.: Addison-Wesley, 2001.

Stroustrup, Bjarne. *The C++ Programming Language.* 3rd ed. Reading, Mass.: Addison-Wesley, 1997.

office automation

The transition from manual to mechanical to electronic processing of information in the office spanned most of the 20th century. In the previous century, the typewriter allowed for the mechanical production of letters and other documents by skilled workers, accommodating (and perhaps encouraging) a growing amount of paperwork. At the turn of the century the card tabulator (see HOLLERITH, HERMAN and PUNCHED CARDS AND PAPER TAPE) began the mechanization of information processing.

During the first half of the 20th century, mechanical or electro-mechanical calculators made by such companies as Burroughs came into more widespread use by bookkeepers and clerks (see CALCULATOR). Meanwhile, one company, International Business Machines (IBM) came to dominate the area of card sorting and tabulating equipment.

When digital computers first came into commercial use in the 1950s, they were too large and expensive to be used in ordinary offices. Bookkeepers and other workers did not deal with computers directly, but were supported by data processing departments or outside service bureaus for what became known as electronic data processing, or EDP.

By the 1970s, the advent of the microprocessor made desk-size information processing systems possible (see MICROPROCESSOR). The first widespread application was the dedicated word processing system, of which the most successful version was developed by An Wang. These systems provided for typing and printing documents and storing them in a file system (see WORD PROCESSING).

During the 1980s, the general-purpose desktop computer (see PERSONAL COMPUTER) became powerful enough to supplant the dedicated word-processing system. Besides providing word-processing functions through ever more versatile versions of programs such as WordPerfect. WordStar, and Microsoft Word, the PC could also run programs to support bookkeeping, accounting, mailing list, and other functions (see DATABASE MANAGEMENT SYSTEM and SPREADSHEET). Gradually, many of these separate programs were merged into office suites such as Microsoft Office (see APPLICATION SUITE). Using a suite meant that information could be easily transferred between word-processing documents, spreadsheets, and database files, facilitating the generation of many kinds of reports and presentations.

Later in the 1980s, two new aspects of office automation began to emerge: communication and collaboration. The use of special hardware and software to connect PCs within an office or throughout the organization (see NETWORK and LOCAL AREA NETWORK) made new applications possible. E-mail began to replace printed memos or phone calls as the preferred way for workers and management to communicate. Programs such as Lotus Notes and Microsoft Outlook added features such as the ability of workers to share a common calendar of tasks, while scheduling software offered more elaborate ways to keep track of large, detailed team projects (see PROJECT MANAGEMENT SOFTWARE).

Today a variety of tools are available for facilitating collaboration. Most word-processing software now offers a feature called revision marking, which lets various editors and reviewers comment on or make revisions to a document. The author can then merge the revisions into a new draft. "Whiteboard" programs let several users on the network work simultaneously on the same virtual screen, drawing diagrams or making outlines.

TRENDS

Even as desk space was being cleared for the first office PCs, pundits began to claim that the "paperless office" was at hand. Actually, the first stages of automation contributed to an increase in the use of paper. On the one hand, word processors and other programs made it easier to generate documents and keep them up to date. On the other hand, the documents were all printed on paper—in part because the ability to share them electronically was nonexistent or rudimentary, and in part because many workers, particularly senior executives, still preferred to work with paper.

The growth of networking made it possible for more people to distribute documents electronically, while higher-resolution video displays made it easier to view pages on the screen. During the 1990s, the inexpensive document scanner (see SCANNER) made it practicable to scan incoming paper documents into text files (see OPTICAL CHARACTER RECOGNITION). While the office is not yet paperless, the tide of paper may now be receding at last.

The ubiquity of the Internet and the use of the HTML format for documents (see HTML and INTRANET) characterize the latest phase in the evolution of office automation. Many corporate procedure manuals and other resources are now being stored on company websites where they can be updated easily and consulted with the aid of search engines. Databases to which workers need shared access are also being hosted through websites. HTML is emerging as a common format for exchanging documents between systems, along with Adobe's Portable Document Format (PDF), which offers a faithful reproduction of the printed page.

Another recent trend is the "mobile office." Many workers can now access the full resources of the office through laptop computers and Internet connections. Workers on the go can also use handheld or palm computers such as the PalmPilot (see PORTABLE COMPUTERS) to access e-mail, calendar, and other information. The growing use of videoconferencing over the Internet using inexpensive cameras and broadband connections is also promoting the "virtual meeting" (see VIDEO CONFERENCING). The economic downturn and concern about air travel following the terrorist attacks of September 2001 may further promote this trend.

Further Reading

Opper, S., and H. Fersko-Weiss. *Technology for Teams: Enhancing Productivity in Networked Organizations.* New York: Van Nostrand Reinhold, 1992.

Ray, C., J. Palmer, and A. Wohl. *Office Automation: A Systems Approach.* Cincinnati, Ohio: South-Western Educational Publishing, 1995.

on-line research

The proliferation of on-line databases, information services (see ON-LINE SERVICES) and websites has made more information accessible to more people than ever before. At the same time, the complexity of the on-line world challenges researchers to develop a new set of skills to cope with it.

It is useful to divide on-line offerings into three broad categories: specialized databases, on-line information services, and the Web as a whole (see WORLD WIDE WEB). Each of these areas requires a somewhat different approach by the on-line researcher.

A common research task is to find and evaluate books or articles on a given subject. Most local libraries have their catalogs on-line, and the world's largest library catalog, that of the Library of Congress (LC), is also available in several forms on the Web.

Newspaper and magazine articles can be found in a number of general-purpose databases such as InfoTrac. These databases can be searched in public libraries: Remote access is generally restricted to the library's cardholders. These records can consist of a bibliographic description only (that is, author, title, periodical, issue date, and so on) or can include an abstract or in many cases the full text of the article. In addition, most major newspapers now offer free access to recent articles on their website, with older articles available for a nominal fee. Magazines, too, frequently offer selected articles or their complete contents on-line.

Using the search facility for an on-line catalog or periodical database is generally simple, particularly if an author or title is known. For subject searching, some familiarity with LC subject headings is helpful. However, the ability of most systems to search for matching words in titles or subjects means that the researcher can be quickly led to the correct subject in most cases.

Another way to get tables of contents, jacket copy, and reviews of books is to browse the on-line catalogs of major booksellers, particularly Amazon.com and BarnesandNoble.com. Publishers' websites are another good way to get information about books, particularly new or forthcoming titles.

Journalists need a broad familiarity with on-line research tools and use computers and on-line services in many facets of their work (see JOURNALISM AND COMPUTERS). Researchers looking for specialized articles in fields such as law or medicine need more rigorous skills.

Most legal research is done using databases such as LexisNexis. These databases are expensive but indispensable to practitioners. However, students and others who can't afford this access can still find U.S. Supreme Court, Court of Appeals, and many state court decisions on-line, thanks to the efforts of organizations such as the Legal Information Institute at Cornell Law School. Because of the complexity of multiple jurisdictions and the need to trace chains of precedent ("shepardizing"), on-line legal research has become an increasingly important paraprofessional task.

Medical research is similarly complex, due to the thousands of precise terms for conditions, procedures, and drugs. The sheer volume of articles (MEDLINE has more than 11 million citations dating back to the 1960s) can make it hard to find and evaluate the most relevant material.

There are a variety of on-line information services that provide organized access to information through forums devoted to particular topics. America On-line and CompuServe have the most extensive offerings.

By far the most extensive information resource today is the World Wide Web with its millions of sites and pages of information. There are two basic approaches to finding material on the Web. The first is to use a search engine by typing in keywords or phrases (see SEARCH ENGINE). Even though search engines such as AltaVista or Google index only a modest fraction of the available pages on the Web, a search on a topic such as "database design" can yield from thousands to millions of possible "hits." Most search engines do attempt to rank results in decreasing order of matching or relevance.

An alternative approach is to browse the categorized list of topics presented by a site such as Yahoo! (www.yahoo.com) or About.com (www.about.com). The advantage of this approach is that the site's researchers have selected the links for each topic that they believe to be the most valuable, and the number of possibilities is likely to be more manageable (see PORTAL).

On-line research remains more an art than a science. The researcher must choose the appropriate tools—bibliographical resources, specialized databases, information services, search engines, and portals—and evaluate and integrate the results so they are useful for a given question or project. Students and researchers now have unprecedented access to information, but sophisticated critical thinking skills must be employed. In particular, it can be difficult to evaluate the background or credentials of the people behind websites that are not associated with recognized media outlets or other organizations.

Further Reading

Cornell Law School. Legal Information Institute. http://www.law.cornell.edu/

Hane, Paula J. *Super Searchers in the News: The On-line Secrets of Journalists and News Researchers.* Medford, N.J.: Information Today, 2000.

LexisNexis. http://www.lexis-nexis.com/default.asp

Library of Congress Catalog. http://catalog.loc.gov/

National Library of Medicine. PubMed. http://www.ncbi.nlm.nih.gov/entrez/query.fcgi

Schlein, Alan M. [and others], eds. *Find It On-line: the Complete Guide to On-line Research.* 2nd ed. Tempe, Az.: Facts on Demand, 2000.

"A Student's Guide to Research with the WWW." http://www.slu.edu/departments/english/research/

on-line services

The ability of PC owners to connect to remote computers (see MODEM) led to the proliferation of both free and commercial on-line information services during the 1980s. At one end of the spectrum were bulletin board systems (BBS), many run by hobbyists on PCs connected to a few phone lines (see BULLETIN BOARD SYSTEMS). They offered users the ability to read and post messages on various topics as well as to download or contribute software (see also SHAREWARE).

The growing number of connect PC owners soon offered entrepreneurs a potential market for a commercial on-line information service. One of the oldest, CompuServe, had actually been started in 1969 as a business time-sharing computer system. In 1979, it launched a service for home computer users, offering e-mail and technical support forums. By the mid-1980s, the service had added an on-line chat service called CB Simulator (see CHAT, ON-LINE) as well as news content. The service's greatest strength, however, remained its forums, which offered technical support for just about every sort of computer hardware or software, together with download libraries containing system patches, drivers, utilities, templates, macros, and other add-ons.

By then, however, the on-line service market had become quite competitive. While CompuServe focused on computer-savvy users, America On-line (AOL), founded in 1985 by Steve Case, targeted the growing legion of new PC users who needed an easy-to-navigate interface. AOL grew steadily, reaching a million customers in 1994 and having more than 22 million today. AOL chat groups became very popular, spawning a vigorous on-line culture while raising controversies about sexual content in some chat "rooms." A third service, Prodigy, also catered to the new user.

Meanwhile, the World Wide Web and the advent of graphical Web browsers such as Netscape and Microsoft Internet Explorer in the mid-1990s led millions of users to connect to the Internet (see INTERNET, WEB BROWSER, and WORLD WIDE WEB). Internet service providers (ISPs) offered direct, no-frills access to the Web. CompuServe and AOL soon offered their users access to the Internet as well. However, accessing the Web through an on-line information service was usually more expensive, and often slower, than using an ISP and a Web browser directly. Additionally, free Web portal services such as Yahoo! began to offer extensive information resources of their own.

The Internet thus threatened to shrink the market for the commercial on-line services. AOL fought back in the late 1990s by cutting its monthly rates to make them competitive with ISPs, flooding the mails with free disks and trial offers, bundling introductory packages with new computer systems, and promoting added-value information services such as stock quotes. In 1998, the market consolidated when AOL bought CompuServe, continuing to run the latter as a subsidiary targeted at more sophisticated users. The same year AOL bought Netscape to gain access to its browser technology. Finally, AOL merged with Time-Warner, creating a massive media conglomerate and providing extensive new content for AOL. AOL continues to sign up large numbers of new users while also creating customized services for different countries, regions, or communities.

The long-term prospects for AOL and other commercial on-line services are uncertain. Many of the advantages these services had until the late 1990s have diminished. For example, the once mutually incompatible e-mail systems of on-line services have been replaced by standard Internet e-mail protocols, so there is little advantage to using a particular service for e-mail. Users can obtain e-mail accounts from a variety of ISPs or through free Web-based services such as hotmail.com. Content such as news, video, and music (see STREAMING) is available from many websites, and most companies now offer extensive on-line technical support for their products. At the same time, attempts to support content-rich sites through either advertising or a subscription model have largely foundered, feeding the "dot-com" collapse of 2000–2001. For services such as AOL, the ultimate question is whether the parts of the service still form a sufficiently compelling whole.

Further Reading

Stauffer, David. *It's a Wired, Wired World: Business the AOL Way.* Dover, N.H.: Capstone, 2000.

Swisher, Kara. *AOL.com: How Steve Case beat Bill Gates, Nailed the Netheads, and Made Millions in the War for the Web.* New York: Crown, 1999.

Wagner, Richard. *Inside CompuServe.* 3rd ed. Indianapolis, Ind.: New Riders, 1995.

open source movement

For a long time programmers have released programs as freeware meaning that users did not have to buy or license the software. There is also "try before you buy" software (see SHAREWARE). However, while freeware sometimes includes not only the executable program but the source code (the actual program instructions), most shareware and virtually all other commercially distributed software does not. As a result, users wishing to fix, modify, or extend the software are generally at the mercy of the company that owns and distributes it.

In university and research computing environments, however, it has been common for programmers to freely share and extend utilities such as program editors. Indeed, much of the necessary software for the earliest minicomputers of the 1960s was created by clever, energetic hackers (see HACKERS AND HACKING). Because the source code (usually on paper tape) was freely distributed, people could easily create and distribute new (and presumably, improved) versions. Having source code also made it possible to "port" software to a newly released machine without having to wait for the relatively ponderous efforts of the official developers.

In particular, although the licensing of the two major versions of the UNIX operating system were controlled by AT&T's Bell Laboratories and the University of California's Berkeley Software Distribution (BSD) respectively, much UNIX software including programming languages (see PERL and PYTHON) and the Web's most popular server, Apache, have been distributed using an *open source* model.

The best-known open source effort is the GNU Project created by Richard Stallman (1953–). GNU, a recursive acronym meaning "GNU's Not UNIX," is a collection of software that provides much of the functionality of AT&T's UNIX without being subject to the latter's licensing fees and restrictions. When creating his own open source version of UNIX (called Linux), Linus Torvalds and his colleagues drew upon the considerable base of software already created by GNU.

According to Stallman and many other advocates, "open source" software is not necessarily free. What is required is that users receive the full source code (or have it readily available for free or at nominal charge). Users are free to modify or expand the source code to create and distribute new versions of the software. Following a legal mechanism that Stallman calls "copyleft," the distributor of open source software must allow subsequent recipients the same freedom to revise and redistribute. However, not all software that is billed as open source follows all of Stallman's requirements, including being copylefted.

Open source software has the potential for providing diversity and alternatives in a world where some categories such as PC operating systems and office software are dominated by one or a few large companies. Indeed, sometimes companies have converted an existing product to open source, as is the case with Sun Microsystems and Star Office, a suite that runs under Linux. Netscape also resorted to open source as part of an unsuccessful attempt to fight off Microsoft for dominance of the browser market in the mid to late 1990s. By making a product open source, a company may hope to tap into the volunteer effort of many talented programmers to improve or expand the program. The company is still free to create proprietary software upon the "base" of a successful open source product. Moderately successful companies such as Linux distributor Red Hat have a business plan based upon providing superior packaging, technical support, and customized solutions around its Linux distribution.

Further Reading

Dibona, Chris. *Open Sources: Voices from the Open Source Revolution.* Sebastopol, Calif.: O'Reilly, 1999.

Free Software Foundation. www.fsf.org

Open Source Initiative. http://www.opensource.org/

"Philosophy of the GNU Project." http://www.gnu.org/philosophy/philosophy.html#AboutFreeSoftware

Raymond, Eric. *The Cathedral and the Bazaar: Musings on Linux and Open Source by an Accidental Revolutionary.* Sebastopol, Calif.: O'Reilly, 2001.

operating system

An operating system is an overarching program that manages the resources of the computer. It runs programs and provides them with access to memory (RAM), input/output devices, a file system, and other services. It provides application programmers with a way to invoke system services, and gives users a way to control programs and organize files.

DEVELOPMENT

The earliest computers were started with a rudimentary "loader" program that could be used to configure the system to run the main application program. Gradually, a more sophisticated way to schedule and load programs, link programs together, and assign system resources to them was developed (see JOB CONTROL LANGUAGE and MAINFRAME).

As systems were developed that could run more than one program at a time (see MULTITASKING), the duties of the operating systems became more complex. Programs had to be assigned individual portions of memory and prevented from accidentally overwriting another program's memory area. A technique called *virtual memory* was developed to enable a disk drive to be treated as an extension of the main memory, with data "swapped" to and from the disk as necessary. This enabled the computer to run more and/or larger applications. The operating system, too, became larger, amounting to millions of bytes worth of code.

During the 1960s, time sharing became popular particularly on new smaller machines such as the DEC PDP series (see MINICOMPUTER), allowing multiple users to run programs and otherwise interact with the same computer. Operating systems such as Multics and its highly successful offshoot UNIX developed ways to assign security levels to files and access levels to users. The UNIX architecture featured a relatively small *kernel* that provides essential process control, memory management, and file system services, while drivers performed the necessary low-level control of devices and a *shell* provided user control. (See UNIX, KERNEL, DEVICE DRIVER, and SHELL.)

Starting in the late 1970s, the development of personal computers recapitulated in many ways the earlier evolution of operating systems in the mainframe world. Early microcomputers had a program loader in read-only memory (ROM) and often rudimentary facilities for entering, running, and debugging assembly language programs.

During the 1980s, more complete operating systems appeared in the form of Apple DOS, CP/M, and MS-DOS for IBM PCs. These operating systems provided such facilities as a file system for floppy or hard disk and a command-line interface for running programs or system utilities. These systems could run only one program at a time (although exploiting a little-known feature of MS-DOS allowed additional small programs to be tucked away in memory).

As PC memory increased from 640K to multiple megabytes, operating systems became more powerful. Apple's Macintosh operating system and Microsoft Windows could manage multiple tasks. Today personal computer operating systems are comparable in sophistication and capability to those used on mainframes. Indeed, PCs can run UNIX variants such as the popular Linux.

COMPONENTS

While the architecture and features of operating systems differ considerably, there are general functions common to almost every system. The "core" functions include "booting" the system and initializing devices, process management (loading programs intro memory assigning them a share of processing time), and allowing processes to communicate with the operating system or one another. Multiprogramming systems often implement not only processes (running programs) but also *threads*, or sections of code within programs that can be controlled separately.

A memory management scheme is used to organize and address memory, handle requests to allocate memory, free up memory no longer being used, and rearrange memory to maximize the useful amount (see MEMORY MANAGEMENT).

There is also a scheme for organizing data created or used by programs into files of various types (see FILE). Most operating systems today have a hierarchical file system that allows for files to be organized into directories or folders that can be further subdivided if necessary. In operating systems such as UNIX, other devices such as the keyboard and screen (console) and printer are also treated like files, providing consistency in programming. The ability to redirect input and output is usually provided. Thus, the output of a program could be directed to the printer, the console, or both.

In connecting devices such as disk drives to application programs, there are often three levels of control. At the top level, the programmer uses a library function to open a file, write data to the file, and close the file. The library itself uses the operating system's lower-level input/output (I/O) calls to transfer blocks of data. These in turn are translated by a *driver* for the particular device into the low-level instructions needed by the processor that controls the device. Thus, the command to write data to a file is ultimately translated into commands for positioning the disk head and writing the data bytes to disk.

Operating systems, particularly those designed for multiple users, must also manage and secure user accounts. The administrator (or sometimes, ultimately, the "super user" or "root") can assign users varying levels of access to programs and files. The owners of files can in turn specify whether and how the files can be read or changed by other users (see DATA SECURITY).

In today's highly networked world most operating systems provide basic support for networking protocols such

Operating System

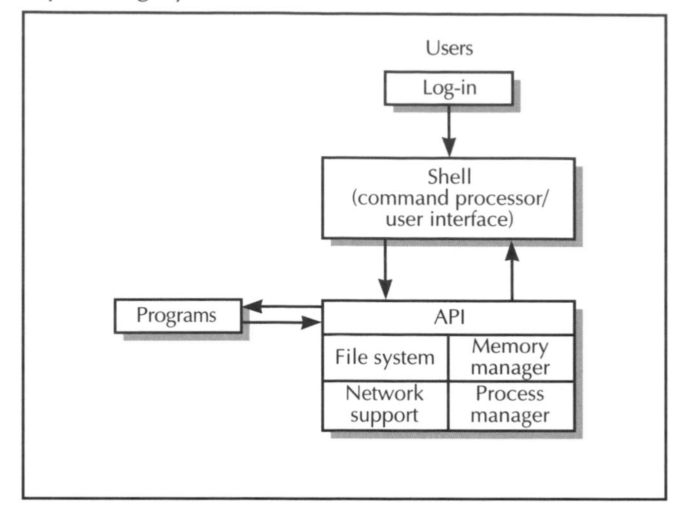

A typical operating system processes user commands or actions using an interface (such as a shell). Both user commands and requests from application programs communicate with the operating system through the application programming interface (API), which provides services such as file, memory, process, and network management.

as TCP/IP. Applications can use this facility to establish network connections and transfer data over the local or remote network (see NETWORK).

The operating system's functions are made available to programmers in the form of program libraries or an application programming interface (API). (See LIBRARY, PROGRAM and APPLICATION PROGRAMMING INTERFACE.)

The user can also interact directly with the operating system. This is done through a program called a *shell* that accepts and responds to user commands. Operating systems such as MS-DOS and early versions of UNIX accepted only typed-in text commands. Systems such as Microsoft Windows and UNIX (through facilities such as XWindows) allow the user to interact with the operating system through icons, menus, and mouse movements. Application programmers can also provide these interface facilities through the API. This means that programs from different developers can have a similar "look and feel," easing the learning curve for users.

ISSUES AND TRENDS

As the tasks demanded of an operating system have become more complex, designers have debated the best overall form of architecture to use. One popular approach, typified by UNIX, is to use a relatively small kernel for the core functions. A community of programmers can then write the utilities needed to manage the system, performing tasks such as listing file directories, editing text, or sending e-mail. New releases of the operating system then incorporate the most useful of these utilities. The user also has a variety of shells (and thus interfaces) available.

The kernel approach makes it relatively easy to port the operating system to a different computer platform and then develop versions of the utilities. (Kernels were also a necessity when system memory was limited and precious, but this consideration is much less important today.)

Designers of modern operating systems face a number of continuing challenges:

- Security, in a world where nearly all computers are networked, often continuously (see COMPUTER CRIME AND SECURITY and FIREWALL)
- The tradeoff between powerful, attractive functions such as scripting and the security vulnerabilities they tend to present
- The need to provide support for new applications such as streaming audio and video (see STREAMING)
- Ease of use in installing new devices (see DEVICE DRIVER and PLUG AND PLAY)
- The continuing development of new user interface concepts, including alternative interfaces for the disabled and for special applications (see USER INTERFACE and DISABLED PERSONS AND COMPUTING)
- The growing use of multiprocessing and multiprogramming, requiring coordination of processors sharing memory and communicating with one another (see MULTIPROCESSING and CONCURRENT PROGRAMMING)
- Distributed systems where server programs, client programs, and data objects can be allocated among many networked computers, and allocations continually adjusted or balanced to reflect demand on the system (see DISTRIBUTED COMPUTING)
- The spread of portable, mobile, and hand-held computers and computers embedded in devices such as engine control systems (see PORTABLE COMPUTERS and EMBEDDED SYSTEMS). Sometimes the choice is between devising a scaled-down version of an existing operating system and designing a new OS that is optimized for devices that may have limited memory and storage capacity.

Further Reading

Bach, Maurice J. *The Design of the UNIX Operating System*. Englewood Cliffs, N.J.: Prentice Hall, 1986.

Ritchie, Dennis M. "The Evolution of the UNIX Time-Sharing System." *Lecture Notes in Computer Science #79: Language Design and Programming Methodology*, New York: Springer-Verlag, 1980. Available on-line at http://cm.bell-labs.com/cm/cs/who/dmr/hist.html

Silberschatz, Abraham, Peter Baer Galvin, and Greg Gagne. *Operating System Concepts.* 6th ed. New York: Wiley, 2001.

optical character recognition

Today it is easy to optically scan text or graphics printed on pages and convert it into a graphical representation for storage in the computer (see SCANNER). However, a shape such as a letter *c* doesn't mean anything in particular as a graphic. Optical character recognition (OCR) is the process of identifying the letter or other document element that corresponds to a given part of the scanned image and converting it to the appropriate character (see CHARACTERS AND STRINGS). If the process is successful, the result is a text document that can be manipulated in a word processor, database, or other program that handles text. Raymond Kurzweil (1948–) marketed the first commercially practicable general-purpose optical character recognition system in 1978.

Once the document page has been scanned into an image format, there are various ways to identify the characters. One method is to use stored *templates* that indicate the pattern of pixels that should correspond to each character. Generally, a threshold of similarity is defined so that an exact match is not necessary to classify a character: The template most similar to the character is chosen. Some systems store a set of templates for each of the fonts most commonly found in printed text. (Recognizing cursive writing is a much more complex process: See HANDWRITING RECOGNITION.)

A more generalized method uses structural features (such as "all t's have a single vertical line and a shorter crossbar line") to classify characters. To analyze a character, the different types of individual features are identified and then compared to a set of rules to determine the character corresponding to that particular combination of features. Sometimes thresholds or "fuzzy logic" are used to decide the *probable* identity of a character.

OCR systems have improved considerably, the process also being speeded up by today's faster processors. Most scanners are sold with OCR software that is perhaps 95 percent accurate, with higher end systems being more accurate still. This is certainly good enough for many purposes, although material that is to be published or used in legal documents should still be proofread by human beings.

Further Reading

More, Shunji, Hirobumi Nishida, and Hiromitsu Yamada, eds. *Optical Character Recognition.* New York: Wiley, 1999.

Rice, S. V., G. Nagy and T. A. Nartker. *Optical Character Recognition: An Illustrated Guide to the Frontier.* Boston: Kluwer, 1999.

optical computing

Light is the fastest thing in the universe, and the science and technology of optics have developed greatly since the invention of the laser in the 1960s. It is not surprising, therefore, that computer designers have explored the possibility of using optics rather than electronics for computation and data storage.

An early idea was to use a grid of laser beams to create logical circuits, exploiting the ability of one laser to be used to "quench" or switch off another one. However, creating a large number of tiny laser beams proved impracticable, as did managing the heat created by the process. However, by the 1980s, experimenters were interacting "microlasers" with semiconductors, exploiting quantum effects. This brought the energy (and heat) problem under control while vastly increasing the potential density of the optical circuitry.

The incredible rate at which conventional silicon-based electronic circuitry continued to increase in density and capacity has limited the incentive to invest in the large-scale research and development that would be needed to develop a complete optical computer with processor and a corresponding optical memory technology.

Instead, current research is exploring the possibility of combining the best features of the optical and electronic system. Silicon chips have a limited surface for connecting data inputs, while light can carry many more channels of data through micro-optics. It may be possible to couple a micro-optic array to the surface of the silicon chip in such a way that the chip could have the equivalent of thousands of connecting pins to transmit data.

The value of optics is more conclusively demonstrated in data transmission and storage technology. Fiber optic cables are being used in many cases to carry large quantities of data with very high capacity (see BANDWIDTH) and may gradually supplant conventional network cable in more applications. The use of lasers to store and read information is seen in CD-ROM and DVD-ROM technology, which has replaced the floppy disk as the ubiquitous carrier of software and handy backup medium (see CD-ROM).

Further Reading

Caulfield, H. John. "Perspectives in Optical Computing." *IEEE Computer,* Feb. 1998, 22–25.

Hecht, Jeff. *Understanding Fiber Optics.* 3rd ed. Upper Saddle River, N.J.: Prentice Hall, 1999.

McCaulay, A. D. *Optical Computer Architectures.* New York: Wiley, 1990.

P

parallel port

There are two basic ways to send data from a computer to a peripheral device such as a printer. A single wire can be used to carry the data one bit at a time (see SERIAL PORT), or multiple parallel wires can be used to send the bits of a data word or byte simultaneously.

Serial ports have the advantage of needing only one line (wire), but sending a byte (eight-bit word) requires waiting for each of the eight bits to arrive in succession at the destination. With a parallel connection, however, the eight bits of the byte are sent simultaneously, each along its own wire, so parallel ports are generally faster than serial ports. Also, since the data is transmitted simultaneously, the protocol for marking the beginning and end of each data byte is simpler. On the other hand, parallel cables are more expensive (since they contain more wires) and are generally limited to a length of 10 feet or so because of electrical interference between the parallel wires.

The original parallel interface for personal computers was designed by Centronics, and a later version of this 36-pin connector remains popular today. Later, IBM designed a 25-pin version. In addition to the wires carrying data, additional wires are used to carry control signals.

Most modern parallel ports use two more advanced interfaces, EPP (Enhanced Parallel Port) or ECP (Extended Capabilities Port). Besides allowing for data transmission up to 10 times faster than the original parallel port, these enhanced ports allow for bi-directional (two-way) communications. This means that a printer can send signals back to the PC indicating that it is low on toner, for example. Printer control software running on the PC can therefore display more information about the status of the printer and the progress of the printing job. Besides printers, the parallel interface has also been used to connect external CD-ROM and other storage devices.

Although early PCs often provided their parallel port connectors on plug-in expansion cards, most PCs today have two parallel connectors built into the motherboard. In recent years the faster and more flexible Universal Serial Bus (see USB) interface has increasingly replaced the parallel port for printers, scanners, digital cameras, external storage drives, and many other devices.

Further Reading

"Parallel Port Central." http://www.lvr.com/parport.htm

White, Ron. *How Computers Work*. Millenium Edition. Indianapolis, Ind.: Que, 1999.

parsing

Just as a speaker or reader of English must be able to recognize the significance of words, phrases, and other components of sentences, a computer program must be able to "understand" the statements, commands, or other input that it is called upon to process.

For example, an interpreter for the BASIC language must be able to recognize that

```
PRINT "End of Run"
```

contains a previously defined command or keyword (PRINT) and that the quote marks enclose a string of characters that are to be interpreted literally rather than standing for something else. Once the type of element or data item is recognized, then the appropriate procedure can be called upon for processing it. (See also COMPILER and INTERPRETER.)

Similarly, a command processor (see SHELL) for an operating system such as UNIX will look at a line of input such as

```
ls -l /bin/MyProgs
```

and recognize that ls is an executable utility program. It will pass the rest of the command line to the ls program, which is then executed. In turn, ls must parse its command line and recognize that –l is a particular option that controls how the directory listing is displayed, and /bin/MyProgs is a pathname that specifies a particular directory location in the file system.

To parse its input, the language or command interpreter begins by looking in the program language or command statement for tokens. (This process is called lexical analysis.) A token is normally defined as a series of one or more characters separated by "whitespace" (blanks, carriage returns, and so on). A token is thus analogous to a word in English.

The series of tokens is then sent to the parser. The parser's job is to identify the significance of each token and to group the tokens into properly formed statements. Generally, the parser first checks the tokens for keywords—words such as "if" or "loop" that have a special meaning in a particular programming language. (In the BASIC example, PRINT is a keyword: In many other languages such functions are external rather than being part of the language itself.) As keywords (and punctuation symbols such as the semicolon used at the end of statements in C and Pascal) are identified, the parser uses a set of rules to determine the overall structure of the statement. For example, a language might define an if statement as follows:

```
If <Boolean-expression> then <statement> else
<statement>
```

This means that when the parser encounters an "if" it will expect to find between that word and "then" an expression that can be tested for being true or false (see BOOLEAN OPERATORS). Following "then," it will expect to find a complete statement. If it finds the optional keyword "else," that word will be followed by an alternative statement. Thus in the statement

```
If Total > Limit Print "Overflow" else Print
Total
```

The elements would be broken down as follows:

If	keyword
Total > Limit	Boolean expression
Print	keyword
"Overflow"	String literal (characters to be printed)
else	keyword
Print	keyword
Total	variable

When writing a parser, the programmer depends on a precise and exhaustive description of the possible legal constructs in the language (see also BACKUS-NAUR FORM). In turn, these rules are turned into procedures by which the parser can construct a representation of the relationships between the tokens. This representation is often

Parsing

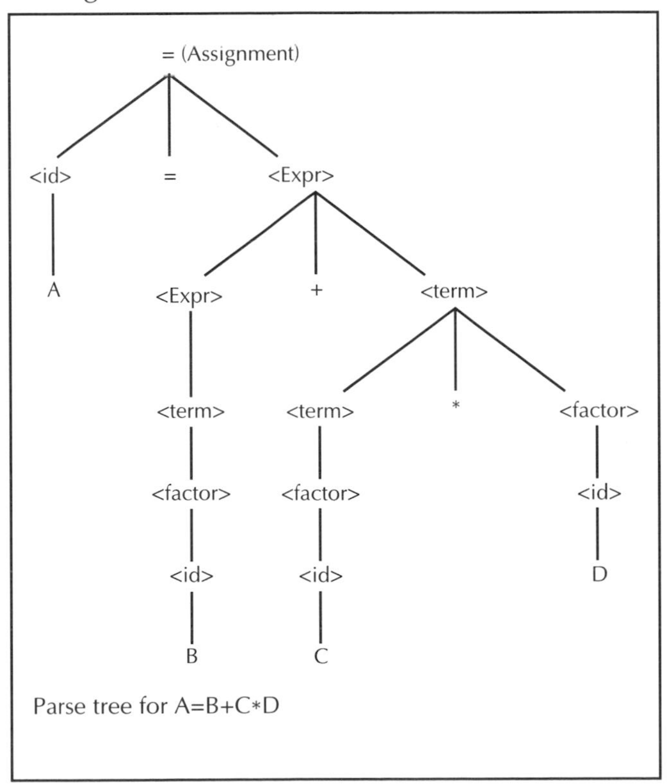

*A parse tree for the statement A = B + C * D. Notice how the expression on the right-hand side of the equals (assignment) sign is eventually parsed into the component identifiers and operators.*

represented as an upside-down tree, rather like the sentence diagrams used in English class.

In general form, an expression, for example, can be diagrammed as consisting of one or more terms (variables, constants, or literal values) or other expressions separated by operators.

Notice that these diagrams are often recursive. That is, the definition of an expression can include expressions. The number of levels that can be "nested" is usually limited by the compiler if not by the definition of the language.

The underlying rules must be constructed in such a way that they are not ambiguous. That is, any given string of tokens must result in one, and only one parse tree.

Once the elements have been extracted and classified, a compiler must also analyze the nonkeyword tokens to make sure they represent valid data types, any variables have been previously defined, and the language's naming conventions have been followed (see COMPILER).

Fortunately, people who are designing command processors, scripting languages, and other applications requiring parsers need not work from scratch. Tools such as YACC (a grammar definition compiler) and BISON and ANTLR (parser generators) are available for UNIX and other platforms.

Further Reading

Brookshear, J. Glenn. *Computer Science: an Overview.* Reading, Mass.: Addison-Wesley, 2000.

Metsker, Steven John. *Building Parsers with Java.* Reading, Mass.: Addison-Wesley, 2001.

Stallman, Richard M., and Charles Donnelly. *The Bison Manual: Using the YACC-Compatible Parser Gernerator for v. 1.29.* Cambridge, Mass.: Free Software Foundation, 1999.

Sun Microsystems. Java Developer Connection. "Parser Technology." http://developer.java.sun.com/developer/technicalArticles/Parser/index.html

Pascal

By the early 1960s, computer scientists had become increasingly concerned with finding ways to better organize or structure programs. Indeed, one language (see ALGOL) had already been developed in part to demonstrate and encourage sound programming practices, including the proper use of control structures (see also LOOP and STRUCTURED PROGRAMMING). However, Algol lacked a full range of data types and other features needed for practical programming, while arguably being too complex and inconsistent to serve as a good teaching language.

Niklaus Wirth at ETH (the Swiss Federal Institute of Technology) worked during the mid-1960s with a committee that was trying to overcome the problems with Algol and make the language more practical and attractive to computer manufacturers and users. However, Wirth gradually became disillusioned with the committee's unwieldy results, and proceeded to develop a new language, Pascal, announcing its specifications in 1970.

Pascal both streamlined Algol and extended it. Besides providing support for character, Boolean, and set data types, Pascal allows users to define new data types by combining the built-in types. This feature is particularly useful for defining a "record" type that, for example, might combine an employee's name and job title (characters), ID number (a long integer), and salary (a floating-point number). The rigorous use of data types also extends to the way procedures are called and defined (see PROCEDURES AND FUNCTIONS).

Pascal attracted much interest among computer scientists and educators by providing a well-defined language in which algorithms could be expressed succinctly. The acceptance of Pascal was also aided by its innovative compiler design. Unlike the machine-specific compilers of the time, the Pascal compiler did not directly create machine code. Rather, its output was "P-code," a sort of abstract machine language (see also PSEUDOCODE). A run-time system written for each computer interprets the P-Code and executes the appropriate machine instructions. This meant that Pascal compilers could be "ported" to a particular model of computer simply by writing a P-Code Interpreter for that machine. This strategy would be used more than two decades later by the creators of a popular language for Web applications (see JAVA).

STRUCTURE OF A PASCAL PROGRAM

The following simple program illustrates the basic structure of a program in Pascal. (The words in bold type are keywords used to structure the program.) The program begins with a Type section that declares user-defined data types. These can include arrays, sets, and records (composite types that can include several different basic types of data). Here an array of up to 10 integers (whole numbers) is defined as a type called IntList.

The Var (variable) section then declares specific variables to be used by the program. Variables can be defined using either the language's built-in types (such as integer) or types previously defined in the Type section. An important characteristic of Pascal is that user-defined types must be defined before they can be used in variable declarations, and variables in turn must be declared before they can be used in the program. Some programmers found this strictness to be confining, but it guards against, for example, a typographical error introducing an undefined variable in place of the one intended. Today most languages enforce the declaration of variables before use.

The word *begin* introduces the executable part of the program. The variables needed for the loop are first initialized by assigning them a value of zero. Note that in

Pascal := (colon and equals sign) is used to assign values. The outer *if* statement is used to ensure that the user does not input an invalid number of items. The *for* loop then reads each input value, assigns it to its place in the array, and keeps a running total. That total is then used to compute the average, which is output by the writeln (write line) statement.

```
program FindAvg (input, output);
  type IntList = array [1 . . 10] of integer;
  var
    Ints: IntList;
    Items, Count, Total, Average: integer;
begin
  Average := 0;
  Total := 0;
  Readln (Items);
  If ((Items > 0) and (Items <= 10)) then
    begin
    for Count := 1 to Items do
      begin
      readln (Ints [Count]);
      Total := Total + Ints [Count]
      end;
    Average := Total / Items;
    Writeln ('The average of the items is:',
      Average)
    end
  else
    Writeln ('Error: Number of items must be
      between 1 and 10')
  end.
```

IMPACT OF PASCAL

Pascal achieved modest commercial success. The P-Code idea was embraced by the UCSD P-System developed by the University of California at San Diego. In the late 1970s and early 1980s, the P-System brought the benefits of Pascal's structured programming to users of computers such as the Apple II, for which the only alternatives had been machine-language or a poorly structured version of BASIC. Later in the 1980s, Borland International came out with Turbo Pascal. This compiler used direct compilation rather than P-Code, sacrificing portability for speed and efficiency. It included an integrated programming environment that made development much cheaper and easier than with existing "bulky" and expensive compilers such as those from Microsoft. Turbo became very popular and eventually included language extensions that supported object-oriented programming. But Pascal became best known in its role as a first language for teaching programming and for expressing algorithms.

However, by 1990 the tide had clearly turned in favor of C and C++. These languages used a more cryptic syntax than Pascal and lacked the latter's rigorous data typing mechanism. Systems programmers in particular preferred C's ability to get "close to the machine" and manipulate memory directly without being confined by type definitions. C had also received a big boost because its developers were also among the key developers of UNIX, a very popular operating system in campus computing environments.

During the 1990s, C, C++, and Java even began to supplant Pascal for computer science instruction. Nevertheless, by encouraging structured programming concepts and helping educate a generation of computer scientists, Pascal made a lasting impact on the computer field. Wirth continued his work with the development of Modula-2 and Oberon, which were confined mainly to the academic world. However, Pascal also was a major influence on the development of Ada, a language endorsed by the U.S. federal government that combines structured programming with object-oriented features and the ability to manage extensive packages of routines (see ADA).

Further Reading

Free Pascal Compiler. http://www.freepascal.org/

Jensen, Kathleen, Niklaus Wirth, and A. Mickel. *Pascal User Manual and Report: ISO Pascal Standard.* 4th ed. New York: Springer-Verlag, 1991.

Koffman, Elliot B. *Turbo Pascal.* 5th update ed. Reading, Mass.: Addison-Wesley, 1997.

Rachele, Warren. *Learn Object Pascal with Delphi.* Plano, Tex.: Wordware Publishing, 2000.

Wirth, Niklaus. *Programming in Modula-2.* 3rd, corr. ed. New York: Springer-Verlag, 1985.

Yue, Tao. "Learn Pascal: the Ultimate Pascal Tutorial." http://web.mit.edu/taoyue/www/tutorials/pascal/

Perl

The explosive growth of the World Wide Web has confronted programmers with the need to find ways to link databases and other existing resources to websites. The specifications for such linkages are found in the Common Gateway Interface (see CGI). However, the early facilities for writing CGI scripts were awkward and often frustrating to use.

Back in 1986, UNIX developer Larry Wall had created a language called Perl (Practical Extraction and Report Language). There were already ways to write scripts for simple data processing (see also SCRIPTING LANGUAGES) as well as a handy pattern-manipulation language (see AWK). However, Wall wanted to provide a greater variety of functions and techniques for finding, extracting, and formatting data. Perl attracted a following within the UNIX community. Since much Web development was being done on UNIX-based systems by the mid- and late-1990s, it was natural that many webmasters and applications programmers would turn to Perl to write their CGI scripts.

As with many UNIX scripting languages, Perl's syntax is broadly similar to C. However, the philosophy behind

C is to provide a sparse core language with most functionality being handled by standard or add-in program libraries. Perl, on the other hand, starts with most of the functionality of UNIX utilities such as sed (stream editor), C shell, and awk, including the powerful regular expressions familiar to UNIX users. The language also includes a "hash" data type (a collection of paired keys and values) that makes it easy for a program to maintain and check lists such as of Internet hosts and their IP addresses (see HASHING).

Wall made it a point to solicit and respond to feedback from Perl users, often by adding features or functions. Wall's approach has been to provide as much practical help for programmers as possible, rather than worrying about the language being well-defined, consistent, and thus easy to learn. For example, in most languages, to make something happen only if a certain condition is not true, one writes something like this:

```
If ! (test for valid data)
  Print Error-Msg;
Else Process_Data;
```

In Perl, however, one can use the "unless" clause. It looks like this:

```
Unless (Test for invalid data) {
  Process_Data;
}
```

Syntactically, the unless clause does not provide anything more than using an If and Else would, and it involves learning a different structure. However, it has the practical benefit of making the program a little easier to read by keeping the emphasis on what the program expects to be doing, not on the possible error. Similarly, Perl offers an "until" loop:

```
Until (Condition is met) {
  Do something;
}
```

In C, one would have to say

```
While (Condition is not met) {
  Do something;
}
```

This "Swiss army knife" approach to providing language features has been criticized by some computer scientists as encouraging undisciplined and hard-to-verify programming. However, Perl's many aficionados see the language as the versatile, essential toolbox for the ever-challenging world of Web programming. As the language evolved through the late 1990s, it also added a full set of object-oriented features (see OBJECT-ORIENTED PROGRAMMING).

SAMPLE PERL PROGRAM

The following very simple code illustrates a Perl program that reads some lines of data from a file and prints them out. The first line tells UNIX to execute the Perl interpreter. The file name data.txt is assigned to the string variable $file. The file is then opened and assigned to the variable INFO. A single statement (not a loop) suffices to assign all the lines in the file to the array @lines. The "foreach" statement is a compact form of For loop that assigns each line in the array to the string variable $line and then prints it to the screen as HTML.

```
#!/usr/local/bin/perl
$file = 'data.txt' ;
open(INFO, "<$file" ) ;
@lines = <INFO> ;
foreach $line (@lines)
{
  print "\n <P> $line </P>" ;
}
# DONE
```

Perl is probably the most widely used scripting language today, particularly in the UNIX environment and especially for Web development. It should be noted, however, that there are also free or low-cost versions of Perl for many other environments including Microsoft Windows.

Further Reading

Wall, Larry, Tom Christiansen, and Jon Orwant. *Programming Perl*. 3rd ed. Sebastopol, Calif.: O'Reilly, 2000.
Marshall, A. D. "Practical Perl Programming." http://www.cs.cf.ac.uk/Dave/PERL/
Schwartz, Randal L., and Tom Phoenix. *Learning Perl*. 3rd ed. Sebastopol, Calif.: O'Reilly, 2001.
"Take 10 Minutes to Learn Perl." http://www.geocities.com/SiliconValley/7331/ten_perl.html#basic

personal computer

The development of the "computer chip" (see MICROPROCESSOR) and the increasing use of integrated circuit technology made it possible by the mid-1970s to begin to think about designing small computers as office machines or consumer devices that could be individually owned or used. In about a decade the personal computer, or PC, would become well established in many businesses and a growing number of homes. After another decade, it became almost as ubiquitous as TV sets and microwaves. Parallel developments in hardware, software, operating systems, and accessory devices made this revolution possible.

The first commercial "personal computer" was the MITS Altair, a microcomputer kit built around an Intel 8080 microprocessor. Building the kit required considerable skill with electronics assembly, but enthusiasts (including a young Bill Gates) were soon writing software and designing add-on modules for the kit (see GATES, WILLIAM). A variety of publications, notably *Byte* magazine, as well as the Homebrew Computer Club gave hobbyists a forum for sharing ideas.

By the late 1970s, personal computing was starting to become accessible to the general public. The Altair enthusiasts had moved on to more powerful systems that offered such amenities as floppy disk drives and an operating system (CP/M, developed by Gary Kildall). Meanwhile, less technically experienced people could also begin to experiment with personal computing, thanks to the complete, ready-to-run PCs being offered by Radio Shack (TRS-80), Commodore (Pet), and in particular, the Apple II.

In order to make serious inroads into the business world, however, the PC needed useful, reliable software. WordStar and later WordPerfect made it possible to replace expensive special-purpose word processing machines (such as those made by Wang) with the more versatile PC. One of the biggest spurs to business use of PCs, however, was an entirely new category of software—the spreadsheet. Dan Bricklin's VisiCalc (see SPREADSHEET) would make the PC attractive to accountants and corporate planners.

The watershed year in personal computing was 1981 because it brought the computer giant IBM into the PC arena (see IBM PC). The IBM PC had a somewhat more powerful processor and could hold more memory than the Apple II, but its main advantage was that it was backed by IBM's decades-long reputation in office machines. Businesses were used to buying IBM products, and conversely, many corporate buyers believed that if IBM was offering desktop computers, then PCs must be useful business machines.

IBM (like Apple) had adopted the idea of open architecture—the ability for third companies to make plug-in cards to add functions to the machine. Thus, the IBM PC became the platform for a burgeoning hardware industry. Further, it turned out that other companies could reverse-engineer the internal code that ran the system hardware (see BIOS) without infringing IBM's legal rights. This meant that companies could make "clones" or IBM-compatible machines that could run the same software as the genuine IBM PC. The first clone manufacturers (such as Compaq) sometimes improved upon IBM such as by offering better graphics or faster processors. However, by the late 1980s the trend was toward companies competing through lower prices for roughly equivalent performance. Facing a declining market share, IBM tried to introduce a new architecture, called *microchannel,* that provided a mainframelike bus architecture for more efficient input/output control. However, whatever technical advantages the new system (called PS/2) might have, the market voted against it by continuing to buy the ever more powerful clones built on the original IBM architecture.

Lower prices and more attractive options led to a growing number of users, which in turn encouraged greater investment in software development. By the mid-1980s, Lotus (headed by Mitch Kapor) dominated the spreadsheet market with its Lotus 1-2-3, while WordPerfect dominated in word processing.

However, Microsoft, whose MS-DOS (or PC-DOS) had become the standard operating system for IBM-compatible PCs, introduced a new operating environment with a graphical user interface (see MICROSOFT WINDOWS). By the mid-1990s, Windows had largely supplanted DOS. Microsoft also committed resources and exploited its intimate knowledge of the operating system to achieve dominance in office software through MS Word, MS Excel (spreadsheet), and MS Access (database).

At the margins Apple's Macintosh (introduced in 1984 and steadily refined) has retained a significant following, particularly in education, publishing, and graphic arts applications (see MACINTOSH). Although Windows now provides a similar user interface, Mac enthusiasts believe their machine is still easier to use (and more stylish), and often see it as a badge for those who "think different."

PC TRENDS

When graphical Web browsing made the Internet widely accessible in the mid-1990s, the demand for PCs increased accordingly. The desire for e-mail, Web browsing, and help with children's homework led many families to purchase their first PCs. By 2000, about two-thirds of American children had access to computers at home, and virtually all schools had at least some PCs in the classroom. Using sophisticated manufacturing and order processing systems, companies such as Dell and Gateway sell PCs directly to consumers and businesses, largely displacing the neighborhood computer store. These efficiencies (and lower prices for memory, processors, and other hardware) have brought the cost for a basic home PC down to around $500, while the capabilities available for those willing to spend $1,500 or so continued to increase. PC users now expect to be able to play CD-based multimedia (and increasingly, the higher-capacity DVDs) while hearing good quality sound.

A number of challenges to the growth of the PC industry have also emerged. As more and more of the activity of PC users began to focus on the Internet, some companies began to host office applications on servers (see APPLICATION SERVICE PROVIDER). Some pundits began to say that with applications being moved to remote servers or offered over the corporate LAN, the PC on the desk could be stripped down considerably. The "network PC" could make do with a slower processor, less memory, and no hard drive, since all data could be stored on the server. Meanwhile, products such as WebTV have tried to replace the PC with an "Internet appliance" that would be as easy to use as a TV remote (see INTERNET APPLIANCE).

Generally, however, the attempts to supplant the full-featured, general-purpose PC have made little progress. One reason is that the cost of complete PC has declined so much that the supposed cost savings of a network PC or Internet appliance have become less significant. Further, privacy issues and the desire of people to have control over their own data are often cited as arguments in favor of the PC.

Ironically, the PC industry's greatest challenge may come from its very success. As more and more households in the United States and other developed countries have PCs, it becomes harder to maintain the sales rate. By 2002, the power of recent PCs has become so great that the desire to upgrade every few years may have become less compelling and the recent economic downturn has hit the computer industry particularly hard. So far it looks like the fastest-growing areas in computer hardware no longer involve the traditional desktop PC, but handheld (palm) computers (see PORTABLE COMPUTERS) and the embedding of more powerful computer capabilities into other machines such as automobiles (see EMBEDDED SYSTEMS).

Further Reading

Freiberger, Paul, and Michael Swaine. *Fire in the Valley: The Making of the Personal Computer.* New York: McGraw-Hill, 1999.

Magid, Larry, and John Grimes (illustrator). *The Little PC Book: Windows XP Edition.* Berkeley, Calif.: Peachpit Press, 2001.

White, Ron. *How Computers Work.* Millennium Ed. Indianapolis, Ind.: Que, 1999.

personal information manager (PIM)

A considerable amount of the working time of most businesspeople is taken up not by primary business tasks but in keeping track of contacts, phone conversations, notes, meetings, deadlines, and other information needed to plan or coordinate activities. Software designers have responded to this reality by creating software to help manage personal information.

Early PC users improvised ways of using available software applications for tracking their activities. For example, a spreadsheet with text fields might be used to record and sort contacts and their associated information such as phone numbers or data could be organized in tables in word processor documents. However, such improvisations can be awkward to use. Loading a full-sized word processor or spreadsheet application takes time (and until Windows and other multitasking solutions came along, only one program could be run at a time). Further, it is hard to integrate information or keep track of the "big picture" with several different kinds of information stored in different formats with different programs.

What was needed was a single application that could integrate the personal information and make it accessible without the user having to shut down the main application program. The first successful PIM was Borland Sidekick, first released in 1984. Although MS-DOS was designed to run only a single program at a time, it had an obscure feature that allowed additional small programs to be loaded into memory where they could be triggered using a key combination. Taking advantage of this feature, Sidekick allowed someone while using, for example, a word processor, to pop up a note-taking window, an address book, calendar, telephone dialer, calculator, or other features. When Microsoft Windows replaced DOS, it became possible to run more than one full-fledged application at a time. PIMs could then become full-fledged applications in their own right, and offer additional features.

As e-mail became more common on local networks in the later 1980s and via the Internet in the 1990s, PIM features began to be integrated with e-mail programs such as Microsoft Outlook and Netscape Navigator's communications facilities. New features included the automatic creation of journal entries from various activities and the creation of "rules" for recognizing and routing e-mail messages with particular senders or subjects. A variety of freeware and shareware PIMs are available for users who want an alternative to the commercial products, and a number of PIMs are available for Macintosh and Linux-based systems.

The growth of handheld (or palm) computers (see PORTABLE COMPUTERS) and more sophisticated cell phones has created a need to provide PIM features for these devices. Since the capacity of handheld devices is limited compared to desktop PCs, there is also a need for software to allow easy transfer of information between portable devices and desktop PCs. This can be done with a serial, USB, or even wireless connection.

In the future, the PIM is likely to become an integrated system that operates on a variety of handheld and desktop devices and seamlessly maintains all information regardless of how it is received. There will also be greater ability to give voice commands (such as to dial a person or to ask for information about a contact), and to have messages read aloud (see SPEECH RECOGNITION AND SYNTHESIS). The software is also likely to include sophisticated "agents" that can be instructed to carry out such tasks as prioritizing messages or returning routine calls (see SOFTWARE AGENT).

Further Reading

Byrne, Randy. *Building Applications with Microsoft Outlook Version 2002.* Redmond, Wash.: Microsoft Press, 2001.

Shareware and Freeware PIMs. www.zdnet.com [search under Downloads]

Microsoft. "Using Outlook." http://www.microsoft.com/office/outlook/using/default.htm

Yahoo! Personal Information Manager Software Page http://dir.yahoo.com/business_and_economy/shopping_and_services/computers/software/personal_information_management/

photography, digital

For more than 150 years photography has depended on the use of film made from light-sensitive chemicals. However, digital photography, first developed in the 1970s, emerged in the late 1990s as a practical, and in some ways superior, alternative to traditional photography.

The basic idea behind digital photography is that light (photons) can create an electrical charge in certain materials. In 1969, engineers at Bell Labs invented a light-sensitive semiconductor that became known as a charge-coupled device (CCD). The original intention of the developers was to use an array of CCDs to make a compact black-and-white video camera for the videophone, a device that did not prove commercially viable. However, astronomers were soon using CCD arrays to capture images too faint for the human eye or even for conventional film.

Digital photography remained confined to such specialized applications until the mid-1990s. By then, the growing use of multimedia and the World Wide Web made digital photography an attractive alternative for getting images on-line quickly, avoiding the need to scan traditional prints or negatives. At the same time, cheaper, more powerful processors and larger capacity memory storage made good quality digital cameras more viable as a consumer product.

A digital camera uses the same type of lenses and optical systems as a conventional camera. Instead of falling upon film, however, the incoming light strikes an array of CCD "photosites." Each photosite represents one picture element, or pixel, which will appear as a tiny dot in the resulting picture. (Camera resolution is typically measured in millions of pixels, or "megapixels.")

The surface of the array contains an abundance of free electrons. As light strikes a photosite, it creates a charge that draws and concentrates nearby electrons. The voltage at a photosite is thus proportional to the intensity of the light striking it. The charge of each row of photosites is transferred to a corresponding read-out register, where it is amplified to facilitate measurement.

The camera uses an analog-to-digital converter (see ANALOG AND DIGITAL) to convert the amplified voltages to digital numeric values. Early consumer digital cameras typically used 8-bit values, limiting the camera to a range of gradated intensity from 0 to 255. However, many cameras today use up to 12 bits, giving a range of 0 to 4096.

The CCD mechanism itself measures only light intensities, not colors. To obtain color, many cameras use a red, green, or blue (RGB) filter at each photosite. (Some manufacturers use cyan, yellow, green, and magenta filters instead.) Since each photosite registers only a single color, interpolation algorithms must be used to estimate the actual color of each pixel by using laws of color optics and comparing the colors and intensity of the adjacent pixels.

New high-end cameras are starting to eschew interpolation in favor of using a complete, separate CCD array for each of the three RGB colors and thus making and combining three complete exposures that directly capture the colors. This produces the best possible color accuracy but is more expensive.

Photography, Digital

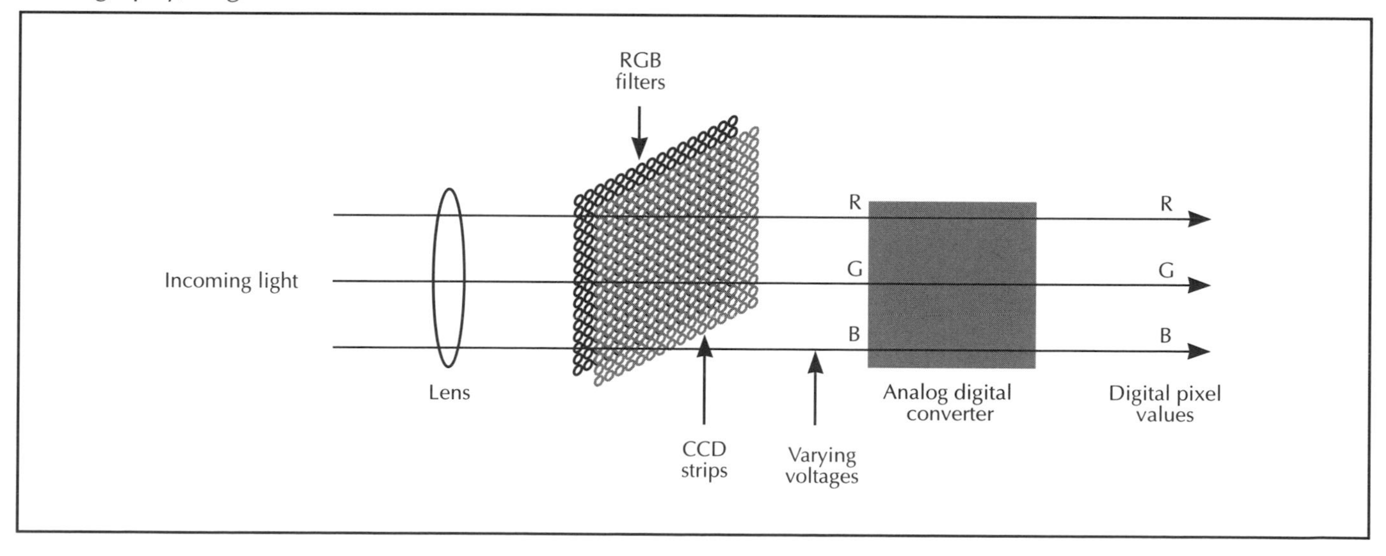

Digital cameras use a charge-coupled device (CCD) to convert incoming light to varying voltages that are digitized to create pixel values. Many more expensive cameras today use separate layers of CCD cells rather than an overlay of RGB filters.

The final image data is stored using a standard file format, usually JPEG (see GRAPHICS FORMATS). The most commonly used storage medium is an insertable "flash" memory module. The major competing memory card standards are CompactFlash, SmartMedia, and Sony Memory Stick. Storage capacities run to 512 MB. As with regular RAM, the cost of flash memory has declined considerably in recent years.

An alternative to memory cards is the use of miniature disk drives. Sony's Mavica line, for example, began with a built-in floppy drive and now offers a rewritable CD drive (see also CD-ROM) storing 150 MB. Although removable disk drives are bulkier than memory sticks, the user can simply insert new disks as needed. As an alternative, a few cameras now offer miniature internal hard drives that can store several gigabytes or more.

In determining the adequacy of the camera's storage capacity, the user must also consider the camera's resolution (number of pixels) and whether images will be compressed before storage (see DATA COMPRESSION). While a certain amount of compression can be achieved without discernable degradation of the image, more drastic "lossy" compression sacrifices image quality for compactness. It should also be noted that as image resolution (and thus file size) increases, the time needed to process and store each image will also increase, limiting how rapidly successive exposures can be made.

Most digital cameras have a USB connector (see USB), making it easy to upload the stored images from the camera to a PC. Once in the PC, images can be edited or otherwise manipulated using the basic photo editing software usually included with the camera or a full-featured professional product such as Adobe Photoshop.

The same trends that have brought more capability per dollar spent on digital cameras have been even more evident in printers (see PRINTER). Using resolutions such up to 2880 dots per inch and special papers, digital camera users can make prints with a quality similar to that produced by traditional photo developers. Computer prints are more subject to color fading over time than are conventional prints, although some printer manufacturers now offer toner that will resist fading for 25 years or more.

FUTURE TRENDS

High-end consumer digital cameras reached the 4-5 megapixel range by 2002, allowing for images that can be "blown up" to 10 by 12 inches or larger while retaining image quality comparable to conventional photos. Professional-grade digital cameras are rated at 8 to 10 megapixels or more. The need for such cameras for professional work arises not only from the higher resolution requirements but also because these cameras have the very high-quality optics used in fine 35 mm cameras, as well as having a greater variety of available specialty lenses. (Consumer digital cameras, however, do offer zoom lenses roughly comparable to those for low-cost 35 mm cameras.) The quality and convenience of digital photography should continue to increase, and it is likely that over the coming decade digital cameras will supplant conventional cameras for most consumer and many professional applications.

Digital video cameras will also become more widely used. Their resolution is generally from about a quarter million pixels to a million pixels—considerably lower than for digital still cameras, but adequate and likely to improve. Digital video cameras are also rated according to lux value, indicating the minimum light level for satisfactory recording. Most digital videos store the captured image to tape (either MiniDV or Hi-8), but some newer cameras use built-in recordable DVD disks instead (see CD-ROM). The ability to digitally edit video direct from the camera is also an important advantage.

Further Reading

Curtin, Dennis P. "Digital Photography: The Textbook. A Complete On-line Course." http://www.photocourse.com/

"Digital Photography Home." http://digital.photography.home.att.net/

"Glossary of Digital Video Terms." http://www.adobe.com/support/techguides/digitalvideo/dv_glossary/main.html

Long, Ben. *Complete Digital Photography.* Hingam, Mass.: Charles River Media, 2001.

NASA. "Charge-Coupled Device (CCD) Imaging Arrays."http://ranier.hq.nasa.gov/Sensors_page/DD/HST&GLL_CCD.html

PL/I

By the early 1960s, two programming languages were in widespread use: FORTRAN for scientific and engineering applications and COBOL for business computing. However, applications were becoming larger and more complex, calling for a wider variety of capabilities. For example, scientific programmers needed to provide data-processing and reporting capabilities as well as computation. Business programmers, in turn, increasingly needed to work with formulas and statistics and needed floating-point and other number formats.

Language developers thus began to look toward a general-purpose language that could be equally at home with words, numbers, and data files. Meanwhile, IBM was preparing to replace its previously separate scientific and business computer systems with the versatile System/360. They and one of their user groups, SHARE, formed a joint committee to develop a new language for this new machine.

At first the designers thought in terms of extending FORTRAN to provide better text and data-processing capabilities, so they designated the new language FORTRAN VI. However, their focus soon changed to designing a completely new language, which was known until 1965 as NPL (New Programming Language). Because this acronym already stood for Britain's National Physical Laboratory, the name of the language was changed to PL/I (Programming Language I).

LANGUAGE FEATURES

PL/I has been described as the "Swiss army knife of languages" because it provides so many features drawn from disparate sources. The basic block structure and control structures (see LOOP and BRANCHING STATEMENT) were adapted from Algol, a relatively small language that had been devised by computer scientists as a model for structured programming (see ALGOL) and is also similar to Pascal (see PASCAL). Blocks can be nested, and variables declared within a block can be accessed only within that block and its nested blocks, unless declared explicitly otherwise.

PL/I includes a particularly rich variety of data types and can specify even the number of digits for numeric data. A PICTURE clause similar to that in COBOL can be used to specify exact layout. However, the language takes a more pragmatic approach than Algol or Pascal; data need not be declared and will be given default characteristics based on context. Input/Output (I/O) is built into the language rather than provided in an external library, and the flexible options include character, streams of characters, blocks, and records with either sequential or random access.

In general, PL/I provides more control over the low-level operation of the machine than Algol or even successors such as C. For example, there is an unusual amount of control over how variables are stored, ranging from STATIC (present throughout the life of the program) to AUTOMATIC (allocated and deallocated as the containing block is entered and exited) to CONTROLLED, where memory must be explicitly allocated and freed. Pointers allow memory locations to be manipulated directly. PL/I also provided more elaborate facilities for handling exceptions (errors) arising from hardware condition, arithmetic, file-handling, or other conditions.

EXAMPLE PROGRAM

The following program executes a DO loop and counts from one to the number of items specified. It then outputs the total of the numbers and their average.

```
COUNTEM: PROCEDURE OPTIONS (MAIN);
DECLARE (ITEMS, COUNTER, SUM, AVG) FIXED;
ITEMS = 10;
SUM = 0;
DO COUNTER = 1 TO ITEMS;
  SUM = SUM + COUNTER;
END;
AVG = SUM / ITEMS;
PUT SKIP LIST ("TOTAL OF ");
PUT ITEMS;
PUT ("ITEMS IS ");
PUT TOTAL;
PUT SKIP LIST ("THE AVERAGE IS: ");
PUT AVG;
END COUNTEM;
```

IMPACT OF THE LANGUAGE

Because of its many practical features and its availability for the popular IBM 360 mainframes, PL/I enjoyed considerable success in the late 1960s and 1970s. The language was later ported to most major platforms and operating systems. When personal computers came along, PL/I became available for IBM's OS/2 operating system as well as for Microsoft's DOS and Windows, although the language never really caught on in those environments.

Computer scientists such as structured programming guru Edsger Dijkstra decried PL/I's lack of a clear, well-defined structure. In his Turing Award Lecture in 1972, Dijkstra opined that "I absolutely fail to see how we can keep our growing programs firmly within our intellectual grip when by its sheer baroqueness the programming language—our basic tool, mind you!—already escapes our intellectual control." (See DIJKSTRA, EDSGER.)

On a practical level the sheer number of features in the language meant that truly mastering it was a lengthy process. A language like C, on the other hand, had a much simpler "core" to master even though it was less versatile. PL/I also tended to retain the mainframe associations from its birth at IBM, while C grew up in the world of mincomputers and the UNIX community and proved more suitable for PCs. Nevertheless, PL/I provided many examples that language designers could use in attempting to design better implementations.

Further Reading

Hughes, Joan Kirby. *PL/I Structured Programming.* 3rd ed. New York: Wiley, 1986.

"PL/I Frequently Asked Questions (FAQ)." http://www.faqs.org/faqs/computer-lang/pli-faq/

"The PL/I Language." http://home.nycap.rr.com/pflass/pli.htm

Sebesta, Robert W. *Concepts of Programming Languages.* Reading, Mass.: Addison-Wesley, 1999.

Plug and Play

In early MS-DOS systems installation of new hardware such as a printer often had to be performed manually by copying files (see DEVICE DRIVER) to the hard drive from floppies and then making specified settings to the system configuration files AUTOEXEC.BAT and CONFIG.SYS. These settings often involved unfamiliar concepts such as interrupts (IRQs) and DMA (direct memory access) channels.

When Windows came along, device manufacturers generally provided an installation program that takes care of copying the files and making the necessary changes to the system registry. However, there was still the problem of ensuring that one had a driver compatible with the version of the operating system in use, and users were sometimes asked to make choices for which they were not prepared (such as choosing which port to use).

By the mid-1990s, Intel was promoting a standard for the automated detection and configuration of devices. Known as Plug and Play (PnP), this standard was incorporated in versions of Microsoft Windows starting with Windows 95 (see MICROSOFT WINDOWS). The required hardware support soon appeared on PC motherboards and expansion cards.

With Plug and Play the user simply connects a printer, scanner, or other device to the PC. Windows detects that a device has been connected and queries it for its official name and other information. If necessary, Windows can then prompt the user for a disk containing the appropriate driver or even search for a driver on a website.

The concept of Plug and Play extends beyond the Windows world, however. In recent years there has been interest in developing a Universal Plug and Play (UPnP) protocol by which a variety of devices could automatically configure themselves with any of a variety of different networks. This would be particularly helpful for home users who are increasingly setting up small networks so they can share broadband Internet connections, as well as the growing number of users who want their desktop PC to work with handheld (palm) computers and other devices. Microsoft is supporting UPnP in recent versions of Windows such as ME and XP.

Further Reading

Bigelow, Stephen J. *The Plug & Play Book.* New York: McGraw Hill, 1999.

Shanley, Tom. *Plug and Play System Architecture.* Reading, Mass.: Addison-Wesley, 1995.

Universal Plug and Play Forum. http://www.upnp.org/about/default.asp

plug-in

A number of applications programs include the ability for third-party developers to write small programs that extend the main program's functionality. For example, thousands of "filters" (algorithms for transforming images) have been written for Adobe Photoshop. These small programs are called *plug-ins* because they are designed to connect to the main program and provide their service whenever it is desired or required.

Perhaps the most commonly encountered plug-ins are those available for Web browsers such as Netscape or Internet Explorer. Here the plug-ins enable the browser to display new types of files (such as multimedia). Many standard programs for particular kinds of files are now provided both as stand-alone applications and as browser plug-ins. Examples include Adobe (PDF document format), Apple QuickTime (graphics, video, and animation), RealPlayer (streaming video and audio), and Macromedia Flash (interactive animation and presentation). These and many other plug-ins are offered free for the downloading, in order to increase the number of potential users for the formats and thus the market for the development packages.

One of the most useful plug-ins found in most browsers is one that allows the browser to run Java applets (see JAVA). In turn, Java is often used to write other plug-ins.

Further Reading

Abacus Development Group. *Netscape Plug-Ins.* Grand Rapids, Mich.: Abacus, 1996.

"Browser Watch Plug-In Plaza." http://browserwatch.internet.com/plug-in.html

Kirk, Cheryl. *Supercharged Web Browsers: A Plug-Ins Field Guide.* Hingham, Mass.: Charles River Media, 1998.

Netscape Plug-ins. http://www.netscape.com/plugins/?cp=dowdep2

pointers and indirection

The memory in a computer is accessed by numbering the successive storage locations (see ADDRESSING). When a programmer declares a variable, the compiler associates its name with a location in available memory (see VARIABLE). If the variable is used in an expression, when the expression is evaluated, the variable's name is replaced by its current value—that is, with the contents of the memory location associated with the variable. Thus, the expression Total + 10 is evaluated as "the contents in the address associated with Total" plus 10.

Sometimes, however, it is useful to have the general capability to access memory locations without assigning explicit variables. This is done through a special type of variable called a *pointer.* The only difference between pointers and regular variables is that the value stored in a pointer is not the data to be ultimately used by the program. Rather, it is the address of that data. Here are some examples from C, a language that famously provides support for pointers:

```
Int MyVar; // Declare a regular variable
Int *MyPtr; // Declare a pointer to an integer
(int) variable
MyVar = 10; // Set the value of MyVar to 10
MyPtr = &MyVar; // Store the address of MyVar in
the pointer
MyPtr
```

In C, an asterisk in front of a variable name indicates that the variable is a pointer to the type declared. In the

Pointers and Indirection (1)

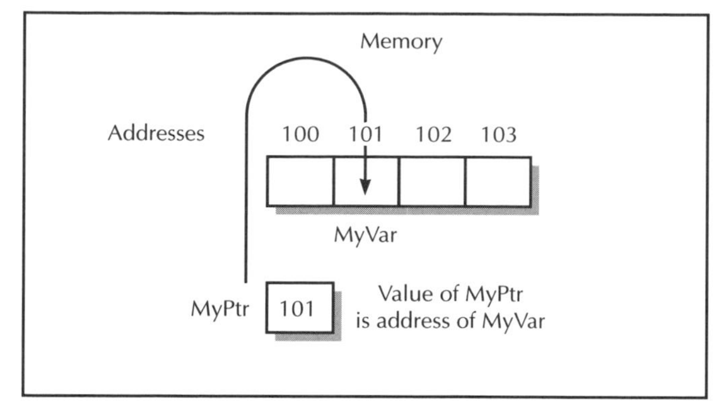

A pointer is a variable whose value is an address location. Here MyPtr holds the address 101.

second line above, therefore, MyPtr is a pointer to an integer variable. This means that the address of any integer variable can be stored in MyPtr. The last line uses the & (ampersand) to represent the address of the variable MyVar. Therefore, it stores that address in MyPtr.

Examining the lines above, one sees that the variable MyVar has the value 10. The pointer variable MyPtr has the value of whatever machine address contains the contents of the variable MyVar. In an expression, putting an asterisk in front of a pointer name "dereferences" the pointer. This means that it returns not the address stored in the pointer, but the value stored at the address stored in the pointer (see the diagram). Therefore if one writes:

```
AnotherVar = * MyPtr;
```

Pointers and Indirection (2)

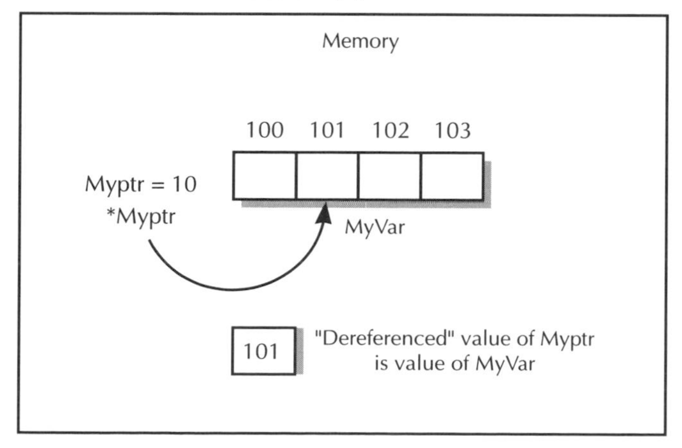

When a pointer is "dereferenced," the value stored in the address in the pointer is retrieved—10 in this case.

What is the value of AnotherVar? The answer is the current value of MyVar (whose address had been stored in MyPtr)—that value, as assigned earlier, is 10.

The general concept of storing the address of another variable in a variable is called indirection, or indirect addressing. It was first used in assembly language to work with index registers—special memory locations in a processor that store memory addresses.

USES FOR POINTERS

Although the concept may seem esoteric, pointers have a number of uses. For example, suppose one has a buffer (perhaps storing video graphics data) and one wants to copy it from one area to another. One could declare the buffer to be an array (see ARRAY) and then reference each element, or memory location and copy it. However, this would be rather awkward. Instead, one can declare a pointer, set it to the starting address of the buffer, and then simply use a loop to increment the pointer, pointing in turn to each location in the buffer.

A similar approach applies to strings in C and related languages. A string of characters in C is declared as an array of char. In an array, the name of the array is actually a pointer to the first data location. It is therefore easy to manipulate strings by getting their starting address by referencing the name and then using one or more pointers to step through the data locations. For example, the following function copies the contents of one string into another:

```
strcpy(char *s1,char *s2)
{
   while (*s2)
     *s1++ = *s2++;
}
```

The function takes two strings, s1 and s2, declared as pointers to char. It then steps (increments) them (using the ++ operator) so that the value in each location in s2 is copied into the corresponding location in s1. The loop exits when the value at s2 is 0 (null), indicating that the end of string marker has been reached.

Another common use for pointers is in memory allocation. Typically, a program requests memory by giving the memory allocation function a pointer and the amount of memory requested. The function allocates the memory and then returns the starting address of the new memory in the pointer, so the program knows how to access that memory.

Pointers are also useful for passing a "bulky" variable such as a data record to a procedure or function. Suppose, for example, a program needs to pass a 65,000 byte record to a procedure for printing a report. If it passes the actual record, the system has to make a copy of the whole record, tying up memory. If, instead, a pointer to the record is used, only the address is passed. The procedure can then access the record at that address without having to make a copy.

In C and some other languages it is even possible to have a pointer that points to another pointer. A common case is an array of strings, such as

```
Char Form [80] [20];
```

representing a form that has 20 lines of 80 characters. Each line is an array of characters and the form as a whole is thus "an array of arrays of characters." Therefore, to dereference (get the value of) a character one would first dereference the line, and then the column.

PROBLEMS WITH POINTERS

Pointers may be useful, but they are also prone to causing programming problems. The simplest one is failing to distinguish between a pointer and its value. For example, suppose one writes:

```
Total = Total + MyPtr;
```

intending to add the value of the variable pointed to by MyPtr to Total. Unfortunately, the asterisk (dereferencing operator) has been inadvertently omitted, so what gets added to Total is the machine address stored in MyPtr!

Another problem comes when a pointer is used to allocate memory, the memory is later deallocated, but the pointer is left pointing to it.

Because pointers can potentially access any location in memory (or at least attempt to), some computer scientists view them as more dangerous than useful. It's true that most things one might want to do with a pointer can be accomplished by alternative means. One attempt to tame pointers is found in C++, which offers the "reference" data type. A reference is essentially a constant pointer that once assigned to a variable always dereferences that variable and cannot be pointed anywhere else. Java has gone even further by not including traditional pointers at all.

Further Reading

Jensen, Ted. "A Tutorial On Pointers and Arrays in C." http://pw1.netcom.com/~tjensen/ptr/pointers.htm

Sebesta, Robert W. *Concepts of Programming Languages.* Reading, Mass.: Addison-Wesley, 1999.

Stroustrup, Bjarne. *The C++ Programming Language.* Special ed. Reading, Mass.: Addison-Wesley, 2000.

popular culture and computing

Computer technology first came to public consciousness with the wartime ENIAC and the first commercial machines such as Univac in the early 1950s. The war had shown the destructive side of new technologies (particularly atomic power), but corporate and government leaders were soon promoting their beneficial prospects. Just as atomic energy advocates promised to provide power that was abundant, cheap, and clean, the computer, or "giant brain" was touted for its ability to solve problems that had been beyond human capabilities.

Science fiction writers, however, had already been considering the ramifications of a related technology, robotics. The term *robot* came from Carl Capek's *R.U.R.* (Rossum's Universal Robots). Although the robot had a human face, it could have inhuman motives and threaten to become Earth's new master, displacing humans. Isaac Asimov offered a more benign vision, thanks to the "laws of robotics" embedded in his machines' very circuitry. The first law states, "A robot shall not harm a human being or, through inaction, cause a human being to come to harm." In the real world, of course, artificial intelligence had no such built-in restrictions (see ARTIFICIAL INTELLIGENCE).

As early as the beginning of the 20th century, writers had been exploring what might happen if some combination of artificial brains and robots offered the possibility of catering to all human needs. In E. M. Forster's "The Machine Stops," published in 1909, people no longer even have to leave their insectlike cells because even their social needs are provided through machine-mediated communication not unlike today's Internet. In the 1930s and 1940s, other writers such as John W. Campbell and Jack Williamson wrote stories in which a worldwide artificial intelligence became the end point of evolution, with humans either becoming extinct or living static, pointless lives.

Science fiction of the "Golden Age" of the pulp magazines had only limited impact on popular culture as a whole. Once actual computers arrived on the scene, however, they became the subject for movies as well as novels. D. F. Jones's novel *Colossus: The Forbin Project* (1966), which became a film in 1970, combined cold war anxiety with fear of artificial intelligence. Joining forces with its Soviet counterpart, Colossus fulfills its orders to prevent war by taking over and instituting a world government. Similarly, Hal in the film *2001: A Space Odyssey* (based on the work of Arthur C. Clarke) puts its own instinct for self-preservation ahead of the frantic commands of the spaceship's crew. However, the artificial can also strive to be human, as in the 2001 movie *A.I.*

As late as the 1970s, computers seemed huge and remote, with little direct connection to people's daily lives. The 1980s and 1990s would bring a different perspective arising from two transformations in computer technology. First, the PC became ubiquitous and millions of people began to directly interact with computers. Second, on-line services and later, the Internet and World Wide Web made computer use a social phenomenon, with people interacting in new ways through e-mail, bulletin boards and even shared fantasy experiences (see also CYBERSPACE AND CYBER CULTURE).

Writers such as William Gibson (*Neuromancer*) and Vernor Vinge (*True Names*) began to explore the world mutually experienced by computer users as a setting where humans could directly link their minds to computer-generated worlds (see VIRTUAL REALITY). A new elite of cyberspace masters were portrayed in a futuristic adaptation of such archetypes as the cowboy gunslinger, samurai, or ninja. Unlike the morally unambiguous world of the old western movies, however, the novels and movies with the new "cyberpunk" sensibility are generally set in a jumbled, fragmented, chaotic world. That world is often dominated by giant corporations (reflecting concerns about economic globalism) and is generally dystopian.

As computers and the Internet continued to become more ubiquitous, they have been increasingly incorporated into more mainstream dramatization. The computer hacker has appeared as both villain and hero (see, for example, the movie *The Net*, 1995). The astonishing rise and crashing fall of entrepreneurs and "dot-coms" is also likely to be fodder for more fiction in years to come. Ultimately, as many aspects of computer technology become inextricably intertwined with daily lives they will simply become part of the background of novels and movies, as necessary and unremarkable as telephones and cars.

Further Reading

Clute, John and Peter Nicholls, editors. *The Encyclopedia of Science Fiction*. New York: St. Martin's Griffin, 1995.

"Cybercinema: An Interactive Site Devoted to the History of Computers and Artificial Intelligence in Film." http://128.174.194.59/cybercinema/main.htm

Gibson, William. *Neuromancer.* New York: Ace Books, 1984.

Science Fiction and Artificial Intelligence. http://www.aaai.org/AITopics/html/scifi.html

Stork, David G. *Hal's Legacy* Cambridge, Mass.: MIT Press, 1996.

Vinge, Vernor, and James Frenkel. *True Name: and the Opening of the Cyberspace Frontier.* New York: Tor Books, 2001.

portable computers

After the microprocessor and integrated circuits made it possible to fit a computer on a desktop, the progress of miniaturization continued (see MICROPROCESSOR). Some designers realized that there might well be a market for a computer compact and light enough to be carried by business travelers, students, and others needing computer access on the go.

The first portable computers were about the size and weight of a large suitcase, making them barely "luggable." The Osborne I (introduced in 1981) and Compaq Portable (1982) were essentially standard computer cases that were "ruggidized" and equipped with handles, built-in CRT monitor, and detachable keyboard.

The Radio Shack Model 100, introduced in 1983, represented a new category of portable computer, the "notebook" PC. About the size of a hardback book and with rather limited memory and display, the Model 100 nevertheless became a favorite with journalists who could write stories on it and file them via the built-in modem.

Meanwhile, the "luggable" computer evolved into the modern laptop during the later 1980s and into the 1990s. These machines generally weighed eight to 10 pounds (their counterparts today are down to around five pounds). These machines feature a somewhat cramped but usable keyboard built into the top of the case. (Since using a mouse would be awkward, a touch pad, trackball, or movable "nub" is used instead.)

A liquid-crystal display (LCD) (see FLAT-PANEL DISPLAY) is built into the inside of the flip-up cover. Data storage started out with floppy disks but soon included an internal hard drive and a CD drive. (Some models have a bay into which any one of several kinds of modular drives can be plugged.) The versatility of laptops was also enhanced by the development of PCMIA, a standard that allows slim cards to be used to plug such devices as modems and network adapters into the laptop.

A portable or "laptop" computer. Notebook computers were originally a smaller, lighter category of computer, but laptops have now "shrunk" to the point where the distinction between laptops and notebooks has become blurred.

A variety of manufacturers make laptops today, including Compaq, Dell, IBM, and Apple, which has laptop versions of the Macintosh called the PowerBook and iBook.

The notebook category still exists, although the distinction between notebooks and regular laptops is diminishing as laptops themselves become smaller and lighter. Designers of laptop and notebook computers have to deal with several problems and tradeoffs. Naturally, laptop users would like to have more powerful processors, higher-capacity hard drives, more storage options, and larger, brighter displays. However, power to run these devices is at a premium. Although the capacity of batteries (usually nickel-cadmium or lithium-ion) has gradually improved, a laptop user on a long flight is still likely to run out of power before reaching his or her destination.

PALMTOP COMPUTERS

By the late 1990s, an even smaller device, the handheld or "palmtop" computer was becoming popular. These computers, sometimes called Personal Digital Assistants (PDAs) are a bit larger than a cell phone. The currently most popular model, the 3COM PalmPilot has a proprietary operating system, but many other PDAs use Windows CE, a stripped-down version of Microsoft's dominant operating system. Since palmtops are too small to have a conventional keyboard, input is performed using a stylus and handwriting recognition software (see HANDWRITING RECOGNITION), or by touching letters on a keyboard display. Scaled-down versions of word processing and other software are available. However, the most popular applications appear to be personal information management (see PERSONAL INFORMATION MANAGER) and accessing the Internet via a limited browsing capability and a wireless connection.

In 2002, Sony announced a line of handheld computers that include much faster processors and higher-resolution displays. Some units even include a built-in digital camera, MP3 player, and a programmable remote control that can be used with TVs and appliances.

A related development is the designing of versatile computer systems built into (or combined with) other devices such as cell phones or watches. There have also been experimental designs for "wearable computers," including displays built into eyeglasses or goggles.

Further Reading

Hansmann, Uwe [and others], ed. *Pervasive Computing Handbook*. New York: Springer-Verlag, 2001.

Lee, Lisa. *Easy iBook*. Indianapolis, Ind.: Que, 1999.

"Mobile Computing related resources on WWW." http://www.cs.purdue.edu/research/cse/scipad/mobicomp.html

"PDA Buzz." http://www.pdabuzz.com/

Proffitt. Brian. *The Practical PDA: Customize Your Palm or Handspring*. Roseville, Calif.: Premier Press, 2001.

portal

The legion of new World Wide Web users who went online in the mid-1990s could easily navigate and "surf' the Web, using browsers such as Netscape and Internet Explorer (see WEB BROWSER). However, the lack of a reliable starting point and a systematic way to find information often led to frustration. Search engines such as AltaVista and Lycos (see SEARCH ENGINE) provided some help, but there was no single guide that could present the most useful information at a glance.

Meanwhile, in 1994, two graduate students, Jerry Yang and David Filo, had begun to circulate an organized listing of their favorite websites by e-mail. When the list proved very popular, they decided they could make a business out of providing a website that could serve as a topical guide to the Web. The result was Yahoo!, the most successful of what would come to be called *Web portals*.

Yahoo! and other portals such as MSN (Microsoft Network), Excite, American On-line (AOL), and Lycos generally provide a listing organized by topic and subtopic. For example, the general topic "Computers and Internet" in Yahoo! is divided into many subtopics such as communications and networking, hardware, software, and so on. Many topics are further subdivided until, at the bottom, there is a list of actual Web links that can be clicked upon to take the user directly to the relevant site.

The advantage of using a portal over using a search engine is that the links on a portal have generally been selected for quality, relevance, and usefulness. The disadvantage is the flip side of that selectivity: The links may reflect the tastes, agenda, or commercial interests of the portal developers and thus exclude important points of view. When seeking to learn more about a subject, many researchers therefore both work "inward" from a portal and "outward" via a search engine (see ON-LINE RESEARCH).

To gain a competitive edge and raise revenue, portals typically include a considerable amount of advertising. Some portals also charge companies for being included or featured in listings or displays. General-purpose portals usually also contain such information as current news, stock prices, weather, and other timely information in an attempt to become their user's default page. Portals (particularly Yahoo!) have also sought to become more attractive (and profitable) by including such services as travel, financial services, games, and auctions.

Some portals emphasize particular approaches to information. For example, About.com goes beyond simply listing links to providing extensive guides to hundreds of subjects in a sort of newsletter format. There are also portals designed to serve particular constituencies, such as professional groups, industries, or hobby or interest groups. Companies can also create "enterprise portals" that can help employees keep in touch with developments and share information. Such portals often serve as the Web-based interface to the corporate local area network (LAN).

As with other information content providers, commercial portal developers have struggled to obtain enough revenue in a slumping Internet sector to keep up with the need to expand and compete in new areas. It is unclear whether the market will support more than a handful of large consumer portals in the long run, but both commercial and specialized portals have become an important part of the way most people access the Web.

Further Reading

About.com. www.about.com

Angel, Karen. *Inside Yahoo! Reinvention for the Road Ahead.* New York: Wiley, 2002.

Microsoft Network (MSN). www.msn.com

"Traffick: the Guide to Portals & Search Engines." http://www.traffick.com/directory/portals/consumer.asp

Yahoo! www.yahoo.com

presentation software

Whether at a business meeting or a scientific conference, the use of slides or transparencies has been largely replaced by software that can create a graphic presentation. Generally, the user creates a series of virtual "slides," which can consist of text (such as bullet points) and charts or other graphics. Often there are templates already structured for various types of presentations, so the user only needs to supply the appropriate text or graphics. There are a variety of options for the general visual style, as well as for transitions (such as dissolves) between slides. Another useful feature is the ability to time the presentation and provide cues for the speaker. Finished presentations can be shown on a standard monitor screen (if the audience is small) or output to a screen projector.

Microsoft PowerPoint is the most widely used presentation program. It includes the ability to import Excel spreadsheets, Word documents, or other items created by Microsoft Office suite applications. The user can switch

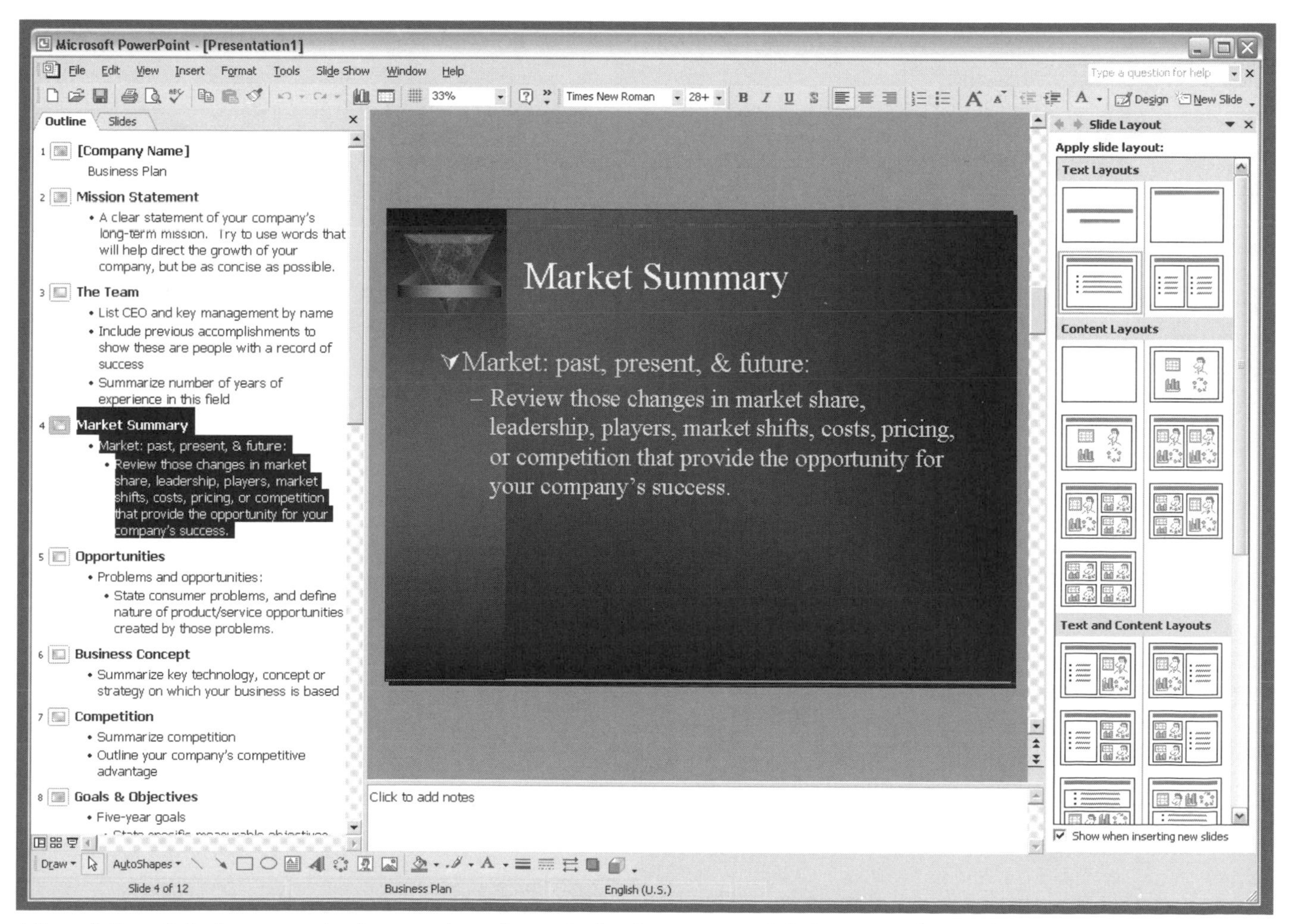

Microsoft PowerPoint is an example of presentation software. Such software uses a "slideshow" metaphor in which screens corresponding to slides can be created and arranged on a timeline for playing. Many types of special effects are also available.

between outline view (which shows the overall structure of the presentation) to viewing individual slides or working with the slides as a collection.

There are a number of alternatives available including Lotus Freelance Graphics and Sun's Star Office, which includes a presentation program comparable to PowerPoint. Another alternative is to use HTML Web-authoring programs to create the presentation in the form of a set of linked webpages. (PowerPoint and other presentation packages can also convert their presentations to HTML.) Although creating presentations in HTML may be more difficult than using a proprietary package and the results may be somewhat less polished, the universality of HTML and the ability to run presentations from a website are strong advantages of that approach.

A number of observers have criticized the general sameness of most business presentations. Some presentation developers opt to use full-fledged animation, created with products such as Macromedia Director.

Further Reading

About.com. "Presentation Design Guide."http://desktoppub.about.com/msubmenu34.htm

Halverson, Margo. *DesignSense for Presentations.* Portland, Me.: Proximity Learning, 1999.

Heid, Jim. "Presenting with HTML: Using the Language of the Web to Build Business Presentations." *Macworld,* April 1998, 93. (Also available on-line through About.com)

Microsoft PowerPoint 2000: Presentation Graphics with Impact. Upper Saddle River, N.J.: Prentice Hall, 1999.

printers

From the earliest days of computing, computer users needed some way to make a permanent record of the machine's output. Although results of a program could be punched onto cards or saved to magnetic tape or some other medium, at some point data has to be readable by human beings. This fact was recognized by the earliest computer and calculator designers: Charles Babbage (see BABBAGE, CHARLES) designed a printing mechanism for his never-finished computing "engine," and Williams Burroughs patented a printing calculator in 1888.

TYPEWRITER-LIKE PRINTERS

The large computers that first became available in the 1950s (see MAINFRAME) used "line printers." These devices have one hammer for each column of the output. A rapidly moving band of type moves under the hammers. Each hammer strikes the band when the correct character passes by. Printing is therefore done line by line, hence the name. Line printers were fast (600 lines per minute or more) but like the mainframes they served, they were bulky and expensive.

The typewriter offered another point of departure for designing printers. A few early computers such as the BINAC (an offshoot of ENIAC) used typewriters rigged with magnetically controlled switches (solenoids). However, a more natural fit was with the Teletype, invented early in the 20th century to print telegraph messages. Since the Teletype is already designed to print from electrically transmitted character codes, it was easy to rig up a circuit to translate the contents of computer data into appropriate codes for printing. (Since the Teletype could send as well as receive messages, it was often used as a control terminal for computer operators or for time-sharing computer users into the 1970s.)

The daisy-wheel printer was another typewriter-like device. It used a movable wheel with the letters embedded in slim "petals" (hence the name). It was slow (about 10 characters a second), noisy, and expensive, but it was the only affordable alternative for early personal computer users who required "letter-quality" output.

Printers (1)

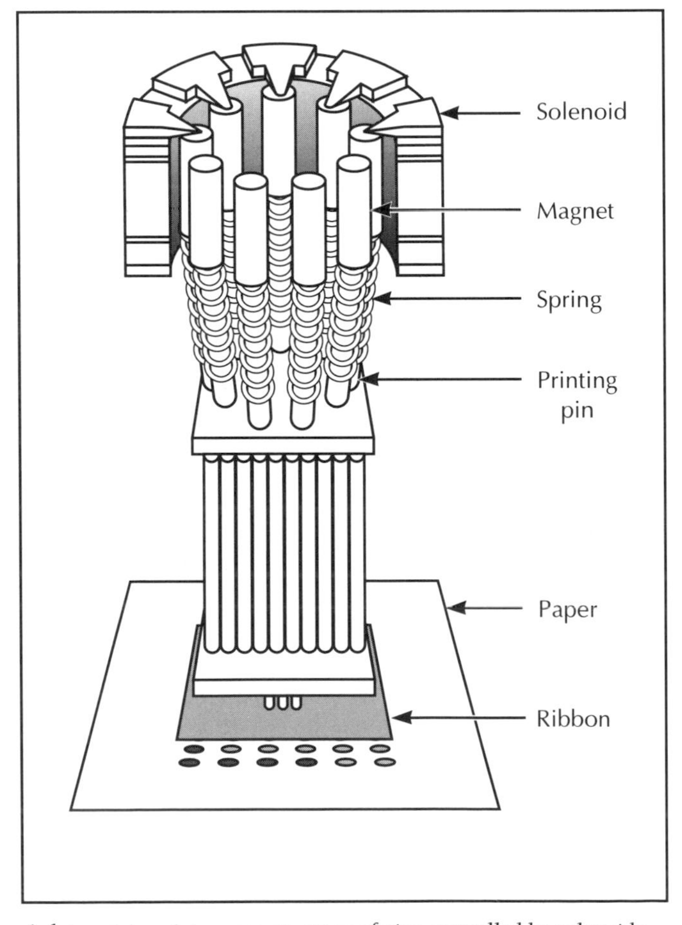

A dot-matrix printer uses an array of pins controlled by solenoids. Each character has a pattern of pins that are pushed against a typewriter-like ribbon to form the character on the paper.

DOT-MATRIX PRINTERS

The dot-matrix printer, which came into common use in the 1980s, uses a different principle of operation than typewriter-style printers. Unlike the latter, the dot-matrix printer does not form solid characters. Instead, it uses an array of magnetically controlled pins (9 pins at first, but 24 on later models). Each character is formed by pressing the appropriate pins into a ribbon that pushes into the paper, leaving a pattern of tiny dots.

Besides being relatively inexpensive, dot-matrix printers are versatile in that a great variety of character styles or fonts can be printed (see FONT), either by loading different sets of bitmaps. Likewise, graphic images can also be printed. However, because the characters are made of tiny dots, they don't have the crisp, solid look of printed type.

LASER AND INKJET PRINTERS

The majority of printers used today use laser or inkjet technology. Both combine the versatility of dot-matrix with the letter quality of typewriter-style printers. Xerox introduced the first laser printer in the 1970s, although the technology was too expensive for most users at first.

The laser printer converts data from the computer into signals that direct the laser beam to hit precise, tiny areas of a revolving drum. The drum is covered with a charged (usually negative) film. The areas hit by the laser, however, gain the opposite charge. As the drum continues to revolve, toner (a black powder) is dispensed. Because the toner is given a charge opposite to the places where the laser hit, the toner sticks to those places. Meanwhile, the paper is drawn into the drum. Because the paper is given the same charge as that produced by the laser beam (but stronger), the toner is pulled from the dots on the drum to the corresponding parts of the paper, forming the characters or graphics. A heating system then fuses the toner to the paper to make the image permanent. Meanwhile, the drum is discharged and the printer is ready for the next sheet of paper.

Color laser printers are also available, although they are still relatively expensive. They work by using four revolutions of the drum for each sheet of paper, depositing appropriate amounts of black, magenta, cyan, and yellow toner.

Laser printers fell in price throughout the 1990s (to $500 or so), but were soon rivaled by a different technology, the inkjet printer.

Printers (2)

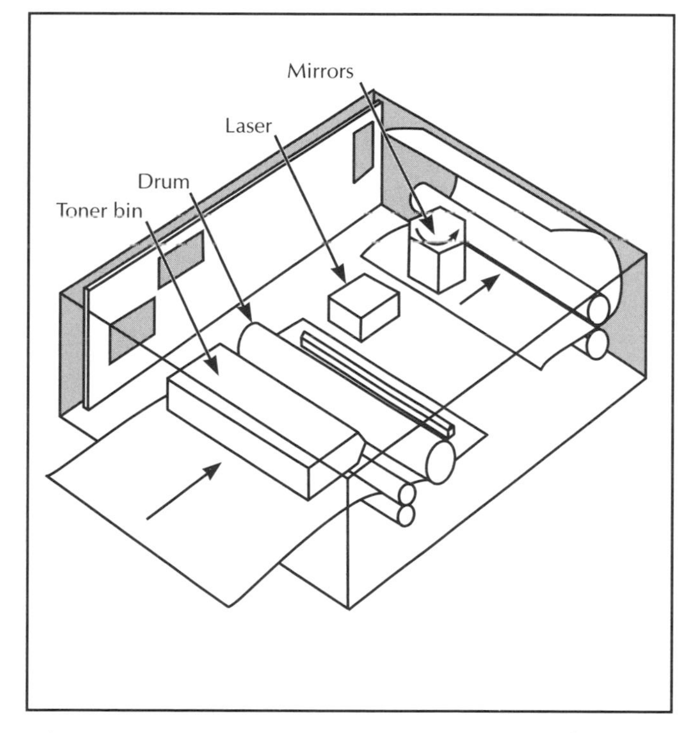

A laser printer uses a mirror-controlled laser beam to strike small spots on a rotating drum (called an OPC or Organic Photoconducting Cartridge) that had been given an electrical charge (usually positive) by a corona wire. The spots where the light beam hits are given an opposite charge (usually negative). The drum is then coated with a powdery toner that is charged opposite to the places where the light hit, so the toner clings to the drum to form the patterns of the characters or graphics. A piece of paper is then given a strong negative charge so it can pull the toner off the drum as it passes under it. Finally, heated rollers called fusers bind the toner to the paper to form the final image.

Printers (3)

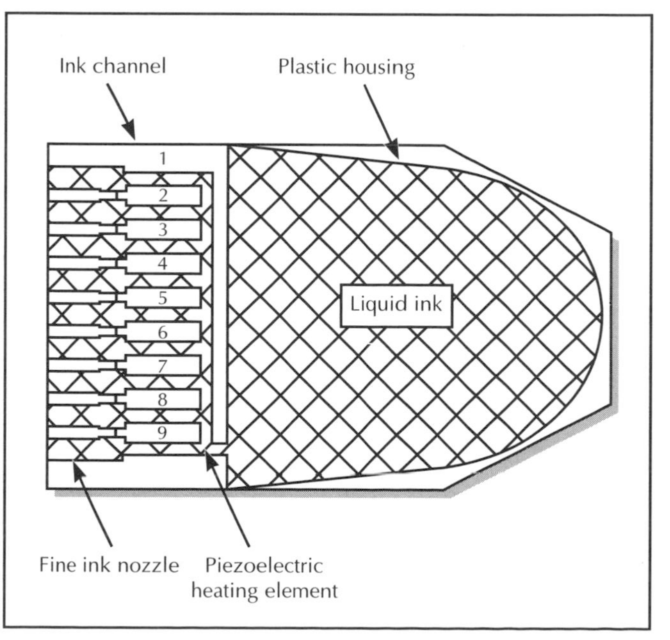

As the name implies, an inkjet printer uses an array of tiny jets or nozzles that are fed from four ink reservoirs (for magenta, cyan, yellow, and black). The appropriate nozzles and colors are activated for each character or graphics area and a heat pulse creates a vapor bubble that pushes a tiny droplet of ink out through the nozzle.

The inkjet printer has a print head that contains an ink cartridge for each primary printing color. Each cartridge has 50 nozzles, each thinner than a human hair. To print, the appropriate nozzles of the appropriate colors are subjected to electric current, which goes through a tiny resistor in the nozzle. An intense heat results for a few microseconds, long enough to create a tiny bubble that in turn forces a droplet of ink onto the page.

Inkjet printers are generally slower than lasers, although fast enough for most purposes. Although the inkjet is like the dot-matrix in producing tiny dots, the dots are much finer. With output at up to 2,880 dots per inch, the resulting characters are virtually indistinguishable from type-printing. Using high resolution and special papers, inkjet printers can now also produce photo prints comparable to those created by traditional processes.

TRENDS

By the end of the 1990s, the inkjet printer was declining steeply in price, and today quite capable units can be purchased for as little as $100 or so. Because of their greater speed, however, lasers are still used for higher-volume printing operations. "Multifunction" units combining printer, scanner, copier, and fax functions are also popular and cost less than a printer alone did only a few years ago.

Advocates of office automation have long predicted the "paperless office," but so far computers and their printers have churned out more paper, not less. However, there are some trends that might eventually reverse this course. Development of practical "electronic books" (page-size displays that can hold thousands of pages of text) may reduce the need for printed output. Another possible replacement for printing is "electronic ink," a sheet of paper with charged ink held in suspension. The text or graphics on the page can be changed electronically, so it can be reused indefinitely. Finally, the ability to access data anywhere on handheld or laptop computers may also reduce the need to make printouts.

Further Reading

PC Magazine. "Buying Guide: Printers." www.pcmag.com. [Navigate to Home > Product Guides > Printers > Buying Guide: Printers]

Rosch, W. L. *Winn L. Rosch's Printer Bible*. New York: MIS Press/Holt, 1996.

White, Ron. *How Computers Work*. Millennium Edition. Indianapolis, Ind.: Queue, 1999.

privacy in the digital age

Quoted in Fred H. Cate's *Privacy in the Information Age*, legal scholar Alan F. Westin has defined privacy as "the claim of individuals, groups, or institutions to determine for themselves when, how, and to what extent information about themselves is communicated to others."

Since the mid-19th century, advances in communications technology have raised new problems for people seeking to protect privacy rights. During the Civil War telegraph lines were tapped by both sides. In 1928, the U.S. Supreme Court in *Olmstead v. U.S.* refused to extend Fourth Amendment privacy protections to prevent federal agents from tapping phone lines without a warrant. Almost 50 years later, the court would revisit the issue in *Katz v. U.S.* and rule that telephone users did have an "expectation of privacy." The decision also acknowledged the need to adapt legal principles to the realities of new technology.

In the second half of the 20th century the growing use of computers would raise two basic kinds of privacy problems: surveillance and misuse of data.

SURVEILLANCE AND ENCRYPTION

Since much sensitive personal and business information is now transmitted between or stored on computers, such information is subject to new forms of surveillance or interception. Keystrokes can be captured using surreptitiously installed software and e-mails can be intercepted from servers or a user's hard drive. Many employers now routinely monitor employees' computer activity at work, including their use of the World Wide Web. When this practice is challenged, courts have generally sided with the employer, accepting the argument that the computers at work exist for business purposes, not private communications, and thus do not carry much of an expectation of privacy. Employers, however, have been encouraged to spell out their employee monitoring or surveillance policies explicitly. Outside the workplace, some protection is offered by the Electronic Privacy Protection Act (ECPA), passed in 1986.

Technology can be used to penetrate privacy, but it can also be used to safeguard it (see ENCRYPTION). Public key encryption programs such as Pretty Good Privacy (PGP) can encode text so that it cannot be read without a very-hard-to-crack key. The U.S. government, whose agencies enjoyed powerful surveillance capabilities, initially fought to suppress the use of encryption, but a combination of unfavorable court decisions and the ability to spread software across the Internet has pretty much decided the battle in favor of encryption users.

Recently, shadowy accounts about a secret system called Echelon have suggested that the National Security Agency has in place a massive system that can intercept worldwide communications ranging from e-mail to cell phone conversations. Apparently, rooms full of supercomputers can sift through this torrent of communication, looking for key words that might indicate a threat to the United States or its allies. (Much communication is in "clear" text; the ability of the government to crack strong encryption is unclear.)

In the aftermath of the terrorist attacks of September 11, 2001 the federal government has pressed for expanded surveillance powers, some of which were granted in the USA PATRIOT Act of 2001. Computerized surveillance and identification systems (see also BIOMETRICS) are also likely to be expanded in airports in other public places as part of the "War on Terrorism."

INFORMATION PRIVACY

Many privacy concerns arise not from the activities of spy or police agencies, but from the potential for the misuse of the many types of personal information now collected by businesses or government agencies. As far back as 1972, the Advisory Committee on Automated Personal Data Systems recommended the following standards to the secretary of the Department of Health, Education, and Welfare:

1. There must be no personal data record-keeping systems whose very existence is secret.
2. There must be a way for an individual to find out what information about him/her is on record and how it is used.
3. There must be a way for an individual to correct or amend a record of identifiable information about him/her.
4. There must be a way for an individual to prevent information about him/her that was obtained for one purpose from being used or made available for other purposes without his/her consent.
5. Any organization creating, maintaining, using, or disseminating records of identifiable personal data must guarantee the reliability of the data for their intended use and must take precautions to prevent misuse of the data.

The Federal Privacy Act of 1974 generally implemented these principles with regard to data maintained by federal agencies. Later, federal laws have attempted to address particular types of information, including school records, medical records, and video rentals.

However, much of the information collected from people results from commercial transactions or other interactions with businesses, particularly via the Internet. Although encrypted processing systems have reduced the chance that a credit card number submitted to a store will be stolen, so-called identity thieves may be able to obtain credit reports under false pretenses or collect enough information about a person from various databases (including Social Security numbers). With that information, the thief can take out credit cards in the person's name and run up huge bills. While the direct financial liability from identity theft is capped, the psychological impact and the effort required for victims to rehabilitate their credit standing can be considerable. In a few cases the same techniques have been used by stalkers, sometimes with tragic consequences.

The ability of websites to track where a visitor clicks by means of small files called "cookies" has also disturbed many people. As with the recording of purchase information by supermarkets and other stores, businesses justify the practice as allowing for targeted marketing that can provide consumers with information likely to be of interest to them. (Many e-mail addresses are also gathered to be sold for use for unsolicited e-mail, or "spam.") An even more intrusive technique involves the surreptitious installation of software on the user's computer for purposes of displaying advertising content or gathering information. In turn, programmers have distributed free utilities for identifying and removing such "adware" or "spyware."

While such consumer tracking is not as dangerous as identity theft, it feels like an invasion of privacy to many people as well as a source of insecurity, particularly because there are as yet few regulations governing such practices. However, in response to such concerns many businesses have put "privacy statements" on their websites, explaining what information about visitors will be collected and how it may be used. Businesses that meet standards for disclosure of their privacy practices can also display the seal of approval of organizations such as TRUSTe.

Many privacy advocates, however, believe that self-regulation is not sufficient to truly protect consumer privacy. They support strong new regulations, including "opt-in" provisions that would require businesses to receive explicit permission from the consumer before collecting information.

Further Reading

Brin, David. *The Transparent Society.* Reading, Mass.: Addison-Wesley, 1998.

Cate, Fred H. *Privacy in the Information Age.* Washington, D.C.: Brookings Institution, 1997.

Electronic Frontier Foundation. www.eff.org

Electronic Privacy Information Center. www.epic.org

"EPIC On-line Guide to Privacy Resources." http://www.epic.org/privacy/privacy_resources_faq.html

Garfinkel, Simson, and Deborah Russell. *Database Nation: the Death of Privacy in the 21st Century.* Sebastopol, Calif.: O'Reilly, 2001.

Henderson, Harry. *Privacy in the Information Age (Library in a Book).* New York: Facts On File, 1999.

Hunter, Richard S. *World Without Secrets: Business, Crime, and Privacy in the Age of Ubiquitous Computing.* New York: Wiley, 2002.

Hyatt, Michael S. *Invasion of Privacy: How to Protect Yourself in the Digital Age.* Washington, D.C.: Regnery Publishing, 2001.

Melanson, Philip H. and Anthony Summers. *Secrecy Wars: National Security, Privacy, and the Public's Right to Know.* Washington, D.C.: Brasseys, 2002.

procedures and functions

From the earliest days of programming, programmers and language designers realized that it would be very useful to organize programs so that each task to be performed by the program had its own discrete section of code. After all, a program will often have to perform the same task, such as sorting or printing data, at several different points in its processing. Instead of writing out the necessary code instructions each time they are needed, why not write the instructions just once and have a mechanism by which they can be called upon as needed? Such callable program sections have been known as procedures, subroutines, or subprograms.

The simplest sort of subroutine is found in assembly languages and early versions of BASIC or FORTRAN. In BASIC, for example, a GOSUB statement contains a line number. When the statement is encountered, execution "jumps" to the statement with that line number, and continues from there until a statement such as RETURN is encountered. For example:

```
10 TOTAL = 10

20 GOSUB 40

30 END

40 PRINT "The total is: ";

50 PRINT TOTAL

60 RETURN
```

Here execution jumps from line 20 to line 40. After lines 40–60 are executed, the program returns to line 30, where it ends.

PROCEDURES WITH PARAMETERS

The simple subroutine mechanism has some disadvantages, however. The subroutine gets the information it needs from the main part of the program implicitly through the global variables that have been defined (see VARIABLE). If it needs to return information, it does it by changing the value of one or more of these global variables. The problem is that many different subroutines may be relying upon the same variables and at the same time changing them, leading to unpredictable results. Modern programming practice therefore generally avoids using global variables as much as possible.

Most high level languages today (including Pascal, C/C++, Java, and modern versions of BASIC) define subprograms as procedures that pass information through specified parameters. For example, a procedure in Pascal might be defined as:

```
Procedure PrintChar (CharNum : integer);
```

Procedure and Functions (1)

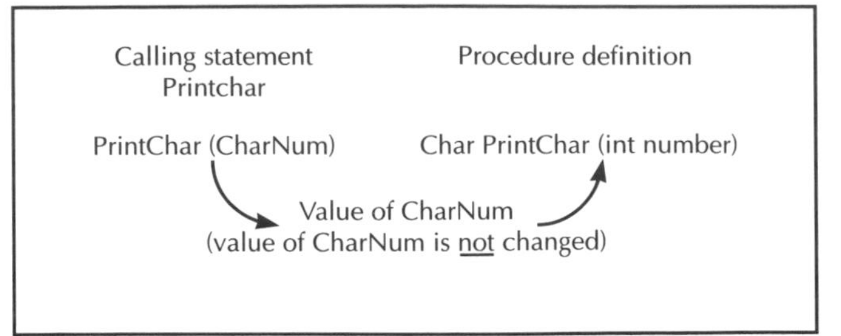

In a "call by value," only the value of the variable parameter (CharNum in this case) is passed from the calling statement to the procedure. The original variable will not be affected.

This procedure has one parameter, an integer that specifies the number of the character to be printed (see CHARACTERS AND STRINGS).

The main program can call the procedure by giving its name and an appropriate character number. For example:

```
PrintChar (32);
```

The code within the procedure does not work with the parameter CharNum directly. Rather, it receives a copy that it can use. Thus, the procedure might include the statements:

```
Writeln ('Character number: ', CharNum );
Writeln (chr (CharNum));
```

The program will print the character number and then print the character itself on the next line (for character number 32 this will actually be a blank).

This typical way of using parameters is called *passing by value*. However, it is possible to pass a parameter to a procedure and have the parameter itself used rather than

Procedure and Functions (2)

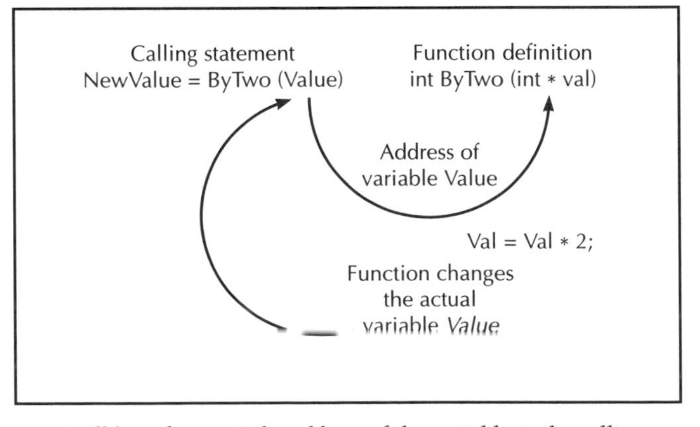

In a "call by reference," the address of the variable in the calling statement is sent to the called function. This means that the latter can change the value of the original variable.

working with a copy. This is called "passing by reference." Pascal uses the var keyword for this purpose, while C passes a pointer to the variable (see POINTERS AND INDIRECTION), and C++ and Java prefix the variable name with an ampersand (&). For example, suppose one has a C function defined as follows:

```
int ByTwo (int * Val)
{
  Val = Val * 2;
}
```

In the following statements in the calling program:

```
Int Value, NewValue;
Value = 10;
NewValue = By Two (Value);
```

NewValue would be set to 20 because the actual variable Value has been multiplied by two inside the ByTwo function.

FUNCTIONS

A function is a procedure that returns its results as a value in place of the function name in the calling statement. For example, a function in C to raise a specified number to a specified power might be defined like this:

```
int Power (int base, int exp)
```

(C and related languages don't use a keyword like Pascal's procedure or function because in C all procedures are functions.)

This definition says that the Power function takes two integer parameters, base and exp, and returns an integer value.

Suppose somewhere in the program there are the following statements:

```
Int Base = 8;

Int Dimensions = 3;
Size = Power (Base, Dimensions);
```

The variable Size will receive the value of Power (8, 3) or 512.

Although the syntax for using procedures or functions varies by language, there are some principles that are generally applicable. The type of data expected by a procedure should be carefully defined (see DATA TYPES). Modern compilers generally catch mismatches between the type of data in the calling statement and what is defined in the procedure declaration. Procedures should also be checked for unwanted "side effects," which they can minimize by not using global variables.

Procedures and functions relating to a particular task are often grouped into separate files (sometimes called units or modules) where they can be compiled and linked into a program that needs to use them (see LIBRARY, PROGRAM).

Object-oriented languages such as C++ think of procedures in a somewhat different way from the examples shown here. While a traditional program sees procedures as blocks of code to be invoked for various purposes, an object-oriented program sees procedures as "methods" or capabilities of the program's various objects (see OBJECT-ORIENTED PROGRAMMING).

Further Reading

Baker, Mark. "Teaching Programming III: Procedures and Functions." http://atschool.eduweb.co.uk/mbaker/prog3.html

Sebesta, Robert W. *Concepts of Programming Languages.* 4th ed. Reading, Mass.: Addison-Wesley, 1999.

programming as a profession

All computer applications depend upon the ability to direct the machine to perform instructions such as fetching or storing data, making logical comparisons, or performing calculations. Although practical electronic computers first began to be built in the 1940s, it took considerable time for programming to emerge as a distinct profession. The first programmers were the computer designers themselves, followed by people (often women) recruited from clerical persons who were good at mathematics. With machines like ENIAC, programming was more like setting up a complicated piece of factory machinery than like writing. Switches or plugboards had to be set, and numeric instruction codes punched on cards to instruct the machine to move each piece of data from one location to another or to perform an arithmetic or logical operation.

Two factors led to greater recognition for the art or craft of programming. First, as more computers were built and put to work for various purposes, more programmers were needed, as well as more attention to their training and management. Second, as programs became larger and more complex, a number of high-level languages such as COBOL and FORTRAN came into use (see PROGRAMMING LANGUAGES). Besides making it easier to write programs, having just a few languages in widespread use made skills more readily transferable from one computer installation to another. And as with any profession, programming developed bodies of knowledge and practice.

At the same time, advances in language development would raise a recurrent question: Are professional programmers really necessary? Since FORTRAN looked a lot like ordinary mathematical notation, couldn't scientists and engineers just write the programs they need without hiring specialists for the job? Similarly, some enthusiasts led managers to think that with COBOL accountants (or even managers) could write their own business programs.

Sometimes part-time or "amateur" programming did prove to be practicable, particularly for scientists who found that writing a quick FORTRAN routine to solve a problem was easier than trying to explain the problem to a professional programmer. However, the professional programmer's job was never really in danger. Businesspeople were less inclined to try to learn COBOL and entrust something like the company's payroll processing to ad hoc efforts. In addition, the programs that controlled the operation of the computer itself, which became known as operating systems, required both arcane knowledge and the ability to design, verify, test, and debug increasingly complex systems (see SOFTWARE ENGINEERING).

DEVELOPMENT OF PRACTICE

In response to this growing complexity, computer scientists approached the improvement of programming practice on several levels. New languages developed in the 1960s and 1970s featured well-defined control structures, data types, and procedure calls (see ALGOL, PASCAL, C, DATA TYPES, LOOP, and STRUCTURED PROGRAMMING.) The management of programming teams and the factors affecting productivity were examined by pioneers such as Frederick Brooks, author of *The Mythical Man-Month,* and IBM sponsored workshops and study groups.

While many mainframe business programmers continued to write and maintain programs written in the older languages (such as COBOL), starting in the 1970s a new generation of systems and applications programmers used C and worked in a different environment—campus minicomputers running UNIX. Unlike the hierarchical, systematic approach of the "mainframe culture," the minicomputer programmers tended toward a decentralized but cooperative approach (see also OPEN SOURCE MOVEMENT, HACKERS AND HACKING).

When the personal computer revolution began to arrive at the end of the 1970s, much of the evolution of programming culture would be recapitulated. Since early microcomputer systems had very limited memory, programmers who wanted to get useful work out of machines such as the Apple II had to work mainly in assembly language or write quick and dirty programs in a limited dialect of BASIC. The hobbyists and early adopters often knew little about the academic world of computer science and software engineering, but they were good at wringing the most out of each clock cycle and byte of memory.

As personal computers gained in power and capability through the 1980s, programmers were able to use higher-level languages such as C. Applications such as word processors, spreadsheets, and graphics programs became more complex, and programmers had to work in larger teams like their mainframe counterparts.

At the same time, the sharp demarcation between programmer and user became less distinct with the personal computer. Many users who were not professional programmers used applications software that included programmable features, such as spreadsheets and simple data bases (see also MACRO and SCRIPTING LANGUAGE). New languages such as Visual Basic let even relatively inexperienced programmers plug in user interfaces and other components and create useful programs (see PROGRAMMING ENVIRONMENT).

Each sector of programming seems to go through a cycle of improvisation and innovation followed by standardization and professionalization. Just as the early ENIAC programmers evolved into the organized hierarchy of corporate programming departments, the individuals and small groups who wrote the first personal computer software evolved into large teams using sophisticated software to track to the modules, versions, and development steps of major programming projects. Similarly, when the explosion of the World Wide Web starting in the mid-1990s brought a new demand for people who could code HTML, CGI, and Java, much of the most interesting work was done by individuals and small companies. But if history repeats itself, the Internet applications field will undergo the same process of professionalization, with increasingly elaborate standards and expectations (see also CERTIFICATION OF COMPUTER PROFESSIONALS).

Throughout the history of programming, visionaries have announced that the time was coming when most if not all programming could be automated. All a person will have to do is give a reasonably coherent description of the desired results and the required program will be coded by some form of artificial intelligence (see EXPERT SYSTEMS, GENETIC ALGORITHMS, and NEURAL NETWORK). But while users have now been given the ability to do many things that formerly required programming, it seems there is still a demand for programmers who can move the bar another step higher. The profession continues to evolve without any signs of impending extinction.

Further Reading

Brooks, Frederick. *The Mythical Man-Month, Anniversary Edition: Essays on Software Engineering.* Reading, Mass.: Addison-Wesley, 1995.

Ceruzzi, Paul. *A History of Modern Computing.* Cambridge, Mass.: MIT Press, 1998.

Henderson, Harry. *Career Opportunities in Computers and Cyberspace.* New York: Facts On File, 1999.

Kohanski, Daniel. *The Philosophical Programmer: Reflections on the Moth in the Machine.* New York: St. Martin's, 1998.

Ullmann, Ellen. *Close to the Machine: Technophilia and its Discontents.* San Francisco: City Lights Books, 1997.

programming environment

The first programmers used pencil and paper to sketch out a series of commands, or punched them directly on cards for input into the machine. But as more computer resources became available, it was a natural thought that programs could be used to help programmers create other programs. The availability of Teletype or early CRT terminals on time-sharing systems by the 1960s encouraged programmers to write simple text editing programs that could be used to create the computer language source code file, which in turn would be fed to the compiler to be turned into an executable program (see TERMINAL and TEXT EDITOR). The assemblers and BASIC language implementations on the first personal computers also included simple editing facilities.

More powerful programming editors soon evolved, particularly in academic settings. One of the best known is EMACS, an editor that contains its own LISP-like language that can be used to write macros to automatically generate program elements (see LISP and MACRO). With the many other utilities available in the UNIX operating system, programmers could now be said to have a programming environment—a set of tools that can be used to write, compile, run, debug, and analyze programs.

More tightly integrated programming environments also appeared. The UCSD "p-system" brought together a program editor, compiler, and other tools for developing Pascal programs. While this system was somewhat cumbersome, in the mid-1980s Borland International released (and steadily improved) Turbo Pascal. This product offered what became known as an "integrated development interface" or IDE. Using a single system of menus and windows, the programmer could edit, compile, run, and debug programs without leaving the main window.

The release of Visual Basic by Microsoft a few years later brought a full graphical user interface (GUI). Visual

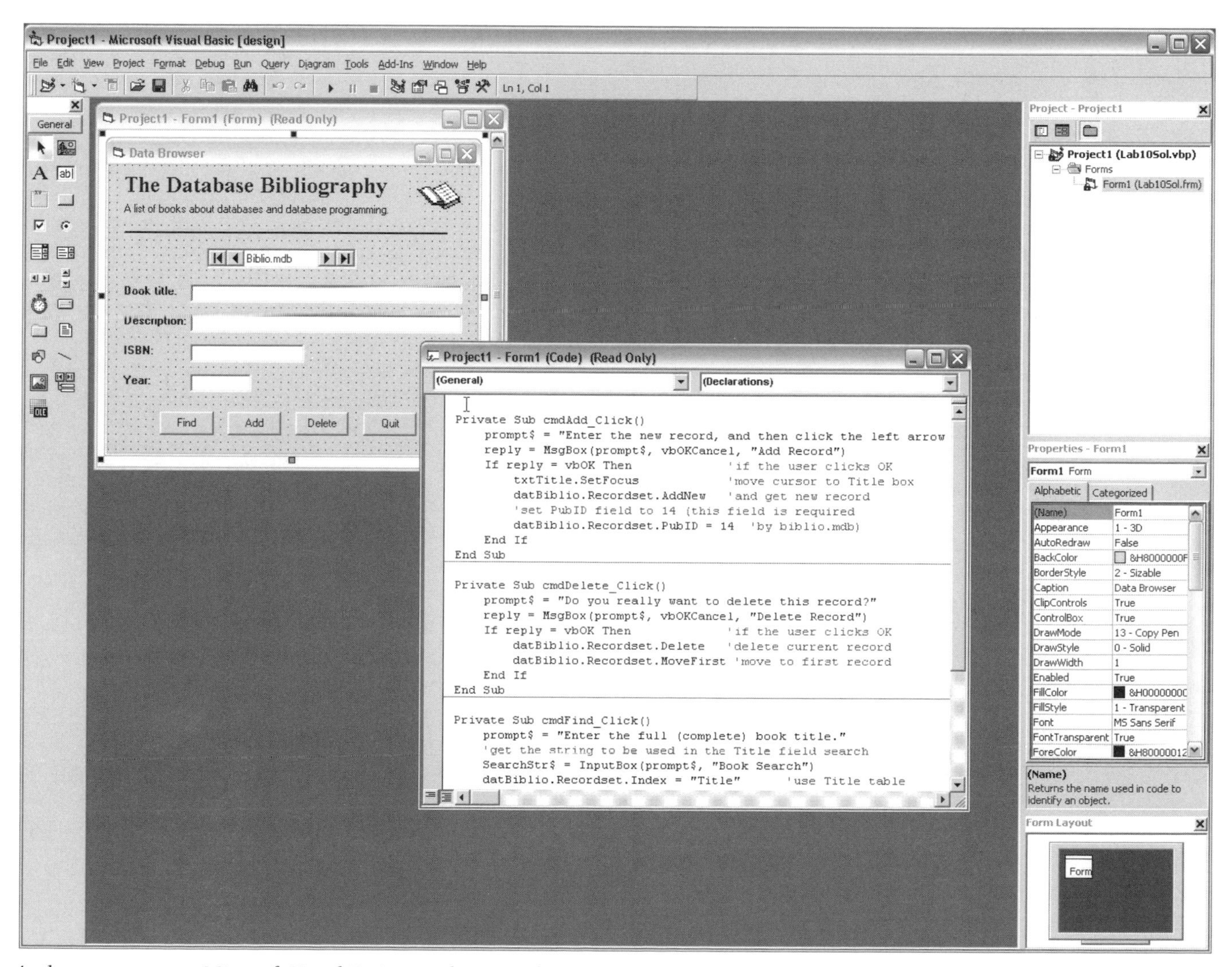

As the name suggests, Microsoft Visual Basic provides a visual programming environment in which the controls that make up a program's user interface can be placed on a form. Various properties (characteristics) of the controls can then be set, and program code is then written and attached to govern how the objects will behave.

Basic not only ran in Windows, it also gave programmers the ability to design programs by arranging user interface elements (such as menus and dialog boxes) on the screen and then attaching code and setting properties to control the behavior of each interface object. This approach was soon extended by Microsoft to development environments for C and C++ (and later, Java) while Borland released Delphi, a visual Pascal development system. Today visual programming environments from a variety of companies are available for most languages.

Modern programming environments help the programmer in a number of ways. While the program is being written, the editor can highlight syntax errors as soon as they're made. Whether arising during editing or after compilation, an error message can be clicked to bring up an explanation, and an extensive on-line help system can provide information about language keywords, built-in functions, data types, or other matters. The debugger lets the programmer trace the flow of execution or examine the value of variables at various points in the program.

Most large programs today actually consists of dozens or even hundreds of separate files, including header files, source code files for different modules, and resources such as icons or graphics. The process of tracking the connections (or dependencies) between all these files, which used to require a list called a *makefile* can now be handled automatically, and relationships between classes in object-oriented programs can be shown visually.

Researchers are working on a variety of imaginative approaches for future programming environments. For example, an interactive graphical display (see VIRTUAL REALITY) might be used to allow the programmer to in effect walk through and interact with various representations of the program.

Further Reading

Halvorson, Michael, ed. *Microsoft Visual Basic 6.0: DeLuxe Learning.* Redmond, Wash.: Microsoft Press, 1998.

Kernighan, Brian W. and Rob Pike. *The UNIX Programming Environment.* Englewood Cliffs, N.J.: Prentice Hall, 1984.

Palmer, I. J., N. Chilton, and R. A. Earnshaw. "Web Programming Environments: Toward a Virtual API." http://www.cs.vu.nl/~eliens/WWW5/papers/VAPIS/

[Program Editors and other tools] www.programfiles.com (click under Programming)

Travers, Michael. "Agent-based Programming Environments." http://xenia.media.mit.edu/~mt/childs-play-pp.html

programming languages

There are many ways to represent instructions to be carried out by a computer. With early machines like ENIAC, programs consisted of a series of detailed machine instructions. The exact movement of data between the processor's internal storage (registers) and internal memory had to be specified, along with the appropriate arithmetic operations. This lowest level, least abstract form of programming languages is hardest for humans to understand and use.

Programming Languages

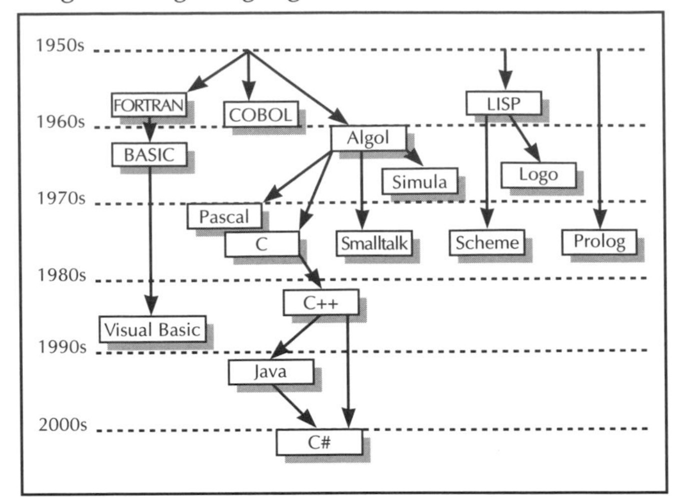

The evolution of a few major programming languages through five decades. There are actually hundreds of different programming languages that have seen at least some use in the past 50 years.

The first step toward a more symbolic form of programming is to use easy-to-remember names for instructions (such as ADD or CMP for "compare") as well as to provide labels for storage locations (variables) and subroutines (see PROCEDURES AND FUNCTIONS). The file of symbolic instructions (called source code) is read by a program called an assembler (see ASSEMBLER), which generates the low-level instructions and actual memory addresses to be used by the program. Because of its ability to closely specify machine operations, assembly language is still used for low-level hardware control or when efficiency is at a premium.

Most languages in use today are higher-level. The mainstream of programming languages consists of languages that are procedural in nature. That is, they specify a main set of instructions that are executed in sequence, although the program can branch off (see BRANCHING STATEMENTS) or repeat a series of statements until a condition is satisfied (see LOOP). A program can also call a set of instructions defined elsewhere in the program. Constant or variable data is declared to be of a certain type such as integer or character (see DATA TYPES) before it is used. There are also rules that determine what parts of a program can access what data (see VARIABLE). For examples of procedural languages see ALGOL, BASIC, C, COBOL, FORTRAN, and PASCAL.

A variant of procedural languages is the object-oriented language (see OBJECT-ORIENTED PROGRAMMING).

Such languages (see C++, JAVA, and SMALLTALK) still use sequential execution and procedures, but the procedures are "packaged" together with relevant data into objects. In order to display a picture, for example, the program will call upon a particular object (created from a class of such objects) to execute its display function with certain parameters such as location and dimensions.

NONPROCEDURAL LANGUAGES

Although the bulk of today's software is written using procedural languages, there are some important languages constructed using quite different paradigms. LISP, for example, is a powerful language used in artificial intelligence applications. LISP is written by putting together layers of functions that carry out the desired processing (see NONPROCEDURAL LANGUAGES, LISP, and FUNCTIONAL LANGUAGES). There are also "logic programming" languages, of which Prolog is best known (see PROLOG). Here a chain of logical steps is constructed such that the program can traverse it to find the solution of a problem.

CONTEXT AND CHANGE IN PROGRAMMING LANGUAGES

Because of the amount of effort it takes to truly master a major programming language, most programmers are fluent in only a few languages and developers tend to standardize on one or two languages. The store of tried-and-true code and lore built up by the programming community tends to make it disadvantageous to radically change languages. Thus, FORTRAN and COBOL, although more than 40 years old, are still in widespread use today. C, which is about 30 years old, has been gradually supplanted by C++ and Java, but the latter languages represent an object-oriented evolution of C, intentionally designed to make it easy for programmers to make the transition. (Smalltalk, which was designed as a "pure" object-oriented language, never achieved widespread use in commercial development.)

Similarly, when programmers had to cope with parallel processing (programs that can have several threads of execution going at the same time), they have tended to favor "parallelized" versions of familiar languages rather than wholly new ones.

While the basic elements of computer languages tend to persist in the same recognizable forms, the way programmers experience their use of languages has changed considerably through the use of modern visual integrated development environments (see PROGRAMMING ENVIRONMENT). A variety of languages have also been designed for tasks such as data management, interfacing webpages, and system administration (see SCRIPTING LANGUAGES, AWK, PERL, and PYTHON).

Further Reading

Sebesta, Robert W. *Concepts of Programming Languages.* Reading, Mass.: Addison-Wesley, 1999.

Bergin, Thomas J., and Richard G. Gibson, eds. *History of Programming Languages-II.* Reading, Mass.: Addison-Wesley/ ACM Press, 1996.

project management software

Whether a project involves only a few people in the same department or thousands of people and several years, there is a variety of software to help managers plan and monitor the status of their projects.

At the simplest level, PIM software (see PERSONAL INFORMATION MANAGER) can be used by an individual to monitor simple personal projects. Such software generally includes the ability to record the description, priority, due date, and reminder date for a task.

Project management software is generally used to plan larger projects involving many persons or teams. A complex project must first be broken down into tasks. (Large projects often have subprojects as an intermediate entity.) Next, dependencies must be taken into account. For example, the user testing program for a software product can't begin until a usable preliminary ("alpha" or "beta") version of the program is available. The various "resources" assigned to a subproject or task must also be tracked, including personnel and number of hours assigned and budget allocations. In tracking personnel assigned to a project, their availability (who is on vacation and who is assigned to what location) must also be considered.

Once the scheduling and priorities are arranged, the inevitable divergences between what was planned and what is actually happening must be monitored. Good project management software provides many tools for the purpose. Available charts and reports often include:

- Gantt charts that use bars to show the duration and percentage of completion of the various overlapping subprojects or tasks.
- PERT (Program Evaluation and Review Technique) charts that show each subproject or task as a rectangular "node" with information about the task. The connections between nodes show the relationships (dependencies) between the items. PERT charts are usually used at the beginning stages of planning.
- Analysis tools that show critical paths and bottlenecks (places where one or more tasks falling behind might threaten large portions of the project). Generally, the more preceding items a task is dependent on, the more likely that task is to fall behind.

- Tools for estimating the probability for completion of a given task based on the probabilities of tasks it is dependent on, as well as other factors such as the likelihood of certain resources becoming available.
- A system of alerts or "stoplights" that show slowdowns, potential problems, or areas where work has stopped completely. These can be set to be triggered when various specified conditions occur.
- Integration between project management and budget reporting so tasks and the project as a whole can be monitored in relation to budget constraints.
- Integration between the project management software and individual schedules kept in PIM software such as Microsoft Outlook or in handheld computers (PDAs) such as the PalmPilot.
- Integration between project management and software for scheduling meetings.

Given the scope and pace of today's business, scientific, and other projects, project management software is often a vital tool. However, using too elaborate a project tracking system for a relatively small and well-defined project may divert time and energy away from the work itself. Fortunately, a wide variety of project management programs are available, ranging from full-fledged products such as Microsoft Project or Primavera Project Planner to simpler shareware or free products.

Further Reading

"A Buyer's Guide: Selecting Project Management Software." http://www.4pm.com/articles/selpmsw.html

The Project Management Center. "Directory of Project Management Software." http://www.infogoal.com/pmc/pmcswr.htm

Lewis, James P. *The Project Manager's Desk Reference.* New York: McGraw Hill Professional Publishing, 1999.

Uyttewaal, Eric. *Dynamic Scheduling with Microsoft Project 2000: The Book by and for Professionals.* New York: International Institute for Learning, 2000.

Prolog

Since the 1950s, researchers have been intrigued by the possibility of automating reasoning behavior, such as logical inference (see ARTIFICIAL INTELLIGENCE). A number of demonstration programs have been written to prove theorems starting from axioms or assumptions. In 1972, French researcher Alain Colmerauer and Robert Kowalski at Edinburgh University created a logic programming language called Prolog (for Programmation en Logique) as a way of making automated reasoning and knowledge representation more generally available.

A conventional procedural program begins by defining various data items, followed by a set of procedures for manipulating the data to achieve the desired result. A Prolog program, on the other hand, begins with a set of facts (axioms) that are assumed to be true. (This is sometimes called declarative programming.)

For example, the fact that Joe is the father of Bill would be written:

```
Father (Joe, Bill).
```

The programmer then defines logical rules that apply to the facts. For example:

```
father (X, Y) :- parent (X, Y), is male (X)
grandfather (X, Y) :- father (X, Z), parent (Z,
Y)
```

Here the first assertion says that a person X is the father of Y if he is the parent of Y and is male. The second assertion says that X is Y's grandfather if he is the father of a person Z who in turn is a parent of Y.

When a program runs, it processes queries, or assertions whose truth is to be proven. Using a process called unification, the Prolog system looks for facts or rules that apply to the query and attempts to create a logical chain leading to proving the query is true. If the chain breaks (because no matching fact or rule can be found), the system "backtracks" by looking for another matching fact or rule from which to attempt another chain.

Prolog aroused considerable interest among artificial intelligence researchers who were hoping to create a powerful alternative to conventional programming languages for automating reasoning. This interest was further spurred by the Japanese Fifth Generation Computer Program of the 1980s, which sought to create logical supercomputers and made Prolog its language of choice. Although some such machines were built, the idea never really caught on. However, Borland International (makers of the highly successful Turbo Pascal) released a Turbo Prolog that made the language more accessible to students using PCs, although it used some nonstandard language extensions.

Despite its commercial success being limited, Prolog has been used in a number of areas of artificial intelligence research. Its rules-based structure is naturally suited for expert systems, knowledge bases, and natural language processing (see EXPERT SYSTEMS and KNOWLEDGE REPRESENTATION). It can also be used as a prototyping language for designing systems that would then be recoded in conventional languages for speed and efficiency.

Further Reading

Bratko, Ivan. *PROLOG Programming for Artificial Intelligence.* 3rd ed. Addison Wesley, 2000.

Brna, Paul. "Prolog Programming: a First Course." http://cbl.leeds.ac.uk/~paul/prologbook/

"Prolog Programming Language." http://www.engin.umd.umich.edu/CIS/course.des/cis400/prolog/prolog.html

Sterling, Leon and Ehud Shapiro, eds. 2nd ed. Boston: MIT Press, 1994.

pseudocode

Because humans generally think on a higher (or more abstract) level than that provided by even relatively high-level programming languages such as BASIC or Pascal, it is sometimes suggested that programmers use some form of pseudocode to express how the program is intended to work. Pseudocode can be described as a language that is more natural and readable than regular programming languages, but sufficiently structured to be unambiguous. For example, the following pseudocode describes how to calculate the cost of wall-to-wall carpet for a room:

```
Get room length (in feet)
Get room width
Multiply length by width to get area (in square
feet)
Get price of carpet per square foot
Multiply price/sq. foot by area to get total
cost.
```

Pseudocode generally includes the basic control structures used in programming languages (see BRANCHING STATEMENTS and LOOP) but is not concerned with small details of syntax. For example, this pseudocode might determine whether to charge sales tax for an on-line purchase:

```
Get customer's state of residence
If state is "CA" then
  Tax = Price * .085
  Total = Price + Tax
End If
```

Once the pseudocode has been written and reviewed, the statements can be recoded in the programming language of choice. For example, the preceding example might look like this in C:

```
If (state == "CA") {
  Tax = Price * .085;
  Total = Price + Tax;
}
```

The term *pseudocode* can also be applied to "intermediate languages" that provide a generic, machine-independent representation of a program. For example, in the UCSD Pascal system the language processor generates a "p-code" that is turned into actual machine language by an interpreter written for each of the different types of computer supported. Today Java takes a similar approach.

Further Reading

Bailey, T. E., and Kris Lundgaard. *Program Design with Pseudocode.* 3rd ed. Pacific Grove, Calif.: Brooks/Cole, 1989.

Daviduck, Brent. "Introduction to Programming in C++: Algorithms, Flowcharts and Pseudocode." http://www.allclearon-line.com/articles/program_intro.pdf

Gilberg, Richard F., and Behrouz A. Forouzan. *Data Structures: a Pseudocode Approach with C++.* Pacific Grove, Calif.: Brooks/Cole, 2001.

Neapolitan, Richard E. *Foundations of Algorithms Using C++ Pseudocode.* 2nd ed. Sudbury, Mass.: Jones and Bartlett, 1998.

punched cards and paper tape

In 1804, the French inventor Joseph-Marie Jacquard invented an automatic weaving loom that used a chain of punched cards to control the pattern in the fabric. A generation later, a British inventor (see BABBAGE, CHARLES) decided that punched cards would be a suitable medium for inputting data into his proposed mechanical computer, the Analytical Engine.

Although Babbage's machine was never built, by 1890 an American inventor was using an electromechanical tabulating machine to process census data punched into cards (see HOLLERITH, HERMAN). Card tabulating machines were improved and marketed by International Business Machines (IBM) throughout the first part of the 20th century. IBM would also create the 80-column standard punched card that would become familiar to a generation of programmers.

Later machines included features such as mechanical sorting, enhanced arithmetic functions, and the ability to group cards by a particular criterion and print subtotals, counts, or other information about each group. Although these machines were not computers, they did introduce the idea of automated data processing.

During the 1930s, a number of companies introduced punch card tabulators that could work with alphanumeric data (that is, letters as well as numbers). With these expanded capabilities, punch card systems could be used to keep track of military recruits, taxpayers, or customers (such as insurance policy holders). IBM emphasized the new machines' features by calling them "accounting machines" instead of tabulators.

While tabulators and calculators using punched cards gave a taste of the power of automated data processing, they had a very limited programming ability. For example, they could not make more than very simple comparisons or decisions, and could not repeat steps under program control (looping). The desire to create a general-purpose data processing system led in the 1940s to the development of the electronic computer.

When the first computers were developed, it was natural to turn to the existing punched cards and their machinery for a medium for inputting data and program instructions into the new machines. Because computers contained working memory, the program could be stored in its entirety during processing, enabling looping, subroutines, and other ways to control processing. Because the amount of available memory or "core" was severely limited, not much data could be stored inside the computer. However, complicated processing could be broken into a series of steps where a program was loaded and run, the input data cards read and processed, and the intermediate results punched onto a set of output cards. The card could then be input to another program to carry out the next phase.

By the 1970s, however, faster and easier to use media such as magnetic tape and disk drives were being employed for program and data storage. Instead of having to use a keypunch machine to create each program statement, programmers could type their commands at a terminal, using a text editor (see also PROGRAMMING ENVIRONMENT). Even the government began to phase out punched cards. Today some "legacy" punch card systems are maintained, and there is sometimes a need to read and convert archival data in punch card form.

Ironically, this workhorse of early data processing would surface again in the U.S. presidential election of 2000, when problems with the interpretation of partly punched "chads" on ballot punch cards would lead to great controversy.

Further Reading

Cardamation Company. http://www.cardamation.com/

Dyson, George. "The Undead: The little secret that haunts Corporate America: A technology that won't go away." *Wired* 7.03 (March 1999), 141–145, 170–172. Also available on-line at http://www.wired.com/wired/archive/7.03/punchcards.html

Philips, N. V. "Everything About Punch Cards." http://www.tno.nl/instit/fel/museum/computer/en/punchcards.html

Province, Charles M. "IBM Punch Cards in the Army." http://www.geocities.com/pattonhq/ibm.html

Python

Created by Guido van Rossum and first released in 1990, Python is a relatively simple but powerful scripting language (see SCRIPTING LANGUAGES and PERL). The name comes from the well-known British comedy group Monty Python.

Python is particularly useful for system administrators, webmasters, and other people who have to link various files, data source, or programs to perform their daily tasks. The language currently has a small but growing (and quite enthusiastic) following.

Python dispenses with much of the traditional syntax used in the C family of languages. For example, the following little program converts a Fahrenheit temperature to its Celsius equivalent:

```
temp = input("Farenheit temperature:")
print (temp-32.0)*5.0/9.0
```

Without the semicolons and braces found in C and related languages, Python looks rather like BASIC. Also note that the type of input data doesn't have to be declared. The runtime mechanism will assume it's numeric from the expression found in the print statement. Python programs thus tend to be shorter and simpler than C, Java, or even Perl programs. The simple syntax and lack of data typing does not mean that Python is not a "serious" language, however. Python contains full facilities for object-oriented programming, for example.

Python programs can be written quickly and easily by trying commands out interactively and then converted the script to bytecode, a machine-independent representation that can be run on an interpreter designed for each machine environment. Alternatively, there are translation programs that can convert a Python script to a C source file that can then be compiled for top speed.

Perl is currently the most popular scripting language for UNIX and Web-related applications. Perl contains a powerful built-in regular expression and pattern-matching mechanism, as well as many other built-in functions likely to be useful for practical scripting. Python, on the other hand, is a more generalized and more cleanly structured language that is likely to be suited for a wider variety of applications, and it is more readily extensible to larger and more complex applications.

Further Reading

Ascher, David, and Alex Martelli, eds. *Python Cookbook*. Sebastopol, Calif.: O'Reilly, 2002.

Lutz, Mark, David Ascher, and Frank Willison. *Learning Python*. Sebastopol, Calif.: O'Reilly, 1999.

[Python Resource Page] http://www.python.org/

van Rossum, Guido. "Comparing Python to Other Languages." http://www.python.org/doc/essays/comparisons.html

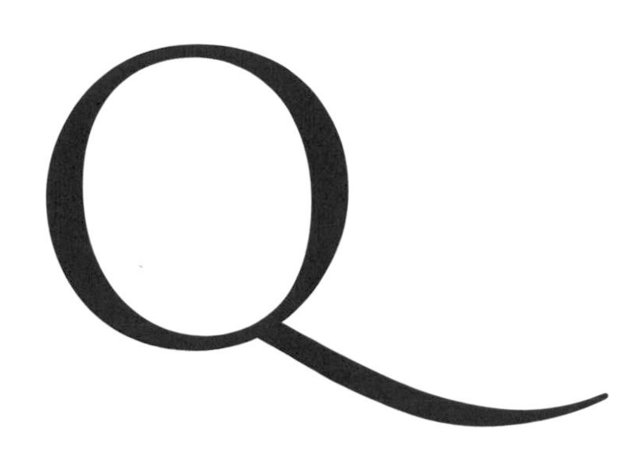

quality assurance, software

Modern software programs are large and complex, and contain many interrelated modules. If a program is not thoroughly tested before it goes into service, it may contain errors that can result in serious consequences (see also RISKS OF COMPUTING).

In the early days of computing, programmers generally tested their code informally and nonsystematically. The assumption was that after the program was given to the users any problems that arose could be fixed through "patches" or replacement versions containing bug fixes. Today, however, it is increasingly recognized that assuring the quality and reliability of software requires a systematic, comprehensive process that begins when software requirements are first specified and continues after the program has been released.

Any program is designed to meet the needs of a specific type of users for specific applications. Therefore, the first step must be to make sure that users are able to communicate their requirements and that the software engineers understand the users' needs and concerns. Detailed written specifications, flowcharts, and other depictions of the program can be reviewed by user representatives (see FLOWCHART and CASE). The specifications can be further explored by creating a prototype or demonstration of the program's features (see also PRESENTATION SOFTWARE). Since a prototype can be dynamic and let users have simulated interactions with the program, it may reveal usability problems that would be hard to spot from charts or documentation. The result of this initial verification process should be that the users agree that the program will do what they need and that they will be comfortable using it.

In moving from design to implementation (writing the actual code), the developers must first choose an appropriate approach (see ALGORITHM) and data representation. Choosing an algorithm that is known to be sound is preferable, but if an algorithm must be modified (or a new one developed), developers may be able to take advantage of mathematical techniques that will suggest, if not totally prove, the algorithm's accuracy and reliability.

As the programmers write the code, they should try to use best practices (see SOFTWARE ENGINEERING). Doing so ensures that the code will be readable and organized in such a way that the source of a problem area can be identified easily, and any "fix" that must be made will be less likely to have unforeseen side effects.

Developers can also include special code that will facilitate testing. This code can include assertions—statements that test specified conditions (such as variable values) at key points in the program, displaying appropriate messages if the values are not within the proper range. Large, complex programs can also include diagnostic modules that give the developers a sort of virtual console that they can use to monitor conditions while the program is running, or "drill into" particular areas for closer inspection (see also BUGS AND DEBUGGING).

Although a certain amount of testing and debugging can and should be done while the code is being written, more extensive and systematic testing is usually performed after a preliminary version of the program has been completed. (This is sometimes called an *alpha version*.) There are two basic approaches to designing the tests. "White box" tests use the developer's knowledge of the code to design test data that will test all of the program's structural features (see PROCEDURES AND FUNCTIONS, BRANCHING STATEMENTS, and LOOP). The testers may be aided by mathematical analysis that identifies "partitions" or ranges within the data that should result in a particular execution path being taken through the program. "Fault coverage" tests can also be designed to test for various specific types of errors (such as input/output, numeric overflow, loss of precision, and so on).

A shortcoming of white box tests is that because the tester knows how the program works, he or she may unconsciously select mainly "reasonable" data or situations. (It has been observed that users are under no such compulsion!) One way to compensate for this bias is to also perform "black box" tests. These tests assume no knowledge of the inner workings of the program. They approach the program from the outside, submitting data (or otherwise interacting with the program) either through the user interface or using an automated process that simulates user input. The tester tries to generate as wide a variety of input data as possible, often by using randomization techniques. The result is that the ability of the program to deal with "unreasonable" data will also be tested, and unforeseen situations may arise and have to be dealt with.

Once this cycle of testing and fixing problems is finished, the program will probably be given to a selected group of users who will operate it under field conditions—that is, in the same sort of environment the program will be used once it is sold or deployed. This process is sometimes called *beta testing*. (Game companies have traditionally relied upon the willingness of gamers to test a new game in exchange for getting to play it sooner.)

The priority (and thus the resources) devoted to testing will vary according to many factors, including

- The complexity of the program (and thus the likelihood of problems)
- The presence of strong competitors who could take advantage of significant problems with the program
- The potential financial impact or legal exposure from bugs or problems
- The ability to "amortize" the costs of developing testing tools and procedures over a number of years as new versions of the program are developed

The Holy Grail for quality assurance would be to develop powerful artificially intelligent automatic testing programs that could analyze a program and develop and execute a variety of thorough tests. However, such a program would itself be very complex, difficult and expensive to develop, and subject to its own bugs. Nevertheless, a number of organizations (notably, IBM) have devoted considerable attention to the problem.

Further Reading

Hower, Rick. "Software QA/Test Resource Center." http://www.softwareqatest.com/

Perry, William E. *Quality Assurance for Information Systems: Methods, Tools, and Techniques*. New York: Wiley, 1991.

Schulmeyer, G. Gordon, and James L. MacManus, eds. *Handbook of Software Quality Assurance*. 3rd ed. Upper Saddle River, N.J.: Prentice Hall, 1999.

quantum computing

The fundamental basis of electronic digital computing is the ability to store a binary value (1 or 0) using an electromagnetic property such as electrical charge or magnetic field.

However, during the first half of the 20th century, physicists discovered the laws of quantum mechanics that apply to the behavior of subatomic particles. An electron or photon, for example, can be said to be in any one of several "quantum states" depending on such characteristics as spin. In 1981, physicist Richard Feynman came up with the provocative idea that if quantum properties could be "read" and set, a computer could use an electron, photon, or other particle to store not just a single 1 or 0, but a number of values simultaneously. The simplest case, storing two values at once, is called a "qubit" (short for "quantum bit"). In 1985, David Deutsch at Oxford University fleshed out Feynman's ideas by creating an actual design for a "quantum computer," including an algorithm to be run on it.

At the time of Feynman's proposal, the techniques for manipulating individual atoms or even particles had not yet been developed (see also NANOTECHNOLOGY), so a practical quantum computer could not be built. However, during the 1990s considerable progress was made, spurred in part by the suggestion of Bell Labs researcher Peter Shor, who outlined a quantum algorithm that might be used for rapid factoring of extremely large integers. Since the security of modern public key cryptography (see ENCRYPTION) depends on the difficulty of such factoring, a working quantum computer would be of great interest to spy agencies.

The reason for the tremendous potential power of quantum computing is that if each qubit can store two values simultaneously, a register with three qubits could store eight values, and in general, for *n* qubits one can operate on 2^n values simultaneously. This means that a single quantum processor might be the equivalent of a huge number of separate processors (see MULTIPROCESSING). Clearly many problems that have been considered not practical to solve (see COMPUTABILITY AND COMPLEXITY) might be tackled with quantum computers.

However, the practical problems involved in designing and assembling a quantum computer are expected to be very formidable. Although scientists during the 1990s achieved the ability to arrange individual atoms, the precise placement of atoms and even individual particles such as photons would be difficult. Furthermore, as more of these components are assembled in very close proximity, it becomes more likely that they will interfere with one another, causing "decoherence," where the superimposed values "break down" to a single 1 or 0, thus causing loss of information. However, some researchers are hopeful that standard mathematical techniques (see ERROR CORRECTION) could be used to keep this problem in check. For example, redundant components could be used so that even if one decoheres, the others could be used to regenerate the information.

Another approach is to use a large number of quantum components to represent each qubit. In 1998, Neil Gershenfeld and Isaac L. Chuang reported successful experiments using a liquid with nuclear magnetic resonance (NMR) technology. Here each atom in a molecule (for example, chloroform), would represent one qubit, and a large number of molecules would be used, for redundancy. Since each "observation" (that is, setting or reading data) affects only a few of the many molecules for each qubit, the stability of the information in the system is not compromised. However, this approach is limited by the number of atoms in the chosen molecule—perhaps to 30 or 40 qubits.

There are many potential applications for quantum computing. While the technology could be used to crack conventional cryptographic keys, researchers have suggested that it could also be used to generate unbreakable keys that depend on the "entanglement" of observers and what they observe. The sheer computational power of a quantum computer might make it possible to develop much better computer models of complex phenomena such as weather, climate, the economy—or of quantum behavior itself.

Further Reading

Benjamin, Simon, and Artur Eckert. "A Short Introduction to Quantum-Scale Computing." http://www.qubit.org/intros/nano/nano.html

Brooks, Michael, ed. *Quantum Computing and Communications.* New York: Springer-Verlag, 1999.

Centre for Quantum Computing. http://www.qubit.org/

Deutsch, David. *The Fabric of Reality.* New York: Allen Lane, 1997.

Quantum Computing

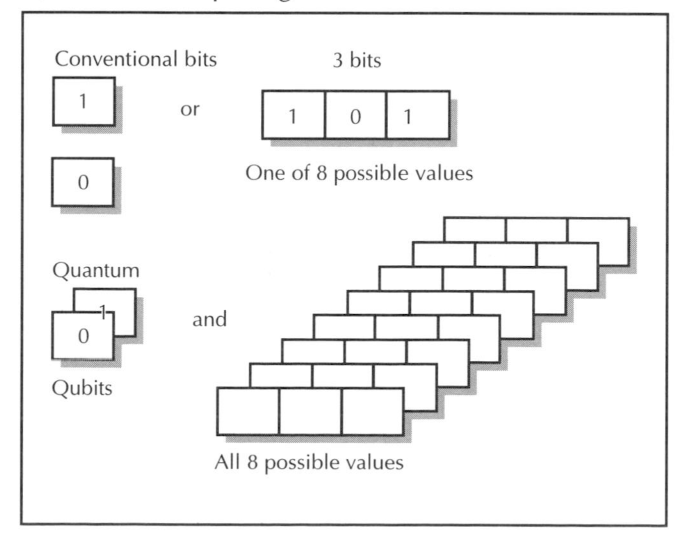

It is often said in computing that it's "always 1 or 0." However, in quantum computing a group of superimposed quantum states can hold ones and zeroes simultaneously. Thus, three quantum bits or "qubits" can hold eight values at once.

queue

A queue is basically a "line" of items arranged according to priority, much like the customers waiting to check out in a supermarket. Many computer applications involve receiving, tracking, and processing requests. For example, an operating system running on a computer with a single processor must keep track of which application should next receive the processor's attention. A print spooler holds documents waiting to be printed. A web or file server must keep track of requests for webpages, files, or other services. Queues provide an orderly way to process such requests. Queues can also be used to efficiently store data in memory until it can be processed by a relatively slow device such as a printer (see BUFFERING).

As a data structure, a queue is a type of list (see LIST PROCESSING). New items are inserted at one end and removed (deleted) from the other end. This contrasts with a stack, where all insertions and deletions are made at the same end (see STACK). Just as the next person served at the supermarket is the one at the head of the line, the end of a queue from which items are removed is called the head or front. And just as new people arriving at the supermarket line join the end of the line, the part of the queue where new items are added is called the tail or rear. Since the first item in line is the first to be removed, a queue is called a FIFO (first in, first out) structure.

To create a queue, a program first allocates a block of memory. It then sets up to pointers (see POINTERS AND

Queue (1)

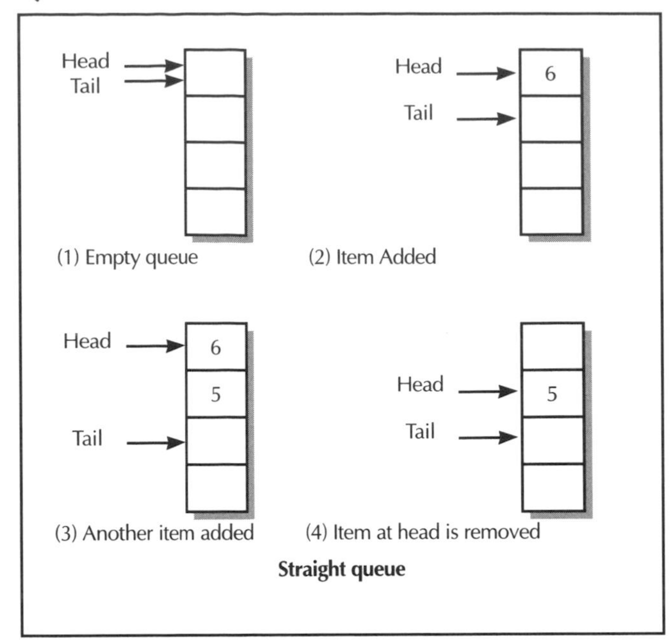

In an empty queue the head and tail pointers point to the first cell in memory. To add a value, it is placed at the cell pointed to by the head pointer, and the tail pointer is moved up one cell. If an item is removed, the head pointer is moved down one place. (Note that items must be added at the tail and removed at the head.)

Queue (2)

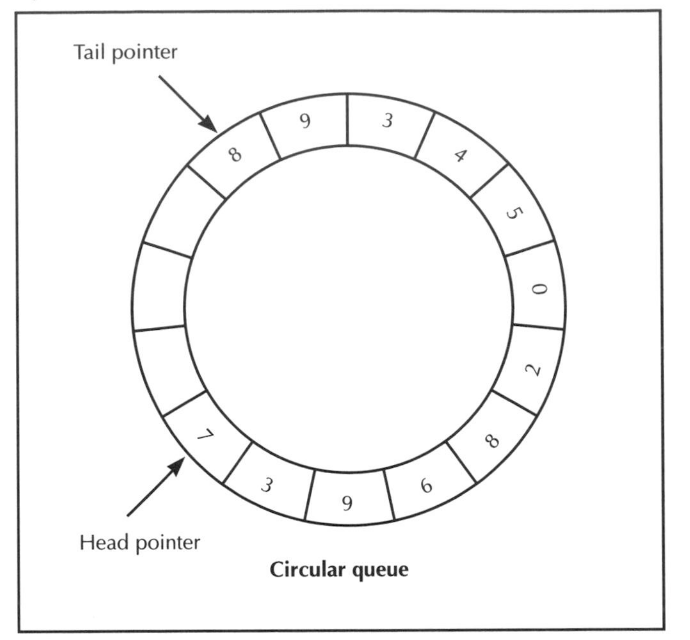

A circular queue works in the same way as a "straight" queue, except that when the last cell in the allotted memory block is reached, the pointer or data "wraps around" to the first cell.

INDIRECTION). One pointer stores the address of the item at the head of the queue; the other has the address of the item at the tail. When the queue starts out, it is empty. This means that both the head and tail pointer start out pointing to the same location.

To add an item, the tail pointer is moved back one location and the item is stored there. To remove an item, the head pointer is simply moved back one location. (The data that had been pointed to by the head pointer can be either retrieved or discarded, depending on the application.)

In actuality it's not quite so simple. As items are added to the queue, the tail pointer keeps moving back in memory with the head pointer trailing behind as items are deleted. If the queue is sufficiently active (many items are being added and removed), the queue will end up "crawling" through memory somewhat like a worm until all the memory is consumed.

In a real line at the supermarket as a customer leaves the checkout stand each of the persons in line moves up one space. In a computer queue this could be accomplished by moving each item up one location whenever an item is removed at the head. However, having to move all the data items each time one is changed would be very inefficient. Instead, one could allow the head of the queue to move only up to some specified location. At that point, the head is moved back to the beginning of the memory block, and thus the space that had been vacated by the tail as it moved up is reutilized. In effect this wraps the memory around into a circle, so this is called a *circular queue.*

Further Reading

Brookshear, J. Glenn. *Computer Science: An Overview.* 6th ed. Reading, Mass.: Addison-Wesley, 2000.

R

random number generation

Computer applications such as simulations, games, and graphics applications often need the ability to generate one or more random numbers (see SIMULATION and COMPUTER GAMES). Random numbers can be defined as numbers that show no consistent pattern, with each number in the series neither affected in any way by the preceding number, nor predictable from it.

One way to get random digits is to simply start with an arbitrary number with a specified number of digits, perhaps 10. This first number is called the seed. Multiply the seed by a constant number of the same length, and take that number of digits off the right end of the product. The result becomes the new seed. Multiply it by the original constant to generate a new product, and repeat as often as desired. The result is a series of digits that appear randomly distributed as though generated by throwing a die or spinning a wheel. This type of algorithm is called a congruential generator.

The quality of a random number generator is proportional to its period, or the number of numbers it can produce before a repeating pattern sets in. The period for a congruential generator is approximately 2^{32}, quite adequate for many applications. However, for applications such as very large-scale simulations, different algorithms (called shift-register and lagged-Fibonacci) can be used, although these also have some drawbacks. Combining two different types of generators produces the best results. The widely used "McGill Random Number Generator Super-Duper" combines a congruential and a shift-register algorithm.

Generating a random number series from a single seed will work fine with most simulations that rely upon generating random events under the control of probabilities (Monte Carlo simulations). However, although the sequence of numbers generated from a given seed is randomly distributed, it is always the same series of numbers for the same seed. Thus, a computer poker game that simply used a given seed would always generate the same hands for each player. What is needed is a large collection of potential seeds from which one can be more or less randomly chosen. If there are enough possible seeds, the odds of ever getting the same series of numbers become vanishingly small.

One way to do this is to read the time (and perhaps date) from the computer's system clock and generate a seed based on that value. Since the clock value is in milliseconds, there are millions of possible values to choose from. Another common technique is to use the interval between the user's keystrokes (in milliseconds). Although they're not perfect, these techniques are quite adequate for games.

Further Reading

Gentle, James E. *Random Number Generation and Monte Carlo Methods.* New York: Springer-Verlag, 1998.

Knuth, Donald E. *The Art of Computer Programming: Volume 2, Seminumerical Algorithms.* 3rd ed. Reading, Mass.: Addison-Wesley, 1997.

real-time processing

There are many computer applications (such as air traffic control or industrial process control) that require that the system respond almost immediately to its inputs.

In designing a real-time system there are always two questions to answer: Will it respond quickly enough most of the time? How much variation in response time can we tolerate? A system that responds to real-time environmental conditions (such as the amount of traction or torque acting on a car's wheels) needs to have a sampling rate and a rate of processing the sampled data that's fast enough so that the system can correct a dangerous condition in time. The responsiveness required of course varies with the situation and with the potential consequences of failure. An air traffic control system may be able to take a few seconds between processing radar samples, but it better get it right in time. Systems like this where real-time response is absolutely crucial are sometimes called "hard real-time systems."

Other systems are less critical. A streaming audio system has to keep its buffer full so it can play in real time, but if it stutters once in a while, no one's life is in danger. (And since download rates over the Internet can vary for many reasons it's not realistic to expect too perfect a level of performance.) A slower "soft real-time system" like a bank's ATM system should be able to respond in tens of seconds, but if it doesn't, the consequences are mainly potential loss of customers and revenue. A fairly wide variation in response time may be acceptable as long waits don't occur often enough to drive away too many customers.

To put together the system, the engineer must look at the inherent speed of the sampling device (such as radar, camera, or simply the keyboard buffer). The speed of the processor(s) and the time it takes to move data to and from memory are also important. The structure, strengths, and weaknesses of the host operating system can also be a factor. Some operating systems (including some versions of UNIX) feature a guaranteed maximum response time for various operating system services. This can be used to help calculate the "worst case scenario"—that combination of inputs and the existing state of the system that should result in the slowest response.

Another approach available in most operating systems is to assign priority to parts of the processing so that the most critical situations are guaranteed to receive the attention of the system. However, things must be carefully tuned so that even lower priority tasks are accomplished in an acceptable length of time.

The design of the data structure or database used to hold information about the process being monitored is also important. In most databases the age of the data is not that important. For a payroll system, for example, it might be sufficient to run a program once in a while to weed out people who are no longer employees. For a nuclear power plant, if data is getting too old such that it's not keeping up with current condition, some sort of alarm or even automatic shutdown might be in order. With a system that has softer constraints (such as an automatic stock trading system), it may be enough to be able to get most trades done within a specified time and to gather data about the performance of the system so the operators can decide whether it needs improvement.

Real-time systems are increasingly important because of the importance of the activities (such as air traffic control and power grids) entrusted to them, and because of their pervasive application in everything from cars to cell phones to medical monitors (see also EMBEDDED SYSTEM and MEDICAL APPLICATIONS OF COMPUTERS). The systems also tend to be increasingly complex because of the increasing interconnection of systems. For example, many real-time systems have to interact with the Internet, with communications services, and with ever more sophisticated multimedia display systems. Further, many real-time systems must use multiple processors (see MULTIPROCESSING), which can increase the robustness and reliability of the system but also the complexity of its architecture, and thus the difficulty in determining and ensuring reliability.

Further Reading

Buttazzo, G. C. *Hard Real-Time Computing Systems.* Boston, Mass.: Kluwer, Academic, 1997.

"Dedicated Systems Encyclopedia." http://www.dedicated-systems.com/encyc/

IEEE Computer Society. "Real-Time Research Repository." http://cs-www.bu.edu/pub/ieee-rts/

Laplante, Phillip A. *Real Time Systems Design and Analysis.* Piscataway, N.J.: IEEE Press, 1996.

recursion

Even beginning programmers are familiar with the idea that a series of program statements can be executed repeatedly as long as (or until) some condition is met (see LOOP). For example, consider this simple function in Pascal. It calculates the factorial of an integer, which is equal to the product of all the integers from 1 to the number. Thus factorial 5, or 5! = 1 * 2 * 3 * 4 * 5 = 120.

```
Function Factorial (n: integer) : integer
Begin
  i: integer;
  For i = 1 to n do
    Factorial := Factorial * i;
End
```

If the main program has the line:

```
Writelin (Factorial (5));
```

The 5 is sent to the function, where the loop simply multiplies the numbers from 1 to 5 and returns 120.

Recursion

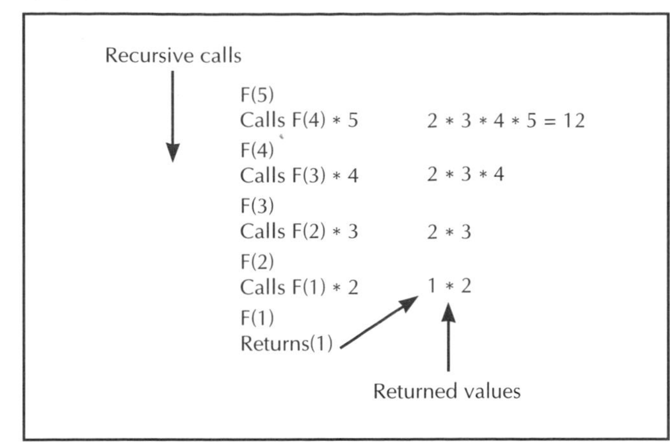

*In recursion a procedure calls itself until some defined condition is met. In this example of a factorial procedure, F(1) is defined to return 1. Once it does, the returned value is plugged into its caller, which then returns the value of 1 * 2 to its caller, and so on.*

However, it is also possible to have a function call *itself* repeatedly until a specified condition is met. This is called recursion, and it allows for some compact but powerful coding. A recursive version of the Factorial function in Pascal might look like this:

```
function Factorial(n:integer):integer
begin
  if (n = 1) then
    Factorial := 1
  else
    Factorial := Factorial(n - 1) * n;
end;
```

Why does this work? An alternative way to define a factorial is to say that the factorial of a number is that number times the factorial of one less than the number. Thus, the factorial of 5 is equal to 5 * 4! or 5 * 4 * 3 * 2 * 1. But in turn the factorial of 4 would be equal to 4 * (3 * 2 * 1), and so on down to the factorial of 1, which is simply 1. Thus, in general terms the factorial of n is equal to n * factorial (n – 1).

What happens if this function is called by the program statement:

```
Writeln ("Factorial of 5 is ") ; Factorial (5)
```

First, the Factorial function is called with the value 5 assigned to n. The If statement checks and sees that n is not 1, so it calls factorial (i. e. itself) with the value of n – 1, or 4. This new instance of the factorial function gets the 4, sees that it is not 1, and calls factorial again with n – 1, which is now 4 – 1 or 3. This continues until n is 2, at which point factorial 1 is called. But this time n *is* 1, so it returns the value of 1 rather than calling itself yet again.

Now the returned value of 1 replaces the call to Factorial (n – 1) in the preceding instance of Factorial (where n had been 2). That 1 is therefore multiplied by 2, and 1 * 2 = 2 is returned to the preceding instance, where n had been 3. Now that 2 gets multiplied by 3 and returned to the instance where n had been 4. This continues until we're back at the first call to factorial 5, where the value of 4 * 3 * 2 * 1 now gets multiplied by that 5, giving 120, or factorial 5. (See the accompanying diagram for help in visualizing this process.)

A RECURSIVE SORTING ALGORITHM

In the preceding example recursion does no more than a simple loop could, but many problems lend themselves more naturally to a recursive formulation. For example, suppose you have an algorithm to merge (combine) two lists of integers that have been sorted into ascending values. The procedure simply takes the smaller of the two numbers at the front of the two lists until one list runs out of numbers (any numbers in the remaining list can then simply be included).

Using an English-like syntax, one can write a recursive procedure to sort a list of numbers by calling itself repeatedly, then using the Merge procedure:

```
Procedure Sort
Begin
  If the list has only one item, return
  Else
    Sort the first half of the list
    Sort the second half of the list
    Merge the two sorted lists
  End If
End (Sort)
```

Sort will call itself until one of the lists has only one item (which by definition is "sorted"), and the Merge procedure will build the sorted list.

To implement recursion, the run-time system for the language must use an area of memory (see STACK) to temporarily store the values associated with each instance of a function as it calls itself. Depending on the implementation, there may be a limit on how many levels of recursion are allowed, or on the size of the stack. (However, the plentiful supply of available memory on most systems today makes this less of an issue.)

The first generation of high-level computer languages (such as FORTRAN and COBOL) did not allow recursion. However, the second generation of procedural languages starting around 1960 with Algol, as well as successors such as Pascal and C do allow recursion. The LISP language (see LISP and FUNCTIONAL LANGUAGES) uses recursive definitions extensively, and recursion turns out to be very useful for processing the grammars for artificial and natural languages (see also PARSING). Recursion can also be used to generate interesting forms of graphics (see also FRACTALS IN COMPUTING).

Further Reading

Hillis, W. Daniel. *The Pattern in the Stone: The Simple Ideas that Make Computers Work.* New York: Basic Books, 1998.

Roberts, E. S. *Thinking Recursively.* New York: Wiley, 1986.

reduced instruction set computer (RISC)

All things being equal, the trend in computer design is to continually add new features. There are several reasons why this is the case with computer processors:

- To create a "family" of upwardly compatible computers (see COMPATABILITY AND PORTABILITY)
- To make a new machine more competitive with existing systems, or to give it a competitive advantage
- To make it easier to write compilers for popular languages
- To allow for more operations to be done with one (or a few) instructions rather than requiring many instructions

There are certainly exceptions to the trend toward complexity. The minicomputer, for example, represented in some ways a simplification of the exiting mainframe design. It didn't have as many ways of working with memory (see ADDRESSING) and lacked the multiple input/output "channels" and their separate processors. But once minicomputers were introduced and achieved success, the same competitive and other pressures led their designers to start adding complexity.

One way processor designers coped with the demand for more complicated instructions was to give the main processor a microprocessor with its own set of simple instructions. When the main processor received one of the complex instructions, it would be executed by being broken down into simpler instructions or "microcode" to be executed by the sub-processor.

This approach gave processor designers greater flexibility. It also made things easier for compiler designers, because the compiler could translate higher-level language statements into fewer, more complex instructions, leaving it to the hardware with its micro engine to break them down into the ultimate machine operations. However, it also meant that the processor had to decode and execute more instructions in every processor cycle, making it less efficient and slower and losing some of the benefits of the faster processors that were becoming available.

In 1975, John Cocke and his colleagues at IBM decided to build a new minicomputer architecture from the ground up. Instead of using complex instructions and decoding them with a micro engine, they would use only simple instructions that could be executed one per cycle. The clock (and thus the cycle time) would be much faster than for existing machines, and the processor would use pipelining so it could decode the next instruction while still executing the previous one. Similarly, in many cases the next item of data needed could be fetched at the same time the data from the previous step was being written (stored). This approach became known as reduced instruction set computing (RISC), because the number of instructions had been reduced compared to exiting systems, which then became known as complex instruction set computing (CISC).

Since the RISC system had only simple instructions, compilers could no longer use many complicated but handy instructions. The compiler would have to take over the job of the micro engine and break all statements down into the basic instructions. It became important that the compiler be able to generate the optimal set of instructions by analyzing how data would have to be moved around in the machine's registers and memory. In other words, RISC hardware gained higher performance through simplification at the hardware level but at the cost of making compilers more complicated. Fortunately, both hardware and software designers were able to meet the challenge and in the process learn how to get the most out of new technology.

RISC would also play a part in the design of the microprocessors that began to power personal computers. For example, the DEC Alpha, a "pure" RISC chip introduced in 1992, provided a level of power that made it suitable for high-performance workstations. Another successful RISC-based development has been the SPARC (Scalable Processor ARChitecture) developed by Sun Microsystems for servers, computer clusters, and workstations.

Perhaps the most interesting development, however, has been the gradual application of RISC principles to mainstream processors such as the Intel 80x86 series used in most personal computers today. Increasingly, the recent Pentium series chips, while supporting their legacy of CISC instructions, are processing them using an inner architecture that uses RISC principles and takes advantage of pipelining, as well as using more registers and a larger data cache. However, the sheer increase in clock cycle speed and performance in the newer chips has made the old tradeoff between complicated and simple instructions less relevant.

Further Reading

Evans, James S., and Richard H. Eckhouse. *Alpha RISC Architecture for Programmers.* Upper Saddle River, N.J.: Prentice Hall, 1999.

Gerritsen, Armin. "CISC vs. RISC: Which one is better?" http://cpusite.examedia.nl/docs/cisc_vs_risc.html

Patterson, Dave, and John Hennessy. *Computer Architecture: a Quantitative Approach.* 2nd ed. San Francisco: Morgan Kaufmann, 1996.

regular expression

Many users of UNIX and the old MS-DOS are familiar with the ability to use "wildcards" to find filenames that match specified patterns. For example, suppose a user wants to list all of the TIF graphics files in a particular directory. Since these files have the extension .tif, a UNIX ls command or a DOS dir command, when given the pattern *.tif, will match and list all the TIF files. (One does have to be aware of whether the operating system in question is case-sensitive. UNIX is, while MS-DOS is not.)

The specification *.tif tells the command "match all files whose names consist of one or more characters and that end with a period followed by the letters tif." It is one of many possible regular expressions. (See the accompanying table for more examples.) The asterisk here is a "metacharacter." This means that it is not treated as a literal character, but as a pattern that will be matched in a specified way.

Most operating systems that have command processors (see SHELL) allow for some form of regular expressions, but don't necessarily implement all of the metacharacters. UNIX provides the most extensive use for regular expressions (see UNIX). UNIX has an operating system facility called *glob* that expands regular expressions (that is, substitutes for them whatever matches) and passes them on to the many UNIX tools or utilities designed to work with regular expressions. These tools include editors such as ex and vi, the character translation utility (tr), the "stream editor" (sed), and the string-searching tool grep. For example, sed can be used to remove all blank lines from a file by specifying

```
sed 's/^$/d' list.txt
```

This command finds all lines with no characters (^$) in the file list.txt and deletes them from the output. Even more extensive use of pattern-matching with regular expressions is found in many scripting languages (see SCRIPTING LANGUAGES, AWK, and PERL).

It is true that most of today's computer users don't enter operating system commands in text form but instead use menus and manipulate icons (see USER INTERFACE and MICROSOFT WINDOWS). If such a user wants to change one word to another throughout a word processing document, he or she is likely to open the Edit menu, select Find, and type the "before" and "after" words into a dialog box. However, even in such cases if the user has some familiarity with regular expressions, more sophisticated substitutions can be accomplished. In Microsoft Word, for example, a variety of wildcards (i.e. metacharacters) can be used for operations that would be hard to accomplish through mouse selections.

Further Reading

Friedl, Jeffrey E. F. *Mastering Regular Expressions: Powerful Techniques for Perl and Other Tools.* 1st ed. Sebastopol, Calif.: O'Reilly, 1997.

Mansour, Steve. "A Tao of Regular Expressions." http://sitescooper.org/tao_regexps.html

research laboratories in computing

The value of creating and maintaining environments for long-term research in computer science and engineering has long been recognized by academic institutions, industry organizations, and corporations.

ACADEMIC RESEARCH INSTITUTIONS

Artificial intelligence and robotics have been the focus of many academic computer science research facilities (see ARTIFICIAL INTELLIGENCE and ROBOTICS). They are examples of areas that show great potential but that demand a substantial investment in long-term research. There are many research organizations in the AI field, but a few stand out as particularly important examples.

The Massachusetts Institute of Technology (MIT) Artificial Intelligence Lab has a wide-ranging program but has emphasized robotics and related fields such as computer vision and language processing.

The MIT Media Lab has become well known for work with new media technologies and the digital and graphi-

METACHARACTERS IN REGULAR EXPRESSIONS

METACHARACTER	MEANING
. (period)	Matches any single character in that position
?	Matches zero or one of any character
*	Matches zero or more of the preceding character (thus * matches any number of characters)
+	Matches one or more of the preceding character (thus 9+ matches 9, 99, 999, etc.)
[ ]	Matches any of the characters enclosed by the brackets
–	Specifies a range of characters. Placing the range in brackets will match any character within the range. For example, [0–9] matches any digit, [A–Z] matches any uppercase character, and [A–Za–z] matches any alphabetic character.
\	"Quotes" the following character. If it is a metacharacter, the following character will be treated as an ordinary character. Thus \? matches an actual question mark.
^	Matches the beginning of a line
$	Matches the end of a line

cal representation of data. However, in recent years it has expanded its focus to the broader area of human-machine interaction and the pervasive presence of intelligent devices in the home and larger environment.

The Stanford Artificial Intelligence Laboratory (SAIL) played an important role in the development of the LISP language (see LISP) and other AI research. Today Stanford's important role in AI is continued by its Robotics Laboratory and the Knowledge Systems Laboratory. Carnegie Mellon University also has a number of influential AI labs and research projects.

On the international scene Japan has had strong research programs in academic and industrial AI, such as the Neural Computing Center at Keio University and the Knowledge-Based Systems Laboratory at Shizuoka University. There are a number of important AI research groups in the United Kingdom, such as at Cambridge, Oxford, King's College, and the University of Edinburgh (where the logic language Prolog was developed).

Some of the most interesting research sometimes emerges from outside the main concerns of an institution. The World Wide Web, for example, was developed by Tim Berners-Lee (see BERNERS-LEE, TIM) while he was working with the coordination of scientific computing at CERN, the giant European particle physics laboratory.

CORPORATE RESEARCH INSTITUTIONS

The challenging nature of computer applications and the competitiveness of the industry have also led a number of major companies to underwrite permanent research institutions. Much corporate-funded research has gone into developing the basic infrastructure of computing rather than to the more esoteric topics pursued by academic departments. However, corporations have also funded "pure" research that may have little short-term application but can ultimately lead to new technologies.

The concept of the industrial laboratory is often attributed to Thomas Edison, whose famous Menlo Park, New Jersey, facility (founded in 1876) put experimentation and development of new inventions on a systematic, continuous basis. Instead of an invention forming the basis for a company, Edison saw invention itself as the core business.

A similar approach motivated the founding of Bell Labs. In the mid-1920s, AT&T, descendant of the original Bell and Western Electric companies, reorganized its various research efforts into one organization. Bell Labs would play a direct role in making modern digital electronics possible when three of its researchers, John Bardeen, Walter H. Brattain, and William B. Shockley invented the transistor in 1947.

On the software side, Bell supported the work of Claude Shannon, whose fundamental theorems of information transmission would become a key to the design of the computer networks (see SHANNON, CLAUDE). The development of the UNIX operating system at Bell in the early 1970s (see RITCHIE, DENNIS and UNIX) would provide much of the infrastructure that would be used for computing at universities and other research institutions and ultimately in the development of the Internet. Similarly, Ritchie and Thompson also developed C, the language that together with its offshoots C++ and Java would become the most widely used general-purpose programming languages for the rest of the century and beyond.

IBM built its first research lab in 1945, beginning a network that would eventually include facilities in Switzerland, Israel, Japan, China, and India. IBM research has generally focused on core hardware and software technologies, including the development of the first hard drive in 1956 and the development of the FORTRAN language by John Backus in 1957 (see FORTRAN). Other IBM innovations have included on-line commerce (the SABRE airline reservation system), the relational database, and the first prototype RISC (reduced instruction set computer).

Xerox is best known for its photocopiers and printers, but in the late 1960s the company decided to try to diversify its products by recasting itself as developer of a comprehensive "architecture of information" in the office. During the 1970s, its Palo Alto Research Center (PARC) worked out of the public eye, where it invented much of the technology (such as the mouse, graphical user interface, and notebook computer) that would become familiar to consumers a decade later in the Macintosh and Microsoft Windows.

In 1991, Microsoft, then a medium-sized company, established its Microsoft Research division, which has since grown to include four laboratories in Redmond, Washington, the San Francisco Bay Area, Cambridge, England, and Beijing. The labs maintain close ties with universities, and their research areas have included data mining and analysis, geographic information systems (Terraserver), natural language processing, and computer conferencing and collaboration.

The role of government agencies in funding computer-related research should not be overlooked. The Internet evolved from a project funded by the Department of Defense's ARPA (Advanced Research Projects Agency) in 1968 (see INTERNET). The network architecture and hardware in turn were developed by a contractor, Bolt, Beranek and Newman (BBN). In the late 1970s, the Defense Department would issue contracts for development of the Ada computer language. Other projects funded by Defense and other government agencies can be found in areas such as robotics, autonomous vehicles, and mapping systems.

COORDINATING RESEARCH

Two large professional organizations for computer scientists and engineers, the Association for Computing

Machinery (ACM) and the Computer Society of the Institute of Electrical and Electronics Engineers (IEEE), serve as clearinghouses and disseminators of research. The Computing Research Association (CRA) brings together more than 190 North American university computer science departments, government-funded research institutions, and corporate research laboratories. Its goal is to improve the opportunities for and quality of research and education in the computer field. (For contact information for these and other selected computer-related organizations, see Appendix 3.)

Further Reading

"Bell Labs Innovations: History." http://www.bell-labs.com/history/
Computer Research Association. www.cra.org
"Computer Science Research." [Google Web Directory] http://directory.google.com/Top/Science/Institutions/Research_Institutes/Computer_Science_ Research/
"IBM Research." http://www.research.ibm.com/
MIT Artificial Intelligence Laboratory. http://www.ai.mit.edu/

risks of computing

Programmers and managers of software development are generally aware of the need for software to properly deal with erroneous data (see ERROR HANDLING). They know that any significant program will have bugs that must be rooted out (see BUGS AND DEBUGGING). Good software engineering practices and a systematic approach to assuring the reliability and quality of software can minimize problems in the finished product (see SOFTWARE ENGINEERING and QUALITY ASSURANCE, SOFTWARE). However, serious bugs are not always caught, and sometimes the consequences can be catastrophic. For example, in the Therac 25 computerized X-ray cancer treatment machine, poorly thought-out command entry routines plus a counter overflow resulted in three patients being killed by massive X-ray overdoses. The overdoses ultimately occurred because the designers had removed a physical interlock mechanism they believed was no longer necessary.

Any computer application is part of a much larger environment of humans and machines, where unforeseen interactions can cause problems ranging from inconvenience to loss of privacy to potential injury or death. Seeing these potential pitfalls requires thinking beyond the specifications and needs of a particular project. For many years the Usenet newsgroup comp.risks (and its collected form, Risks Digest) have chronicled what amounts to an ongoing symposium where knowledgeable programmers, engineers, and others have pointed out potential risks in new technology and suggested ways to minimize them.

UNEXPECTED SITUATIONS

A common source of risks arises from designers of control systems failing to anticipate extreme or unusual environmental conditions (or interactions between conditions). This is a particular problem for mobile robots, which unlike their tethered industrial counterparts must share elevators, corridors, and other places with human beings. For example, a hospital robot was not designed to recognize when it was blocking an elevator door—a situation that could have blocked a patient being rushed into surgery. A basic principle of coping with unexpected situations is to try to design a fail-safe mode that does not make the situation worse. For example, an automatic door should be designed so that if it fails it can be opened manually rather than trapping people in a fire or other disaster.

UNANTICIPATED INTERACTIONS

The more systems there are that can respond to external inputs, the greater the risk that a spurious input might trigger an unexpected and dangerous response. For example, the growing number of radio-controlled (wireless) devices have great potential for unexpected interactions between different devices. In one case reported to the Risks Forum, a Swedish policeman's handheld radio inadvertently activated his car's airbag, which slammed the radio into him. Several military helicopters have crashed because of radio interference. Banning the use of electronic devices at certain times and places (for example, aboard an aircraft that is taking off or landing) can help minimize interference with the most safety-critical systems.

At the same time, regulations themselves introduce the risk that people will engage in other forms of risky behavior in an attempt to either follow or circumvent the rule. For example, the Japanese bullet train system imposed a stiff penalty for operators who failed to wear a hat. In one case an operator left the train cabin to retrieve his hat while the train kept running unsupervised. This minor incident actually conceals two additional sorts of risks—that of automating a system so much that humans no longer pay attention, and the inability of the system to sense the lack of human supervision.

UNANTICIPATED USE OF DATA

The growing number of different databases that track even the intimate details of individual lives has raised many privacy issues (see PRIVACY IN THE DIGITAL AGE). Designers and maintainers of such databases had some awareness of the threat of unauthorized persons breaking into systems and stealing such data (see COMPUTER CRIME AND SECURITY). However, most people were surprised and alarmed by the new crime of identity theft, which began to surface in significant numbers in the mid- to late-1990s.

It turned out that while a given database (such as customer records, bank information, illicitly obtained DMV records, and so on) usually did not have enough information to allow someone to successfully impersonate another's identity, it was not difficult to use several of these sources together to obtain, for example, the information needed to apply for credit in another's name. In particular, while most people guarded their credit card numbers, they tended not to worry as much about Social Security numbers (SSN). However, since many institutions use the SSN to index their records, the number has become a key for unlocking personal data.

Further, as more organizations put their records online and make them Web-accessible, the ability of hackers, private investigators (legitimate or not), and "data brokerage" services to quickly assemble a dossier of sensitive information on any individual was greatly increased. Here we have a case where a powerful tool for productivity (the Internet) also becomes a facilitator for using the vulnerabilities in any one system to compromise others.

In an increasingly networked and technologically-dependent world, the anticipation and prevention of computer risks has become very important. To the extent companies may be legally liable for the more direct forms of risk, there is more incentive for them to devote resources to risk amelioration. However, many computer-related risks are at least as much social as technological in nature, and are beyond the scope of concern of any one company or organization. Social risks ultimately demand a broader social response.

Technology itself can be used to help ameliorate technological risks. Artificial intelligence techniques (see also EXPERT SYSTEMS and NEURAL NETWORK) might be used improve the ability of a system to adapt to unusual conditions. However, any such programming then becomes prone to bugs and risks itself.

So far, the most successful way to deal with the broad range of computer risks has been through human collaboration as facilitated by the Internet. Through venues such as the Risks Forum computer-mediated communications and collaboration allows for the pooling of human intelligence in the face of the growing complexity of human inventiveness.

Further Reading

Association for Computing Machinery. Committee on Computers and Public Policy. "The Risks Digest: Forum On Risks To the Public in Computers and Related Systems." http://catless.ncl.ac.uk/Risks/ Also available in the newsgroup comp.risks.

Fox, John, and Subrata Das. *Safe and Sound. Artificial Intelligence in Hazardous Operations*. Boston: MIT Press, 2000.

Neumann, Peter G. *Computer-Related Risks*. Reading, Mass.: Addison-Wesley, 1995.

———. "Illustrative Risks to the Public in the Use of Computer Systems and Related Technology." http://www.csl.sri.com/users/neumann/illustrative.html

Ritchie, Dennis

(1941–)
American
Computer Scientist

Together with Ken Thompson, Dennis Ritchie developed the UNIX operating system and the C programming language—two tools that have had a tremendous impact on the world of computing for three decades.

Ritchie was born on September 9, 1941, in Bronxville, New York. He was exposed to communications technology and electronics from an early age because his father was director of the Switching Systems Engineering Laboratory at Bell Laboratories. (Switching theory is closely akin to computer logic design.) Ritchie attended Harvard University and graduated with a B.S. in physics. However, by then his interests had shifted to applied mathematics and in particular, the mathematics of computation, which he later described as "the theory of what machines can possibly do." (See COMPUTABILITY AND COMPLEXITY). For his doctoral thesis he wrote about

Dennis Ritchie codeveloped the UNIX operating system and the C programming language, two of the most important developments in the history of computing. (PHOTO COURTESY OF LUCENT TECHNOLOGIES' BELL LABS)

recursive functions (see RECURSION). This topic was proving to be important for the definition of new computer languages in the 1960s (see ALGOL).

In 1967, however, Ritchie decided that he had had enough of the academic world. Without finishing the requirements for his doctorate, he started work at Bell Labs, his father's employer. Bell Labs is an institution that has made a number of key contributions to communications and information theory (see RESEARCH LABORATORIES IN COMPUTING).

By the late 1960s, computer operating systems had become increasingly complex and unwieldy. As typified by the commercially successful IBM System/360, the operating system was proprietary, had many hardware-specific functions and tradeoffs in order to support a family of upwardly-compatible computer models, and was designed with a top-down approach.

During his graduate studies, however, Ritchie had encountered a different approach to designing an operating system. A new system called Multics was being designed jointly by Bell Labs, MIT, and General Electric. Multics was quite different from the batch-processing world of mainframes: It was intended to allow many users to share a computer. He had also done some work with MIT's Project Mac. The MIT computer students, the original "hackers" (in the positive meaning of the term), emphasized a cooperative approach to designing tools for writing programs. This, too, was quite different from IBM's highly structured and centralized approach.

Unfortunately, the Multics project itself grew increasingly unwieldy. Bell Labs withdrew from the Multics project in 1969. Ritchie and his colleague Ken Thompson then decided to apply many of the same principles to creating their own operating system. Bell Labs wasn't in a mood to support another operating system project, but they eventually let Ritchie and Thompson use a DEC PDP-7 minicomputer. Although small and already obsolete, the machine did have a graphics display and a Teletype terminal that made it suitable for the kind of interactive programming they preferred. They decided to call their system UNIX, punning on Multics by suggesting something that was simpler and better integrated.

Instead of designing from the top down, Ritchie and Thompson worked from the bottom up. They designed a way to store data on the machine's disk drive (see FILE), and gradually wrote the necessary utility programs for listing, copying, and otherwise working with the files. Thompson did the bulk of the work on writing the operating system, but Ritchie did make key contributions such as the idea that devices (such as the keyboard and printer) would be treated the same way as other files. Later, he reconceived data connections as "streams" that could connect not only files and devices but applications and data being sent using different protocols. The ability to flexibly assign input and output, as well as to direct data from one program to another, would become hallmarks of UNIX.

When Ritchie and Thompson successfully demonstrated UNIX, Bell Labs adopted the system for its internal use. UNIX turned out to be ideal for exploiting the capabilities of the new PDP-11 minicomputer. As Bell licensed UNIX to outside users, a unique community of user-programmers began to contribute their own UNIX utilities (see also OPEN SOURCE MOVEMENT).

In the early 1970s, Ritchie also collaborated with Thompson in creating C, a streamlined version of the earlier BCPL and CPL languages. C would be a "small" language that was independent of any one machine but could be linked to many kinds of hardware thanks to its ability to directly manipulate the contents of memory. C became tremendously successful in the 1980s. Since then, C and its offshoots C++ and Java became the dominant languages used for most programming today.

Ritchie and Thompson still work at Bell Labs' Computing Sciences Research Center. (When AT&T spun off many of its divisions, Bell Labs became part of Lucent Technologies.) Ritchie has developed an experimental operating system called Plan 9 (named for a cult sci-fi movie). Plan 9 attempts to take the UNIX philosophy of decentralization and flexibility even further, and is designed especially for networks where computing resources are distributed.

Ritchie has received numerous awards, often given jointly to Thompson. These include the ACM Turing Award (1985), the IEEE Hamming Medal (1990), the Tsutomu Kanai Award (1999), and the National Medal of Technology (also 1999).

Further Reading

[Dennis Ritchie Home Page] http://www.cs.bell-labs.com/who/dmr/

Kernighan, B. W., and Dennis M. Ritchie. *The C Programming Language.* Upper Saddle River, N.J., Prentice Hall, 1978. (A second edition was published in 1989)

Lohr, Steve. *Go To.* New York: Basic Books, 2001.

Ritchie, Dennis M., and Ken Thompson. "The UNIX Time-Sharing System." *Communications of the ACM,* vol. 17, no. 7, 1974, 365–375.

Slater, Robert. *Portraits in Silicon.* Cambridge, Mass.: MIT Press, 1987.

robotics

The idea of the automaton—the lifelike machine that performs intricate tasks by itself—is very old. Simple automatons were known to the ancient world. By the 18th century, royal courts were being entertained by intricate humanlike automatons that could play music, draw pictures, or dance. A little later came the "Turk," a chess-playing automaton that could beat most human players.

However, things aren't always what they seem. The true automatons, controlled by gears and cams, could play only whatever actions had been designed into them. They could not be reprogrammed and did not respond to changes in their environment. The chess-playing automaton held a concealed human player.

True robotics began in the mid-20th century and has continued to move between two poles: the pedestrian but useful industrial robots and the intriguing but tentative creations of the artificial intelligence laboratories.

INDUSTRIAL ROBOTS

In 1921, the Czech playwright Karel Capek wrote a play called *R.U.R.* or *Rossum's Universal Robots*. Robot is a Czech word that has been translated as work(er), serf, or slave. In the play the robots, which are built by factories to work in other factories, eventually revolt against their human masters.

During the 1960s, real robots began to appear in factory settings. They were an outgrowth of earlier machine tools that had been programmed by cams and other mechanisms. An industrial robot is basically a movable arm that ends in a "hand" called an end effector. The arm and hand can be moved by some combination of hydraulic, pneumatic, electrical, or mechanical means.

An experimental NASA robot arm. (NASA PHOTO)

Typical applications include assembling parts, welding, and painting. The robot is programmed for a task either by giving it a detailed set of commands to move to, grasp, and manipulate objects, or by "training" the robot by moving its arm, hand, and effectors through the required motions, which are then stored in the robot's memory. By the early 1970s, Unimation, Inc. had created a profitable business from selling its Unimate robots to factories.

The early industrial robots had very little ability to respond to variations in the environment, such as the "work piece" that the robot was supposed to grasp being slightly out of position. However, later models have more sophisticated sensors to enable them to adjust to variations and still accomplish the task. The more sophisticated computer programs that control newer robots have internal representations or "frames of reference" to keep track of both the robot's internal parameters (angles, pressures, and so on) and external locations in the work area.

MOBILE ROBOTS AND SERVICE ROBOTS

Industrial robots work in an extremely restricted environment, so their world representation can be quite simple. However, robots that can move about in the environment have also been developed. Military programs have developed automatic guided vehicles (AGVs) with wheels or tracks, capable of navigating a battlefield and scouting or attacking the enemy (see also MILITARY APPLICATIONS OF COMPUTERS). Space-going robots including the Sojourner Mars rover also have considerable onboard "intelligence," although their overall tasks are programmed by remote commands.

Indeed, the extent to which mobile robots are truly autonomous varies considerably. At one end is the "robot" that is steered and otherwise controlled by its human operator, such as law enforcement robots that can be sent into dangerous hostage situations. (Another example is the robots that fight in arena combat in the popular "Robot Wars" shows.)

Moving toward greater autonomy, we have the "service robots" that have begun to show up in some institutions such as hospitals and laboratories. These mobile robots are often used to deliver supplies. For example, the HelpMate robot can travel around a hospital by itself, navigating using an internal map. It can even take an elevator to go to another floor.

Service robots have had only modest market penetration, however. They are relatively expensive and limited in function, and if relatively low-wage more versatile human labor is available, it is generally preferred. For now mobile robots and service robots are most likely to turn up in specialized applications in environments too dangerous for human workers, such as in the military, law enforcement, handling of hazardous materials, and so on.

SMART ROBOTS

Robotics has always had great fascination for artificial intelligence researchers (see ARTIFICIAL INTELLIGENCE). After all, the ability to function convincingly in a real-world environment would go a long way toward demonstrating the viability of true artificial intelligence.

Building a smart, more humanlike robot involves several interrelated challenges, all quite difficult. These include developing a system for seeing and interpreting the environment (see COMPUTER VISION) as well a way to represent the environment internally so as to be able to navigate around obstacles and perform tasks.

One of the earliest AI robots was "Shakey," built at the Stanford Research Institute (SRI) in 1969. Shakey could navigate only in a rather simplified environment. However, the "Stanford Cart," built by Hans Moravec in the late 1970s could navigate around the nearby campus without getting into too much trouble.

Today it's possible to buy an AI robot for one's home, in the form of toys such as Sony's AIBO robot dog, which can emulate various doggy behaviors such as chasing things and communicating by body language. Some robot toys not only have an extensive repertoire of behavior and vocalizations, but also can learn to some extent (see NEURAL NETWORK.)

It's also possible to experiment with robotics at home or school, thanks to kits such as the Logo-Lego, which combines a popular building set with a versatile educational programming language (see LOGO).

FUTURE APPLICATIONS

A true humanoid robot with the kind of capabilities written about by Isaac Asimov and other science fiction writers is not in sight yet. However, there are many interesting applications of robots that are being explored today. These include the use of remote robots for such tasks as performing surgery (see also TELEPRESENCE) and the application of robotics principles to the design of better prosthetic arms and legs for humans (bionics). Farther afield is the possibility of creating artificial robotic "life" that can self-reproduce (see ARTIFICIAL LIFE).

Further Reading

Arkin, Ronald C. *Behavior-Based Robotics.* Cambridge, Mass.: MIT Press, 1998.

Brooks, Rodney A. *Flesh and Machines: How Robots will Change Us.* New York: Pantheon Books, 2002.

Fuller, J. L. *Robotics: Introduction, Programming, and Projects.* Upper Saddle River, N.J.: Prentice Hall, 1998.

Jeffries, David. *Artificial Intelligence: Robotics and Machine Evolution.* New York: Crabtree, 1999.

Kortenkamp, David, R. Peter Bonasso, and Robin Murphy, eds. *Artificial Intelligence and Mobile Robots: Case Studies of Successful Robot Systems.* Menlo Park, Calif.: AAAI Press and MIT Press, 1998.

Moravec, Hans. *ROBOT: Mere Machines to Transcendent Minds.* New York: Oxford University Press, 1998.

Nof, Shimon Y. *Handbook of Industrial Robotics.* 2nd ed. New York: Wiley, 1999.

"Robotics Internet Resources." http://www-robotics.cs.umass.edu/robotics.html

RPG

Many business computer programs written for mainframe computers involved reading data from files, performing relatively simple procedures, and outputting printed reports. During the 1960s, some people believed that COBOL, a general-purpose (but business-oriented) computer language, would be easy enough for nonprogrammers to use (see COBOL). Although this turned out not to be the case, IBM did succeed in creating RPG (Report Program Generator), a language designed to make it easier for programmers (including beginners) to generate business reports.

Most COBOL programs read data, perform tests and calculations, and print the results. RPG, first released in 1964 for use with the new System/360 mainframe and the smaller System/3, simplifies this process and eliminates most writing of program code statements.

An RPG program is built around the "RPG Cycle," consisting of three stages. During the input stage, the input device(s), file type, access specifications, and data record structure are specified. (These specifications can be quite elaborate.) The heart of the program specifies calculations to be performed with the various data fields, while the output section specifies how the results will be laid out in report form, including such things as headers, footers, and sections.

Subsequent versions of RPG added more features. RPG-IV, released in 1994, includes the ability to define subroutines, for example. IBM has also released VisualAge RPG, which allows for the creation and running of RPG programs in the Microsoft Windows environment. There are also tools for interfacing RPG programs with various database systems and to use RPG for writing Web-based (CGI) programs.

Further Reading

Cozzi, Robert, Jr. *The Modern RPG IV Language.* 2nd ed. Carlsbad, Calif.: Midrange Computing, 1999.

Meyers, Bryan, and Jeff Sutherland. *VisualAge for RPG by Example.* Loveland, Colo.: Duke Press, 1998.

Stone, Bradley V. *e-RPG: Building AS/400 Web Applications with RPG.* Carlsbad, Calif.: Midrange Computing, 2000.

S

scanner

In order for a computer to work with information, the information must be digitized—converted to data that application programs can recognize and manipulate (see CHARACTERS AND STRINGS). Computer users have thus been confronted with the task of converting millions of pages of printed words or graphics into machine-readable form. Since it is expensive to re-key text (and impractical to redraw images), some way is needed to automatically convert the varying shades or colors of the text or images into a digitized graphics image that can be stored in a file.

This is what a scanner does. The scanner head contains a charge-coupled device (CCD) like that used in digital cameras (see PHOTOGRAPHY, DIGITAL). The CCD contains thousands or millions of tiny regions that can convert incoming light into a voltage level. Each of these voltage levels, when amplified, will correspond to one pixel of the scanned image. (A color scanner uses three different diodes for each pixel, each receiving light through a red, green, or blue filter.)

The operation of the head depends on the type of scanner. In the most common type, the flatbed scanner, a motor moves the head back and forth across the paper, which lies facedown on a glass window. In a sheet-fed scanner, the head remains stationery and the paper is fed past it by a set of rollers. Finally, there are handheld scanners, where the job of moving the scanner head is performed by the user moving the scanner back and forth over the page.

The resolution of a scanner depends on the number of pixels into which it can break the image. The color depth depends on how many bits of information that it can store per pixel (more information means more gradations of color or gray). Resolutions of 1200 dots per inch (dpi) or more are now common, with up to 36 bit color depth, allowing for about 68.7 billion colors or gradations (see also COLOR IN COMPUTING).

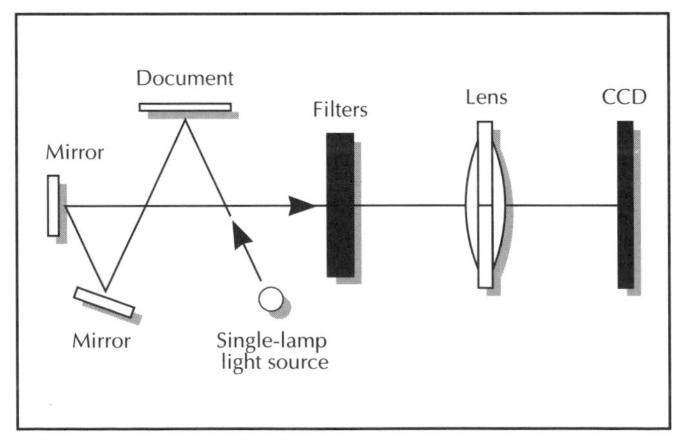

Like a digital camera, a scanner uses a charge-coupled device (CCD) array to read light. In the case of a scanner, the light is reflected from a lamp off the surface of the document being scanned. The resulting signals are converted to digital values to be stored as a graphics image.

Besides considerations of resolution and color depth, the quality of a scanned image depends on the quality of the scanner's optics as well as on how the page or other object reflects light. As anyone who has browsed eBay listings knows, the quality of scans can vary considerably. Most scanners come with software that allows for the scanner to be controlled and adjusted from the PC, and image-editing software can be used to further adjust the scanned image.

Even if the input is a sheet of text, the scanner's output is simply a graphical image. Special software must be used to interpret scanned images of text and identify which characters and other features are present (see OPTICAL CHARACTER RECOGNITION). Since such software is not 100 percent accurate, human proofreaders may have to inspect the resulting documents.

Like printers, scanners have become quite inexpensive in recent years. Quite serviceable units are available for around $100 or so. (Popular multifunction devices often include scanner, copier, fax, and printer capabilities. A scanner can be used as a copier or fax by sending its output to the appropriate mechanism.)

Many home users now use scanners to digitize images for use in personal webpages, on-line auctions, and other venues. Since sheet-fed scanners can only process individual sheets (not books, magazines, or objects) they are now less popular. Handheld scanners are somewhat tedious to use and require a steady hand, so they are generally used only in special circumstances where a flatbed scanner is not available. For capturing images of three-dimensional objects it is often easier to use a digital camera than a scanner.

Further Reading

Fulton, Wayne. "A Few Scanning Tips." http://www.scantips.com/

Gilbert, Jim. *How to Do Everything with Your Scanner.* Berkeley, Calif.: Osborne McGraw-Hill, 2001.

PC Tech Guide. "Scanners." http://www.pctechguide.com/18scanners.htm

scheduling and prioritization

Often in computing, a fixed resource must be parceled out among a number of competing users. The most obvious example is the operating system's scheduling the running of programs. Most computers have a single central processor (CPU) to execute programs. However, today virtually all operating systems (except for certain dedicated applications—see EMBEDDED SYSTEM) are expected to have many programs available simultaneously. For example, a Microsoft Windows user might have a word processor, spreadsheet, e-mail program, and Web browser all open at the same time. Not only might all of these programs be carrying out tasks or waiting for the user's input, but dozens of "hidden" system programs are also running in the background, providing services such as network support, virus protection, and printing services (see also MULTITASKING).

In this environment each executing program (or "process") will be in one of three possible states. It may be actively executing (that is, its code is being run by the CPU). It may be ready to execute—that is, "wanting" to perform some activity but needing access to the CPU. Or, the program may be "blocked"—that is, not executing and unable to execute until some external condition is met. Blockage is usually caused by an input/output (I/O) operation. An example would be a program that's waiting for data to finish loading from a file.

In this sort of single-processor multiple-program system, the simplest arrangement is to have the operating system dole out fixed amounts of execution time to each program. Each program that indicates that it's ready to run gets placed in a list (see QUEUE) and given its turn. When the amount of time fixed for a turn has passed, the operating system saves the program's "state" in the processor—the contents of the registers, address pointed to by the pointer to the next instruction to be executed, and so on. This stored information can be considered to be a "virtual processor." When the program's turn comes around again, the processor is reloaded with the contents of the virtual processor and execution continues where it had left off.

USE OF PRIORITY

The above scheme assumes that all programs should have equal priority. In other words, that the timely completion of one program is not more important than that of another, or that no program should be "bumped up" in the queue for some reason. In reality, however, most operating systems do give some programs preference over others.

For example, suppose the word processor has just received a user's mouse click on a menu. The next program in the queue for execution, however, is an antivirus program that's checking all the files on the hard drive for possible viruses. The latter program is important, but since the user is not waiting for it to finish, a delay in its execution won't cause a significant problem. The user, on the other hand, is expecting the menu just clicked on to open almost instantly, and will become irritated with even a short delay. Therefore, it makes sense for the operating system to give a program that's responding to immediate user activity a higher priority than a program that's carrying out tasks that don't require user intervention.

There are other times when a program must (or should) be given a higher priority. A program may be required to complete a task within a guaranteed time frame (for example, to dispatch emergency services personnel). The operating system must therefore provide a way that the program can request priority execution.

In general, an operating system that supports real-time applications or that requires great attention to efficiency in using valuable devices may need a much more sophisticated scheduling algorithm that factors in the availability of key devices or services and adjusts program priorities in order to minimize bottlenecks and guarantee that the system's response will be within required parameters. Indeed, the method used for assigning priorities may actually be changed in response to changes in the various "loads" on the system. Sophisticated systems may also include programs that can predict the likely future load on the system in order to adjust for it as quickly as possible.

SCHEDULING MULTIPROCESSOR SYSTEMS

These general principles also apply to systems where more than one processor is available, but there is the added complication of deciding where the scheduling program will be run. In a multiprocessing system that has one "master" and many "slave" processors, the scheduling program runs on the master processor. This arrangement is simple, but it means that when a slave processor wants to schedule a program it must wait until the scheduling program gets its next time-slice on the master processor.

One alternative is to allow any processor that has free time to run the scheduling algorithm. This is harder to set up because it requires a mechanism to make sure two processors don't try to run the scheduling program at the same time, but it smoothes out the bottleneck that would arise from relying on a single processor.

A variant of this approach is "distributed scheduling." Here each processor runs its own scheduling program. All the schedulers share the same set of information about the status and queuing of processes on the system, and a locking mechanism is used to prevent two processors from changing the same information at the same time. This approach is easiest to "scale up" since added processors can come with their own scheduling programs.

Two trends in recent years have changed the emphasis in scheduling algorithms. One is the continuing drop in price per unit of processing power and memory. This means that maximum efficiency in using the hardware can often give way in favor of catering to the user's convenience and perceptions by giving more priority to interaction with the user. The other development is the growing use of systems where much of the burden of graphics and interactivity is placed on the user's desktop, thus simplifying the complexity of scheduling for the server (see CLIENT-SERVER COMPUTING).

Principles of scheduling and priority can be applied in areas other than computer operating systems. Scheduling human activities (such as factory work) adds further complications such as the dependence of one task upon the prior performance of one or more other tasks (see also PROJECT MANAGEMENT SOFTWARE) and the "just-in-time" scheduling for minimizing the investment in materials or inventory.

Further Reading

Brucker, Peter. *Scheduling Algorithms.* New York: Springer-Verlag, 2001.

Pinedo, M. *Scheduling: Theory, Algorithms, and Systems.* 2nd ed. Upper Saddle River, N.J.: Prentice Hall, 2001.

scientific computing applications

From microbiology to plasma physics, modern science would be impossible without the computer. This is not because the computer has replaced the scientific method of observation, hypothesis, and experiment. Modern scientists essentially follow the same intellectual procedures at did Galileo, Newton, Darwin, and Einstein. Rather, understanding of the layered systems that make up the universe has now reached so complex and detailed a level that there is too much data for an individual human mind to grasp. Further, the calculations necessary to process the data usually can't be performed by unaided humans in any reasonable length of time. This can be caused either by the inherent complexity of the calculation (see COMPUTABILITY AND COMPLEXITY) or the sheer amount of data (as in DNA sequencing).

INSTRUMENTATION

Some apparatus such as particle accelerators are complicated enough to make it expedient to control the operation by computer. It is simply more convenient to have instruments such as spectrocopes process samples automatically under computer control and produce printed results.

Most instruments for gathering data use electronics to turn physical measurements into numeric representations (see ANALOG AND DIGITAL and DATA ACQUISITION). The modern instrument's built-in processor and software performs preliminary processing that used to have to be done later in the lab. This can include scaling the data to an appropriate range of values, eliminating "noise" data, and providing an appropriate time framework for interpreting the data. Use of electronics also enables the data to be transmitted from a remote location (telemetry). See also SPACE EXPLORATION AND COMPUTERS.

DATA ANALYSIS

The analysis of data to obtain theoretical understanding of the processes of nature also greatly benefits from the power of computers ranging from ordinary PCs to high-performance scientific workstations to large supercomputers. The possible significance of variables can be determined by statistical techniques (see also STATISTICS AND COMPUTING).

The fundamental task in understanding any system is to isolate the significant variables and determine how they affect one another. In many cases this can be done by solving differential equations, where a dependent variable changes as a result of changes in one or more independent variables. For example, the classical Maxwell theory of wave behavior is a system of differential equations that could be used to understand, for example, how radar waves will bounce off an object with a given shape and reflectivity. However, real-world objects have complicating factors: A given problem may include aspects of wave behavior, electromagnetic interaction, deformation of material, and so on. While the great scientists of the late 19th to mid-20th century could develop elegant formulas showing key relationships in nature, the interaction of many different phenomena often requires much more formidable computation that must be applied to many individual components.

It might be considered fortunate that the computer came along at about the time that it was required for further scientific progress. However, another way to look at it is to note that much of the pressure that led to investment in the development of computers came from that very need for computational resources, albeit primarily for wartime projects.

Computer processing of photographic or scanned data can provide detailed information about the environment. In this NASA test project aerial and satellite imagery is analyzed to yield information about the ripeness of grapes in a vineyard, as well as moisture, soil conditions, and plant disease. (NASA PHOTO)

SIMULATION AND VISUALIZATION

Even if scientists have a basic understanding of a system, it may be hard to determine what the overall results of the interaction of the many particles (or other elements) in the system will be. This is true, for example, in the analysis of events taking place in nuclear reactors. Fortunately computers can apply the laws of the system to each of many particles and determine the resulting actions from their aggregate behavior (see SIMULATION). Simulation is particularly important in fields where actual experiments are not possible because of distance or time. Thus, a hypothesis about the formation of the universe can be tested by applying it to a set of initial conditions believed to reflect those at or near the time of the Big Bang.

However, even the most skilled scientists have trouble relating numbers to the shape and interaction of real-world objects. Computers have greatly aided in making it possible to visualize structures and phenomena using high-resolution 3D color graphics (see COMPUTER GRAPHICS). Features of interest can be enhanced, and arbitrary ("false") colors can be used to visually show such things as temperature or blood flow. These techniques can also be used to create interactive models where scientists can, for example, combine molecules in new ways and have the computer calculate the likely properties of the result. Finally, computer visualization and modeling can be used both to teach science and to give the general public some visceral grasp of the meaning of scientific theories and discoveries.

Further Reading

Computing in Science and Engineering. http://computer.org/ciseportal/

Heath, Michael T. *Scientific Computing: An Introductory Survey.* 2nd ed. New York: McGraw-Hill, 2002.

Langtangen, Hans Petter, Are Magnus Bruaset, and Ewald Quak, eds. *Advances in Software Tools for Scientific Computing.* New York: Springer-Verlag, 2000.

National Center for Supercomputing Applications. http://www.ncsa.uiuc.edu/About/NCSA/

Scientific Computing World. http://www.scientific-computing.com/main.html

scripting languages

There are several different levels at which someone can give commands to a computer. At one end, an applications programmer writes program code that ultimately results in instructions to the machine to carry out specified processing (see PROGRAMMING LANGUAGES and COMPILER). The result is an application that users can control in order to get their work done.

At the other end, the ordinary user of the application uses menus, icons, keystrokes, or other means to select program features in order to format a document, calculate a spreadsheet, create a drawing, or perform some other task. Today most users also control the operating system by using a graphical user interface to, for example, copy files.

However, there is an intermediate realm where text commands can be used to work with features of the operating system, to process data through various utility programs, and to create simple reports. For example, a system administrator may want to log the number of users on the system at various times, the amount of disk capacity being used, the number of hits on various pages on a Web server, and so on. (See SYSTEM ADMINISTRATOR.) It would be expensive and time-consuming to write and compile full-fledged application programs for such tasks, particularly if changing needs will dictate frequent changes in the processing.

The use of the operating system shell and shell scripting (see SHELL) has traditionally been the way to deal with automating routine tasks, especially with systems running UNIX. However, the complexity of modern networks and in particular, the Internet, has driven administrators and programmers to seek languages that would combine the quick, interactive nature of shells, the structural features of full-fledged programming languages, and the convenience of built-in facilities for pattern-matching, text processing, data extraction, and other tasks. The result has been the development of a number of popular scripting languages (see also AWK, PERL, and PYTHON).

WORKING WITH SCRIPTING LANGUAGES

Although the various scripting languages differ in syntax and features, they are all intended to be used in a similar way. Unlike languages such as C++, scripting languages are interpreted, not compiled (see INTERPRETER). Typically, a script consists of a number of lines of text in a file. When the file is invoked (such as by someone typing the name of the language followed by the name of the script file at the command prompt), the script language processor parses each statement (see PARSER). If the statement includes a reference to one of the language's internal features (such as an arithmetic operator or a print command), the appropriate function is carried out. Most languages include the basic types of control structures (see BRANCHING STATEMENTS and LOOP) to test various variables and direct execution accordingly.

The trend in higher-level languages has been to require that all variables be declared to be used for particular kinds of data such as integer, floating-point number, or character string (see DATA TYPES). Scripting languages, however, are designed to be easy to use and scripts are relatively simple and easy to debug. Since the consequences of errors involving data types are less likely to be severe, scripting languages don't require that variables be declared before they are used. The language processor will make "common sense" assumptions about data. Thus if an integer such as 23 and a floating-point number like 17.5 must be added together, the integer will be converted to floating point and the result will be expressed as the floating-point value 40.5.

Similarly, scripting languages take a relaxed view about scope, or the parts of a program from which a variable's value can be accessed. Scripting languages do provide for some form of subroutine or procedure to be declared (see PROCEDURES AND FUNCTIONS). Generally, variables used within a subroutine will be considered to be "local" to that subroutine, and variables declared outside of any subroutine will be treated as global.

With compilers for regular programming languages, a great deal of attention must be paid to creating fast, efficient code. A scientific program may need to optimize calculations so that it can tackle cutting-edge problems in physics or engineering. A commercial application such as a word processor must implement many features to be competitive, and yet be able to respond immediately to the user and complete tasks quickly.

Scripting languages, on the other hand, are typically used to perform housekeeping tasks that don't place much demand on the processor, and that often don't need to be finished quickly. Because of this, the relative inefficiency of on-the-fly interpretation instead of optimized compilation is not a problem. Indeed, by making it easy for users to write and test programs quickly, the interpreter makes it much easier for administrators and others to create simple but useful tools for monitoring the system and extracting necessary information. Scripting languages can also be used to quickly create a prototype version of a program that will be later recoded in a language such as C++ for efficiency.

Scripting languages were originally written for operating systems that process text commands. However, with the popularity of Microsoft Windows, Macintosh, and various UNIX-based graphical user interfaces, many users and even system administrators now prefer a visual scripting environment. For example, Microsoft Visual Basic for Windows (and the related Visual Basic for Applications and VBScript) allow users to write simple programs that can harness the features of the Windows operating system and user interface and take advantage of prepackaged functionality available in ActiveX controls (see BASIC). In these visual environments the tasks that had been performed by script files can be automated by setting up and linking appropriate objects and adding code as necessary.

BRIEF SURVEY OF SCRIPTING LANGUAGES

Here is a brief characterization of the leading scripting languages. Note that awk, Perl, and Python have their own entries in this book, which include more extensive discussion of features and some sample scripts.

Awk was named for its developers (Aho, Kernighan, and Weinberger) and introduced in the late 1970s was the earliest major UNIX scripting language. It was designed primarily for editing and formatting text and for extracting data from simple text databases. Its strength is in its built-in text pattern-matching and processing facilities. However, awk has been mostly superseded by the more powerful Perl.

JavaScript (a sort of subset of the full Java language) was developed by Netscape in the mid-1990s and allows for scripts to be embedded in the HTML code of webpages. When a script is triggered on a page being viewed, it is interpreted by the JavaScript interpreter in the viewer's Web browser. Although it can be used to enhance webpages such as by creating textual effects and animation, its main use has been in processing on-line forms.

Perl, an acronym for "Practical Extraction and Report Language," was developed by Larry Wall in the late 1980s. Just as awk resulted from frustration with the limitations of the scripting supported by the various UNIX shells, Perl began as an attempt to extend the strengths of awk in text-processing to provide a language that could deal better with the many sources of data in a modern networked (and Internet) environment. Perl thus has full facilities for connecting to processes and network sockets. Perl's data-processing capabilities have been extended through additional list-processing functions and the facility for working with hashes (arrays in which each data item is associated with a key value). The latest version of Perl (1999) adds support for object-oriented programming.

Python is one of the more recent scripting languages, developed by Guido van Rossum in the early 1990s. Unlike the "Swiss army knife" approach of Perl, Python is tightly designed and object-oriented "from the ground up." Although it was first conceived as a scripting language, Python is now used as a general-purpose alternative to C/C++ and Java for developing moderate-size Web applications, including CGI and other database functions.

Tcl/Tk, or Tool Command Language/Toolkit, developed in the late 1980s by John Ousterhout, takes a somewhat different approach to scripting. It's actually a general-purpose "toolkit" that can be added to an application to provide a way for its functions to be made available to outside scripts. (Thus, Tcl provides a functionality roughly similar to that available in Windows through ActiveX objects or Object-Linking and Embedding [OLE].) Tcl is useful for creating extended user interfaces for applications. Because of the language's consistent structure, interfaces created with Tcl will be similar from one application to another, aiding learning.

Further Reading

"JavaScript tutorial." http://www.w3schools.com/js/default.asp

Ousterhout, J. *Tcl and the Tk Toolkit*. Reading, Mass.: Addison-Wesley, 1994.

TCL Developer Xchange.http://tcl.activestate.com:8002/resource/

Weltner, Tobias. *Windows Scripting Secrets*. New York: Hungry Minds, 2000.

"Scripting Languages: Automating the Web." *World Wide Web Journal*. no. 2, Spring 1997. Available on-line at http://www.w3j.com/6/

search engine

By the mid-1990s, many thousands of pages were being added to the World Wide Web each day (see WORLD WIDE WEB). The availability of graphical browsing programs such as Mosaic, Netscape, and Microsoft Internet Explorer (see WEB BROWSER) made it easy for ordinary PC users to view webpages and to navigate from one page to another. However, people who wanted to use the Web for any sort of systematic research found they needed better tools for finding the desired information.

There are basically three approaches to exploring the Web: casual "surfing," portals, and search engines. A user might find (or hear about) an interesting webpage devoted to a business or other organization or perhaps a particular topic. The page includes a number of featured links to other pages. The user can follow any of those links to reach other pages that might be relevant. Those pages are likely to have other interesting links that can be followed, and so on. Most Web users have surfed in this way: It can be fun and it can certainly lead to "finds" that can be bookmarked for later reference. However, this approach is not systematic, comprehensive, or efficient.

Alternatively, the user can visit a site such as the famous Yahoo! started by Jerry Yang and David Filo (see PORTAL). These sites specialize in selecting what their editors believe to be the best and most useful sites for each topic, and organizing them into a multilevel topical index. The portal approach has several advantages: The work of sifting through the Web has already been done, the index is easy to use, and the sites featured are likely to be of good quality. However, even Yahoo!'s busy staff can examine only a tiny portion of the estimated 4 billion or so webpages being presented on about 120 million different websites (as of 2001). Also, the sites selected and featured by portals are subject both to editorial discretion (or bias) and in some cases to commercial interest.

ANATOMY OF A SEARCH ENGINE

Search engines such as Lycos and AltaVista were introduced at about the same time as portals. Although there is some variation, all search engines follow the same basic approach. On the host computer the search engine runs automatic Web searching programs (sometimes called "spiders" or "Web crawlers"). These programs systematically visit websites and following the links to other sites and so on through many layers. Usually, several such programs are run simultaneously, from different starting points or using different approaches in an attempt to cover as much of the Web as possible. When a Web crawler reaches a site, it records the address (URL) and compiles a list of significant words. The Web crawlers give the results of their searches to the search engine's indexing program, which adds the URLs to the associated keywords, compiling a very large word index to the Web.

Search engines can also receive information directly from websites. It is possible for page designers to add a special HTML "metatag" that includes keywords for use by search engines. However, this facility can be misused by some commercial sites to add popular words that aren't actually relevant to the site, in the hope of attracting more hits.

To use a search engine, the user simply navigates to the search engine's home page with his or her Web browser. (Many browsers can also add selected search engines to a special "search pane" or menu item for easier access.) The user then types in a word or phrase. Most search engines accept logical specifiers (see BOOLEAN OPERATORS) such as and, or, or not. Thus, a search for "internet and statistics" will find only pages that have both words. Some engines also allow for phrases to be put in quote marks so they will be searched for as a whole. A search for "internet statistics" will match only pages that have these two words next to each other.

Because of the huge size of the Web, even seemingly esoteric search words can yield thousands of "hits" (results). Therefore, most search engines rank the results by analyzing how relevant they are likely to be. This can be done in a simple way by comparing the frequency with which the search terms appear on the various pages. More sophisticated search engines such as Google can determine how relevant a word or phrase seems to be because of its placement or presence in a heading or how often a site is referred to from other sites. Some search engines also offer the ability to "refine" searches by adding further words and performing a new match against the set of results.

LIMITATIONS AND FUTURE OF SEARCH ENGINES

Search engines do provide many useful "hits" for both casual and professional researchers, but the current technology does have a number of limitations. Even the most comprehensive search engines now reach and index only a small fraction of the total available webpages. One way to maximize the number of pages searched is to use a "metasearch" program such as Copernic, which submits a user's search to many different search engines. It then collates the results, removing duplicates and attempting to rank them in relevance.

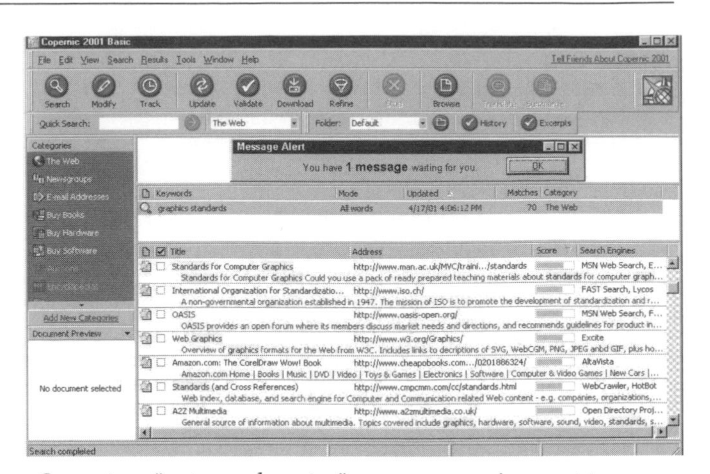

Copernic, a "metasearch engine" can pass a user's request to many different search engines and then prioritize and collate the results, weeding out duplicates.

Even with "relevancy" algorithms, searches for broad, general topics are likely to retrieve many less than useful hits. Also, current search engines have difficulty finding image and sound (music) files, which are among the most sought-after Web content. This is because the search engine cannot recognize graphics or sound as such, only file names or extensions or text descriptions. Search engines also vary considerably in their ability to read and index files in proprietary text formats such as Microsoft Word or Adobe PDF.

In the future artificial intelligence techniques may make it possible for search engines to recognize types of images or sounds through pattern recognition. Search engines may be able to respond more appropriately to "natural language" queries such as "How many pages are there on the Web?" and find the answer, or at least webpages that are likely to have the answer. (Current services of this type such as AskJeeves.com tend to give hit-and-miss results.)

For now, search engines remain a useful tool, but systematic researchers should complement their results with links from portals and recommendations from authoritative sites.

Further Reading

Ackermann, Ernest, and Karen Hartman. *The Information Searcher's Guide to Searching and Researching on the Internet and World Wide Web.* 2nd ed. Wilsonville, Oreg.: ABF Content, 2000.

Belew, Richard K., and C. J. van Rijsbergen. *Finding Out About: A Cognitive Perspective on Search Engine Technology and the WWW.* New York: Cambridge University Press, 2001.

"Meta-Search Engines." http://www.lib.berkeley.edu/TeachingLib/Guides/Internet/MetaSearch.html

"Search Engine Showdown: The User's Guide to Web Searching." http://www.searchengineshowdown.com/

serial port

There are basically two ways to move data from a computer to or from a peripheral device such as a printer or modem. A byte (8 bits) of data can be moved all at once, with each bit traveling along its own wire (see PARALLEL PORT). Alternatively, a single wire can be used to carry the data one bit at a time. Such a connection is called a *serial port.*

The serial port receives data a full byte at a time from the computer bus and uses a UART (Universal Asynchronous Receiver-Transmitter) to extract the bits one at a time and send them through the port. A corresponding circuit at the other end accumulates the incoming bits and reassembles them into data bytes.

The data bits for each byte are preceded by a start-bit to signal the beginning of the data and terminated by an stop-bit. Depending on the application, an additional bit may be used for parity (see ERROR CORRECTION). Devices connected by a serial port must "negotiate" by requesting a particular connection speed and parity setting. Failure to agree results in gibberish being received.

The official standard for serial transmission is called RS-232C. It defines various additional pins to which wires are connected, such as for synchronization (specifying when the device is ready to send or receive data) and ground. Physically, the old-style connectors are called DB-25 because they contain 25 pins (many of which are not used). Most newer PCs have DB-9 (i.e. nine pin) connectors. A "gender changer" can be used in cases where two devices both have male connectors (with pins) or female connectors (with corresponding sockets).

Because they use a single data transmission line and include error-correction, serial cables can be longer than parallel cables (25 feet or more, as opposed to 10–12 feet). Serial transmission is generally slower (at up to 115,200 bits/second) than parallel transmission. Serial connections have generally been used for such devices as modems (whose speed is already limited by phone line characteristics), keyboards, mice, and some older printers. Today the faster and more flexible USB (see UNIVERSAL SERIAL BUS) is replacing serial connections for many devices including even keyboards.

Further Reading

"The Complete Pinout Guide to Parallel-Serial Port, Network and Monitor Cables." http://www.indiacam.net/pinout/

White, Ron. *How Computers Work.* Millennium Edition. Indianapolis, Ind.: Que, 1999.

Shannon, Claude E.

(1916–2001)
American
Mathematician, Computer Scientist

The information age would not have been possible without a fundamental understanding of how information could be encoded and transmitted electronically. Claude Elwood Shannon developed the theoretical underpinnings for modern information and communications technology and then went on to make important contributions to the young discipline of artificial intelligence (AI).

Shannon was born in Gaylord, Michigan, on April 30, 1916. He received bachelor's degrees in both mathematics and electrical engineering at the University of Michigan in 1936. He went on to MIT, where he earned a master's degree in electrical engineering and a Ph.D. in mathematics, both in 1940. Shannon's background thus well equipped him to relate mathematical concepts to practical engineering issues.

While a graduate student at MIT, Shannon was in charge of programming an elaborate analog computer called the Differential Analyzer that had been built by Vannevar Bush (see ANALOG COMPUTER and BUSH, VANNEVAR). Actually "programming" is not quite the right word: To solve a differential equation with the Differential Analyzer, it had to be translated into a variety of physical settings and arrangements of the machine's intricate electromechanical parts.

The Differential Analyzer was driven by electrical relay and switching circuits. Shannon became interested in the underlying mathematics of these control circuits. He realized that their fundamental operations corresponded to the Boolean algebra he had studied in undergraduate mathematics classes (see BOOLEAN OPERATORS). It turned out that the seemingly abstract Boolean AND, OR

Claude Shannon developed the fundamental theory underlying modern data communications, and made contributions to the development of artificial intelligence as well. (PHOTO COURTESY OF LUCENT TECHNOLOGIES' BELL LABS)

and NOT operations had a practical engineering use. Shannon used the results of his research in his 1938 M.S. thesis, titled "A Symbolic Analysis of Relay and Switching Circuits." This work was honored with the Alfred Nobel prize of the combined engineering societies (this is not the same as the more famous Nobel Prize).

Along with the work of Alan Turing and John von Neumann (see TURING, ALAN and VON NEUMANN JOHN), Shannon's logical analysis of switching circuits would become essential to the inventors who would build the first digital computers in just a few years. (Demonstrating the breadth of his interests, Shannon's Ph.D. thesis would be in an entirely different application—the algebraic analysis of problems in genetics.)

In 1941, Shannon joined Bell Laboratories, perhaps America's foremost industrial research organization. The world's largest phone company had become increasingly concerned with how to "scale up" the burgeoning telephone system and still ensure reliability. The coming of war also highlighted the importance of cryptography—securing one's own transmissions while finding ways to break opponents' codes. Shannon's existing interests in both data transmission and cryptography neatly dovetailed with these needs.

Shannon's paper titled "A Mathematical Theory of Cryptography" would be published after the war. But Shannon's most lasting contribution would be to the fundamental theory of communication. His formulation would explain what happens when information is transmitted from a sender to a receiver—in particular, how the reliability of such transmission could be analyzed (see also INFORMATION THEORY).

Shannon's 1948 paper, "A Mathematical Theory of Communication" was published in *The Bell System Technical Journal*. Shannon identified the fundamental unit of information (the binary digit, or "bit" that would become familiar to computer users). He showed how to measure the redundancy (duplication) within a stream of data in relation to the transmitting channel's capacity, or bandwidth. Finally, he showed methods that could be used to automatically find and fix errors in the transmission. In essence, Shannon founded modern information theory, which would become vital for technologies as diverse as computer networks, broadcasting, data compression, and data storage on media such as disks and CDs.

One of the unique strengths of Bell Labs is that it did not limit its researchers to topics that were directly related to telephone systems or even data transmission in general. Like Alan Turing, Shannon became interested after the war in the question of whether computers could be taught to perform tasks that are believed to require true intelligence (see ARTIFICIAL INTELLIGENCE). He developed algorithms to enable a computer to play chess and published an article on computer chess in *Scientific American* in 1950. He also became interested in other aspects of machine learning, and in 1952 he demonstrated a mechanical "mouse" that could solve mazes with the aid of a circuit of electrical relays.

The mid-1950s would prove to be a very fertile intellectual period for AI research. In 1956, Shannon and AI pioneer John McCarthy (see MCCARTHY, JOHN) put out a collection of papers titled "Automata Studies." The volume included contributions by two other seminal thinkers, John von Neumann and Marvin Minsky (see MINSKY, MARVIN).

Although he continued to do research, by the late 1950s Shannon had changed his emphasis to teaching. As Donner Professor of Science at MIT (1958–1978) his lectures inspired a new generation of AI researchers. During the same period Shannon also explored the social impact of automation and computer technology as a Fellow at the Center for the Study of Behavioral Sciences in Palo Alto, California.

Shannon received numerous prestigious awards, including the IEEE Medal of Honor and the National Medal of Technology (both in 1966). Shannon died on February 26, 2001 in Murray Hill, New Jersey.

Further Reading

Shannon, Claude Elwood. "A Chess-Playing Machine." *Scientific American,* February 1950, 48–51.

———. *Collected Papers.* Ed. N. J. A. Sloane and Aaron D. Wyner. New York: IEEE Press, 1993.

———. "A Mathematical Theory of Communication." *Bell System Technical Journal* 27; July and October, 1948, 379–423, 623–656. Available for download at http://cm.bell-labs.com/cm/ms/what/shannonday/paper.html

Waldrop, M. Michael. "Claude Shannon: Reluctant Father of the Digital Age." *Technology Review,* July/Aug. 2001, n.p. Available for purchase on-line at http://www.techreview.com/articles/waldrop0701.asp

shareware

The early users of personal computers generally had considerable technical skill and a desire to write their own programs. This was partly by necessity: If one wanted to get an Apple, Atari, Commodore, or Radio Shack machine to perform some particular task, chances were one would have to write the software oneself. Commercial software was scarce and relatively expensive. However, given enough time, it was possible for hobbyists to write programs using the machine's built-in BASIC language or (with more effort) assembly language.

Programs such as utilities and games were often freely shared at gatherings of PC enthusiasts (see USER GROUPS). Many talented amateur programmers considered trying to turn their avocation into a business. However, a utility to provide better file listings or a colorful graphics program that creates kaleidoscopic images was unlikely to interest the commercial software companies who developed large programs in-house for marketing primarily to business.

In 1982, Andrew Fuegelman created a program called PC-Talk. This program provided a better way for users with modems to connect to the many bulletin board systems that were starting to spring up. Fluegelman was familiar with the common practice of public radio and TV broadcasters of soliciting pledge payments to help support their "free" service. He decided to do something similar with his program. He distributed it to many bulletin boards, where users could download it for free. However, he asked users who liked the program and wanted to continue to use it to pay him $25.

Fluegelman dubbed his method of software distribution "freeware" (because it cost nothing to try out the program). Other programmers began to use the same method with their own software. This included Jim Knopf, author of the PC-File database program, and Bob Wallace, who offered PC-Write as a full-featured alternative to expensive commercial word processing program. Because Fluegelman had trademarked the term *freeware,* these other authors began to call their offerings *shareware.*

Today freeware means software that can be downloaded at no cost and for which there is no charge for continued use. The program may be redistributed by users as long as they don't charge for it.

Shareware, on the other hand, follows Fluegelman's original concept. The software can be downloaded for free. The user is allowed to try the program for a limited period (either a length of time such as 30 days, or a maximum number of times that the program can be run). After the trial period, the user is expected to pay the author the specified fee of continued use. (Today this is usually done through the author's website or a service that can accept secure credit card payments on-line.) Once the user pays, he or she receives either an unrestricted version of the software or frequently, an alphanumeric key that can be typed into the program to remove all restrictions. At this point the program is said to be "registered."

Users can be encouraged (or forced) to pay in various ways. Some programs keep working after the trial period, but display continual "nag" messages or remove some functionality, such as the ability to print or save one's work. ("Demos" of commercial games or other programs also have limited functionality, but cannot be registered or upgraded. They are there simply to entice consumers to buy the commercial product.)

Alternatively, some shareware authors prefer to entice their users to register by offering bonuses, such as additional features, free upgrades, or additional technical support. Sometimes (as with the RealPlayer streaming sound and video player and the Eudora e-mail program) a useful but limited "lite" version is offered as freeware, but users are encouraged to upgrade to a more full-featured "professional" version.

Shareware has been a moderately successful business for a number of program authors. For example, Phil Katz's PKZip file compression and packaging program is so useful that it has found its way onto millions of PCs, and enough users paid for the program to keep Katz in business. (PKZip and its cousin WinZip are examples of shareware programs that became so popular that they spawned commercially packaged versions.)

Shareware and freeware should be distinguished from public domain and open source software (see OPEN SOURCE MOVEMENT). Public domain software is not only free (as with freeware), but the author has given up all rights including copyright, and users are free to alter the program's code or to use it as part of a new program. Open source software, on the other hand, allows users free access to the software and its source code, but with certain restrictions—notably, that it not be used in some other product for which access will be restricted.

Today tens of thousands of shareware and freeware programs are available on the Internet via ftp archives, author's websites, and giant on-line libraries maintained by zdnet.com, cnet.com, tucows.com, and others.

Further Reading

Association of Shareware Professionals. http://www.asp-shareware.org/pad/

Lehnert, Wendy G. *The Web Wizard's Guide to Freeware and Shareware.* Boston: Addison-Wesley, 2002.

shell

During the 1950s, using a computer generally meant that operators submitted batch-processing command cards (see JOB CONTROL LANGUAGE) that controlled how each program would use the computer's resources. One program ran at a time, and interaction with the user was minimal. However, when time-sharing computers began to appear in the 1960s, users gained the ability to control the computer interactively from terminals. The operating system therefore needed to have a facility that would interpret and execute the commands being typed in by the users, such as a request to list the files in a directory or to send a file to the printer. This command interpreter is called a *shell.*

To see a simple shell in action, a Windows user need only bring up an MS-DOS prompt, type the word *dir,* and press Enter. A shell called command.com provides the user interface for users of IBM PC-compatible systems running MS-DOS. The command processor displays a prompt on the screen. It then interprets (see PARSING) the user's commands. If the command involves one of the shell's internal operations (such as "dir" to list a file directory), it simply executes that routine. For example the command:

```
dir temp /p
```

would be interpreted as a call to execute the dir function, passing it the name "temp" (a directory) and the /p, which dir interprets as a "switch" or instruction telling it to pause the directory listing after each screenful of text. If the command is an external MS-DOS utility such as "xcopy" (a file copying program), the shell runs that program, passing it the information (mainly file names) from the command line. Finally, the shell can run any other executable program on the system. It is then that program's responsibility to interpret and act upon any additional information that was provided.

MS-DOS also has the ability for the command.com shell to read a series of commands stored in a text file called a batch file, and having the *.bat (batch) extension. This allowed for rudimentary scripting of system housekeeping operations or other routine tasks (see SCRIPTING LANGUAGES).

UNIX SHELLS

MS-DOS largely faded away in the 1990s as more users switched to Microsoft Windows and begun to use a graphical user interface to control their machines. However, shells have achieved their greatest proliferation and elaboration with UNIX, the operating system developed by Ken Thompson and Dennis Ritchie starting in 1969 and widely used for academic, scientific, engineering, and Web applications.

UNIX shells serve the same basic purposes as the MS-DOS shell: interactive control of the operating system and the ability to run stored command scripts. However, the UNIX shells have considerably more complex syntax and capabilities.

Part of the design philosophy of the UNIX system was to place the core operating system functions in the kernel (see KERNEL and UNIX). This modular design meant that UNIX, unlike most other operating systems, did not have to commit itself to a particular form of user interface or command processor. Accordingly, a number of such processors (shells) have been developed, reflecting the programming style preferences of their originators.

The first shell to be developed was the Bourne Shell, named for its creator, Steven R. Bourne, who developed it at Bell Labs, the original home of UNIX. The Bourne shell implemented some basic ideas that are characteristic of UNIX: the ability to redirect input and output to and from files, devices or other sources (using the < and > characters), and the ability to use "pipes" (the | character) to connect the output of one command to the input of another.

The next major development was the C shell (csh). The Bourne shell used a relatively simple and clean syntax devised by its creator. As the name suggests, the C shell (developed at the University of California, Berkeley) takes its syntax from the C programming language, which was by far the most commonly used language on UNIX systems. One logical reason for this choice was that C programmers could quickly learn to write scripts with the C shell. The C shell also added support for job control (that is, moving processes between foreground and background operation) and in general was easier to use for interactively controlling programs from the command line.

UNIX users sometimes used both shells, since the simpler and more consistent syntax of the Bourne shell is generally thought to be better for writing scripts. (The two shells also reflected the split in the UNIX world between the version of the operating system provided by AT&T and the variant developed at UC Berkeley.)

David Korn at AT&T then decided to combine the best features of both shells. His Korn shell (Ksh) kept the better scripting language features from the Bourne shell but added job-control and other features from the C shell. He also added the programming language concept of functions (see PROCEDURES AND FUNCTIONS), allowing for cleaner organization of code.

Another popular shell, BASH (Bourne Again Shell) was developed by the Free Software Foundation for GNU, an open-source version of UNIX. BASH and Ksh share most features and both are compatible with POSIX, a standard specification for connecting programs to the UNIX operating system.

SHELL SCRIPTS

Regardless of the version of the shell used, shell scripts work in the same basic way. A shell script is a text file containing commands to the shell. The commands can use control statements (see BRANCHING STATEMENTS and LOOP) and invoke both the shell's internal features and the many hundreds of utility programs that are available on UNIX systems.

Once the script is written, there are two ways to execute it. One way is to type the name of the shell at the command prompt, followed by the name of the script file, as in:

```
$ sh MyScript
```

Alternatively, the chmod (change mode) command can be used to mark the script's file type as executable, and the first line of the script then contains a statement that invokes the shell, which will parse the rest of the script. The script can now be executed simply by typing its name at the command prompt (or it can be included as a command in another script).

Here is a simple example of a shell script that prints out various items of information about the user and the current session on a UNIX system:

```
#! /sbin/sh

echo My username: 'whoami'
echo My current directory: 'pwd'
echo
echo My disk usage:
du -k
echo
echo System status:
uptime
if test -f log.txt; then
  cat log.txt
else echo Log file not found
fi
```

The first line tells UNIX which shell to use to interpret the script (in this case the Bourne shell, sh, will be executed). The echo command simply outputs the text that follows it to the screen. "whoami" is a UNIX command that prints the user's name. The script takes advantage of an interesting UNIX feature: The whoami command is put in "backquotes" (' '). This inserts the output of the whoami command (the user name) in place of that command, and the resulting text is output by the echo command.

The du command gives the user's disk usage, while the uptime command gives some statistics about how many users are on the system and how long the system has been running. Finally, the if statement at the end of the script tests for the presence of the file log.txt. If the file exists, its contents are displayed by the "cat" command.

When "myinfo" is typed at the UNIX prompt, the output might look like the following:

```
$ myinfo
My username: hrh
My current directory: /home/h/r/hrh

My disk usage:
132 ./.nn
4 ./Mail
48 ./.elm
296 .

System status:
  7:34pm up 56 day(s), 20:39, 73 users, load
  average: 3.62, 3.45, 3.49

This is a test file.
```

Further Reading

Bourne, Steve. "An Introduction to the UNIX Shell." http://www.ling.helsinki.fi/users/reriksso/unix/shell.html

"How to write a shell script." http://www.hsrl.rutgers.edu/ug/shell_help.html

Kochan, Stephen G., and Patrick H. Wood. *UNIX Shell Programming.* 2nd ed. Carmel, Ind.: Hayden Books, 1990.

Rosenblatt, Bill, and Arnold Robbins. *Learning the Korn Shell.* 2nd ed. Sebastopol, Calif.: O'Reilly, 2002.

Simula

One of the most interesting applications of computers is the simulation of systems in which many separate actions or events are happening simultaneously (see SIMULATION). During the 1950s, Norwegian computer scientist Kristen Nygaard began to develop a more formal way of describing and designing simulations. A typical simulation consists of a number of "objects," such as cars in a traffic flow or customers waiting in a bank line. In a bank simulation, for example, the objects (customers) would demand service from particular serving objects (teller windows). They would move in a queue and their motion would be captured at various points of time.

Nygaard used his ideas to create symbols and flow diagrams to represent the events going on in a simulation. However, existing computer languages such as Algol 60 were designed to carry out procedures sequentially and one at a time, not simultaneously. This made it difficult to write a program representing a situation in which many cars or customers were moving simultaneously.

In the early 1960s, Nygaard was joined by Ole-Johan Dahl, who had more experience with systems programming and computer language design. They worked together to create a new language that they called Simula, reflecting their emphasis on simulation programming. In designing Simula, the authors sought to create a data structure that was better suited to simultaneous actions or events. For example, in a simulation of automobile traffic, each car would be an "object" with data such as its location and speed as well as actions or capabilities such as changing speed or direction. The data for each object must be maintained separately and updated frequently.

The Algol 60 language already had a way to define code "blocks" (see PROCEDURES AND FUNCTIONS) that could contain their own local data as well as actions to be performed. Further, such blocks could be called repeatedly such that many copies could be "open" at the same time. However, these calls were still essentially sequential, not simultaneous. In their new Simula 1 language (introduced in 1965), Dahl and Nygaard created a way to simulate simultaneous processing. Even though the computer would (probably) only have a single processor such that only one copy of a block of code could be executing at a given time, Simula set up special variables for keeping track of simulated time. Control would "jump" from one instance of a block to another such that all blocks would, for example, have their actions for the time 20:15 executed, then actions for 20:16 would be executed, and so on. A list kept track of processes in time order. Thus, Simula 1 kept all the features of Algol but made it more suitable for modeling simultaneous events (see also MULTIPROCESSING).

Simula 1 was quite successful as a simulation language, but the authors soon realized that the ability to use separate invocations of a procedure to create individual "objects" had a more general application to representing data in applications other than simulations. In creating Simula 67 (the version of the language still used today), they therefore introduced the formal concept of the class as a specification that could be used to create objects of that type. They also introduced the key idea of inheritance (where one class can be derived from an earlier class), as well as a way that a derived class could redefine a procedure that it had inherited from the original (base) class (see OBJECT-ORIENTED PROGRAMMING, and CLASS).

Although Simula 67 would continue to be used primarily for simulations rather than as a general-purpose programming language, its object-oriented ideas would prove to be very influential. The designers of Smalltalk and Ada would look to Simula for structural ideas, and the popular C++ language began with an effort to create a "C with classes" language along the lines of Simula. (See SMALLTALK, ADA, and C++.)

Further Reading

Holmevik, Jan-Rune. "Compiling Simula: a Study in the History of Computing and the Construction of the SIMULA Programming Languages." *STS Report* (Trondheim). Available on-line at http://lingua.utdallas.edu/jan/simula.html

Nygaard, K., and O.-J. Dahl. "The Development of the Simula Languages" in *The History of Programming Languages*, R. L. Wexelblat, ed. New York: Academic Press, 1981.

Pooley, R. *Introduction to Programming with Simula.* Oxfordshire, U.K.: Alfred Waller, 1987.

Sebesta, Robert W. *Concepts of Programming Languages.* 4th ed. Reading, Mass.: Addison-Wesley, 1999.

simulation

A simulation is a simplified (but adequate) model that represents how a system works. The system can be an existing, real-world one, such as a stock market or a human heart, or a proposed design for a system, such as a new factory or even a space colony.

If a system is simple enough (a cannonball falling from a height, for example), it is possible to use formulas such as those provided by Newton to get an exact answer. However, many real-world systems involve many discrete entities with complex interactions that cannot be captured with a single equation. During the 1940s, scientists encountered just this problem in attempting to understand what would happen under various conditions in a nuclear reaction.

Together with physicist Enrico Fermi, two mathematicians, John von Neumann (see VON NEUMANN, JOHN) and Stanislaw Ulam, devised a new way to simulate complex systems. Instead of trying fruitlessly to come up with some huge formula to "solve" the whole system, they applied probability formulas to each of a number of particles—in effect, "rolling the dice" for each one and then observing their resulting distribution and behavior. Because of its analogy to gambling, this became known as the Monte Carlo method. It turned out to be widely useful not only for simulating nuclear reactions and particle physics but for many other activities (such as bombing raids or the spread of disease) where many separate things behave according to probabilities.

A number of other models and techniques have made important contributions to simulation. For example, the attempt to simulate the operation of neurons in the brain has led to a powerful technique for performing tasks such as pattern recognition (see NEURAL NETWORK). The application of simple rules to many individual objects can result in beautiful and dynamic patterns (see CELLULAR AUTOMATA), as well as ways to model behavior (see ARTIFICIAL LIFE). Here, instead of a system being simplified into a simulation, a simulation can be created in order to see what sort of systems might emerge.

SOFTWARE IMPLEMENTATION

Because of the number of calculations (repeated for a single object and/or applied to many objects) required for an accurate simulation, it is obviously useful for the simulation designer to have as much computer power as possible. Similarly, having many processors or a network of separate computers not only increases the available computing power, but may make it more natural to represent different objects or parts of a system by assigning each to its own processor. (This naturalness goes the other way, too: Simulation techniques can be very important in modeling or predicting the performance of computer networks including the Internet.)

However, it is also important to have programming languages and techniques that are suited for representing the simultaneous changes to objects (see also MULTIPROCESSING). Using object-oriented languages such as Simula or Smalltalk makes it easier to package and manage the data and operations for each object (see OBJECT-ORIENTED PROGRAMMING, SIMULA, and SMALLTALK).

APPLICATIONS

Simulations and simulation techniques are used for a tremendous range of applications today. Besides helping with the understanding of natural systems in physics, chemistry, biology, or engineering, simulation techniques are also applied to human behavior. For example, the behavior of consumers or traders in a stock market can be explored with a simulation based on game theory concepts. Artificial intelligence techniques (such as expert systems) can be used to give the individual "actors" in a simulation more realistic behavior.

Simulations are often used in training. A modern flight simulator, for example, not only simulates the aerodynamics of a plane and its response to the environment and to control inputs, but detailed graphics (and simulated physical motion) can make such training simulations feel very realistic, if not quite to *Star Trek* holodeck standards. Whether for flight, military exercises, or stock trading, simulations can provide a much wider range of experiences in a relatively short time than would be feasible (or safe) using the real-world activity. Simulations can also play an important part in testing software or systems or in predicating the results of business decisions or strategies.

Simulations are also frequently sold as entertainment. Many commercial strategy and role-playing games as well as vehicle simulators contain surprisingly complex simulations that make the games both absorbing and challenging (see COMPUTER GAMES). Such games can also have considerable educational value.

Further Reading

Arsham, Hossein. "Systems Simulation: The Shortest Path from Learning to Applications." http://ubmail.ubalt.edu/~harsham/simulation/sim.htm

"A Collection of Modeling and Simulation Resources on the Internet." http://www.idsia.ch/~andrea/simtools.html

Evans, James R., and David Louis Olson. *Introduction to Simulation and Risk Analysis.* Upper Saddle River, N.J.: Prentice Hall, 2001.

Law, A., and W. Kelton. *Simulation Modeling and Analysis.* 3rd ed. New York: McGraw-Hill, 1999.

Smalltalk

Working during the 1970s at the Xerox Palo Alto Research Laboratory (PARC), computer scientist Alan Kay created many ideas and devices that have found their way into today's personal computers. While designing a proposed notebook computer called the Dynabook, Kay decided to take a new approach to creating its operating system. The result would be a language (and system) called *Smalltalk.*

In developing Smalltalk, Kay built upon two important ideas. The first was that people could master the power of the computer most easily by being able to create, test, and revise programs interactively rather than having to go through the cumbersome process of traditional compilation. Seymour Papert had already created Logo, an interactive, graphics-rich language that proved especially good for teaching children surprisingly sophisticated computer science concepts. The name *Smalltalk* reflects how the first implementation of this language was also designed to be a simple, child-friendly language.

The other key idea Kay used in Smalltalk was object-oriented programming, which had first been developed in the language Simula 67 (see SIMULA and OBJECT-ORIENTED PROGRAMMING). However, instead of simply adding classes and objects to existing language features, Kay designed Smalltalk to be object-oriented from the ground up. Even the data types (such as integer and character) that are used to declare variables in traditional languages become objects in Smalltalk. Users can define new classes that are treated just like the "built-in" ones. There is no need to worry about having to declare variables to be of a certain type before they can be used; in Smalltalk variables can be associated with any object.

To get a program to perform an action, a "message" is sent to an object, which invokes one of the object's defined capabilities (methods). For a very simple example, consider the BASIC assignment statement:

```
Total = Total + 1
```

In a traditional language like BASIC, this is conceptualized as "add 1 to the value stored at the location labeled Total and store the result back in that location." In the object-oriented Smalltalk language, however, the equivalent statement would be:

```
Total <- Total + 1
```

This means "send the message + 1 to the object that is referenced by the variable called Total." This message references the + method, one of the methods that numeric objects "understand." The object therefore adds 1 to its value, and returns that value as a new object, which in turn is now referenced by the variable Total.

A "program" in Smalltalk is simply a collection of objects with the capabilities to carry out whatever processes are required. The objects and their associated variables make up the "workspace," which can be saved to disk periodically.

For the Smalltalk programmer there is no distinction between Smalltalk and the host computer's operating system. The operating system's capabilities (such as file handling) are provided within the Smalltalk system as pre-defined objects. Kay envisaged Smalltalk as a complete environment that could be extended by users who were not necessarily experienced programmers, and he designed its pioneering graphical user interface as a way to make it easy for users to work with the system.

Smalltalk includes a "virtual machine," whose instructions are then implemented in specific code for each major type of computer system. Because of Smalltalk's consistent structure and ability to build everything up from objects, almost all of the Smalltalk system is written in Smalltalk itself, making it easy to transplant to a new computer once the machine-specific details are provided.

Because of its elegance and consistency and its availability on personal computers, by the 1980s Smalltalk had aroused considerable interest. The language has not been widely used for mainstream applications, in part because the mechanisms needed to kept track of classes and inheritance of methods are hard to implement as efficiently as the simpler mechanisms used in traditional languages. The approach of building object-oriented features onto existing languages (as with developing C++ from C) had greater appeal to many because of efficiency and a less steep learning curve.

Nevertheless, the conceptual power of Smalltalk has made it attractive for certain AI and complex simulation projects, and it appeals to those who want a pure object-oriented approach where an application can cleanly mirror a real-world situation. Smalltalk also remains a good choice for teaching programming to children (and others). A version called *Squeak* provides a rich environment of graphics and other functions. Squeak and a number of other Smalltalk implementations are available for free download for a number of different computer systems.

Further Reading

Kamath, Monty. "Monty Kamath's Good Start" [Smalltalk resources] http://www.goodstart.com/

Korienek, Gene, Tom Wrensch, and Doug Dechow. *Squeak: a Quick Trip to ObjectLand.* Boston: Addison-Wesley, 2002.

Lewis, Simon. *The Art and Science of Smalltalk.* Upper Saddle River, N.J.: Prentice Hall, 1995.

"Welcome to Squeak." http://squeak.org/index.html

"www. Smalltalk.org" http://www.smalltalk.org/

smart card

The smart card is the next generation of transaction devices. Magnetically coded credit, debit, and ATM cards have been in use for many years. These cards contain a magnetic strip encoded with a small amount of fixed data to identify the account. All the actual data (such as account balances) is kept in a central server, which is why credit cards must be validated and transactions approved through a phone (modem) link. Some magnetic strip cards such as those used in rapid transit systems are rewritable, so that, for example, the fare for the current ride can be deducted. Telephone cards work the same way. Nevertheless, these cards are essentially passive tokens containing a small amount of data. They have little flexibility.

However, since the mid-1970s it has been possible to put a microprocessor and rewritable memory into a card the size of a standard credit card. These smart cards can store a hundred or more times the data of a magnetic strip card. Further, because they have an onboard computer (see also EMBEDDED SYSTEM), they can interact with a computer at the point of service, exchanging and updating information.

Magnetic strip cards have no way to verify whether they're being used by their legitimate owner, and it is relatively easy for criminals to obtain the equipment for creating counterfeits. With a smart card, the user's PIN can be stored on the card and the terminal can require that the user type in that number to authorize a transaction. Again, the PIN can be validated without reference to a remote server.

Smart Card

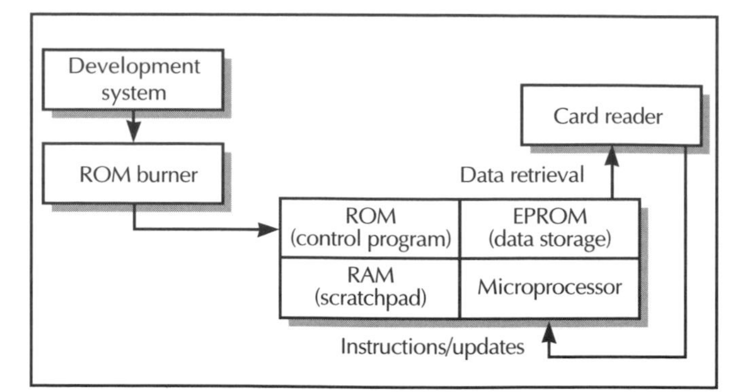

A smart card is "smart" because it doesn't just hold and update data, but has an embedded program and the ability to respond to a variety of requests.

HARDWARE AND PROGRAMMING

Besides the microprocessor and associated circuitry, the smart card contains a small amount of RAM (random access memory) to hold "scratch" data during processing, as well as up to 64K of ROM (read-only memory) containing the card's programming instructions. The program is created on a desktop computer and written to the ROM that is embedded in the card. Finally, the card includes up to 64 KB of EEPROM (Electrically Erasable Programmable Read Only Memory) for holding account balances and other data. This memory is nonvolatile (meaning that no power is needed to maintain it), and can be erased and rewritten by the card reader.

"Contact" cards must be swiped through the reader and are most commonly used in retail, phone, pay TV, or health care applications. "Contactless" cards need only be brought into the proximity of the reader, which communicates with it via radio signals or a low-powered laser beam. Contactless cards are more practical for applications such as collecting bridge tolls.

The card reader (or terminal) at the point of sale contains its own computer, which runs software that requests particular services from the card's program, including providing identifying information and balances, updating balances, and so on.

Microsoft and some other companies have introduced the PC/SC standard for programming smart cards from Windows-based systems. Another standard, Open Card, promises to be compatible with a wide range of platforms and languages, including Java. (Java, after all, descended from a project to develop a language for programming embedded systems.) However, the first commercially available Java-based smart card programming system is based on another standard called *JavaCard*.

APPLICATIONS

The same smart card might also be programmed to handle several different types of transactions, and could function as a combination phone card, ATM card, credit card, and even medical insurance card. Europe has been well ahead of the United States in adopting smart card technology, with both France and Germany beginning during the 1980s to use smart cards for their phone systems. During the 1990s, they began to develop infrastructure for universal use of smart cards for their national health care systems. In 2002, Ontario, Canada, began to replace citizenship papers with a smart card, as well as creating a health services card. Other innovative uses for smart cards include London's city pass for tourists, which can be programmed to provide not only prepaid access but also various bonuses and promotions.

The packing of many services and the associated information onto a smart card raises greater concern that the information might be illicitly captured and abused (see also PRIVACY IN THE DIGITAL AGE). Smart chips about the size of a grain of rice can be implanted beneath the skin. When scanned by hospital personnel, the patient's entire medical record can be retrieved, which can be vital for deciding which drugs to administer in an emergency when the patient is unable to communicate. However, the chips might be surreptitiously scanned by, for example, employers seeking to screen out workers with expensive medical conditions.

Smart cards (such as for digital TV access) have been counterfeited with the aid of sophisticated programs and intrusion equipment. Card makers try to design the card's circuits so that it resists intrusion and tampering and rejects programming attempts from unauthorized equipment.

Another way to prevent unauthorized use is to have the card store identifying information that can be verified through fingerprint scanners or other means (see also BIOMETRICS). Smart ID and access cards are being deployed by more U.S. government agencies to control access to sensitive areas in the wake of the September 11, 2001, terrorist attacks. The newest smart cards, such as one called the Ultra Card, can hold 20 MB of information, allowing the use of much more extensive biometric data. The controversial "national ID card," if implemented, is likely to be a smart card.

A service called GSM (Global System of Mobile Communications) is gradually being adopted. Through the use of a smart card "subscriber identity module," it allows wireless phone users in any participating country to make calls and have the appropriate fees deducted. Further, the GSM can route calls to a person's number automatically to that person's handset, regardless of the country of origin and destination.

There is a very large investment in the current credit card technology, but the flexibility and potential security of smart credit and debit cards is attractive. Already a few issuers have released credit cards with smart chip technology.

Further Reading

Fancher, Carol H. "Smart Cards: as potential applications grow, computers in the wallet are making unobtrusive inroads." *Scientific American,* vol. 275, no. 2, August 1996, 40 ff. Also available on-line at http://www.sciam.com/0896issue/0896fancher.html

Rankl, W. *Smart Card Handbook, 2nd ed.* New York: Wiley, 2000.

Smart Card Alliance. http://www.scia.org/

Wilson, Chuck. *Get Smart: The Emergence of Smart Cards in the United States and their Pivotal Role in Internet Commerce.* Richardson, Tex.: Mullaney Publication Group, 2001.

social impact of computing

In 2001, the Computer Professionals for Social Responsibility (CPSR) held a conference titled "Nurturing the Cybercommons, 1981–2021." Speakers looked back at the amazing explosion in computing and computer-mediated communications in the last two decades of the 20th century. They then turned to the next 20 years, discussing how computing technology offered both the potential for a more robust democracy and the threat that control of information by the few could disenfranchise the many. Their challenge was to create a "cybercommons"—a way in which the benefits of technology could be shared more equitably.

It is sobering to realize just how much happened in only two decades. The computer went from being an esoteric possession of large institutions to a ubiquitous companion of daily work and home life. At the same time, the

One of the earliest hints that computers might have a broader impact on society came in 1952 when Univac's prediction of an Eisenhower election victory was relayed by anchor Walter Cronkite. (PHOTO COURTESY COMPUTER MUSEUM HISTORY CENTER)

Internet, which in 1981 had been a tool for a small number of campus computing departments and government-funded researchers, has burgeoned to a medium that is fast changing the way people buy, learn, and socialize.

The use of computing for specific applications generally brings risks along with benefits (see RISKS OF COMPUTING). Sometimes risks can go beyond a specific program into the interaction between that program and other systems. In the broadest sense, however, computer use as a human activity affects all other human activities. The ultimate infrastructure is not the computer, the software program, or even the entire Internet. Rather, it is society as a whole. There are a several dimensions along which both positive and negative possibilities can be seen.

STRATIFICATION VS. OPPORTUNITY

In the past 30 years, the computer has created millions of new jobs, ranging from webmaster to support technician to Internet café proprietor (see EMPLOYMENT IN THE COMPUTER FIELD). Millions of other jobs have been redefined: The typist has become the word processor, for example. Many other jobs have disappeared or are in the process of disappearing—such as travel agents, who have found themselves under pressure both from do-it-yourself Internet booking and the airlines deciding that they no longer needed to give agents incentives for booking.

In a rapidly changing technological and economic landscape, there are always emerging opportunities. The primacy of computer skills in the job market has, however, exacerbated a trend that was seen throughout the 20th century. New, well-paid jobs increasingly require technical training and skills—expanding the definition of "functional literacy." Throughout the second half of the century, the traditional blue-collar factory jobs that could assure a comfortable living for persons with only a high school education have become increasingly scarce. This has been the result both of increasingly competitive (and lower-priced) overseas labor and factory automation (see ROBOTICS) at home. Essentially, the well-paid tech sector and the low-paid service sector have grown rapidly, while the ground in between has eroded.

Sometimes jobs don't disappear, but are "dumbed down," becoming low-skill and low-paid. Fifty years ago, a store clerk had to be able to count up from the cash register total to the amount of money presented by the customer. Today, computerized cash registers tell the clerk exactly how much change to give (and often dispense the coins automatically). Old-style clerks had to know about prices, discounts, and special offers. Today these are handled automatically by bar codes and smart cards. Although the supermarket clerk still is moderately well paid, the ultimate end of the process is seen in the fast food clerk, who often needs only push buttons with pictures of food on them. He or she is likely to be paid little more than minimum wage. The impact of technology on jobs can even go through several stages. For example, skilled photo technicians have been replaced by the use of automated photo processing equipment. In turn, however, the growing use of digital cameras is reducing the use of film-based photography in general.

The result of these trends may well be increased social stratification. The best jobs in the information age require skills such as programming, systems analysis, or the ability to create multimedia content. However, the opportunity to acquire such skills varies and is not evenly distributed through all groups in the population (see DIGITAL DIVIDE). Although minority groups are now catching up in terms of access to computers at home and in school, disparities in the quality of education will only be magnified as technical skills increasingly correlate with good pay and benefits.

At the same time the computer offers powerful new tools for education (see EDUCATION AND COMPUTERS). Potentially, this could overcome much of the disadvantages of poverty because once the threshold of access is met, the poor person's Internet is much the same as that available to the privileged. However, mastering the necessary skills requires both provision of adequate resources and that prevailing cultural attitudes support intellectual achievement.

DEPENDENCY VS. EMPOWERMENT

Computers have made people more dependent in some ways while empowering them in others. Society is increasingly dependent on computers to operate the systems that provide transportation, power, and communications infrastructure. The "Y2K" scare at the end of the century proved to be unfounded, but it did give people a chance to consider what a major, prolonged failure in the information infrastructure would mean for maintaining the physical necessities of life, the viability of the economy, and the cohesion of society itself (see Y2K PROBLEM). The terrorist attacks of September 11, 2001, brought to greater public awareness the concerns about "cyberterrorism" that experts had been debating since the late 1990s.

At the same time, computers—and particularly the Internet—have give individuals a greater feeling of empowerment in many respects. The savvy Web user now has numerous ways to shop for everything from airline tickets to Viagra pills at prices that reflect disintermediation—the elimination of the middleman. Many people are less inclined to take the word of traditional authority figures (such as doctors) and instead are tapping into the sort of information that had been previously been accessible only to professionals. However, access to information is not the same thing as having the necessary background and skills to evaluate that information. Whether falling victim to an outright scam or simply not fully understanding the consequences of a decision, the Web user finds little in the way of a regulatory safety net. The tension between the high degree of regulation now existing in much of our society and the frontierlike qualities of cyberspace will no doubt be a major theme in the next few decades.

CENTRALIZATION VS. DEMOCRACY

Many observers feel that cyberspace is no longer like the Wild West of the 1870s, but perhaps not yet 1893, when historian Frederick Jackson Turner wrote about the closing of the frontier. It is true that anyone with a PC and a connection can post a webpage that can potentially be viewed by 400 million or so people worldwide. However, the consolidation of media companies (such as AOL-Time Warner) has continued, and it is these companies that provide not only most of the content on TV but the most-viewed websites as well. It is feared that such concentration will diminish the diversity of viewpoints that the average Web user will encounter, making the new medium more like broadcast and cable TV. There may still be many "channels," but only the most dedicated and curious would be willing to seek them out.

There are many activist groups such as the Electronic Frontier Foundation and the Center for Democracy and Technology that seek to protect citizens' privacy on-line while requiring greater disclosure of how corporations intend to use information they gather from website visitors (see PRIVACY IN THE DIGITAL AGE). Web users also have access to sites that monitor campaign contributions and other exercises of political influence.

ISOLATION VS. COMMUNITY

There are many on-line facilities that allow individuals and groups to maintain an ongoing dialog (see CHAT, ON-LINE and CONFERENCING SYSTEMS). Students at a school in Iowa can now collaborate with their counterparts in Kenya or Thailand on projects such as measuring global environmental conditions. Senior citizens who have become isolated from family members and lack access to transportation can find social outlets on-line.

However, critics such as Clifford Stoll believe that the growth of on-line communication (see also VIRTUAL COMMUNITY) may be leading to a further erosion of physical communities and a sense of neighborhood. For many years, it has been observed that people in suburbia often don't know their neighbors: The car and the phone let them form relationships and "communities" without much regard to geography. It is possible that the growth in on-line communities will accelerate this effect. Further, with people being able to order an increasing array of goods and services on-line, might the market plaza and its modern counterpart the mega mall become less of a meeting place? Even the proposal to allow people to vote on-line might promote democracy at the expense of the contact between citizens and the shared rituals that give people a stake in the larger community.

Thus, computer technology offers many opposing prospects and visions. The social changes that are cascading from information and communications technology are likely to be at least as pervasive in the early 21st century as the those wrought by the telephone, automobile, and television were in the 20th.

Further Reading

Association for Computing Machinery. Special Interest Group on Computers and Society. http://www.acm.org/sigcas/

Baase, S. *A Gift of Fire: Social, Legal and Ethical Issues of Computing*. Upper Saddle River, N.J.: Prentice Hall, 1997.

Center for Democracy and Technology. http://www.cdt.org

Computer Professionals for Social Responsibility (CPSR). http://www.cpsr.org/

Electronic Frontier Foundation. http://www.eff.org

Johnson, D. *Computer Ethics*. Upper Saddle River, N.J.: Prentice Hall, 1994.

Miller, S. *Civilizing Cyberspace: Policy, Power and the Information Superhighway*. New York: ACM Press, 1996.

Stoll, C. Silicon *Snake Oil: Second Thoughts on the Information Highway*. New York: Doubleday, 1995.

software agent

Most software is operated by users giving it commands to perform specific, short-duration tasks. For example, a user might have a word processor change a word's typestyle to bold, or reformat a page with narrower margins. On the other hand, a person might give a human assistant higher-level instructions for an ongoing activity: for example, "Start a clippings file on the new global trade treaty and how it affects our industry."

In recent years, however, computer scientists and developers have created software that can follow instructions more like those given to the human assistant than those given to the word processor. These programs are variously called software agents, intelligent agents, or bots (short for "robots"). Some consumers have already used software agents to comb the Web for them, looking, for example, for the best on-line price for a certain model of digital camera. Agent programs can also assist with on-line auctions, travel planning and booking, and filtering e-mail to remove unwanted "spam" or to direct inquiries to appropriate sales or technical support personnel.

Practical agents or bots can be quite effective, but they are relatively inflexible and able to cope only with narrowly defined tasks. A travel planning agent may be able to interface with on-line reservations systems and book airline tickets, for example. However, the agent is unlikely to be able to recognize that a recent upsurge in civil strife suggests that travel to that particular country is not advisable.

Researchers have, however, been working on a variety of more open-ended agents that, while not demonstrably "intelligent," do appear to behave intelligently. The first program that was able to create a humanlike conversation was ELIZA. Written in the mid-1960s by Joseph Weizenbaum, ELIZA simulated a conversation with a "nondirective" psychotherapist. More recently, Internet "chatterbots" such as one called Julia have been able to carry on apparently intelligent conversations in IRC (Internet Relay Chat) rooms, complete with flirting. Other "social bots" have served as players in on-line games.

Chatterbots are effective because they can mirror human social conventions and because much of casual human conversation contains stereotyped phrases or clichés that can be easily imitated. Ideally, however, one would want bots to be able to combine the ability to carry out practical tasks with a more general intelligence and a more "sociable" interface. This requires that the bot have an extensive knowledge base (see KNOWLEDGE REPRESENTATION) and a greater ability to understand human language (see LINGUISTICS AND COMPUTING). Small strides have been made in providing on-line help systems that can deal with natural language questions, as well as being able to interactively help users step through a particular tasks.

Agents or bots have also suggested a new paradigm for organizing programs. Currently, the most widely accepted paradigm treats a program as a collection of objects with defined capabilities that respond to "messages" asking for services (see OBJECT-ORIENTED PROGRAMMING). A move to "agent-oriented programming" would carry this evolution a step further. Such a program would not simply have objects that wait passively for requests. Rather, it would have multiple agents that are given ongoing tasks, priorities, or goals. One approach is to allow the agents to negotiate with one another or to put tasks "up for bid," letting agents that have the appropriate ability contract to perform the task. With each task having a certain amount of "money" (ultimately representing resources) available, the negotiation model would ideally result in the most efficient utilization of resources.

If Marvin Minsky's (see MINSKY, MARVIN) "society of mind" theory is correct and the human brain actually contains many cooperating "agents," then it is possible that systems of competing and/or cooperating agents might eventually allow for the emergence of a true artificial intelligence.

In the future, agents are likely to become more capable of understanding and carrying out high-level requests while enjoying a great deal of autonomy. Some possible application areas include data mining, marketing and survey research, intelligent Web searching, security, and intelligence gathering. However, autonomy may cause problems if agents get out of control or exhibit viruslike behavior.

Further Reading

"Bot Spot." http://www.botspot.com/

Davis, R., and R. G. Smith. "Negotiation as a Metaphor for Distributed Problem Solving." *Artificial Intelligence,* vol. 20, 1983.

Janson, Sverker. "Agent-Based Systems: Web Resources" Swedish Institute of Computer Science. http://www.agentbase.com/survey-resources.html

Leonard, Andrew. *Bots: The Origin of a New Species.* San Francisco: Hardwired, 1997.

Weiss, Gerhard, ed. *Multiagent Systems: A Modern Approach to Distributed Artificial Intelligence.* Cambridge, Mass.: MIT Press, 1999.

software engineering

By the late 1960s, large computer programs (such as the operating systems for mainframe computers) consisted of thousands of lines of computer code. In what became known as "the software crisis," managers of software development were facing great uncertainty about both program development schedules and the reliability of the resulting programs.

Programming had started out in the 1940s as an offshoot of mathematics, just as the building of computers was an offshoot of electrical or electronic engineering. Increasingly, however, programmers were searching for a new professional identity. What paradigm was truly appropriate? Should programmers strive to be more like mathematicians, seeking to rigorously prove the correctness of their programs? On the other hand, many programmers thought of their work as a craft, performed using individual experience and intuition, and not easily subject to standardization. Between the two poles of mathematics (or science) and craft came another possibility: engineering.

The concept of software engineering proved to be attractive. Mathematics (and science in general) are usually carried on without being immediately and directly applied to creating a particular device or process. Outside of research programs, however, computer applications were written to perform real-world tasks (such as flight control) that have real-world consequences. Thus, although the notation of a computer program resembles that of mathematics, the operation of a program more nearly resembles that of complex mechanical systems created by engineers. By attaching the label of engineering to what programmers do, advocates of software engineering hoped to develop a body of practices and standards comparable in some way to those used in engineering. Some critics, however, believe that this paradigm is inappropriate, either because they believe one should strive for the greater rigor of science or out of a preference for individual craft over standardization.

PROGRAMMING PRACTICES

One of the most pervasive contributions to software engineering has been in computer language design and coding practices. At about the same time that the concept of software engineering was being promulgated, computer scientists were advocating better facilities for defining and structuring programs (see STRUCTURED PROGRAMMING). These included well-defined control structures (see BRANCHING STATEMENTS and LOOP), use of built-in and user-defined kinds of data (see DATA TYPES), and the breaking of programs into more manageable modules (see PROCEDURES AND FUNCTIONS).

The next paradigm came in the late 1970s and had taken hold by the late 1980s (see OBJECT-ORIENTED PROGRAMMING). The ability to "hide" details of function within objects that mirrored those in the real world provided a further way to make complex programs easier to understand and maintain. The growing use of well-tested collections of procedures or objects (see LIBRARY, PROGRAM) has been essential for keeping up with the growing complexity of application programs.

Software engineers are also concerned with developing tools that will better manage the programming process and help ensure that standards are being followed (see PROGRAMMING ENVIRONMENT). The use of CASE (Computer-Aided Software Engineering) tools such as sophisticated program editors, documentation generators, class diagrammers, and version control systems has also steadily increased. Today many of these tools are available even on modest desktop computing environments (see CASE).

THE PROGRAM DEVELOPMENT PROCESS

Perhaps the most important task for software engineering has been seeking to define and improve the process by which programs are developed. In general, the overall steps in developing a program are:

- Detailed specification of what the program will be required to do. This can include developing a prototype and getting user's reaction to it.
- Creation of a suitable program architecture—algorithm(s) and the data types, objects, or other structures needed to implement them (see ALGORITHM).
- Coding—writing the program language statements that implement the structure.
- Verification and testing of the program using realistic data and field testing (see QUALITY ASSURANCE, SOFTWARE).
- Maintenance, or the correction of errors and adding of requested minor features (short of creating a new version of the program).

There are a number of competing ways in which to view this software development cycle. The "iterative" or "evolutionary" approach sees software development as a linear process of progress through the above steps.

The "spiral" approach, on the other hand, sees the steps of planning, risk analysis, development, and evaluation being applied repeatedly, until the risk analysis and evaluation phases result in a go/no go to finish the project.

The most commonly used approach is called *waterfall*. In it the results (output) of each stage become the input of the next stage. This approach is easiest for scheduling (see PROJECT MANAGEMENT SOFTWARE), since each stage is strictly dependent on its predecessor. However, some advocates of this approach have included the ability for a given stage to feed back to the preceding stage if necessary. For example, a problem found in implementation (coding) may require revisiting the preceding design phase.

DEVELOPING SOFTWARE ENGINEERING STANDARDS

Two organizations have become prominent in the effort to promote software engineering. The federally funded Software Engineering Institute (SEI) at Carnegie Mellon

Software Engineering (1)

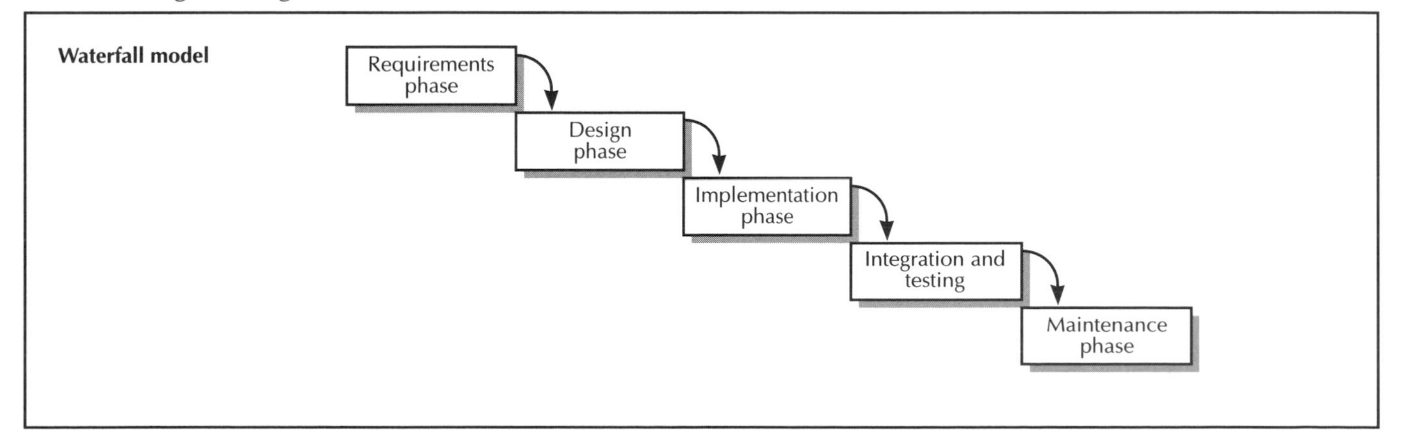

The Spiral Model visualizes software development as a process of planning, risk analysis, development, and evaluation. The cycle repeats until the project is developed to its full scope.

Software Engineering (2)

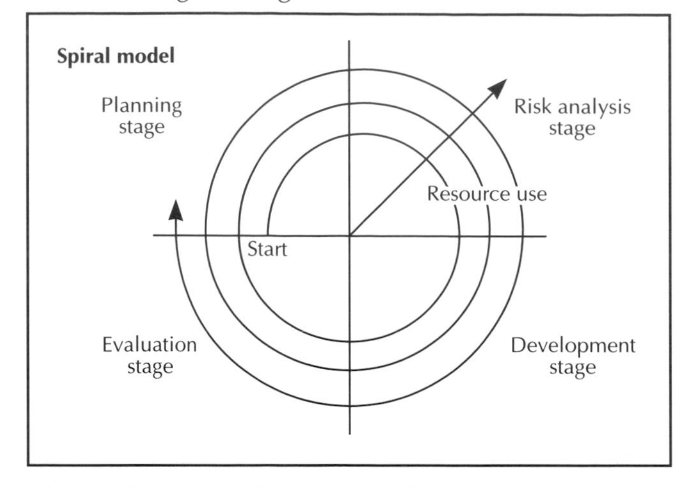

The Waterfall or Cascade model sees software development as a more linear process going through the requirements, design, implementation, integration and testing, and maintenance phases. The results of each phase cascade down into the next.

University was established in 1984. Its mission statement is to:

1. Accelerate the introduction and widespread use of high-payoff software engineering practices and technology by identifying, evaluating, and maturing promising or underused technology and practices.
2. Maintain a long-term competency in software engineering and technology transition.
3. Enable industry and government organizations to make measured improvements in their software engineering practices by working with them directly.
4. Foster the adoption and sustained use of standards of excellence for software engineering practice.

Since 1993, the IEEE Computer Society and ACM Steering Committee for the Establishment of Software Engineering as a Profession has been pursuing a set of goals that are largely complementary to those of the SEI:

1. Adopt Standard Definitions
2. Define Required Body of Knowledge and Recommended Practices (In electrical engineering, for example, electromagnetic theory is part of the body of knowledge while the National Electrical Safety Code is a recommended practice.)
3. Define Ethical Standards
4. Define Educational Curricula for (a) undergraduate, (b) graduate (MS), and (c) continuing education (for retraining and migration).

Further Reading

Brooks, Frederick. *The Mythical Man-Month: Essays on Software Engineering*. Reading, Mass.: Addison-Wesley, 1995.

Carnegie Mellon Software Engineering Institute. http://www.sei.cmu.edu/

Joint IEEE Computer Society and ACM Steering Committee for the Establishment of Software Engineering as a Profession. http://www.computer.org/tab/seprof/index.htm

Shaw, M., and D. Garlan. *Software Architecture: Perspectives on an Emerging Discipline*. Upper Saddle River, N.J.: Prentice Hall, 1996.

Sommerville, Ian. *Software Engineering*. 6th ed. Reading, Mass.: Addison-Wesley, 2000.

sorting and searching

Because they are so fundamental to maintaining databases, the operations of sorting (putting data records in order) and searching (finding a desired record) have received extensive attention from computer scientists. A variety of different and quite interesting sorting methods have been devised (see ALGORITHM).

Any application that involves keeping track of a significant number of data records will have to keep them sorted in some way. After all, if records are simply inserted as they arrive without any attempt at order, the time it will take to find a given record will, on the average, be the time it would take to search through half the records in the database. While this might not matter for a few hundred records on a fast modern computer, it would be quite unacceptable for databases that might have millions of records.

SORTING CONSIDERATIONS

While some sorting algorithms are better than others in almost all cases, there are basic considerations for choosing an approach to sorting. The most obvious is how fast the algorithm can sort the number of records the application is likely to encounter. However, it is also necessary to consider whether the speed of the sort increases steadily (linearly) as the number of records increases, or it becomes proportionately worse. That is, if an algorithm can sort a thousand records in two seconds, will it take 20 seconds for 10,000 records, or perhaps five minutes?

In most cases one assumes that the records to be sorted are in more or less random order, but what happens if the records to be sorted are already partly sorted . . . or almost completely sorted? Some algorithms can take advantage of the partial sorting and complete the job far more quickly than otherwise. Other algorithms may slow down drastically or even produce errors under those conditions.

The range or variation in the key (the data field by which records are being sorted) may also play a role. In some cases if the keys are close together, some algorithms may be able to take advantage of that fact.

Finally, the available computer resources must be considered. Today many desktop PCs have 128 MB or more of main memory (RAM), while servers or mainframes may have several gigabytes (GB). If the database is small enough that it can be entirely kept in main memory, sorting is fast because any record can be accessed in the same amount of time at electronic speeds. If, however, part of the database must be kept in secondary storage (such as hard drives), the sorting program will have to be designed so that it reads a number of records from the hard drive in a single reading operation, in order to avoid the overhead of repeated disk operations. Most likely the individual batches will be read from the disk, sorted in memory, written back to disk, and then merged to sort the whole database.

SORTING ALGORITHMS

There are numerous sorting algorithms ranging from the easy-to-understand to the commonly used to the exotic and quirky. Only the highlights can be covered here; see Further Reading for sources for more detailed discussions.

Sorting and Searching (1)

9	**7**	3	8	6	2
7	**9**	3	8	6	2

7	**9**	**3**	8	6	2
7	**3**	**9**	8	6	2

7	3	**9**	**8**	6	2
7	3	**8**	**9**	6	2

7	3	8	**9**	**6**	2
7	3	8	**6**	**9**	6

7	3	8	9	**6**	**2**
7	3	8	9	**2**	**6**

Boldface indicates two numbers being compared
Bubble sort

In a bubble sort, pairs of adjacent numbers are compared and switched if they are out of order. Eventually, the lowest values (such as 2 in this case) will "bubble up" to the front of the list.

SELECTION SORT

The simplest and least efficient kind of sort is called the *selection sort.* Rather like a bridge player organizing a hand, the selection sort involves finding the record with the lowest key and swapping it with the first record, then scanning back through for the next lowest key and swapping it with the second record, and so on until all the records are sorted. While this uses memory very efficiently (since the records are sorted in place), it is not only slow, but also gets worse fast. That is, the time taken to sort n records is proportional to n^2.

The selection approach suffers because on each pass the sort determines not only the record with the lowest key but the one with the next lowest key. However, that information is not retained. The heapsort, invented by John Williams in 1964, uses a binary tree to store a heap of sorted records (see TREE and HEAP). Once the heap is built, the tree nodes can be used to store record numbers in a corresponding array that will represent the sorted database. The heapsort is efficient because no records are physically moved, and the only memory needed is for the heap and array. The heapsort is generally considered the fastest and most reliable general-purpose sorting algorithm, with a maximum running time of log n.

BUBBLE SORT

The bubble sort is based on making comparisons and swaps. It makes the most convenient comparison possible: each record with its neighbor. The algorithm looks at the first two records. If the second has a lower key than the first, the records are swapped. The procedure continues with the second and third records, then the third and fourth, and so on through all the records, swapping pairs of adjacent records whenever they are out of order. After one pass the record with highest key will have "bubbled up to" the end of the list. The procedure is then repeated for all but the last record until the two highest records are at the end, and so on until all the records are sorted. Unfortunately, the number of comparisons and swaps that must be made makes the bubble sort as slow as the selection sort.

QUICKSORT

The quicksort improves on the basic bubble sort by first choosing a record with a key approximately midway between the lowest and highest. This key is called the *pivot.* The records are then moved to the left of the pivot if they are lower than it, and to the right if higher (that is, the records are divided into two *partitions*). The process is then repeated to split the left side with a new pivot, and then the right side likewise. This is continued until the partition size is one, and the records are now all sorted. (Because of this repeated partitioning, quicksort is usually implemented using a procedure that calls itself repeatedly—see RECURSION.)

Devised by C. A. R. Hoare in 1962, quicksort is much faster than the bubble sort because records are moved over greater distances in a single operation rather than simply being exchanged with their neighbors. Assuming an appropriate initial pivot value is chosen, running time is proportional to the logarithm of n rather than to the square of n. The difference becomes dramatic as the size of the database increases.

Sorting and Searching (2)

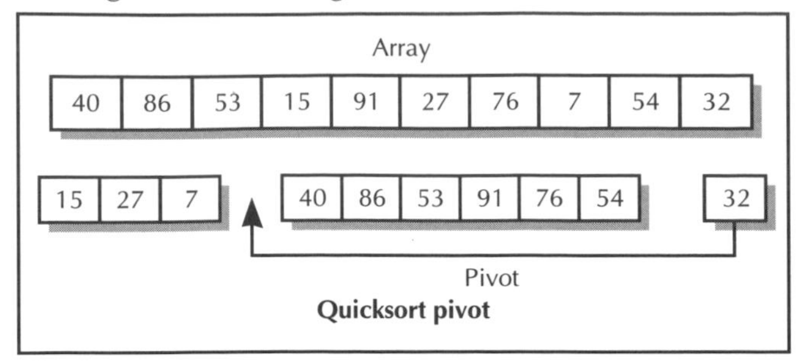

The Quicksort uses a value called the pivot to partition the list into two smaller lists. This process is repeated until the list has been divided and "conquered" (sorted).

INSERTION SORT

The bubble sort and quicksort are designed to work with records that are in random order. However, in many applications a database grows slowly over time. At any given time the existing database is already sorted, so it hardly makes sense to have to resort the whole database each time a new record is added.

Instead, an *insertion sort* can be used. In its simplest form, the algorithm looks sequentially through the sorted records until it finds the first record whose key is higher than that of the new record. The new record can then be inserted just before that record, much like the way a bridge player might organize the cards in a hand. (Since inserting a record and physically moving all the higher records up in memory can be time-consuming, a linked list of key values and associated record number is often used instead. (See LIST PROCESSING.) That way only the links need to be changed rather than any records being moved.

The insertion sort was improved by Donald L. Shell in 1959. His "shellsort" takes a recursive approach (like that in the quicksort), and applies the insertion sort procedure to successively smaller partitions.

Another improvement on the insertion sort is the mergesort. As the name implies, this approach begins by creating two small lists of sorted records (using a simple comparison algorithm), then merging the lists into longer lists. Merging is accomplished by looking at the two keys on the top of two lists and taking whichever is lowest until the lists are exhausted. The merge sort also lends itself to a recursive approach, and it is comparable in speed and stability to the heapsort.

HASH SORTS

All of the sorting algorithms discussed so far rely upon some form of comparison. However, it also possible to sort records by calculating their relative positions or distribution (see HASHING). In its simplest form, an array can be created whose range of indexes is equal to 1 to the maximum possible key value. Each key is then stored in the index position equal to its value (that is, a record with a key of 2314 would be stored in the array at position Array[2314]. This procedure works well, but only if the keys are all integers, the range is small enough to fit in memory, and there are no duplicate keys (since a duplicate would in effect overwrite the record already stored in that position).

A more practical approach is to use a formula (hash function) that should create a unique hash value for each key. The function must be chosen to minimize "collisions" where two keys end up with the same hash value, which creates the same problem as with duplicate keys. A hash sort is quite efficient within those constraints.

SEARCHING

Once one has a database (sorted or not), the next question is how to search for records in it. As with sorting, there are a variety of approaches to searching. The simplest and least efficient is the linear search. Like the selection sort, the linear search simply goes through the database records sequentially until it finds a matching key or reaches the end without a "hit." If there is indeed a matching record, on the average it will be found in half the time needed to process the whole database.

In most real applications the database will have been sorted using one of the methods discussed earlier. Here, the basic approach is to do a binary search. First the key in the middle record in the database is examined. The key is compared with the search key. If the search key is smaller, then any matching key must be in the first half of the database. Otherwise, it must be in the second half (unless, of course, it happens to *be* the matching key). The process is then repeated. That is, if the key is somewhere in the first half, that portion of the list is in turn split in half and its middle value is examined, and the comparison to the search key is made. Thus, the area in which the matching key must be found is progressively cut in half until either the matching key is found or there are no more records to check. Because of the power of successive division, the binary search is very quick, and doubling the size of the database means adding only one more comparison on the average.

Sometimes knowledge about the distribution of keys in the database can be used to improve even the binary search. For example, if keys are alphabetical and the search key begins with S, it is likely to be faster to pick a starting point near the end of the list rather than from the middle. A binary tree (see TREE) can be constructed from the keys in a database in order to analyze the most likely starting points for a search.

Finally, hashing (as previously discussed) can be used to quickly calculate the expected location of the desired record, provided there are no collisions.

Further Reading

Cormen, Thomas H., Charles E. Leiserson, and Ronald L. Rivest. *Introduction to Algorithms.* 2nd ed. Cambridge, Mass.: MIT Press, 2001.

Gosling, James, Jason Harrison, and Jim Boritz. Animated Sorting Algorithms. http://www.cs.ubc.ca/spider/harrison/Java/sorting-demo.html

Knuth, Donald E. *The Art of Computer Programming, vol. III: Sorting and Searching.* 2nd ed. Reading, Mass.: Addison-Wesley, 1997.

sound file formats

There are a number of ways that sound can be sampled, stored, or generated digitally (see MUSIC, COMPUTER). Here we will look at some of the most popular sound file formats.

WAV

The WAV (wave) file format is specific to Microsoft Windows. It essentially stores the raw sample data that represents the digitized audio content, including information about the sampling rate (which in turns affects the sound quality). Since WAV files are not compressed, they can consume considerable disk space.

AIFF

AIFF stands for Audio Interchange File Format, and is specific to the Apple Macintosh and to Silicon Graphics (SGI) platforms. Like WAV, it stores actual sound sample data. A variant, AIFF-C, can store compressed sound.

AU

The AU (audio) file format was developed by Sun Microsystems and is used mainly on UNIX systems, and also in Java programming.

MIDI

MIDI stands for Musical Instrument Digital Interface. Unlike most other sound formats, MIDI files don't represent sampled sound data. Rather, they represent virtual musical instruments that synthesize sound according to complex algorithms that attempt to mirror the acoustic characteristics of real pianos, guitars, or other instruments. Since MIDI is like a "score" for the virtual instruments rather than storing the sounds, it is much more compact than sampled sound formats. MIDI is generally used for music composition rather than casual listening.

MP3

MP3 is actually a component of the MPEG (Moving Picture Expert Group) multimedia standard, and stands for MPEG-1 Audio Layer 3. It is now the most popular sound format, using compression to provide a balance of sound quality and compactness that is comparable to that of standard audio CDs and suitable for most listeners. The compression algorithm relies upon psychoacoustics (the study of how people perceive the components of sound) to identify frequencies that humans can't hear, and thus may be safely discarded. The digitized sound on a CD is compressed up to 1/12 or less of its original size, so a 630 MB CD becomes about 50 MB in MP3 files.

Since most PC users now have hard drives rated in the tens of gigabytes (GB), it is easy to store an extensive music library in MP3 form. Most PCs now come with software that can play MP3 files (such as Windows Media Player), and there are also free and shareware programs from a variety of sources, as well as plug-ins for playing sound files directly from the Web browser.

Since MP3 is much more compact than "raw" CD format, users with inexpensive CD-RW drives can "burn" large amounts of music in MP3 form onto a single CD. This is typically done using software that "rips" the raw tracks from an audio CD and converts them to an MP3 file, which can then be stored on the PC's hard drive.

MP3 CDs can't be played by regular CD players, but special portable MP3 CD players are available. Alternatively, one can obtain a portable MP3 player that stores the MP3 data in memory rather than using CDs. These players provide for music on the go without the skipping caused by jiggling conventional portable CD players. It is easy to download MP3 files from the PC to the portable player. However, memory capacity (and thus the number of songs that can be stored) may be more limited.

The ease with which CDs can be turned into MP3 files has spawned widespread unlicensed distribution of copyrighted music, such as by file-sharing services like Napster, which has been largely shut down by legal action. This doesn't affect the usefulness of the MP3 format itself, however, and record companies are in the process of developing ways to distribute licensed MP3s.

Further Reading

Johnson, Dave, and Rick Broida. *How To Do Everything with MP3 and Digital Music.* New York: McGraw Hill Professional, 2001.

"Sound Files." http://www.webteacher.org/winexp/sounds/sounds.html

"The Ultimate MP3 Source." http://users.bestweb.net/~warrior/what_mp3.html

Young, Robert. *The MIDI Files.* 2nd ed. New York: Prentice Hall, 2001.

"ZD Webopedia." http://www.webopedia.com

space exploration and computers

It might have been barely possible to put a satellite (or person) in orbit without the use of computers, but any more extensive exploration of space requires many types of computer applications.

HUMAN SPACE EXPLORATION

Flying to the Moon required precisely calculated and controlled "burns" to inject the *Apollo* spacecraft from orbit into its arcing trajectory to the Moon. The detachable Lunar Excursion Module (LEM) also had a computer on board (roughly comparable in power to something found in today's programmable calculators). Although the pilot controlled the final landing manually, the computer interpreted radar data to fix the lander's position, monitored fuel consumption, and provided other key data.

The Space Shuttle, the most complex vehicle ever built by human beings, has five onboard computer systems that control flight maneuvers (including rendezvous and docking operations), monitor and control environmental conditions, keep track of fuel, batteries, life support, and other consumables, and provide many other functions to support the crew's tasks and experiments.

AUTOMATED SPACE EXPLORATION

Thus far, human explorers have flown no farther than the Moon. However, in the last 40 years an extensive survey of most of the solar system has been carried out by robot (that is to say, computerized) probes and landers. These probes have landed on Mars and visited every planet except Pluto, as well as making close approaches to asteroids and comets.

The control computer aboard a space probe has several jobs. It must keep the probe oriented in such a way that its solar panels can receive energy from the Sun, as well as keeping an antenna pointed toward Earth so it can receive commands and return data from the probe's scientific instruments.

Starting with *Voyager* 2 (a probe that is still returning data from more than 7 billion miles from Earth), space probe computers have been more autonomous, able to make attitude corrections and course corrections as needed. The onboard computer can even be reprogrammed with new instructions sent from Earth. Space probes have returned incredibly detailed pictures of the surface of the Moon and planets, preparing the way for human missions or robot landers.

Landers reach a fixed point on a planetary surface and transmit photographs, temperature, radiation, and other readings. Probes can survive only for minutes on the hostile surface of Venus, but have functioned for many months on Mars. In a remarkably ambitious mission beginning in 1976, the two Viking Mars landers were able to carry out experiments on soil samples in an unsuccessful attempt to find evidence of life while a third probe mapped the planet's surface from orbit. Besides demonstrating remarkable reliability (*Viking* 2 was still operating in 1982 when it was accidentally turned off by a remote command), the mission also demonstrated the ability to coordinate surface and orbital exploration.

In July 1997, the *Mars Pathfinder* probe landed on the red planet, rolling and bouncing to a stop inside a sort of giant airbag. After deflating, the *Pathfinder* base station deployed the *Sojourner* mobile robot. This vehicle (see ROBOTICS) was controlled by operators on Earth, but because of the 10–15-minute time delay in signals arriving from Earth, the *Sojourner* had some autonomous ability to avoid collisions or other hazards. The onboard computer also had to compress and transmit images and other data. Much larger, longer-range rovers are scheduled to arrive later in the 2000 decade.

The need to build compact computers and other electronics for space exploration helped spur the development of techniques now found in garden-variety consumer electronics. Space computers are also important for demonstrating the reliability and robustness that is necessary for applications on Earth (such as in the military). Space electronics must be shielded and "hardened" to withstand the intense solar radiation, extreme changes in temperature, and electromagnetic fluxes or surges. Redundancy can be used where possible, but weight is always at a high premium. With the exception of certain satellites and the *Hubble Space Telescope*, space computers cannot receive on-site service visits.

Because of the high cost and risk of maintaining human life for long periods in space, it is likely that robotic probes and rovers will remain the main means for space exploration in the early 21st century.

Further Reading

Freudenrich, Craig. "How Space Shuttles Work." http://www.howstuffworks.com/space-shuttle.htm

Space probes and computers have now explored all of the solar system except for Pluto. This computer-enhanced image shows the fine structure of Saturn's rings. (NASA PHOTO)

Graham, John F. *Space Exploration: from Talisman of the Past to Gateway for the Future.* Available on-line at http://www.space.edu/projects/book/
Hall, Eldon. C. *Journey to the Moon: the Story of the Apollo Guidance Computer.* Reston, Va.: American Institute of Aeronautics, 1996.
Matloff, Gregory L. *Deep-Space Probes.* New York: Springer-Verlag, 2000.
"Mars Pathfinder Project Information." http://nssdc.gsfc.nasa.gov/planetary/mesur.html

speech recognition and synthesis

The possibility that computers could use spoken language entered popular culture with Hal 2001, the self-aware talking computer in the film *2001: A Space Odyssey.* On a practical level, the ability of users to communicate using speech rather than a keyboard would bring many advantages, such as mobile, hands-free computing and greater independence for disabled persons. Considerable progress has been made in this technology since Hal "talked" in 1968.

Speech recognition begins with digitizing the speech sounds and converting them into a standard, compact representation. The analysis can be based on matching the input sounds to one of about 200 "spectral equivalence classes" from which the representation can be created. Alternatively, algorithms can use data based on modeling how the human vocal tract produces speech sounds, and extract key features that then become the speech representation. Neural networks can also be "trained" to recognize speech features (see NEURAL NETWORK). The latter two approaches are potentially more flexible but also considerably more difficult, and tend to be used in research rather than in commercial voice recognition systems.

Whichever form of representation is used, it must then be matched to the characteristics of particular words or phonemes, usually with the aid of sophisticated statistical and time-fitting techniques. The simplest systems work on a word level, which may suffice if the system is restricted to a simple vocabulary and the user speaks slowly and distinctly enough. Such systems usually require that the user "train" the system by speaking selected words and phrases. The user can then control the system with a set of voice commands.

Creating a system that can handle the full range of language is much more difficult. This kind of system breaks the language down into phonemes, its basic sound constituents (English has about 40 phonemes). The system includes a stored dictionary of phoneme sequences and the corresponding words. However, "understanding" which words are being spoken is more than a matter of matching phoneme sequences to a dictionary. For one thing, the sound of the first or last phoneme in a word can change depending on the phoneme in an adjacent word.

Once the speech has been recognized, it can be converted to character data (see CHARACTERS AND STRINGS) and treated as though the text had been entered from the keyboard. This means, for example, that a user could dictate text to be placed in a word processor document as well as using voice commands to perform tasks such as formatting text. (Special words can be used to introduce and end commands.)

Voice control and dictation have been offered commercially by such companies as Dragon Systems and Kurzweil. Microsoft now includes speech recognition and synthesis facilities in the latest version of its popular office suite, Office XP.

VOICE SYNTHESIS

The other part of the speech equation is the ability to have the computer turn character codes into spoken words. The most primitive approach is to digitally record appropriate spoken words or phrases, which can then be replayed when speech is desired. Naturally, what is spoken is limited to what is available in the recorded library, although the words and phrases can be combined in various ways. Since the combinations lack the natural transitions that speakers use, the result sounds "mechanical." Common applications include automated announcements in train stations or in prompts for voicemail systems.

To produce a synthesizer that can "speak" any natural language text, the system must have a dictionary that gives the phonemes found in each word. The 40 or so different phonemes can then be digitally recorded and the system would then identify the phonemes in each word and play them to create speech. While this solves the limited vocabulary problem, the synthesized speech is rather unnatural and hard to understand. This is because, as noted earlier, the way phonemes are sounded changes under the influence of adjacent phonemes, and these nuances are lacking in a simple phoneme playback.

More sophisticated voice synthesis systems record natural speech and identify all the possible combinations of half of a phoneme and half of an adjacent phoneme. That way the possible transition sounds are also recorded, and the resulting speech sounds considerably more natural. The drawback is that more memory and processing power are required, but these commodities are becoming increasingly cheaper.

Speech recognition and synthesis technology has made only slow inroads into the computing mainstream, such as office applications. Given the costs of hardware, software, and training, the keyboard remains more productive and cost-effective for most applications. However, voice technology does have a growing number of specialty uses, including security and access systems, speech synthesis for disabled persons who cannot see or speak, and enabling service robots to interact with people in the environment. Speech technology has also been a long-standing topic in artificial intelligence and robotics research.

Further Reading

"Commercial Speech Recognition." http://www.tiac.net/users/rwilcox/speech.html

Usenet Groups: comp.speech.users, comp.speech.research

Huang, Xuedong [and others]. *Spoken Language Processing: A Guide to Theory, Algorithm and System Development.* Upper Saddle River, N.J.: Prentice Hall, 2001.

Morgan, Nelson, and Ben Gold. *Speech and Audio Signal Processing: Processing and Perception of Speech and Music.* New York: Wiley, 1999.

Web-based speech synthesis demo. http://wwwtios.cs.utwente.nl/say/form/

spreadsheet

With the possible exception of word processing, no personal computer application caught the imagination of the business world as quickly as did the spreadsheet, which first appeared as Daniel Bricklin's *VisiCalc* in 1979. VisiCalc quickly became the "killer app"—the application that could justify corporate purchases of Apple II computers. When the IBM PC began to dominate the office computing industry in the mid-1980s, it had a new spreadsheet, Lotus 1-2-3. By the end of the decade, however, Microsoft's Excel spreadsheet had come to the forefront, running on Microsoft Windows. It remains the market leader today.

HOW SPREADSHEETS WORK

A spreadsheet is basically a tabular arrangement of rows and columns that define many individual cells. Typically, the columns are lettered (A to Z, then AA, AB, and so on) while the rows are numbered. A particular cell is referenced using its column and row coordinates; thus A1 is the cell in the upper left corner of the spreadsheet.

Any cell can contain a numeric value, a formula, or a label (such as for giving a title to the spreadsheet or some section of it). Formulas reference the values in other cell locations. For example, if the formula =SUM (A1:B1) is inserted into cell C1, when the spreadsheet is calculated the sum of the contents of cells A1 and B1 will be inserted into C1. Modern spreadsheets let users select from a variety of functions (predefined formulas) for such things as interest or rates of return. Instead of having to type the individual coordinates of cells to be used in a formula, he or she can simply click on or drag across the cells to select them. Formulas can also include conditional evaluation (similar to the If statements found in programming languages—see branching statements).

Spreadsheets provide a variety of "housekeeping" commands that can be used for functions such as copying or moving a range of cells or "cloning" a cell's value into a range of cells. Large spreadsheets can be broken down into multiple linked spreadsheets to make it easier to understand and maintain.

Macros offer a powerful way to simplify and automate spreadsheet operations. A macro is essentially a set of programmed instructions to be carried out by the spreadsheet (see MACRO). One use of macros is to carry out complicated procedures by taking advantage of features similar to those found in programming languages such as Visual Basic. Macros can also be used to automate data entry into the spreadsheet and validate the data. Depending on their complexity, macros can either be typed in as a series of statements or recorded as the user takes appropriate menu and mouse actions. "Solver" utilities can also simplify the process of tweaking input variables in order to achieve a defined goal. Although spreadsheets can certainly solve many types of algebraic equations, symbolic manipulation is better handled by programs such as Mathematica (see MATHEMATICS SOFTWARE).

Besides having extensive graphics and charting capabilities, modern spreadsheets are often part of integrated office programs (see APPLICATION SUITE). Thus, a Microsoft Excel spreadsheet could obtain data from an Access database and create charts suitable for webpages or PowerPoint presentations.

Further Reading

Benninga, Simon, and Benjamin Czaczkes. *Financial Modeling.* 2nd ed. Cambridge, Mass.: MIT Press, 2000.

Harvey, Greg. *Excel 2000 for Windows for Dummies.* New York: Hungry Minds, 1999.

Ragsdale, Cliff. *Spreadsheet Modeling and Decision Analysis W/Excel.* 3rd ed. Cincinnati, Ohio: South-Western College Pub., 2001.

"Spreadsheets, Mathematics, Science, and Statistics Education." http://sunsite.univie.ac.at/spreadsite/

SQL

Structured query language was originally developed in the early 1970s as a command interface for IBM mainframe databases. Today, however, SQL has become the lingua franca for relational database systems (see DATABASE MANAGEMENT SYSTEM).

A relational database (such as Oracle, Sybase, IBM DB2, and Microsoft Access) stores data in tables called *relations.* The columns in the table describe the characteristics of an entity (corresponding to data fields). For example, in a customer database the Customer table might include attributes such as customer number, First_name, Last_Name, Street, City, Phone_number, and so on. The rows in the table (sometimes called *tuples*) represent the data records for the various customers.

Many database systems have more than one table. For example, a store's database might contain a Customers

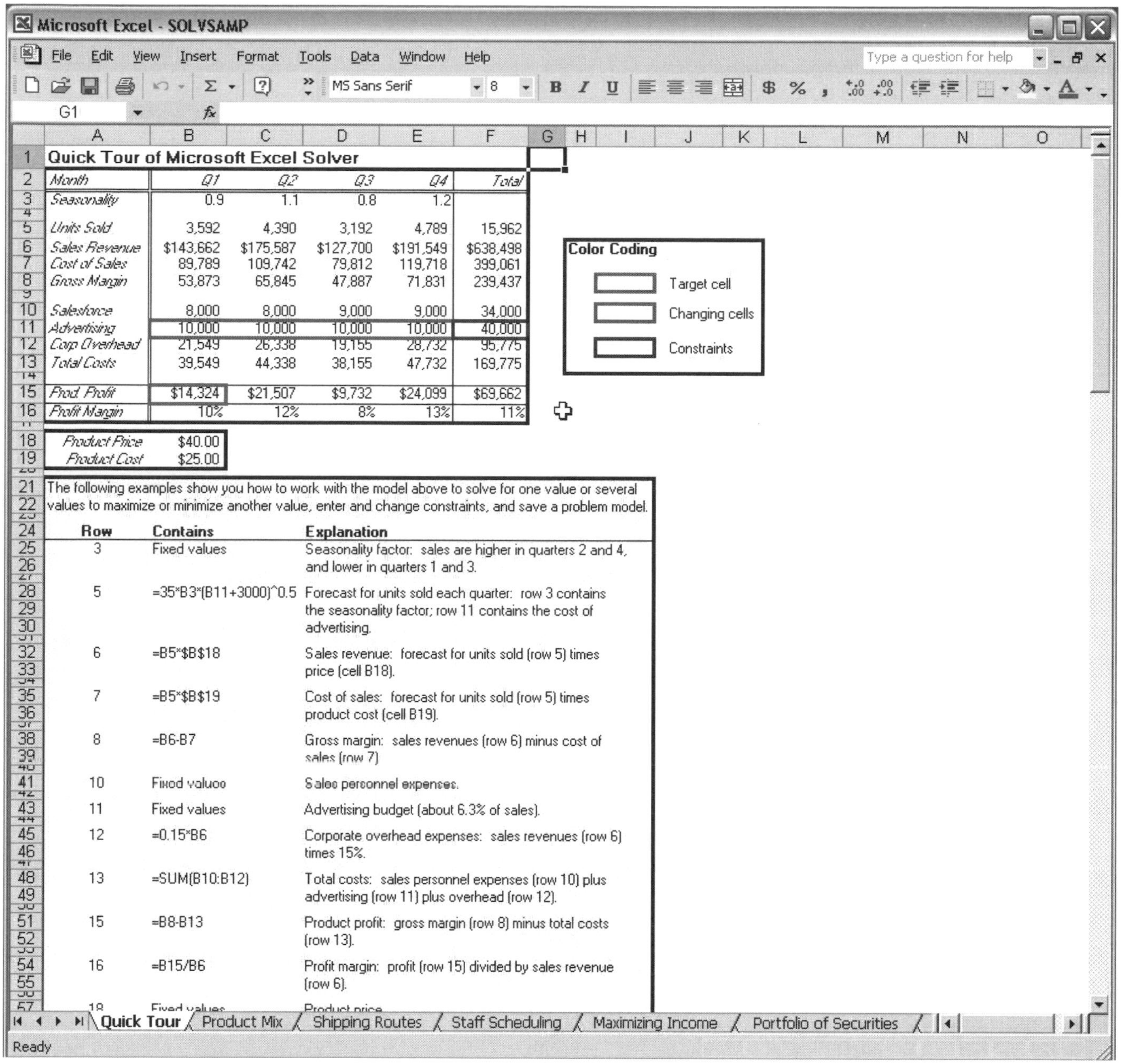

Microsoft Excel - SOLVSAMP

	A	B	C	D	E	F
1	Quick Tour of Microsoft Excel Solver					
2	Month	Q1	Q2	Q3	Q4	Total
3	Seasonality	0.9	1.1	0.8	1.2	
5	Units Sold	3,592	4,390	3,192	4,789	15,962
6	Sales Revenue	$143,662	$175,587	$127,700	$191,549	$638,498
7	Cost of Sales	89,789	109,742	79,812	119,718	399,061
8	Gross Margin	53,873	65,845	47,887	71,831	239,437
10	Salesforce	8,000	8,000	9,000	9,000	34,000
11	Advertising	10,000	10,000	10,000	10,000	40,000
12	Corp Overhead	21,549	26,338	19,155	28,732	95,775
13	Total Costs	39,549	44,338	38,155	47,732	169,775
15	Prod. Profit	$14,324	$21,507	$9,732	$24,099	$69,662
16	Profit Margin	10%	12%	8%	13%	11%
18	Product Price	$40.00				
19	Product Cost	$25.00				

Color Coding
- Target cell
- Changing cells
- Constraints

The following examples show you how to work with the model above to solve for one value or several values to maximize or minimize another value, enter and change constraints, and save a problem model.

Row	Contains	Explanation
3	Fixed values	Seasonality factor: sales are higher in quarters 2 and 4, and lower in quarters 1 and 3.
5	=35*B3*(B11+3000)^0.5	Forecast for units sold each quarter: row 3 contains the seasonality factor; row 11 contains the cost of advertising.
6	=B5*B18	Sales revenue: forecast for units sold (row 5) times price (cell B18).
7	=B5*B19	Cost of sales: forecast for units sold (row 5) times product cost (cell B19).
8	=B6-B7	Gross margin: sales revenues (row 6) minus cost of sales (row 7).
10	Fixed values	Sales personnel expenses.
11	Fixed values	Advertising budget (about 6.3% of sales).
12	=0.15*B6	Corporate overhead expenses: sales revenues (row 6) times 15%.
13	=SUM(B10:B12)	Total costs: sales personnel expenses (row 10) plus advertising (row 11) plus overhead (row 12).
15	=B8-B13	Product profit: gross margin (row 8) minus total costs (row 13).
16	=B15/B6	Profit margin: profit (row 15) divided by sales revenue (row 6).
18	Fixed values	Product price

Quick Tour / Product Mix / Shipping Routes / Staff Scheduling / Maximizing Income / Portfolio of Securities

Ready

Modern spreadsheets have many sophisticated features. Microsoft Excel, for example, has a "Solver" module that can be used to solve for particular values or to maximize or minimize specified values.

table (for information identifying a customer), an Item table (giving characteristics of an item, such as price and number in stock), and a Transaction table (whose characteristics might be customer number, date, item bought, and so on). Notice that the Transaction record contains both a customer number and an item number. It thus serves as a sort of bridge or link between the Customer and Item tables.

SQL provides commands that can be used to specify and access components of a database. For example, the INSERT and DELETE commands can be used to add or remove rows (records) from tables.

To query a database means to give criteria for selecting certain records from a table. For example, the query

```
SELECT * FROM CUSTOMERS WHERE LAST_NAME =
"Howard"
```

would return the complete records for all customers whose last name is Howard. If only selected fields are desired, they can be specified like this:

```
SELECT NUMBER, NAME, PRICE FROM ITEMS WHERE
PRICE > = 50.00
```

This query will display the Number, Name, and Price fields for all items whose price is greater than or equal to $50.00.

SQL includes many commands to further refine data processing and reporting. There are built-in mathematical functions as well as a GROUP BY command for further breaking down a report by a particular field name or value.

SQL can be used interactively by typing commands at a prompt, but database applications designed for less technical users often provide a user-friendly query form (and perhaps menus or buttons). After the user selects the appropriate fields and values, the program will then generate the necessary SQL statements and send them to the internal "database engine" for processing. The results will then be displayed for the user.

SQL procedures can be stored and managed as part of a database. SQL can also be "embedded" within a more complete programming language environment so that, for example, a Java program can perform SQL operations while using Java for processing that cannot be specified in SQL. In the mid-1990s an object-oriented version of SQL called OQL (object query language), allowing the use of that popular paradigm for database operations (see OBJECT-ORIENTED PROGRAMMING).

Further Reading

Bowman, Judith S., Sandra L. Emerson, and Mark Donovsky. *The Practical SQL Handbook: Using Structured Query Language.* Reading, Mass.: Addison-Wesley, 1996.

Brockwood, Ted. "Simple SQL." http://www.webdevelopersjournal.com/articles/sql.html

Elsmari, R., and S. Navathe. *Fundamentals of Database Systems.* Redwood City, Calif.: Benjamin Cummings, 2000.

stack

Often a temporary storage data area is needed during processing. For example, a program that calls a procedure (see PROCEDURES AND FUNCTIONS) usually needs to pass one or more data items to the procedure. These items are specified as arguments that will be matched to the procedure's defined parameters. For example, the procedure call

```
Square (50, 50, 20)
```

could draw a square whose upper left corner is at the screen coordinates 50, 50 and whose length per side is 20 pixels.

When the compiler generates the machine code for this statement, that code will probably instruct the processor to store the numbers 50, 50, and 20 onto a stack. A stack is simply a list that represents successive locations in memory into which data can be inserted. The operation of a stack can be visualized as being rather like the spring-loaded platform onto which dishes are stacked for washing in some restaurants. As each dish (number) is added, the stack is "pushed." Because only the item "on top" (the last one added) can be removed ("popped") at any given time, a stack is described as a LIFO (last in, first out) structure. (Note that this is different from a queue, where items can be added or removed from either end ([see QUEUE].)

Stacks are useful whenever nested items must be tracked. For example, a procedure might call a procedure

SQL

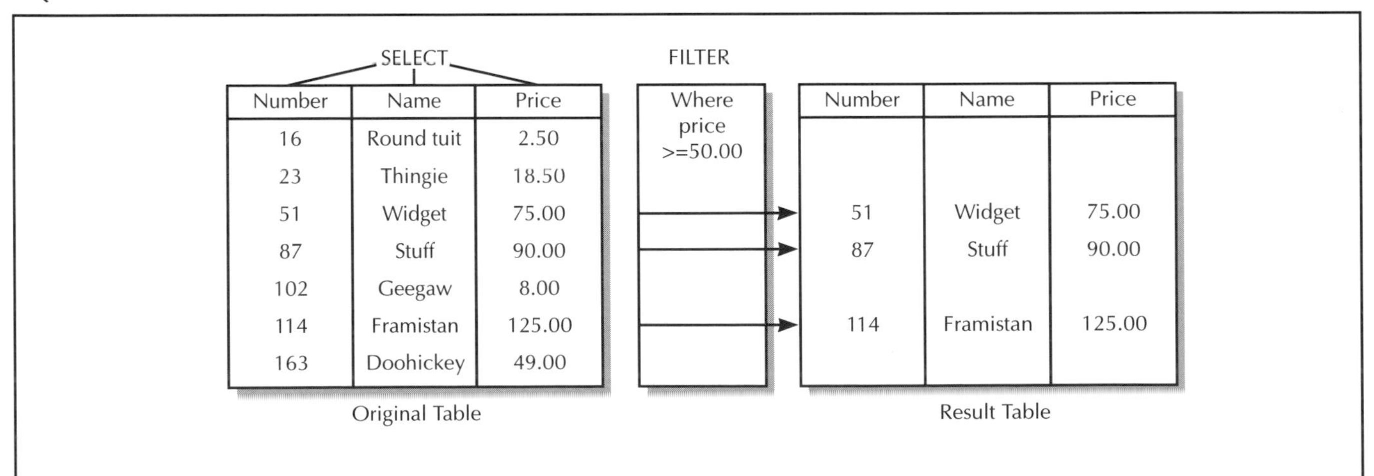

Structured Query Language (SQL) is a standardized way to query and manipulate databases. Here the statement SELECT NUMBER, NAME, PRICE WHERE PRICE >= 50.00 extracts only the records meeting that criterion.

Stack

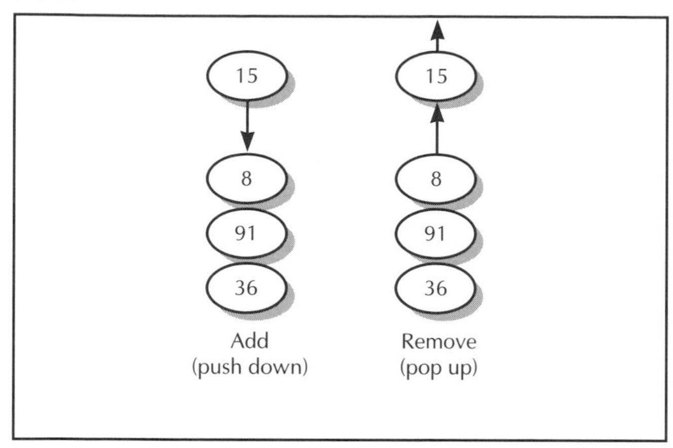

A stack can be visualized like the spring-loaded stack of plates in a restaurant kitchen. Values can only be added (pushed) or removed (popped) from the top of the stack.

that in turn calls another procedure. The stack can keep track of the parameters (as well as the calling address) for each pending procedure.

Stacks can also be used to evaluate nested arithmetic expressions. For example, the expression that we write in conventional (prefix) notation as

```
7 * 5 + 2
```

can be represented internally in postfix form as:

```
* + 5 7 2
```

Here one stack can be used to hold the operators (* +) and one the operands (5 7 2). The evaluation then proceeds in the following steps:

```
Pop the * from the operator stack

Since * is a binary operator (one that needs two
  operands), pop the 5 and 7 from the operand
  stack

Multiply 5 and 7 to get 35.

Pop the + from the operator stack.

Pop the 35 (which is now on the top of the
  operand stack) and the 2

Add 35 and 2 to get 37.
```

An interesting programming language uses this stack mechanism for all processing (see FORTH). In working with stacks, it may be necessary to keep in mind any limitations on the amount of memory allocated to the stack, although a stack can also be implemented dynamically as a linked list (see LIST PROCESSING).

Further Reading

Knuth, Donald E. *The Art of Computer Programming,* vol. 1. 3rd ed. Reading, Mass.: Addison-Wesley, 1997.

Morris, John. "Data Structures and Algorithms. Stacks." http://ciips.ee.uwa.edu.au/~morris/Year2/PLDS210/stacks.html

Stallman, Richard

(1953–)
American
Computer Scientist

Richard Stallman created superb software tools—the programs that help programmers with their work. He went on to spearhead the open source movement, a new way to develop software.

Stallman was born on March 16, 1953, in New York City. He quickly showed prodigious talent for mathematics and was exploring calculus by the age of eight. Not much later, his summer camp reading included a manual for the IBM 7094 mainframe belonging to one of the counselors. Fascinated with the idea of programming languages, young Richard began writing simple programs, even though he had no access to a computer.

Fortunately, a high school honors program let him obtain some time on a mainframe, and his programming talents led to a summer job with IBM. While studying for his B.A. in physics at Harvard (which he received in 1970), Stallman found himself sneaking across town to the MIT Artificial Intelligence Lab. There he developed Emacs, a powerful text editor that could be programmed with a language modeled after LISP, the favorite language of AI researchers. While working on Emacs and other system software for the AI Lab, Stallman participated in the unique MIT "hacker culture." (During the 1970s, "hacker" still meant a creative computing virtuoso, not a cyber-criminal.)

Stallman's experience in the freewheeling, competitive yet cooperative atmosphere at MIT led him to decide in 1984 to start the Free Software Foundation, which would become his life's work. Stallman and his colleagues at the FSF worked through the 1980s to develop GNU. At the time, UNIX, the operating system of choice for most campuses and researchers, required an expensive license from Bell Laboratories. GNU (a recursive acronym for "GNU's Not UNIX") was intended to include all the functionality of UNIX but with code that owed nothing to Bell Labs. Stallman's key contributions to the project included the GNU C compiler and debugger, as well as his management of a cooperative effort in which many talented programmers would coordinate their efforts over the Internet.

By the early 1990s, most of GNU was complete except for a key component: the kernel containing the essential functions of the operating system. A Finnish programmer named Linus Torvalds decided to write the kernel and integrate it with much of the existing GNU software. The result would become known as Linux, and today it is a popular operating system that runs on many servers and workstations. While acknowledging Torvalds's efforts, Stallman insists that the operating system is more properly called GNU Linux, to reflect the large amount of GNU code it employs.

Stallman has received a number of important awards, including the ACM Grace Hopper Award (1990), Electronic Frontier Foundation Pioneer Award (1998), and a MacArthur Foundation fellowship (1990).

Further Reading

DiBona, Chris, Sam Ockman, and Mark Stone, eds. *Open Sources: Voices from the Open Source Revolution*. Sebastopol, Calif.: O'Reilly, 1999.

Free Software Foundation / GNU. http://www.fsf.org

"Richard Stallman's Personal Homepage." http://www.stallman.org

Williams, Sam. *Free as in Freedom: Richard Stallman's Crusade for Free Software*. Sebastopol, Calif.: O'Reilly, 2002.

standards in computing

One hallmark of the maturity of a technology is the development of a variety of kinds of standards that are accepted by a majority of practitioners. There are several reasons why standards develop.

MARKETPLACE STANDARDS

In many cases, a particular product gains a prominent position in an emerging market, and would-be competitors adopt its interface and specifications. For example, the parallel port printer interface (and plug) developed by Centronics for its printers was adopted by virtually all printer manufacturers. Since it would be impracticable for computer manufacturers to provide many different parallel connectors on their machines, there was a clear market advantage in setting a standard. When a particular product (Centronics in this case) becomes that standard, it is mainly a matter of timing.

Once a marketplace standard is established, manufacturers and consumers will generally not want products that are incompatible with it. When the IBM PC and its ISA expansion card became the standard followed by many "clone" manufacturers, IBM discovered that even Big Blue flouted the standard at its peril. When IBM came out with its MCA (Microchannel Architecture) in the late 1980s, the new machines, although possessing some technical advances, did not sell as well as expected. Most people stayed with the existing IBM standard and built upwardly compatible machines upon it.

OFFICIAL STANDARDS

Some standards are developed by official bodies. For example, the International Standards Organization (ISO) has an elaborate formal process where panels of experts develop standards for a huge variety of technologies, including many relating to computing. In an increasingly global economy, international standards allow equipment (or software) from one country to be used with that from another. For example, credit cards, phone cards, and "smart cards" around the world have a common format established by ISO standards. (Standards specific to electrical and electronic engineering are developed by a similar body, the International Electrotechnical Commission, or IEC.) Standards that have become widely accepted but are not yet official ISO standards take the form of Publicly Available Specifications, or PAS. Government contracts often specify ISO standards as well as a variety of other standards developed by various government agencies. The ISO 9001 standards apply specifically to computer systems, software, and its development.

EVOLUTION OF STANDARDS

The extent of standardization within the broad information technology (IT) industry varies widely among applications. Generally, things that have been established for a long time (meaning, in computing terms, a couple decades or so) are likely to be well standardized. An example is the standards for character sets.

For areas in which new applications are emerging, practitioners tend to have less interest (or patience) with the idea of standards. For example, the World Wide Web is still relatively new, and standards for the operation of websites are emerging only slowly. In this case, it is mainly concern about such matters as privacy protection that has encouraged the adoption of standards for matters such as the secure transmission of credit card information on-line or privacy policies regarding the use of information obtained from Web users. The potential threat of government regulation often encourages the development of marketplace standards as an alternative.

Technical societies such as the Institute for Electrical and Electronic Engineering (IEEE) and the World Wide Web Consortium are an important forum for the discussion and development of standards.

Further Reading

Goetsch, David L., and Stanley Davis. *Understanding and Implementing ISO 9000 and Other ISO Standards*. Upper Saddle River, N.J.: Prentice Hall, 2001.

IEEE Computer Society. "Standards Activities." http://www.computer.org/standards/

International Organization for Standardization (ISO). http://www.iso.ch/iso/en/ISOOn-line.frontpage

Radice, Ronald A. *ISO 9001: Interpreted for Software Organizations*. Andover, Mass.: Paradoxicon, 1995.

statistics and computing

The application of computing technology to the collection and analysis of statistics is as old as computing itself.

Indeed, Charles Babbage was an early proponent of the collection of social and economic statistics in order to understand how society was being changed by the Industrial Revolution in the early 19th century. By the end of that century, Herman Hollerith had come to the rescue of the U.S. Census Bureau by providing his card tabulation machines for the 1890 Census. (See BABBAGE, CHARLES and HOLLERITH, HERMAN.)

In the era of the mainframe, performing statistical analysis with a computer generally required writing a customized program (although the development of FORTRAN around 1960 gradually led the accumulation of an extensive library of subroutines that could be employed to perform statistical functions). Programs generally run in a batch mode, with data supplied from punched cards or tape.

When the personal computer arrived, it wasn't yet powerful enough for much statistical work, although a program such as VisiCalc (see SPREADSHEET) could be used for simple operations. Gradually, spreadsheets grew more powerful, but statisticians truly rejoiced when software packages specifically designed for statistical work began to appear.

Today there are hundreds of statistical packages available, of which the best known one for personal computers is SPSS. Most packages can be used to perform the standard forms of statistical analysis, including analysis of variance, regression analysis, discrete data analysis, time series analysis, and cluster analysis. There are also packages for specialized applications. Moving in the direction of greater generality, mathematical software such as Mathematica and MATLAB can also be used for statistical applications (see MATHEMATICS SOFTWARE). This category of software experiences steady growth because the ability to analyze data quickly and interactively is increasingly important given the growing pace of human activity, whether one is confronted with a rapidly spreading disease or a volatile economy.

Other areas related to statistical computing include the extraction of useful correlations from existing data bases (see DATA MINING) and the development of dynamic models based on probability and statistics (see SIMULATION).

Further Reading

American Statistical Association. http://www.amstat.org/

Green, Samuel B. [and others]. *Using SPSS for Windows: Analyzing and Understanding Data.* Upper Saddle River, N.J.: Prentice Hall, 1999.

"Rainier's Web Site for Statisticians." http://ourworld.compuserve.com/homepages/Rainer_Wuerlaender/stathome.htm

"Statistics and Computing. Statistics on the Web." http://www.vanderbilt.edu/~rtucker/methods/econometrics/

Venables, William N., and Brian D. Ripley. *Modern Applied Statistics with S-Plus.* New York: Springer-Verlag, 1999.

streaming

Web users increasingly have access to such content as news broadcasts, songs, and even full-length videos. The problem is that the user must receive the content in real time at a steady pace, not in sputters or jerks. However, factors such as load on the Web server and network congestion between the server and user can cause delays in transmission. One way to reduce the problem would be to compress the data (see DATA COMPRESSION). However, excessive compression would compromise audio or picture quality to an unacceptable extent. Fortunately, a technology called streaming offers a way to smooth out the transmission of large amounts audio or video content (see also MULTIMEDIA).

When a user clicks on an audio or video link, the player software (or Web browser plug-in) is loaded and the transmission begins. Typically, the player stores a few seconds of the transmission (see BUFFERING), so any momentary delays in the transmission of data packets will not appear as the data starts to play. Assuming the rate of transmission remains sufficient, enough data remains in the buffer so that data can be "fed" to the playing software at a steady pace. If, however, there is too much delay due to network congestion, the playback will pause while the player refills its buffer.

The most popular media players for PCs (such as WinAmp, RealPlayer, and Windows Media Player) provide for streaming data. Despite streaming, connections of less than about 56K bps are likely to result in occasional interruption of content. Together with the use of streaming, the gradual move to faster cable or DSL connections (see BROADBAND) is improving the multimedia experience for Web users. However, the delivery of full-length videos in real time is problematic except over very fast dedicated networks.

Further Reading

Mack, Steve. *Streaming Media Bible.* New York: Hungry Minds, 2002.

RealNetworks. http://www.real.com/realone/?src=realaudio

Streaming

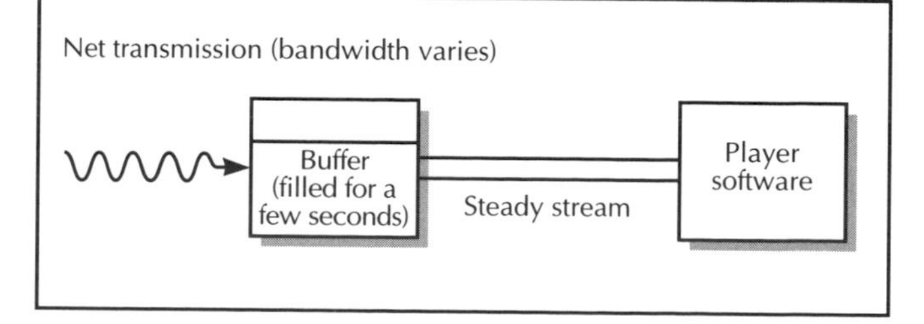

Streaming evens out the vagaries of Internet traffic by buffering audio or video data as it's received and then feeding it to the browser or player at a steady rate.

"Video on the World Wide Web." http://www.videonics.com/videos/about-web-video.html
"Video Over the Internet." http://www.rad.com/networks/1996/video/video.htm

Stroustrup, Bjarne

(1950–)
Danish
Computer Scientist

Bjarne Stroustrup created C++, an object-oriented successor to the popular C language that has now largely supplanted the original language.

Stroustrup was born on December 30, 1950, in Aarhus, Denmark. As a student at the University of Aarhus his interests were far from limited to computing (indeed, he found programming classes to be rather dull). However, unlike literature and philosophy, programming did offer a practical job skill, and Stroustrup began to do contract programming for Burroughs, an American mainframe computer company. To do this work, Stroustrup had to pay attention to both the needs of application users and the limitations of the machine, on which programs had to be written in assembly language to take optimal advantage of the memory available.

By the time Stroustrup received his master's degree in computer science from the University of Aarhus, he was an experienced programmer, but he soon turned toward the frontiers of computer science. He became interested in distributed computing (writing programs that run on multiple computers at the same time) and developed such programs at the Computing Laboratory at Cambridge University in England, where he earned his Ph.D. in 1979.

The 1970s was an important decade in computing. It saw the rise of a more methodical approach to programming and programming languages (see STRUCTURED PROGRAMMING). It also saw the development of a powerful and versatile new computing environment: the UNIX operating system and C programming language developed by Dennis Ritchie (see RITCHIE, DENNIS) and Ken Thompson and Bell Laboratories. Soon after getting his doctorate, Stroustrup moved to Bell Labs, where he became part of that effort.

As Stroustrup continued to work on distributed computing, he decided that he needed a language that was better than C at working with the various modules running on the different computers. He studied an early object-oriented language (see OBJECT-ORIENTED PROGRAMMING and SIMULA). Simula had a number of key concepts including the organization of a program into classes, entities that combined data structures and associated capabilities (methods). Classes and the objects created from them offered a better way to organize large programs, and

In the 1980s, Bjarne Stroustrup created the object-oriented C++ language that became the most popular language for general applications programming. (PHOTO COURTESY OF BJARNE STROUSTRUP)

was particularly suited for distributed computing and parallel programming where there were many separate entities running at the same time.

However, Simula was fairly obscure, and it was unlikely that the large community of systems programmers who were using C would switch to a totally different language. Instead, starting in the early 1980s, Stroustrup decided to add object-oriented features (such as classes with member functions, user-defined operators, and inheritance) to C. At first he gave the language the rather unwieldy name of "C with Classes" However, in 1985 he changed the name to C++. (The ++ is a reference to an operator in C that adds one to its operand, thus C++ is "C with added features.")

At first some critics criticized C++ for retaining most of the non-object oriented features of C (unlike pure object languages such as Smalltalk), while others complained that the overhead required in processing classes made C++ slower than C. During the 1990s, however, C++ became increasingly popular, aided by its relatively smooth learning curve for C programmers and the development or more efficient compilers. C++ is now the most widely used general purpose computer language.

Stroustrup has continued to work at AT&T Labs (one of the two successors to the original Bell Labs). He received the 1993 ACM Grace Hopper Award for his work on C++ and is an AT&T Fellow

Further Reading

"Bjarne Stroustrup's Homepage." http://www.research.att.com/~bs/homepage.html
Gribble, Cheryl. "History of C++." http://www.hitmill.com/programming/cpp/cppHistory.html

Stroustrup, Bjarne. *The C++ Programming Language*. 3rd ed. Reading, Mass.: Addison-Wesley, 2000.
———. *The Design and Evolution of C++*. Reading, Mass.: Addison-Wesley, 1995.

structured programming

As programs grew longer and more complex during the 1960s, computer scientists began to pay more attention to the ways in which programs were organized. Most programming languages had a statement called "GOTO" or its equivalent. This statement transfers control to some arbitrary other point in the program, as identified by a label or line number.

In 1968, computer scientist Edsger Dijkstra (see DIJKSTRA, EDSGER) sent a letter to the editor of the *Proceedings of the ACM* with the title "GO TO Statement Considered Harmful." In it he pointed out that the more such jumps programs made from place to place, the harder it was for someone to understand the logic of the program's operation.

The following year, Dijkstra introduced the term *structured programming* to refer to a set of principles for writing well-organized programs that could be more easily shown to be correct. One of these principles is statements such as If . . . Then . . . Else be used to organize a choice between two or more alternatives (see BRANCHING STATEMENTS) and that statements such as While be used to control repetition or iteration of a statement (see LOOP).

Other computer scientists added further principles, such as modularization (breaking down a program into separate procedures, such as for data input, different stages of processing, and output or printing). Modularization makes it easier to figure out which part of a program may be causing a problem, and to fix part of a problem without affecting other parts. A related principle, information hiding, keeps the data used by a procedure "hidden" in that procedure so that it can't be changed from some other part of the program.

Structured programming also encourages stepwise refinement, a program design process described by Niklaus Wirth, creator of Pascal. This is a top-down approach in which the stages of processing are first described in high-level terms (see also PSEUDOCODE), and then gradually fleshed out in their details, much like the writing of an outline for a book.

The principles of structured programming were soon embodied in a new generation of programming languages (see ALGOL, PASCAL, and C). Although use of well-structured language didn't guarantee good structured programming practice, it at least made the tools available.

The ideas of structured programming form a solid basis for programming style today. They have been supplemented rather than replaced by a new paradigm developed in the 1970s and 1980s (see OBJECT-ORIENTED PROGRAMMING).

Further Reading

Dhal, Ole-Johan, Edsger W. Dijkstra, and C. A. R. Hoare, eds. *Structured Programming*. New York: Academic Press, 1972.
Dijkstra, Edsger. *A Discipline of Programming*. Englewood Cliffs, N.J.: Prentice Hall, 1976.
———. "Go To Statement Considered Harmful." *Communications of the ACM* 11, no. 3, 1968, 147–148.
Orr, Kenneth T. *Structured Systems Development*. New York: Yourdon Press, 1977.

supercomputer

The term *supercomputer* is not really an absolute term describing a unique type of computer. Rather, it has been used through successive generations of computer design to describe the fastest, most powerful computers available at a given time. However, what makes these machines the fastest is usually their adoption of a new technology or computer architecture that later finds its way into standard computers.

The first supercomputer is generally considered to be the Control Data CDC 6600, designed by Seymour Cray in 1964. The speed of this machine came from its use of the new, faster silicon (rather than germanium) transistors and its ability to run at a clock speed of 10 MHz (a speed that would be achieved by personal computers by the mid-1980s). Even with transistors, these machines generated so much heat that they had to be cooled by a Freon-based refrigeration system.

Cray then left CDC to form Cray Research. He designed the Cray 1 in 1976, the first of a highly successful series of supercomputers. The Cray 1 took advantage

A Cray 190 A supercomputer. Seymour Cray's leading-edge machines defined supercomputing for many years. (NASA PHOTO)

of a new technology, integrated circuits, and new architecture: vector processing, in which a single instruction can be applied to an entire series (or array) of numbers simultaneously. This innovation marked the use of parallel processing as one of the distinguishing features of supercomputers. The machine's monolithic appearance gave it a definite air of science fiction, and the first one built was installed at the secretive Los Alamos National Laboratory.

The next generation, the Cray X-MP, carried parallelism further by incorporating multiple processors (the successor, Cray Y-MP, had 8 processors, which together could perform a billion floating-point operations per second [1 gigaflop]).

Soon Cray no longer had the supercomputer field to itself, and other companies (particularly the Japanese manufacturers NEC and Fujitsu) entered the market. The number of processors in supercomputers increased to as many as 1,024 (in the 1998 Cray SV1), which can exceed 1 *trillion* floating-point operations per second (1 teraflop).

Meanwhile, processors for desktop computers (such as the Intel Pentium) also continued to increase in power, and it became possible to build supercomputers by combining large numbers of these readily available (and relatively low-cost) processors.

The ultimate in multiprocessing is the series of Connection Machines built by Thinking Machines Inc. (TMI) and designed by Daniel Hillis. These machines have up to 65,000 very simple processors that run simultaneously, and can form connections dynamically, somewhat like the process in the human brain. These "massively parallel" machines are thus attractive for artificial intelligence research. It is also possible to achieve supercomputerlike power by having many computers on a network divide the work of, for example, cracking a code or analyzing radio telescope data for signs of intelligent signals.

Programs for supercomputers must be written using special languages (or libraries for standard languages) that are designed to provide for many processes to run at the same time and that allow for communication and coordination between processing (see MULTIPROCESSING).

APPLICATIONS

Supercomputers are always more expensive and somewhat less reliable than standard computers, so they are used only when necessary. As the power of standard computers continues to grow, applications that formerly required a multimillion-dollar supercomputer can now run on a desktop workstation (a good example is the creation of detailed 3D graphics).

On the other hand, there are always applications that will soak up whatever computing power can be brought to bear on them. These include analysis of new aircraft designs, weather and climate models, the study of nuclear reactions, and the creation of models for the synthesis of proteins. The never-ending battle of organizations such as the National Security Agency (NSA) to monitor worldwide communications and crack ever-tougher encryption also demands the fastest available supercomputers (see also QUANTUM COMPUTING).

The future of supercomputing seems to involve two competing paradigms. One is typified by "Q," a supercomputer at the Los Alamos National Laboratory that is intended to create 3D simulations so accurate that they will eliminate the need to physically test nuclear weapons. This machine, expected to be one of the two most powerful in the world, is rated at up to 30 Teraflops (trillions of floating-point operations per second). However this performance is achieved at the cost of very high energy consumption (up to 5 megawatts) and the need for a nuclear power plant-style cooling system.

An alternative is to connect together modest "off the shelf" processors similar to those found in today's desktop machines. One such architecture is the Beowulf Cluster, which packs together its processors into a refrigerator-sized space. While less powerful (measured in mere gigaflops rather than teraflops), these machines also consume far less power and provide considerably more processing power per dollar spent.

In the future it is likely that there will be a few super-supercomputers like Q for the highest priority applications, while the less expensive cluster-type supercomputers will increasingly bring supercomputer power to many other applications.

Further Reading

Hord, R. Michael. *Understanding Parallel Supercomputing.* Piscataway, N.J.: IEEE Press, 1999.

Johnson, George. "At Los Alamos, Two Visions of Supercomputing." *The New York Times,* June 25, 2002. Available on-line at www.nytimes.com.

Murray, C. J. *The Supermen: The Story of Seymour Cray and the Technical Wizards behind the Supercomputer.* New York: Wiley, 1995.

National Center for Atmospheric Research. Scientific Computing Division. "Supercomputer Gallery." http://www.scd.ucar.edu/computers/gallery/

National Center for Supercomputing Applications (NCSA). http://www.ncsa.uiuc.edu/

system administrator

A system administrator is the person responsible for managing the operations of a computer facility to ensure that it runs properly, meets user needs, and protects the integrity of users' data. Such facilities range from offices with just a few users to large campus or corporate facilities that may be served by a large staff of administrators.

The system administrator's responsibilities often include:

- Setting up accounts for new users
- Allocating computing resources (such as server space) among users
- Configuring the file, database, or local area network (LAN) servers
- Installing new or upgraded software on users' workstations
- Keeping up with new versions of the operating system and networking software
- Using various tools to monitor the performance of the system and to identify potential problems such as device "bottlenecks" or a shortage of disk space
- Ensuring that regular backups are made
- Configuring network services such as e-mail, Internet access, and the intranet (local TCP/IP network)
- Using tools such as firewalls and virus scanners to protect the system from viruses, hacker attacks, and other security threats (see also COMPUTER CRIME AND SECURITY)
- Providing user orientation and training
- Creating and documenting policies and procedures

System administrators often write scripts to automate many of the above tasks (see SCRIPTING LANGUAGES). Because of the complexity of modern computing environments, an administrator usually specializes in a particular operating system such as UNIX or Windows 2000.

A good system administrator needs not only technical understanding of the many components of the system, but also the ability to communicate well with users—good "people skills." Larger organizations are more likely to have separate network and database administrators, while the administrator of a small facility must be a jack (or jill) of all trades.

Further Reading

Frisch, Aeleen. *Essential System Administration.* 2nd ed. Sebastopol, Calif.: O'Reilly, 1996.

Nemeth, Evi [and others], *UNIX System Administration Handbook.* 3rd ed. Upper Saddle River, N.J.: Prentice Hall, 2000.

"UNIX System Administration." http://darkwing.uoregon.edu/~hak/unix.html

Willis, Will, David Watts, and Tillman Strahan. *Windows 2000 System Administration Handbook.* Upper Saddle River, N.J.: Prentice Hall, 2000.

systems analyst

The systems analyst serves as the bridge between the needs of the user and the capabilities of the computer system. The systems analyst goes into action when users request that some new application or function be provided (usually in a corporate computing environment).

The first step is to define the user's requirements and to prepare precise specifications for the program. In doing so, the systems analyst is aided by methodologies developed by computer scientists over the last several decades (see STRUCTURED PROGRAMMING and OBJECT-ORIENTED PROGRAMMING). Often flowcharts or other aids are used to help visualize the operation of the program (see also CASE).

After communicating with the user, the systems analyst must then communicate with the programmers, helping them understand what is needed and reviewing their work as they begin to design the program. Although the systems analyst may do little actual programming, he or she must be familiar with programming tools and practices. This may make it possible to suggest existing software or components that could be adapted instead of undertaking the cost and time involved with creating a new program. As a program is developed, systems analysts are often responsible for designing tests to ensure that the software works properly (see QUALITY ASSURANCE, SOFTWARE).

Depending on the organizational structure, all or part of the analysis function may be included in the job description "programmer-analyst" or included as part of the duties of a senior software engineer or manager of program development. Experienced systems analysts are likely to be called upon to participate in the evaluation of possible investments in new software or hardware, and other aspects of long-term planning for computing facilities.

Further Reading

Lejk, M., and D. Deeks. *Introduction to Systems Analysis Techniques.* Upper Saddle River, NJ: Prentice-Hall, 1998.

Robertson, James, and Suzanne Robertson. *Complete Systems Analysis: The Workbook, the Textbook, the Answers.* New York: Dorset House, 1998.

"Systems Analysis Web Sites." http://www.umsl.edu/~sauter/analysis/analysis_links.html

Whitten, J. L., L. D. Bentley, and K. C. Dittman. *Systems Analysis and Design Methods.* Homewood, IL: Richard D. Irwin, 1997.

systems programming

Applications programmers write programs to help users work better, while systems programmers write programs to help the computer itself work better (see OPERATING SYSTEM). Systems programmers generally work for companies in the computer industry that develop operating systems, network facilities, program language compilers and other software development tools, utilities, and device drivers. However, systems programmers can also work for applications developers to help them interface their programs to the operating system or to devices (see DEVICE DRIVER and APPLICATIONS PROGRAMMING INTERFACE).

Modern operating systems are highly complex, so systems programmers tend to specialize in particular areas. These might include device drivers, software development tools, program language libraries, applications programming interfaces (APIs), and utilities for monitoring system conditions and resources. Systems programmers develop the infrastructure needed for networking, as well as multiple-processor computers and distributed computing systems. Systems programmers also play a key role when an application program must be "ported" to a different platform or simply modified to run under a new version of the operating system.

Generally, an application programmer works at a fairly high level, using language functions and APIs to have the program ask the operating system for services such as loading or saving files, printing, and so on. The systems programmer, on the other hand, must be concerned with the internal architecture of the system (such as the buffers allocated to hold various kinds of temporary data) and with how commands are constructed for disks and other devices. Generally, the systems programmer must also have a more thorough knowledge of data structures and how they are physically represented in the machine as well as the comparative efficiency of various algorithms. Because it determines how efficiently the system's resources can be used, systems programming must often be "tight" and optimized for peak performance. Thus, although lower-level assembly language is no longer used for much applications programming, it can still be found in systems programming.

Further Reading

Beck, Leland L. *System Software: an Introduction to Systems Programming.* 3rd ed. Reading, Mass.: Addison-Wesley, 1998.

Hart, Johnson M. *Win32 System Programming, Second Edition. A Windows 2000 Application Developer's Guide.* Reading, Mass.: Addison-Wesley, 2000.

Stevens, W. Richard. *Advanced Programming in the UNIX Environment.* Reading, Mass.: Addison-Wesley, 1992.

T

tape drives

Anyone who has seen computers in old movies is familiar with the row of large, freestanding tape cabinets with their spinning reels of tape. The visual cue that the computer was running consisted of the reels thrashing back and forth vigorously while rows of lights flashed on the computer console. Magnetic tape was indeed the mainstay for data storage in most large computers (see MAINFRAME) in the 1950s through the 1970s.

In early mainframes the main memory (corresponding to today's RAM chips) consisted of "core"—thousands of tiny magnetized rings crisscrossed with wires by which they could be set or read. Because core memory was limited to a few thousand bytes (KB), it was used only to hold the program instructions (see also PUNCHED CARDS AND PAPER TAPE) and to store temporary working data while the program was running.

The source data to be processed by the program was read from a reel of tape on the drive. If the program updated the data (rather than just reporting on it), it would generally write a new tape with the revised data. In large facilities a person called a tape librarian was in charge of keeping the reels of tape organized and providing them to the computer operators as needed.

OPERATION

A mainframe tape drive had two reels, the supply reel and the take-up reel. Because each reel had its own motor, they could be spun at different speeds. This allowed a specified length of tape to be suspended between the two reels, serving as sort of a buffer and allowing the take-up reel to accelerate at the start of a read or write operation without danger of breaking the tape. The "buffer" tape was actually suspended in a partial vacuum, which both kept the tape taut enough to prevent snarling and allowed for air pressure sensors to activate the appropriate motor when the amount of tape in the buffer went above or below preset points.

Data was read or written by the read and write heads respectively, in units called frames. In addition to the 1 or 0 data bits, each frame included parity bits (see ERROR CORRECTION). The frames were combined into blocks, with each block having a header in front of the data frames and one or more frames of check (parity) bits following the data.

The two predominant tape formats were the IBM format, which used variable-length data blocks (and thus could not be rewritten) and the DEC format, which used fixed-length blocks, allowing data to be rewritten in place, albeit at some cost in speed and efficiency.

During the 1960s, magnetic disks (see HARD DISK) increasingly came into use, and more of the temporary data being used by programs began to be stored on disk rather than on tape. Eventually, tapes were relegated to storing very large data sets or archiving old data.

However, when the first desktop microcomputers (such as the Apple II and Radio Shack TRS-80) came along in the late 1970s and early 1980s, they, like the first mainframes, had very limited main memory and disk drives were unavailable or expensive. As a result, programs (such as Bill Gates's Microsoft Basic) often came on tape cassettes, and the computer included an interface allowing it to be connected to an ordinary audio cassette recorder. However, this use of tapes was quite short-lived, and was soon replaced by the floppy disk drive and later, hard drives and CD-ROM drives.

TAPES AS BACKUP DEVICES

By the 1990s, PC users generally used tapes only for making backups. A typical backup tape drive uses DAT (digital audio tape) cartridges that hold from hundreds of megabytes to several gigabytes of data. Most drives use a rotating assembly of four heads (two read and two write) that verify data as it's being written. As a backup medium, tape has a far lower cost per gigabyte than disk devices. It is easy to use and can be set up to run unattended (except for periodically changing cartridges).

However, since tapes are written and read sequentially, they are not convenient for restoring selected files (see also BACKUP AND ARCHIVE SYSTEMS). Many smaller installations now prefer using a second ("mirror") hard drive as backup, using disk arrays (see DISK ARRAY) or using recordable CDs or optical drives for smaller amounts of data (see CD-ROM).

Many large companies and government agencies have thousands of reels of tape stored away in their vaults since the 1960s, including data returned from early NASA space missions. As time passes, it becomes increasingly difficult to guarantee that this archived data can be successfully read. This is due both to gradual deterioration of the medium and the older data formats becoming obsolete.

The Storagetek 4400 Tape Silo uses multiple robot devices to store and mount tapes under remote control. This installation at the NAS (NASA Advanced Supercomputing) center can store up to 3,300 terabytes (trillions of bytes). (NASA PHOTO)

Further Reading

Rothenberg, J. "Ensuring the Longevity of Digital Documents." *Scientific American*. 272, no. 1, 42–47.

White, Ron. *How Computers Work*. Millennium Ed. Indianapolis, Ind.: Que, 1999.

TCP/IP

Contrary to popular perception, the Internet is not e-mail, chat rooms, or even the World Wide Web. It is a system by which computers connected to various kinds of networks and with different kinds of hardware can exchange data according to agreed rules, or protocols. All the applications mentioned (and many others) then use this infrastructure to communicate.

TCP/IP (Transmission Control Protocol/Internet Protocol) provides the rules for transmitting data on the Internet. It consists of two parts. The IP (Internet Protocol) routes packets of data. The header information also includes:

- The total length of the packet. In theory packets can be as large as 65K bytes; in practice they are limited to a smaller maximum.
- An identification number that can be used if a packet is broken into smaller pieces for efficiency in transmission. This allows the packet to be reassembled at the destination.
- A "time to live" value that specifies how many hops (movements from one intermediate host to another) the packet will be allowed to take. This is reduced by 1 for each hop. If it reaches 0, the packet is assumed to have gotten "lost" or stale, and is discarded.
- A protocol number (the protocol is usually TCP, see below).
- A checksum for checking the integrity of the header itself (not the data in the packet).
- The source and destination addresses.

The source and destination are given as IP addresses, which are 32 bits long and typically written as four sets of up to three numbers each—for example, 208.162.106.17

A NETWORK OF NETWORKS

As the name implies, the Internet is a network that connects many local networks. The IP address includes an ID for each network (called a subnet) and each host computer on the network. The arrangement and meaning of these fields differs somewhat among five classes of IP addresses. The first three classes are designed for different

TCP/IP (1)

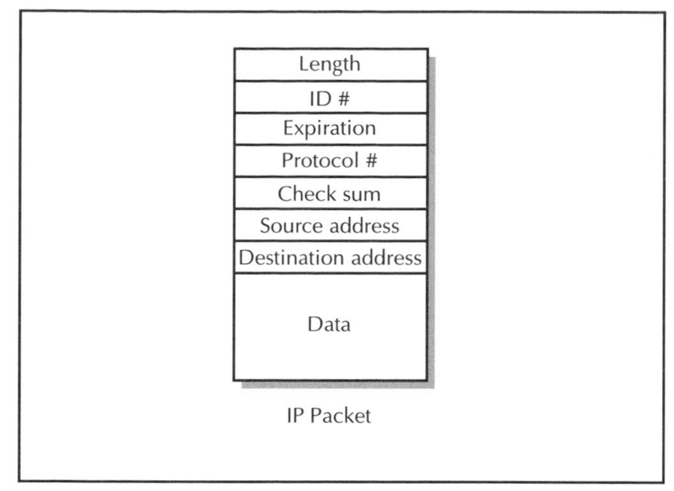

The header fields and data for an IP (Internet Protocol) packet. The packets can travel over different routes to the destination address and then be reassembled in the correct order.

sizes of networks, and the latter two are used for special purposes such as "multicasting" where the same data packet is sent to multiple hosts.

Many Internet users (at home as well as in offices) are part of a local network (see LOCAL AREA NETWORK). Typically, all users on the local network share a single Internet connection, such as a DSL or cable line. This sharing is enabled by having one computer (or a hardware device

TCP/IP (2)

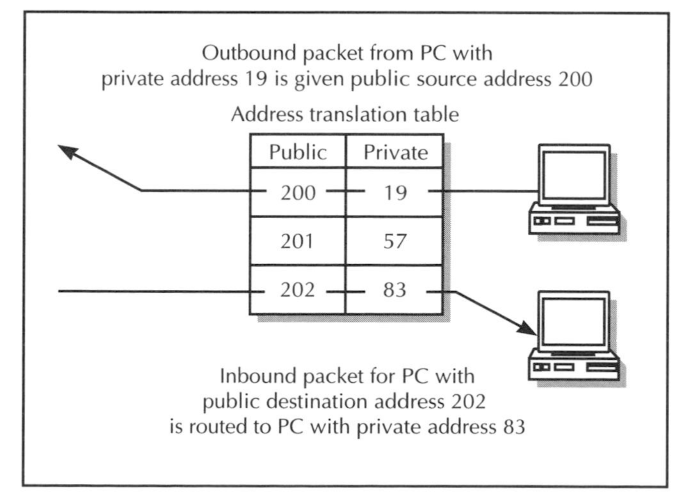

Network Address Translation (NAT) can protect computers on a local network by giving outgoing packets an arbitrary public source address in place of a computer's actual (private) address. Incoming packets addressed to that public address are then routed to the correct private address using a table.

called a router) connected to the Internet, serving as the link between the local network and the rest of the world. A facility called Network Address Translation (NAT) assigns a private IP address to each computer on the network. When a computer wants to make an Internet connection, its outgoing packet is assigned a public IP address from a pool. When packets replying to that public address are received, they are converted back to the private address and thus routed to the appropriate user.

NAT has the benefit of providing some security against intrusion, since from the outside only the single public IP address is visible, not the private addresses of the various machines on the network. However, using NAT (and a similar scheme called PAT that allows difference hosts to use the same IP address by being assigned different port numbers) causes some slowdown because of the translation process.

Another facility, Dynamic Host Configuration Protocol (DHCP), is used to assign an arbitrary available public IP address to each host when it connects to the network. This system is now used by most DSL and cable systems, and it reduces the danger of running out of IP numbers (each network is assigned a range of numbers, and is thus limited to that many IP addresses).

DOMAIN NAME SYSTEM

Internet users typically don't have to worry about IP numbers, except perhaps when configuring their software. Instead they use alphabetic addresses, such as http://www.factsonfile.com. The Domain Name System (DNS) sets up a correspondence between the names (which include domains such as .com for commercial or .edu for educational institutions) and the IP numbers (see DOMAIN NAME SYSTEM).

TRANSMISSION CONTROL PROTOCOL

The Transmission Control Protocol (the TCP part of TCP/IP) controls the flow of packets that have been structured as described above. To use TCP, the sending computer opens a special file called a socket, which is identified by the computer's IP number plus a port number. Standard port numbers are used for the various protocols such as www (Web) and ftp (File Transfer Protocol). The receiving computer connects using a corresponding socket. TCP includes basic flow control and error-checking features similar to those used for most data transmissions. For some applications (such as connecting to the domain name server) error control isn't needed, so a simpler protocol called the User Datagram Protocol is used.

THE BIG PICTURE

How does TCP/IP fit into the use of the Internet? When an application such as an e-mail program, Web browser, or ftp client makes a connection, IP packets using TCP flow control carry the requests from the client to the server and the server's response back to the client. Each application has its own protocol to specify these requests (such as for a webpage). For e-mail the protocol is SMTP (Simple Mail Transfer Protocol); Web servers and browsers use HTTP (Hypertext Transfer Protocol); and for file transfers it is FTP (File Transfer Protocol). (See also E-MAIL, HTML, HYPERTEXT AND HYPERMEDIA, and FILE TRANSFER PROTOCOLS.)

Further Reading

Comer, D. *Internetworking with TCP/IP.* 3 vols., various editions. Upper Saddle River, N.J.: Prentice Hall, 1998–2000.

Kessler, Gary C. "An Overview of TCP/IP Protocols and the Internet." http://www.garykessler.net/library/tcpip.html

Stevens, W. Richard. *TCP/IP Illustrated: vol. 1, The Protocols.* Reading, Mass.: Addison-Wesley, 1994.

technical support

Competition and user demand have led to modern software becoming increasingly complex and often stuffed with esoteric features. Despite improvement in programs' own built-in help systems (see HELP SYSTEMS), users will often have questions about how to perform particular tasks. There will also be times when a program doesn't perform as the user expects because the user misunderstands some feature of the program, the program has an internal flaw (see BUGS AND DEBUGGING), or there is a problem in interaction between the application program, the user's operating system, or the user's hardware (see DEVICE DRIVER).

To get help when problems arise, users often turn to the technical support facility, often called a help desk. This facility can either be internal to an organization (helping the organization's computer users with a wide range of problems), or belong to the maker of the software (and available in varying degrees to all licensed users of that software).

Large help desks often have two or more levels or tiers of assistants. The first tier assistant can respond to the simplest (and usually most common) situations. For example, a first-tier support person for a cable or DSL Internet Service Provider could tell a caller whether service has been interrupted in their area and if not, take the caller through a set of steps to reset a "hung" modem. If the situation is more complex (or the basic steps do not resolve it), the call will be "escalated" to the next tier, where a more experienced technician can address detailed software configuration issues.

Advanced technical support representatives can use tools such as remote operation software that lets them take over control of the user's PC in order to see exactly what is going on. They can also submit detailed problem reports to engineers in cases were a modification (patch) to the software might be needed.

SUPPORT ALTERNATIVES

Users who are dissatisfied with the wait for phone support or dealing with poorly trained support personnel may be able to take advantage of alternative sources of information and support. Most software companies now have websites that include a support section that offers services such as

- Frequently Asked Questions (FAQ) files with answers to common problems.
- A searchable "knowledge base" of articles relating to various aspects of the software, including compatibility with other products, operating system issues, and so on.
- Forms or e-mail links that can be used to submit questions to the company. Typically questions are answered in one or two working days.
- A bulletin board where users can share solutions and tips relating to the software.

Websites for publications such as *PC Magazine* and ZDNet also offer articles and other resources for working with the various versions of Microsoft Windows and popular applications.

TECHNICAL SUPPORT ISSUES

As with many other aspects of the computer industry, the changing economic climate has had an impact on technical support practices. Many companies are hoping that providing more extensive Web-based technical support will reduce the need for help desk representatives. Companies that don't want to create their own support websites can turn to consultants such as Expertcity.com or PCSupport.com to create and manage such services for a fee.

Another way companies have sought to reduce help desk costs is to outsource their technical support operations. Most software companies are in areas with a relatively high cost of labor. With modern communications and network services, there is no need for the help desk personnel to be at the company headquarters. Workers in less expensive parts of the United States or even in countries such as India that have a large pool of technically trained, English-speaking persons can sometimes offer help services at a lower cost than running an in-house help desk, even when the cost of training and phone line charges are taken into account.

Poor technical support can lead customers to switch to competing products. While this may not be much of a concern in a rapidly expanding industry (where new customers seem to be available in abundance), the situation is different in stagnant or contracting economic conditions. Trying to reduce technical support costs may bring some short-term help to the bottom line, but in the longer run the result might be fewer customers and less revenue. An alternative approach is to consider technical support to be part of a broad effort to maintain customer loyalty; this is often called Customer Relationship Management (CRM). With regard to technical support, CRM is implemented by using software to better track the resolution of customer's problems as well as to use information obtained in the support process to offer the customer additional products or services custom-tailored to individual situations. With such an approach the effort to provide better technical support is seen not simply as a necessary business expense but as an investment with an expected (though hard to measure) return.

Further Reading

CRM Forum. http://www.crm-forum.com/
Czegel, Barbara. *Running an Effective Help Desk*. 2nd ed. New York: Wiley, 1998.
Help Desk Institute. http://www.helpdeskinst.com/
McBride, Dione. *A Guide to Help Desk Technology, Tools & Techniques*. Boston: Course Technology, 2000.
"Supportgate." [Portal for technical support professionals] http://www.supportgate.com/home.html
Wooten, Bob. *Building & Managing a World Class IT Help Desk*. New York: McGraw Hill, 2001.

technical writing

Users of complex systems require a variety of instructional and reference materials, which are produced by technical writers and editors. (It should be noted that technical writing covers many areas other than computer software and systems. However, it is the latter that fall within the scope of this book.)

The traditional products produced by technical writers in the computer industry can be divided into three broad categories: software manuals, trade books, and in-house documentation for developers.

SOFTWARE MANUALS

Until the mid-1990s, just about every significant software product came with a manual (or a set of manuals). A typical manual might include an overview of the program, an introductory tutorial, and a complete, detailed reference guide to all commands or functions.

In theory, staff technical writers (or sometimes contractors) develop the manuals during the time the program is being written. They have access to the programmers for asking questions about the program's operation, and they receive updates from the developers that describe changes or added features. In practice, however, writers may not be assigned to a project until the program is almost done. The programmers, who are under deadline pressure, may not be very communicative, and the writers may have to make their best guess about some matters. The result can be a manual that is no longer "in synch" with the program's actual feature set.

Technical writers often work in a publications department with other professionals including editors, desktop publishers, and graphics specialists. While manuals can be written using an ordinary word processing program, many departments use programs such as Framework that are designed for the production and management of complex documents.

In recent years, many software manufacturers have stopped including printed user manuals with their packages, or include only slim "Getting Started" manuals. As a money-saving measure the traditional documentation is often replaced by a PDF (Adobe Portable Document Format) document on the CD. There is also a greater reliance on extensive on-line help, using either a Windows or Macintosh-specific format or the HTML format that is the lingua franca of the World Wide Web. (See also DOCUMENTATION, USER.)

Technical writers have thus had to learn how to construct Help files in these various formats (see also AUTHORING SYSTEMS, HELP SYSTEMS, and HTML). Creation of interactive tutorials also requires knowledge of multimedia formats and even animation (such as Flash).

TRADE BOOKS

As millions of people became new computer users during the 1980s, a thriving computer book publishing industry offered users a more user-friendly approach than that usually provided in the manuals issued by the software companies. The "Dummies" books, offering bite-sized servings of information written in a breezy style and accompanied by cartoons, eventually spread beyond computers into hundreds of other fields and the format was then copied by other publishers. Publishers such as Sams, Coriolis, and particularly O'Reilly have aimed their offerings at more experienced users, programmers, and multimedia developers.

Computer trade books are often written by experienced developers and systems programmers who can offer advanced knowledge and "tips and tricks" to their less experienced colleagues. Since many technical "gurus" are not experienced writers, the best results often come from collaboration between the expert and an experienced technical writer and/or editor who can review the material for completeness, organization, and clarity.

In recent years there has been some contraction in the computer book industry. This has arisen from several sources: improved on-line help included in products; the dominance of many applications areas by a handful of products; and fewer people needing beginner-level instruction.

IN-HOUSE DOCUMENTATION

Many technical writers work within software companies or in the information systems departments of other corporations, universities, or government agencies. Their work is generally more highly structured than that of the manual or book writer. As part of a development team, a technical writer may be in charge of creating documentation describing the data structures, classes, and functions within the program. This task is aided by a variety of tools including facilities for extracting such information automatically from C++ or Java programs. The writer may also be responsible for maintaining logs that show each change or addition made to the program during each compiled version or "build."

This type of technical writing requires detailed knowledge of operating systems, programming languages, software development tools, and software engineering methodology. It also requires the ability to work well as part of a team, often under conditions of high pressure.

TECHNICAL WRITING AS A PROFESSION

Until the 1980s, few institutions offered degrees in technical writing. Programmers with an interest in writing or writers with a technical bent entered the field informally. During the 1980s, the number of degree offerings increased, and people began to specifically prepare for the field, often by earning a computer science degree with a specialization in technical writing. Organizations such as the Society for Technical Communication have offered technical writers and editors a forum for discussing their profession, including issues relating to certification.

Further Reading

Bremer, Michael. *The User Manual Manual: How to Research, Write, Edit, Test & Produce a Software Manual.* Concord, Calif.: Untechnical Press, 1999.

Lindsell-Roberts, Sheryl. *Technical Writing for Dummies.* New York: Hungry Minds, 2001.

Society for Technical Communication. http://www.stc.org/

telecommunications

Since its birth in the mid-20th century, the digital computer and the telephone have had a close mutual relationship. Many of the first programmable calculators and computers built in the early 1940s used relays and other components that were being manufactured for the increasingly automated phone system (see AIKEN, HOWARD). The phone industry contributed ideas as well as hardware. Scientists at the Bell Laboratories carried out fundamental research into information theory that would soon be applied to data communications (see SHANNON, CLAUDE).

As computers became more capable in the 1950s and 1960s, they began to return the favor, making possible increasing automation for the phone system. Meanwhile, computers were starting to be hooked up to telephone lines (see MODEM) so they could exchange data and allow their users to communicate (see NETWORK).

The development of a global network (see INTERNET) and its growth through the 1980s provided a universal platform for data communications. At first, the Internet was used mostly by academics and engineers, but the advent of the World Wide Web and in particular, graphical Web browsers made Internet access ubiquitous among small businesses and home users by the late 1990s.

Institutional Internet users often had fast access through dedicated phone lines (designated T-1, and so on), while homes, small businesses, and schools were limited to much slower dial-up access. This began to change in the late 1990s as alternatives to POTS ("plain old telephone service") emerged in the form of DSL (a much faster service running over regular phone lines) and cable modems that used the infrastructure that already brought TV to millions of homes.

IMPACT OF DEREGULATION

Prior to the court-ordered breakup of AT&T in 1984, the phone industry functioned in a monolithic way and was not very responsive to the needs of the growing computer networking industry.

The breakup of AT&T led to growing competition, providing a wider variety of telecommunications equipment and lower phone rates just as PC users were starting to buy modems and sign up with on-line services and bulletin boards. The growing deregulation movement in the 1990s (culminating in the Telecommunications Act of 1996) furthered this process by opening cable and broadcast television, radio, and other wireless communication to competition.

Although as of the early 2000s relatively few consumers actually had a choice between cable and DSL at their location, these technologies, along with wireless and satellite, are expected to become increasingly available. The use of fiber optic networks also continues to grow, and new buildings and even homes are being built from the ground up to incorporate network outlets for computers and "smart appliances."

CONVERGENCE AND THE FUTURE

The ability of the Internet to transmit any sort of data virtually anywhere at relatively low cost has created new alternatives to traditional communications technologies. For example, sending digitized voice telephone calls as packets over the Internet can provide a lower-cost alternative to conventional long distance calling (see INTERNET TELEPHONY). At the same time, previously separate functions are converging into "smarter" devices. Thus, the handheld computer and the cell phone seem to be converging into a single device that can provide data management (see PERSONAL INFORMATION MANAGEMENT), Web browsing, and communications in a single package.

Computers and communications technology will continue to grow more intertwined. Today it is increasingly hard to distinguish information technology, media content, and communications technology as being distinctive sectors. After all, a consumer can watch a movie in the theater or later on broadcast, cable, or satellite TV, rent it on commercial videotape or DVD disk (playable on PCs as well as dedicated players), or even view it as a streaming file direct from the Internet. Although these technologies have differing technical constraints, their end products are the same for the computer.

This multiplicity of function means that the competitive environment is increasingly hard to predict, since there are so many possible players. The companies offering content through this variety of technologies are also increasingly intertwined, as typified by the merger of publishing giant Time-Warner with the premier on-line service to form AOL-Time Warner. Is this a media company or a communications company, or something new?

For analysts, studying any technology requires awareness of the many possible alternatives, while studying any application means considering the many possible technological implementations. For policy makers and regulators, the challenge is to provide for such public goods as equal access, privacy, and protection of intellectual property in a communications infrastructure that is truly global in scope and evolving at a pace that frequently outdistances the political process.

Further Reading

Benjamin, Stuart Minor, Douglas Lichtman, and Howard A. Shelanski. *Telecommunications Law and Policy.* Durham, N.C.: Carolina Academic Press, 2001.

Green, James Harry. *The Irwin Handbook of Telecommunications.* 4th ed. New York: McGraw-Hill Professional Publishing, 2000.

National Academy of Sciences. Computer Science and Telecommunications Board. http://www4.nationalacademies.org/cpsma/cstb.nsf

———. *The Changing Nature of Telecommunications/Information Infrastructure* (1995). Available on-line at http://www.nap.edu/books/030905091X/html/index.html

———. *The Evolution of Untethered Communications* (1997). Available on-line at http://stills.nap.edu/html/evolution/index.html#contents

———. *Keeping the U.S. Computer and Communications Industry Competitive: Convergence of Computing, Communications and Entertainment* (1995). Available on-line at http://www.nap.edu/books/0309050898/html/index.html

Pecar, Joseph A., and David A. Garbin. *The New McGraw-Hill Telecom Factbook.* 2nd ed. New York: McGraw-Hill, 2000.

telecommuting

According to a July 2000 survey by the Behavioral Research Center, about 9.3 million Americans work from home at least one day per week. Although about a 10th of these are self-employed persons, most hold regular corporate positions but are able to do some of their work from home.

Telecommuting was made possible by the growing capabilities of home computers and the availability of network connections that allow the worker at home to have access to most of the people and facilities that would be available if the worker were on site. Workers and companies that promote telecommuting often cite the following advantages:

- Elimination of stressful, time-wasting commutes
- Workers may be more productive because they have fewer office distractions, unnecessary meetings, etc.
- Reduction of traffic and air pollution
- Greater flexibility in working hours
- The ability of working parents with small children to combine child care and work to some extent
- Reduction of costs associated with office facilities

However, telecommuting has its critics in management. Some of the problems or disadvantages cited include:

- Worker productivity may decrease due to lack of sufficient discipline and workers becoming distracted at home.
- Managers may have trouble keeping track of or evaluating the activities of workers who are not physically present.
- Telecommuters may miss critical information and go "out of the loop."
- Possible legal liabilities and application of OSHA rules to home working situations.

It is true that telecommuting is suitable mainly for jobs that involve information processing rather than person-to-person contact, such as service jobs. However, the use of videoconferencing or Web conferencing technology increasingly makes it possible for suitably equipped telecommuters to participate in meetings almost as directly as if they were physically present (see CONFERENCING SYSTEMS and TELEPRESENCE).

In some cases involving videoconferencing or other activities that require high-powered computer systems and high bandwidth connections, telecommuters physically commute to a "satellite work center" near their home that has the appropriate equipment. This can provide some of the advantages of telecommuting such as flexibility and lower commute and office costs.

A number of issues must be worked out between workers and management for any telecommuting program, including:

- Who will pay for the equipment used by the telecommuter

- Procedures for monitoring the work
- How telecommuters will participate in meetings (either remotely or in person)
- The portion of the worker's hours involving telecommuting, and the portion requiring attendance in-house

TRENDS

Telecommuting was touted in the mid-1990s as the wave of the future. In reality, the statistics given earlier suggest while it is a viable option for a significant minority of workers, telecommuting is not growing as rapidly as had been predicted. The growing power of desktop PCs and the availability of broadband (DSL or cable) network connections should help facilitate telecommuting. In the longer term new technologies may make the distinction between telecommuters and physically present workers much less important (see TELEPRESENCE and VIRTUAL REALITY).

Further Reading

American Telecommuting Association. http://www.knowledgetree.com/ata.html

Dziak, Michael J. and Gil Jordan. *Telecommuting Success: A Practical Guide for Staying in the Loop While Working Away from the Office.* Indianapolis, Ind.: Park Avenue, 2001.

Johnson, Nancy J., ed. *Telecommuting and Virtual Offices: Issues and Opportunities.* Hershey, Pa.: Idea Group Publishing, 2001.

Network World. Net Worker Research: Telecommuting. http://www.nwfusion.com/net.worker/research/telecommuting.html

telepresence

An old phone company slogan asserted that "long distance is the next best thing to being there." Today technology has made the ability to "be there" a much more complete experience. It is now quite common for businesspersons to "attend" a meeting in a distant city using video cameras to see and be seen, with images and voice traveling over special leased lines or high speed Internet connections (see VIDEOCONFERENCING).

While videoconferencing and suitable software allows remote interaction and collaboration (such as being able to build a spreadsheet or diagram together), the remote participant has little ability to physically interact with the environment. He or she can't walk freely around, perhaps joining other meeting participants in an adjacent room while they have pastries and coffee. The remote participant also cannot handle physical objects such as models.

There are two basic approaches to letting persons have an unconstrained experience in a remote environment. The first is to use technology to create a virtual presence where a person can experience a simulated environment from many different angles and move freely through it while grasping and manipulating objects (see VIRTUAL REALITY). In a virtual environment each participant can be represented by an "avatar" body that can be programmed to move in response to head trackers, gloves, and other devices.

However, a virtual reality is an artificial representation of the world. A group of people having a meeting in a physical space can't interact with someone who is in virtual space except in the most rudimentary ways. To be on an equal footing, all participants would have to be in either physical or virtual space.

TELEROBOTICS

The alternative is to connect the remote participant to a mobile robot (this is sometimes called *telerobotics*). Such robots already exist, although their capabilities are limited and they are not yet widely used for meetings. Rodney Brooks, Director of the MIT Artificial Intelligence Laboratory, foresees a not very distant future in which such robots will be commonplace.

The robot will have considerable built-in capabilities, so the person who has "roboted in" to it won't need to worry about the mechanics of walking, avoiding obstacles, or focusing vision on particular objects. Seeing and acting through the robot, the person will be able to move around an environment as freely as persons who are physically present. The operator can give general commands amounting to "walk over there" or "pick up this object" or perform more delicate manipulations by using his or her hands to manipulate gloves connected to a force-feedback mechanism.

Brooks sees numerous applications for robotic telepresence. For example, someone at work could "robot in" to his or her household robot and do things such as checking to make sure appliances are on or off, respond to a burglar alarm, or even refill the cat's food dish. Robotic telepresence could also be used to bring expertise (such as that of a surgeon) to any site around the world without the time and expense of physical travel. Indeed, robots may be the only way (for the foreseeable future) that humans are likely to explore environments far beyond Earth (see SPACE EXPLORATION AND COMPUTERS).

Further Reading

Brooks, Rodney A. *Flesh and Machines: How Robots Will Change Us.* New York: Pantheon Books, 2002.

Sheridan, T. B. *Telerobotics, Automation and Human Supervisory Control.* Cambridge, Mass.: MIT Press, 1992.

The Telegarden. http://telegarden.aec.at

template

The term *template* is used in a several contexts in computing, but they all refer to a general pattern that can be customized to create particular products such as documents.

In a word processing program such as Microsoft Word, a template (sometimes called a style sheet) is a document that comes with a particular set of styles for various elements such as titles, headings, first and subsequent paragraphs, lists, and so on. Each style in turn consists of various characteristics such as type font, type style (such as bold), and spacing. The template also includes properties of the document as a whole, such as margins, header, and footer.

To create a new document, the user can select one of several built-in templates for different types of documents such as letters, faxes, and reports, or design a custom template by defining appropriate styles and properties. Special sequences of programmed actions can also be attached to a template (see MACRO).

Templates can be created and used for applications other than word processing. A spreadsheet template consists of appropriate macros and formulas in an otherwise blank spreadsheet. When it is run, the template prompts the user to enter the appropriate values and then the calculations are performed. A database program can have input forms that serve in effect as templates for creating new records by inputting the necessary data.

CLASS TEMPLATES

Some programming languages use the term *template* to refer to an abstract definition that can be used to create a variety of similar classes for handling different types of data, which in turn are used to create actual objects. For example, once the programmer defines the following template:

```
template<class ANY_TYPE>
ANY_TYPE maximum(ANY_TYPE a, ANY_TYPE b)
{
  return (a > b) ? a : b;
}
```

This template provides any class with the maximum function, which can compare any two objects of that class and return the larger one. (See C++, CLASS, and OBJECT-ORIENTED PROGRAMMING.)

Further Reading

Stroustrup, Bjarne. *The C++ Programming Language.* 3rd ed. Reading, Mass.: Addison-Wesley, 1997.

"Tech. Talk About C++ Templates." http://www.comeaucomputing.com/techtalk/templates/

Template Central. http://www.templatecentral.com/index.asp

terminal

Throughout the 1950s, operators interacted with computers primarily by punching instructions onto cards that were then fed into the machine (see PUNCHED CARDS AND PAPER TAPE). Although this noninteractive batch processing procedure would continue to be used with mainframe computers during the next two decades, another way to use computers began to be seen in the 1960s.

With the beginning of time-sharing computer systems, several users could run programs on the computer at the same time. The users communicated with the machine by typing commands into a Teletype or similar device. Such a device is called a *terminal.*

The simple early terminals did little more than accept lines of text commands from the user and print responses or lines of output coming from the computer. However, a newer type of terminal began to replace the Teletype. It consisted of a keyboard attached to a televisionlike cathode ray tube (CRT) display. User still typed commands, but the computer's output could now be displayed on the screen.

Gradually, CRT terminals gained additional capabilities. The text being entered was now stored in a memory buffer that corresponded to the screen and the user could use special control commands or keys to move the input cursor anywhere on the screen when creating a text file. This made it much easier for users to revise their input (see TEXT EDITING). These "smart terminals" had their own small processor and ran software that provided these functions.

During the 1970s, the UNIX operating system developed a sophisticated way to support the growing variety of terminals. It provided a library of cursor-control routines (called *curses*) and a database of terminal characteristics (called *termcap*).

When the personal computer came along, it had a keyboard, a processor, the ability to run software, and a connection for a TV or monitor. The PC thus had all the ingredients to become a smart terminal. Indeed, a modern PC is a terminal, but users don't usually have to think in those terms. The exception is when the user runs a communications program to connect to a remote computer (perhaps a bulletin board) with a modem. These programs, such as the Hyperterminal program that comes with Windows, allow the PC to emulate (work like) one of the standard terminal types such as VT-100. This ability to emulate a standard terminal means that any software that supports that physical terminal should also work remotely with a PC.

Today most interaction with remote programs is through a Web browser, although protocols such as telnet are still used to provide terminallike access to remote programs. Many commands previously entered as text lines in a terminal are now given using the mouse with menus and icons (see USER INTERFACE).

The relationship between a terminal and remote computer is analogous to that between a workstation (or desktop PC) and the network server in that the burden of processing is divided between the two devices in various ways (see CLIENT-SERVER COMPUTING). A "thin client" PC performs relatively little processing with the server doing most of the work.

Specialized terminals are still used for many applications. An ATM, for example, is a special-purpose banking terminal driven by a keypad and touch screen.

Further Reading

"Archive of Video Terminal Information." http://www.cs.utk.edu/~shuford/terminal_index.html

Free Software Foundation. "The Termcap Library." http://www.gnu.org/manual/termcap-1.3/termcap.html

Overview of How Terminals Work in Linux. http://www.linuxdoc.org/HOWTO/Text-Terminal-HOWTO-5.html

text editor

As noted in the previous article (see TERMINAL), an alternative to batch-processing punch card driven computer operations emerged in the 1960s in the form of text commands typed at an interactive console or terminal. At first text could be typed only a line at a time and there was no way to correct a mistake in a previous line.

Soon, however, programmers began to create text editing programs. The first editors were still line-oriented, but they stored the lines for the current file in memory. To display a previous line, the user might simply type its number. To correct a word in the line the user might type something like

```
c/fot/for
```

to change the typo "fot" to the word "for" in the current line.

Starting in the early 1970s, the UNIX system provided both a line editor (ed or ex) and a "visual editor" (vi). The latter editor works with terminals that can display full screens of text and allow the cursor to be moved anywhere on the screen. This type of editor is also called a screen editor.

Most ordinary PC users use word processors rather than text editors to create documents. Unlike a text editor, a word processor's features are designed to create output that looks as much like a printed document as possible. This includes the ability to specify text fonts and styles. However, most systems also include a simpler text editor that can be useful for making quick notes (in Windows this program is indeed called Notepad).

The primary use of text editors today is to create programs and scripts. These must generally be created using only standard ASCII characters (see CHARACTERS AND STRINGS), without all the embedded formatting commands and graphics found in word processing documents. Programmer's text editors can be very sophisticated in their own right, providing features such as built in syntax checking and formatting or (as with the Emacs editor) the ability to program the editor itself. Ultimately, however, program editors must create a source code file that can be processed by the compiler.

Text editors are also useful for writing quick, short scripts (see SCRIPTING LANGUAGES) and can be handy for writing HTML code for the Web. However, many webpages are now designed using word processor–like programs that convert the WYSIWYG (what you see is what you get) formatting into appropriate HTML codes automatically.

Further Reading

Finseth, C. A. *The Craft of Text Editing: Emacs for the Modern World.* New York: Springer-Verlag, 1991.

Lamb, L. *Learning the vi Editor.* 5th ed. Sebastopol, Calif.: O'Reilly, 1990.

Smith, P. D. *An Introduction to Text Processing.* Cambridge, Mass.: MIT Press, 1990.

"Shareware Text Editors." http://www.blockdev.net/Community/Editors/shareware.shtml

Yahoo! Text Editors page. http://dir.yahoo.com/Computers_and_Internet/Software/Text_Editors/

Torvalds, Linus

(1969–)
Finnish
Software Developer

Linus Torvalds developed Linux, a free version of the UNIX operating system that has become the most popular alternative to proprietary operating systems.

Torvalds was born on December 28, 1969, in Helsinki, Finland. His childhood coincided with the microprocessor revolution and the beginnings of personal computing. At the age of 10, he received a Commodore PC from his grandfather, a mathematician. He learned to write his own software to make the most out of the relatively primitive machine.

In 1988, Torvalds enrolled in the University of Helsinki to study computer science. There he encountered UNIX, a powerful and flexible operating system that was a delight for programmers who liked to tinker with their computing environment. Having experienced UNIX, Torvalds could no longer be satisfied with the operating systems that ran on most PCs, such as MS-DOS, which lacked the powerful command shell and hundreds of utilities that UNIX users took for granted.

Torvalds's problem was that the UNIX copyright was owned by AT&T, which charged $5,000 for a license to run UNIX. To make matters worse, most PCs weren't powerful enough to run UNIX anyway.

At the time there was already a project called GNU underway (see OPEN SOURCE and STALLMAN, RICHARD). The Free Software Foundation was attempting to replicate all the functions of UNIX without using any of AT&T's proprietary code. This would mean that the AT&T copyright would not apply, and the functional equivalent of UNIX could be given away for free. Stallman and the FSF had already provided key tools such as the C compiler and the Emacs program editor. However, they had not yet created the heart of the operating system (see KERNEL). The kernel contains the essential functions needed for the operating system to control the computer's hardware, such as creating and managing files on the hard drive.

In 1991, Torvalds wrote his own kernel and put it together with the various GNU utilities to create what soon became known as Linux. Torvalds adopted the open source license (GPL) pioneered by Stallman and the FSF, allowing Linux to be distributed freely. The software soon spread through ftp sites on the Internet, where hundreds of enthusiastic users (mainly at universities) helped to improve Linux, adding features and writing drivers to enable it to work with more kinds of hardware.

By the mid-1990s, the free and reliable Linux had become the operating system of choice for many website developers. Torvalds, who still worked at the University of Helsinki as a researcher, faced an ever-increasing burden of coordinating Linux development and deciding when to release successive versions. As companies sprang up to market software for Linux, they offered Torvalds very attractive salaries, but he did not want to be locked into one particular Linux package (distribution).

Instead, in 1997 Torvalds moved to California's Silicon Valley, where he became a key software engineer at Transmeta, a company that makes Crusoe, a processor designed for mobile computing. However, Torvalds continues to keep strong ties to the Linux community.

Further Reading

Dibona, Chris. *Open Sources: Voices from the Open Source Revolution.* Sebastopol, Calif.: O'Reilly, 1999.

Moody, Glenn. *Rebel Code: Linux and the Open Source Revolution.* New York: Allen Lane, 2001.

Torvalds, Linus. *Just for Fun: The Story of an Accidental Revolutionary.* New York: HarperBusiness, 2001.

transaction processing

Many computer applications involve the arrival of a set of data that must be processed in a specified way. For example, a bank's ATM system receives a customer's request to deposit money together with identification of the account and the amount to be deposited. The system must accept the deposit, update the account balance, and return a receipt to the customer. This is an example of real-time transaction processing.

Some applications process transactions in batches. For example, a company may run a program once a month that generates paychecks and withholding stubs from employee records that include hours worked, number of dependents claimed, and so on. Indeed, in the ATM example, the account balance is typically not updated during the on-line transaction, but instead a batch transaction is stored. Overnight that transaction will be processed together with other transactions affecting that account (such as checks), and the balance will then be officially updated. (The program module that keeps track of the progress of transactions is called a transaction monitor.)

There are several considerations that are important in designing transaction systems. While some transactions may simply involve a request for information and do not update any files, many transactions may require that several files or database records be updated. For example, a transfer of funds from a saving account to a checking account will require that both accounts be updated: the first with a debit and the second with a credit. What happens if the computer performs the savings debit but then goes down before the checking account can be credited? The result would be an upset customer whose money seems to have disappeared.

The solution to this problem is to design a process where the various updates are done not to the actual databases but to associated temporary databases. Once these potential transactions are posted, the system issues a "commit" command to the databases. Each database must send a reply acknowledging that it's ready to perform the actual update. Only if all databases reply affirmatively is the commit command given, which updates all the databases simultaneously.

Further Reading

Bernstein, P. A., and E. Newcomer. *Principles of Transaction Processing.* New York: Academic Press/Morgan Kaufmann, 1997.

Gray, J., and A. Reuter. *Transaction Processing: Concepts and Techniques.* New York: Academic Press/Morgan Kaufmann, 1992.

tree

The tree is a data structure that consists of individual intersections called *nodes*. The tree normally starts with a single root node. (Unlike real trees, data trees have their root at the top and branch downward.) The root connects to one or more nodes, which in turn branch into additional nodes, often through a number of levels. (A node that branches downward from another node is called that node's *child* node.) A node at the bottom that does not branch any further is called a terminal node or sometimes a leaf.

Trees are useful for expressing many sorts of hierarchical structures such as file systems where the root of a disk holds folders that in turn can hold files or additional folders, and so on down many levels. (A corporate organization chart is a noncomputer example of a hierarchical tree.)

The most common type of tree used as a data structure is the binary tree. A binary tree is a tree in which no node has more than two child nodes. To move through data stored in a binary tree, a program can use two pointers, one to the current node's left child and one to its right child (see POINTERS AND INDIRECTION). The pointers can then be used to trace the paths through the nodes. If the tree represents a file that has been sorted (see SORTING AND SEARCHING), comparing nodes to the desired value and branching accordingly quickly leads to the desired record.

Alternatively, the data can be stored directly in contiguous memory locations corresponding to the successive numbers of the nodes. This method is faster than having to trace through successive pointers, and a binary search algorithm can be applied directly to the stored data. On the other hand it is easier to insert new items into a linked list (see LIST PROCESSING).

A common solution is to combine the two structures, storing the linked list in a contiguous range of memory by storing its root in the middle of the range, its left child at the beginning of the range, its right child at the end, and then repeatedly splitting each portion of the range to store each level of children. Intuitively, one can see that algorithms for processing such stored trees will take a recursive approach (see RECURSION).

For efficiency it's important to keep all branches of a tree approximately the same length. A B-tree (balanced tree) is designed to automatically optimize itself in this way.

Trees lend themselves to game programs where a series of moves and their possible replies must be explored to varying levels. A chess program will typically create a tree from the current position, but use various criteria to determine which moves should be explored beyond just a few levels, thus "pruning" the game tree.

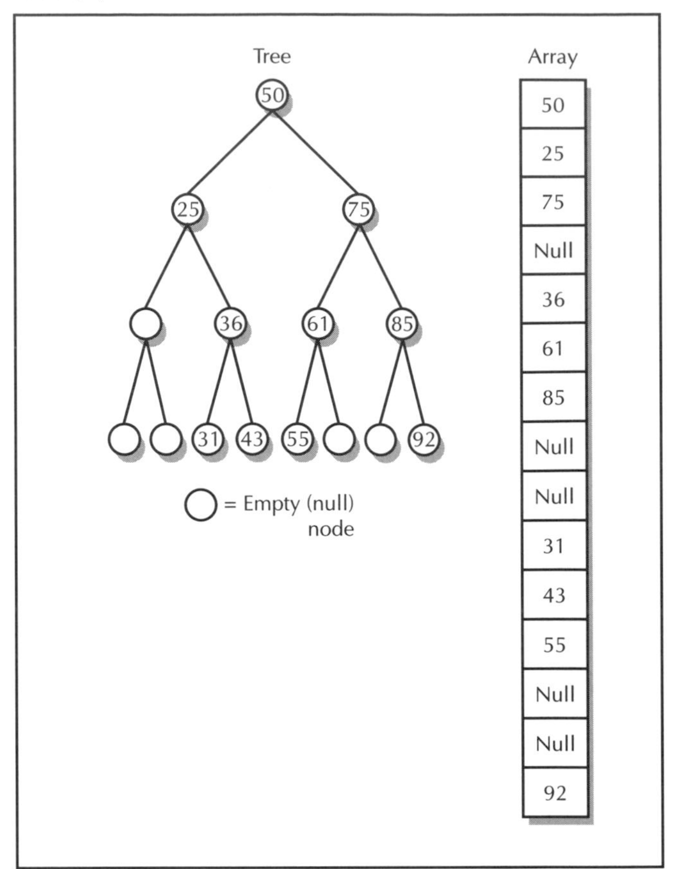

In a binary tree each node either has two branch (child) nodes or is a terminal node. Here a binary tree is shown with the equivalent representation in an array of memory locations. Notice that the numbers are stored level by level (50 in the top level, 25 and 75 in the second level, and so on). Null would be a special value (such as –1) representing an empty node.

Tree (1)

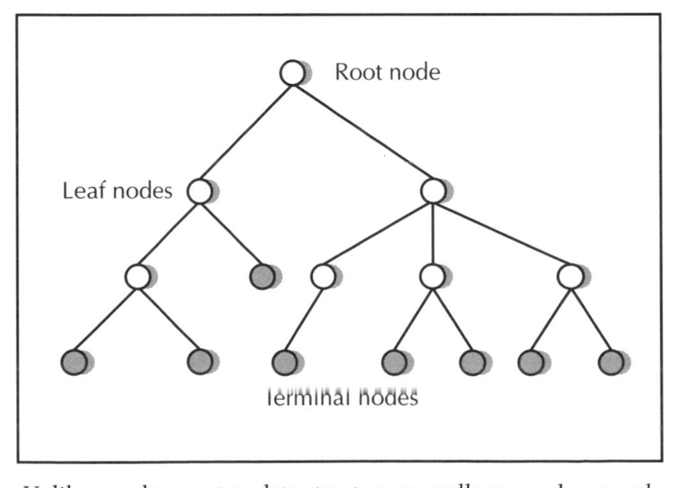

Unlike a real tree, a tree data structure generally grows downward from a root node to leaf nodes. A node that does not branch further is called a terminal node.

Further Reading

Brookshear, J. Glenn. *Computer Science: An Overview.* 6th ed. Reading, Mass.: Addison-Wesley, 2000.

Neubauer, Peter. "B-Trees: Balanced Data Structures." http://www.public.asu.edu/~peterjn/btree/

Turing, Alan Mathison

(1912–1954)

British

Mathematician, Computer Scientist

Alan Turing's broad range of thought pioneered many branches of computer science, ranging from the fundamental theory of computability to the question of what might constitute true artificial intelligence.

Turing was born in London on June 23, 1912. His father worked in the Indian (colonial) Civil Service, while his mother came from a family that had produced a number of distinguished scientists. As a youth Turing showed great interest and aptitude in both physical science and mathematics. When he entered King's College, Cambridge, in 1931, his first great interest was in probability, where he wrote a well-regarded thesis on the Central Limit Theorem.

Turing's interest then turned to the question of what problems could be solved through computation (see COMPUTABILITY AND COMPLEXITY). Instead of pursuing conventional mathematical strategies, he re-imagined the problem by creating the Turing Machine, an abstract "computer" that performs only two kinds of operations: writing or not writing a symbol on its imaginary tape, and possibly moving one space on the tape to the left or right. Turing showed that from this simple set of states (see FINITE STATE MACHINE) any possible type of calculation could be constructed. His 1936 paper "On Computable Numbers" together with another researcher's different approach (see CHURCH, ALONZO) defined the theory of computability. Turing then came to America, studied at Princeton University, and received his Ph.D. in 1938.

Alan Turing made vital contributions to the theory of computability and then raised many of the key questions that would later shape the field of artificial intelligence. (PHOTO © PHOTOGRAPHER, SCIENCE SOURCE/PHOTO RESEARCHERS)

Turing did not remain in the abstract realm, however, but began to think about how actual machines could perform sequences of logical operations. When World War II erupted, Turing returned to Britain and went into service with the government's Bletchley Park code-breaking facility. He was able to combine his previous work on probability and his new insights into computing devices to help analyze cryptosystems such as the German Enigma cipher machine and to design specialized code-breaking machines.

As the war drew to an end, Turing's imagination brought together what he had seen of the possibilities of automatic computation, and particularly the faster machines that would be made possible by harnessing electronics rather than electromechanical relays. In 1946, he received a British government grant to build the ACE (Automatic Computing Engine). This machine's design incorporated advanced programming concepts such as the storing of all instructions in the form of programs in memory without the mechanical setup steps required for machines such as the ENIAC. Another important idea of Turing was that programs could modify themselves by treating their own instructions just like other data in memory. However, the engineering of the advanced memory system led to delays, and Turing left the project in 1948 (it would be completed in 1950). Turing also continued his interest in pure mathematics and developed a new interest in a completely different field, biochemistry.

Turing's last and perhaps greatest impact would come in the new field of artificial intelligence. Working at the University of Manchester, Turing devised a concept that became known as the Turing Test. In its best-known variation, the test involves a human being communicating via a Teletype with an unknown party that might be either another person or a computer. If a computer at the other end is sufficiently able to respond in a humanlike way, it may fool the human into thinking it is another person. This achievement could in turn be considered strong evidence that the computer is truly intelligent. Since Turing's 1950 article computer programs such as ELIZA and Web "chatterbots" have been able to temporarily fool people they encounter, but no computer program has yet been able to pass the Turing Test when subjected to extensive probing questions by a knowledgeable person.

Alan Turing had a secret that was very dangerous in that time and place: He was gay. In 1952, the socially awkward Turing stumbled into a set of circumstances that led to his being arrested for homosexual activity, which was illegal and heavily punished at the time. The effect of his trial and forced medical "treatment" suggested that his death from cyanide poisoning on June 7, 1954, was probably a suicide.

Alan Turing's many contributions to computer science were honored by his being elected a Fellow of the British Royal Society in 1951 and by the creation of the prestigious Turing Award by the Association for Computing Machinery, given every year since 1966 for outstanding contributions to computer science.

Further Reading

Henderson, Harry. *Modern Mathematicians.* New York: Facts On File, 1996.

Herken, R. *The Universal Turing Machine.* 2nd ed. London: Oxford University Press, 1988.

Hodges, A. *Alan Turing: The Enigma.* New York: Simon & Schuster, 1983. Reprinted New York: Walker, 2000.

Turing, Alan M. "Computing Machinery and Intelligence." *Mind,* vol. 49, 1950, 433–460.

———. "On Computable Numbers, with an Application to the Entscheidungsproblem." *Proceedings of the London Mathematical Society,* vol. 2, no. 42, 1936–1937, 230–265.

———. "Proposed Electronic Calculator." In Carpenter, B. E. and R. W. Doran, eds. *A. M. Turing's ACE Report of 1946 and other Papers.* Charles Babbage Institute Reprint Series in the History of Computing, vol. 10. Cambridge, Mass.: MIT Press, 1986.

Turing website. http://www.turing.org.uk/turing/

U

UNIX

By the 1970s, time-sharing computer systems were in use at many universities and engineering and research organizations. Such systems, often running on computers such as the PDP series (see MINICOMPUTER), required a new kind of operating system that could manage the resources for each user as well as the running of multiple programs (see MULTITASKING).

An elaborate project called Multics had been begun in the 1960s in an attempt to create such an operating system. However, as the project began to bog down, two of its participants, Ken Thompson and Dennis Ritchie (see RITCHIE, DENNIS) decided to create a simple, more practical operating system for their PDP-7. The result would become UNIX, an operating system that today is a widely used alternative to proprietary operating systems such as those from IBM and Microsoft.

ARCHITECTURE

The essential core of the UNIX system is the kernel, which provides facilities to organize and access files (see KERNEL and FILE), move data to and from devices, and control the running of programs (processes). In designing UNIX, Thompson deliberately kept the kernel small, noting that he wanted maximum flexibility for users. Since the kernel was the only part of the system that could not be reconfigured or replaced by the user, he limited it to those functions that reliability and efficiency dictated be handled at the system level.

Another way in which the UNIX kernel was kept simple was through device independence. This meant that instead of including specific instructions for operating particular models of terminal, printers, or plotters within the kernel, generic facilities were provided. These could then be interfaced with device drivers and configuration files to control the particular devices.

A UNIX system typically has many users, each of whom may be running a number of programs. The interface that processes user commands is called the *shell*. It is important to note that in UNIX a shell is just another program, so there can be (and are) many different shells reflecting varying tastes and purposes (see SHELL). Traditional UNIX shells include the Bourne shell (sh), C shell (csh), and Korn shell (ksh). Modern UNIX systems can also have graphical user interfaces similar to those found on Windows and Macintosh personal computers (see USER INTERFACE).

WORKING WITH COMMANDS

UNIX systems come with hundreds of utility programs that have been developed over the years by researchers working at Bell Labs and campuses such as the University of California at Berkeley (UCB). These range from simple commands for working with files and directories (such as cd to set a current directory and ls to list the files in a directory) to language compilers, editors, and text-processing utilities.

Whatever shell is used, UNIX provides several key features for constructing commands. A powerful system of patterns (see REGULAR EXPRESSION) can be used to find files that match various criteria. For a very simple example, the command

```
% ls *.doc
```

will list all files in the current directory that end in .doc. (The % represents the command prompt given by the shell.)

Most earlier operating systems used special syntax to refer to devices such as the user's terminal and the printer. UNIX, however, treats devices just like other files. This means that a program can receive its input by opening a terminal file and send its output to another file. For example:

```
% cat > note
This is a note.
^D
```

The cat (short for concatenate) command adds the user's input to a file called note. The ^D stands for Control-D, the special character that marks end-of-file. Once the command finishes, there is a file called note on the disk, which can be listed by the ls command:

```
% ls -l note
-rw——- 1 hrh well 16 Mar 25 20:16 note
```

The contents of the file can be checked by issuing another cat command:

```
% cat note
This is a note
```

Many commands default to taking keyboard input if no input file is specified. For example, one can type sort followed by a list of words to sort:

```
% sort
apple
pear
orange
tangerine
lemon
^D
```

Once the input is finished, the sort command outputs the sorted list:

```
apple
lemon
orange
pear
tangerine
```

One of the things that makes UNIX attractive to its users is the ability to combine a set of commands in order to perform a task. For example, suppose a user on a timesharing system wants to know which other users are logged on. The who command provides this information, but it includes a lot of details that may not be of interest. Suppose one just wants the names of the current users. One way to do this is to connect the output of the who command to awk, a scripting language (see AWK and SCRIPTING LANGUAGES).

```
% who | awk ' { print $1 }'
```

Here the vertical bar (called a *pipe*) connects the output of the first command to the input of the second. Thus, the awk command receives the output of the who command. The statement print $1 tells awk to output the first column from who's output, which is just the names of the users. The first part of the list looks like this:

```
mnemonic
bernie
kryan
nanlev
goddessj
brady
demaris
techgirl
```

This is fine, but the output might be better if it were sorted. All that's needed is to add one more pipe to connect the output of the awk command to the sort command:

```
% who | awk ' {print $1}' | sort
aarong
aimee
almanac
amicus
autumn
biscuit
bradburn
brian
```

The ability to redirect input and output and to use pipes to connect commands makes it easy for UNIX users to create mini-programs called scripts to perform tasks that would require full-fledged compiled programs on other systems. For example, the preceding command could be put into a file called users, and the file could be set to be executable. Once this is done, all the user has to do to get the user list is to type users at a shell prompt. Today UNIX users have a wide choice of powerful scripting languages (see PERL and PYTHON).

UNIX THEN AND NOW

The versatility of UNIX quickly made it the operating system of choice for most campuses and laboratories, as well as for many software developers. When PCs came along in the late 1970s and 1980s, they generally lacked the resources to run UNIX, but developers of PC operating systems such as CP/M and MS-DOS were influenced by UNIX ideas including the hierarchical file system with its levels of directories, the use of a command-processing shell, and wildcards for matching filenames.

Besides hardware requirements, another barrier to the use of UNIX by home and business users was that the operating system was copyrighted by Bell Labs and a UNIX license often cost more than the PC to run it on. However, a combination of the efforts of the Free Software Foundation (see STALLMAN, RICHARD) and a single inspired programmer (see TORVALDS, LINUS) resulted in the release of Linux, an operating system that is fully functionally compatible with UNIX but uses no AT&T code and is thus free of licensing fees.

The maturation of Linux coincided with the explosion of interest in Web development in the mid- to late-1990s, and Linux proved to be a solid, reliable operating system for Web servers and Web development. At first Linux was somewhat intimidating to users used to operating systems such as Microsoft Windows. However, modern Linux distributions such as those from Red Hat, Debian, Caldera, and others come with an installation program that takes care of scary details such as disk partitioning and installs a graphical user interface (such as KDE or Gnome) that is much like that found in Windows or on the Macintosh. (Meanwhile, Apple computer recently replaced its longstanding Macintosh operating system with OS X, a system based on UNIX.)

The final criticism of Linux was that it lacked a full-featured office software suite comparable to Microsoft Office. Today, however, office suites such as Star Office are compatible with and comparable to MS Office and are available for free.

Barring some external event (such as a legally mandated breakup of Microsoft and a change in the way it bundles Windows), it is unlikely that Linux will displace Microsoft as the operating system leader for PCs. However, UNIX in several flavors has been one of the most successful operating systems in the history of computing, running on everything from pocket PCs to supercomputers.

Further Reading

Bach, M. J. *The Design of the UNIX Operating System.* Upper Saddle River, N.J.: Prentice Hall, 1986.

"A Basic UNIX Tutorial." http://www.ee.byu.edu/support/computer_tutorial/

Kernighan, B. W., and R. Pike. *The UNIX Programming Environment.* Upper Saddle River, N.J.: Prentice Hall, 1984.

"Linux Office Apps." http://www.linux-office.net/

Salus, P. A. *A Quarter Century of UNIX.* Reading, Mass.: Addison-Wesley, 1994.

Torvalds, Linus. *Just for Fun: The Story of an Accidental Revolutionary.* New York: HarperBusiness, 2001.

USB

The traditional ways to connect a computer to peripheral devices such as printers are via parallel and serial connections (see PARALLEL PORT and SERIAL PORT). Both methods are standardized and reliable, but by the mid-1990s people wanted to connect many more data-hungry devices to their PCs, including scanners, digital cameras, and external storage drives. Besides wanting faster data transfers, system designers looked for a way to connect more devices without having to add more ports to the motherboard. It would also be convenient to be able to plug or unplug devices without having to reboot the PC. The Universal Serial Bus (USB) has all these features: It is relatively fast and quite flexible.

Introduced in 1996, USB uses a four-wire cable with small rectangular connectors. Devices can be connected directly to the host USB hub built into the computer. Alternatively, a second hub can be connected to the host hub, allowing for several devices to share the same connection. (Often for convenience, monitors and other devices now include built-in USB hubs for ease in connecting other devices on the desktop.)

Two of the four wires carry power from the PCs power supply (or from a secondary powered hub) to the connected devices. The other two wires carry data. The 1s and 0s in the data are signaled by the difference in voltage between the two wires. (This tends to reduce the effects of outside electromagnetic interference, since if both wires are affected similarly, the difference between them won't change.)

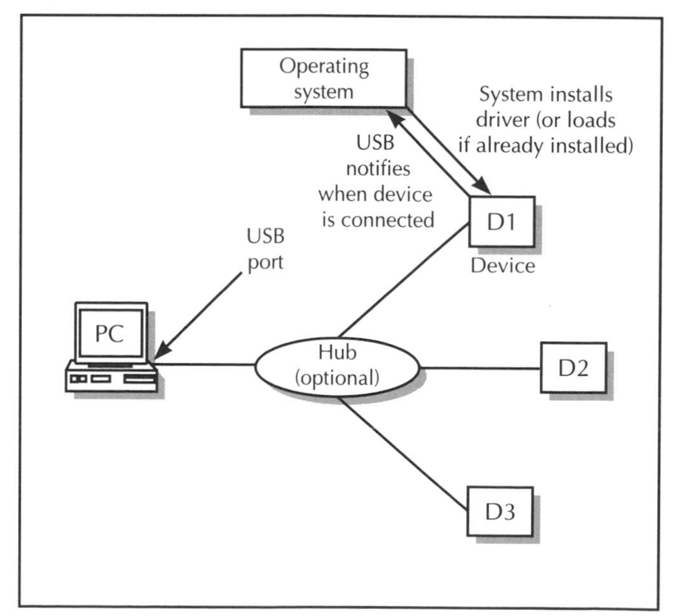

Most PCs today have one or more USB ports, and one can always buy a hub to provide additional connections. When a USB device is first connected, the operating system can automatically load its device driver or, if necessary, prompt the user to provide it on disk or CD.

When a USB device is connected, it creates a voltage change that causes the USB system in the PC to query it for identifying information. If the information indicates that the device has not been installed, the operating system begins an installation procedure that can be carried out either automatically or with a little help from the user (see PLUG AND PLAY).

Once a device is installed, its identifying information tells the USB system what data rate it can handle. (Some devices, such as keyboards, don't need to be very fast, while others, such as CD drives, place a premium on speed.) The USB system assigns each device an address. The system functions like a miniature token-ring network, sending queries or commands with tokens identifying the appropriate device. The devices respond to requests that have their token and in turn send requests when they have data to transmit.

The USB system can assign priorities to devices according to their need for an uninterrupted flow of data. A recordable CD (CD-RW) drive, for example, is sensitive to interruptions in the flow of data, so it is given a high priority. A keyboard sends only a tiny bit of data at a time, and it can get by with a low priority, requesting service as needed. Other devices such as scanners and printers may handle a large flow of data, but are not very sensitive to interruptions and can have a medium priority.

The original USB specification allowed for up to 12 MB/sec data transfers. However, as of 2002 the USB 2.0 specification coming into use allows speeds up to 480 MB/sec, enough to easily handle a digital camera, scanner, and CD-RW drive simultaneously.

Further Reading

USB.org. http://www.usb.org/

White, Ron. *How Computers Work*. Millennium Ed. Indianapolis, Ind.: Que, 1999.

user groups

Computer users have always had an interest in finding and sharing information about the systems they are trying to use. As early as 1955, users of the IBM 701 mainframe banded together, in this case to try to influence IBM's decisions about new software. Later, users of minicomputers made by Digital Equipment Corporation formed DECUS.

By the mid-1970s, microcomputer experimenters had organized several groups, of which the most influential was probably the Homebrew Computer Club, meeting first in a garage in Menlo Park, California, 1975. The group soon was filling an auditorium at Stanford University. Members demonstrated and explained their hand-built computer systems, argued the merits of kits such as the Altair, and later, witnessed Steve Wozniak's prototype Apple I computer.

At the other end of the scale, users of UNIX on university computer systems had formed USENIX, the UNIX user's group. A growing system of newsgroups called USENET (see NETNEWS) would soon extend beyond UNIX concerns to hundreds of other topics.

Early PC users had great need for user groups. Technical support was primitive and the variety of computer books limited, so the best way to get quirky hardware or balky software to work was often to ask fellow users, read user group newsletters, or skim through the great variety of small publications that catered to users of particular systems. Users could also meet to swap public domain software disks. User groups could be formed around software as well as hardware. Thus, users could swap spreadsheet templates or discuss Photoshop techniques.

User groups have gradually become less important, or perhaps it is better to say that they have changed their mode of existence. Starting in the mid-1980s, the modem and bulletin board, on-line services such as CompuServe and later, websites offered more convenient access to information and software without the need to attend meetings. At the same time, the quality and reliability of hardware and software has steadily improved, even though there is always a new crop of problems.

User groups played a key role in the adoption of new technology, much as they had in earlier movements such as amateur radio. Today it might be said that every user has the opportunity to join numerous virtual user groups, although the sense of fellowship and mutual exploration may be somewhat lacking.

Further Reading

Association of Personal Computer User Groups. http://www.apcug.org/

Linux User Groups. http://www.linux.org/groups/

Usenix. http://www.usenix.org/

User Group Network. http://www.user-groups.net/

user interface

All computer designers are faced with the question of how users are going to communicate with the machine in order to get it to do what they want it to do. User interfaces have evolved considerably in 50 years of computing.

The user interface for ENIAC and other early computers consisted of switches or plugs for configuring the machine for a particular problem, followed by loading instructions from punch cards. The mainframes of the 1950s and 1960s had control consoles from which text commands could be entered (see also JOB CONTROL LANGUAGE).

The time-sharing computers that became popular starting in the 1960s still used only text commands, but they were more interactive. Users could type commands to examine directories and files, and run utilities and other programs. Starting in the 1970s, UNIX provided a powerful and flexible way to combine commands to carry out a variety of tasks interactively or through batch processing (see UNIX and SHELL).

The first graphical user interfaces (GUIs) resulted from experimental work at the Xerox Palo Alto Research Center (PARC) during the 1970s. Instead of typing commands at a prompt, GUI users can use a mouse to open menus and select commands, and click on icons to open programs and files. For operations that require detailed specifications, a standard dialog box can be presented, using controls such as check boxes, buttons, text boxes, and sliders.

GUIs entered the mainstream thanks to Apple's Macintosh and Microsoft Windows for IBM-compatible PCs. By the mid-1990s, the GUI had supplanted text-based operating systems such as MS-DOS for most PC users. The strength of the GUI is that it can visually model the way users work with objects in the real world. For example, a file can be deleted by dragging it to a trash can icon and dropping it in. Dragging a slider control to adjust the volume for a sound card is directly analogous to moving a slider on a home stereo system.

Because a system like Windows or the Macintosh provides developers with standardized interface objects and conventions, users are able to learn the basics of operating a new application more quickly. Whereas in the old days different programs might use slightly different keystrokes or commands for saving a file, Windows users know that in virtually any application they can open the File menu and select Save, or press Ctrl-S.

With the growth of the World Wide Web, interface design has extended to webpages. Generally, webpages use similar elements to desktop GUIs, but there are some special considerations such as browser compatibility, response at differing connection speeds, and the integration of text and interactive elements.

GUIs do have some general drawbacks. An experienced user of a text-based operating system might be able to type a precise command that could find all files of a given type on the system and copy them to a backup directory. The GUI counterpart might involve opening the Search menu, typing a file specification, and making further selections and menu choices to perform the copy. Command-driven systems also provide for powerful scripting capabilities. GUI systems often allow for the recording of keystrokes or menu selections, but this is less powerful and versatile.

Designers of user interfaces have to consider whether the elements of the system are intuitively understandable and consistent and whether they can be manipulated in efficient yet natural ways (see also ERGONOMICS OF COMPUTING).

ALTERNATIVE AND FUTURE INTERFACES

The marketplace has spoken, and the desktop GUI is now the mainstream interface for most ordinary PC users. However, there are a variety of other interfaces that are used for particular circumstances or applications, such as:

- Touchscreens (as with ATMs)
- Handwriting or written "gesture" recognition, such as on handheld computers (see HANDWRITING RECOGNITION) or for drawing tablets
- Voice controlled systems (see SPEECH RECOGNITION AND SYNTHESIS)
- Trackballs, joysticks, and touchpads (used as mouse alternatives)
- Virtual reality interfaces using head-mounted systems, sensor gloves, and so on (see VIRTUAL REALITY)

Because much interaction with computers is now away from the desktop and taking place on laptops, handheld, or palm computers, and even in cars, there is likely to be continuing experimentation with user interface design.

Further Reading

Johnson, Jeff. *GUI Bloopers: Do's and Don'ts for Software Developers and Web Designers.* San Francisco: Morgan Kaufmann, 2000.

Raskin, Jeff. *The Humane Interface: New Directions for Designing Interactive Systems.* Reading, Mass.: Addison-Wesley, 2000.

Schneiderman, B. *Designing the User Interface: Strategies for Effective Human-Computer Interaction.* 2nd ed. Reading, Mass.: Addison-Wesley, 1992.

Stephenson, Neil. *In the Beginning Was the Command Line.* New York: Avon Books, 1999.

variable

Virtually all computer programs must keep track of a variety of items of information that can change as a result of processing. Such values might include totals or subtotals, screen coordinates, the current record in a database, or any number of other things. A *variable* is a name given to such a changeable quantity, and it actually represents the area of computer memory that holds the relevant data.

Consider the following statement in the C language:

```
int Total = 0;
```

Variables have several attributes. First, every variable has a name—Total in this case. Although this name actually refers to an address in memory, in most cases the programmer can use the much more readable name instead of the actual address.

It is possible to have more than one name for the same variable by having another variable point to the first variable's contents, or by declaring a "reference" variable (see POINTERS AND INDIRECTION).

Each variable has a data type, which might be number, character, string, a collection (such as an array), a data record, or some special type defined by the programmer (see DATA TYPES). With some exceptions (see SCRIPTING LANGUAGES) most modern programming languages require that the programmer declare each variable before it is used. The declaration specifies the variable's type—in the current example, the type is int (integer, or whole number).

A variable is usually given an initial value by using an assignment statement; in the example above the variable Total is given an initial value of 0, and the assignment is combined with the declaration. (Some languages automatically assign a default value such as 0 for a number or a null character for a string, but with other languages failure to assign a value results in the variable having as its value whatever happens to be currently stored in the memory address associated with the variable. An explicit assignment is thus always safer and more readable.)

When exactly do variables get set up, and when do they get their values? This varies with the programming language (see BINDING). With C and similar languages, a variable receives its data type when the program is compiled (compile time). The type in turn determines the range of values that the variable can hold (physically based on the number of bytes of memory allocated to it). The variable's value is actually stored in that location when the program is executed (run time).

A few languages such as APL and LISP use dynamic binding, meaning that a data type is not associated with a variable until run time. This makes for flexibility in programming, but at some cost in efficiency of storage and execution speed.

During processing, a variable's value can change through the use of operators in expressions (see OPERATORS AND EXPRESSIONS). Thus, the example value Total might be changed by a statement such as:

```
Total = Total + Subtotal;
```

When this statement is executed, the following happens:

The value of the memory location labeled "Subtotal" is obtained.

The value of the memory location labeled "Total" is obtained.

The two values are added together.

The result is stored in the location labeled "Total," replacing its former value.

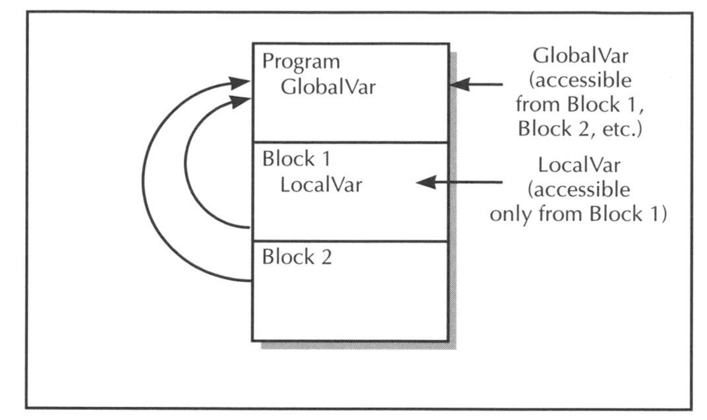

In modern programming languages, local variables can be accessed only within the procedure of block in which they are used. Global variables, which can be accessed (and changed) from anywhere in the program, are seldom used because it is easy for them to unexpectedly change. Instead, procedures communicate by receiving and passing values through their defined parameters.

SCOPE OF VARIABLES

In early programming languages variables were generally global, meaning that they could be accessed and changed from any part of the program. While this practice is convenient, it became riskier as programs became larger and more complex. One part of a program might be using a variable called Total or Subtotal to keep track of some quantity. Later, another part is written to deal with some other calculation, and uses the same names. The programmer may think of the second Total and Subtotal as being quite separate from the first, but in reality they refer to the same memory locations and any change affects both of them. Thus, it's easy to create unwanted "side effects" when using global variables.

Starting in the 1960s and more systematically during the 1970s, there was great interest in designing computer languages that could better manage the structure and complexity of large programs (see SRUCTURED PROGRAMMING). One way to do this is to break programs up into more manageable modules that each deal with some specific task (see PROCEDURES AND FUNCTIONS). Unless explicitly declared to be global, variables within a procedure or function are local to that unit of code. This means that if two procedures both have a variable called Total, changes to one Total do not affect the other.

Generally, in block-structured languages such as Pascal a variable is by default local to the block of code in which it is defined. This means it can be accessed only within that block. (Its visibility is said to be limited to that block.) The variable will also be accessible to any block that is nested within the defining block, unless another variable with the same name is declared in the inner block. In that case the inner variable supersedes the outer one, which will not be visible in the inner block.

Some languages such as APL and early versions of LISP define scope differently. Since these languages are not block structured, scope is determined not by the relationship of blocks of code but by the sequence in which functions are called. At run time each variable's definition is searched for first in the code where it is first invoked, then in whatever function called that code, then in the function that called *that* function, and so on. As with dynamic binding, dynamic scooping offers flexibility but at a considerable price. In this case, the price is that the program's effects on variables will be hard to understand, and the search mechanism slows down program execution. Dynamic scoping is thus not often used today, even in LISP.

Global variables were convenient because they allowed information generated by one part of a program to be accessed by any other. However, such accessibility can be provided in a safer, more controlled form by explicitly passing variables or their values to a procedure or function when it is called (see PROCEDURES AND FUNCTIONS).

Object-oriented languages provide another way to control or encapsulate information. Variables describing data used within a class are generally declared to be private (accessible only within the functions used by the class). Public (i.e. global) variables are used sparingly. The idea is that if another part of the program wants data belonging to a class, it will call a member function of the class, which will provide the data without giving unnecessary access to the class's internal variables.

A final concept that is important for understanding variables is that of lifetime, that is, how long the definition of a variable remains valid. For efficiency, the runtime environment must deallocate memory for variables when they can no longer be used by the program (that is go "out of scope"). Generally, a variable exists (and can be accessed) only while the block of code in which it was defined is being executed (including any procedures or functions called from that block). In the case of a variable

Variable (2)

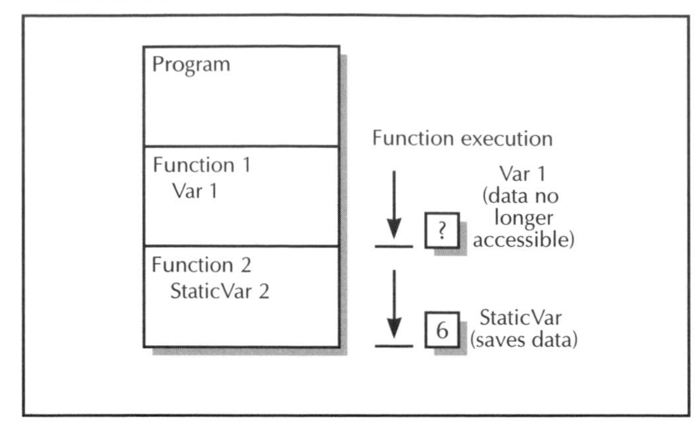

In the C language, a regular variable within a function or procedure exists (and is defined) only so long as the function or procedure is executing. However, a "static" variable retains its value between procedure calls, so it can be used as a counter or status flag.

declared in the main program, this will be until the program as a whole reaches its end statement. For variables within procedures or functions, however, the lifetime lasts only until the procedure or function ends and control is returned to the calling statement. However, languages such as C allow the special keyword *static* to be used for a variable that is to remain in existence as long as the program is running. This can be useful when a procedure needs to "remember" some information between one call and the next, such as an accumulating total.

Further Reading

Kernighan, Brian W., and Dennis Ritchie. *The C Programming Language*. 2nd ed. Englewood Cliffs, N.J.: Prentice Hall, 1988.

Sebesta, Robert W. *Concepts of Programming Languages*. Reading, Mass.: Addison-Wesley, 1999.

Stroustrup, Bjarne. *The C++ Programming Language*. 3rd ed. Reading, Mass.: Addison-Wesley, 1997.

videoconferencing

The growth of the global economy has meant that many companies have operations in many locations around the world. The time and expense involved in travel have encouraged the search for alternatives to face-to-face meetings (see TELEPRESENCE). The added discomfort and uncertainty following the September 11, 2001, terrorist attacks is likely to further spur this movement.

Basic videoconferencing is carried out by using video cameras and microphones to carry the image and voice of each person so that it can be seen by all participants. The video and sound data is digitized and transmitted between the participants' locations, using some existing communications link. Although direct satellite technology can be used, it is very expensive. A more practicable alternative is the use of a proprietary system over special phone lines (such as ISDN or DSL). Increasingly, however, broadband connections to the general Internet are used (see also INTERNET TELEPHONY). This is relatively inexpensive and flexible, but sometimes less reliable because of the effects of network congestion.

The quality of imagery depends on the system. High-end systems, which can cost tens of thousands of dollars, use large, high-definition screens or even special projection equipment that can give a 3D look to peoples' faces. Although high-end videoconferencing software and hardware can be expensive, there are now a variety of alternatives for small businesses and individual users. (As of 2002 the printing store chain Kinko's is offering videoconferencing through some of its stores for $450/hr.)

For smaller, less formal meetings there are more affordable alternatives. Products such as Microsoft NetMeeting, CuSeeMe, and Yahoo Messenger set up user accounts and a directory that makes it easy for users to connect. Other than the Internet connection, the only hardware needed is a microphone and an inexpensive camera (see WEB CAM).

Business videoconferencing systems often include the ability for participants to view and interact with software applications. This makes it possible not only to view slide shows or other presentations (see PRESENTATION SOFTWARE) but to collaborate on creating documents. An "electronic whiteboard" can be used to display not only computer text and graphics but also handwritten notes created by participants using electronic drawing pads. The system can also create a hardcopy record of documents developed during the meeting.

Besides business meetings and conferences and product roll-outs, videoconferencing can also be used for a variety of other applications including sales presentations and for conducting focus groups for market research.

Videoconferencing is also being used increasingly in education. For K-12 classes, a videoconferencing field trip can take children to a museum or science laboratory that would otherwise be too far to visit. Both docent and students can see and hear one another, as well as being able to see exhibits or experiments close up. For college students and adults, it is possible to attend classes given by eminent lecturers and participate fully just as though they were enrolled on campus (see also EDUCATION AND COMPUTERS).

Further Reading

Rhodes, John D., and Brad Caldwell. *Videoconferencing for the Real World*. Boston: Focal Press, 2001.

Digital Horizon [Newsletter], vol. 1, no. 1. "Desktop Video Conferencing (DVC)." http://www.broadband-guide.com/news/dhnewssept96/dh1.html

Freed, Les. "Videoconferencing Software." *PC Magazine*, March 31, 2000. Available on-line at http://www.zdnet.com/products/stories/reviews/0,4161,2470119,00.html
"Internet Conferencing." http://netconference.miningco.com/
Pachnowski, Lynn M. "Virtual Field Trips Through Teleconferencing." *Learning & Leading with Technology* 29, no. 6, March 2002, 10 ff.
Prencipe, Loretta W. "Management Briefing: Do You Know the Rules and Mannners of an Effective Virtual Meeting?" *InfoWorld* 23, no. 18, April 30, 2001, 46.

video editing, digital

When videotape first became available in the 1950s, recorders cost thousands of dollars and could only be afforded by TV studios. Today the VCR is inexpensive and ubiquitous. However, it is hard to edit videotape. Tape is a linear medium, meaning that to find a given piece of video the tape has to be moved to that spot. Removing or adding something involves either physically splicing the tape (as is done with film) or more commonly, feeding in tape from two or more recorders onto a destination tape. Besides being tedious and limited in capabilities, "linear editing" by copying loses a bit of quality with each copying operation.

Today, however, it is easy to shoot video in digital form (see PHOTOGRAPHY, DIGITAL) or to convert analog video into digital form. Digital video is a stream of data that represents sampling of the source signal, such as from the charge-coupled device (CCD) that turns light photons into electron flow in a digital camera or digital camcorder. This process involves either software or hardware compression for storage and decompression for viewing and editing (such a scheme is called a CODEC for "compression/decompression"). The most widely used formats include DV (Digital Video) and MPEG (Motion Picture Expert Group), which has versions that vary in the amount of compression and thus fidelity.

In a turnkey system, the input source is automatically digitized and stored. In desktop video using a PC, a video capture card must be installed. The card turns the analog video signal into a digital stream. The most commonly used interface to bring video into a PC is IEE1394, better known as FireWire, which has the high bandwidth needed to transfer video data.

Once the video is captured, it can be stored in frame buffers in memory and edited in various ways using a variety of software. Expensive turnkey systems come with advanced software, while desktop video users can choose from products such as Media Studio Pro or Adobe Premiere. The editing interface usually has a timeline and thumbnails showing the location of key frames in the sequence. Individual clips can be extracted and tweaked with motion and transition effects; a variety of filters (see also PLUG-IN) can be applied to the video. The accompanying sound track(s) can also be edited. Once things look right, the software is told to render (create) the finished video and save it to disk.

The ever-increasing processing power and disk capacity of today's PC is likely to make real-time video editing more feasible. This means that video can be played back directly from the edited timeline without transitions or effects having to be rendered first. Digital video cameras are also likely to increase in picture quality (see PHOTOGRAPHY, DIGITAL). Already desktop video is proving to be an affordable, viable alternative to expensive turnkey systems for many applications.

Further Reading

Bovik, Al, ed. *Handbook of Image and Video Processing*. San Diego, Calif.: Academic Press, 2000.
"Desktop Video." http://desktopvideo.about.com/
"Desktop Video Handbook On-line." http://www.videoguys.com/dtvhome.html
"Digital Video Editing: Professional Resources." http://www.digitalvideoediting.com/Htm/DVEditHomeSet1.htm
"Glossary of Digital Video Terms." http://www.adobe.com/support/techguides/digitalvideo/dv_glossary/main.html
Long, Ben, and Sonja Schenk. *The Digital Filmmaking Book (with CD-ROM)*. Hingham, Mass.: Charles River Media, 2000.

virtual community

Back in the mid-19th century, a number of technical professionals began to "chat" on-line without meeting physically—they were telegraph operators who relayed messages across the growing web that one author has called "The Victorian Internet." When computer networking began to grow in the 1970s, its own pioneers used facilities such as newsgroups (see NETNEWS AND NEWSGROUPS) to discuss a variety of topics. By the early 1980s, users were interacting on-line in complex fantasy games called MUDs (Multi-user Dungeons, or Dimensions) or MOOs (Muds, Object-Oriented). A little later, bulletin boards and especially systems such as the Well (Whole Earth 'Lectronic Link) based in the San Francisco Bay Area (see BULLETIN BOARD and CONFERENCING SYSTEMS) provided long-term outlets for people to share information and interact on-line.

Looking at the Well, a writer named Howard Rheingold introduced the term *virtual community* in a 1993 book. He explored the ways in which a sufficiently compelling and versatile technology encouraged people to form long-term contacts, form personal relationships, and carry out feuds. When on-line, participants experience such a venue as the Well as a place that becomes almost as tangible (and often as "real") as a physical place such as a small town or corner bar.

Virtual community members who live in the same geographical area sometimes do get together physically (the Well has had picniclike "Well Office Parties" for many years). Members can band together to support a colleague who faces a crisis such as the life-threatening illness of a son (on the Well, blank postings called *beams* are often used as an expression of sympathy). The virtual community can also serve as a rallying point following a physical disaster such as the 1989 earthquake in the San Francisco Bay Area. On a daily basis, virtual communities can often provide help or advice from a remarkable variety of highly qualified experts.

Virtual communities have their share of human foibles and worse. A virtual world that is compelling enough to immerse participants for hours on end is also powerful enough to engage emotions and expose vulnerabilities. For example, in a MUD called LambdaMOO one participant used descriptive language to have his game character "rape" a female character created by another participant, inflicting genuine distress. Like physical communities, virtual communities must evolve rules of governance, and actions in a virtual community can have real-world legal consequences.

Critics such as Clifford Stoll have argued that virtual communities are not only not a substitute for "true" physical community, but also may be further fragmenting neighborhoods and isolating people. (On the other hand, people who are already physically isolated, such as rural folk and the elderly or disabled, may find an outlet for their social needs in a virtual community.) Certainly the "bandwidth" in terms of human experience is less in a virtual community than in a physical community. Ideally, individuals should cultivate a mixture of virtual and physical community relationships.

Further Reading

Barnes, S. B. *On-line Connections: Internet Personal Relationships.* Cresskill, N.J.: Hampton Press, 2000.

Dibbel, J. "Rape in Cyberspace: How an Evil Clown, a Haitian Trickster Spirit, Two Wizards and a Cast of Dozens Turned a Database into a Society." *The Village Voice,* Dec. 21, 1993, 39.

———. *My Tiny Life: Crime and Passion in a Virtual World.* New York: Henry Holt, 1998.

Rheingold, Howard. *The Virtual Community: Homesteading on the Electronic Frontier.* 2nd ed. Reading, Mass.: Addison-Wesley, 1993.

———. "Howard Rheingold Home Page." http://www.rheingold.com/

Turkle, Sherry. *Life on the Screen: Identity in the Age of the Internet.* New York: Simon & Schuster, 1995.

virtual reality

As the graphics and processing capabilities of computers grew increasingly powerful starting in the 1980s, it became possible to think in terms of creating a 3D environment that would not only appear to be highly realistic to the user, but also would respond to the user's natural motions in realistic ways.

This idea is not that new in itself. Starting as early as the 1930s, the military built mechanical flight trainers or simulators that could create a somewhat realistic experience of what a pilot would see and feel during flight. More sophisticated versions of these mechanical simulators helped the United States train the tens of thousands of pilots it needed during World War II while reducing the resources needed for actual flight hours. Today the military continues to pioneer the use of realistic computerized simulators to train tank crews and even individual soldiers in the field (see MILITARY APPLICATIONS OF COMPUTERS).

Early simulators used "canned" graphics and could not respond very smoothly to control inputs (such as a pilot moving stick or rudder). Modern virtual reality, however, depends on the ability to smoothly and quickly generate realistic 3D graphics. At first such graphics could only be generated on powerful workstations such as those made by Sun or Silicon Graphics. However, as anyone who has recently played a computer game or simulation knows, there has been great improvement in the graphics available on ordinary desktop PCs since the mid-1990s.

A variety of software and programming tools can be used to generate 3D worlds on a PC (see COMPUTER GRAPHICS). First released in 1995, a facility called VRML (Virtual Reality Modeling Language) is now supported by many Web browsers. There are also programming extensions for Java (Java 3D).

Modern computer games thus embody aspects of virtual reality in terms of graphics and responsiveness. But true VR is generally considered to involve a near total immersion. Instead of a screen, a head-mounted display (HMD) is generally used to display the virtual world to the user while shutting out environmental distractions.

A NASA researcher wearing an early virtual reality (VR) outfit, including head-mounted display and gloves whose position can be tracked. (NASA PHOTO)

Typically, slightly different images are presented to the left and right eyes to create a 3D stereo effect.

The other half of the VR equation is the way in which the user interacts with the virtual objects. Head-tracking sensors are used to tell the system where the user is looking so the graphics can be adjusted accordingly. Other sensors can be placed in gloves worn by the user. The system can thus tell where the user's hand is within the virtual world, and if the user "grasps" with the glove, the user's hand in the virtual world will grasp or otherwise interact with the virtual object. More elaborate systems involve a full-body suit studded with sensors.

To make interaction realistic, VR researchers have had to study both the operation of human senses and that of the skeleton and muscles. For a truly realistic experience, the user must be able to feel the resistance of objects (which can be implemented by a force-feedback system). Sound can be handled easily, but as of yet not much has been done with the senses of smell and taste.

In designing a VR system, there are a number of important considerations. Will the user be physically immersed (such as with an HMD), or, as in some military applications, will the user be seeing both a virtual and the actual physical world? How important is graphic realism vs. real-time responsiveness? (Opting too much for computationally intensive realism might cause unacceptable latency, or delay between a user action and the environment's response.)

APPLICATIONS

Besides military training, currently the most viable application for VR seems to be entertainment. VR techniques have been used to create immersive experiences in elaborate facilities at venues such as Disneyland and Universal Studios, and to some extent even in local arcades. VR that is accompanied by convincing physical sensations has allowed for the creation of a new generation of roller coasters that if built physically would be too expensive, too dangerous, or even physically impossible.

However, there are other significant emerging applications for VR. When combined with telerobotic technology (see TELEPRESENCE), VR techniques are already being used to allow surgeons to perform operations in new ways. VR technology can also be used to make remote conferencing more realistic and satisfactory for participants. Clearly the potential uses for VR for education and training in many different fields are endless. VR technology combined with robotics could also be used to give disabled persons much greater ability to carry out the tasks of daily life.

In the ultimate VR system, users will be networked and able to simultaneously experience the environment, interacting both with it and one another. The technical resources and programming challenges are also much greater for such applications. The result, however, might well be the sort of environment depicted by science fiction writers such as William Gibson (see CYBERSPACE AND CYBER CULTURE).

Further Reading

Durlach, N., and A. Mavor, eds. *Virtual Reality: Scientific and Technical Challenges.* Washington, D.C.: National Academy Press, 1995.

Heim, Michael. *Virtual Realism.* New York: Oxford University Press, 1998.

Vince, John. *Essential Virtual Reality Fast: How to Understand the Techniques and Potential of Virtual Reality.* New York: Springer-Verlag, 1998.

Virtual Reality Society. http://www.vrs.org.uk/

Yahoo! Virtual Reality Page. http://dir.yahoo.com/Computers_and_Internet/Multimedia/Virtual_Reality/

von Neumann, John

(1903–1957)
Hungarian–American
Mathematician, Computer Scientist

John von Neumann made wide-ranging contributions in fields as diverse as pure logic, simulation, game theory, and quantum physics. He also developed many of the key concepts for the architecture of the modern digital computer and helped design some of the first successful machines.

Von Neumann was born on December 28, 1903, in Budapest, Hungary, to a family with banking interests. As a youth he showed a prodigious talent for calculation and interest in mathematics, but his father opposed his pursuing a career in pure mathematics. Therefore, when von Neumann entered the University of Berlin in 1921 and the Technische Hochschule in 1923, he earned his Ph.D. in chemical engineering. However, in 1926 he went back to Budapest and earned a Ph.D. in mathematics with a dissertation on set theory. He would then serve as privatdozcent, or lecturer, at Berlin and the University of Hamburg.

During the mid-1920s, two competing mathematical descriptions of the behavior of atomic particles were being offered by Erwin Schrödinger's wave equations and Werner Heisneberg's matrix approach. Von Neumann showed that the two theories were mathematically equivalent. His 1932 book, *The Mathematical Foundations of Quantum Mechanics,* remains a standard textbook to this day. Von Neumann also developed a new form of algebra where "rings of operators" could be used to describe the kind of dimensional space encountered in quantum mechanics.

Meanwhile, von Neumann had become interested in the mathematics of games, and developed the discipline that would later be called game theory. His "minimax theorem" described a class of two-person games in which

John von Neumann developed automata theory as well as fundamental concepts of computer architecture such as storing programs in memory along with the data. He also did seminal work in logic, quantum physics, simulation, and game theory. (PHOTO COURTESY OF COMPUTER MUSEUM HISTORY CENTER)

both players could minimize their maximum risk by following a specific strategy.

COMPUTATION AND COMPUTER ARCHITECTURE

In 1930, von Neumann immigrated to the United States, where he would become a naturalized citizen and spend the rest of his career. He was made a Fellow at the new Institute for Advanced Study at Princeton at its founding in 1933, and would serve in various capacities there and as a consultant for the U.S. government.

In the late 1930s, interest had begun to turn to the construction of programmable calculators or computers (see also CHURCH, ALONZO and TURING, ALAN). Just before and during World War II, von Neumann worked on a variety of problems in ballistics, aerodynamics, and later, the design of nuclear weapons. All of these problems cried out for machine assistance, and von Neumann became acquainted both with British research in calculators and the massive Harvard Mark I programmable calculator (see AIKEN, HOWARD).

A little later, von Neumann learned that two engineers were working on a new kind of machine: an electronic digital computer called ENIAC that used vacuum tubes for its switching and memory, making it about a thousand times faster than the Mark I. Although the first version of ENIAC had already been built by the time von Neumann came on board, he served as a consultant to the project at the University of Pennsylvania's Moore School.

The earliest computers (such as the Mark I) read instructions from cards or tape, discarding each instruction as it was performed. This meant, for example, that to program a loop, an actual loop of tape would have to be mounted and controlled so that instructions could be repeated. The electronic ENIAC was too fast for tape readers to keep up, so it had to be programmed by setting thousands of switches to store instructions and constant values. This tedious procedure meant that it wasn't practicable to use the machine for anything other than massive problems that would run for many days.

In his 1945 "First Draft of a Report on the EDVAC" and his more comprehensive 1946 "Preliminary Discussion of the Logical Design of an Electronic Computing Instrument," von Neumann established the basic architecture and design principles of the modern electronic digital computer.

Von Neumann declared that in future computers the machine's internal memory would be used to store constant data and all instructions. With programs in memory, looping or other decision making can be accomplished simply by "jumping" from one memory location to another. Computers would have two forms of memory: relatively fast memory for holding instructions, and a slower form of storage that could hold large amounts of data and the results of processing. (In today's PCs these functions are provided by the random access memory [RAM] and hard drive respectively.) The storage of programs in memory also meant that a program could treat its own instructions like data and change them in response to changing conditions.

In general, von Neumann took the hybrid design of ENIAC and conceived of a design that would be all-electronic in its internal operations and store data in the most natural form possible for an electronic machine—binary, with 1 and 0 representing the on and off switching states and, in memory, two possible "marks" indicated by magnetism, voltage levels, or some other phenomenon. The logical design would be consistent and largely independent of the vagaries of hardware.

Eckert and Mauchly (see ECKERT, J. PRESPER and MAUCHLY, JOHN WILLIAM) and some of their supporters would later claim that they had already conceived of the idea of storing programs in memory, and in fact they had already designed a form of internal memory called a mercury delay line. Whatever the truth in this assertion, it remains that von Neumann provided the comprehensive theoretical architecture for the modern computer, which would become known as the von Neumann architecture.

Von Neumann's reports would be distributed widely and would guide the beginnings of computer science research in many parts of the world.

Looking beyond EDVAC, von Neumann, together with Herman Goldstine and Arthur Burks, designed a new computer for the Institute for Advanced Study that would embody the von Neumann principles. The IAS machine's design would in turn lead to the development of research computers for RAND Corporation, the Los Alamos National Laboratory, and in several countries including Australia, Israel, and even the Soviet Union. The design would eventually be commercialized by IBM in the form of the IBM 701.

In his later years, von Neumann continued to explore the theory of computing. He studied ways to make computers that could automatically maintain reliability despite the loss of certain components, and he conceived of an abstract self-reproducing automaton (see CELLULAR AUTOMATA).

Von Neumann's career was crowned with many awards reflecting his diverse contributions to American science technology. These include the Distinguished Civilian Service Award (1947), Presidential Medal of Freedom (1956), and the Enrico Fermi Award (1956). Von Neumann died on February 8, 1957 in Washington, D.C.

Further Reading

Aspray, William. *John von Neumann and the Origins of Modern Computing.* Cambridge, Mass.: MIT Press, 1990.

Heims, S. J. *John von Neumann and Norbert Wiener: From Mathematics to the Technologies of Life and Death.* Cambridge, Mass.: MIT Press, 1980.

"John Louis von Neumann" [biography]. http://ei.cs.vt.edu/~history/VonNeumann.html

von Neumann, John. *The Computer and the Brain.* New Haven, Conn.: Yale University Press, 1958.

———. *Theory of Self-Reproducing Automata.* Edited and compiled by Arthur W. Burks. Urbana: University of Illinois Press, 1966.

W

Web browser

The World Wide Web consists of millions of sites (see WORLD WIDE WEB and WEB SERVER) that provide hypertext documents (see HTML and WEBPAGE DESIGN) that can include not only text but still images, video, and sound. To access these pages, the user runs a Web-browsing program.

The basic function of a Web browser is to request a page by specifying its address (URL, uniform [or universal] resource locator). This request resolves to a request (HTTP, HyperText Transport Protocol) that is processed by the relevant Web server. The server sends the HTML document to the browser, which then displays it for the user. Typically, the browser stores recently requested documents and files in a local cache on the user's PCs. Use of the cache reduces the amount of data that must be resent over the Internet. However, sufficiently skilled snoopers can examine the cache to find details of a user's recent Web surfing. (Caching is also used by Internet Service Providers so they can provide frequently requested pages from their own server rather than having to fetch them from the hosting sites.)

When the Web was first created in the early 1990s (see BERNERS-LEE, TIM) it consisted only of text pages, although there were a few experimental graphical Web extensions developed by various researchers. The first graphical Web browser to achieve widespread use was Mosaic created by Marc Andreessen, developed at the National Center for Supercomputing Applications (NCSA). By 1993, Mosaic was available for free download and had become the browser of choice for PC users.

Andreessen left NCSA in 1994 to found Netscape Corporation. The Netscape Navigator browser improved Mosaic in several ways, making the graphics faster and more attractive. Netscape included a facility called Secure Sockets Layer (SSL) for carrying out encrypted commercial transactions on-line (see E-COMMERCE).

Microsoft, which had been a latecomer to the Internet boom, entered the fray with its Microsoft Internet Explorer. At first the program was inferior to Netscape, but it was steadily improved. Aided by Microsoft's controversial tactic of bundling the free browser starting with Windows 95, Internet Explorer has taken over the leading browser position with about a 75 percent market share by 2001.

Some typical features of a modern Web browser include

- Navigation buttons to move forward and back through recently visited pages
- A "history" panel allowing return to pages visited in recent days
- A search button that brings up the default search engine (which can be chosen by the user)
- The ability to save page as "favorites" or "bookmarks" for easy retrieval

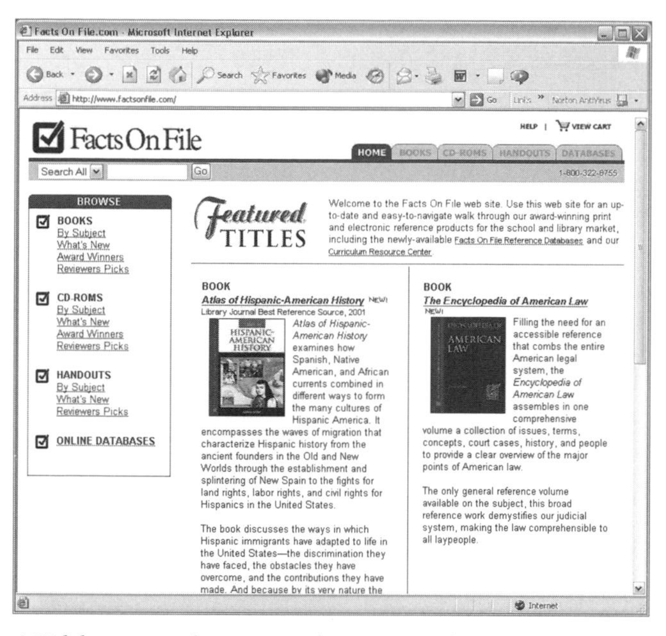

A Web browser such as Microsoft Internet Explorer or Netscape Navigator makes it easy to find and move between linked webpages. Browser users can record or "bookmark" favorite pages. Browser plug-ins provide support for services such as streaming video and audio.

THE BROWSER AS PLATFORM

Today a Web user can view a live news broadcast, listen to music from a radio station, or view a document formatted to near-print quality. All these activities are made possible by "helper" software (see PLUG-IN) that gives the Web browser the capability to load and display or play files in special formats. Examples include the Adobe PDF (Portable Document Format) reader, the Windows Media Player, and RealPlayer for playing video and audio content (see also STREAMING).

What makes the browser even more versatile is the ability to load and run programs from websites (see JAVA). Java was highly touted starting in the mid-1990s, and some observers believed that by making Web browsers into platforms capable of running any sort of software, there would be less need for proprietary operating systems such as Microsoft Windows. Microsoft has responded by trying to shift developers' emphasis from Java to its proprietary technology called .NET. Meanwhile, the tools for making webpages more versatile and interactive continue to proliferate, including later versions of HTML and XML (see WEBPAGE DESIGN). This proliferation, as well as use of proprietary extensions can cause problems in accessing websites from older or less-known browsers.

The growing numbers of handheld or palm computers (see PORTABLE COMPUTERS) are accompanied by scaled-down Web browsers. These are generally controlled by touch and have a limited display size, but can provide information useful to travelers such as driving directions, weather forecasts, and capsule news or stock summaries.

Further Reading

"Browser Watch." http://browserwatch.internet.com/

Lowe, Doug. *Microsoft Internet Explorer 6 for Dummies.* New York: Hungry Minds, 2001.

Microsoft Internet Explorer Home Page. http://www.microsoft.com/windows/ie/default.asp

Netscape Communications Corporation. "Browser Central." http://browsers.netscape.com/browsers/main.tmpl

Yahoo! Web browser page. http://dir.yahoo.com/Computers_and_Internet/software/internet/world_wide_web/browsers/

Web cam

Thousands of real-time views of the world are available on the Web. These include everything from the prosaic (a coffee machine at MIT) to the international (a view of downtown Paris or Tokyo) to the sublime (a Rocky Mountain sunset). All of these views are made possible thanks to the availability of inexpensive digital cameras (see PHOTOGRAPHY, DIGITAL).

To create a basic Web cam, the user connects a digital camera to a PC, usually via a USB cable. A program controls the camera, taking a picture at frequent intervals (perhaps every 30 seconds or minute). The picture is received from the camera as a JPG (JPEG) file. The program then uploads the picture to the user's webpage (usually using file transfer protocol, or ftp), replacing the previous picture. Users connected to the website can click to see the latest picture. Alternatively, a script running on the server can update the picture automatically.

HISTORY AND APPLICATIONS

One of the earliest and most famous Web cams was created by Quentin Stafford-Fraser in 1991. He later recalled that he and his fellow "coffee club" members were tired to making the long trek to the coffee room at the Cambridge University computer laboratory. It seemed that more often than not the life-giving brew so necessary to computer science had already been consumed. So they rigged a video camera, connected it to a video capture card, and fed the image into the building's local network. Now researchers working anywhere in the building could get an updated image of the coffee machine three times a minute. This wasn't technically a Web cam. At the time the Web was just being developed by Tim Berners-Lee. However, the camera was put on the Web in 1993, where it resided until 2001 when the laboratory housing the now famous coffee machine was moved.

The Web cam became a social phenomenon in 1996 when a college student named Jennifer Ringley started Jennicam, a Web cam set up to make a continuing record of her daily life available on the Web. There were soon many imitators. Apparently this use of Web cams taps into humans' intense curiosity about the details of each other's lives—a curiosity that to some critics tips over into voyeurism and obsession. The popularity of such social Web cams may have contributed to the "reality TV" phenomenon at the turn of the new century.

Web cams have many practical applications, however. People on the road can log into the Web and check to make sure everything's OK at home. A Web cam also makes an inexpensive monitor for checking on infants or toddlers in another room, or checking on the behavior of a babysitter ("Nannycam").

Web cams can also serve an educational purpose. They can take viewers to remote volcanoes or the interior of an Amazon rain forest. In a sense viewers who saw the pictures of the Martian surface and the explorations of the Sojourner rover were using the farthest-reaching Web cam of all.

Further Reading

"A Birdseye View of . . . Webcams." http://www.abirdseyeviewof.com/WIE.html

Brain, Marshall. "How Webcams Work." http://www.howstuffworks.com/webcam.htm

Jennicam. http://www.jennicam.org/index.html

MIT Technology Review. "Trailing Edge. Coffee Cam." June 2001. Available on-line at http://www.techreview.com/articles/trailing0601.asp

Parker, Elisabeth, and John Grimes. *The Little Web Cam Book.* Berkeley, Calif.: Peachpit Press, 1999.

Web filter

Listings of the most frequent requests typed into Web search engines usually begin with the word *sex.* Although sensational journalism of the mid-1990s sometimes unfairly portrayed the World Wide Web as nothing more than an electronic red light district, it is indisputable that there are many websites that feature material that most people would agree is not suitable for young people. Many parents as well as some schools, libraries, and workplaces have installed Web filter programs, marketed under names such as SurfWatch or NetNanny. Popular Internet security programs (such as those from Norton/Symantec) also include Web filter modules.

The Web filter examines requests made by a Web user (see WORLD WIDE WEB and WEB BROWSER) and blocks those associated with sites deemed by the filter user to be objectionable. There are two basic mechanisms for determining whether a site is unsuitable. The first is to check the site's address (URL) against a list and reject a request for any site on the list. (Most filter programs come with default lists; the filter user can add other sites as desired. Generally, the filter is installed with a password so only the authorized user [such as a parent] can change the filter's behavior.)

The other filtering method relies on a list of keywords associated with objectionable activities (such as pornography). When the user requests a site, the filter checks the page for words on the keyword list. If a matching word or phrase is found, the site is blocked and not shown to the user.

Each method has its drawbacks: Using a site list will miss new sites that appear between list updates, while using keywords can result in appropriate sites also being blocked. For example, a keyword filter that blocks sites with the word *breast* will probably also block a site devoted to breast cancer research, a fact often pointed out by opponents of laws requiring the use of Web filters. The list and keyword methods can be combined.

Besides protecting children from inappropriate material at home or in a school or library, Web filters are also used in workplaces. Besides wanting to keep workers from becoming distracted, employers are concerned that allowing Internet pornography in the workplace may make them liable for creating a "hostile work environment" under sexual harassment laws.

However, civil liberties groups such as the ACLU object to the use of Web filters in public libraries on First Amendment grounds and have vigorously fought such legislation in the courts. The 1996 Communications Decency Act was declared unconstitutional by the U.S. Supreme Court, and a later law, the 1998 Child On-line Protection Act (which requires that users of adult websites provide proof of age) was set aside by the 3rd U.S. District Court pending a Supreme Court ruling later in 2002.

Meanwhile, the 2002 Children's Internet Protection Act attempts to sidestep the First Amendment issue by merely denying certain types of federal funding to libraries that do not install Web filters. However, this law is currently being challenged by the American Library Association, ACLU, and a group of public libraries.

Critics of Web filters suggest that rather using technical tools to block access to the Internet, parents and teachers should talk to children about their use of the Internet and supervise it if necessary. Another approach is to focus on websites that are designed especially for kids.

Further Reading

Aftab, Perry. *The Parent's Guide to Protecting Your Children in Cyberspace.* New York: McGraw-Hill, 1999.

Coursey, David. "Why the Feds CAN'T Protect Kids from Internet Porn." *ZDNet Anchor Desk,* April 2, 2002. http://www.zdnet.com/anchordesk/stories/story/0,10730,2059542,00.html

McDonald, Tim. "ACLU Seeks to Overturn Filter Law." NewsFactor Network. March 20, 2001. http://www.osopinion.com/perl/story/8310.html

"Web Filters." http://www.teachersfirst.com/tutorial/filters2.shtml

webmaster

There are many on-line services (including some free ones) that will provide users with personal webpages. There are also programs such as Microsoft FrontPage that allow users to design webpages by arranging objects visually on the screen and setting their properties. However, creating and maintaining a complete website with its many linked pages, interactive forms and interfaces to databases and other services is a complicated affair. For most moderate to large-size organizations, it requires the services of a new category of IT professional: the webmaster.

Although the mixture of tasks and responsibilities will vary with the extent and purpose of the website, the skill set for a webmaster can include the following:

DEVELOPING AND EXTENDING THE WEBSITE

- Understanding how the website responds to and manages requests (see WEB SERVER)
- Fluency in the basic formatting of text and other page content and the use of frames and other tools for organizing and presenting text (see HTML)
- Extended formatting and content organization facilities such as Cascading Style Sheets (CSS), Dynamic HTML (DHTML), and Extensible Markup Language (see XML)
- Use of graphics formats and graphics and animation programs (such as Photoshop, Flash, and DreamWeaver)
- Extending the interactivity of webpages through writing scripts using tools such as JavaScript and Perl (see CGI, PERL, and PYTHON)
- Dealing with platform and compatibility issues, including browser compatibility

It is hard to draw a bright line between advanced tasks for webmasters and full-blown applications designed to run on servers or Web browsers. Some additional tools for extending Web capabilities include:

- Programming languages suitable for Web-based applications, such as Java, Visual Basic, and the new C# (pronounced C-sharp) (see JAVA)
- Web server and browser plug-ins
- Active X controls and the Microsoft .NET framework (for Windows-based systems)

ADMINISTRATIVE TASKS

- Obtaining, organizing, and updating the content for webpages (this may be delegated to writers, editors, or graphics specialists)
- Monitoring the performance of the Web server
- Ensuring site availability and response time
- Recommending acquisition of new hardware or software as necessary
- Using tools to gather information about how the site is being used, what parts are being visited, the effectiveness of advertising, and so on. This is particularly relevant to commercial sites, and can raise privacy issues.
- Setting up and managing facilities for on-line shopping (see E-COMMERCE)
- Installing and using security tools (particularly important for commercial and sensitive government sites)
- Working with major search engine providers to ensure that the site is presented to relevant searches
- Fielding queries from users about the operation of the site
- Relating the website operation to other concerns such as marketing, technical support, or the legal department
- Developing policies for website use
- Integrating the website operations into the overall corporate planning and budgeting process

The mixture of technical professional and administrator that is the webmaster makes for an always interesting and challenging career. In larger organizations there may be further differentiation of roles, with the webmaster mainly charged with operation and maintenance of the site, with the development and extension of the site handled by content providers and programmers. However, even in such cases the webmaster will need to have a general understanding of how the various features of the website interact and of the tools used to create and maintain them. People with webmaster skills can also work as independent consultants to set up and run websites for smaller businesses, schools, and nonprofit organizations.

Webmaster skills are now taught in high school, community college, vocational school, and as part of university information technology programs. However, as of 2002, the situation with regard to certification remains chaotic, with a variety of proprietary and multivendor certifications competing for attention.

Despite the economic downturn in the information technology sector, the long-term outlook for qualified webmasters is still good. Many organizations have made a fundamental commitment to use of the Web for business functions, and webmasters are needed to manage this effort.

Further Reading

Gerend, Jason, and Stephen L. Nelson. *New Webmaster's Guide to Dreamweaver 4: The Seven Steps for Designing, Building and Managing Dreamweaver 4 Web Sites*. Redmond, Wash.: Redmond Technology Press, 2001.

———. *New Webmaster's Guide to FrontPage 2002: The Eight Steps for Designing, Building and Managing FrontPage 2002 Web Sites*. Redmond, Wash.: Redmond Technology Press, 2002.

Spainhour, Stephen, and Robert Eckstein. *Webmaster in a Nutshell*. 2nd ed. Sebasatpol, Calif.: O'Reilly, 1999.

World Organization of Webmasters. http://www.joinwow.org/careerpreso/index.asp

webpage design

The World Wide Web has existed for only about a decade, so it's not surprising that the principles and practices for the design of attractive and effective webpages are still emerging. As seen in the preceding entry (see WEBMASTER), creating webpages involves the integration of many skills. In addition to the basic art of writing, many skills that had belonged to separate professions in the print world now often must be exercised by the same individual. These include typography (the selection and use of type and type styles), composition (the arrangement of text on the page), and graphics. To this mix must be added nontraditional skills such as designing interactive features and forms, interfacing with other facilities (such as databases), and perhaps the incorporation of features such as animation or streaming audio or video.

However new the technology, the design process still begins with the traditional questions any writer must ask: What is the purpose of this work? Who am I writing for? What are the needs of this audience? A website that is designed to provide background information and contact for a university department is likely to have a printlike format and a restrained style. Nevertheless, the designer of such a site may be able to imaginatively extend it beyond the traditional bounds—for example, by including streaming video interviews that introduce faculty members.

A site for an on-line store is likely to have more graphics and other attention-getting features than an academic or government site. However, despite the pressure to "grab eyeballs," the designer must resist making the site so cluttered with animations, pop-up windows, and other features that it becomes hard for readers to search for and read about the products they want.

A site intended for an organization's own use (see INTRANET) should not be visually unattractive, but the emphasis is not on grabbing users' attention, since the users are already committed to using the system. Rather, the emphasis will be on providing speedy access to the information people need to do their job, and in keeping information accurate and up to date.

Once the general approach is settled on, the design must be implemented. The most basic tool is HTML, which has undergone periodic revisions and expansions (see HTML). Even on today's large, high-resolution monitors a screen of text is not the same as a page in a printed book or magazine. There are many ways text can be organized (see HYPERTEXT AND HYPERMEDIA). A page that is presenting a manual or other lengthy document can mimic a printed book by having a table of contents. Clicking on a chapter takes the reader there. Shorter presentations (such as product descriptions) might be shown in a frame with buttons for the reader to select different aspects such as features and pricing. Frames (independently scrollable regions on a page) can turn a page into a "window" into many kinds of information without the user having to navigate from page to page, but there can be browser compatibility issues. Tables are another important tool for page designers. Setting up a table and inserting text into it allows pages to be formatted automatically.

Many sites include several different navigation systems including buttons, links, and perhaps menus. This can be good if it provides different types of access to serve different needs, but the most common failing in Web design is probably the tendency to clutter pages with features to the point that they are confusing and actually harder to use.

Although the Web is a new medium, much of the traditional typographic wisdom still applies. Just as many people who first encountered the variety of Windows or Macintosh fonts in the 1980s filled their documents with a variety of often bizarre typefaces, beginning webpage designers sometimes choose fonts that they think are "edgy" or cool, but may be hard to read—especially when shown against a purple background!

Today it is quite possible to create attractive webpages without extensive knowledge of HTML. Programs such as FrontPage and DreamWeaver mimic the operation of a word processor and take a WYSIWYG (what you see is what you get) approach. Users can build pages by selecting and arranging structural elements, while choosing styles for headers and other text as in a word processor. These programs also provide "themes" that help keep the visual and textual elements of the page consistent. Of course, designing pages in this way can be criticized as leading to a "canned" product. People who want more distinctive pages may choose instead to learn the necessary skills or hire a professional webpage designer. A feature called Cascading Style Sheets (CSS) allows designers to precisely control the appearance of webpages while defining consistent styles for elements such as headings and different types of text.

Most webpages include graphics, and this raises an additional set of issues. Some users now have fast Internet connections (see BROADBAND), but the majority are still limited to slower dial-up speeds. One way to deal with this situation is to display relatively small, lower-resolution graphics (usually 72 pixels per inch), but to allow the user to click on or near the picture to view a higher-resolution version. Page designers must also make sure that the graphics they are using are created in-house, are public domain, or are used by permission.

Animated graphics (animated GIFs or more elaborate presentations created with software) can raise performance and compatibility issues. Generally, if a site offers, for example, Flash animations, it also offers users an alternative presentation to accommodate those with slower connections or without the necessary browser plug-ins.

Further Reading

Lynch, Patrick J., and Sarah Horton. *Web Style Guide: Basic Design Principles for Creating Web Sites.* New Haven, Conn.: Yale University Press, 1999. Also available on-line at http://www.med.yale.edu/caim/manual/contents.html

Meyer, Eric A. *Cascading Style Sheets: The Definitive Guide.* Sebastopol, Calif.: O'Reilly, 2000.

Niederst, Jennifer. *Web Design in a Nutshell.* 2nd ed. Sebastopol, Calif.: O'Reilly, 2001.

On-line Training Solutions, Inc. *FrontPage Version 2002 Step by Step.* Redmond, Wash.: Microsoft Press, 2001.

University of Nebraska-Lincoln. "Internet Tools and Helper Applications." http://www.unl.edu/websat/tools.html#helper

Web Browser Guide. http://www.webreview.com/browsers/index.shtml

Weinman, Lynda, and Garo Green. *Dreamweaver 3 Hands On Training.* Berkeley, Calif.: Peachpit Press, 2000.

Web server

Most Web users aren't aware of exactly how the information they click for is delivered, but the providers of information on the Web must be able to understand and use the Web server. In simple terms, a Web server is a program running on a networked computer (see INTERNET). The server's job is to deliver the information and services that are requested by Web users.

When a user types in (or clicks on) a link in the browser window, the browser sends a HTTP request (see HTTP and WEB BROWSER). To construct the request, the browser first looks at the address (URL) in the user request. An address such as http://www.well.com/conferencing.html consists of three parts:

- The protocol, specifying the type of request. For webpages this is normally http. In many cases this part can be omitted and the browser will assume that it is meant.
- The name of the server—in this case, www.well.com. The www indicates that it is a World Wide Web server. The rest of the server name gives the organization and the domain (.com, or commercial).
- The specific page being requested. A webpage is simply a file stored on the server, and has the extension htm or html to indicate that it is an HTML-formatted page. If no page is specified, the server will normally provide a default page such as index.html.

In order to direct the browser's request to the appropriate host and server, the browser sends the URL to a name server (see DOMAIN NAME SYSTEM). The name server provides the appropriate numeric IP address (see TCP/IP). The browser then sends an HTTP "get" request to the server's IP address.

Assuming the page requested is valid, the server sends the HTML file to the browser. The browser in turn interprets the formatting and display instructions in the HTML file and "renders" the text and graphics appropriately. It is remarkable that this whole process from user click to displayed page usually takes only a few seconds, even if the website is thousands of miles away and requests must be relayed through many intervening computers.

WEB SERVER FEATURES

Web servers would be simple if webpages consisted only of static text and graphics. However, webpages today are dynamic: They can display animations, sound, and video. They also interact with the user, responding to menus and other controls, presenting and processing forms, and retrieving data from linked databases. To do these things, the server can't simply serve up a preformatted page, it must dynamically generate a unique page that responds to the user's actions.

This interactivity requires that the server be able to run programs (scripts) embedded in webpages. The Common Gateway Interface (CGI) is the basic mechanism for this, though many webpage developers can now work at a higher level to create their page's interaction through scripts in languages such JavaScript. (See CGI and SCRIPTING LANGUAGES). The task of interfacing webpages with database facilities is often accomplished using powerful data-management languages (see PERL and PYTHON).

Windows-based servers use ASP (Active Server Pages), a facility that links the Web server to Windows ActiveX controls to access databases. The interaction is usually scripted in VB Script or JScript.

Modern Web server software also contains modules for monitoring and security—an increasingly important consideration as websites become essential to business and the delivery of goods and services.

One of the most popular and reliable Web servers in use today is Apache, developed in 1995 and freely distributed with Linux and other UNIX systems (there is also a Windows version). The name is a pun on "a patchy server," meaning that it was developed by adding a series of "software patches" to existing NCSA server code. Microsoft also provides its own line of Web server software that is specific to Windows.

The future should see an increasingly seamless integration between Web servers, browsers, and other applications. Microsoft has been promoting .NET, an initiative that is designed to build Internet access and interoperability into all applications, providing operating system extensions and programming frameworks.

Beyond Microsoft's mainly proprietary efforts, another source of integration is the growing use of the Extensible Markup Language (see XML) and its offshoot SOAP (Simple Object Access Protocol). The goal is to give Web documents and other objects the ability to "communicate" their content and structure to other programs, and to allow programs to freely request and provide services to one another regardless of vendor, platform, or location. As this trend progresses, the Web server starts to "disappear" as a separate entity and the provision of Web services becomes a distributed, cooperative effort.

Further Reading

"Apache Server Frequently Asked Questions." http://httpd.apache.org/docs/misc/FAQ.html

Apache Software Foundation. http://www.apache.org/

Butler, Jason. "ASP Data Access for Beginners." http://www.15seconds.com/Issue/001025.htm

Graham, Steve [and others]. *Building Web Services with Java: Making Sense of XML, SOAP, WSDL and UDDI.* Indianapolis, Ind.: Sams, 2001.

Stanek, William Robert. *Microsoft Windows 2000 and IIS 5.0 Administrator's Pocket Consultant.* Redmond, Wash.: Microsoft Press, 2001.

Ullman, Chris [and others]. *Beginning Active Server Pages 3.0.* Birmingham, UK: Wrox Press, 2000.

Walther, Stephen. *ASP .NET Unleashed.* Indianapolis, Ind.: Sams, 2001.

Wiener, Norbert

(1894–1964)
American
Mathematician, Philosopher

Norbert Wiener developed the theory of cybernetics, or the process of communication and control in both machines and living things. His work has had an important impact both on philosophy and on design principles.

Wiener was born on November 26, 1894, in Columbia, Missouri. His father was a linguist at Harvard University, and spurred an interest in communication which the boy combined with an avid pursuit of mathematics and science (particularly biology). A child prodigy, Wiener started reading at age three, entered Tufts University at age 11, and earned his B.A. in 1909 at the age of 14, after concluding that his lack of manual dexterity made biological work too frustrating. He earned his M.A. in mathematics from Harvard only three years later, and his Harvard Ph.D. in mathematical logic just a year later in 1913. He then traveled to Europe, where he met leading mathematicians such as Bertrand Russell, G. H. Hardy, Alfred North Whitehead, and David Hilbert. When the United States entered World War I, Wiener served at Aberdeen Proving Ground, where he designed artillery firing tables.

After the war, Wiener was appointed as an instructor at MIT, where he would serve until his retirement in 1960. However, he continued to travel widely, serving as a Guggenheim Fellow at Copenhagen and Göttingen in 1926, and a visiting lecturer at Cambridge (1931–1932) and Tsing-Hua University in Beijing (1935–1936). Wiener's scientific interests proved to be as wide as his travels, including research into stochastic and random processes (such as the Brownian motion of microscopic particles) where he sought more general mathematical tools for the analysis of irregularity.

During the 1930s, Wiener began to work more closely with MIT electrical engineers who were building mechanical computers (see BUSH, VANNEVAR and ANALOG COMPUTER). He learned about feedback controls and servomechanisms that enabled machines to respond to forces in the environment.

During World War II, he did secret military research with an engineer, Julian Bigelow, on antiaircraft gun control mechanisms, including methods for predicting the future position of an aircraft based upon limited and possibly erroneous information.

Wiener became particularly interested in the feedback loop—the process by which an adjustment is made on the basis of information (such as from radar) to a predicted new position, a new reading is taken and a new adjustment made, and so on. (He had first encountered these concepts at MIT with his friend and colleague Harold Hazen.) The use of "negative feedback" made it possible to design systems that would progressively adjust themselves such as by intercepting a target. More generally, it suggested mechanisms by which a machine (perhaps a robot) could progressively work toward a goal.

Wiener's continuing interest in biology led him always to relate what he was learning about control and feedback mechanisms to the behavior of living organisms. He had followed the work of Arturo Rosenbleuth, a Mexican physiologist who was studying neurological conditions that appeared to result from excessive or inaccurate feedback. (Unlike the helpful negative feedback, positive feedback in effect amplifies errors and sends a system swinging out of control.)

By the end of World War II, Wiener, Rosenbleuth, the neuropsychiatrist Warren McCulloch, and the logician Walter Pitts were working together toward a mathematical description of neurological processes such as the firing of neurons in the brain. This research, which started out with the relatively simple analogy of electromechanical relays (as in the telephone system) would eventually result in the development of neural network theory (see NEURAL NETWORK and MINSKY, MARVIN). More generally, these scientists and others (see VON NEUMANN, JOHN) had begun to develop a new discipline for which Wiener in 1947 gave the name *cybernetics*. This word is from a Greek word referring to the steersman of a ship, suggesting the control of a system in response to its environment.

The field of cybernetics attempted to draw from many sources, including biology, neurology, logic, and what would later become robotics and computer science. Wiener's 1948 book, *Cybernetics or Control and Communication in the Animal and the Machine,* was as much philosophical as scientific, suggesting that cybernetic principles could be applied not only to scientific research and engineering but also to the better governance of society. (On a more practical level Wiener also worked with Jerome Wiesner on designing prosthetics to replace missing limbs.)

Although Wiener did not work much directly with computers, the ideas of cybernetics would indirectly influence the new disciplines of artificial intelligence (AI) and robotics. However, in his 1950 book, *The Human Use of Human Beings,* Wiener warned against the possible misuse of computers to rigidly control or regiment people, as was the experience in Stalin's Soviet Union. Wiener became increasingly involved in writing these and other popular works to bring his ideas to a general audience.

Wiener received the National Medal of Technology from President Johnson in 1964. The accompanying citation praised his "marvelously versatile contributions, profoundly original, ranging within pure and applied mathematics, and penetrating boldly into the engineering and biological sciences." He died on March 18, 1964, in Stockholm, Sweden.

Further Reading

Heims, Steve J. *John von Neumann and Norbert Wiener: From Mathematics to the Technologies of Life and Death.* Cambridge, Mass.: MIT Press, 1980.

Rosenblith, Walter, and Jerome Wiesner. "A Memoir: from Philosophy to Mathematics to Biology." http://ic.media.mit.edu/JBW/ARTICLES/WIENER/WIENER1.HTM

Wiener, Norbert. *Cybernetics, or Control and Communication in the Animal and Machine.* Cambridge, Mass.: MIT Press, 1950 (2nd ed. 1961).

———. *The Human Use of Human Beings: Cybernetics and Society.* Boston: Houghton Mifflin, 1950. (2nd ed., Avon Books, 1970).

———. *Invention: The Care and Feeding of Ideas.* Cambridge, Mass.: MIT Press, 1993.

wireless computing

Using suitable radio frequencies to carry data among computers on a local network has several advantages. The trouble and expense of running cables (such as for Ethernet) in older buildings and homes can be avoided. With a wireless LAN (WLAN) a user could work with a laptop on the deck or patio while still having access to a high-speed Internet connection.

Typically, a wireless LAN uses a frequency band with each unit on a slightly different frequency, thus allowing all units to communicate without interference. (Although radio frequency is now most popular, wireless LANs can also use microwave links, which are sometimes used as an alternative to Ethernet cable in large facilities.)

Usually there is a network access point, a PC that contains a transceiver and serves as the network hub (it may also serve as a bridge between the wireless network and a wired LAN). The hub computer can also be connected to a high-speed Internet service via DSL or cable. It has an antenna allowing it to communicate with wireless PCs up to several hundred feet away, depending on building configuration.

Each computer on the wireless network has an adapter with a transceiver so it can communicate with the access point. The adapter can be built-in (as is the case with some handheld computers), or mounted on a PC card (for laptops) or an ISA card (for desktop PCs) or connected to a USB port.

Simple home wireless LANs can be set up as a "peer network" where any two units can communicate directly with each other without going through an access point or hub. Applications needing Internet access (such as e-mail and Web browsers) can connect to the PC that has the Internet cable or DSL connection.

A wireless LAN can make it easier for workers who have to move around within the building to do their jobs. Examples might include physicians or nurses entering patient data in a hospital or store workers checking shelf inventory.

PROTOCOLS

Several protocols or standards have been developed for wireless LANs. The most common today is IEEE 802.11b, also called WiFi (or FireWire), with speeds up to 11 mbps (megabits per second) transmitting on 2.4 GHz (gigahertz) band. Although that would seem to be fast enough for most applications, a new alternative, 802.11a, can offer speeds up to 54 mbps. Because it uses the unlicensed 5 GHz frequency range it is not susceptible to interference from other devices.

The question of security for 802.11 wireless networks has been somewhat controversial. Obviously, wireless data can be intercepted in the same way that cell phone or other radio transmissions can. The networks come with a security feature called Wired Equivalent Privacy (WEP), but many users neglect to enable it and even the 128-bit version is vulnerable to certain types of attack. Users can obtain greater security by reducing emissions outside the building, changing default passwords and device IDs, and disabling DHCP to make it harder for snoopers to obtain a valid IP address for the network. Users can also add another layer of encryption and possibly isolate the wireless network from the wired network by using a more secure Virtual Private Network (VPN). Many of these measures do involve a tradeoff between the cost of software and administration on the one hand and greater security on the other. However, the growing popularity of wireless access should spur the development of improved built-in security.

Another wireless protocol called Bluetooth has been embedded in a variety of handheld computers, appliances, and other devices. It provides a wireless connection at speeds up to 1 MB/second.

MOBILE WIRELESS NETWORKING

Wireless connections can also keep computer users in touch with the Internet and their home office while they travel. Increasingly, more devices are becoming wireless capable while at the same time the functions of handheld computers, cell phones, and other devices are being merged (see also PORTABLE COMPUTERS). A new initiative called *3G* (third generation) involves the establishment of ubiquitous wireless services that can connect users to the Internet (and thus to one another) from a growing number of locations.

Currently, the 3G agenda is further advanced in Europe than in the United States. One problem is that a standard protocol has not yet emerged. The leading candidates appear to be GSM (Global System for Mobile Communications), which is used by European cell phone networks, and CDMA (Code Division Multiplexing Access).

3G has different speeds ranging from 144 bps for vehicular connections to 384 kbps for personal handheld devices to 2 bps for indoor installations. All providers, ranging from cell phones to packet (IP) telephony would using a standard billing format and database so that users could operate across many sorts of services seamlessly. The Federal Communications Commission should designate the frequency bands for this service by the end of 2002. Verizon is already offering a 3G network in limited geographical areas and other major telecommunications carriers will soon be following suit.

Ultimately, 3G may bring a *Star Trek*–like world, with handheld devices that include not only e-mail and Web browsing capability but a "smart phone," MP3 music player, and even a digital camera.

Further Reading

Derfler, Frank J., Jr., and Les Freed. "Wireless LANs." *PC Magazine,* March 2000. Available on-line: http://www.zdnet.com/products/stories/reviews/0,4161,2470130,00.html

Dornan, Andy. *The Essential Guide to Wireless Communications Applications: From Cellular Systems to WAP and M-Commerce.* Upper Saddle River, N.J.: Prentice Hall, 2000.

Ellison, Craig. "Wireless LANs at Risk." *PC Magazine,* April 9, 2002, 66–68.

Federal Communications Commission. "Third Generation ("3G") Wireless." http://www.fcc.gov/3G/

Wireless LAN Association. http://www.wlana.com/

Wirth, Niklaus

(1934–)
Swiss
Computer Scientist

Niklaus Wirth created new programming languages such as Pascal that helped change the way computer scientists and programmers thought about their work. His work influenced later languages and ways of organizing program resources.

Wirth was born on February 15, 1934, in Winterhur, Switzerland. He received a degree in electrical engineering at the Swiss Federal Institute of Technology (ETH) in 1959, then earned his M.S. at Canada's Laval University. He went to the University of California, Berkeley, where he received his Ph.D. in 1963 and taught in the newly founded Computer Science Department at nearby Stanford University. By then he had become involved with computer science and the design of programming languages.

Wirth returned to the ETH in Zurich in 1968, where he was appointed a full professor of computer science. He had been part of an effort to improve Algol. Although Algol offered better program structures than earlier languages such as FORTRAN, the committee revising the language had become bogged down in adding many new features to the language that would become Algol-68 (see ALGOL).

Wirth believed that adding several ways to do the same thing did not improve a language but simply made it harder to understand and less reliable. Between 1968 and 1970, Wirth therefore crafted a new language, Pascal, named after the 17th-century mathematician who had built an early calculating machine.

Pascal required that data be properly defined (see DATA TYPES) and allowed users to define new types of data such as records (similar to those used in databases). It provided all the necessary control structures (see LOOP and BRANCHING STATEMENTS). Following the new thinking about structured programming (see DIJKSTRA, EDSGER) Pascal retained the "unsafe" GOTO statement but discouraged its use.

Pascal became the most popular language for teaching programming. By the 1980s, versions such as UCSD Pascal and later, Borland's Turbo Pascal were bringing the benefits of structured programming to desktop computer users. Meanwhile, Wirth was working on a new language, Modula-2. As the name suggested, the language featured the use of modules, packages of program code that could be linked to programs to extend their data types and functions. Wirth also designed a computer workstation called *Lilith.* This powerful machine not only ran Modula-2; its operating system, device drivers and all other facilities were also implemented in Modula-2 and could be seamlessly integrated, essentially removing the distinction between operating system and application programs. Wirth also helped design Modula-3, an object-oriented extension of Modula-2, as well as another language, Oberon, which was originally intended to run in built-in computers (see EMBEDDED SYSTEMS).

Looking back at the development of object-oriented programming (OOP), the next paradigm that captured the attention of computer scientists and developers after structured programming, Wirth has noted that OOP isn't all that new. Its ideas (such as encapsulation of data) are largely implicit in structured procedural programming, even if it shifted the emphasis to binding functions into objects and allowing new objects to extend (inherit from) earlier ones. But he believes the fundamentals of good programming haven't really changed in 30 years. In a 1997 interview Wirth noted that "the woes of Software Engineering are not due to lack of tools, or proper management, but largely due to lack of sufficient technical competence. A good designer must rely on experience, on precise, logical thinking; and on pedantic exactness. No magic will do."

Wirth has received numerous honors, including the ACM Turing Award (1984) and the IEEE Computer Pioneer Award (1987).

Further Reading

Pescio, Carlo. "A Few Words with Niklaus Wirth." *Software Development,* vol. 5, no. 6, June 1997. Available on-line at http://www.eptacom.net/pubblicazioni/pub_eng/wirth.html

Wirth, Niklaus. *Algorithms + Data Structures = Programs.* Englewood Cliffs, N.J.: Prentice Hall, 1976.

———, and Kathy Jensen. *PASCAL User Manual and Report.* 4th ed. New York: Springer-Verlag, 1991.

———. *Project Oberon: The Design of an Operating System and Compiler.* Reading, Mass.: Addison-Wesley, 1992.

———. "Recollections about the Development of Pascal." In Bergin, Thomas J., and Richard G. Gibson, eds., *History of Programming Languages-II,* 97–111. New York: ACM Press; Reading, Mass.: Addison-Wesley, 1996.

———. *Systematic Programming: An Introduction.* Reading, Mass.: Addison-Wesley, 1973.

women in computing

Although computer science and technology would come to be considered male-dominated professions, women were part of the computer story from its very beginnings. Augusta Byron, Lady Lovelace not only served as the world's first computer technical writer in helping popularize the Analytical Engine (see BABBAGE, CHARLES), she also was the first programmer, giving instructions for the machine to carry out various calculations.

During World War II, a number of women with mathematical backgrounds received a mysterious summons from the government. When they arrived at the restricted facility, women such as Jean Bartik and Betty Holberton discovered that they had been assigned the task of programming ENIAC, generally considered the first functioning electronic digital computer (see ECKERT, J. PRESPER and MAUCHLY, JOHN). Meanwhile, Grace Hopper (see HOPPER, GRACE MURRAY), who would become probably the most famous woman in computer science, programmed the Harvard Mark I calculator (see AIKEN, HOWARD). After the war she joined Eckert and Mauchly to develop the first automatic program compilers. During the 1950s, Hopper went on to create Flow-Matic, the first business-oriented programming language, and then played a key role on the committee that designed COBOL. (Three other women, Holberton, Jean E. Sammet, and Mary K. Hawes also served on the committee.)

Women would continue to play an important if little-known role in computing, particularly in military and other government facilities. However, as computer science became more formalized as a profession in the 1960s and 1970s, most practitioners came from mathe-

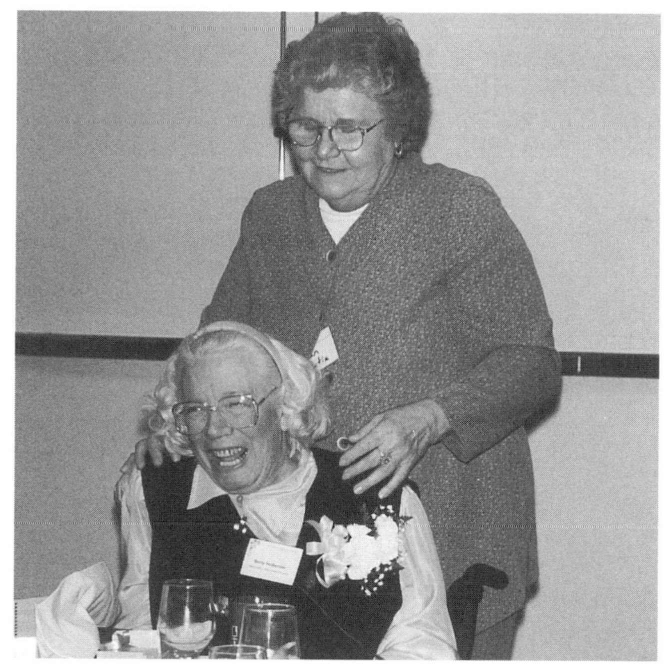

Jean Bartik (standing) and Betty Holberton answered a call for "computers" during World War II. At the time, that was the name for a clerical person who performed calculations. These two computer pioneers, shown here at a reunion, would go on to develop important programming techniques for the ENIAC and later machines. (COURTESY OF THE ASSOCIATION FOR WOMEN IN COMPUTING)

matics and engineering, two professions traditionally dominated by men. The microcomputer and PC revolution of the 1970s and 1980s, too, sprung from a male culture of garage tinkerers and gadget-happy "nerds."

Even today only about 30 percent of undergraduate computer science degrees in the United States are awarded to women, and women account for only 12 percent of doctorates. Indeed, the percentage seemed to actually decline as the computer field became more popular and erstwhile nerds became multimillionaire CEOs.

However, the growth of the Internet and in particular the World Wide Web may be helping to reverse this trend. The ubiquitous Web is exposing girls and boys equally to the possibilities of computing, even if boys still play more videogames. Designers of games and educational software have also been paying more attention to the needs and interests of girls. Finally, the continuing broad social trend toward greater opportunity and acceptance for women in all professions should also have its effect on the computing field.

Today many organizations are helping to encourage more female participation in computer science and technology. These include the ACM Committee on Women in Computing and the Association for Women in Computing (see Appendix IV for more contact information and for more names of organizations involved with women in computing). In recent years there have also been a series of national conferences entitled The Grace Hopper Celebration of Women in Computing.

Further Reading

The Ada Project. http://tap.mills.edu

Gürer, Denise. "Pioneering Women in Computer Science." *Communications of the ACM* 38, no. 1, 1995, 45–54.

Institute for Women and Technology [sponsors the Grace Hopper Celebration of Women in Computing Conference].http://www.iwt.org/home.html

Margolis, Jane, and Allan Fisher. *Unlocking the Clubhouse.* Cambridge, Mass.: MIT Press, 2001.

Seminerio, Maria. "The Secret Mission of the First [Women] Computer Programmers." *ZDNet News,* October 19, 1998. http://zdnet.com.com/2100-11-512403.html?legacy=zdnn

Toole, Betty Alexandra, ed. *Ada, the Enchantress of Numbers: A Selection from the Letters of Lord Byron's Daughter and Her Description of the First Computer.* Mill Valley, Calif.: Strawberry Press, 1992.

Yount, Lisa. *A–Z of Women in Science and Mathematics.* New York: Facts On File, 1999.

word processing

Although computers are most often associated with numbers and calculation, creating text documents is probably the most ubiquitous application for desktop PCs.

The term *word processor* was actually coined by IBM in the 1960s to refer to a system consisting of a Selectric typewriter with magnetic tape storage. This allowed the typist to record keystrokes (and some data such as margin settings) on tape. Material could be corrected by being re-recorded. The tape could then be used to print as many perfect copies of the document as required. A version using magnetic cards instead of tape appeared in 1969.

The first modern-style word processor was marketed by Lexitron and Linolex. It also used magnetic tape, but it added a video display screen. Now the writer could see and correct text without having to print it first. A few years later, a new invention, the floppy disk, became the standard storage medium for dedicated word processing systems.

The word-processing systems developed by Wang, Digital Equipment Corporation, Data General, and others became a feature in large offices in the late 1970s. These systems were essentially minicomputers with screens, keyboards, and printers and running a specialized software program. Because these systems were expensive (ranging from about $8,000 to $20,000 or more), they were not affordable by smaller businesses. Typically, they were operated by specially trained personnel (who became known also as "word processors") to whom documents were funneled for processing, as with the old "typing pool."

PC WORD PROCESSING

The first microcomputer systems had very limited memory and storage capacity. However, by the late 1970s various systems using the S-100 bus and running CP/M had word-processing programs, as did the Apple II and other first-generation PCs. However, it took the entry of the IBM PC into the market in 1981 to make the PC a word-processing alternative for mainstream businesses. The machine had more memory and storage than earlier machines, and the IBM name provided reassurance to business.

A number of word-processing programs were written for the IBM PC running MS-DOS, but the market leaders were WordStar and WordPerfect. Both programs offered basic text editing and formatting, including the ability to embed commands to mark text for boldface, italic, and so on. The programs came with drivers for the more popular printers.

In 1984, the Macintosh offered a new face for word processing and other applications. Using bitmapped fonts, the Mac could show a good representation of the fonts and typestyles that would be in the printed document. This "what you see is what you get" (see WYSIWYG) approach, together with the graphical user interface with mouse-driven menus meant that users did not have to learn the often obscure command key sequences used in WordStar or WordPerfect.

Microsoft then developed Windows as a graphical user interface alternative to MS-DOS for IBM-compatible

PCs. By 1990, Windows was rapidly replacing DOS as the operating system of choice, and Microsoft Word was winning the battle against WordPerfect, whose Windows version was rather flawed at first.

In addition to being able to visually show fonts and formatting, Word and other modern word processors are packed with features. Some typical features today include:

- Different views of the document, including an outline showing headings down to a user-specified level
- Automatic table of contents and index generation
- Tables and multicolumn text
- Automatic formatting of bulleted and numbered lists
- Built-in and user-defined styles for headings, paragraphs, and so on.
- The ability to use built-in or user-defined templates to provide starting settings for new documents (see TEMPLATE)
- The ability to record or otherwise specify a series of commands to be performed automatically (see MACRO)
- Spelling and grammar checkers
- The ability to incorporate a variety of graphics image formats in the document
- Automatic formatting and linking of Web hyperlinks within documents
- The ability to import and export documents in a variety of formats, including Web documents (see HTML)
- An extensive on-line help system including "wizards" to guide the user step-by-step through various tasks

As word processors become more extensive in their capabilities, it has become harder to distinguish them from programs designed to create precise copy for publication (see DESKTOP PUBLISHING). However, copy prepared by writers with a word processor must generally be further processed through a desktop publishing or in-house computerized typography system.

At the other end of the spectrum many users find that word processors are "overkill" for making simple notes. A variety of programs for entering simple text are available, including the Notepad program that comes with Windows. There are also applications for which plain text must be produced, without the formatting codes added by word processors. In particular, programmers often use specialized editing programs to create source code (see TEXT EDITOR).

TRENDS

Today word processing programs are generally part of an office software suite such as Microsoft Office, Corel Office, or Star Office. Documents created by other components of the suite can be embedded in word processing documents. (In Windows, object linking and embedding [OLE] is a system that allows for embedded documents to be automatically updated and to be edited using the functions of the host program. Thus, an Excel spreadsheet embedded in a Word document can be worked in place using the standard Excel interface.)

There are also features that can facilitate collaboration between workers in a networked office, such as by keeping track of revisions made by various people working on the same document.

As with other applications, word processors are increasingly being integrated with the Web, and include the ability to create HTML documents. In turn, the programs specifically designed for creating HTML documents now have many word-processor features including templates, styles, and the visual representation of the page.

Further Reading

Acklen, Laura. *Special Edition Using Corel WordPerfect 10.* Indianapolis, Ind.: Que, 2001.

Kunde, Brian. "A Brief History of Word Processing (through 1986)." http://www.stanford.edu/~bkunde/fb-press/articles/wdprhist.html

Millhollon, Mary, and Katherine Murray. *Microsoft Word Version 2002 Inside Out.* Redmond, Wash.: Microsoft Press, 2001.

"A Potted History of WordStar." http://www.petrie.u-net.com/wordstar/history/history.htm

workstation

Like minicomputer, *workstation* is a rather slippery term whose meaning and significance has changed somewhat with the growing power of desktop PCs.

In the late 1960s and 1970s, most "personal" computing was done by individuals connected to time-sharing mainframes or minicomputers by terminals. Generally, the terminals could only display text, not graphics.

However, researchers at the Xerox Palo Alto Research Center (PARC) began to develop a more powerful computer for individual use (see ENGLEBART, DOUGLAS and KAY, ALAN). The Xerox Alto had a high-resolution bitmapped graphics display and a mouse-controlled graphical user interface. While it was expensive and not very successful commercially, the Alto set the stage for the Macintosh in 1984 and for Microsoft Windows.

Although the desktop PCs of the 1980s such as the IBM PC had some graphics capabilities, the machines lacked the capacity for graphics-intensive applications such as engineering design and the generation of movie effects. Led by Sun and Silicon Graphics (SGI), the high-performance graphics workstation emerged as a distinctive product category. These machines used relatively powerful microprocessors (such as the Sun SPARC and the MIPS) with instruction sets optimized for speed (see RISC). These systems generally ran UNIX as their operating system.

However, by the late 1990s, ordinary desktop PCs were catching up to dedicated workstations in terms of processing power and graphics features. By 2002, a desktop PC costing about $2,000 offered a 2 GB processor, 256 MB of RAM, 120 GB hard drive, and an optimized 3D graphics card that can drive displays up to 1600 by 1200 pixels or more. These systems can run Windows NT or XP, or, for users preferring UNIX, Linux offers a robust and inexpensive operating system. This sort of system rivals the capabilities of a dedicated workstation while offering all of the versatility of a general-purpose PC. As a result, the term *workstation* today refers more to a way of using a computer than to a specific class of hardware. Machines are thought of as workstations if they emphasize graphics performance and are dedicated to particular activities such as science, imaging, engineering, design (see also COMPUTER-AIDED DESIGN AND MANUFACTURING), or video editing.

Further Reading

Miller, Michael J. "Do You Need a Workstation?" *ZDNet AnchorDesk*. August 20, 2000. http://www.zdnet.com/products/stories/reviews/0,4161,2616686,00.html

Overview of SGI Workstations. http://www.sgi.com/workstations/

Sun Microsystems. *The New User's Guide to Sun Workstation*. New York: Springer-Verlag, 1991.

World Wide Web

In little more than a decade the World Wide Web has become nearly as ubiquitous as the telephone and has become for many a preferred medium for shopping, news, entertainment, and education. Some cultural observers believe that this vast system of linked information may be having an impact on society as great as that of the invention of the printing press more than five centuries earlier.

By the beginning of the 1990s, the Internet had become well established as a means of communication between relatively advanced computer users, particularly scientists, engineers, and computer science students—primarily using UNIX-based systems (see UNIX). A number of services used the Internet protocol (see TCP/IP) to carry messages or data. These included e-mail, file transfer protocol (see FTP) and newsgroups (see NETNEWS AND NEWSGROUPS). A Wide Area Information Service (WAIS) even provided a protocol for users to retrieve information from databases on remote hosts. Another interesting service, Gopher, was developed at the University of Minnesota in 1991. It used a system of nested menus to organize documents at host sites so they could be browsed and retrieved by remote users.

Gopher was quite popular for a few years, but it would soon be overshadowed by a rather different kind of networked information service. A physicist/programmer (see BERNERS-LEE, TIM) working at CERN, the European particle physics laboratory in Switzerland had devised in 1989 a system that he eventually called the World Wide Web (sometimes called WWW or W3). By 1990, he was running a prototype system and demonstrating it for CERN researchers and a few outside participants.

USING THE WEB

The Web consists essentially of three parts. Berners-Lee devised a markup language: that is, a system for indicating document elements (such as headers), text characteristics, and so on (see HTML). Any document could be linked to another (see HYPERTEXT AND HYPERMEDIA) by specifying that document's unique address (called a Uniform Resource Locator or URL) in a request. Berners-Lee defined the HyperText Transport Protocol, or HTTP, to handle the details needed to retrieve documents. (Although HTTP is most often used to retrieve HTML-formatted Web documents, it can also be used to specify documents using other protocols, such as ftp, news, or Gopher.)

A program (see WEB SERVER) responds to requests for documents sent over the network (usually the Internet, that is, TCP/IP). The requests are issued by a client program as a result of the user clicking on highlighted links or buttons or specifying addresses (see WEB BROWSER). The browser in turn interprets the HTML codes on the page to display it correctly on the user's screen.

At first the Web had only text documents. However, thanks to Berners-Lee's flexible design (see CLIENT-SERVER COMPUTING) new, improved Web browsers could be created and used with the Web as long as they followed the rules for HTTP. The most successful of these new browsers was Mosaic, created by Marc Andreesen at the National Center for Supercomputing Applications. NCSA Mosaic was available for free download and could run on Windows, Macintosh, and UNIX-based systems. Mosaic not only dispensed with the text commands used by most of the first browsers, it also had the ability to display graphics and play sound files. With Mosaic the text-only hypertext of the early Web rapidly became a richer hypermedia experience. And thanks to the ability of browsers to accept modules to handle new kinds of files (see PLUG-IN), the Web could also accommodate real-time sound and video transmissions (see STREAMING).

In 1994, Andreessen left NCSA and co-founded a company called Netscape Communications, which improved and commercialized Mosaic. Microsoft soon entered with a competitor, Internet Explorer; today these two browsers dominate the market with Microsoft having taken the lead. Together with relatively low-cost Internet access (see MODEM and INTERNET SERVICE PROVIDER) these user-friendly Web browsers brought the Web (and thus the underlying Internet) to the masses. Schools and libraries began to offer Web access while workplaces began to use internal webs to organize information and organize operations (see INTRANET). Meanwhile, companies such as the on-line bookseller Amazon.com demonstrated new ways to deliver traditional products, while the on-line auction site eBay took advantage of the unique characteristics of the on-line medium to redefine the auction.

The burgeoning Web was soon offering millions of pages, especially as entrepreneurs began to find additional business opportunities in the new medium (see E-COMMERCE). Two services emerged to help Web users make sense of the flood of information. Today users can search for words or phrases (see SEARCH ENGINE) or browse through structured topical listings (see PORTAL). Estimates from various sources suggest that as of 2002 approximately 500 million people worldwide access the Web.

IMPACT AND TRENDS

The Web is gradually emerging as an important news medium (see JOURNALISM AND THE COMPUTER INDUSTRY). The medium combines the ability of broadcasting to reach many people from one point with the ability to customize content to each person's preferences. Traditional broadcasting and publishing are constrained by limited resources and the need for profitability, and thus the range and diversity of views made available tend to be limited. With the Web, anyone with a PC and a connection to a service provider can put up a website and say just about anything. Millions of people now display aspects of their lives and interests on their personal webpages, and even publish on-line journals called *weblogs* that chronicle their day-to-day thoughts. The Web has also provided a fertile medium for the creation of on-line communities (see VIRTUAL COMMUNITY) while contributing to significant issues (see PRIVACY IN THE DIGITAL AGE).

The setback in Web commerce at the beginning of the new century should therefore be taken in context: The growth in use of the Web continues, as does its importance in daily life and the continuing development of new applications and activities. Although the technology has matured to some extent, its long-term social and economic impact will be hard to assess for some time.

Further Reading

Berners-Lee, Tim. *Weaving the Web: the Original Design and Ultimate Destiny of the World Wide Web.* New York: Harperbusiness, 2000.

CyberAtlas [Internet Statistics].http://cyberatlas.internet.com/

"Navigating the World Wide Web." http://www.imaginarylandscape.com/helpweb/www/www.html

Rosenfeld, Louis and Peter Morville. *Information Architecture for the World Wide Web.* Sebastopol, Calif.: O'Reilly, 1998.

Webreference.com. http://www.webreference.com/

Wozniak, Steven

(1950–)
American
Computer Inventor and Engineer

Steve Wozniak, often known as "Woz," co-founded Apple computer and designed the Apple II, one of the first popular personal computers.

Born on August 11, 1950, in San Jose, California, Wozniak grew up to be a classic "electronics whiz." He built a working electronic calculator when he was 13, winning the local science fair. After graduating from Homestead High School, Wozniak tried community college but quit to work with a local computer company. Although he then enrolled in the University of California, Berkeley, to study electronic engineering and computer science, he dropped out in 1971 to go to work again, this time as an engineer at Hewlett-Packard, at that time one of the most successful companies in the young Silicon Valley.

By the mid-1970s, Wozniak was in the midst of a technical revolution in which hobbyists explored the possibilities of the newly available microprocessor or "computer on a chip." A regular attendee at meetings of the Homebrew Computer Club, Wozniak and other enthusiasts were excited when the MITS Altair, the first complete microcomputer kit, came on the market in 1975. The Altair, however, had a tiny amount of memory, had to be programmed by toggling switches to input hexadecimal codes (rather like the ENIAC), and had very primitive input/output capabilities. Wozniak decided to build a computer that would be much easier to use—and more useful.

Wozniak's prototype machine, the Apple I, had a keyboard and could be connected to a TV screen to provide a video display. He demonstrated it at the Homebrew Computer Club and among the interested spectators was his friend Steve Jobs. Jobs had a more entrepreneurial interest than Wozniak, and spurred him to set up a business to manufacture and sell the machines. Together they founded Apple Computer in June 1976. Their "factory" was Jobs's parents' garage, and the first machines were assembled by hand.

Wozniak designed most of the key parts of the Apple, including its video display and later, its floppy disk interface, which is considered a model of elegant engineering to this day. He also created the built-in operating system and BASIC interpreter, which were stored in read-only memory (ROM) chips so the computer could function as soon as it was turned on.

In 1981, just as the Apple II was reaching the peak of its success, Wozniak was almost killed in a plane crash. He took a sabbatical from Apple to recover, get married, and return to UC Berkeley (under an assumed name!) to finish his B.S. in electrical engineering and computer science.

Wozniak's life changes affected him in other ways. As Apple grew and became embroiled in the problems of large companies, "Woz" sold large amounts of his Apple stock and gave the money to Apple employees that he thought had not been properly rewarded for their work. Later in the 1980s, he produced two rock festivals that lost $25 million, which he paid out of his own money. He was quoted as saying, "I'd rather be liked than rich." He left Apple for good in 1985 and founded Cloud Nine, an unsuccessful company that designed remote control and "smart appliance" hardware.

During the 1990s, Wozniak organized a number of charitable and educational programs, including cooperative activities with people in the former Soviet Union. He particularly enjoyed classroom teaching, bringing the excitement of technology to young people. In 1985, Wozniak received the National Medal of Technology.

Further Reading

Cringely, Robert X. *Accidental Empires: How the Boys of Silicon Valley Make Their Millions, Battle Foreign Competition, and Still Can't Get a Date.* Reading, Mass.: Addison-Wesley, 1992.

Freiberger, Paul, and Michael Swaine. *Fire in the Valley: the Making of the Personal Computer.* 2nd ed. New York: McGraw-Hill, 1999.

Kendall, Martha E. *Steve Wozniak, Inventor of the Apple Computer.* 2nd rev. ed. Los Gatos, Calif.: Highland Publishing, 2001.

Woz.org [Steve Wozniak's Home Page]. www.woz.org

WYSIWYG

With early computer applications there was little resemblance between the user interface used to create a product and the appearance of the product itself. For example, a user of an early PC word-processing program such as WordStar could indicate that certain words be put in boldface or italic by starting and ending the word with hidden code characters (such as Ctrl-B or Ctrl-I). However, the actual appearance of the word on the screen did not change or at most appeared in highlighted text.

Powerful and elaborate formatting programs for UNIX such as nroff and troff could specify not only typestyles but font, size, character spacing, margins, and many other characteristics. However, this was done by specifying special commands (beginning with a dot) throughout the otherwise plain text.

During the 1980s, however, the growing processing and graphics capabilities of PCs together with a visual philosophy at Apple and Microsoft led to a new philosophy, "what you see is what you get" (WYSIWYG). (This is probably a reference to Flip Wilson's "Geraldine" character in 1970s TV.)

On a Macintosh, for example, a document could be shown as bitmapped graphics rather than text characters. This meant that actual fonts, typestyles, spacing, and other effects could be shown on the screen in a way that reasonably accurately corresponds to the appearance the document would have when printed on paper. Microsoft then adopted the same approach in its Windows operating system and applications followed suit.

Today virtually all word processors use WYSIWYG, although plain-text editors are still used for programming and scripting (see TEXT EDITOR). Graphics, too, can be shown as actual pictures rather than just markers. Typically there are different "views" that can be selected by the user. For example, there might be a normal or draft view that shows only text with markers for graphics, and a "page view" that shows a representation of the printed page. Usually text can be entered and edited in either mode, but the nongraphical mode may respond more fluidly.

Although WYSIWYG has become the dominant user interface mode for most applications that create documents, there are many reasons why the correspondence between screen and printed output may be less than perfect. In word processing, there may be fonts that are available as screen fonts but not as printer fonts (or vice versa), and the system must therefore either choose the most similar font to the missing font or approximate it in some way. With graphics there is frequently a difference between the way in which the screen and printer renders a particular color. Photoshop users, for example, must go through an elaborate process of color correction if they want the most accurate results. Critics sometimes refer to less than perfect WYSIWYG as WYSIAWYG (what you see is almost what you get).

Some critics also find that WYSIWYG has become so much the expected norm that application designers use it even when it may be inappropriate or inefficient. When combined with a related idea, the GUI (see GRAPHICAL USER INTERFACE), the results may be slow performance on older systems and the use of repetitive, tedious mouse movements instead of concise commands. Many experienced Web designers, for example, prefer coding their own HTML (often with the aid of templates or macros) rather than using a GUI/WYSIWYG applications such as FrontPage. Advocates of using markup commands rather than WYSIWYG also point to the ability to ensure a consistent style and to automate more of the process of creating documents.

However, by providing immediate visual feedback, WYSIWYG doubtless makes it easier for users to learn how to create the desired documents in word processing or desktop publishing programs. As part of the broad movement toward "visual computing," WYSIWYG is here to stay.

Further Reading

Calore, Michael. "WYSIWYG Editor Shootout 2001." *Webmonkey,* September 28, 2001. Available on-line at http://hotwired.lycos.com/webmonkey/01/39/index4a.html

Stephenson, Neal. *In the Beginning . . . Was the Command Line.* New York: Avon Books, 1999.

Taylor, Conrad. "What has WYSIWYG Done to Us?" *Seybold Report on Publishing Systems* 26, no. 2, September 30, 1996. Available on-line at http://www.ideography.co.uk/library/seybold/WYS_intro.html

XML

Several markup languages have been devised for specifying the organization or format of documents. Today the most commonly known markup language is the Hypertext Markup Language (see HTML), which is the organizational "glue" of the Web (see WORLD WIDE WEB).

HTML is primarily concerned with rendering (displaying) documents. It describes structural features of documents (such as headers, sections, tables, and frames), but it doesn't really convey the structure of the information within the document. Further, HTML is not extensible—that is, one can't define one's own tags and use them as part of the language. XML, or Extensible Markup Language, is designed to meet both of these needs. In effect, while HTML is a descriptive coding scheme, XML is a scheme for creating data definitions and manipulating data within documents. (XML can be viewed as a subset of the powerful and generalized SGML, or Standard Generalized Markup Language.)

The basic building block of XML is the element, which can be used to define an entity (rather like a database record). For example, the following statement:

```
<team name="New York Yankees">
   <players>
      <player name="1">Babe Ruth</player>
      <player name="2">Lou Gehrig</player>
   </players>
</team>
```

XML text is bracketed by tags as with HTML. The "team" element has an attribute called "name" that is assigned the value "New York Yankees." (Attribute values must be enclosed in quote marks.) It also contains a nested element called *players,* which in turn defines player names, Babe Ruth and Lou Gehrig. The elements are defined at the beginning of the XML document by a DTD (Document Type Definition), or such a definition can be "included" from another file.

XML is currently supported by the leading Web browsers. In effect, it includes HTML as a subset, or more accurately XHTML (HTML conformed to XML 1.0 standards). Thus, XML documents can be properly rendered by browsers, while applications that are XML-enabled (or that use XML-aware ActiveX controls or similar Java facilities, for example) can parse the XML and identify the data structures and elements in the document. Together with programming languages such as Java and facilities such as SOAP (Simple Object Access Protocol), XML can be used to create applications that connect servers and documents across the Internet.

XML can be viewed as part of a trend to make data "self-describing." The ability to encode not just the structure but the logical content of documents promises a growing ability for automated agents or "bots" to take over much of the work of sifting through the Web for desired information, bringing the Web closer to the intentions of its inventor, Tim Berners-Lee.

Further Reading

Clark, Scott. "The Webdeveloper.com Guide to XML." http://www.webdeveloper.com/html/html_xml_1.html

Jones, Jeff. "XML 101." http://www.swynk.com/friends/jones/articles/xml_101.asp

Ray, Erik T. *Learning XML*. Sebastopol, Calif.: O'Reilly, 2001.

"The XML Cover Pages." http://www.oasis-open.org/cover/sgml-xml.html

Y

Y2K problem

Sherlock Holmes once referred to a dog barking in the night. Watson, puzzled as usual, replied that no dog had barked. Holmes replied that it was the nonbarking that was significant. The same can be said about the growing concern toward the end of the 1990s that the year 2000 might bring massive, disastrous failures to many of the computer systems on which society now depended for its well-being.

Most programs written in the 1960s and 1970s (see MAINFRAME and COBOL) saved expensive memory space by storing only the second two digits of year dates. After all, dates could be understood to begin with "19" for many years to come (although some farsighted computer scientists did warn of future trouble). Eventually the century began to draw to an end.

Although much computing activity had moved onto newer systems by the 1990s, many large government and corporate computer systems were still running the original applications or their descendents. If such a program were run in the year 2000, it would have no way to distinguish a date in that year from a date in 1900. While the prospect of a centenarian being suddenly treated as a newborn was likely to be more amusing than significant, what would happen to a 30-year mortgage that was written in 1975 and intended to come due in 2005? Would people be billed based on a -70-year term? Many observers feared that some systems would actually crash because they would begin to generate nonsensical data. What, for example, might happen to an air traffic control system or automated power grid system that used dates and times to track events?

No one really knew. One problem was that there were millions of lines of code, often written by programmers who had long since retired. Nor was it simply a matter of looking for references to date fields (such as in decision statements), because of the many ways programmers could express such statements. In addition to mainframe applications, there were also the computers hardwired into devices of all kinds including cars and airplanes (see EMBEDDED SYSTEM). As with the early mainframes, these systems were often designed with limited available memory, and thus their programmers, too, may have been tempted to save bytes by lopping off the century years.

As the fateful date approached, government agencies and businesses began to invest billions of dollars and hire expensive consultants to check code for "Y2K compatibility." In the end, Y2K problems were found and fixed in the most critical systems, and the year 2000 dawned without significant mishaps. (It turned out that the virtually all the embedded systems did not in fact have Y2K problems, mostly because they didn't even track year dates.)

But although the "dog didn't bark" and in retrospect some of the hype about Y2K seems excessive, it did lead to improvement in a great deal of software. Further, it increased awareness of dependence on computers for so many aspects of life—a dependence that has been cast in a harsh new light by the terrorist events of September 11, 2001 (see RISKS OF COMPUTING).

Further Reading

Crawford, Walt. "Y2K: Lessons from a Non-Event." *Online,* vol. 25, issue 2, March 2001, 73.

Finkelstein, Anthony. "Y2K: a Retrospective View." *Computing and Control Engineering Journal,* vol. 11, no. 4, August 2000, 156–159. Available on-line at http://www.cs.ucl.ac.uk/staff/A.Finkelstein/papers/y2kpiece.pdf

Yourdon, Edward and Jennifer Yourdon. *Time Bomb 200!: What the Year 2000 Computer Crisis Means to You.* Upper Saddle River, N.J.: Prentice Hall, 1997.

Z

Zuse, Konrad

(1910–1995)
German
Engineer, Inventor

Great inventions seldom have a single parent. Although popular history credits Alexander Graham Bell with the telephone, the almost forgotten Elisha Gray invented the device at almost the same time. And although the ENIAC is widely considered to be the first practical electronic digital computer (see ECKERT, J. PRESPER and MAUCHLY, JOHN) another American inventor built a smaller machine on somewhat different principles that also has a claim to being "first" (see ATANASOFF, JOHN). Least known of all is Konrad Zuse, perhaps because he did most of his work in a nation that was plunging the world into war.

Zuse was born on June 22, 1910, in Berlin. He studied civil engineering at the Technische Hochschule Berlin-Charlottenburg, receiving his degree in 1935. One of his tasks in engineering was performing calculations of the stress on structures such as bridges. At the time these calculations were carried out by going through a series of steps on a form over and over again, plugging in the data and calculating by hand or using an electromechanical calculator. Like other inventors before him, Zuse began to wonder whether he could build a machine that could carry out these repetitive steps automatically.

Zuse was unaware of the nearly forgotten work of Charles Babbage and that of other inventors in America and Britain who were beginning to think along the same lines (see BABBAGE, CHARLES). With financial help from his parents (and the loan of their living room), Zuse began to assemble his first machine from scrounged parts. His first machine, the Z1, was completed in 1938. The machine used slotted metal plates with holes and pins that could slide to carry out binary addition and other operations (in using the simpler binary system rather than decimal, Zuse was departing from other calculator designers).

The Z1 had trouble storing and retrieving numbers and never worked well. Undeterred, Zuse began to develop a new machine that used electromechanical telephone relays (a ubiquitous component that was also favored by Howard Aiken [see AIKEN, HOWARD]). The new machine worked much better, and Zuse successfully demonstrated it at the German Aerodynamics Research Institute in 1939.

With World War II under way, Zuse was able to obtain funding for his Z3, which was able to carry out automatic sequences from instructions (Zuse used discarded movie film instead of punched tape). The machine used 22-bit words and had 600 relays in the calculating unit and 1,800 for the memory. However, the machine could not do branching or looping the way modern computers can. It was destroyed in a bombing raid in 1944. Meanwhile, Zuse used spare time from his military duties at the Henschel aircraft company to work on the Z4, which was completed in 1949. This machine was more fully programmable and was comparable to Howard Aiken's Mark I.

By that time, however, Zuse's electromechanical technology had been surpassed by the fully electronic vacuum tube computers such as the ENIAC and its successors. (Zuse had considered vacuum tubes but had rejected them, believing that their inherent unreliability and the large numbers needed would make them impracticable for a large-scale machine.) During the 1950s and 1960s, Zuse ran a computer company, ZUSE KG, which eventually produced electronic vacuum tube computers.

Zuse's most interesting contribution to computer science would not be his hardware but a programming language called *Plankalkül* or "programming calculus." Although the language was never implemented, it was far in advance of its time in many ways. It started with the radically simple concept of grouping individual bits to form whatever data structures were desired. It also included program modules that could operate on input variables and store their results in output variables (see PROCEDURES AND FUNCTIONS). Programs were written using a notation similar to mathematical matrices.

Zuse labored in obscurity even within the computer science fraternity. However, toward the end of his life his work began to be publicized. He received numerous honorary degrees from European universities as well as awards and memberships in scientific and engineering academies. Zuse also took up abstract painting in his later years. He died on December 18, 1995.

Further Reading

Bauer, F. L., and H. Wössner. "The Plankalkül of Konrad Zuse: A Forerunner of Today's Programming Languages." *Communications of the ACM*, vol. 15, 1972, 678–685.

Lee, J. A. N. *Computer Pioneers*. Los Alamitos, Calif.: IEEE Computer Society Press, 1995.

Zuse, Konrad. *The Computer—My Life*. New York: Springer-Verlag, 1993.

APPENDIX I
BIBLIOGRAPHIES AND WEB RESOURCES

PRINTED BIBLIOGRAPHIES

The following listing is generally limited to works published 1990 or later, except for a few works of historical interest.

Aangeenbrug, Robert T. *A Bibliographic Analysis of Statewide Geographic Information Systems, 1970–1990.* Chicago: Council of Planning Librarians, 1992.

Bane, Adele F. *Technology and Adult Learning: a Selected Bibliography.* Englewood Cliffs, N.J.: Educational Technology Publications, c1994.

Buchenrieder, Klaus and Jerzy Rosenblit, eds. *Codesign: Computer Aided Software/Hardware Engineering.* Piscataway, N.J.: IEEE Press, 1995.

Buchenrieder, Klaus. *Hardware/Software Co-Design: An Annotated Bibliography: 350 Bibliographical References.* Chicago: IT Press, 1994.

Chalmers, Lex. *Expert Systems In Geography And Environmental Studies: An Annotated Review of Recent Work in the Field.* Waterloo, Ont.: Dept. of Geography, University of Waterloo, 1990.

Ciampi, Costantino and Roberta Nannucci, eds. *Information Technology and the Law: An International Bibliography.* Boston: Kluwer Academic, 1998.

Computer Abstracts. Bethesda, Md.: Jointly published by Cambridge Scientific Abstracts, and Engineering Information, Inc., 1993–.

Cortada, James W., ed. *Archives of Data-processing History: A Guide to Major U.S. Collections.* New York: Greenwood Press, 1990.

———. *A Bibliographic Guide to The History Of Computer Applications, 1950–1990.* Westport, Conn.: Greenwood Press, 1996.

———. *A Bibliographic Guide to The History Of Computing, Computers, and The Information Processing Industry.* New York: Greenwood Press, 1990.

Devonport, C. C. *A Selected GIS Bibliography.* Canberra: Australia Government Publishing Service, 1992.

Drew, Wilfred, ed. *Key Guide to Electronic Resources. Agriculture.* Medford, N.J.: Learned Information, 1995.

Gelfand, Julia and Locke Morrisey. *Selective Guide to Literature on Artificial Intelligence and Expert Systems.* Washington, D.C.: American Society for Engineering Education, Engineering Libraries Division, 1992.

Goodrich, Anna Rose. *Computer Science In Health Sciences: Index Of New Information With Authors, Subjects, And References.* Washington, D.C.: Abbe Publishers Association of Washington, D.C., 1994.

Gray, Elaine. *Automation in Local Government: a Partially Annotated Bibliography.* Chicago: Council of Planning Librarians, 1990.

Hancox, Peter J., William J. Mills, and Bruce J. Reid. *Keyguide to Information Sources in Artificial Intelligence/Expert Systems.* Lawrence, Kan.: Ergosyst Associates, 1990.

Hawkes, Lory, Christina Murphy, and Joe Law. *The Theory and Criticism of Virtual Texts: An Annotated Bibliography, 1988–1999.* Westport, Conn.: Greenwood Publishing, 2000.

Internet and Personal Computing Abstracts. Oxford, U.K.: Cambridge Scientific Abstracts, 1989–.

Johansson, Stig and Anna-Brita Stenström. *English Computer Corpora: Selected Papers and Research Guide.* Berlin; New York: Mouton de Gruyter, 1991.

Knee, Michael and Steven D. Atkinson. *Hypertext/Hypermedia: An Annotated Bibliography.* New York: Greenwood Press, 1990.

Linten, Allison J. *The Internet: A Bibliography with Indexes.* Huntington, N.Y.: Nova Science Publishers, 2000.

Mashack, Thea. *Bibliography Of Computer Science Reports, 1963–1992.* Stanford, Calif.: Dept. of Computer Science, Stanford University, 1993.

MSC/NASTRAN *bibliography.* 2nd ed. Los Angeles: MacNeal-Schwendler Corp., 1993.

Murphy, Donal P., ed. *Bibliography Of Civil Engineering Computer Applications 1984-1990.* New York: CITIS, 1991.

Palmer, Marlene A. *Expert Systems And Related Topics: Selected Bibliography and Guide to Information Sources.* Harrisburg, Pa.: Idea Group, 1990.

Rickman, Tamara J., and Allen H. Miller. *A Categorized Bibliography of Coastal Applications of Geographic Information Systems.* Madison: University of Wisconsin-Madison, Sea Grant Institute, c1995.

Romiszowski, A. J. *Computer Mediated Communication: A Selected Bibliography.* Englewood Cliffs, N.J.: Educational Technology Publications, 1992.

Sabourin, Conrad. *Computational Character Processing: Character Coding, Input, Output, Synthesis, Ordering, Conversion, Text Compression, Encryption, Display Hashing, Literate Programming: Bibliography.* Montréal: Infolingua, 1994.

———. *Computational Lexicology And Lexicography: Dictionaries, Thesauri, Term Banks, Analysis, Transfer And Generation Dictionaries, Machine Readable Dictionaries, Lexical Semantics, Lexicon Grammars: Bibliography.* Montréal: Infolingua, 1994.

———. *Computational Linguistics In Information Science: Information Retrieval (Full-Text Or Conceptual), Automatic Indexing, Text Abstraction, Content Analysis, Information Extraction, Query Languages: Bibliography.* Montréal: Infolingua, 1994.

———. *Computational Morphology: Morphological Analysis And Generation Lemmatization: Bibliography.* Montréal: Infolingua, c1994.

———. *Computational Parsing: Syntactic Analysis, Semantic Interpretation, Parsing Algorithms, Parsing Strategies: Bibliography.* Montréal: Infolingua, 1994.

———. *Computational Speech Processing: Speech Analysis, Recognition, Understanding, Compression, Transmission, Coding, Synthesis, Text to Speech Systems, Speech to Tactile Displays, Speaker Identification, Prosody Processing: Bibliography.* Montréal: Infolingua, 1994.

———. *Computational Text Generation: Generation from Data or Linguistic Structure, Text Planning, Sentence Generation, Explanation Generation: Bibliography.* Montréal: Infolingua, 1994.

———. *Computational Text Understanding: Natural Language Programming, Argument Analysis: Bibliography.* Montréal: Infolingua, 1994.

——— with Elca Tarrab. *Computer Assisted Language Teaching: Teaching Vocabulary, Grammar, Spelling, Writing, Composition, Listening, Speaking, Translation, Foreign Languages, Text Composition Aids, Error Detection And Correction, Readability Analysis: Bibliography.* Montréal: Infolingua, 1994.

——— with Rolande M. Lamarche. *Computer Mediated Communication: Computer Conferencing, Electronic Mail, Electronic Publishing, Computer Interviewing, Interactive Text Reading, Group Decision Support Systems, Idea Generation Support Systems, Human Machine Communication, Multi-Media Communication, Hypertext, Hypermedia, Linguistic Games: Bibliography.* Montréal: Infolingua, 1994.

——— with Rolande M. Lamarche. *Electronic Document Processing: Document Editing, Formatting, Typesetting, Mark-Up, Storing, Interchanging, Managing: Bibliography.* Montréal: Infolingua, 1994.

———. *Literary Computing: Style Analysis, Author Identification, Text Collation, Literary Criticism: Bibliography.* Montréal: Infolingua, 1994.

——— with Laurent R. Bourbeau. *Machine Translation: Aids to Translation, Speech Translation: Bibliography.* Montréal: Infolingua, 1994.

———. *Mathematical and Formal Linguistics: Grammar Formalisms, Grammar Testing, Logics, Quantifiers: Bibliography.* Montréal: Infolingua, 1994.

———. *Natural Language Interfaces: Interfaces To Databases, To Expert Systems, To Robots, To Operating Systems, And Question-Answering Systems: Bibliography.* Montréal: Infolingua, 1994.

———. *Optical Character Recognition And Document Segmentation: Character Preprocessing, Thinning, Isolation, Segmentation, Feature Extraction, Cursive and Multi-Font Recognition, Writer/Scriptor Identification: Bibliography.* Montréal: Infolingua, 1994.

———. *Quantitative And Statistical Linguistics: Frequencies Of Characters, Phonemes, Words, Grammatical Categories, Syntactic Structures, Lexical Richness, Word Collocations, Entropy, Word Length, Sentence Length: Bibliography.* Montréal: Infolingua, 1994.

Salisbury, Lutishoor. *Artificial Intelligence, 1986–1989: A Select Bibliography.* Monticello, Ill.: Vance Bibliographies, 1990.

Wasserman, Philip D. and Roberta M. Oetzel. *Neuralsource: The Bibliographic Guide To Artificial Neural Networks.* New York: Van Nostrand Reinhold, 1990.

White, Anthony G. *A List Of Personal Computer Programs for Basic Architectural Design: A Selected Bibliography.* Monticello, Ill.: Vance Bibliographies, 1990.

———. *A Selected, Annotated Source List of Personal Computer Programs for School District Public Administrators.* Monticello, Ill.: Vance Bibliographies, 1990.

Wick, Robert L. *Electronic and Computer Music.* Westport, Conn.: Greenwood Press, 1997.

Wilson, Michael A. *Geographic Information Systems: A Partially Annotated Bibliography.* Chicago: Council of Planning Librarians, 1990.

Zaffarano, Mark A. and Anna Gnadt. *Court Technology: A Selected Bibliography.* Monticello, Ill.: Vance Bibliographies, 1990.

Web Resources

About.com. "Computing and Technology." http://about.com/compute/

Academic Info: Computer Science & Computer Engineering. http://www.academicinfo.net/compsci.html

ACM Digital Library. http://portal.acm.org/dl.cfm?coll=portal&dl=ACM&CFID=2406658&CFTOKEN=91312318

Collection of Computer Science Bibliographies. ftp://ftp.cs.umanitoba.ca/pub/bibliographies/index.html

Computer Dictionaries, Acronyms, and Glossaries. http://www.compinfo-center.com/tpdict-t.htm

Computer Industry Almanac. http://www.c-i-a.com/

Computer Museum History Center. http://www.computer.org/annals/

Computer.Org Digital Library (IEEE Computer Society). http://portal.acm.org/dl.cfm?coll=portal&dl=ACM&CFID=2406 658&CFTOKEN=91312318

Computing Reviews. http://www.reviews.com/home.cfm

CyberAtlas. http://cyberatlas.internet.com/

How Stuff Works: Computers. http://www.howstuffworks.com/category.htm?cat=comp

IEEE Annals of the History of Computing. http://www.computer.org/annals/

Internet Electronic Library Project. http://elib.cs.sfu.ca/

PC Guide. http://www.pcguide.com/index.htm

Programming Language Critiques. http://www.people.virginia.edu/~sdm7g/LangCrit/index.html

Research Language Overviews. http://www-2.cs.cmu.edu/afs/cs.cmu.edu/user/mleone/web/language/overviews.html

Resources for Programming Language Research. http://www-2.cs.cmu.edu/afs/cs.cmu.edu/user/mleone/web/languageresearch.html

Science and Technology Sources on the Internet: Guide to Computer Science Internet Resources. [compiled by Michael Knee] http://www.library.ucsb.edu/istl/97-summer/internet2.html

Understanding the Computer Culture [reviews and guides], Danielle Bernstein. http://www.kean.edu/~dbernste/cult.html

Virtual Museum of Computing. http://vlmp.museophile.com/computing.html

Yahoo! Science: Computer Science. http://dir.yahoo.com/Science/Computer_Science/

ZD Webopedia. http://www.webopedia.com/

APPENDIX II
A CHRONOLOGY OF COMPUTING

The following chronology lists some significant events in the history of computing. Although the first CALCULATORS (i.e. the abacus) were known in ancient times, the chronology begins with the development of modern mathematics and the first calculators in the 17th century.

1617
John Napier published an explanation of "Napier's bones," a manual aid to calculation based on logarithms, and the ancestor to the slide rule.

1624
William Schickard invented a mechanical CALCULATOR that can perform automatic carrying during addition and subtraction. It can also multiply and divide by repeated additions or subtractions.

1642
Blaise Pascal invented a CALCULATOR that he calls the Pascaline. Its improved carry mechanism used a weight to allow it to carry several places. A small batch of the machines was made, but it did not see widespread use.

1673
Gottfried Wilhelm Leibniz (co-inventor with Isaac Newton of the calculus) invented a CALCULATOR called the Leibniz Wheel. He also wrote about the binary number system that eventually became the basis for modern computation.

1786
J. H. Muller invented a "difference engine," a machine that can solve polynomials by repeated addition or subtraction.

1822
Charles BABBAGE designed and partially built a much more elaborate difference engine.

1832
Babbage sketched out a detailed design for the Analytical Machine. This machine was to have been programmed by PUNCHED CARDS, storing data in a mechanical MEMORY, and even including a PRINTER. Although it was not built during his lifetime, Babbage's machine embodied most of the concepts used in modern computers.

1843
Ada Lovelace provided extensive commentary on a book by BABBAGE's Italian supporter Menabrea. Besides being the first technical writer, Lovelace also wrote what might be considered the world's first computer program.

1844
Samuel Morse demonstrated the electromagnetic telegraph by sending a message from Washington to Baltimore. The telegraph inaugurated both electric data transmission and the use of a binary character code (dots and dashes).

1850
Amedee Mannheim created the first modern slide rule. It will become an essential accessory for engineers and scientists until the inexpensive electronic CALCULATOR arrived in the 1970s.

1854
George Boole's book *The Laws of Thought* described what is now called Boolean algebra. BOOLEAN OPERATORS are essential for the BRANCHING STATEMENTS and LOOPS that control the operation of computer programs.

1884
W. S. Burroughs marketed his first adding machine, beginning what will become an important CALCULATOR (and later, computer) business.

1890
Herman HOLLERITH's PUNCHED CARD tabulator enabled the U.S. government to complete the 1890 census in record time.

1896
Hollerith founded the Tabulating Machine Company, which will become the Computing, Tabulating, and Recording company (CTR) in 1911. In 1924, it will become International Business Machines (IBM).

1904

J. A. Fleming invented the diode vacuum tube. Together with Lee de Forest's invention of the triode two years later, this development defined the beginnings of electronics, offering a switching mechanism much faster than mechanical relays.

1919

The "flip-flop" circuit was invented by two American physicists, W. H. Eccles and R. W. Jordan. The ability of the circuit to switch smoothly between two (binary) states would form the basis for computer arithmetic logic units.

1921

Karl Capek's play *R.U.R.* introduced the term *robot.* Robots will become a staple of science fiction "pulps" starting in the 1930s.

1930

Vannevar BUSH's elaborate ANALOG COMPUTER, the Differential Analyzer, went into service.

1936

Alonzo CHURCH developed the lambda calculus, which can be used to demonstrate the COMPUTABILITY of mathematical problems.

Konrad ZUSE built his first computer, a mechanical machine based on the binary system.

1937

Alan TURING provided an alternative (an equivalent) demonstration of COMPUTABILITY through his Turing Machine, an imaginary computer that can reduce any computable problem to a series of simple operations performed on an endless tape.

Bell Laboratories mathematician George Stibitz created the first circuit that could perform addition by combining BOOLEAN OPERATORS.

1938

In a key development in ROBOTICS, Doug T. Ross, an American engineer, created a robot that can store its experience in MEMORY and "learn" to navigate a maze.

G. A. Philbrick developed an electronic version of the ANALOG COMPUTER.

Working in a garage near Stanford University, William Hewlett and David Packard began to build audio oscillators. They called their business the Hewlett-Packard Company. Fifty years later, the garage would be preserved as a historical landmark.

1939

John ATANASOFF and Clifford Berry built a small electronic binary computer called the Atanasoff-Berry Computer (ABC). A 1973, court decision would give this machine precedence over ENIAC as the first electronic digital computer.

George Stibitz built the Complex Number Calculator, which is controlled by a keyboard and uses relays.

1940

Claude SHANNON introduced the fundamental concepts of DATA COMMUNICATIONS theory.

George Stibitz demonstrated remote computing by controlling his Complex Number Calculator in New York from a Teletype TERMINAL at Dartmouth College in New Hampshire.

1941

Working in isolation in wartime Germany, Konrad ZUSE completed the Z3. Although still mechanical rather than electronic, the machine used sophisticated floating-point NUMERIC DATA.

1943

The British built Colossus, an electronic (vacuum tube) special-purpose computer that can rapidly analyze permutations to crack the German Enigma cipher.

1944

Howard AIKEN completed the Harvard Mark I, a large programmable calculator (or computer) using electromechanical relays.

John VON NEUMANN and Stanislaw Ulam developed the Monte Carlo method of probabilistic SIMULATION, a tool that would find widespread use as computer power becomes available.

1945

ZUSE continued computer development and created a sophisticated matrix-based programming language called Plankalkül.

Vannevar BUSH envisioned HYPERTEXT and knowledge linking and retrieval in his article "As We May Think."

Alan TURING developed the concept of using PROCEDURES AND FUNCTIONS (subroutines) called with parameters. His team also developed the Pilot ACE (Automatic Computing Engine), which would help the development of a British computer industry.

1946

ENIAC went into service. Developed by J. Presper ECKERT and John MAUCHLY, the machine is widely considered to be the first large-scale electronic digital computer. It used 18,000 vacuum tubes.

In the "Princeton Reports" based upon the ENIAC work, John VON NEUMANN, together with Arthur W. Burks and Herman Goldstine described the fundamental operations of modern computers including the stored program concept—the holding of all program instructions in memory, where they can be referred to repeatedly and even manipulated like other data.

1947

The Association for Computing Machinery (ACM) was founded.

ECKERT and MAUCHLY formed the Eckert-Mauchly Corporation for commercial marketing of computers based on the ENIAC design.

John VON NEUMANN began development of the EDVAC (Electronic Discrete Variable Automatic Calculator) for the U.S. government's Ballistic Research Laboratory. This machine, completed in 1952, would be the first to use programs completely stored in memory and able to be changed without physically changing the hardware.

Richard Hamming developed ERROR CORRECTION algorithms.

Alan TURING'S paper on "Intelligent Machinery" began laying the groundwork for ARTIFICIAL INTELLIGENCE research.

In Britain, Manchester University built the first electronic computer that can store a full program in MEMORY. It was called "baby" because it was a small test version of a planned larger machine. For its main memory it used a CRT-like tube invented by F. C. Williams.

IBM under Thomas J. Watson, Sr. decided to enter the new computer field in a big way by beginning to develop the Selective Sequence Electronic Calculator (SSEC) as a competitor to ENIAC and the Harvard Mark I. The huge machine used thousands of both vacuum tubes and relays.

Tom Kilburn and M. H. A. Newman invented the index register, which would be used to keep track of the current location in memory of instructions or data.

The transistor was invented at Bell Labs by John Bardeen, Walter Brattain, and William Shockley. The solid-state device could potentially replicate all the functionality of the vacuum tube with much less size and power consumption. It would be some time before it was inexpensive enough to be used in computers, however.

Norbert WIENER coined the term *cybernetics* to refer to control and feedback systems.

Claude SHANNON formally introduced statistical information theory.

1949

The Cambridge EDSAC demonstrated versatile stored-program computing. Meanwhile, ECKERT and MAUCHLY work on BINAC, a successor/spinoff of ENIAC for Northrop Aircraft Corporation.

Frank Rosenblatt developed the perceptron, the first form of NEURAL NETWORK, for solving pattern-matching problems.

An Wang patented "core memory," using an array of magnetized rings and wires, which would become the main memory (RAM) for many MAINFRAMES in the 1950s.

1950

Alan TURING proposed the Turing Test as a way to demonstrate ARTIFICIAL INTELLIGENCE.

Development began of the high-speed computers Whirlwind and SAGE for the U.S. military. The military also began to use computers to run war games or simulations.

Claude SHANNON outlined the algorithms for a chess-playing program that could evaluate positions and perform heuristic calculations. He would build a chess-playing computer called Caissac.

Japan began development of electronic computers under the leadership of Hideo Yamashita, who would build the Tokyo Automatic Calculator.

Approximately 60 electronic or electromechanical computers were in operation worldwide. Each was built "by hand" as there were no production models yet.

1951

ECKERT and MAUCHLY marketed Univac I, generally considered the first commercial computer (although the Ferranti Mark I is sometimes given co-honors).

An Wang founded Wang Laboratories, which would become a major computer manufacturer through the 1970s.

Grace HOPPER at Remington Rand coined the word *COMPILER* and began developing automatic systems for creating machine codes from higher-level instructions.

1952

Alick Glennie developed autocode, generally considered to be the first true high-level PROGRAMMING LANGUAGE.

Magnetic core MEMORY began to come into use.

On election night a Univac I predicted that Dwight D. Eisenhower would win the 1952 U.S. presidential election. It made its prediction an hour after the polls closed, but its findings were not released at first because news analysts insisted the race was closer.

MANIAC was developed to do secret nuclear research in Los Alamos.

The IBM 701 went into production. It was one of the first computers to use magnetic TAPE DRIVES as primary means of data storage.

IBM was accused of violating the Sherman Antitrust Act in its computer business. Litigation in one form or another would drag on until 1982.

John VON NEUMANN described self-reproducing automata.

The symbolic ASSEMBLER was introduced by Nathaniel Rochester.

IBM and Remington Rand (Univac) dominated the young computer industry.

1954

The IBM 650 was marketed. It was the first truly mass-produced computer, and relatively affordable by businesses and industries. It used a magnetic drum memory.

In Britain, the Lyons Electronic Office (LEO) became the first integrated computer system for use for BUSINESS APPLICATIONS, primarily accounting and payroll.

1955

Grace HOPPER created Flow-matic, the first high-level language designed for BUSINESS APPLICATIONS OF COMPUTERS.

The Computer Usage Company (CUC) was founded by John W. Sheldon and Elmer C. Kubie. It is considered to be the first company devoted entirely to developing computer software rather than hardware.

Bendix marketed the G-15, its competitor to the IBM 650 in the "small" business computer market.

Users of the new IBM 704 MAINFRAME, frustrated at the lack of technical supported, formed the first computer USER GROUP, called SHARE.

The large IBM 705 MAINFRAME is marketed by IBM. It uses magnetic core MEMORY.

1956

The IBM 704 and Univac 1103 introduced a new generation of commercial MAINFRAMES with magnetic core storage.

John MCCARTHY coined the term ARTIFICIAL INTELLIGENCE, or AI.

The Dartmouth AI conference brought together leading researchers such as McCarthy, Marvin MINSKY, Herbert Simon, and Allen Newell. It would set the agenda for the field.

Newell, Shaw, and Simon developed Logic Theorist, the first program that can prove theorems.

A. I. Dumey described HASHING, a procedure for quickly sorting or retrieving data by assigning calculated values.

The infant transistor industry began to grow as companies such as IBM began to build transistorized calculators.

IBM signed a consent decree ending the 1952 antitrust complaint by restricting some of its business practices in selling mainframe computers.

1957

John Backus and his team released FORTRAN, which would become the most widely used language for SCIENTIFIC COMPUTING APPLICATIONS.

Digital Equipment Corporation (DEC) was founded by Ken Olsen and Harlan Anderson. The company's agenda involved the development of a new class of smaller computer, the MINICOMPUTER.

Minicomputer development would be inspired by the MIT TX-0 computer. While not yet a "mini," the machine was the first fully transistorized computer.

The hard drive came into service in IBM's 305 RAMAC.

IBM developed the first dot matrix PRINTER.

1958

The I/O interrupt used by devices to signal their needs to the CPU was developed by IBM. It would be used later in personal computers.

China began to build computers based on Soviet designs, which in turn had been based upon American and British machines.

Sperry Rand introduced the Univac II, a huge, powerful, and surprisingly reliable computer that used 5,200 vacuum tubes, 18,000 crystal diodes, and 184,000 magnetic cores.

Jack Kilby of Texas Instruments built the first integrated circuit, fitting five components onto a half-inch piece of germanium.

As the cold war continued, the U.S. Air Force brought SAGE on-line. This integrated air defense system featured REAL-TIME PROCESSING and graphics displays.

1959

John MCCARTHY developed LISP, a language based on Alonzo CHURCH's lambda calculus and including extensive facilities for LIST PROCESSING. It would become the favorite language for ARTIFICIAL INTELLIGENCE research.

COBOL was introduced, with much of the key work and inspiration coming from Grace HOPPER.

IBM marketed the 7090 MAINFRAME, a large transistorized machine that could perform 229,000 additions a second. The smaller IBM 1401 would prove to be even more popular. IBM also introduced a high-speed PRINTER using type chains.

Robert Noyce of Fairchild Semiconductor built a different type of integrated circuit, using aluminum traces and layers deposited on a silicon substrate.

1960

Digital Equipment Corporation (DEC) marketed the PDP-1, generally considered the first commercial MINICOMPUTER.

Control Data Corporation (CDC) impressed the industry with its CDC 1604, designed by Seymour Cray. It offered high speed at considerably lower prices than IBM and the other major companies.

The ALGOL language demonstrated block structure for better organization of programs. The report on the language introduced BNF (BACKUS-NAUR FORM) as a systematic description of computer language grammar.

Donald Blitzer introduced PLATO, the first large-scale interactive COMPUTER-AIDED INSTRUCTION system. It would later be marketed extensively by Control Data Corporation (CDC).

Paul Baran of RAND developed the idea of packet-switching to allow for decentralized information NETWORKS; the idea would soon attract the attention of the U.S. Defense Department.

In an advance in practical ROBOTICS, the remote-operated "Handyman" robot arm and hand was put to work in a nuclear power plant.

The U.S. Navy began to develop the Naval Tactical Data System (NTDS) to track targets and the status of ships in a combat zone.

1961

Time-sharing computer systems came into use at MIT and other facilities. Among other things, they encouraged the efforts of the first HACKERS to find clever things to do with the computers.

Leonard Kleinrock's paper "Information Flow in Large Communication Nets" was the first description of the packet-switching message transfer system that would underlie the INTERNET.

Arthur Samuel's ongoing research into COMPUTER GAMES design culminated in his checkers program reaching master level. The program includes learning algorithms that can improve its play.

The IBM STRETCH (IBM 7030) is installed at Los Alamos National Laboratory. Its advanced "pipeline" architecture allowed new instructions to begin to be processed while preceding ones were being finished. It and Univac's LARC are sometimes considered to be the first SUPERCOMPUTERS.

IBM made a major move into scientific computing with its modular 7040 and 7044 computers, which can be used together with the 1401 to build a "scalable" installation for tackling complex problems.

Unimation introduced the industrial robot (the Unimate).

Fairchild Semiconductor marketed the first commercial integrated circuit.

1962

The discipline of COMPUTER SCIENCE began to emerge with the first departments established at Purdue and Stanford.

MIT students created *Spacewar,* the first video COMPUTER GAME, on the PDP-1.

On a more practical level, MIT programmers Richard Greenblatt and D. Murphy develop TECO, one of the first TEXT EDITORS.

J. C. R. Licklider described the "Intergalactic Network," a universal information exchange system that would help inspire the development of the INTERNET.

Douglas Engelbart invented the computer MOUSE at SRI.

IBM developed the SABRE on-line ticket reservation system for American Airlines. The system will soon be adopted by other carriers and demonstrate the use of networked computer systems to facilitate commerce. Meanwhile, IBM earned $1 billion from its computer business, which by then had overtaken its traditional office machines as the company's leading source of revenue.

1963

Joseph Weizenbaum's *Eliza* program carried on natural-sounding conversations in the manner of a psychotherapist.

Ivan Sutherland developed *Sketchpad,* the first computer drawing system.

Reliable Metal Oxide Semiconductor (MOS) integrated circuits were perfected, and would become the basis for many electronic devices in years to come, including computers for space exploration.

1964

The IBM System/360 was announced. It would become the most successful MAINFRAME in history, with its successors dominating business computing for the next two decades.

IBM introduced the MT/ST (Magnetic Tape/Selectric Typewriter), considered to be the first dedicated WORD PROCESSING system. While rudimentary, it allowed text to be corrected before printing.

Seymour Cray's Control Data CDC 6600 is announced. When completed, it ran about three times faster than IBM's STRETCH, irritating Thomas Watson, head of the far larger IBM.

J. Kemeny and T. Kurtz developed BASIC to allow students to program on the Dartmouth time-sharing system.

At the other end of the scale, IBM introduced the complex, feature-filled PL/1 (Programming Language 1) for use with its System/360.

The American National Standards Institute (ANSI) officially adopted the ASCII (American Standard Code for Information Interchange) character code.

Paul Baran of SRI wrote a paper, "On Distributed Communication Networks," further describing the implementation of packet-switched network that could route around disruptions. The work began to attract the attention of military planners concerned with air defense and missile control systems surviving nuclear attack.

Jean Sammet and her colleagues developed the first computer program that can do algebra.

Gordon Moore (a founder of Fairchild Semiconductor and later, of Intel Corporation) stated that the power of CPUs would continue to double every 18 to 24 months. "Moore's Law" proved to be remarkably accurate.

1965

IBM introduced the FLOPPY DISK (or diskette) for use with its mainframes.

Edsgar Dijkstra devised the semaphore, a variable that two processes can use to synchronize their operations and aiding the development of CONCURRENT PROGRAMMING.

The APL language developed by Kenneth Iverson provided a powerful, compact, but perhaps cryptic way to formulate calculations.

The SIMULA language introduced what will become known as OBJECT-ORIENTED PROGRAMMING.

The DEC PDP-8 became the first mass-produced minicomputer, with over 50,000 systems being sold. The machine brings computing power to thousands of universities, research labs, and businesses that could not afford mainframes. Designed by Edson deCastro and engineered by Gordon Bell, the PDP-8 design marked an important milestone on the road to the desktop PC.

NASA uses an IBM onboard computer to guide Gemini astronauts in their first rendezvous in space.

The potential of the EXPERT SYSTEM was demonstrated by *Dendral,* a specialized medical diagnostic program that began development by Edward Feigenbaum, Joshua Lederberg, and Bruce Buchanan.

The U.S. Defense Department's ARPA (Advanced Research Projects Agency) sponsored a study of a "co-operative network of time-sharing computers." A testbed network was begun by connecting a TX-2 minicomputer at MIT via phone line to a computer at System Development Corporation in Santa Monica, Calif.

Ted Nelson's influential vision of universal knowledge sharing through computers introduced the term HYPERTEXT.

1966

In the first federal case involving COMPUTER CRIME (*U.S. v. Bennett*), a bank programmer is convicted of altering a bank program to allow him to overdraw his account.

The first ACM Turing Award is given to Alan Perlis.

The New York Stock Exchange automated much of its trading operations.

1967

The memory CACHE (a small amount of fast memory used for instructions or data that are likely to be needed) was introduced in the IBM 360/85 series.

IBM developed the first FLOPPY DISK drive.

Seymour Papert introduced LOGO, a Lisp-like language that would be used to teach children programming concepts intuitively.

A chess program written by Richard Greenblatt of MIT, Mac Hack IV, achieved the playing skill of a strong amateur human player.

Fred Brooks did early experiments in computer-mediated sense perception, laying groundwork for VIRTUAL REALITY.

1968

Edsger DIJKSTRA'S little letter entitled "GO TO Considered Harmful" argued that the GOTO or "jump" statement made programs hard to read and more prone to error. The resulting discussion gave impetus to the STRUCTURED PROGRAMMING movement. Another aspect of this movement was the introduction of the term *SOFTWARE ENGINEERING.*

Robert Noyce, Andrew Grove, and Gordon Moore founded Intel, the company that would come to dominate the microprocessor industry by the early 1980s.

IBM introduced the System/3, a lower-cost computer system designed for small businesses.

Bolt, Beranek and Newman (BBN) was awarded a government contract to build "interface message processors" or IMPs to translate data between computers linked over packet-switched networks.

Alan Kay prototyped the Dynabook, a concept that led toward both the PORTABLE COMPUTER and the graphical USER INTERFACE.

Stanley Kubrick's movie *2001* introduced Hal 9000, the self-aware (but paranoid) computer that kills members of a deep-space exploration crew.

1969

Ken Thompson and Dennis RITCHIE began work on the UNIX OPERATING SYSTEM. It will feature a small KERNEL that can be used with many different command SHELLS, and will eventually incorporate hundreds of utility programs that can be linked to perform tasks.

Edgar F. Codd introduced the concept of the relational system that would form the foundation for most modern DATABASE MANAGEMENT SYSTEM.

IBM was sued by the U.S. Department of Justice for antitrust violations. The voluminous case would finally be dropped in 1982. However, government pressure may have led the computer giant to finally allow its users to buy software from third parties, giving a major boost to the software industry.

ARPANET is officially launched. The first four nodes of the ARPANET came on-line, prototyping what would eventually become the INTERNET.

SRI researchers developed Shakey, the first mobile robot that could "see" and respond to its environment. The actual control computer was separate, however, and controlled the robot through a radio link.

Neil Armstrong and Edwin Aldrin successfully made the first human landing on the Moon, despite problems with the onboard Apollo Guidance Computer.

The first automatic teller machine (ATM) was put in service.

1970

Gene Amdahl left IBM to found Amdahl Corporation, which would compete with IBM in the mainframe "clone" market.

An Intel Corporation team led by Marcian E. Hoff began to develop the Intel 4004 MICROPROCESSOR.

Digital Equipment Corporation announced the PDP-11, the beginning of a series of 16-bit minicomputers that will support time-sharing computing in many universities.

John Conway's "Game of Life" popularized CELLULAR AUTOMATA.

The ACM held its first all-computer chess tournament in New York. Northeastern University's Chess 3.0 topped the field of six programs competing.

Charles Moore began writing programs to demonstrate the versatility of his programming language FORTH.

Xerox Corporation established the Palo Alto Research Center (PARC). This laboratory will create many innovations in interactive computing and the graphical USER INTERFACE.

1971

Niklaus WIRTH formally announced PASCAL, a small, well-structured language that will become the most popular language for teaching COMPUTER SCIENCE for the next two decades.

The IEEE Computer Society was founded.

The IBM System/370 series ushered in a new generation of mainframes using densely packed integrated circuits for both CPU and memory.

1972

Dennis RITCHIE and Brian Kernighan developed C, a compact language that would become a favorite for SYSTEMS PROGRAMMING, particularly in UNIX.

The creation of an e-mail program for the ARPANET included the decision to use the at (@) key as part of email addresses.

Alan KAY developed SMALLTALK, building upon SIMULA to create a powerful, seamless OBJECT ORIENTED PROGRAMMING language and operating system. The language would eventually be influential although not widely used. Kay also prototyped the Dynabook, a notebook computer, but Xerox officials showed little interest.

Seymour CRAY left CDC and founded Cray Research to develop new SUPERCOMPUTER.

Intel introduced the 8008, the first commercially available 8-bit MICROPROCESSOR.

The 5.25-inch diskette first appeared. It would become a mainstay of personal computing until it was replaced by the more compact 3.5-inch diskette in the 1990s.

Nolan Bushnell's Atari Corp. had the first commercial COMPUTER GAME hit, *Pong*. It and its beeping cousins would soon become an inescapable part of every parent's experience.

1973

Alain Colmerauer and Philippe Roussel at the University of Marseilles developed PROLOG (Programming in Logic), a language that could be used to reason based upon a stored base of knowledge. The language would become popular for EXPERT SYSTEMS development.

Bell Laboratories established a group to support and promulgate the UNIX OPERATING SYSTEM.

The Ethernet protocol for LANs (LOCAL AREA NETWORKS) was developed by Robert Metcalfe.

In a San Francisco hotel lobby Vinton CERF sketched the architecture for an Internet gateway on a napkin.

Don Lancaster published his "TV Typewriter" design in *Radio Electronics*. It would enable hobbyists to build displays for the soon-to-be available microcomputer.

The Boston Computer Society (BCS) was founded. It became one of the premier computer user groups.

Gary Kildall founded Digital Research, whose CP/M OPERATING SYSTEM would be an early leader in the microcomputer field.

A federal court declared that the Eckert-Mauchly ENIAC patents were invalid because John ATANASOFF had the same ideas earlier in his ABC computer.

1974

The Alto graphical workstation was developed by Alan KAY and others at Xerox PARC. It did not achieve commercial success, but a decade later something very much like it would appear in the form of the Apple MACINTOSH.

An international computer chess tournament is won by the Russian KAISSA program, which crushed the American favorite Chess 4.0.

Computerized product scanners were introduced in an Ohio supermarket.

Intel released the 8080, a MICROPROCESSOR that had 6,000 transistors, could execute 640,000 instructions per second, was able to access 64K of memory, and ran at a clock rate of 2 MHz.

David Ahl's *Creative Computing* magazine began to offer an emphasis on using small computers for education and other human-centered tasks.

Vinton CERF and Robert Kahn began to publicize their TCP/IP INTERNET protocol.

A group at the University of California, Berkeley, began to develop their own version of the UNIX OPERATING SYSTEM.

The 1974 Privacy Act began the process of trying to protect individual PRIVACY IN THE DIGITAL AGE.

1975

Fred Brooks published the influential book *The Mythical Man-Month*. It explained the factors that bog down software development and focused more attention on SOFTWARE ENGINEERING and its management.

Electronics hobbyists were intrigued by the announcement of the MITS Altair, the first complete microcomputer system available in the form of a kit. While the basic kit cost only $395, the keyboard, display, and other peripherals were extra.

MITS founder Ed Roberts also coined the term PERSONAL COMPUTER. Hundreds of hobbyists built the kits and yearned for more capable machines. Many hobbyists flocked to meetings of the Homebrew Computer Club in Menlo Park, California.

IBM introduced the first commercially available laser PRINTER. The very fast, heavy-duty machine was suitable only for very large businesses.

The first ARPANET discussion mail list was created. The most popular topic for early mail lists was science fiction.

In Los Angeles, Dick Heiser opened what is believed to be the first retail store to sell computers to "ordinary people."

1976

Seymour CRAY'S sleek, monolithlike Cray 1 set a new standard for SUPERCOMPUTERS.

Whitfield Diffie and Martin Hellman announced a public key ENCRYPTION system that allowed users to securely send information without previously exchanging keys.

IBM developed the first (relatively crude) inkjet PRINTER for printing address labels.

Shugart Associates offered a FLOPPY DISK drive to microcomputer builders. It cost $390.

Steve Wozniak proposed that Hewlett-Packard fund the creation of a PERSONAL COMPUTER, while his friend Steve Jobs made a similar proposal to Atari Corp. Both proposals were rejected, so the two friends started Apple Computer Company.

Chuck Peddle of MOS Technology developed the 6502 MICROPROCESSOR, which would be used in the Apple, Atari, and some other early personal computers.

Bill GATES complained about software piracy in his "Open Letter to Hobbyists." People were illicitly copying his BASIC language tapes. COPY PROTECTION would soon be used in an attempt to prevent copying of commercial programs for personal computers.

Computer enthusiasts found an erudite forum in the magazine *Dr. Dobb's Journal of Computer Calisthenics and Orthodontia: Running Light Without Overbyte*. The more mainstream *Byte* magazine also became a widely known forum for describing new projects and selling components.

William Crowther and Don Woods at Stanford University developed the first interactive COMPUTER GAME involving an adventure with monsters and other obstacles. University administrators would soon complain that the game was wasting too much computer time.

1977

Benoit Mandelbrot's book on FRACTALS IN COMPUTING popularized a mathematical phenomenon that would find uses in computer graphics, data compression, and other areas.

The Data ENCRYPTION Standard (DES) was announced. Critics charged that it was too weak and probably already compromised by spy agencies.

Vinton CERF demonstrated the versatility and extent of the Internet Protocol (IP) by sending a message around the world via radio, land line, and satellite links.

The Charles Babbage Institute was founded. It would become an important resource for the study of computing history.

Bill GATES and Paul Allen found a tiny company called Microsoft. Its first product was a BASIC INTERPRETER for the newly emerging PERSONAL COMPUTER systems.

Radio Shack began selling its TRS-80 Model 1 personal computer.

The Apple II was released. It will become the most successful of the early (pre-IBM) personal computers.

1978

Diablo Systems marketed the first daisy wheel PRINTER.

Atari announced its Atari 400 and Atari 800 personal computers. They offered superior graphics (for the time).

Daniel Bricklin's VisiCalc SPREADSHEET is announced. It will become the first software "hit" for the Apple II, leading businesses to consider using PERSONAL COMPUTERS.

Ward Christiansen and Randy Suess developed the first software for BULLETIN BOARD SYSTEMS (BBS).

The first West Coast Computer Faire was organized in San Francisco. The annual event became a showcase for innovation and a meeting forum for the first decade of personal computing.

The BSD (Berkeley Software Distribution) version of UNIX was released by the group at the University of California, Berkeley, under the leadership of Bill Joy.

The AWK (named for Aho, Weinberger, and Kernighan) SCRIPTING LANGUAGE appeared.

1979

MEDICAL APPLICATIONS OF COMPUTING were highlighted when Allan M. Cormack and Godfrey N. Hounsfield received the Nobel Prize in medicine for the development of computerized tomography (CAT), creating a revolutionary way to examine the structure of the human body.

The Ashton-Tate company began to market dBase II, a DATABASE MANAGEMENT SYSTEM that becomes the leader in personal computer databases during the coming decade.

Intel's new 16-bit processors, the 8086 and 8088, began to dominate the market.

Hayes marketed the first MODEM, and the CompuServe ONLINE SERVICE and early bulletin boards gave a growing number of users something to connect to.

UNIX users Tom Truscott, Jim Ellis, and Steve Bellovin developed a program to exchange news in the form of files copied between the Duke University and University of North Carolina computer systems. This gradually grew into USENET (or NETNEWS), providing thousands of topical newsgroups.

The first networked computer fantasy game, MUD (Multi-User Dungeon), was developed.

The first COMDEX was held in Las Vegas. It would become the PC industry's premier trade show.

Boston's Computer Museum was founded. This perhaps signaled the computing field's consciousness of coming of age.

1980

ADA, a modular descendent of PASCAL, was announced. The language was part of efforts by the U.S. Defense Department to modernize its software development process.

RISC (reduced instruction set computer) microprocessor architecture was introduced.

Apple's initial public offering of 4.6 million shares at $22 per share sold out immediately. It was the largest IPO since that of Ford Motor Company in 1956. Apple founders, Steve Jobs and Steve Wozniak, became the first multimillionaires of the microcomputer generation.

XENIX, a version of UNIX for PERSONAL COMPUTERS, was offered. It met with limited success.

Shugart Associates announced a HARD DISK drive for personal computers. The disk stored a whopping 5 megabytes.

1981

The IBM PC was announced. Apple "welcomed" its competitor in ads, but the IBM machine would soon surpass its competitors as the personal computer of choice for business. Its success is aided by a version of the VisiCalc SPREADSHEET that sells more than 200,000 copies.

Osborne introduced the PORTABLE (sort of) COMPUTER, a machine with the size and weight of a heavy suitcase.

Apple tried to market the Apple III as a more powerful desktop computer for business, but the machine was plagued with technical problems and did not sell well.

Digital Equipment Corporation introduced its DECmate dedicated word-processing system.

Xerox PARC displayed the Star, a successor to the Alto with 512K of RAM. It was intended for use in an Ethernet NETWORK.

A network called BITNET ("Because it's Time Network") began to link academic institutions worldwide.

Tracy Kidder's best-selling *The Soul of a New Machine* recounted the intense Silicon Valley working culture as seen in the development of Data General's latest workstation, the Eclipse.

Japan announced a 10-year effort to create "Fifth Generation" computing based on application of ARTIFICIAL INTELLIGENCE.

1982

Sun Microsystems was founded. It would specialize in high-performance WORKSTATIONS.

AT&T began marketing UNIX (System III) as a commercial product.

Compaq became one of the most successful makers of "clones" or IBM PC-compatible computers, introducing a portable (luggable) machine.

The AutoCad program brought CAD (COMPUTER-AIDED DESIGN AND MANUFACTURING) to the desktop.

The *Time* magazine "man of the year" wasn't a person at all—it was the personal computer!

1983

Business use of personal computers continued to grow. WORD PROCESSING leaders WordStar and WordPerfect were joined by the first version of Microsoft Word. Lotus 1-2-3 became the new SPREADSHEET leader.

Borland International introduced Turbo Pascal, a speedy, easy to use programming environment for personal computers.

An industry pundit introduced the term *vaporware* to refer to much-hyped but never-released software, such as a product called Ovation for IBM PCs.

IBM tried to market the PC Jr., a less-expensive PC for home and school users. It failed to gain a foothold in the market.

More successfully, IBM offered the PC XT, the first PERSONAL COMPUTER that had a built-in hard drive.

Radio Shack introduced the Model 100, the first practical notebook computer.

Apple introduced the Lisa, a $10,000 computer with a graphical USER INTERFACE. Its high price and slow performance made it a flop, but its ideas would be more successfully implemented the following year in the MACINTOSH.

John Sculley became president of Apple Computer, beginning a bitter struggle with Apple co-founder Steve Jobs.

Richard Stallman began the GNU (GNU's not UNIX) project to create a version of UNIX that would not be subject to AT&T licensing.

The movie *War Games* portrayed teenage HACKERS taking control of nuclear missile facilities.

1984

A classic Super Bowl commercial introduced the Apple MACINTOSH, the computer "for the rest of us." Based largely on Alan KAY's earlier work at Xerox PARC, the "Mac" used menus, icons, and a mouse instead of the cryptic text commands required by MS-DOS.

Meanwhile, IBM introduced a more powerful personal computer, the PC/AT with the Intel 80286 chip.

Steve Jobs leaves Apple Computer to found a company called NeXT.

Microsoft CEO Bill GATES was featured on a *Time* magazine cover.

The DOMAIN NAME SYSTEM began. It allows Internet users to connect to remote machines by name without having to specify an exact network path.

British institutions develop JANET, the Joint Academic Network.

Science fiction writer William Gibson coined the word CYBERSPACE in his novel *Neuromancer*. It began a new SF genre called *cyberpunk*, featuring a harsh, violent, immersive high-tech world.

1985

Desktop publishing was fueled by several developments including John Warnock's PostScript page description language and the Aldus PageMaker page layout program. The MACINTOSH'S graphical interface gave it the early lead in this application.

MICROSOFT WINDOWS 1.0 was released, using many of the same features as the Macintosh, although not nearly as well.

There was increasing effort to unify the two versions of UNIX (AT&T and BSD), with guidelines including the System V Interface Standard and POSIX.

Commodore introduces the Amiga, a machine with a sophisticated OPERATING SYSTEM and powerful color graphics. The machine had many die-hard fans but ultimately could not survive in the marketplace.

IBM marketed the IBM 3090, a large, powerful MAINFRAME that cost $9.3 million.

The Cray 2 SUPERCOMPUTER broke the 1-billion-instructions-a-second barrier.

A CONFERENCING SYSTEM called the Whole Earth 'Lectronic Link (Well) was founded. Its earliest users are largely drawn from Grateful Dead fans and assorted techies.

1986

The National Science Foundation funded NSFNET, which provides high-speed Internet connections to link universities and research institutions.

Borland released a PROLOG compiler, making the ARTIFICIAL INTELLIGENCE language accessible to PC users. A PC version of SMALLTALK also appeared from another company.

Apple beefed up the relatively anemic Macintosh with the Macintosh Plus, which has more MEMORY.

1987

Bjarne STROUSTRUP'S C++ language offered OBJECT-ORIENTED PROGRAMMING in a form that was palatable to the legions of C programmers. The language would surpass its predecessor in the coming decade.

Sun marked its first WORKSTATION based on RISC (REDUCED INSTRUCTION SET COMPUTING) technology.

Apple sold its one millionth MACINTOSH. Apple also brought out a new line of Macs (the Macintosh SE and Macintosh II) that, unlike the original Macs, were expandable by plugging in cards.

Apple also introduced Hypercard, a simple HYPERTEXT AUTHORING SYSTEM that became popular with educators.

IBM introduced a new line of personal computers called the PS/2. It featured a more efficient BUS called the Microchannel and some other innovations, but it sold only modestly. Most of the industry continued to further develop standards based upon the IBM PC AT.

The Thinking Machines Corporation's Connection Machine introduced massive parallel processing. It contained 64,000 MICROPROCESSORS that could collectively perform 2 billion instructions per second.

1988

Robert Morris Jr.'s "worm" accidentally ran out of control on the Internet, bringing concerns about COMPUTER CRIME AND SECURITY to public attention. The Computer Emergency Response Team (CERT) was formed in response.

Wolfram's Mathematica program was a milestone in mathematical computing, allowing users to not merely calculate but also to solve symbolic equations automatically.

Cray introduced the Cray Y-MP SUPERCOMPUTER. It could process 2 billion operations per second.

IBM announced a new midrange MAINFRAME, the AS/400.

Sandia National Laboratory began to build a massively parallel "hypercomputer" that would have 1,024 processors working in tandem.

A consortium called the Open Software Foundation was established to promote OPEN SOURCE shared software development.

1989

The Internet now had more than 100,000 host computers.

Deep Thought defeated Danish chess grandmaster Bent Larsen, marking the first time a grandmaster had been defeated by a computer.

Intel announced the 80486 CPU, a chip with over a million transistors.

Astronomer Clifford Stoll's book *The Cuckoo's Egg* recounted his pursuit of German hackers who were seeking military secrets. Stoll soon became a well-known critic of computer technology and the Internet.

The ARPANET officially ends, having been succeeded by the NSFNET.

1990

MICROSOFT WINDOWS became truly successful with version 3.0, diminishing the user interface advantages of the MACINTOSH.

At Sun Microsystems, James Gosling developed the Oak language to control EMBEDDED SYSTEMS. After the original project was canceled, Gosling redesigned the language as JAVA.

IBM announced the System/390 MAINFRAME.

IBM and Microsoft developed OS/2, an operating system intended to replace MS-DOS. Microsoft withdrew in favor of Windows, and despite considerable technical merits, OS/2 never really takes hold.

Secret Service agents raided computer systems and bulletin boards, seeking evidence of illegal copying of a BellSouth manual, disrupting an innocent game company. In response, Mitch Kapor founded the Electronic Frontier Foundation to advocate for civil liberties of computer users. Another group, the Computer Professionals for Social Responsibility, filed a Freedom of Information Act (FOIA) request for FBI records involving alleged government surveillance of BULLETIN BOARD SYSTEMS.

1991

The Science Museum in London exhibited a reconstruction of Charles BABBAGE'S never-built difference engine.

A Finnish student named Linus TORVALDS found that he couldn't afford a UNIX license, so he wrote his own UNIX KERNEL and combined it with GNU utilities. The result would eventually become the popular Linux operating system.

Developers at the University of Minnesota created Gopher, a system for providing documents over the Internet using linked menus. However, it was soon to be surpassed by the WORLD WIDE WEB, created by Tim BERNERS-LEE at the CERN physics laboratory in Geneva, Switzerland.

Advanced Micro Devices began to compete with Intel by making IBM PC-compatible CPU chips.

Apple and IBM signed a joint agreement to develop technology in areas that include object-oriented operating systems, multimedia, and interoperability between Macintosh and IBM networks.

1992

Reports of the Michelangelo COMPUTER VIRUS frightened computer users. Although the virus did little damage, it spurred more users to practice "safe computing" and install antivirus software.

Motorola announced the Power PC, a 32-bit RISC MICROPROCESSOR that contains 28 million transistors.

An estimated 1 million host computers were on the Internet. The Internet Society is founded to serve as a coordinator of future development of the network.

1993

Apple's Newton handheld computer created a new category of machine called the PDA, or Personal Digital Assistant.

MICROSOFT WINDOWS NT was announced. It is a version of the operating system designed especially for network servers.

Steve JOBS announced that his NeXT company would abandon its hardware efforts and concentrate on marketing its innovative operating system and development software.

Leonard Adleman demonstrated MOLECULAR COMPUTING by using DNA molecules to solve the Traveling Salesman problem.

The Cray 3 SUPERCOMPUTER continued the evolution of that line. It could be scaled up to a 16-processor system.

The Mosaic graphical WEB BROWSER popularized the WORLD WIDE WEB.

The Clinton administration announced plans to develop a national "Information Superhighway" based on the Internet. Volunteer "Net Day" programs would begin to connect schools to the network.

The White House established its website, www.whitehouse.gov.

1994

Mosaic's developer, Marc ANDRESSEN, left NCSA and joined Jim Clark to found Netscape. Netscape soon released an improved browser called Netscape Navigator.

Apple announced that it would license the Mac operating system to other companies to make Macintosh "clones." Few companies would take them up on it, and Apple would soon withdraw the licensing offer.

Intel Corporation was forced to recall millions of dollars worth of its new Pentium CHIPS when a mathematical flaw was discovered in the floating-point routines.

Marc Andreesen and Jim Clark founded Netscape and developed a new WEB BROWSER, Netscape Navigator. It would become the leading Web browser for several years.

Red Hat released a commercial distribution of Linux 1.0.

Search engines such as Lycos and Alta Vista started helping users find webpages. Meanwhile, a graduate student named Jerry Yang started compiling an on-line list of his favorite websites. That list would eventually become Yahoo!

Advertising in the form of banner ads began to appear on websites.

1995

MICROSOFT WINDOWS 95 gave a new look to the operating system and provided better support for devices, including PLUG AND PLAY device configuration.

Microsoft began its own on-line service, the Microsoft Network (MSN). Despite its startup icon being placed on the Windows 95 desktop, the network would trail industry leader America On-line, which had overtaken CompuServe and Prodigy.

Jeff Bezos's on-line bookstore, Amazon.com, opened for business. It would become the largest e-commerce retailer.

The major on-line services began major promotion of access to the WORLD WIDE WEB.

NSFNET retired from direct operation of the Internet, which had now been fully privatized. The agency then focused on providing new BROADBAND connections between SUPERCOMPUTER sites.

Sun announced the JAVA language. It would become one of the most popular languages for developing applications for the World Wide Web.

Motorola announced the Power PC-602, a 64-bit CPU chip.

Compaq ranked first in personal computer sales in the United States, followed by Apple.

Physicists Peter Fromherz and Alfred Stett of the Max Planck Institute of Biochemistry in Munich, Germany, demonstrated the direct stimulation of a specific nerve cell in a leech by a computer probe. This conjured visions of the "jacked-in" neural implants foreseen by science fiction writers such as William Gibson.

The next generation of Cray supercomputers, the T90 series, could be scaled up to a rate of 60 billion instructions per second.

STREAMING (real-time video and audio) began to become popular on the Web.

Computer generated imagery (CGI) was featured by Hollywood in the movie *Toy Story*.

1996

A product called Web TV attempted to bring the WORLD WIDE WEB to home consumers without the complexity of full-fledged computers. The product achieved only modest success as the price of personal computers continued to decline.

The U.S. postal service issued a stamp honoring the 50th anniversary of ENIAC.

The Boston Computer Society, one of the oldest computer USER GROUPS, disbanded.

World chess champion Garry Kasparov won his first match against IBM's Deep Blue chess computer, but said the match had been unexpectedly tough.

Yahoo! offered its stock to the public, running up the second-highest first-day gain in NASDAQ history.

Seymour CRAY's Cray Research (a developer of SUPERCOMPUTERS) was acquired by Silicon Graphics.

Pierre Omidyar turned a small hobby auction site into eBay and was soon attracting thousands of eager sellers and buyers to the site.

In one of its infrequent ventures into hardware, Microsoft announced the NetPC, a stripped-down diskless PC that would run software from a network. Such "network computers" never really caught on, being overtaken by the ever-declining price for complete PCs.

1997

The chess world was shocked when world champion Garry Kasparov was defeated in a rematch with Deep Blue.

A single INTERNET domain name, business.com, was sold for $150,000.

Amazon.com had a successful Initial Public Offering (IPO).

A technology called "push" began to be hyped. It involved websites continually feeding "channels" of news or entertainment to user's desktops. However, the idea would fail to make much headway.

Internet users banded together to demonstrate DISTRIBUTED COMPUTING by cracking a 56-bit DES cipher in 140 days.

The Association for Computing Machinery (ACM) celebrated its 50th anniversary.

1998

MICROSOFT WINDOWS 98 provided an incremental improvement in the operating system.

Apple announced the iMac, a stylish machine that rejuvenated the MACINTOSH line.

eBay's IPO was wildly successful, making Pierre Omidyar, Meg Whitman, and other eBay executives instant millionaires.

Merger-mania hit the on-line service industry, with America On-line buying CompuServe's on-line service (spinning off the network facilities to WorldCom). AOL then acquired Netscape and its Web hosting technology.

In another significant merger, Compaq acquired Digital Equipment Corporation (DEC).

1999

Federal Judge Thomas Penfield Jackson found that Microsoft violated antitrust laws. The case dragged on with appeals, with the process of crafting a remedy (such as possibly the split-up of the company) still unresolved in 2002.

Another virus, Melissa, panicked computer users.

Some companies began to offer "free" computers to people who agreed to sign up for long-term, relatively expensive INTERNET service.

Computer scientists and industry pundits debated the possibility of widespread computer disasters due to the Y2K PROBLEM. Companies spent millions of dollars trying to find and fix old computer code that used only two digits to store year dates.

Apple released OS X, a new UNIX-based operating system for the MACINTOSH.

2000

New Year's Day found the world to be continuing much as before, with only a few scattered Y2K PROBLEMS.

Unknown hackers, however, brought down some commercial websites with denial-of-service (DOS) attacks.

AOL merged with Time-Warner, creating the world's largest media company. Critics worried about the affects of growing corporate concentration on the diversity of the INTERNET.

MICROSOFT WINDOWS 2000 began the process of merging the consumer Windows and Windows NT lines into a single family of operating systems that would no longer use any of the underlying MS-DOS code.

The WORLD WIDE WEB was estimated to have about 1 billion pages on-line.

Tech stocks (and particularly E-COMMERCE companies) began to sharply decline as investors became increasingly skeptical about profitability.

A growing number of Web users were beginning to switch to much faster BROADBAND connections using DSL or cable lines.

2001

The decline in E-COMMERCE stocks continued, with tens of thousands of jobs lost. One of the many failures was Webvan, the Internet grocery service. Amazon.com suffered losses but continued trying to expand into profitable niches. Only eBay among the major e-commerce companies continued to be profitable.

MICROSOFT WINDOWS XP offered consumer and "professional" versions of Windows on the same code base.

IBM researchers created a seven "qubit" quantum computer to execute Shor's algorithm, a radical approach to factoring that could potentially revolutionize cryptography.

Among the specters raised in the wake of the September 11 terrorist attacks was "cyberterrorism" having the potential to disrupt vital infrastructure, services, and the economy. BIOMETRICS and more sophisticated database techniques were enlisted in the war on terrorism while civil liberties groups voiced concerns.

2002

Wireless networking using the faster 802.11 standard became increasingly popular as an alternative to cabled or phone line networks for homes and small offices.

Consumer digital cameras began to approach "professional" quality.

The U.S. Supreme Court ruled that "virtual" child pornography (in which no actual children were used) was protected by the First Amendment.

Continuing stock market declines threaten growth in the computer and Internet sectors.

The music-sharing service Napster goes out of business, when it is forced to stop distributing copyrighted music.

APPENDIX III
SOME SIGNIFICANT AWARDS

This appendix describes some of the major awards in computer science and technology and lists recipients as of 2001. The last names of persons with entries in this book are given in SMALL CAPITAL LETTERS.

ASSOCIATION FOR COMPUTING MACHINERY (ACM)

ACM TURING AWARD

The ACM Turing Award "is given to an individual selected for contributions of a technical nature made to the computing community. The contributions should be of lasting and major technical importance to the computer field."

ANNUAL RECIPIENTS
(A few years have joint recipients.)

1966 A. J. Perlis: "For his influence in the area of advanced programming techniques and compiler construction."

1967 Maurice V. Wilkes: "Professor Wilkes is best known as the builder and designer of the EDSAC, the first computer with an internally stored program. Built in 1949, the EDSAC used a mercury delay line memory. He is also known as the author, with Wheeler and Gill, of a volume on *'Preparation of Programs for Electronic Digital Computers'* in 1951, in which program libraries were effectively introduced."

1968 Richard Hamming: "For his work on numerical methods, automatic coding systems, and error-detecting and error-correcting codes."

1969 Marvin MINSKY [Citation not listed by ACM. However, Minsky was a key pioneer in artificial intelligence research, including neural networks, robotics, and cognitive psychology.]

1970 J. H. Wilkinson: "For his research in numerical analysis to facilitate the use of the high-speed digital computer, having received special recognition for his work in computations in linear algebra and 'backward' error analysis."

1971 John MCCARTHY: "Dr. McCarthy's lecture 'The Present State of Research on Artificial Intellegence' is a topic that covers the area in which he has achieved considerable recognition for his work."

1972 E. W. DIJKSTRA.: "Edsger Dijkstra was a principal contributor in the late 1950's to the development of the ALGOL, a high-level programming language which has become a model of clarity and mathematical rigor. He is one of the principal exponents of the science and art of programming languages in general, and has greatly contributed to our understanding of their structure, representation, and implementation. His fifteen years of publications extend from theoretical articles on graph theory to basic manuals, expository texts, and philosophical contemplations in the field of programming languages."

1973 Charles W. Bachman: "For his outstanding contributions to database technology."

1974 Donald E. KNUTH: "For his major contributions to the analysis of algorithms and the design of programming languages, and in particular for his contributions to the 'art of computer programming' through his well-known books in a continuous series by this title."

1975 Alan Newell and Herbert A. Simon: "In joint scientific efforts extending over twenty years, initially in collaboration with J. C. Shaw at the RAND Corporation, and subsequentially with numerous faculty and student collegues at Carnegie-Mellon University, they have made basic contributions to artificial intelligence, the psychology of human cognition, and list processing."

1976 Michael O. Rabin and Dana S. Scott: "For their joint paper 'Finite Automata and Their Decision Problem,' which introduced the idea of nondeterministic machines, which has proved to be an enormously valuable concept. Their [Scott & Rabin] classic paper has been a continuous source of inspiration for subsequent work in this field."

1977 John Backus: "For profound, influential, and lasting contributions to the design of practical high-level programming systems, notably through his work on FORTRAN, and for seminal publication of formal procedures for the specification of programming languages."

1978 Robert W. Floyd: "For having a clear influence on methodologies for the creation of efficient and reliable software, and for helping to found the following important subfields of computer science: the theory of parsing, the semantics of programming languages, automatic program verification, automatic program synthesis, and analysis of algorithms."

1979 Kenneth E. Iverson: "For his pioneering effort in programming languages and mathematical notation resulting in what the computing field now knows as APL, for his contributions to the implementation of interactive systems, to educational uses of APL, and to programming language theory and practice."

1980 C. Anthony R. Hoare: "For his fundamental contributions to the definition and design of programming languages."

1981 Edgar F. Codd: "For his fundamental and continuing contributions to the theory and practice of database management systems. He originated the relational approach to database management in a series of research papers published commencing in 1970. His paper 'A Relational Model of Data for Large Shared Data Banks' was a seminal paper, in a continuing and carefully developed series of papers. Dr. Codd built upon this space and in doing so has provided the impetus for widespread research into numerous related areas, including database languages, query subsystems, database semantics, locking and recovery, and inferential subsystems."

1982 Stephen A. Cook: "For his advancement of our understanding of the complexity of computation in a significant and profound way. His seminal paper, 'The Complexity of Theorem Proving Procedures,' presented at the 1971 ACM SIGACT Symposium on the Theory of Computing, laid the foundations for the theory of NP-Completeness. The ensuing exploration of the boundaries and nature of NP-complete class of problems has been one of the most active and important research activities in computer science for the last decade."

1983 Ken Thompson and Dennis RITCHIE: "For their development of generic operating systems theory and specifically for the implementation of the UNIX operating system."

1984 Niklaus Wirth: "For developing a sequence of innovative computer languages, EULER, ALGOL-W, MODULA and PASCAL. PASCAL has become pedagogically significant and has provided a foundation for future computer language, systems, and architectural research."

1985 Richard M. Karp: "For his continuing contributions to the theory of algorithms including the development of efficient algorithms for network flow and other combinatorial optimization problems, the identification of polynomial-time computability with the intuitive notion of algorithmic efficiency, and, most notably, contributions to the theory of NP-completeness. Karp introduced the now standard methodology for proving problems to be NP-complete which has led to the identification of many theoretical and practical problems as being computationally difficult."

1986 John Hopcroft and Robert Tarjan: "For fundamental achievements in the design and analysis of algorithms and data structures."

1987 John Cocke: "For significant contributions in the design and theory of compilers, the architecture of large systems and the development of reduced instruction set computers (RISC); for discovering and systematizing many fundamental transformations now used in optimizing compilers including reduction of operator strength, elimination of common subexpressions, register allocation, constant propagation, and dead code elimination."

1988 Ivan Sutherland: "For his pioneering and visionary contributions to computer graphics, starting with Sketchpad, and continuing after. Sketchpad, though written twenty-five years ago, introduced many techniques still important today. These include a display file for screen refresh, a recursively traversed hierarchical structure for modeling graphical objects, recursive methods for geometric transformations, and an object oriented programming style. Later innovations include a 'Lorgnette' for viewing stereo or colored images, and elegant algorithms for registering digitized views, clipping polygons, and representing surfaces with hidden lines."

1989 William (Velvel) Kahan: "For his fundamental contributions to numerical analysis. One of the foremost experts on floating-point computations. Kahan has dedicated himself to 'making the world safe for numerical computations.'"

1990 Fernando J. Corbato: "For his pioneering work organizing the concepts and leading the development of the general-purpose, large-scale, time-sharing and resource-sharing computer systems, CTSS and Multics."

1991 Robin Milner: "For three distinct and complete achievements: 1) LCF, the mechanization of Scott's Logic of Computable Functions, probably the first theoretically based yet practical tool for machine assisted proof construction; 2) ML, the first language to include polymorphic type inference together with a type-safe exception-handling mechanism; 3) CCS, a general theory of concurrency. In addition, he formulated and strongly advanced full abstraction, the study of the relationship between operational and denotational semantics."

1992 Butler W. Lampson: "For contributions to the development of distributed, personal computing environments and the technology for their implementation: workstations, networks, operating systems, programming systems, displays, security and document publishing."

1993 Juris Harmanis and Richard E. Stearns: "In recognition of their seminal paper which established the foundations for the field of computational complexity theory."

1994 Edward FEIGENBAUM and Raj Reddy: "For pioneering the design and construction of large-scale artificial intelligence systems, demonstrating the practical importance and potential commercial impact of artificial intelligence technology."

1995 Manuel Blum: "In recognition of his contributions to the foundations of computational complexity theory and its application to cryptography and program checking."

1996 Amir Pneueli: "For seminal work introducing temporal logic into computing science and for outstanding contributions to program and systems verification."

1997 Douglas ENGELBART: "For an inspiring vision of the future of interactive computing and the invention of key technologies to help realize this vision."

1998 James Gray: "For seminal contributions to database and transaction processing research and technical leadership in system implementation."

1999 Frederick P. Brooks, Jr.: "For landmark contributions to computer architecture, operating systems, and software engineering."

2000 Andrew Chi-Chih Yao: "In recognition of his fundamental contributions to the theory of computation, including the complexity-based theory of pseudorandom number generation, cryptography, and communication complexity."

2001 Ole-Johan Dahl and Kristen Nygaard: "For ideas fundamental to the emergence of object oriented programming, through their design of the programming languages Simula I and Simula 67."

ECKERT-MAUCHLY AWARD

Administered jointly by ACM and IEEE Computer Society and "given for contributions to computer and digital systems architecture where the field of computer architecture is considered at present to encompass the combined hardware-software design and analysis of computing and digital systems."

Annual Recipients

1979 Robert S. Barton: "For his outstanding contributions in basing the design of computing systems on the hierarchical nature of programs and their data."

1980 Maurice V. Wilkes: "For major contributions to computer architecture over three decades including notable achievements in developing a working stored-program computer, formulation of the basic principles of microprogramming, early research on cache memories, and recent studies in distributed computation."

1981 Wesley A. Clark: "For contributions to the early development of the minicomputer and the multiprocessor, and for continued contributions over 25 years that have found their way into computer networks, modular computers, and personal computers."

1982 C. Gordon Bell: "For his contributions to designing and understanding computer systems: for his contributions in the formation of the minicomputer; for the creation of the first commercial, interactive timesharing computer; for pioneering work in the field of hardware description languages; for co-authoring classic computer books and co-founding a computer museum."

1983 Tom Kilburn: "For major seminal contributions to computer architecture spanning a period of three decades. For establishing a tradition of collaboration between university and industry which demands the mutual understanding of electronics technology and abstract programming concepts."

1984 Jack B. Dennis: "For contributions to the advancement of combined hardware and software design through innovations in data flow architectures."

1985 John Cocke: "For contributions to high performance computer architecture through lookahead, parallelism and pipeline utilization, and to reduced instruction set computer architecture through the exploitation of hardware-software tradeoffs and compiler optimization."

1986 Harvey G. Cragon: "For major contributions to computer architecture and for pioneering the application of integrated circuits for computer purposes. For serving as architect of the Texas Instruments scientific computer and for playing a leading role in many other computing developments in that company."

1987 Gene M. Amdahl: "For outstanding innovations in computer architecture, including pipelining, instruction look-ahead, and cache memory."

1988 Daniel P. Siewiorek: "For outstanding contributions in parallel computer architecture, reliability, and computer architecture education."

1989 Seymour CRAY: "For a career of achievements that have advanced supercomputing design."

1990 Kenneth E. Batcher: "For contributions to parallel computer architecture, both for pioneering theories in interconnection networks and for the pioneering implementations of parallel computers."

1991 Burton J. Smith: "For pioneering work in the design and implementation of scalable shared memory multiprocessors."

1992 Michael J. Flynn: "For his important and seminal contributions to processor organization and classification, computer arithmetic and performance evaluation."

1993 David Kuck: "For his impact on the field of supercomputing, including his work in shared memory multiprocessing, clustered memory hierarchies, compiler technology, and application/library tuning."

1994 James E. Thornton: "For his pioneering work on high-performance processors; for inventing the 'scoreboard' for instruction issue; and for fundamental contributions to vector supercomputing."

1995 John Crawford: "In recognition of your impact on the computer industry through your development of microprocessor technology."

1996 Yale N. Patt: "For important contributions to instruction level parallelism and superscalar processor design."

1997 Robert Tomasulo: "For the ingenious Tomasulo's algorithm, which enabled out-of-order execution processors to be implemented."

1998 T. Watanabe [Citation not available, but NEC notes that Watanabe "was a chief architect for NEC's first supercomputer, the SX-2, and is recognized for his significant contributions to the architectural design of supercomputers having multiple, parallel vector pipelines and programmable vector caches."]

1999 James E. Smith: "For fundamental contributions to high-performance microarchitecture, including saturating counters for branch prediction, reorder buffers for precise exceptions, decoupled access/execute architectures, and vector supercomputer organization, memory, and interconnects."

2000 Edward Davidson: "For his seminal contributions to the design, implementation, and performance evaluation of high-performance pipelines and multiprocessor systems."

2001 John Hennessy: "For being the founder and chief architect of the MIPS Computer Systems and contributing to the development of the landmark MIPS R2000 microprocessor."

GRACE MURRAY HOPPER AWARD

The ACM gives this award for "the outstanding young computer professional of the year . . . selected on the basis of a single recent major technical or service contribution."

Annual Recipients

Note: this award has not been given every year.

1971 Donald E. KNUTH: "For the publication in 1968 (at age 30) of Volume I of his monumental treatise 'The Art of Computer Programming.'"

1972 Paul E. Dirksen and Paul H. Kress: "For the creation of WATFOR Compiler, the first member of a powerful new family of diagnostic and educational programming tools."

1973 Lawrence M. Breed, Richard Lathwell, and Roger Moore: "For their work in the design and implementation of APL/360, setting new standards in simplicity, efficiency, reliability and response time for interactive systems."

1974 George N. Baird: "For his successful development and implementation of the Navy's COBOL Compiler Validation System."

1975 Allen L. Scherr: "For his pioneering study in quantitative computer performance analysis."

1976 Edward H. Shortliffe: "For his pioneering research which is embodied in the MYCIN program. MYCIN is a program which consults with physicians about the diagnosis and treatment of infections. In creating MYCIN, Shortliffe employed his background of medicine, together with his research in knowledge-based systems design, to produce an integrated package which is easy for expert physicians to use and extend. Shortliffe's work formed the basis for a research program supported by NIH, and has been widely studied and drawn upon by others in the field of knowledge-based systems."

1978 Raymond C. Kurzweil: "For his development of a unique reading machine for the blind, a computer-based device that reads printed pages aloud. The Kurzweil machine is an 80-pound device that shoots a beam of light across each printed page, converts the reflected light across each printed page, converts the reflected light into digital data that is analyzed by its built-in computer, and then transformed into synthetic speech. It is expected to make reading of all printed material possible for blind people, whose reading was previously limited to material translated into Braille. The machine would not have been possible without another achievement by Kurzweil, that is, a set of rules embodied in the mini-computer program by which printed characters of a wide variety of sizes and shapes are reliably and automatically recognized."

1979 Steven WOZNIAK: "For his many contributions to the rapidly growing field of personal computing and, in particular, to the hardware and software for the Apple Computer."

1980 Robert M. Metcalfe: "For his work in the development of local networks, specifically the Ethernet."

1981 Daniel S. Bricklin: "For his contributions to personal computing and, in particular, to the design of VisCalc. Bricklin's efforts in the development of the 'Visual Calculator' provide the excellence and elegance that ACM seeks to sustain through such activities as the Awards program."

1982 Brian K. Reid: "For his contributions in the area of computerized text-production and typesetting systems, specifically Scribe which represents a major advance in this area. It embodies several innovations based on computer science research in programming language design, knowledge-based systems, computer document processing, and typography. The impact of Scribe has been substantial due to the excellent documentation and Reid's efforts to spread the system."

1984 Daniel H. H. Ingalls, Jr.: "For his work at the Xerox Palo Alto Research Center, where he was a major force, both technical and inspirational, in the development of the SMALLTALK language and its graphics facilities. He is the designer of the BITBLT primitive that is now widely used for generating images on raster-scan displays. The combination of a good idea, a good design, and very effective and careful implementation has led to BITBLT's wide acceptance in the computing community. Mr. Ingalls' research has also directly and dramatically affected the computing industry's view of what people should have in the way of accessible computing."

1985 Cordell Green: "For establishing several key aspects of the theoretical basis for logic programming and providing a resolution theorem prover to carry out a programming task by constructing the result which the computer program is to compute. For proving the constructive technique correct and for presenting an effective method for constructing the answer; these contributions providing an early theoretical basis for Prolog and logic programming."

1986 William N. JOY: "For his work on the Berkeley UNIX Operating System as a designer, integrator, and implementor of many of its advanced features including Virtual Memory, the C-shell, the vi Screen editor, and Networking."

1987 John K. Ousterhout. "For his contribution to very large scale integrated circuit computer aided design. His systems, Caesar and Magic, have demonstrated that effective CAD systems need not be expensive, hard to learn, or slow."

1988 Guy L. Steele: "For his general contributions to the development of Higher Order Symbolic Programming, principally for his advancement of lexical scoping in LISP."

1989 W. Daniel Hillis: "For his basic research on data parallel algorithms and for the conception, design, implementation and commercialization of the Connection Machine."

1990 Richard STALLMAN: "For pioneering work in the development of the extensible editor EMACS (Editing Macros)."

1991 Feng-hsuing Hsu: "For contributions in architecture and algorithms for chess machines. His work led to the creation of the Deep Thought Chess Machine, which led to the first chess playing computer to defeat Grandmasters in tournament play and the first to achieve a certified Grandmaster level rating."

1993 Bjarne STROUSTRUP: "For his early work laying the foundations for the C++ programming language. Based on the foundations and Dr. Stroustrup's continuing efforts, C++ has become one of the most influential programming languages in the history of computing."

1996 Shafrira Goldwasser: "For her early work relating computation, randomness, knowledge committee and proofs, which has shaped the foundations of probabilistic computation theory, computational number theory, and cryptography. This work is a continuing influence in design and certification of secure communications protocols, with practical applications to development of secure networks and computer systems."

1999 Wen-mei Hwu: "For the design and implementation of the IMPACT compiler infrastructure which has been used extensively both by the microprocessor industry as a baseline for product development and by academia as a basis for advanced research and development in computer architecture and compiler design."

2000 Lydia Kavraki: "For her seminal work on the probabilistic roadmap approach which has caused a paradigm shift in the area of path planning, and has many applications in robotics, manufacturing, nanotechnology and computational biology."

2001 George Necula: "For his seminal work on the concept and implementation of Proof Carrying Code, which has had a great impact on the field of programming languages and compilers and has given a new direction to applications of theorem proving to program correctness, such as safety of mobile code and component-based software."

Electronic Frontier Foundation (EFF)

PIONEER AWARDS

The EFF gives annual "Pioneer Awards" to leaders in "expanding knowledge, freedom, efficiency, and utility."

1992 Douglas C. ENGELBART, Robert Kahn, Jim Warren, Tom Jennings, and Andrzej Smereczynski.

1993 Paul Baran, Vinton CERF, Ward Christensen, Dave Hughes, and the USENET software developers, represented by the software's originators Tom Truscott and Jim Ellis.

1994 Ivan Sutherland, Whitfield Diffie and Martin Hellman, Murray Turoff and Starr Roxanne Hiltz, Lee Felsenstein, Bill Atkinson, and the Well.

1995 Philip Zimmermann, Anita Borg, and Willis Ware.

1996 Robert Metcalfe, Peter Neumann, Shabbir Safdar, and Matthew Blaze.

1997 Marc Rotenberg, Johan "Julf" Helsingius, and (special honorees) Hedy Lamarr and George Antheil.

1998 Richard STALLMAN, Linus TORVALDS, and Barbara Simons.

1999 Jon Postel, Drazen Panic, and Simon Davies.

2000 Tim BERNERS-LEE, Phil Agre, and "Librarians Everywhere."

2001 Seth Finkelstein, Stephanie Perrin, and Bruce Ennis.

IEEE Computer Society

COMPUTER PIONEER AWARD

The IEEE Computer Society presents the Computer Pioneer Award "for significant contributions to concepts and developments in the electronic computer field which have clearly advanced the state of the art in computing." The award is given a minimum of 15 years after the achievement being awarded.

Charter Recipients

Howard H. AIKEN
Samuel N. Alexander

Gene M. Amdahl
John W. Backus
Robert S. Barton
C. Gordon Bell
Frederick P. Brooks, Jr.
Wesley A. Clark
Fernando J. Corbato
Seymour R. CRAY
Edsgar W. DIJKSTRA
J. Presper ECKERT
Jay W. Forrester
Herman H. Goldstine
Richard W. Hamming
Grace M. HOPPER
Alston S. Householder
David A. Huffman
Kenneth E. Iverson
Tom Kilburn
Donald E. KNUTH
Herman Lukoff
John W. MAUCHLY
Gordon E. Moore
Allen Newell
Robert N. Noyce
Lawrence G. Roberts
George R. Stibitz
Shmuel Winograd
Maurice V. Wilkes
Konrad ZUSE

ANNUAL RECIPIENTS

(With year and achievement as cited by the Computer Society.)

1981 Jeffrey Chuan Chu: "For his early work in electronic computer logic design"

1982 Harry D. Huskey: "For the first parallel computer SWAC"

1982 Arthur Burks: "For his early work in electronic computer logic design"

1984 John Vincent ATANASOFF: "For the first electronic computer with serial memory"

1984 Jerrier A. Haddad: "For his part in the lead IBM 701 design team"

1984 Nicholas C. Metropolis: "For the first solved atomic energy problems on ENIAC"

1984 Nathaniel Rochester: "For the architecture of IBM 702 electronic data processing machines"

1984 Willem L. van der Poel: "For the serial computer ZEBRA"

1985 John G. Kemeny: "For BASIC"

1985 John MCCARTHY: "For LISP and artificial intelligence"

1985 Alan Perlis: "For computer language translation"

1985 Ivan Sutherland: "For the graphics SKETCHPAD"

1985 David J. Wheeler: "For assembly language programming"

1985 Heniz Zemanek: "For computer and computer languages—MAILUEFTERL"

1986 Cuthbert C. Hurd: "For contributions to early computing"

1986 Peter Naur: "For computer language development"

1986 James H. Pomerene: "For IAS and Harvest computers"

1986 Adriann van Wijngaarden: "For ALGOL 68"

1987 Robert E. Everett: "For Whirlwind"

1987 Reynold B. Johnson: "For RAMAC"

1987 Arthur L. Samuel: "For Adaptive non-numeric processing"

1987 Nicklaus E. WIRTH: "For PASCAL"

1988 Freidrich L. Bauer: "For computer stacks"

1988 Marcian E. Hoff, Jr.: "For microprocessor on a chip"

1989 John Cocke: "For instruction pipelining and RISC concepts"

1989 James A. Weidenhammer: "For high speed I/O mechanisms"

1989 Ralph L. Palmer: "For the IBM 604 electronic calculator"

1989 Mina S. Rees: "For the ONR Computer R&D development beginning in 1946"

1989 Marshall C. Yovits: "For the ONR Computer R&D development beginning in 1946"

1989 F. Joachim Weyl: "For the ONR Computer R&D development beginning in 1946"

1989 Gordon D. Goldstein: "For his work with the Office of Naval Research and computer R&R beginning in 1946"

1990 Werner Buchholz: "For computer architecture"

1990 C. A. R. Hoare: "For programming languages definitions"

1991 Bob O. Evans: "For compatable computers"

1991 Robert W. Floyd: "For early compilers"

1991 Thomas E. Kurtz: "For BASIC"

1992 Stephen W. Dunwell: "For project stretch"

1992 Douglas C. ENGELBART: "For human computer interaction"

1993 Erich Bloch: "For high speed computing"

1993 Jack S. Kilby: "For co-inventing the integrated circuit"

1993 Willis H. Ware: "For the design of IAS and Johnniac computers"

1994 Gerrit A. Blaauw: "In recognition of your contributions to the IBM System/360 Series of computers"

1994 Harlan B. Mills: "In recognition of contributions to Structured Programming"

1994 Dennis M. RITCHIE: "In recognition of contributions to the development of UNIX"

1994 Ken L. Thompson: "For his work with UNIX"

1995 Gerald Estrin: "For significant developments on early computers"

1995 David Evans: "For seminal work on computer graphics"

1995 Butler Lampson: "For early concepts and developments of the PC"

1995 Marvin MINSKY: "For conceptual development of artificial intelligence"

1995 Kenneth Olsen: "For concepts and development of minicomputers"

1996 Angel Angelov: "For computer science technologies in Bulgaria"

1996 Richard F. Clippinger: "For computing laboratory staff member, Aberdeen Proving Ground, who converted the ENIAC to a stored program"

1996 Edgar Frank Codd: "For the invention of the first abstract model for database management"

1996 Norber Fristacky: "For pioneering digital devices"

1996 Victor M. Glushkov: "For digital automation of computer architecture"

1996 Jozef Gruska: "For the development of computer science in former Czechoslovakia with fundamental contributions to the theory of computing and extraordinary organizational activities"

1996 Jiri Horejs: "For informatics and computer science"

1996 Lubomir Georgiev Iliev: "A founder and influential leader of computing in Bulgaria; leader of the team that developed the first Bulgarian computer; made fundamental and continuing contributions to abstract mathematics and software"

1996 Robert E. Kahn: "For the co-invention of the TCP/IP protocols and for originating the Internet program"

1996 Laszlo Kalmar: "For recognition as the developer of a 1956 logical machine and the design of the MIR computer in Hungary"

1996 Antoni Kilinski: "For pioneering work in the construction of the first commercial computers in Poland, and for the development of university curriculum in computer science"

1996 Laszlo Kozma: "For development of the 1930 relay machines, and going on to build early computers in post-war Hungary"

1996 Sergey A. Lebedev: "For the first computer in the Soviet Union"

1996 Alexej A. Lyuponov: "For Soviet cybernetics and programming"

1996 Romuald W. Marczynski: "For pioneering work in the construction of the first Polish digital computers and contributions to fundamental research in computer architecture"

1996 Grigore C. Moisil: "For polyvalent logic switching circuits"

1996 Ivan Plander: "For the introduction of computer hardware technology into Slovakia and the development of the first control computer"

1996 Arnols Reitsakas: "For contributions to Estonia's computer age"

1996 Antonin Svoboda: "For the pioneering work leading to the development of computer research in Czechoslovakia and the design and construction of the SAPO and EPOS computers"

1997 Homer (Barney) Oldfield: "For pioneering work in the development of banking applications through the implementation of ERMA, and the introduction of computer manufacturing to GE"

1997 Francis Elizabeth (Betty) Snyder-Holberton: "For the development of the first sort-merge generator for the Univac which inspired the first ideas about compilation"

1998 Irving John (Jack) Good: "For significant contributions to the field of computing as a cryptologist and statistician during World War II at Bletchley Park, as an early worker and developer of the Colossus at Bletchley Park and on the University of Manchester Mark I, the world's first stored program computer"

1999 Herbert Freeman: "For pioneering work on the first computer built by the Sperry Corporation, the SPEEDAC, and for subsequent contributions to the areas of computer graphics and image processing"

2000 Harold W. Lawson: "For inventing the pointer variable and introducing this concept into PL/I, thus providing for the first time, the capability to flexibly treat linked lists in a general-purpose high level language"

2000 Gennady Stolyarov: "For pioneering development in 'Minsk' series computers' software, of the information systems' software and applications and for data processing and data base management systems concepts dissemination and promotion"

2000 Georgy Lopato: "For pioneering development in Belarus of the 'Minsk' series computers' hardware, of the multicomputer complexes and of the 'RV' family of mobile computers for heavy field conditions"

2001 Vernon Schatz: "For the development of Electronics Funds Transfer which made possible computer to computer commercial transactions via the banking system"

2001 William H. Bridge: "For the marrying of computer and communications technology in the GE DATANET 30, putting terminals on peoples' desks to communicate with and timeshare a computer, leading directly to the development of the personal computer, computer networking and the internet"

2002 Per Brinch Hansen: "For pioneering development in operating systems and concurrent programming, exemplified by work on the RC4000 multiprogramming system, monitors, and Concurrent Pascal"

ECKERT-MAUCHLY AWARD

The Eckert-Mauchly Award is given by the IEEE Computer Society "for outstanding contributions to the field of computer and digital systems architecture."

Annual Recipients

(With year and achievement as cited by the Computer Society)

1979 Robert S. Barton

1980 Maurice V. Wilkes

1981 Wesley A. Clark

1982 C. Gordon Bell

1983 Tom Kilburn

1984 Jack B. Dennis

1985 John Cocke: "For contributions to high-performance computer architecture through lookahead, parallelism and pipeline utilization, and to reduced instruction set computer architecture through the exploitation of hardware-software tradeoffs and compiler optimization."

1986 Harvey G. Cragon: "For major contributions to computer architecture and for pioneering the application of integrated circuits for computer purposes and for serving as architect of the Texas Instruments scientific computer and for playing a leading role in many other computing developments in that company."

1987 Gene M. Amdahl: "For outstanding innovations in computer architecture, including pipelining, instruction look-ahead and cache memory."

1988 Daniel P. Siewiorek: "For outstanding contributions in parallel computer architecture, reliability, and computer architecture education."

1989 Seymour CRAY: "For a career of achievements that have advanced supercomputer design."

1990 Kenneth E. Batcher: "For contributions to parallel computer architecture, both for pioneering theories in interconnection networks and for the pioneering implementations of parallel computers."

1991 Burton J. Smith: "For pioneering work in the design and implementation of scalable shared memory multiprocessors."

1992 Michael J. Flynn: "For his important and seminal contributions to processor organization and classification, computer arithmetic and performance evaluation."

1993 David Kuck: "For his impact on the field of supercomputing, including his work in shared memory multiprocessing, clustered memory hierarchies, compiler technology, and application/library tuning."

1994 James E. Thornton: "For his pioneering work on high-performance processors; for inventing the 'scoreboard' for instruction issue; and for fundamental contributions to vector supercomputing."

1995 John H. Crawford: "In recognition of your impact on the computer industry through your development of microprocessor technology."

1996 Yale N. Patt: "For important contributions to instruction-level parallelism and superscalar processor design."

1997 Robert Tomasulo: "For the ingenious Tomasulo's algorithm, which enabled out-of-order execution processors to be implemented."

1998 Tadashi Watanabe: "For contributions to the architectural design of supercomputers with multiple/parallel vector pipelines and programmable vector caches."

1999 James E. Smith: "For fundamental contributions to high-performance micrarchitecture, including saturating counters for branch prediction, reorder buffers for precise exceptions, decoupled access/execute architectures, and vector supercomputer organization memory, and interconnects."

2000 Edward S. Davidson: "For seminal contributions to the design, implementation, and performance evaluation of high-performance pipelines and multiprocessor systems."

2001 John L. Hennessy: "For being the founder and chief architect of the MIPS Computer Systems and contributing to the development of the landmark MIPS R2000 microprocessor."

National Medal of Technology

Given by the President of the United States, the National Medal of Technology is "the highest honor bestowed by the President of the United States to America's leading innovators."

Computer-Related Recipients

1985

AT&T Bell Laboratories: "For contribution over decades to modern communication systems."

Frederick P. Brooks, Jr., Erich Bloch, and Bob O. Evans, International Business Machines Corp.: "For their contributions to the development of the hardware, architecture and systems engineering associated with the IBM System/360, a computer system and technologies which revolutionized the data processing industry and which helped to make the United States dominant in computer technology for many years."

Steven P. Jobs and Steven Wozniak, Apple Computer, Inc.: "For their development and introduction of the personal computer which has sparked the birth of a new industry extending the power of the computer to individual users."

John T. Parsons and Frank L. Stulen, John T. Parsons Company: "For their development and successful demonstration of the numerically-controlled machine tool for the production of three-dimensional shapes, which has been essential for the production of commercial airliners and which is seminal for the growth of the robotics, CAD-CAM, and automated manufacturing industries."

1986

Bernard Gordon, Analogic Corp.: "Father of high-speed analog-to-digital conversion which has been applied to medical, analytical, computer and communications products; founder of two companies with over 2,000 employees and over $100 million in annual sales and creator of a new master's level institute located in Massachusetts to teach engineering leadership and project engineering to engineers."

Reynold B. Johnson, International Business Machines Corp.: "Introduction and development of magnetic disk storage for computers that provided access to virtually unlimited amounts of information in fractions of a second and is the basis for time sharing systems and storage of millions of records. Over $10 billion in annual sales and over 100,000 jobs arose from this development."

William C. Norris, Control Data Corp.: "Advancement of micro electronics and computer technology and creation of one of the Fortune 500—Control Data Corporation—which has over $5 billion in annual sales and over 50,000 employees."

1987

Robert N. Noyce, Intel Corp.: "For his inventions in the field of semiconductor integrated circuits, for his leading role in the establishment of the microprocessor which has led to much wider use of more powerful computers, and for his leadership of research and development in these areas, all of which have had profound consequences both in the United States and throughout the world."

1988

Robert H. Dennard, IBM T.J. Watson Research Center: "For invention of the basic one-transistor dynamic memory cell used worldwide in virtually all modern computers."

David Packard, Hewlett-Packard Company: "For extraordinary and unselfish leadership in both industry and government, particularly in widely diversified technological fields which strengthened the competitiveness and defense capabilities of the United States."

1989

Jay W. Forrester, Massachusetts Institute of Technology and Robert R. Everett, The MITRE Corp.: "For their creative work in developing the technologies and applying computers to real-time applications. Their important contributions proved vital to national and free world defense and opened a new era of world business."

1990

John V. Atanasoff, Iowa State University (Ret.): "For his invention of the electronic digital computer and for contributions toward the development of a technically trained U.S. work force."

Jack St. Clair Kilby, Jack Kilby Co.: "For his invention and contributions to the commercialization of the integrated circuit and the silicon thermal print-head; for his contributions to the development of the first computer using integrated circuits; and for the invention of the hand-held calculator, and gate array."

John S. Mayo, AT&T Bell Laboratories: "For providing the technological foundation for information-age communications, and for overseeing the conversion of the national switched telephone network from analog to a digital-based technology for virtually all long-distance calls both nationwide and between continents."

Gordon E. Moore, Intel Corp.: "For his seminal leadership in bringing American industry the two major postwar innovations in microelectronics large scale integrated memory and the microprocessor—that have fueled the information revolution."

1991

C. Gordon Bell, Stardent Computers: "For his continuing intellectual and industrial achievements in the field of computer design; and for his leading role in establishing cost-effective, powerful computers which serve as a significant tool for engineering, science and industry."

John Cocke, International Business Machines Corp.: "For his development and implementation of Reduced Instruction Set Computer (RISC) architecture that significantly increased the speed and efficiency of computers, thereby enhancing U.S. technological competitiveness."

Grace Murray HOPPER, U.S. Navy (Ret.)/Digital Equipment Corp.: "For her pioneering accomplishments in the development of computer programming languages that simplified computer technology and opened the door to a significantly larger universe of users."

1992

William H. GATES III, Microsoft Corp.: "For his early vision of universal computing at home and in the office; for his technical and business management skills in creating a world-wide technology company; and for his contribution to the development of the personal computer industry."

1993

Kenneth H. Olsen, Digital Equipment Corp.: "For his contributions to the development and use of computer technology; and for his entrepreneurial contribution to American business."

1994

[No computer-related recipients]

1995

Edward R. McCracken, Silicon Graphics, Inc.: "For his groundbreaking work in the areas of affordable 3D visual computing and super computing technologies; and for his technical and leadership skills in building Silicon Graphics, Inc., into a global advanced technology company."

IBM Team: Praveen Chaudhari, IBM TJ Watson Research Center; Jerome J. Cuomo, North Carolina State University (formerly with IBM); and Richard J. Gambino, State University of New York at Stony Brook (formerly with IBM): "For the discovery and development of a new class of materials—the amorphous magnetic materials—that are the basis of erasable, read-write, optical storage technology, now the foundation of the worldwide magnetic-optic disk industry."

1996

James C. Morgan, Applied Materials, Inc.: "For his leadership of 20 years developing the U.S. semiconductor manufacturing equipment industry, and for his vision in building Applied Materials, Inc. into the leading equipment company in the world, a major exporter and a global technology pioneer which helps enable Information Age technologies for the benefit of society."

1997

Vinton Gray CERF, MCI and Robert E. Kahn, Corporation for National Research Initiatives: "For creating and sustaining development of Internet Protocols and continuing to provide leadership in the emerging industry of internetworking."

1998

Kenneth L. Thompson, Bell Laboratories, and Dennis M. RITCHIE, Lucent Technologies: "For their invention of UNIX® operating system and the C programming language, which together have led to enormous growth of an entire industry, thereby enhancing American leadership in the Information Age."

1999

Raymond Kurzweil, founder, chairman, and chief executive officer, Kurzweil Technologies, Inc.: "For pioneering and innovative achievements in computer science such as voice recognition, which have overcome many barriers and enriched the lives of disabled persons and all Americans."

Robert Taylor (Ret.): "For visionary leadership in the development of modern computing technology, including computer networks, the personal computer and the graphical user interface."

2000

Douglas C. ENGELBART, director, Bootstrap Institute: "For creating the foundations of personal computing including continuous, real-time interaction based on cathode-ray tube displays and the mouse, hypertext linking, text editing, on-line journals, shared-screen teleconferencing, and remote collaborative work. More than any other person, he created the personal computing component of the computer revolution."

The IBM Corporation: "For 40 years of innovations in the technology of hard disk drives and information storage products. IBM is widely recognized as the world's leader in basic data storage technologies, and holds over 2,000 U.S. patents. IBM is a top innovator of component technologies—such as flying magnetic heads (thin film heads, and magneto resistive heads), film disks, head accessing systems, digital signal processing and coding, as well as innovative hard disk drive systems. Some specific IBM inventions are used in every modern hard drive today: thin film inductive heads, MR and GMR heads, rotary actuators, sector servos and advanced disk designs. These advances outran foreign hard disk technology and enabled the U.S. industry to maintain the lead it holds today."

APPENDIX IV
COMPUTER-RELATED ORGANIZATIONS

Following is a list of some important computer-related organizations, including contact information.

GENERAL COMPUTER SCIENCE ORGANIZATIONS

American Society for Information Science (http:www.asis.org/) 8720 Georgia Avenue, Suite 501, Silver Spring, MD 20910. Telephone: (301) 495-0810 e-mail: asis@asis.org

Association for Computing Machinery [The oldest computer science organization, with extensive member services and on-line resources.] (http://www.acm.org/) One Astor Plaza, 1515 Broadway, New York, NY 10036-5701 Telephone: (212) 869-7440 e-mail: ACMHELP@acm.org

Computing Research Association (http://www.cra.org) 1100 Seventeenth Street, NW, Suite 507, Washington, DC 20036-4632 Telephone: (202) 234-2111 e-mail: info@cra.org

IEEE Computer Society [A branch of the Institute for Electrical and Electronics Engineering; offers extensive member services and on-line resources.] 1730 Massachusetts Avenue, NW, Washington, DC 20036-1992 Telephone: (202) 371-0101 e-mail: membership@computer.org

Software Engineering Institute (http://sei.cmu.edu) 4500 Fifth Avenue, Pittsburgh, PA 15213-3890 Telephone: (412) 268-5800 e-mail: customer-relations@sei.cmu.edu

APPLICATION AND INDUSTRY SPECIFIC GROUPS

American Association for Artificial Intelligence (http://www.aaai.org/) 445 Burgess Drive, Menlo Park, CA 94025-3442 Telephone: (650) 328-3123 e-mail: info@aaai.org

American Congress on Surveying and Mapping (http://www.landsurveyor.com/acsm/) 5410 Grosvenor Lane, Suite 100, Bethesda, MD 20814-2144 Telephone: (301) 493-0200 e-mail: infoacsm@mindspring.com

American Design Drafting Association (http://www.adda.org) 5522 Norbeck Road, Rockville, MD 20853 Telephone: (301) 460-6875 e-mail: national@adda.org

American Electronics Association (http://www.aeanet.org/) P.O. Box 54990, Santa Clara, CA 54990 Telephone: (800) 284-4232

American Society for Photogrammetry and Remote Sensing (http://www.asprs.org) 5410 Grosvenor Lane, Suite 210, Bethesda, MD 20814-2160 Telephone: (301) 493-0290

American Statistical Association (http://ww.amstat.org) 1429 Duke Street, Alexandria, VA 22314-3415 Telephone: (703) 684-1221

Association for Library and Information Science Education (http://www.alise.org) P.O. Box 7640, Arlington, VA 22207 Telephone: (703) 243-8040 e-mail: sroger7@ibm.net

Association for Multimedia Communication (http://www.amcomm.org) P.O. Box 10645, Chicago, IL 60610 Telephone: (312) 409-1032

Association of American Geographers (http://www.aag.org) 1710 Sixteenth Street, NW, Washington, DC 20009-3198 Telephone: (202) 234-1450 e-mail: gaia@aag.org

Association of Internet Professionals (http://www.association.org/index.html) 9200 Sunset Boulevard, Suite 710, Los Angeles, CA 90069 Telephone: (800) JOIN-AIP or 310-724-6636 e-mail: info@association.org

Computer-Aided Manufacturing International (http://www.cam-i.org/) 3301 Airport Freeway, Suite 324, Bedford, TX 76021 Telephone: (817) 860-1654

Computing Technology Industry Association (CompTIA) (http://www.comptia.org) 1815 S. Meyers Road, Suite 300, Oakbrook Terrace, IL 60181-5228 Telephone: (630) 268-1818 e-mail: info@comptia.org

Computer Law Association (http://www.cla.org) 3028 Javier Road, Suite 402, Fairfax, VA 22031 Telephone: (703) 560-7747 e-mail: clanet@aol.com

Digital Library Federation (http://www.clir.org/diglib/dlfhomepage.htm) e-mail: dwaters@clir.org

Electronics Industries Association (http://www.eia.org/) 2500 Wilson Boulevard, Arlington, VA 22201 Telephone: (703) 907-7500

Information Industry Association (http://www.infoindustry.org/) 1625 Massachusetts Avenue, NW, Suite 700, Washington, DC 20036 Telephone: (202) 986-0280

Information Technology Association of America (http://www.itaa.org) 1616 North Fort Myer Drive, Suite 1300, Arlington, VA 22209 Telephone: (703) 522-5055

Interactive Media Alliance (http://www.tima.org/) GCATT Building, 250 14th Street, NW, 4th floor, Atlanta, GA 30318-5394

International Game Developer's Association (http://www.igdn.org/index.html) 1030 East El Camino Real, 210 Sunnyvale, CA 94087 e-mail: webmeister@igdn.org

International Society for Technology in Education (http://www.iston-line.uoregon.edu) 1787 Agate Street, Eugene, OR 97403 Telephone: (503) 346-4414

International Webmasters Association. (http://www.irwa.org/) 119 East Union Street, Suite F, Pasadena, CA 91103 Telephone: (626) 449-3709

Libraries for the Future (http://www.lff.org) 121 West 27th Street, Suite 1102, New York, NY 10001 Telephone: (212) 352-2330 e-mail: lff@lff.org

Library and Information Technology Association (http://www.lita.org) American Library Association, 50 East Huron Street, Chicago, IL 60611-2795 Telephone: (800) 545-2433, ext. 4270 e-mail: lita@ala.org

National Multimedia Association of America (http://www.digizen.net/nmaa/) 4290 Niagara Road, 3rd floor, College Park, MD 20740 Telephone: (800) 819-1335

Office Automation Society International (http://www.pstcc.cc.tn.us/ost/oasi.html) 5170 Meadow Wood Boulevard, Lyndhurst, OH 44124 Telephone: (216) 461-4803 e-mail: JBDYKE@aol.com

Resource Center for Cyberculture Studies (http://otal.umd.edu/~rccs/) David Silver, 2123 Taliaferro Hall, University of Maryland, College Park, MD 20742-7725 Telephone: (301) 405-1354

Robotics Industries Association (http://www.robotics.org) 900 Victors Way, P.O. Box 3724, Ann Arbor, MI 48106 Telephone: (734) 994-6088 e-mail: ria@robotics.org

SIGGRAPH [Graphics special interest group of the Association for Computing Machinery] (http://www.siggraph.org)

Society for Computer Simulation International (http://www.scs.org) P.O. Box 17900, San Diego, CA 92177-7900 Telephone: (619) 277-3888 e-mail: info@scs.org

Society for Information Management (http://www.simnet.org/) 401 North Michigan Avenue, Chicago, IL 60611-4267 Telephone: (312) 644-6610 e-mail: info@simnet.org

Society for Technical Communication (http://www.stc.org/) 901 North Stuart Street, Suite 904, Arlington, VA 22203-1854 Telephone: (703) 522-4114

Telecommunications Industry Association (http://www.tiaonline.org) 2500 Wilson Boulevard, Suite 300, Arlington, VA 22201-3834 Telephone: (703) 907-7700

Government, Standards and Security Organizations

American National Standards Institute (ANSI) (http://ansi.org/default.htm) 11 West 42nd Street, New York, NY 10036 Telephone: (212) 642-4900

Computer Emergency Response Team (CERT) (http://www.cert.org) CERT Coordination Center, Software Engineering Institute, Carnegie Mellon University, Pittsburgh, PA 15213-3890 Telephone: (412) 268-7090 e-mail: cert@cert.org

Computer Security Institute (http://www.gocsi.com/) 600 Harrison Street, San Francisco, CA 94107 Telephone: (415) 905-2626

Information Systems Security Association (http://www.uh.edu/~bmw/issa/) 1926 Waukegan Road, Suite 1, Glenview, IL 60025-1770 Telephone: (847) 657-6746

Institute for the Certification of Computing Professionals (http://www.iccp.org) 2200 East Devon Avenue, Suite 268, Des Plaines, IL 60018 Telephone: (708) 299-4227

International Organization for Standardization (ISO) (http://www.iso.org)

Internet Society (http://www.isoc.org/) 12020 Sunrise Valley Drive, Suite 210, Reston, VA 20191-3429 Telephone: (703) 648-9888

National Center for Supercomputing Applications (NCSA) (http://www.ncsa.uiuc.edu) 605 East Springfield Avenue, Champaign, IL 61820-5518. Telephone: (217) 244-0072

National Telecommunications and Information Administration (http://www.ntia.doc.gov/) U.S. Dept. of Commerce, 14th & Constitution, NW, Washington, DC 20230.

Quality Assurance Institute (http://www.qaiusa.com/index.html) 7575 Dr. Phillips Boulevard, Suite 350, Orlando, FL 32819 Telephone: (407) 363-1111

Urban and Regional Information Systems Association (http://www.urisa.org) 1460 Renaissance Drive, Suite 305, Park Ridge, IL 60068 Telephone: (847) 824-6300 e-mail: info@urisa.org

World Wide Web Consortium (www.w3c.org) [The premier body for planning and coordination for the Internet.] Massachusetts Institute of Technology, Laboratory for Computer Science, 200 Technology Square, Cambridge, MA 02139 Telephone: (617) 253-2613

Advocacy Groups

Association for Women in Computing (http://www.awc.org) 41 Sutter Street, Suite 1006, San Francisco, CA 94104 Telephone: (415) 905-4663

Black Data Processing Associates (http://www.bdpa.org) P.O. Box 2420, Washington, DC 20013

Center for Democracy and Technology (http://www.cdt.org) 1634 I Street, NW, Washington, DC 20006 Telephone: (202) 637-9800

Computer Professionals for Social Responsibility (http://www.cpsr.org) P.O. Box 717, Palo Alto, CA 94302 e-mail: cpsr@cpsr.org

Electronic Frontier Foundation (www.eff.org) P.O. Box 170190, San Francisco, CA 94117

Electronic Privacy Information Center (http://www.epic.org) 1718 Connecticut Avenue, NW, Suite 200, Washington, DC 20009 Telephone: (202) 483-1140 e-mail: info@epic.org

Women in Technology (http://www.wiatlanta.org) P.O. Box 88247, Atlanta, GA 30356 Telephone: (404) 872-1WIT

INDEX

Boldface page numbers denote main entries; *italic* page numbers indicate illustrations.

Q

R